Intellectual Property

Law:

Cases & Materials

Lydia Pallas Loren
Henry J. Casey Professor of Law
Lewis & Clark Law School

Joseph Scott Miller
Professor of Law
University of Georgia Law School

Edition 6.0 © 2018 Lydia Pallas Loren & Joseph Scott Miller
ISBN 978-1-943689-05-7

www. semaphorepress.com

For Mom and Dad, who taught me to think outside the box.
-LPL

For Harlan and Ginger, who take me on their own survey.
-JSM

Intellectual Property Law: Cases & Materials
Lydia Pallas Loren & Joseph Scott Miller

Copyright and Your Rights

The authors retain the copyright in this book. By downloading a copy of this book from the Semaphore Press website, you have made an authorized copy of the book from the website for your personal use. If you lose it, or your computer crashes or is stolen, don't worry. Come back to the Semaphore Press website and download a replacement copy, and don't worry about having to pay again. Just to be clear, Semaphore Press and the Authors of this casebook are *not* granting you permission to reproduce the material and books available on our website, except to the extent needed for your personal use. We are not granting you permission to distribute copies either. Indeed, we ask that you not resell or give away your copy. Instead, please direct people who are interested in obtaining a copy to the Semaphore Press website, www.semphorepress.com, where they can download their own copies. The resale market in the traditional casebook publishing world is part of what drives casebook prices above $200: When a publisher prices a book at $220, it is factoring in the competition and lost opportunities that the resold books embody. Things are different at Semaphore Press: Because anyone can get his or her own copy of a Semaphore Press book at a reasonable price, we ask that you help us keep legal casebook materials available at reasonable prices by directing anyone interested in this book to our website.

Printing a Paper Copy

If you would like to have a printed copy of the book in addition to the electronic copy, you are welcome to print out a copy of any part, or all, of the book. We anticipate that students may wish to carry only portions of the book at a time. Alternatively, a print version of this book is available on Amazon.com.

Finding Aids and Annotations

Finally, please note that the book does not include an index, a table of cases, or other finding aids that are conventional for printed books. This is because a Semaphore Press book, in pdf form, can be searched electronically for any word or phrase in which you are interested. With the book open in Adobe Reader, simply hit control-f (or select the "find" option in the "Edit" pull-down menu) and enter the search term you want to find. Additionally, the pdf version has been bookmarked for easy "jumping" to sections, or even specific cases, in the book.

We also enable Adobe Reader's commenting features in our pdf books, so you can highlight text, insert comments, and personally annotate your copy in other ways you find helpful. If your copy of Reader does not appear to permit these commenting features, please check to make sure you have the most recent version; any version numbered "8" or higher should permit you to annotate a Semaphore Press book.

TABLE OF CONTENTS

EDITORS' NOTE

In editing cases, we do not indicate omitted footnotes or citations; we do, however, retain footnote numbering from the cases as reported. Where we omit substantive discussion, we note the omission with an ellipsis.

CHAPTER 1: INTRODUCTION

The study of intellectual property law is a study in the protection of intangible assets. A formula for a new drug, information about a process that makes a product more durable, the words that make up a poem or a novel, a symbol that signifies to consumers the source of a product, all these information intangibles can be protected through a variety of intellectual property laws. One can, of course, embody these intangibles in physical items, such as pills, production lines, books, sheet music, and product tags. Coping with this distinction between intangible information and its tangible embodiments is a key part of what sets intellectual property law apart from the real and personal property law doctrines you studied as a new law student. The basic laws of intellectual property protection are the federal Patent Act, the federal Copyright Act, state and federal trademark and trade secret laws, and a variety of statutory and common law protections for what are sometimes referred to as "allied rights," such as the right of publicity. The laws that you will learn about in this book define the rights that individuals and entities have in new and creative products of the mind.

A. INTELLECTUAL PROPERTY LAW'S BASIC POLICIES

Law has a tremendous ability to influence how we invest our resources, including our time. Some laws encourage us to engage in certain activities by rewarding us for those activities. One common way to reward people for an activity, a way with which you are already familiar, is to give them the right to exclude others from sharing in the activity's fruits without permission. Many aspects of intellectual property law encourage creative activity in just this way, giving creators exclusive rights to the fruits of their productivity. Specifically, the law gives creators rights to exclude others from taking newly created information without permission, or from marketing competing embodiments of the new information (at least for a time). Relying on an analogy to real property law, we believe these rights will translate into an incentive to engage in creative endeavors because the rights will have value for the creators. Some laws are designed to keep us from engaging in certain activities by defining them as unlawful and subjecting those who engage in that activity to a variety of sanctions, ranging from monetary payments to loss of freedom through incarceration. Intellectual property laws include a range of sanctions for violating the rights of others.

Like every analogy, the analogy to real property law, which animates many of the doctrines in Anglo-American intellectual property law, has limits and can lead one astray. Information isn't land. We should take care when using what we know about legal doctrines for land, streams, and cows to craft legal doctrines for melodies, logos, and inventions.

Two characteristics that distinguish tangible objects from intangible items are known as excludability and rivalrous consumption. Tangible objects, by their nature, are "excludable"—they can be kept by their owners and, to varying degrees, can be kept away from others. For example, an apple is highly excludable. The person who possesses an apple can lock it away in a box and make it very difficult for anyone else to obtain that apple without permission. An intangible product of the mind, on the other hand, is not excludable like the apple. Once an author releases a poem to the public, the author cannot lock the poem away in a box to keep it from others. While the author can lock her physical copy of the poem in a box, anyone who heard or read the poem before it was locked away can recite it for themselves or others, and can type the words into a computer and put the poem on their website or blog. The intangible asset of the poem lacks the characteristic of excludability.

In addition to non-excludability, intangible assets are non-rivalrous in consumption. Consumption of tangible assets is rivalrous: If Jack eats his apple, Jill cannot eat and enjoy the same apple. Jack and Jill, however, can both enjoy the same poem and can both use it on their websites or recite it to anyone they desire to share it with. One person's use of an intangible asset does not interfere with another's ability to use that same intangible asset.

Thomas Jefferson, in a famous letter to Isaac McPherson in August 1813, talked about both excludability and rivalrousness, although he didn't use those words. According to Jefferson,

> [i]f nature has made any one thing less susceptible than all others of exclusive property, it is the action of the thinking power called an idea, which an individual may exclusively possess as long as he keeps it to himself; but the moment it is divulged, it forces itself into the possession of every one, and the receiver cannot dispossess himself of it. Its peculiar character, too, is that no one possesses the less, because every other possesses the whole of it. He who receives an idea from me, receives instruction himself without lessening mine; as he who lights his taper at mine, receives light without darkening me.[*]

This passage shows that the founders were well aware of these differentiating characteristics of information.

These characteristics of non-excludability and non-rivalrous consumption can pose challenges for the creators of intangible assets. Economists call non-excludable, non-rivalrous goods "public goods," and providing an adequate supply of a public good calls for a proper mechanism. The provision of national defense, for example, is a public good. All citizens are protected by having a national defense, none can be excluded from it, and the consumption of that defense is not rivalrous (practically speaking). To get it, we raise taxes to pay for it. We do the same thing to encourage the creation of new ideas, with grants from the National Institutes of

---

[*] Available at http://press-pubs.uchicago.edu/founders/documents/a1_8_8s12.html.

Health and the National Endowment for the Humanities. But direct government payment for the creation of intangible assets is only a tiny sliver of the compensation that individuals can receive for their creative activities.

One might ask, with respect to products of the mind, why not simply rely on a private market aided by basic contract, property, and tort law? In first year property class you learned about the tragedy of the commons, the idea that property held in common may be harmed by overuse. The problem for intangible assets is distinctly different. With public goods, the risk is one of underproduction, not overuse. Consider an innovative company that has created a groundbreaking new device that significantly increases gas mileage in a traditional car. Assume that the creation of the device took many years and much experimentation. Assume, also, that the only legal rules are the simple contract, property, and tort rules. Once the creator begins selling the device, other companies are going to purchase the device, examine it, and quickly begin marketing their own versions of the device; there isn't any intellectual property law to stop them. To attract consumers, the other companies will likely offer the device at a lower price than the creator. They can afford to do so, given that they—unlike the creator—have little or no research and development costs to recoup. The competitors also benefit by focusing only on technologies that are proven to work, without having funded any duds. While the innovative company may have been able to charge supra-competitive prices initially, an aspect called "lead-time advantage," once the competitors come along the innovative company will need to lower its price to stay in the game. In the end, the price-cutting cycle that competition provides will drive the unit cost of the device down to the marginal cost of producing it. At this competitive price, which takes no account of the creator's research and development costs (except in the atypical case where the lead-time is prolonged), the innovative company will find it hard to recoup its initial R&D investment, much less make a return. For the consumer, this price reduction would appear to be a positive result. Knowing how the story ends, however, the innovative company may decide not to engage in the initial research in the first place, and the new technology is never developed. If that occurs, the end result for the consumer may be stagnation in innovation. This simple story reveals an important reason for providing the creator a legal right to exclude others from competing by direct imitation, at least for a time. It also shows the tradeoff between higher consumer prices in the short run (from the lack of imitative competition) and more innovation in the long run (which can benefit consumers greatly).

The bulk of compensation for the creation of intellectual property assets—inventions, creative works, and trademarked goods—comes from the market, a market that intellectual property law helps to create by providing producers with a right to exclude. In large measure, intellectual property law is designed with a utilitarian purpose: to facilitate market transactions in intangible assets. Intellectual property law provides intangible assets with the characteristic of excludability by granting exclusive rights to the creators and making those rights eligible for market transactions, for example through sales and licenses. This excludability reduces the risk that once an intellectual creation is shared with others the creator will be undercut in the market by a merely imitative competitor who makes no substantial R&D investment in creation. Intellectual property law assists the creator or subsequent owner of that asset by providing a way to deter and prevent "infringement"—unauthorized use of the intangible asset by others.

The risk of underproduction for intellectual assets stems from potential undercompensation for creators. However, it is important to recognize that not all compensation is monetary. Indeed, individuals create for a variety of reasons. Some desire riches, but others desire recognition, affirmation, or advancement in their careers. Some would happily continue creating if their basic investment were covered—they do not need the superprofits that come with a blockbuster hit. Thus, in some areas of intellectual property, the animating policies are not solely utilitarian. Additionally, Lockean labor theory, positing that property rights come into being because of the investment of labor, also plays a role in the desirability of protecting intangible "property."

Consider this very page that you are reading. It is not contained in the standard law school casebook. Instead, you had the choice of obtaining this material from a website. The authors of the material and the company providing the website did not first insist upon compensation in return for you obtaining a copy. Yet, it did take time and effort to create this material, and it takes resources to maintain the website. What might have motivated the authors to create these materials and to provide them in this manner? What may motivate Semaphore Press to provide these materials if it is not receiving $200 from each student using the material? Did you pay for the materials, as the Semaphore Press website encouraged you to do? Why or why not? As you learn more about intellectual property law think about what motivates the choices made by the authors of this material, by Semaphore Press, by your professor who selected this book, and by you.

Finally, while providing rights to exclude may have the effect of creating incentives to create new and useful items, providing too many rights in intangibles also has a risk. Granting the first person to come up with a basic or general idea the right to stop others from using that idea can create too large a bottleneck against competition. The right to exclude can prevent follow-on creativity, and block new inventions that build on past discoveries. Depending on what kinds of activities the right holder can prohibit and for how long, intellectual property rights can lead to monopolistic stagnation. Additionally, much of what is protected by intellectual property law involves expression, and allowing the creator of the expression to prevent the speech activities of others creates tensions with our First Amendment values.

B. OVERVIEW OF INTELLECTUAL PROPERTY RIGHTS

These materials are designed to guide a student new to intellectual property law through the core areas in the field: trade secret law, patent law, copyright law, and trademark law. As you build your knowledge base, the material will begin to make comparisons among the different fields of protection. The remainder of this Chapter considers the common law doctrine of misappropriation as an introduction to some fundamental concepts in intellectual property.

Chapter 2 explores trade secret protection. Keeping an intellectual creation a secret is one way to assure that others do not copy that creation. However, exploiting the value of the creation often requires some disclosure of the secret, perhaps to an employee or a trusted business partner. Trade secret law provides protection for valuable business information, so long as the trade secret owner has made reasonable efforts to maintain the information's secrecy. The protection provided is not absolute, so those who figure out the information on their own (often referred to as "independent creation") or figure out the information by lawfully studying the products sold by the trade secret owner (often referred to as "reverse engineering") are not considered to be "misappropriating" the trade secret. The trade secret law materials in Chapter 2 focus on understanding what information can qualify, what constitutes reasonable efforts, and what are the lawful ways to obtain trade secret information. Trade secret protection, the foundations of which are in state common law, has been codified in state and, more recently, federal statutes.

Governed by a federal statute, patent law provides a right to exclude others from "making, using, selling, offering to sell, or importing" a patented invention. 35 U.S.C. § 271. As explored in Chapter 3, the protection offered by patent law is broad and thick, but its duration is relatively short: protection expires, at most, 20 years after the date an inventor files for patent protection. A patent is obtained through an application and examination process overseen by the federal Patent and Trademark Office (PTO). A wide array of information can be eligible for patent protection, including everything from drug formulae to the design of industrial equipment, but the invention must be both novel and nonobvious, terms of art that the patent law materials explore.

Also governed by federal statute, copyright law provides a right to exclude others from reproducing the copyrighted expression, publicly performing or displaying the expression, distributing copies of the expression, and from creating derivative works based on the expression. 17 U.S.C. § 106. Today, obtaining copyright protection does not require registration with the Copyright Office, a division of the Library of Congress; nor does it require a notice of copyright (for example, the "©" often placed on copyrighted works), although there are benefits to both registration and notice. To obtain protection, a work need only be original and "fixed" in a tangible medium of expression. Copyright protection has many different limits to the rights granted to copyright owners, including the important doctrine of fair use. Copyright protection is not as robust as patent protection, but it lasts for far longer. Copyright protection in a work created today lasts for the life of the author plus 70 years. Chapter 4 explores the basics of copyright law.

Trademark law protects the signals that individuals and entities use to identify the source of their products or services for consumers. Trademark protection stems from both state statutes and from the federal trademark statute, known as the Lanham Act. Chapter 5 focuses on federal protection for trademarks. Obtaining a federally registered trademark from the Patent and Trademark Office is not necessary for federal protection, although such registration provides many benefits. Any word, name, symbol, or device can serve as a trademark so long as it is capable of identifying and distinguishing the goods or services as coming from a particular source. 15 U.S.C. § 1127. Trademark law makes it unlawful for someone to use another's mark in a way that is likely to create consumer confusion about the source of a good or service. More recent developments in federal trademark law also protect certain famous marks from dilution of the signaling power of the marks. The trademark materials also introduce the basics of the prohibition against dilution.

Building on the solid foundation laid in the chapters exploring the four principal types of intellectual property protection (trade secret, patent, copyright, and trademark*), Chapter 6 examines some overlapping concerns relating to the limits of protection. One important source of limits is the supremacy of national law over conflicting state law, the focus of the doctrine known as federal preemption. Another important limitation comes from the conflicts that arise when an intangible asset protected by intellectual property law is embodied in a tangible object subject to the rules concerning personal property ownership. Those limits manifest in doctrines referred to as "exhaustion" or "first sale." Finally, Chapter 7 explores the remedies available for violations of intellectual property rights, ranging from injunctive relief to monetary damages.

Legal protection for intellectual property is bound to the nation or state that creates it. For example, the Patent Act provides that a U.S. Patent confers a right "throughout the United States." 35 U.S.C. § 154(a)(1). As an introduction to the field of intellectual property law, these materials focus on the basics of protection under U.S. law. Personal and business relations, however, frequently flow across national boundaries. Communications and trade increasingly take place on a global scale and may implicate the intellectual property laws of more than one nation. Enforcement of the rights of an intellectual property owner will largely come down to questions of personal jurisdiction over the infringing party. In the past two decades, the role of intellectual property in global economics has taken on increased significance. Several areas of intellectual property protection have international treaties that bind countries to provide a minimum level of protection for intellectual property and employ the fundamental principal of national treatment. National treatment is best understood by the simple axiom that the rights afforded to a foreign national should be identical to those afforded to a home national. We leave full exploration of international intellectual property for a subsequent course.

C. An Example of Intellectual Property Protection: Misappropriation

In many ways, intellectual property protection is a form of trade regulation. While intellectual property law can define the rights of the "property" owner, what these laws attempt to do is balance a desire for free competition (and all the benefits such competition provides) with a recognition that the underproduction risk is real in the area of intellectual endeavors.

---

* Available as a separate supplemental chapter on the Semaphore Press website, we offer coverage of the right of publicity, which some consider to be a type of intellectual property protection.

If someone has the bright idea that a particular intersection is a
good location for a gas station, and builds one at one corner of the
intersection, he cannot prevent someone else from appropriating
his idea by building a gas station at the opposite corner. A fundamental principle of American law is that competition is not a tort,
that is, an invasion of a legally protected right. Freedom to imitate, to copy, is a cornerstone of competition and operates to minimize monopoly profits.

William M. Landes & Richard A. Posner, *The Economic Structure of Intellectual
Property Law* 23 (2003).

Today, almost all of intellectual property law is based in statute—either federal
or state. The one area in which the common law continues to play a major role in
defining rights is misappropriation law. As a common law doctrine, misappropriation has its roots in a famous Supreme Court opinion, *International News Service
v. Associated Press*, 248 U.S. 215 (1918). In that case, the Court draws on the authority of the federal courts, prior to *Erie R. Co. v. Tompkins,* 304 U.S. 64 (1938),
to create general federal common law. Thus, after *Erie*, the law articulated in the
INS opinion is no longer legally binding except to the extent that state courts have
incorporated *INS* into state law. Nonetheless, the opinion is the starting point for
understanding misappropriation.

International News Service v. Associated Press
248 U.S. 215 (1918)

[The Associated Press (AP), a cooperative of newspapers, sued International
News Service (INS), which was owned by William Randolph Hearst. Hearst also
owned many newspapers throughout the country. AP's reporters wrote dispatches that were then sold to non-Hearst as well as Hearst papers. AP competed
against other news services, including INS, in gathering news to be published in
newspapers. INS was barred during much of World War I by British and French
censors from sending war dispatches to the United States, because Hearst had
offended the British and French by siding with Germany at the outset of the war.
To circumvent this boycott, INS paraphrased AP's published war dispatches and
provided those paraphrased reports in its own dispatches. AP did not assert a
claim for copyright infringement, presumably because, as discussed in more detail
in Chapter 4, copyright protects only expression, not the facts or ideas expressed.
Additionally, AP had not registered copyrights in its dispatches, a requirement for
bringing a copyright infringement action.]

PITNEY, JUSTICE:

The parties are competitors in the gathering and distribution of news and its
publication for profit in newspapers throughout the United States. [AP] . . . is a
co-operative organization . . . its members being individuals who are either propri-
etors or representatives of about 950 daily newspapers published in all parts of the
United States. . . . [AP] gathers in all parts of the world, by means of various instrumentalities of its own, by exchange with its members, and by other appropriate

means, news and intelligence of current and recent events of interest to newspaper readers and distributes it daily to its members for publication in their newspapers. The cost of the service, amounting approximately to $3,500,000 per annum, is assessed upon the members and becomes a part of their costs of operation, to be recouped, presumably with profit, through the publication of their several newspapers. . . .

. . . .

. . . [INS's] business is the gathering and selling of news to its customers and clients, consisting of newspapers published throughout the United States, under contracts by which they pay certain amounts at stated times for defendant's service. It has widespread news-gathering agencies; the cost of its operations amounts, it is said, to more than $2,000,000 per annum; and it serves about 400 newspapers . . . .

. . . .

The only matter that has been argued before us is whether defendant [INS] may lawfully be restrained from appropriating news taken from bulletins issued by complainant [AP] or any of its members, or from newspapers published by them, for the purpose of selling it to defendant's clients. . . .

. . . .

. . . The peculiar value of news is in the spreading of it while it is fresh; and it is evident that a valuable property interest in the news, as news, cannot be maintained by keeping it secret. Besides, except for matters improperly disclosed, or published in breach of trust or confidence, or in violation of law, none of which is involved in this branch of the case, the news of current events may be regarded as common property. What we are concerned with is the business of making it known to the world, in which both parties to the present suit are engaged. That business consists in maintaining a prompt, sure, steady, and reliable service designed to place the daily events of the world at the breakfast table of the millions at a price that, while of trifling moment to each reader, is sufficient in the aggregate to afford compensation for the cost of gathering and distributing it, with the added profit so necessary as an incentive to effective action in the commercial world. The service thus performed for newspaper readers is not only innocent but extremely useful in itself, and indubitably constitutes a legitimate business. The parties are competitors in this field; and, on fundamental principles, applicable here as elsewhere, when the rights or privileges of the one are liable to conflict with those of the other, each party is under a duty so to conduct its own business as not unnecessarily or unfairly to injure that of the other.

Obviously, the question of what is unfair competition in business must be determined with particular reference to the character and circumstances of the business. The question here is not so much the rights of either party as against the public but their rights as between themselves. And, although we may and do assume that neither party has any remaining property interest as against the public in uncopyrighted news matter after the moment of its first publication, it by no means follows that there is no remaining property interest in it as between themselves. For, to both of them alike, news matter, however little susceptible of ownership or dominion in the absolute sense, is stock in trade, to be gathered at the cost of enterprise, organization, skill, labor, and money, and to be distributed and sold to those who will pay money for it, as for any other merchandise. Regarding the news, therefore, as but the material out of which both parties are seeking to make profits at the same time and in the same field, we hardly can fail to recognize that for this purpose, and as between them, it must be regarded as *quasi* property, irrespective of the rights of either as against the public.

. . . .

The peculiar features of the case arise from the fact that, while novelty and freshness form so important an element in the success of the business, the very processes of distribution and publication necessarily occupy a good deal of time. [AP]'s service, as well as [INS]'s, is a daily service to daily newspapers; most of the foreign news reaches this country at the Atlantic seaboard, principally at the City of New York, and because of this, and of time differentials due to the earth's rotation, the distribution of news matter throughout the country is principally from east to west; and, since in speed the telegraph and telephone easily outstrip the rotation of the earth, it is a simple matter for [INS] to take [AP]'s news from bulletins or early editions of [AP]'s members in the eastern cities and at the mere cost of telegraphic transmission cause it to be published in western papers issued at least as early as those served by [AP]. Besides this, and irrespective of time differentials, irregularities in telegraphic transmission on different lines, and the normal consumption of time in printing and distributing the newspaper, result in permitting pirated news to be placed in the hands of [INS]'s readers sometimes simultaneously with the service of competing [AP] papers, occasionally even earlier.

[INS] insists that when, with the sanction and approval of [AP], and as the result of the use of its news for the very purpose for which it is distributed, a portion of [AP]'s members communicate it to the general public by posting it upon bulletin boards so that all may read, or by issuing it to newspapers and distributing it indiscriminately, [AP] no longer has the right to control the use to be made of it; that when it thus reaches the light of day it becomes the common possession of all to whom it is accessible; and that any purchaser of a newspaper has the right to communicate the intelligence which it contains to anybody and for any purpose, even for the purpose of selling it for profit to newspapers published for profit in competition with [AP]'s members.

The fault in the reasoning lies in applying as a test the right of [AP] as against the public, instead of considering the rights of [AP] and [INS], competitors in business, as between themselves. The right of the purchaser of a single newspaper to spread knowledge of its contents gratuitously, for any legitimate purpose not unreasonably interfering with [AP]'s right to make merchandise of it, may be admitted; but to transmit that news for commercial use, in competition with [AP]—which is what [INS] has done and seeks to justify—is a very different matter. In doing this [INS], by its very act, admits that it is taking material that has been acquired by [AP] as the result of organization and the expenditure of labor, skill, and money, and which is salable by [AP] for money, and that [INS] in appropriating it and selling it as its own is endeavoring to reap where it has not sown, and by disposing of it to newspapers that are competitors of [AP]'s members is appropriating to itself the harvest of those who have sown. Stripped of all disguises, the process amounts to an unauthorized interference with the normal operation of [AP]'s legitimate business precisely at the point where the profit is to be reaped, in order to divert a material portion of the profit from those who have earned it to those who have not; with special advantage to [INS] in the competition because of the fact that it is not burdened with any part of the expense of gathering the news. The transaction speaks for itself and a court of equity ought not to hesitate long in characterizing it as unfair competition in business.

. . . .

The contention that the news is abandoned to the public for all purposes when published in the first newspaper is untenable. Abandonment is a question of intent, and the entire organization of [AP] negatives such a purpose. The cost of the service would be prohibited if the reward were to be so limited. No single newspaper, no small group of newspapers, could sustain the expenditure. Indeed, it is one of the most obvious results of [INS]'s theory that, by permitting indiscriminate publication by anybody and everybody for purposes of profit in competition with the news-gatherer, it would render publication profitless, or so little profitable as in effect to cut off the service by rendering the cost prohibitive in comparison with the return. . . . [P]ublication by each member must be deemed not by any means an abandonment of the news to the world for any and all purposes, but a publication for limited purposes; for the benefit of the readers of the bulletin or the newspaper as such; not for the purpose of making merchandise of it as news, with the result of depriving [AP]'s other members of their reasonable opportunity to obtain just returns for their expenditures.

It is to be observed that the view we adopt does not result in giving to complainant the right to monopolize either the gathering or the distribution of the news, or, without complying with the copyright act, to prevent the reproduction of its news articles, but only postpones participation by [AP]'s competitor in the processes of distribution and reproduction of news that it has not gathered, and only to the extent necessary to prevent that competitor from reaping the fruits of [AP]'s efforts and expenditure . . . .

It is said that the elements of unfair competition are lacking because there is no attempt by [INS] to palm off its goods as those of [AP], characteristic of the most familiar, if not the most typical, cases of unfair competition. But we cannot concede that the right to equitable relief is confined to that class of cases. In the present case the fraud upon [AP]'s rights is more direct and obvious. Regarding news matter as the mere material from which these two competing parties are endeavoring to make money, and treating it, therefore, as quasi property for the purposes of their business because they are both selling it as such, [INS]'s conduct differs from the ordinary case of unfair competition in trade principally in this that, instead of selling its own goods as those of [AP], it substitutes misappropriation in the place of misrepresentation, and sells [AP]'s goods as its own.

Besides the misappropriation, there are elements of imitation, of false pretense, in [INS]'s practices. The device of rewriting complainant's news articles, frequently resorted to, carries its own comment. The habitual failure to give credit to [AP] for that which is taken is significant. Indeed, the entire system of appropriating [AP]'s news and transmitting it as a commercial product to [INS]'s clients and patrons amounts to a false representation to them and to their newspaper readers that the news transmitted is the result of [INS]'s own investigation in the field. But these elements, although accentuating the wrong, are not the essence of it. It is something more than the advantage of celebrity of which complainant is being deprived.

. . . .

HOLMES, JUSTICE, DISSENTING (WITH JUSTICE MCKENNA):

When an uncopyrighted combination of words is published there is no general right to forbid other people repeating them—in other words there is no property in the combination or in the thoughts or facts that the words express. Property, a creation of law, does not arise from value, although exchangeable—a matter of fact. Many exchangeable values may be destroyed intentionally without compensation. Property depends upon exclusion by law from interference, and a person is not excluded from using any combination of words merely because some one has used it before, even if it took labor and genius to make it. If a given person is to be prohibited from making the use of words that his neighbors are free to make some other ground must be found. One such ground is vaguely expressed in the phrase unfair trade. This means that the words are repeated by a competitor in business in such a way as to convey a misrepresentation that materially injures the person who first used them, by appropriating credit of some kind which the first user has earned. The ordinary case is a representation by device, appearance, or other indirection that the defendant's goods come from the plaintiff. But the only reason why it is actionable to make such a representation is that it tends to give the defendant an advantage in his competition with the plaintiff and that it is thought undesirable that an advantage should be gained in that way. Apart from that the defendant may use such unpatented devices and uncopyrighted combinations of words as he likes. The ordinary case, I say, is palming off the defendant's product as the plaintiff's but the same evil may follow from the opposite falsehood—from saying whether in words or by implication that the plaintiff's product is the defendant's, and that, it seems to me, is what has happened here.

Fresh news is got only by enterprise and expense. To produce such news as it is produced by the defendant represents by implication that it has been acquired by the defendant's enterprise and at its expense. When it comes from one of the great news collecting agencies like the Associated Press, the source generally is indicated, plainly importing that credit; and that such a representation is implied may be inferred with some confidence from the unwillingness of the defendant to give the credit and tell the truth. If the plaintiff produces the news at the same time that the defendant does, the defendant's presentation impliedly denies to the plaintiff the credit of collecting the facts and assumes that credit to the defendant.

If the plaintiff is later in Western cities it naturally will be supposed to have obtained its information from the defendant. The falsehood is a little more subtle, the injury, a little more indirect, than in ordinary cases of unfair trade, but I think that the principle that condemns the one condemns the other. It is a question of how strong an infusion of fraud is necessary to turn a flavor into a poison. The dose seems to me strong enough here to need a remedy from the law. But as, in my view, the only ground of complaint that can be recognized without legislation is the implied misstatement, it can be corrected by stating the truth; and a suitable acknowledgment of the source is all that the plaintiff can require. I think that within the limits recognized by the decision of the Court [INS] should be enjoined from publishing news obtained from [AP] for hours after publication by the plaintiff unless it gives express credit to [AP]; the number of hours and the form of acknowledgment to be settled by the District Court.

BRANDEIS, JUSTICE, DISSENTING:

. . . .

No question of statutory copyright is involved. The sole question for our consideration is this: Was [INS] properly enjoined from using, or causing to be used gainfully, news of which it acquired knowledge by lawful means (namely, by reading publicly posted bulletins or papers purchased by it in the open market) merely because the news had been originally gathered by [AP] and continued to be of value to some of its members, or because it did not reveal the source from which it was acquired?

. . . .

News is a report of recent occurrences. The business of the news agency is to gather systematically knowledge of such occurrences of interest and to distribute reports thereof. [AP] contended that knowledge so acquired is property, because it costs money and labor to produce and because it has value for which those who have it not are ready to pay; that it remains property and is entitled to protection as long as it has commercial value as news; and that to protect it effectively, the defendant must be enjoined from making, or causing to be made, any gainful use of it while it retains such value. An essential element of individual property is the legal right to exclude others from enjoying it. If the property is private, the right of exclusion may be absolute; if the property is affected with a public interest, the right of exclusion is qualified. But the fact that a product of the mind has cost its producer money and labor, and has a value for which others are willing to pay, is not sufficient to ensure to it this legal attribute of property. The general rule of law is, that the noblest of human productions—knowledge, truths ascertained, conceptions, and ideas—become, after voluntary communication to others, free as the air to common use. Upon these incorporeal productions the attribute of property is continued after such communication only in certain classes of cases where public policy has seemed to demand it. These exceptions are confined to productions which, in some degree, involve creation, invention, or discovery. But by no means all such are endowed with this attribute of property. The creations which are recognized as property by the common law are literary, dramatic, musical, and other artistic creations; and these have also protection under the copyright statutes. The inventions and discoveries upon which this attribute of property is conferred only by statute, are the few comprised within the patent law. . . .

The knowledge for which protection is sought in the case at bar is not of a kind upon which the law has heretofore conferred the attributes of property; nor is the manner of its acquisition or use nor the purpose to which it is applied, such as has heretofore been recognized as entitling a plaintiff to relief.

. . . .

. . . [AP] further contended that [INS]'s practice constitutes unfair competition, because there is "appropriation without cost to itself of values created by" [AP]; and it is upon this ground that the decision of this court appears to be based. To appropriate and use for profit, knowledge and ideas produced by other men, without making compensation or even acknowledgment, may be inconsistent with a finer sense of propriety; but, with the exceptions indicated above, the law has heretofore sanctioned the practice. Thus it was held that one may ordinarily make and sell anything in any form, may copy with exactness that which another has produced, or may otherwise use his ideas without his consent and without the payment of compensation, and yet not inflict a legal injury; and that ordinarily one is at perfect liberty to find out, if he can by lawful means, trade secrets of another, however valuable, and then use the knowledge so acquired gainfully, although it cost the original owner much in effort and in money to collect or produce.

Such taking and gainful use of a product of another which, for reasons of public policy, the law has refused to endow with the attributes of property, does not become unlawful because the product happens to have been taken from a rival and is used in competition with him. The unfairness in competition which hitherto has been recognized by the law as a basis for relief, lay in the manner or means of conducting the business; and the manner or means held legally unfair, involves either fraud or force or the doing of acts otherwise prohibited by law. In the "passing off" cases (the typical and most common case of unfair competition), the wrong consists in fraudulently representing by word or act that defendant's goods are those of plaintiff. In the other cases, the diversion of trade was effected through physical or moral coercion, or by inducing breaches of contract or of trust or by enticing away employes. In some others, called cases of simulated competition, relief was granted because defendant's purpose was unlawful; namely, not competition but deliberate and wanton destruction of plaintiff's business.

That competition is not unfair in a legal sense, merely because the profits gained are unearned, even if made at the expense of a rival, is shown by many cases . . . . He who follows the pioneer into a new market, or who engages in the manufacture of an article newly introduced by another, seeks profits due largely to the labor and expense of the first adventurer; but the law sanctions, indeed encourages, the pursuit. He who makes a city known through his product, must submit to sharing the resultant trade with others who, perhaps for that reason, locate there later. . . .

The means by which [INS] obtains news gathered by [AP] is also clearly unobjectionable. It is taken from papers bought in the open market or from bulletins
publicly posted. No breach of contract . . . or of trust . . . [occurred]; and neither
fraud nor force is involved. The manner of use is likewise unobjectionable. No
reference is made by word or by act to [AP], either in transmitting the news to
subscribers or by them in publishing it in their papers. Neither [INS] nor its subscribers is gaining or seeking to gain in its business a benefit from the reputation
of [AP]. They are merely using its product without making compensation. That,
they have a legal right to do; because the product is not property, and they do not
stand in any relation to [AP], either of contract or of trust, which otherwise precludes such use. The argument is not advanced by characterizing such taking and
use a misappropriation.

It is also suggested that the fact that defendant does not refer to [AP] as the
source of the news may furnish a basis for the relief. But the defendant and its
subscribers, unlike members of [AP], were under no contractual obligation to disclose the source of the news; and there is no rule of law requiring acknowledgment
to be made where uncopyrighted matter is reproduced. [INS] is said to mislead its
subscribers into believing that the news transmitted was originally gathered by it
and that they in turn mislead their readers. There is, in fact, no representation by
either of any kind. Sources of information are sometimes given because required
by contract; sometimes because naming the source gives authority to an otherwise
incredible statement; and sometimes the source is named because the agency does
not wish to take the responsibility itself of giving currency to the news. But no
representation can properly be implied from omission to mention the source of
information except that [INS] is transmitting news which it believes to be credible.

   . . . .

  . . . The great development of agencies now furnishing country-wide distribution of news, the vastness of our territory, and improvements in the means of
transmitting intelligence, have made it possible for a news agency or newspapers
to obtain, without paying compensation, the fruit of another's efforts and to use
news so obtained gainfully in competition with the original collector. The injustice
of such action is obvious. But to give relief against it would involve more than the
application of existing rules of law to new facts. It would require the making of a
new rule in analogy to existing ones. The unwritten law possesses capacity for
growth; and has often satisfied new demands for justice by invoking analogies or
by expanding a rule or principle. This process has been in the main wisely applied
and should not be discontinued. Where the problem is relatively simple, as it is
apt to be when private interests only are involved, it generally proves adequate.
But with the increasing complexity of society, the public interest tends to become
omnipresent; and the problems presented by new demands for justice cease to be
simple. Then the creation or recognition by courts of a new private right may work
serious injury to the general public, unless the boundaries of the right are definitely established and wisely guarded. In order to reconcile the new private right
with the public interest, it may be necessary to prescribe limitations and rules for
its enjoyment; and also to provide administrative machinery for enforcing the

rules. It is largely for this reason that, in the effort to meet the many new demands for justice incident to a rapidly changing civilization, resort to legislation has latterly been had with increasing frequency.

The rule for which the plaintiff contends would effect an important extension of property rights and a corresponding curtailment of the free use of knowledge and of ideas; and the facts of this case admonish us of the danger involved in recognizing such a property right in news, without imposing upon news-gatherers corresponding obligations. . . .

. . . .

. . . [L]egislators dealing with the subject might conclude, that the right to news values should be protected to the extent of permitting recovery of damages for any unauthorized use, but that protection by injunction should be denied, just as courts of equity ordinarily refuse (perhaps in the interest of free speech) to restrain actionable libels, and for other reasons decline to protect by injunction mere political rights; and as Congress has prohibited courts from enjoining the illegal assessment or collection of federal taxes. If a Legislature concluded to recognize property in published news to the extent of permitting recovery at law, it might, with a view to making the remedy more certain and adequate, provide a fixed measure of damages, as in the case of copyright infringement.

Or again, a legislature might conclude that it was unwise to recognize even so limited a property right in published news as that above indicated; but that a news agency should, on some conditions, be given full protection of its business; and to that end a remedy by injunction as well as one for damages should be granted, where news collected by it is gainfully used without permission. If a legislature concluded . . . that under certain circumstances news-gathering is a business affected with a public interest; it might declare that, in such cases, news should be protected against appropriation, only if the gatherer assumed the obligation of supplying it at reasonable rates and without discrimination, to all papers which applied therefor. If legislators reached that conclusion, they would probably go further, and prescribe the conditions under which and the extent to which the protection should be afforded; and they might also provide the administrative machinery necessary for insuring to the public, the press, and the news agencies, full enjoyment of the rights so conferred.

Courts are ill-equipped to make the investigations which should precede a determination of the limitations which should be set upon any property right in news or of the circumstances under which news gathered by a private agency should be deemed affected with a public interest. Courts would be powerless to prescribe the detailed regulations essential to full enjoyment of the rights conferred or to introduce the machinery required for enforcement of such regulations. Considerations such as these should lead us to decline to establish a new rule of law in the effort to redress a newly disclosed wrong, although the propriety of some remedy appears to be clear.

Notes & Questions

1. Is the intellectual property field properly classified as a part of "property" law? Is it possible to have a property interest "between" two people, but not have a property interest as against the world? Is that really a property interest? Justice Pitney calls the right in this case *quasi* property. Is it the "exchange value" that makes it property for Justice Pitney? Is everything that is valuable therefore property? Scholars debate the use of the phrase "intellectual property" for the field that you are studying. Might the use of the term "property" skew thinking in the field? The phrase "intellectual property" did not come into vogue until the last few decades of the Twentieth Century. An alternative phrase is "exclusive rights," because intellectual property protection typically provides one party with an exclusive right to exploit a creation, be it an invention, a work of authorship, or a trademark.

2. How long does it take for hot news to cool down? Is a competitor's interest in publishing the material, even after a delay, proof that the material hasn't cooled down enough yet?

3. Is Justice Holmes correct that, absent a statutory right providing some other relief, *credit* for the news is what matters in this case, with the real, and perhaps only, problem being that INS failed to credit AP as the source of the stories? Do you think that AP would have sued even if INS *had* indicated AP as the source of the information contained in INS' stories?

4. Justice Brandeis states an axiomatic "general rule of law": knowledge and ideas, once voluntarily communicated to others, are "free as the air to common use." At the root of all intellectual property law is a struggle to balance that general rule of law with the desire, and perhaps the need, to provide exclusive rights to some creators to encourage creative behavior in society. Did the majority in *INS* get the balance right?

5. Much lore has built up around the *INS v. AP* case. In *The Story of* INS v. AP: *Property, Natural Monopoly, and the Uneasy Legacy of a Concocted Controversy* in INTELLECTUAL PROPERTY STORIES 9, 10 (Ginsburg and Dreyfuss eds. Foundation Press 2006), Douglas G. Baird provides a full account of the parties and the facts that led to the litigation, debunking much of the misinformation that has come to be associated with the case and asserting that the case "has more to do with the regulation of a natural monopoly than with property rights." Hearst was both the owner of INS and the most important member of the AP because AP was a membership based organization and Hearst owned more newspapers than any other member. Baird argues that Hearst "had no interest in an outcome that undercut his own ability to prevent rival newspapers from competing with his AP papers" and thus had little reason to have INS' lawyer challenge AP's legal arguments. *Id.* at 28.

6. Justice Brandeis is clearly concerned with the role courts play in a system that recognizes the legislature's sweeping power to establish, or to forebear from establishing, rights and remedies. Did Justice Brandeis' arguments convince you? His concern is a recurring theme in intellectual property law.

The interplay of the common law claim of misappropriation and federal statutory regime of copyright manifests itself as an inquiry into whether the protection granted under a federal statute "preempts" a particular state common law form of protection. Chapter 6 explores the specific tests used to determine whether a state law claim is preempted by federal intellectual property law. In the next case the court discusses the continuing viability of an *INS*-style misappropriation claim.

Barclays Capital Inc. v. Theflyonthewall.Com, Inc.
650 F.3d 876 (2d Cir. 2011)

SACK, JUDGE:

The plaintiffs-appellees—Barclays Capital Inc. ("Barclays"); Merrill Lynch, Pierce, Fenner & Smith Inc. ("Merrill Lynch"); and Morgan Stanley & Co. Inc. ("Morgan Stanley") (collectively, the "Firms")—are major financial institutions that, among many other things, provide securities brokerage services to members of the public. Largely in that connection, they engage in extensive research about the business and prospects of publicly traded companies, the securities of those companies, and the industries in which those companies are engaged. The results of the research are summarized by the Firms in reports, which customarily contain recommendations as to the wisdom of purchasing, holding, or selling securities of the subject companies. Although the recommendations and the research underlying them in the reports are inextricably related, it is the alleged misappropriation of the recommendations, each typically contained in a single sentence, that is at the heart of the district court's decision and the appeal here.

Each morning before the principal U.S. securities markets open, each Firm circulates its reports and recommendations for that day to clients and prospective clients. The recipients thus gain an informational advantage over non-recipients with respect to possible trading in the securities of the subject companies both by learning before the world at large does the contents of the reports and, crucially for present purposes, the fact that the recommendations are being made by the Firm. The existence of that fact alone is likely to result in purchases or sales of the securities in question by client and non-client alike, and a corresponding short-term increase or decrease in the securities' market prices. The Firms and similar businesses, under their historic and present business models, profit from the preparation and circulation of the reports and recommendations principally insofar as they earn brokerage commissions when a recipient of a report and recommendation turns to the firm to execute a trade in the shares of the company being reported upon.

The defendant-appellant is the proprietor of a news service distributed electronically, for a price, to subscribers. In recent years and by various means, the defendant has obtained information about the Firms' recommendations before the Firms have purposely made them available to the general public and before exchanges for trading in those shares open for the day. Doing so tends to remove the informational and attendant trading advantage of the Firms' clients and prospective clients who are authorized recipients of the reports and recommendations. The recipients of the information are, in turn, less likely to buy or sell the securities using the brokerage services of the reporting and recommending Firms, thereby reducing the incentive for the Firms to create such reports and recommendations in the first place. This, the Firms assert, will destroy their business models and have a severely deleterious impact on their ability to engage in further research and to create further reports and recommendations.

In an attempt to preserve their business models, the Firms have increasingly taken measures to seek to prevent or curtail such pre-market—and therefore, from their point of view, premature—public dissemination of their recommendations. As the district court reported in *Barclays Capital Inc. v. Theflyonthewall.com* ("*Fly I*"), 700 F. Supp. 2d 310 (S.D.N.Y. 2010), the Firms have, for example: "communicated to their employees that the unauthorized dissemination of their equity research or its contents is a breach of loyalty to the Firm, undermines the Firm's creation of revenue, and can result in discipline, including firing," *id.* at 319-20; included in their licensing agreements with third-party distributors and in the reports themselves provisions prohibiting redistribution of their content, *id.* at 320; adopted policies limiting public dissemination of the reports and the information they contain, *id.*; and employed emerging Internet technology by which the Firms can seek to find the source of such "leaks" and to "plug" them, *id.* It is not clear from the record the extent to which these efforts are currently effective, but no concern has been expressed to us as to their legality or legitimacy.

The Firms instituted this litigation as part of the same endeavor. The first of their two sets of claims against the defendant sounds in copyright and is based on allegations of verbatim copying and dissemination of portions of the Firms' reports by the defendant. The Firms have been entirely successful on these copyright claims. *See Fly I*, 700 F. Supp. 2d at 328 ("Fly no longer disputes * * * that it infringed the copyrights in [seventeen of the Firms' reports]* * *. [J]udgment shall [therefore] be entered for the [Firms] on their claims of copyright infringement."). Although the extent to which the Firms' success on the copyright claims has alleviated their overall concerns is not clear, their victory on these claims is secure: Fly has not challenged the resulting injunction on appeal.

What remains before us, then, is the second set of claims by the Firms, alleging that Fly's early republication of the securities recommendations that the Firms create—their "hot news"—is tortious under the New York State law of misappropriation. The district court agreed and granted carefully measured injunctive relief. It is to the misappropriation cause of action that this appeal and therefore this opinion is devoted.

Background

We find little to take issue with in the district court's careful findings of facts, to which we must in any event defer. We therefore borrow freely from them.[3]

The Firms and their Research Reports

The Firms are multinational financial entities that provide a variety of asset management, sales and trading, investment banking, and brokerage services to institutional investors, businesses of various sizes, and individuals. Among their many activities, the Firms compile research reports on specific companies whose securities are publicly traded, on industries, and on economic conditions generally. They disseminate such reports and accompanying trading recommendations to clients, such as hedge funds, private equity firms, pension funds, endowments, and individual investors. The reports, which vary in format, range from a single page to hundreds of pages in length. They typically include data analysis, qualitative discussion, and the recommendation. In the process of producing and disseminating the reports, the Firms employ hundreds of research analysts and spend hundreds of millions of dollars annually.

In preparing a company report, an analyst will gather data related to its business, and may visit its physical facilities, converse with industry experts or company executives, and construct financial or operational models. The analyst then uses that information in light of his or her expertise, experience, and judgment to arrive at formal projections and recommendations regarding the value of the company's securities.

This litigation concerns the trading "Recommendations," a term which the district court defined as "actionable reports," i.e., Firm research reports "likely to spur any investor into making an immediate trading decision.[4] Recommendations upgrade or downgrade a security; begin research coverage of a company's security (an event known as an 'initiation'); or predict a change in the security's target price." *Fly I*, 700 F. Supp. 2d at 316. The better known and more respected an analyst is, the more likely that a recommendation for which he or she is primarily responsible will significantly affect the market price of a security.

Most Recommendations are issued sometime between midnight and 7 a.m. Eastern Time, allowing stock purchases to be made on the market based on the reports and Recommendations upon the market opening at 9:30 a.m. Timely receipt of a Recommendations affords an investor the opportunity to execute a trade in the subject security before the market has absorbed and responded to it.

---

[3] The irony of doing so in the context of a copyright-infringement and "hot news"-misappropriation case is not lost on us.

[4] We refer to Recommendations by the Firms, as opposed to others who make recommendations but are not party to this litigation, with a capital "R."

The Firms typically provide complimentary copies of the reports and Recommendations to their institutional and individual clients using a variety of methods.[6] The Firms then conduct an orchestrated sales campaign in which members of their sales forces contact the clients the Firms think most likely to execute a trade based upon the Recommendation, with the understanding that continued receipt of reports and Recommendations may be made contingent on the generation of a certain level of trading commissions paid to the Firm.[7]

The Firms contend that clients are much more likely to place a trade with a Firm if they learn of the Recommendation directly from that Firm rather than elsewhere, and estimate that more than sixty percent of all trades result from Firm solicitations, including those highlighting Recommendations. It is from the commissions on those trades that Firms profit from the creation and dissemination of their reports and Recommendations. They assert that the timely, exclusive delivery of research and Recommendations therefore is a key to what they frequently refer to as their "business model."

Theflyonthewall.com

The defendant-appellant Theflyonthewall.com, Inc. ("Fly") is, among other things, a news "aggregator." For present purposes, "[a]n aggregator is a website that collects headlines and snippets of news stories from other websites. Examples include Google News and the Huffington Post."

Understanding that investors not authorized by the Firms to receive the reports and Recommendations are interested in and willing to pay for early access to the information contained in them—especially the Recommendations, which are particularly likely to affect securities prices—several aggregators compile securities-firm recommendations, including the Recommendations of the Firms, sometimes with the associated reports or summaries thereof, and timely provide the information to their own subscribers for a fee. Fly is one such company. It employs twenty-eight persons, about half of whom are devoted to content production. It does not itself provide brokerage, trading, or investment-advisory services beyond supplying that information.

---

[6] The Firms distribute reports directly to some of their clients via, inter alia, online platforms that the Firms maintain which provide authorized individuals with access to such research. The Firms also grant licenses to third-party distributors such as Bloomberg, Thomson Reuters, FactSet, and Capital IQ to distribute the reports and Recommendations on their respective platforms.

The universe of authorized report recipients is strikingly large. Morgan Stanley estimates that it distributes its research reports to 7,000 institutional clients and 100,000 individual investors. Each institutional client may in turn identify multiple employees to receive reports. Morgan Stanley estimates that in aggregate approximately 225,000 separate people are authorized to receive its reports.

[7] Each of the Firms conducts a daily morning meeting at roughly 7:15 a.m. During this meeting, analysts will describe to the sales force interesting or important Recommendations issued the previous night. Starting around 8:00 a.m., the sales staff will in turn call, e-mail, and instant message clients to draw their attention to the report and Recommendation, in the hopes that a client will decide to place a trade with the Firm as a result of this contact, earning the firm a commission.

Typical clients of the Firms are hedge funds, private equity firms, pension funds, endowments, and wealthy individual investors. By contrast, Fly's subscribers are predominately individual investors, institutional investors, brokers, and day traders. These customers purchase one of three content packages on Fly's website, paying between $25 and $50 monthly for unlimited access to the site.

In addition to maintaining its website, Fly distributes its content through third-party distributors and trading platforms, including some, such as Bloomberg and Thomson Reuters, that also separately provide authorized dissemination of the Firms' Recommendations. Fly has about 3,300 direct subscribers through its website, and another 2,000 subscribers who use third-party platforms to receive the service.

Fly characterizes itself as a source for breaking financial news, claiming to be the "fastest news feed on the web." *Fly I*, 700 F. Supp. 2d at 322. It advertises that its "quick to the point news is a valuable resource for any investment decision." *Id.* Fly has emphasized its access to analyst research, saying that its newsfeed is a "one-stop solution for accessing analyst comments," and brags that it posts "breaking analyst comments as they are being disseminated by Wall Street trading desks, consistently beating the news wires." *Id.* at 322-23.

The cornerstone of Fly's offerings is its online newsfeed, which it continually updates between 5:00 a.m. and 7:00 p.m. during days on which the New York Stock Exchange is open. The newsfeed typically streams more than 600 headlines a day in ten different categories, including "hot stocks," "rumors," "technical analysis," and "earnings." One such category is "recommendations." There, Fly posts the recommendations (but not the underlying research reports or supporting analysis) produced by sixty-five investment firms' analysts, including those at the plaintiff Firms. A typical Recommendation headline from 2009, for example, reads "EQIX: Equinox initiated with a Buy at BofA/Merrill. Target $110."

. . . .

Fly publishes most of its recommendation headlines before the New York Stock Exchange opens each business day at 9:30 a.m. Fly estimates that the Firms' Recommendation headlines currently comprise approximately 2.5% of Fly's total content, down from 7% in 2005.

According to Fly, over time it has changed the way in which it obtains information about recommendations. Some investment firms, such as Wells Fargo's investment services, will send Fly research reports directly as soon as they are released. Others, including the plaintiff Firms, do not. Until 2005, for recommendations of firms that do not, including the plaintiff Firms, Fly relied on employees at the investment firms (without the firms' authorization) to e-mail the research reports to Fly as they were released. Fly staff would summarize a recommendation as a headline (e.g., "EQIX initiated with a Buy at BofA/Merrill. Target $110."). Sometimes Fly would include in a published item an extended passage taken verbatim from the underlying report.

Fly maintains that because of threats of litigation in 2005, it no longer obtains recommendations directly from such investment firms. Instead, it gathers them using a combination of other news outlets, chat rooms, "blast IMs" sent by people in the investment community to hundreds of recipients, and conversations with traders, money managers, and its other contacts involved in the securities markets. Fly also represents that it no longer publishes excerpts from the research reports themselves, and now disseminates only the Recommendations, typically summarizing only the rating and price target for a particular stock.

The Firms' Response to the Threat Posed by Fly and Other Aggregators

Because the value of the reports and Recommendations to an investor with early access to a Recommendation is in significant part derived from the informational advantage an early recipient may have over others in the marketplace, most of the trading the Firms generate based on their reports and Recommendations occurs in the initial hours of trading after the principal U.S. securities markets have opened. Such sales activity typically slackens by midday. The Firms' ability to generate revenue from the reports and Recommendations therefore directly relates to the informational advantage they can provide to their clients. This in turn is related to the Firms' ability to control the distribution of the reports and Recommendations so that the Firms' clients have access to and can take action on the reports and Recommendations before the general public can.

The Firms have employed a variety of measures in an attempt to stem the early dissemination of Recommendations to non-clients. Most of them have either been instituted or augmented relatively recently in response to the increasing availability of Recommendations from Fly and competing aggregators and news services. The Firms describe these steps as follows:

The Firms have made a "very substantial and costly effort to study the unauthorized dissemination of their research reports and ... to plug the leaks they have found." Merrill Lynch, for example, has: (a) worked with third-party vendors to limit access to Merrill Lynch clients; (b) employed an internal security program to detect breaches of security; (c) investigated Merrill Lynch employees, including a review of cell phones, for leaks to third parties; (d) internalized Merrill Lynch's email subscription system; (e) identified and blacklisted websites that seek to post links to Merrill Lynch content; and (f) created unique signature URLs when links to research are sent to clients so that clients' usage can be monitored and abuse tracked. . . .

. . . .

The Complaint and Pre-Trial District Court Proceedings

. . . .

. . . The gravamen of the [Firm's "hot news" misappropriation] claim is that the aggregate widespread, unauthorized reporting of Recommendations by Fly and other financial news providers—including better known, better financed, more broadly accessed outlets—has threatened the viability of the Firms' equity research operations. The Firms allege that this unauthorized distribution allows clients and prospective clients to learn of Recommendations from sources other than the Firms before the Firms' sales staff can reach out to them to solicit their business, thereby reducing the ability of research to drive commission revenue. This, they assert, seriously threatens their ability to justify the expense of maintaining their extensive research operations.

On August 16, 2006, Fly answered, raising several affirmative defenses, including "fair use" and protections purportedly afforded to it and its dissemination of news by the First Amendment.

. . . .

The Trial and The District Court Decision

. . . .

On March 18, 2010, the district court issued its Opinion and Order, deciding for the plaintiffs on both the copyright-infringement and the "hot news" misappropriation claims. It awarded the plaintiffs statutory damages and attorney's fees related to the copyright infringement claim. . . .

. . . .

Having concluded that the Firms had established the tort of "hot news" misappropriation, the district court entered a permanent injunction barring Fly from reporting a Recommendation until either (a) half an hour after the market opens, if the report containing the recommendation was released before 9:30 a.m., or (b) two hours after release, if the report was released after 9:30 a.m. . . .

. . . .

On appeal, Fly argues principally that (1) the district court erred in finding that the plaintiffs established "hot news" misappropriation under New York law, specifically in that the plaintiffs failed to prove time-sensitivity, free-riding, direct competition, and reduced incentives; (2) that the district court's injunction violates Fly's free-speech rights under the First Amendment; [and] (3) that the district court's finding of "hot news" misappropriation violates the Copyright Clause of the Constitution and the Copyright Act . . . .

Discussion

II. Viability of the "Hot News" Misappropriation Tort

Amici Google, Inc. and Twitter, Inc., referring to the "hot news" misappropriation tort as an "end-run" around the Constitution's Copyright Clause and Supreme Court precedent, and arguing that their position is supported by "[i]mportant public policy concerns," urge us to "repudiate the tort." Brief for Google, Inc. and Twitter, Inc. as Amici Curiae Supporting Reversal at 3.

We need not address the viability *vel non* of a "hot news" misappropriation tort under New York law. Were we to do so, though, plainly we would be bound by the conclusion of the previous Second Circuit panel in [*National Basketball Association v. Motorola, Inc.*, 105 F.3d 841 (2d Cir. 1997) ("*NBA*")] that the tort survives. We are therefore without the authority to "repudiate" that view.

Were we indeed called upon to consider the continued viability of the tort under New York law, perhaps we would certify that issue to the New York Court of

Appeals. The issue we address, however, is federal preemption. As a federal court, we answer that question ourselves.

III. Copyright Act Preemption

. . . .

INS itself is no longer good law. Purporting to establish a principal of federal common law, the law established by *INS* was abolished by *Erie Railroad Co. v. Tompkins*, 304 U.S. 64 (1938), which largely abandoned federal common law. But, as the *NBA* panel pointed out, "[b]ased on legislative history of the 1976 [Copyright Act amendments], it is generally agreed that a 'hot-news' *INS*-like claim survives preemption." *NBA*, 105 F.3d at 845 (citing H.R. Rep. No. 94-1476 at 132).

The House of Representatives Report with respect to the preemption provisions of the 1976 Copyright Act amendments commented in this regard:

> "Misappropriation" is not necessarily synonymous with copyright infringement, and thus a cause of action labeled as "misappropriation" is not preempted if it is in fact based neither on a right within the general scope of copyright as specified by [17 U.S.C. §] 106 [specifying the general scope of copyright] nor on a right equivalent thereto. For example, state law should have the flexibility to afford a remedy (under traditional principles of equity) against a consistent pattern of unauthorized appropriation by a competitor of the facts (i.e., not the literary expression) constituting "hot" news, whether in the traditional mold of [*INS*], or in the newer form of data updates from scientific, business, or financial data bases.

H.R. Rep. No. 94-1476 at 132, reprinted in 1976 U.S.C.C.A.N. at 5748 (footnote omitted). The House Report thus anticipated that *INS*-like state-law torts would survive preemption. It did not itself create such a cause of action or recognize the existence of one under federal law. It allowed instead for the survival of such a state-law claim.

The *NBA* Court . . . used *INS* as a description of the type of claims—"*INS*-like"—that, Congress has said, are not necessarily preempted by federal copyright law. Some seventy-five years after its death under *Erie*, *INS* thus maintains a ghostly presence as a description of a tort theory, not as precedential establishment of a tort cause of action.

ii. Moral Dimensions

One source of confusion in addressing these misappropriation cases is that *INS* itself was a case brought in equity to enjoin INS from copying AP's uncopyrightable news. In that context, the *INS* Court emphasized the unfairness of INS's practice of pirating AP's stories. It condemned, in what sounded biblical in tone, the defendant's "reap[ing] where it ha[d] not sown." *INS*, 248 U.S. at 239. The Court said:

> This defendant * * * admits that it is taking material that has been acquired by complainant as the result of organization and the expenditure of labor, skill, and money, and which is salable by complainant for money, and that defendant in appropriating it and selling it as its own is endeavoring to reap where it has not sown, and by disposing of it to newspapers that are competitors of complainant's members is appropriating to itself the harvest of those who have sown. Stripped of all disguises, the process amounts to an unauthorized interference with the normal operation of complainant's legitimate business precisely at the point where the profit is to be reaped, in order to divert a material portion of the profit from those who have earned it to those who have not; with special advantage to defendant in the competition because of the fact that it is not burdened with any part of the expense of gathering the news. The transaction speaks for itself, and *a court of equity ought not to hesitate long in characterizing it as unfair competition* in business.

Id. at 239-40 (emphasis added). This dicta has been absorbed by New York misappropriation law:

> New York courts have noted the incalculable variety of illegal practices falling within the unfair competition rubric, calling it a broad and flexible doctrine that depends more upon the facts set forth than in most causes of action. It has been broadly described as encompassing any form of commercial immorality, or simply as endeavoring to reap where one has not sown; it is taking the skill, expenditures and labors of a competitor, and misappropriating for the commercial advantage of one person a benefit or property right belonging to another. The tort is adaptable and capacious.

Roy Exp. Co. Establishment of Vaduz, Liech. v. Columbia Broad. Sys., Inc., 672 F.2d 1095, 1105 (2d Cir. 1982). And it has been reflected in the rhetoric of federal district courts applying New York law. *See, e.g., Fly I*, 700 F. Supp. 2d at 336 (quoting *INS*).

   The *NBA* Court also noted that the district court whose decision it was reviewing had "described New York misappropriation law as standing for the 'broader principle that property rights of commercial value are to be and will be protected from any form of commercial immorality'; that misappropriation law developed 'to deal with business malpractices offensive to the ethics of [] society'; and that the doctrine is 'broad and flexible.'" *NBA*, 105 F.3d at 851 (brackets in original). But Judge Winter explicitly rejected the notion that "hot news" misappropriation cases based on the disapproval of the perceived unethical nature of a defendant's ostensibly piratical acts survive preemption. The Court concluded that "such concepts are virtually synonymous [with] wrongful copying and are in no meaningful fashion distinguishable from infringement of a copyright. The broad misappropriation doctrine relied upon by the district court is, therefore, the equivalent of exclusive rights in copyright law." *NBA*, 105 F.3d at 851.

No matter how "unfair" . . . Fly's use of the fact of the Firms' Recommendations may be to the Firms . . . then, such unfairness alone is immaterial to a determination whether a cause of action for misappropriation has been preempted by the Copyright Act. The adoption of new technology that injures or destroys present business models is commonplace. Whether fair or not,[29] that cannot, without more, be prevented by application of the misappropriation tort. Indeed, because the Copyright Act itself provides a remedy for wrongful copying, such unfairness may be seen as supporting a finding that the Act preempts the tort.

iii. Narrowness of the Preemption Exception

The *NBA* panel repeatedly emphasized the "narrowness" of the "hot news" tort exception from preemption. *See id.* at 843, 848, 851, 852 (using the word "narrow" or "narrowness" five times). . . .

    . . . .

Indeed, central to the principle of preemption generally is the value of providing for legal uniformity where Congress has acted nationally.

This is a pressing concern when considering the "narrow" "hot news" misappropriation exemption from preemption. The broader the exemption, the greater the likelihood that protection of works within the "general scope" of the copyright and of the type of works protected by the Act will receive disparate treatment depending on where the alleged tort occurs and which state's law is found to be applicable.

The problem may be illustrated by reference to a recent case in the Southern District of New York. In *Associated Press v. All Headline News Corp.*, 608 F. Supp. 2d 454 (S.D.N.Y. 2009), the court sought to determine whether there was a difference between New York and Florida "hot news" misappropriation law in order for it to analyze, under choice-of-law principles, which state's law applied. Judge

---

[29] It is in the public interest to encourage and protect the Firms' continued incentive to research and report on enterprises whose securities are publicly traded, the businesses and industries in which they are engaged, and the value of their securities. But under the Firms' business models, that research is funded in part by commissions paid by authorized recipients of Recommendations trading not only with the benefit of the Firms' research, but on the bare fact that, for whatever reason, the Recommendation has been (or is about to be) issued. If construed broadly, the "hot news" misappropriation tort applied to the Recommendations alone could provide some measure of protection for the Firms' ability to engage in such research and reporting. But concomitantly, it would ensure that the authorized recipients of the Recommendations would in significant part be profiting because of their knowledge of the fact of a market-moving Recommendation before other traders learn of that fact. In that circumstance, the authorized recipient upon whose commissions the Firms depend to pay for their research activities would literally be profiting at the expense of persons from whom such knowledge has been withheld who also trade in the shares in question ignorant of the Recommendation.

None of this affects our analysis, nor do we offer a view of its legal implications, if any. We note nonetheless that the Firms seem to be asking us to use state tort law and judicial injunction to enable one class of traders to profit at the expense of another class based on their court-enforced unequal access to knowledge of a fact—the fact of the Firm's Recommendation.

Castel observed that "[n]o authority has been cited to show that Florida recognizes a cause of action for hot news misappropriation. Then again, defendants have not persuasively demonstrated that Florida would not recognize such a claim." *Id.* at 459-60.

It appears, then, that the alleged "hot news" misappropriation in *All Headline News Corp.* might have been permissible in New York but not in Florida. The same could have been said for the aggregation and publication of basketball statistics in *NBA*, and the same may be said as to the aggregation and publication of Recommendations in the case at bar. To the extent that "hot news" misappropriation causes of action are not preempted, the aggregators' actions may have different legal significance from state to state—permitted, at least to some extent, in some; prohibited, at least to some extent, in others. It is this sort of patchwork protection that the drafters of the Copyright Act preemption provisions sought to minimize, and that counsels in favor of locating only a "narrow" exception to Copyright Act preemption.

[Looking to Second Circuit precedent, the Court engaged in an extended discussion of the appropriate test for determining whether a state law claim, such as a claim for misappropriation, is preempted. The details of the preemption doctrine are discussed at length in Chapter 6. When it came to applying the doctrine to the claim asserted in this case, the Court provided a lengthy discussion of the element of "free-riding" that precedent had indicated could save a hot-news claim from preemption.]

. . . .

B. Preemption and This Appeal

. . . .

. . . [T]he Firms' claim is not a so-called *INS*-type non-preempted claim because Fly is not, under *NBA*'s analysis, "free-riding." It is collecting, collating and disseminating factual information—the facts that Firms and others in the securities business have made recommendations with respect to the value of and the wisdom of purchasing or selling securities—and attributing the information to its source. The Firms are making the news; Fly, despite the Firms' understandable desire to protect their business model, is breaking it.[36] As the *INS* Court explained, long before it would have occurred to the Court to cite the First Amendment for the proposition:

---

[36] For purposes of evaluating its behavior, at least, INS was not "breaking" news in this sense. It was not reporting on news AP was making by itself reporting news—e.g., "The Associated Press and major news networks reported late Sunday that President Obama plans to nominate Solicitor General Elena Kagan to replace retiring Supreme Court Justice John Paul Stevens." Maureen Hoch, Reports: President Obama to Name Elena Kagan as Supreme Court Pick, PBS Newshour (May 9, 2010, 11:08 PM) available at http:// www.pbs.org/newshour/rundown/2010/05/reports-obama-to-name-elena-kagan-as-supreme-court-pick.html (latest visit Mar. 7, 2011)—let alone making news—e.g., "Tamer Fakahany, an assistant managing editor at the AP's Nerve Center in New York, has been named deputy managing editor overseeing the center at AP headquarters." Tamer Fakahany Named AP Deputy Managing Editor, Associated Press, Feb. 8, 2011, available at

> [T]he news element—the information respecting current events contained in the literary production—is not the creation of the writer, but is a report of matters that ordinarily are *publici juris*; it is the history of the day. It is not to be supposed that the framers of the Constitution, when they empowered Congress "to promote the progress of science and useful arts, by securing for limited times to authors and inventors the exclusive right to their respective writings and discoveries" (Const., Art. I, § 8, par. 8), intended to confer upon one who might happen to be the first to report a historic event the exclusive right for any period to spread the knowledge of it.

INS, 248 U.S. at 234.

The use of the term "free-riding" in recent "hot news" misappropriation jurisprudence exacerbates difficulties in addressing these issues. Unfair use of another's "labor, skill, and money, and which is salable by complainant for money," *INS*, 248 U.S. at 239, sounds like the very essence of "free-riding," and the term "free-riding" in turn seems clearly to connote acts that are quintessentially unfair.

It must be recalled, however, that the term free-riding refers explicitly to a requirement for a cause of action as described by *INS*. As explained by the *NBA* Court, "[a]n indispensable element of an *INS* 'hot news' claim is free-riding by a defendant on a plaintiff's product." *NBA*, 105 F.3d at 854.

The practice of what *NBA* referred to as "free-riding" was further described by *INS*. The *INS* Court defined the "hot news" tort in part as "taking material that has been acquired by complainant as the result of organization and the expenditure of labor, skill, and money, and which is salable by complainant for money, and * * * appropriating it and selling it as [the defendant's] own. * * * " *INS*, 248 U.S. at 239. That definition fits the facts of *INS*: The defendant was taking news gathered and in the process of dissemination by the Associated Press and selling that news as though the defendant itself had gathered it. But it does not describe the practices of Fly. The Firms here may be "acquiring material" in the course of preparing their reports, but that is not the focus of this lawsuit. In pressing a "hot news" claim against Fly, the Firms seek only to protect their Recommendations, something they create using their expertise and experience rather than acquire through efforts akin to reporting.

Moreover, Fly, having obtained news of a Recommendation, is hardly selling the Recommendation "as its own," *INS*, 248 U.S. at 239. It is selling the information with specific attribution to the issuing Firm. Indeed, for Fly to sell, for example, a Morgan Stanley Recommendation "as its own," as INS sold the news it cribbed from AP to INS subscribers, would be of little value to either Fly or its customers. If, for example, Morgan Stanley were to issue a Recommendation of

---

http://www.cnbc.com/id/41478155 (latest visit Mar. 7, 2011). By significant contrast, in *INS*, AP broke news, and INS repackaged that news as though it were "breaking" news of its own.

Boeing common stock changing it from a "hold" to a "sell," it hardly seems likely that Fly would profit significantly from disseminating an item reporting that "Fly has changed its rating of Boeing from a hold to a sell." It is not the identity of Fly and its reputation as a financial analyst that carries the authority and weight sufficient to affect the market. It is Fly's accurate attribution of the Recommendation to the creator that gives this news its value.

We do not perceive a meaningful difference between (a) Fly's taking material that a Firm has created (not "acquired") as the result of organization and the expenditure of labor, skill, and money, and which is (presumably) salable by a Firm for money,[37] and selling it by ascribing the material to its creator Firm and author (not selling it as Fly's own), and (b) what appears to be unexceptional and easily recognized behavior by members of the traditional news media—to report on, say, winners of Tony Awards or, indeed, scores of NBA games with proper attribution of the material to its creator.[38] *INS* did not purport to address either.

It is also noteworthy, if not determinative, that *INS* referred to INS's tortious behavior as "amount[ing] to an unauthorized interference with the normal operation of complainant's legitimate business *precisely at the point where the profit is to be reaped*, in order to *divert a material portion of the profit* from those who have earned it to those who have not. * * * " *Id.* at 240 (emphases added). As we have seen, the point at which the Firms principally reap their profit is upon the execution of sales or purchases of securities. It is at least arguable that Fly's interference with the "normal operation" of the Firms' business is indeed at a "point" where the Firms' profits are reaped. But it is not at all clear that that profit is being in any substantial sense "diverted" to Fly by its publication of Recommendations news. The lost commissions are, we would think, diverted to whatever broker happens to execute a trade placed by the recipient of news of the Recommendation from Fly.

. . . .

. . . [W]e see nothing in the district court's opinion or in the record to indicate that . . . a significant portion of the diversion of profits to which the Firms object is lost to brokers in league with Fly or its competitors. Firm clients are, moreover, free to employ their authorized knowledge of a Recommendation to make a trade with a discount broker for a smaller fee. And, as we understand the record, the Firms channel fees to their brokerage operations using a good deal more than their

---

[37] The Firms do not sell their Recommendations for money. We understand this to be in keeping with their business model, under which the Firms are compensated through commissions for executing trades for clients. But we assume that the Firms could sell the Recommendations, were they so inclined.

[38] Another analogue that comes readily to mind is the regular practice of members of the news media—traditional and otherwise—to report on political endorsements by the editorial boards of competitors. The fact that the New York Times endorses a particular candidate seems to us to be news. When the newspaper publishes its endorsement, that fact is widely reported, without controversy so far as we know, by other news outlets. *See, e.g., Major Newspapers Reveal Their Favorite Candidates*, L.A. Times, Oct. 23, 2000, at A14 (describing and quoting from various major newspapers' endorsements during the 2000 U.S. Presidential election).

Recommendations alone. A non-public Firm report, quite apart from the attached Recommendation—by virtue of the otherwise non-public information the report contains, including general news about the state of the markets, securities, and economic conditions—seems likely to play a substantial part in the Firms' ability to obtain trading business through their research efforts. It is difficult on this record for us to characterize Fly's publication of Recommendations as an unauthorized interference with the normal operation of Firms' legitimate business precisely at the point where the profit is to be reaped which, directly or indirectly, diverts a material portion of the Firms' profits from the Firms to Fly and others engaged in similar practices. *See INS*, 248 U.S. at 240.

. . . .

. . . We think that the *NBA* panel's decision that the absence of "free riding" was fatal to the plaintiff's claim in that case is binding upon us on the facts presented here. In other words, even were we to conclude, hypothetically . . . , that there was indeed direct competition between the Firms and Fly with respect to the Recommendations, we would nonetheless be bound to reverse the judgment of the district court based on our reading of *NBA*. The presence or absence of direct competition is thus not determinative and is therefore a matter we are not called upon to decide here.

Conclusion

We conclude that in this case, a Firm's ability to make news—by issuing a Recommendation that is likely to affect the market price of a security—does not give rise to a right for it to control who breaks that news and how. We therefore reverse the judgment of the district court to that extent and remand with instructions to dismiss the Firms' misappropriation claim.

Notes & Questions

1. Throughout intellectual property law, we confront the need to balance free competition with the necessary incentives for creation. Did the court get that balance right?

2. From the news of distant wars, to sports scores, to fast-paced financial information, the common law tort of misappropriation may provide some limited protection to those who invest in the creation of information. When it does provide protection, the creators of that information may then control *who* can access that information, *when* they can access it, and at what *price*. In the age of the internet some declare that "information wants to be free," but the creators of that information may have different plans! In some ways, the state-law claim of misappropriation may be a claim of last resort: where no other type of intellectual property is available, the only chance of stopping a competitor may be a claim for unfair competition or misappropriation. As a common law claim, misappropriation has some malleability. How far should courts being willing to push the doctrine in the name of preventing "free-riding?"

3. Assertions of "hot news" misappropriation made something of a comeback in the web era, where websites routinely aggregate headlines and news snippets

from other news publishers. For example, the Associated Press defeated a motion to dismiss in its misappropriation case against a news aggregator. *Associated Press v. All Headline News Corp.*, 608 F. Supp. 2d 454 (S.D.N.Y. 2009). The case was subsequently settled, with the defendants paying an unspecified sum and agreeing to "not make competitive use of content or expression from AP stories." *See* Joint Press Release, 7/13/2009, http://www.ap.org/pages/about/pressreleases/pr_ 071309a.html. The *Flyonthewall* case demonstrates another attempt to use the "hot news" misappropriation claim. This time, however, the plaintiffs were unsuccessful. On the other hand, the plaintiffs *did* succeed on their copyright infringement claim. As you will see when you study copyright law, the balance contained in the federal Copyright Act allows for free copying, even by competitors, of underlying facts and of ideas, so long as the particular expression is not copied. Flyonthewall had infringed 17 specific reports issued by the plaintiffs by copying more than just the facts, a finding that Fly did not appeal.

4. In *INS v. AP* Justice Pitney observed that an injunction against misappropriation "postpones participation by [AP]'s competitor in the processes of distribution and reproduction of news that it has not gathered . . . only to the extent necessary to prevent that competitor from reaping the fruits of [AP]'s efforts and expenditure." Is that accomplished with a day's delay in republication? Twelve hours? A week? In the final paragraph of the majority's opinion, the court acknowledged the indefinite nature of the language of the injunction:

> There is some criticism of the injunction that was directed by the District Court upon the going down of the mandate from the Circuit Court of Appeals. In brief, it restrains any taking or gainfully using of the complainant's news, either bodily or in substance from bulletins issued by the complainant or any of its members, or from editions of their newspapers, '*until its commercial value as news to the complainant and all of its members has passed away.*' The part complained of is the clause we have italicized; but if this be indefinite, it is no more so than the criticism. Perhaps it would be better that the terms of the injunction be made specific, and so framed as to confine the restraint to an extent consistent with the reasonable protection of complainant's newspapers, each in its own area and for a specified time after its publication, against the competitive use of pirated news by defendant's customers. But the case presents practical difficulties; and we have not the materials, either in the way of a definite suggestion of amendment, or in the way of proofs, upon which to frame a specific injunction; hence, while not expressing approval of the form adopted by the District Court, we decline to modify it at this preliminary stage of the case, and will leave that court to deal with the matter upon appropriate application made to it for the purpose.

After the decision the parties soon settled out of court, and there are no further opinions in the case. *See* Douglas G. Baird, *The Story of* INS v. AP: *Property, Natural Monopoly, and the Uneasy Legacy of a Concocted Controversy*, in INTELLECTUAL PROPERTY STORIES 9, 10 (Ginsburg and Dreyfuss eds., 2006). Did the injunction the district court had entered in *Barclays Capital*, while it was in effect, provide the parties with better guidance? How do you think this type of remedy might affect the decision of parties in the position of Barclays Capital to pursue a misappropriation claim? How might this remedy affect the decision to invest in the creation of this type of information in the first place?

5. It is said that information is power. In *Barclays Capital*, information translates into money, and those with more money to spend on it, *i.e.*, Barclays Capital's clients, are the ones who obtained access to that information first. As the case illustrates, the distribution of wealth in society is influenced by the distribution of information, which in turn can entrench that distribution of wealth. Of course, information can also disrupt existing patterns of wealth. Should a court take into account the societal impact of permitting a misappropriation claim when determining whether a defendant is "free-riding"? Intellectual property protection, whether in the form of the tort of misappropriation or the other forms of protection explored in the subsequent chapters of this book, is a set of legal mechanisms that profoundly affect who has what information, when, and on what terms.

CHAPTER 2: TRADE SECRET LAW

I. DEFINING A TRADE SECRET

One way to protect competitively valuable information is to keep it secret. Secrecy, if one can preserve it, may allow the secret-holder to prevent others from exploiting the information unless they, too, invest in developing the information themselves. Absolute secrecy, however, makes it difficult for the information holder to exploit the information's commercial value; the possessors of secrets often need to share those secrets with others to exploit the secret effectively or efficiently. Legal protection for what are known as "trade secrets" allows holders of secrets to share the secrets in a controlled way with confidence that the law will protect the holders from invasive behavior by others, called "misappropriation."

Keep in mind, too, that this legal protection for trade secrets exists against the backdrop of market competition, where others may try to develop the same, or similar, information for themselves by using their own acumen and resources. As Landes & Posner observe, "competition is not a tort," and "[f]reedom to imitate, to copy, is a cornerstone of competition."[*]

Having created trade secret protection, we must maintain a framework for distinguishing a wrongful invasion of another's secrecy from lawful competition. We know at the outset that, no matter how we define misappropriation—a definition we explore in Section II of this chapter—it is inevitable that we will judge cases correctly in two ways and incorrectly in two ways:

ERRORS IN ADJUDICATING A TRADE SECRET MISAPPROPRIATION CLAIM

The defendant …	is held …	
	to have misappropriated,	*not* to have misappropriated,
and did misappropriate.	**correct**	**error**
and did *not* misappropriate.	**error**	**correct**

We want to judge cases correctly, of course, both to encourage people to develop valuable business information and to discourage misappropriation. At the same time, we want to avoid incorrect rulings that may deter people from relying on trade secret protection, and that may increase baseless suits in which a plaintiff claims misappropriation merely to harass a competitor. In short, we want to minimize errors, at reasonable cost. Given our baseline of free competition, we probably also incline, at least informally, toward concluding that false negatives (*i.e.*, holding that the defendant did not misappropriate when, in fact, it did) are less

---

[*] William Landes & Richard Posner, *The Economic Structure of Intellectual Property Law* 23 (2003).

harmful to competition than are false positives (*i.e.*, holding that the defendant misappropriated when, in fact, it did not).

How can we reduce errors in sorting wrongful invasions of secrecy from rightful vigorous—even imitative—competition? Part of the answer is found in how we test whether a complaining party's purported trade secret is, in fact, a secret. Section 1 of the Uniform Trade Secret Act (UTSA), a version of which has been adopted in 48 states,[*] Puerto Rico, and the District of Columbia, contains the following definition of a trade secret:

> "Trade Secret" means information, including a formula, pattern, compilation, program, device, method, technique or process, that:
> (1) Derives independent economic value, actual or potential, from not being generally known to, and not being readily ascertainable by proper means by, other persons who can obtain economic value from its disclosure or use; and
> (2) Is the subject of efforts that are reasonable under the circumstances to maintain its secrecy.

UTSA § 1. A new federal trade secret statute—the Defend Trade Secrets Act of 2016 ("DTSA"), Pub. L. No. 114 153 —uses a nearly identical definition. *See* 18 U.S.C. § 1839(3) (codifying the federal definition).

A statutory supplement for this Chapter that includes Kentucky's adopted version of the UTSA and the federal trade secret act is available through Semaphore Press' website: https://www.semaphorepress.com/downloads/IP/TradeSecretSupp2016.pdf

The Uniform Trade Secret Act, with comments, is available from NCCUSL: http://www.uniformlaws.org/shared/docs/trade secrets/utsa final 85.pdf

---

The two elements required to qualify information as a trade secret, set out in subsections (1) and (2), help reduce the errors described above. The second requirement focuses on whether the purported owner took steps to keep the information a secret. The steps an owner took to maintain secrecy can indicate at least three things. First, the objective evidence of the steps taken is some evidence that the trade secret owner subjectively believed the information was valuable as a secret and therefore worth protecting. That subjective belief, in turn, is evidence that the information did, in fact, have value as a secret. Second, the intensity of the owner's effort is evidence of the secret's relative value. Third, the more effort the owner put into keeping the information a secret, the more likely the information actually stayed a secret. As a result, if someone else knows of and is using that very information, it is more likely than it would otherwise be (absent such efforts) that he or she obtained it in an improper way.

---

[*] The remaining states that have *not* adopted the UTSA are Massachusetts and New York.

The first element directly considers how easy it would be for someone to discover the information through means we think are proper. Specifically, if the disputed information is simple, it is at least plausible that an accused misappropriator either independently discovered or reverse-engineered the information on its own—both of which are considered "proper" ways to obtain the information. Moreover, the *simpler* the information is, the *less* likely the accused's independent efforts will have left a clear record of exonerating documents or other evidence. As a result, *when the purported trade secret is relatively simple information, we face a higher risk of wrongly mistaking lawful competition for misappropriation.* To better guard against such an erroneous liability finding, the UTSA requires consideration of an objective test of secrecy that focuses on how readily people in the trade secret plaintiff's business community can obtain the disputed information by proper means (*i.e.*, without resorting to misappropriation). If others can readily obtain the disputed information properly, perhaps we should deny the information trade secret status in the first place. Error balancing considerations such as these pervade legal doctrine. Here, they help us develop a system for protecting trade secrets amid thriving competition.

The following cases further explore this initial topic: *what qualifies for trade secret protection?* As you read these cases pay close attention to the source of the legal protection and the definition of a trade secret. What types of items qualify for protection? What qualities must be present in order for something to be protected as a trade secret? How does one prove the existence of those qualities?

A. When is a Secret a Secret?

Amoco Production Co. v. Laird
622 N.E.2d 912 (Ind. 1993)

maps were a trade secret

Dickson, Justice:

. . . .

In early 1991, Amoco, a corporation engaged in the identification and production of new sources of oil in the continental United States, became interested in a large area of the Northeast Central United States. Amoco was hopeful of locating oil reserves of 50 million barrels in this area where such a significant discovery would be regarded as quite surprising. In April, 1991, Amoco authorized the formation of a team of experts to more specifically determine the oil reserve potential of the project area. The nucleus of this team consisted of a geologist and a geophysicist deployed on a full-time basis under the supervision of Amoco's exploration manager, Christopher Williams. In its quest for oil reserves, the team first reviewed published literature from the United States Geologic Survey and examined substantial territorial documentation kept by Amoco in a confidential library. The team also interviewed Amoco personnel to learn of previous Amoco co-sponsored experience in the region. Twenty-four possible production locations were identified through this process.

Williams, after further statistical evaluation by the team, narrowed the endeavor to four sites. Based upon additional assessment of geological fault lines and fracture swarms, the team thereafter was authorized to refine its focus on a 13,000-square-mile area in southern Michigan and northern portions of Ohio and Indiana known as the Trenton Black River formation. The team subsequently suggested that microwave radar, a means by which micro-emissions associated with large concentrations of underground hydrocarbons are detected by radar return to an over-flying aircraft, be considered in Amoco's intensified assessment of the formation. John Clendenning, an Amoco geologist consulted as an expert in microwave radar, recommended that Amoco contract with Airborne Petroleum, Inc. ("Airborne") to conduct a microwave radar survey of the area. Supplied by Amoco with a navigational grid designed by the team to locate trending geological fault patterns, Airborne conducted the microwave radar study at a cost to Amoco of $150,000.00. Data retrieved by Airborne was then forwarded by Amoco to QC Data, Inc. ("QC"), which digitalized the information and incorporated it into corresponding maps. In addition to its own internal security measures, Amoco took steps to preserve the confidentiality of the survey results through contractual agreements with Airborne and QC.

After its evaluation of the digitalized maps, Amoco authorized further microwave radar study in the Indiana counties of Fulton, Marshall, and Kosciusko. Subsequent analysis directed Amoco's interest to two sites lying primarily in Fulton and Marshall Counties with an estimated yield of 22 to 23 million barrels of oil. Amoco sent a senior land negotiator to Fulton County to view the area and gather information relevant to the obtaining of oil and gas leases in the vicinity. On October 29, 1991, Williams met with the team to further assess the two suspected oil fields. Because estimated production fell below the original 50-million-barrel objective, the group recommended deferral of site development pending future study of the reserve locations.

Clendenning, long disappointed by his perception of Amoco's reluctance to rely upon microwave radar findings in favor of more conventional, established technologies, was convinced that Amoco once again would disregard the promise of microwave radar. On November 9, 1991, he sent to Texan William Laird, a former neighbor, oil wildcatter, and oil exploration financier, a facsimile transmission of a page from a road atlas upon which he had drawn circles accurately defining the location of the potential reserve sites. Within nine days, Laird traveled to Fulton County where he employed a douser to more definitively establish the contours of the prospective fields and eventually obtained oil and gas exploration leases for a significant portion of the reserve locations.

In November of 1991, an Amoco vice-president in charge of North American exploration overruled the team's recommended postponement of site development and directed that development should proceed. Surprised at this unexpected turn of events, Clendenning contacted Laird but failed in his attempt to terminate Laird's leasing efforts. Amoco again sent its land negotiator to Indiana whereupon he discovered the extensive lease purchases within the charted reserve areas. This information was reported to Amoco security which later determined that Clendenning was the source of the information provided to Laird.

On January 24, 1992, Amoco brought an action against Laird for a temporary restraining order, a preliminary injunction, damages, and attorney's fees. The trial court granted Amoco's request for a preliminary injunction which prohibited Laird from pursuing or developing oil and gas leases in areas identified by the map and from using or disclosing any other information gained in the discovery or litigation of this case. From the issuance of the preliminary injunction Laird brought the present interlocutory appeal.

. . . .

. . . Our review in this case focuses upon the trial court's finding that there was sufficient evidence to support the conclusion that information on the Clendenning map is protectable as a "trade secret," defined in relevant part by Ind. Code § 24-2-3-2 as

> information, including a formula, pattern, compilation, program, device, method, technique or process, that:
> (1) Derives independent economic value, actual or potential, from not being generally known to, and *not being readily ascertainable by proper means* by, other persons who can obtain economic value from its disclosure or use; and
> (2) Is the subject of efforts that are reasonable under the circumstances to maintain its secrecy.

Ind. Code § 24-2-3-2 (emphasis added).

. . . .

Amoco contends that the information appearing on Clendenning's map, the product of an extended and costly exploration effort, is not readily ascertainable and thus constitutes a trade secret under Ind. Code § 24-2-3-2. Conversely, Laird argues that because the information provided by the map could have been duplicated by other economically feasible means, it is readily ascertainable and therefore not a trade secret. We thus recognize the essence of the parties' dispute to be the construction of "not being readily ascertainable," a phrase which, due to its apparent susceptibility to more than one interpretation, is ambiguous.

. . . .

In this case, the Court of Appeals . . . concluded that because the trial court did not find that it was "economically infeasible" for Laird to identify the location of the oil reserves by proper means other than Clendennings map, the information was therefore readily ascertainable and thus not a trade secret. We disagree.

. . . .

. . . [T]he economic infeasibility standard derives no support from the plain language of either the Indiana Uniform Trade Secrets Act or the UTSA. Our examination of these provisions reveals that there exists neither explicit mention nor implication that the economic infeasibility of obtaining information by other proper means is a factor to be considered in determining what information merits trade secret protection. A defendant's economic capacity to obtain information by other proper means is thus a notion extraneous to either statute. In fact, the Comment to § 1, the UTSA's definitional component adopted in its entirety as Ind. Code § 24-2-3-2, states in pertinent part, only that "[i]nformation is readily ascertainable if it is available in trade journals, reference books, or other published materials." Thus, the overlay of an economically infeasible standard upon the UTSA's readily ascertainable standard is clearly inconsistent with the definitional elements of "trade secret" contained in model statute endorsed *in toto* by our legislature. Economic infeasibility thus would alter Indiana's Uniform Trade Secrets Act by having it mean what it does not say.

Beyond its lack of support in statutory language, the economic infeasibility standard is inconsistent with apparent legislative intent. We note that our General Assembly has declared within the Indiana Trade Secrets Acts that

> [t]his chapter shall be applied and construed to effectuate its general purpose to make uniform law with respect to the subject matter of this chapter among states enacting the provisions of this chapter.

Ind. Code § 24-2-3-1(b).

We also observe that in 1982, Indiana adopted the UTSA substantially as promulgated by the National Conference of Commissioners on Uniform State Laws. The Prefatory Note to the UTSA declares not only that the Act is a substitution of unitary definitions of trade secrets, but that it "also codifies the results of the better reasoned cases concerning the remedies for trade secret misappropriation." 14 *Uniform Laws Annotated,* Uniform Trade Secrets Act With 1985 Amendments, Prefatory Note, 435 (1990); *see also* Jager, *supra*, § 3.04. It is thus apparent that Indiana Legislators, adopting the UTSA, sought the uniform application of UTSA definitions of trade secret consistent with the application of the act in other adopting jurisdictions. Therefore, case law from other UTSA jurisdictions becomes relevant authority for construction of trade secret law in Indiana.

. . . .

In determining what information is "readily ascertainable," other jurisdictions have frequently looked to the degree of time, effort, and expense required of a defendant to acquire or reproduce the alleged trade secret information by other proper means. Occasionally, courts have invoked the common law predecessor to the UTSA, *The Restatement of Torts,* § 757 (1939), to assist in the definitional task. "Although all of the Restatement's factors no longer are required to find a trade secret, those factors still provide helpful guidance to determine whether the information in a given case constitutes 'trade secrets' within the definition of the statute." *Optic Graphics, Inc. v. Agee* (1991), 87 Md. App. 770, 784.

Of particular relevance to the case before us is the sixth factor enumerated in the Restatement: "the ease or difficulty with which the information could be properly acquired or duplicated by others." *Restatement of Torts,* § 757(b) (1939). Variation exists as to how much time, effort, and funding must be provided in duplicating information before such information is eligible for trade secret protection. Illinois, for example, has found that a trade secret may exist where the duplication of effort was time-consuming, relatively time-consuming, or producible "after significant time, effort, and expense." *Gillis Associated Indus., Inc. v. Cari-All, Inc.* (1990), 206 Ill. App. 3d 184, 190. Similarly,

> [i]f the information can be readily duplicated without involving considerable time, effort or expense, then it is not secret. Conversely, information which can be duplicated only by an expensive time-consuming method of reverse engineering, for instance, could be secret, and the ability to duplicate it would not constitute a defense.

Hamer Holding Group, Inc. v. Elmore (1990), 202 Ill. App. 3d 994, 1011-12. When information can be derived from research, "there must be at least a fair degree of inconvenient or difficult research required before trade secret status will be recognized." Thad G. Long, *Alabama Trade Secrets Act,* 18 Cumb. L. Rev. 557, 564 (1987). Even if information potentially could have been duplicated by other proper means, it is "no defense to claim that one's product could have been developed independently of plaintiff's, if in fact it was developed by using plaintiff's proprietary designs." *Television Communication,* 119 Ill. Dec. at 506. That is, the mere availability of other proper means will not excuse a trade secret misappropriation. Jager, *supra,* § 5.04.

Alabama has held that information may not be protected as a trade secret where there is not a "substantial research investment" required to obtain such information. *Public Sys., Inc. v. Towry* (1991), Ala., 587 So. 2d 969, 972. Similarly, under New York law, it has been recognized that information is a trade secret only when "discoverable with great effort . . . through the expenditure of considerable time and money . . . ." *Webcraft Technologies, Inc. v. McCaw* (S.D.N.Y. 1987), 674 F. Supp. 1039, 1045. Thus, "[t]he more difficult, time consuming and costly it would be to develop the information, the less likely it is 'readily' ascertainable." Peter J. Couture, *Independent Derivation and Reverse Engineering,* Trade Secret Protection and Litigation 615, 623 (1992).

The requirement that there be some measure of time, effort, and funding expended in the duplication or acquisition of information appears calculated to assure not "that trade secrets be unascertainable at all by proper means, but only that they not be *readily* or *quickly* ascertainable by such means." Gerald B. Buechler, Jr., *Revealing Nebraska's Trade Secrets Act,* 23 Creighton L.Rev. 323, 340 (1990) (emphasis in original).

Although the standard utilized by other jurisdictions to determine "not being readily ascertainable" varies, we find no case holding that "not being readily ascertainable" adheres when measures required to duplicate or acquire information are so prohibitively burdensome as to be "economically infeasible". An economic infeasibility standard in trade secrets law not only would be unique to Indiana but also would be singularly extreme in its demand that the effort required to duplicate or acquire alleged trade secret information be not merely considerable or significant but so burdensome as to be a virtual economic impossibility.

We thus find that, consistent with the interpretation of the UTSA in other jurisdictions, where the duplication or acquisition of alleged trade secret information requires a substantial investment of time, expense, or effort, such information may be found "not being readily ascertainable" so as to qualify for protection under the Indiana Uniform Trade Secrets Act. Therefore, the trial court's finding that methods of acquiring the information pertaining to the location of the Indiana oil reserve sites "were not simple or easy to accomplish, and are expensive to develop," is sufficient to support its conclusion that such information was not readily ascertainable and thus entitled to trade secret protection.

We also address Laird's assertion that the alleged trade secret information is readily ascertainable because the discovery of the Indiana reserve sites was facilitated by an already-known process rather than by a procedure exclusively developed, owned, or controlled by Amoco. Laird contends that anyone could have contracted Airborne and hired it to perform the microwave radar survey. Laird also asserts that Amoco's claim must fail because Amoco failed to show that any other form of prospecting, such as satellite imaging, could not have generated the same information.

Other jurisdictions have held that a trade secret often may include elements which by themselves may be readily ascertainable in the public domain, but when viewed together may still qualify for trade secret protection.

> A trade secrets plaintiff need not prove that every element of an information compilation is unavailable elsewhere. Such a burden would be insurmountable since trade secrets frequently contain elements that by themselves may be in the public domain but taken together qualify as trade secrets.

Boeing [Co. v. Sierracin Corp.], 108 Wash. 2d [38] at 50 [(1987)] (citations omitted).

> [I]t is a well settled principle 'that a trade secret can exist in a combination of characteristics and components, each of which, by itself, is in the public domain, but the unified process and operation of which, in unique combination, affords a competitive advantage and is a protectable secret.'

Trandes Corp. v. Guy F. Atkinson Co. (D. Md. 1992), 798 F. Supp. 284, 288, *quoting Q-Co. Indus., Inc. v. Hoffman* (S.D.N.Y 1985), 625 F. Supp. 608, 617. Thus, "The effort of compiling useful information is, of itself, entitled to protection even if the information is otherwise generally known." *ISC-Bunker Ramo Corp. v. Altach, Inc.* (N.D. Ill. 1990), 765 F. Supp. 1310, 1322.

Applying the foregoing to the facts of the present case, the Record reveals that, in its exploratory quest, Amoco utilized easily accessible information already within the public domain. United States Geologic Survey reports, as well as other data pertaining to geological fault lines and fracture swarms, are examples of information generally known, at least to those interested in the oil production marketplace, and utilized by Amoco in which Amoco cannot claim a proprietary interest. Similarly, Laird correctly asserts that microwave radar detection of potential oil reserve sites is not a technology which Amoco itself developed or to which it could claim ownership or exclusive access.

Notwithstanding Amoco's use of some information and technology residing in the public domain, Amoco's exploratory effort was nevertheless a unique undertaking. Amoco engaged in considerable outlay of resources of time, effort, and funding preliminary to and culminating in its decision to utilize microwave radar and contact Airborne. Amoco initially conceived of the exploratory project in an area where 50-million-barrel production capacity was considered surprising. Prior to Laird's receipt of the map, Amoco had invested nearly seven months in project development, including the formation and supervision of an exploration team which included the full-time services of two highly trained employees as well as those efforts of other individuals assigned various tasks related to the project.

Amoco utilized its own proprietary, confidentially-stored information. The team consulted with other Amoco personnel who had previous geological experience with the territory under exploratory consideration. The team engaged in repeated statistical assessments of extensive areas, enabling it to narrow its exploratory focus. Only after allocation of considerable resources did Amoco contract with Airborne for $150,000 to conduct microwave radar surveys of the 13,000-square-mile area. Airborne was provided with an Amoco-generated grid pattern to guide Airborne's flyover procedure.

While some tools leading to Amoco's site discoveries were easily accessible within the public domain, such as U.S. Geological Survey data and the existence of microwave technology, we find that, taken together, the integration of pertinent site information and resultant projections as to potential oil reserves constitutes a unique compilation of information not previously known in the marketplace. We thus agree with the trial court's conclusion that the information generated by Amoco, later appearing on Clendenning's map, was not readily ascertainable.

As to the applicable burden of proof in a trade secrets case, "[t]he burden is on the one asserting the trade secret . . . to show that it is included or embodied in the categories [of protectable trade secrets information] listed [in the trade secrets statute]." *Public Systems*, 587 So. 2d at 971. Hence, a plaintiff who seeks relief for misappropriation of trade secrets must identify the trade secrets and carry the burden of showing they exist.

In the present case, Amoco, by providing evidence sufficient to demonstrate that duplication of its trade secret information would require a substantial investment of time, expense, and effort, established that its geographical information was not readily ascertainable. In meeting the "not being readily ascertainable" element of "trade secret," the only definitional component of Indiana's Uniform Trade Secrets Act at issue in this case, Amoco has met its burden of showing that the geographical information displayed on Clendenning's map is entitled to trade secret protection.

Our decision is consistent with the policies underlying trade secret law which have evolved to promote "[t]he maintenance and standards of commercial ethics and the encouragement of invention . . . ." *Kewanee Oil Co. v. Bicron Corp.* (1974), 416 U.S. 470, 481.

> It seems only fair that one should be able to keep and enjoy the fruits of his labor. If a businessman has worked hard, has used his imagination, and has taken bold steps to gain an advantage over his competitors, he should be able to profit from his efforts. Because a commercial advantage can vanish once the competition learns of it, the law should protect the businessman's efforts to keep his achievements a secret.

Metallurgical Indus. Inc. v. Fortek, Inc., (5th Cir. 1986), 790 F.2d 1195, 1201. In protecting individual property rights in trade secrets, the purpose of trade secrets law also has been recognized as the promotion of the development of new products and technology. Without the availability of such protection, particularly with respect to exploration for subterranean natural resources, "corporations and individual investors would not risk the large sums of money for geophysical exploration, an expensive but only infrequently rewarding venture." 1 Roger Milgrim, *On Trade Secrets,* § 2.09 (1993).

The initial identification of significant oil reserve locations though not strictly a new invention, product, or technology, represents the unique discovery of previously unknown deposits of valuable natural resource. As such, we find that protection of such a discovery is consistent with the policies underpinning trade secrets law.

Notes & Questions

1. If trade secret protection were not available for the oil field information at issue in this case, what other means might a firm like Amoco use to protect its investment in developing the information?

2. In *Amoco Production Company*, the Indiana Supreme Court looks to the precursor to the UTSA, the Restatement of Torts § 757. Comment b to § 757 provides:

> Some factors to be considered in determining whether given information is one's trade secret are:
> 1. the extent to which the information is known outside of his business;
> 2. the extent to which it is known by employees and others involved in his business;
> 3. the extent of measures taken by him to guard the secrecy of the information;
> 4. the value of the information to him and to his competitors;
> 5. the amount of effort or money expended by him in developing the information;
> 6. the ease or difficulty with which the information could be properly acquired or duplicated by others.

The court in *Amoco* relied on factor #6. However, all these factors can be relevant and helpful in determining whether information qualifies as a trade secret.

3. The *Amoco* court's construction of the Indiana trade secret provision is an authoritative statement of Indiana state law. Imagine a company, tomorrow, files suit under the new federal DTSA in the Northern District of Indiana. The DTSA defines "trade secret" to mean:

> all forms and types of financial, business, scientific, technical, economic, or engineering information, including patterns, plans, compilations, program devices, formulas, designs, prototypes, methods, techniques, processes, procedures, programs, or codes, whether tangible or intangible, and whether or how stored, compiled, or memorialized physically, electronically, graphically, photographically, or in writing if—
>
> (A) the owner thereof has taken reasonable measures to keep such information secret; and
>
> (B) the information derives independent economic value, actual or potential, from not being generally known to, and not being readily ascertainable through proper means by, another person who can obtain economic value from the disclosure or use of the information.

18. U.S.C. 1839(3). Should the federal trial judge interpret the federal requirements for trade secret protection in the same way that *Amoco* did under the Indiana statute? Although the *Erie* doctrine (*Erie R. Co. v. Tompkins*, 304 U.S. 64 (1938)) is not formally applicable to a federal DTSA claim, do the principles that animate it—especially the avoidance of incentives to forum shop—speak to this scenario? What if the plaintiff not only pleads a claim under the DTSA, but also pleads a supplemental claim (28 U.S.C. § 1367) under the Indiana statute—should the fact that the plaintiff asserts *both* state and federal misappropriation claims affect the analysis?

4. Notice who the defendant was in this suit. In Section II of this chapter, we will explore what acts constitute misappropriation of trade secrets and are therefore actionable. Assuming that Mr. Clendenning engaged in some wrong that was actionable, why do you think Amoco chose not to sue him?

5. The valuable oil field information in the case includes not only those places that are profitable to drill, but also those that are *not* profitable to drill. Sometimes, the most valuable information a firm has is information about the options that *don't work*, that are *blind alleys*—information by which a competitor could profit, by sparing itself the expense of learning it the hard way, *i.e.*, at full cost. The UTSA definition of "trade secret" covers this information as well. *See* UTSA § 1 Comment ("The definition includes information that has commercial value from a negative viewpoint, for example the results of lengthy and expensive research which proves that a certain process will *not* work could be of great value to a competitor."). How would you square this conclusion with the statutory definition?

6. Does it matter how available (or expensive) the other proper means are? The court states that "the mere availability of other proper means will not excuse a trade secret misappropriation." But if, in a given case, the information the plaintiff asserts is a trade secret can quickly be discovered through simple research or near-

costless reverse engineering, wouldn't that make the information "readily ascertainable by proper means"? And if the information truly is readily ascertainable, and thus *not* protected by trade secret law, why should it matter how the defendant obtained the information? Is the court ignoring the statute here?

7. The court recognizes that a combination of various pieces of information, each of which is in the public domain, can be a trade secret. Trade secret protection encourages companies to make the effort to combine into a new secret whole what are otherwise disparate bits of public information, in valuable ways. But, be careful: effort in combining information is not enough, by itself, to obtain trade secret protection. The additional requirements for trade secret protection must still be met, including taking reasonable steps to keep the combination of information secret, addressed in subsection B below.

CDI Energy Services, Inc. v. West River Pumps, Inc.
567 F.3d 398 (8th Cir. 2009)

Melloy, Judge:

CDI Energy Services, Inc. ("CDI"), sells and services equipment for use in the oilfield industry. CDI maintained a field office in Dickinson, North Dakota, with three employees, John Martinson, Dale Roller, and Kent Heinle. CDI alleges that these men started a competing company, West River Pumps, Inc. ("West River"), stole proprietary information, and solicited business from CDI's clients while still employed by CDI. Upon discovering its employees' actions, CDI filed the present diversity action asserting, among other claims, . . . trade-secret misappropriation . . . . CDI obtained an initial, ex parte temporary restraining order and moved for preliminary injunctive relief. After the parties briefed the matter and submitted affidavits, the district court denied the motion for a preliminary injunction and dissolved the temporary restraining order. CDI appeals, and we affirm.

I. Background

In February 2000, CDI entered the market for selling and servicing oilfield equipment in Dickinson, North Dakota, by hiring Martinson and Roller to open a CDI field office. At that time, Martinson and Roller were experienced in the industry, had client contacts in the area, and were working for one of CDI's competitors. While the men still were working for their previous employer, CDI encouraged them to solicit business from the competitor's clients and bring those clients with them to CDI. Roller and Martinson did so. One of the clients Roller brought with him to CDI subsequently accounted for approximately half of CDI's business.

At CDI, Martinson served as the district manager in North Dakota, and Roller served as the sales and service representative. Martinson hired defendant Kent Heinle in 2007 to work for CDI as a service technician.

In 2007, while employed by CDI, the three men formed West River and contacted CDI's clients. They informed CDI's clients of their plan to commence operations as a separate business and asked those clients to do business with West River. They also secured permission from several clients to move the clients' equipment from CDI's shop to West River's new location. On October 16, 2007, the three men resigned from CDI.

Martinson, Roller and Heinle were CDI's only employees in the area, and when they left CDI, CDI no longer had a presence in the local market. CDI's nearest field office was over 140 miles away in eastern Montana.

CDI argues that it made substantial investments to train Martinson, Roller, and Heinle, develop business in the Dickinson area, and develop trade-secret information. CDI argues that these investments and efforts support its . . . trade-secret claim[] because the investments demonstrate the value that the defendants took from CDI. For example, CDI asserts that it provided extensive training to the men regarding marketing strategies and several aspects of CDI's business operations. The defendants argue that CDI provided no education or formal training but that CDI hired them specifically because they already had experience in the industry and, in fact, had customers they could bring to CDI.

CDI argues that it took efforts reasonable under the circumstances to protect certain information as trade secrets. CDI identifies customer lists, customer contact information, business strategies, customer repair and purchase histories, and CDI pricing information as trade secrets. CDI argues that it had policies in place informing employees that information was confidential and that employees were to maintain the information as confidential. The defendants assert that CDI's Dickinson office was open, the purportedly trade-secret materials were unguarded and unmarked, and CDI made no substantial efforts to ensure that the materials were kept confidential.

The defendants admit that Martinson, Roller, and Heinle took limited records with them when they left CDI. They provided the district court with a description of all the documents that they took, however, and claim that they returned those documents to CDI. They assert none of the materials are trade secrets.

. . . The court found CDI had not made a showing sufficient to demonstrate that any of the materials the defendants took constituted trade secrets under North Dakota law. Accordingly, the court found CDI had not established a likelihood of success on the merits of the trade-secret claim. . . .

II. Discussion

. . . .

We find no error in the district court's conclusion that CDI failed to meet its burden to prove that the defendants had taken trade-secret information or that CDI itself had taken reasonable steps to protect any purported trade secrets. The information at issue was of the type that may, in some industries, be treated as trade-secret information (customer names, contact information, pricing information, etc.). CDI, however, failed to show that any of the information in this case actually was a trade secret, *i.e.*, information that has economic value by virtue of having been kept secret and that cannot be "ascertain[ed] by proper means." *See* N.D. Cent. Code § 47-25.1-01(4) ("'Trade secret' means information * * * that: a. Derives independent economic value, actual or potential, from not being generally known to, and not being readily ascertainable by proper means by, other persons who can obtain economic value from its disclosure or use; and b. Is the subject of efforts that are reasonable under the circumstances to maintain its secrecy.").

It appears undisputed that the potential customers for CDI and West River in the area surrounding Dickinson are a small collection of easily identifiable, locally operating oilfield companies. Information about these companies would be easily obtainable, if not already known, by relevant actors in the local oilfield service and equipment industry. Also, the record shows little effort by CDI to conceal data as trade secrets, and the defendants' affidavits contest CDI's assertions regarding those efforts. Because we review the district court's assessment of the factual record only for clear error, we find no basis to disturb the district court's conclusion regarding the trade-secret claim.

With no likelihood of success on the merits, there is little justification for granting a preliminary injunction regarding the trade-secret claim. . . .

Notes & Questions

1. *Amoco* suggested the potential for liability where a third party avoids the work of reverse engineering the information and instead obtains that information in an inappropriate way. How can this be, if the information is truly *ineligible* for protection? In *CDI*, the defendants did not need to make a substantial investment in developing client information; they knew much of it from a prior job, and the local market they served had just a few customers. In both *Amoco* and *CDI*, then, it seems clear that if the information were to be found eligible for trade secret protection, the defendants would be liable for misappropriation. Thus, the defendants' argument in both cases is that the information is *ineligible* for protection. In the introductory material we described the cost of error calculation that is common in shaping legal doctrine. In these two cases you can see the court policing the boundaries of the definition of "trade secret" in cases where it is clear how the defendants came to know the alleged trade secret information.

2. Customer information certainly can, as a general matter, qualify as protectable trade secret information. *See* 1 Roger M. Milgrim & Eric E. Bensen, *Milgrim on Trade Secrets* § 1.09[7] (collecting and discussing cases). Protection depends on the facts of a given case. What, exactly was the problem in *CDI*? How might a plaintiff prove that information has economic value by virtue of having been kept secret?

In most states that have adopted the UTSA, the first requirement is phrased as it is in the UTSA: "A 'Trade Secret' means information . . . that: (1) Derives independent economic value, actual or potential, from not being generally known to, and *not being readily ascertainable by proper means* by, other persons who can obtain economic value from its disclosure or use . . . ." UTSA §1 (emphasis added). Some states, however, omit the italicized language. *See, e.g.,* Oregon Rev. Stat. § 646.461(4); California Civ. Code § 3426.1(d). The states that include the italicized language contained in the uniform law make it harder for a trade secret owner to prove the information at issue is a trade secret. On what policy ground should a state preserve, or delete, this aspect of the model definition of "trade secret"? Consider: How, if at all, does including this phrase in the definition of "trade secret" reduce the number of baseless, or ill-conceived, claims of trade secret misappropriation? The significance of that phrase requires understanding what are "proper means" by which one can obtain a trade secret. The UTSA does not define that phrase in the statute, but the commentary contains some guidance.

> Proper means include:
> 1. Discovery by independent invention;
> 2. Discovery by "reverse engineering," that is, by starting with the known product and working backward to find the method by which it was developed. The acquisition of the known product must, of course, also be by a fair and honest means, such as purchase of the item on the open market for reverse engineering to be lawful;
> 3. Discovery under a license from the owner of the trade secret;
> 4. Observation of the item in public use or on public display;
> 5. Obtaining the trade secret from published literature.

UTSA §1 Comment. Here we focus on one very important proper means: reverse engineering.

Pamela Samuelson & Suzanne Scotchmer, *The Law and Economics of Reverse Engineering,* 111 YALE L.J. 1575, 1582-84 (2002)

Reverse engineering has always been a lawful way to acquire a trade secret, as long as "acquisition of the known product . . . [is] by a fair and honest means, such as purchase of the item on the open market."[26] As the Restatement of Unfair Competition points out, "The owner of a trade secret does not have an exclusive right to possession or use of the secret information. Protection is available only against a wrongful acquisition, use, or disclosure of the trade secret," as when the use or disclosure breaches an implicit or explicit agreement between the parties or when improper means, such as trespass or deceit, are used to obtain the secret. Even when a firm has misappropriated another firm's trade secret, injunctive relief may be limited in duration based in part on the court's estimation of how long it would take a reverse engineer to discover the secret lawfully.

The legal right to reverse-engineer a trade secret is so well-established that courts and commentators have rarely perceived a need to explain the rationale for this doctrine. A rare exception is the 1989 U.S. Supreme Court decision, *Bonito Boats, Inc. v. Thunder Craft Boats, Inc.,* [489 U.S. 141], which characterized reverse engineering as "an essential part of innovation," likely to yield variations on the product that "may lead to significant advances in the field." Moreover, "the competitive reality of reverse engineering may act as a spur to the inventor" to develop patentable ideas. Even when reverse engineering does not lead to additional innovation, the *Bonito Boats* decision suggests it may still promote consumer welfare by providing consumers with a competing product at a lower price.

Further justification for the law's recognition of a right to reverse-engineer likely derives from the fact that the product is purchased in the open market, which confers

---

[26] Unif. Trade Secrets Act § 1, cmt. 2, 14 U.L.A. 437, 438 (1990).

on its owner personal property rights, including the right to take the purchased product apart, measure it, subject it to testing, and the like. The time, money, and energy that reverse engineers invest in analyzing products may also be a way of "earning" rights to the information they learn thereby. Still another justification stems from treating the sale of a product in the open market as a kind of publication of innovations it embodies. This publication dedicates these innovations to the public domain unless the creator has obtained patent protection for them.

Courts have also treated reverse engineering as an important factor in maintaining balance in intellectual property law. Federal patent law allows innovators up to twenty years of exclusive rights to make, use, and sell an invention, but only in exchange for disclosure of significant details about their invention to the public. This deal is attractive in part because if an innovator chooses to protect its invention as a trade secret, such protection may be short-lived if it can be reverse-engineered. If state legislatures tried to make trade secrets immune from reverse engineering, this would undermine federal patent policy because it would "convert the . . . trade secret into a state-conferred monopoly akin to the absolute protection that a federal patent affords."[36] Reverse engineering, then, is an important part of the balance implicit in trade secret law.

Cemen Tech Inc. v. Three D Industries, L.L.C.
753 N.W.2d 1 (Iowa 2008)

Larson, Judge:

When the defendants in this case began to manufacture a cement mixer similar to one manufactured by Cemen Tech (CTI), CTI sued them, alleging [among other claims] misappropriation of trade secrets . . . . The district court granted the defendants' motion for summary judgment on virtually all of the plaintiff's claims, and the plaintiff appealed. We affirm in part, reverse in part, and remand for further proceedings.

I.   Facts and Prior Proceedings.

[Cemen Tech Inc. ("CTI")] is a manufacturer of mobile volumetric concrete mixers—machines designed to mix concrete components at job sites. Defendants Dean Longnecker and David Enos, through their business, Three D Company, L.L.C., were interested in purchasing CTI and, on October 25, 1999, sent a letter of intent to CTI requesting information regarding the business. After a number of letters of intent and nondisclosure and confidentiality agreements, CTI provided Longnecker and Enos with business information, including organizational charts; employee handbooks; a strategic plan; and information on customer deposits, assets, accounts payable, accounts receivable, financial statements, and lists of customers and suppliers.

By the spring of 2001, it became clear that Longnecker and Enos were not going to purchase CTI. On June 5, 2001, CTI terminated Three D Company, L.L.C.'s latest letter of intent. . . .

In July 2001 Brad Luhrs, an employee of CTI, contacted Longnecker about the possibility of leaving CTI and going to work with Longnecker and Enos to start

---

[36] Chicago Lock Co. v. Fanberg, 676 F.2d 400, 405 (9th Cir. 1981). . . .

their own mobile mixer business. By the end of 2001, Dan Jones, Brad Luhrs, Mark Dorman, Dan Pothast, and Scott Longnecker resigned from CTI and began working for Three D Industries developing mobile volumetric concrete mixers in direct competition with CTI.

In January 2002 the defendants exhibited a prototype cement mixer at the World of Concrete show closely resembling CTI's mixer. CTI sued Three D Industries, L.L.C. and eight individual defendants for breach of contract, misappropriation of trade secrets, unfair competition, breach of fiduciary duty, and tortious interference with contract. . . . The district court granted the defendants' motion for summary judgment in part and denied it in part, and CTI appealed. . . .

. . . .

IV.   The Claim of Misappropriation of Trade Secrets.

The principal issue in this case is whether CTI generated a genuine issue of material fact on its claim that the defendants misappropriated CTI's trade secrets, including information regarding the construction of its machines and general information about their manufacture and sale. . . .

A.   Legal Principles. Iowa Code chapter 550 (2001) is Iowa's version of the Uniform Trade Secrets Act. . . . A trade secret is defined as

> information, including but not limited to a formula, pattern, compilation, program, device, method, technique, or process that is both of the following:
> a. Derives independent economic value, actual or potential, from not being generally known to, and not being readily ascertainable by proper means by a person able to obtain economic value from its disclosure or use.
> b. Is the subject of efforts that are reasonable under the circumstances to maintain its secrecy.

Iowa Code § 550.2(4).

The Restatement (Third) of Unfair Competition (1995) [hereinafter Restatement] is intended to be consistent with the Uniform Trade Secrets Act, such as our Code chapter 550, and we rely on it here. See Restatement § 39 cmt. d. . . .

. . . .

B.   Evidence Presented in Resistance to the Motion for Summary Judgment. In this case, CTI contends that its cement volumetric mixer, component parts, manufacturing processes, and supplier and customer information constitute trade secrets. A CTI officer identified numerous processes and design features developed by CTI that were unique, features that CTI had sought to protect by its confidentiality agreements and which it claimed as trade secrets. This evidence is sufficient to generate a fact question regarding the economic value of the information CTI argues is trade-secret protected.

. . . .

The defendants argue that information regarding CTI's volumetric mixer is not a trade secret because it is "readily ascertainable." This argument is based, in large part, on the concept of reverse engineering. ("'Reverse engineering is the process by which a completed process is systematically broken down into its component parts to discover the properties of the product with the goal of gaining the expertise to reproduce the product.'" *Revere Transducers, Inc. v. Deere & Co.*, 595 N.W.2d 751, 775 n. 8 (Iowa 1999) (quoting *Christianson v. Colt Indus. Operating Corp.*, 870 F.2d 1292, 1295 n. 4 (7th Cir. 1989)). However, the fact that information may be obtained by lawful means, including reverse engineering, is not necessarily dispositive of the trade-secret issue.

> The theoretical ability of others to ascertain the information through proper means does not necessarily preclude protection as a trade secret. Trade secret protection remains available unless the information is readily ascertainable by such means. *Thus, if acquisition of the information through an examination of a competitor's product would be difficult, costly, or time-consuming, the trade secret owner retains protection* against an improper acquisition, disclosure, or use.

[Restatement (Third) of Unfair Competition (1995)] § 39 cmt. f (emphasis added).

In CTI's resistance to the defendants' motion for summary judgment, it provided a report by Dr. Bruce Johnson, who holds a Ph.D. in mechanical engineering and who has considerable experience in design and development engineering. Dr. Johnson evaluated the defendants' reverse engineering claim—he inspected the defendants' prototype machine; he inspected CTI's, as well as the defendants', manufacturing facilities; he inspected the drawings of both CTI and the defendants; and he reviewed witness depositions.

Dr. Johnson stated that a representative of the defendant company told him they did not start construction on their machine until November 2001, and the defendants exhibited their fully functioning machine at a trade show in March 2002. Dr. Johnson concluded:

> It would be impossible for anyone, even a trained and experienced engineer, to complete the design process of a complex piece of machinery such as involved in this instance, to the point of having a perfectly functioning prototype, in less than six months without having at his disposal a vast amount of information. In the context of this case, it is my opinion that it would have been impossible for the Defendants collectively to complete the process without access to the proprietary information of Cemen Tech, regardless of whether that information was in written or other form.

Dr. Johnson's report is strong evidence that, though CTI's machine and component parts could be reverse engineered, the difficulty in doing so may not defeat CTI's trade-secret claim.

Even though CTI conceded that the defendants, if given sufficient time, could produce a machine similar to CTI's based on reverse engineering, a fact finder could reasonably conclude that the delay in production of the machine would give CTI a temporal advantage. Even if the period of that advantage would be short, this would not deny trade-secret protection to CTI because neither Iowa Code section 550.2(4) nor section 1(4) of the Uniform Trade Secrets Act "include[s] any requirement relating to the duration of the information's economic value." Restatement § 39 cmt. d. Presumably, the extent of this temporal advantage would be reflected in any damage award.

Finally, the defendants' means of obtaining the information at issue lends credence to CTI's claim that the information was intended to be confidential. Generally,

> [i]nformation that is readily ascertainable by proper means is not protectable as a trade secret, and the acquisition of such information even by improper means is therefore not actionable. However, the accessibility of information, and hence its status as a trade secret, is evaluated in light of the difficulty and cost of acquiring the information by proper means. In some circumstances the actor's decision to employ improper means of acquisition is itself evidence that the information is not readily ascertainable through proper means and is thus protectable as a trade secret. Because of the public interest in deterring the acquisition of information by improper means, doubts regarding the status of information as a trade secret are likely to be resolved in favor of protection when the means of acquisition are clearly improper.

Restatement § 43 cmt. d (citations omitted).

. . . [A] jury could find that Dan Jones and Brad Luhrs gave Three D Industries information they gathered from CTI in the course of strategy and production meetings, including new product ideas, CTI's engineering data, computations, calculations, and certain innovations. In fact, evidence in the record suggests that Luhrs and Jones copied files from a CTI computer onto CDs that were not returned to CTI upon their resignations. The actions of these defendants could reasonably be considered to be improper and, therefore, "evidence that the information is not readily ascertainable through proper means and * * * thus protectable as a trade secret." Restatement § 43 cmt. d. Viewing the evidence in the light most favorable to CTI, we believe there was sufficient evidence to generate a fact question on the trade-secret issue . . . .

. . . .

In summary, we conclude it was error to enter summary judgment on CTI's trade-secret claim . . . . We remand for further proceedings on this issue.

. . . .

Notes & Questions

1. As Samuelson and Scotchmer explain, the ability to reverse engineering contributes in important ways to technological advances. Trade secret protection thus leaves breathing room for reverse engineering in two ways. First, the ability to reverse engineer information can affect whether the law grants trade secret protection for that information in the first place. Second, if someone obtains the information through reverse engineering it does not qualify as a misappropriation.

Some state statutes include reverse engineering within the definition of trade secrets, *see*, *e.g.*, N.C. Gen. Stat. § 66–152(3) ("'Trade secret' means business or
technical information . . . that . . . derives independent actual or potential commercial value from not being generally known or readily ascertainable through independent development or lawful reverse engineering"), while others expressly
exclude reverse engineering as a type of misappropriation. *See*, *e.g.*, Calif. Civil
Code § 3426.1(a) ("[r]everse engineering or independent derivation alone shall
not be considered improper means" of acquiring a trade secret). Be sure to keep
straight these two separate ways in which the reverse-engineerability of the information at issue can be relevant to the trade secret claim.

2. The court in *Cemen Tech* quotes from the Restatement § 43 cmt. d: "[i]nfor-
mation that is readily ascertainable by proper means is not protectable as a trade
secret, and the acquisition of such information even by improper means is therefore not actionable." In what way is the cost-of-errors analysis being employed by
the court?

3. If reverse engineering fuels advances in technology and knowledge, should
entities be able to restrict the freedom to engage in reverse engineering through
contract? Imagine Alpha sells to Beta a product that embodies Alpha's trade secret. The written sales agreement between them states "Beta agrees, as part of the
bargained-for-price of the product, to waive any right to reverse-engineer any
trade secret the product embodies." Beta later reverse engineers Alpha's secret,
using the product. If Alpha sues for breach of contract, should the court enforce
the contract? Would it matter whether Alpha & Beta negotiated the agreement or
if the agreement was a standard form sales agreement that Alpha insists upon for
all its sales?

In both the *Amoco* and *CDI Energy Services* cases the defendants were potential
competitors seeking to use the alleged trade secret information to their advantage.
A different type of trade secret case involves a "spoiler" who desires to destroy
the trade secret's secrecy, but not because of a competitive advantage to be gained.
In the more traditional trade secret case the defendant usually stands to gain by
using the trade secret while also keeping it secret from *other* competitors. Spoiler
cases have a different dynamic at play, one that is often far more damaging for the
trade secret owner. Moreover, the internet has significantly empowered spoilers:

DVD Copy Control Ass'n v. Bunner
10 Cal. Rptr. 3d 185 (Cal. Ct. App. 2004)

PRIMO, JUDGE:

Plaintiff DVD Copy Control Association, Inc. (DVD CCA) sued defendant
Andrew Bunner (Bunner) and others under California's Uniform Trade Secrets
Act (UTSA) (Civ. Code, § 3426 et seq.), seeking an injunction to prevent defend-
ants from using or publishing "DeCSS," a computer program allegedly containing
DVD CCA's trade secrets.

. . . .

I. FACTUAL AND PROCEDURAL BACKGROUND

A. Introduction

. . . Unlike motion pictures on videocassettes, motion pictures contained on DVDs may be copied without perceptible loss of video or audio quality. This aspect of the DVD format makes it particularly susceptible to piracy. For this reason, motion pictures stored on DVDs have been protected from unauthorized use by a content scrambling system referred to as CSS. Simply put, CSS scrambles the data on the disk and then unscrambles it when the disk is played on a compliant DVD player or computer. CSS does not allow the content on the DVD to be copied.

For obvious reasons, the motion picture industry desired to keep the CSS technology a secret. But to make DVD players and computer DVD drives that can unscramble and play a CSS-protected DVD, the manufacturers had to have the CSS "master keys" and an understanding of how the technology works. In an attempt to keep CSS from becoming generally known, the industries agreed upon a restrictive licensing scheme and formed DVD CCA to be the sole licensing entity for CSS. Under the CSS licensing scheme, each licensee receives a different master key to incorporate into its equipment and sufficient technical know how to permit the manufacture of a DVD-compliant device. All licensees must agree to maintain the confidentiality of CSS.

In spite of these efforts to maintain the secrecy of CSS, DeCSS appeared on the Internet sometime in October 1999 and rapidly spread to other Web sites, including those of the defendants. According to DVD CCA, DeCSS incorporates trade secret information that was obtained by reverse engineering CSS in breach of a license agreement. DVD CCA alleges that DeCSS allows users to illegally pirate the copyrighted motion pictures contained on DVDs, "activity which is fatal to the DVD video format and the hundreds of computer and consumer electronics companies whose businesses rely on the viability of this digital format."

DVD CCA filed the instant complaint for injunctive relief on December 27, 1999, alleging that Bunner and the other defendants had misappropriated trade secrets by posting DeCSS or links to DeCSS on their Web sites, knowing that DeCSS had been created by improper means. The requested injunctive relief sought to prevent defendants from using DeCSS, from disclosing DeCSS or other proprietary CSS technology on their Web sites or elsewhere, and from linking their Web sites to other Web sites that disclosed DeCSS or other CSS technology.

. . . .

B. The Factual Record

. . . John Hoy, President of DVD CCA explained that DeCSS first appeared on the Internet on October 6, 1999. That first posting was in machine-readable form referred to as object code.[5] The DeCSS source code was posted about three weeks later, on or around October 25, 1999. Hoy declared that both postings contain CSS

---

[5] To oversimplify, object code is a set of instructions comprised of strings of 1's and 0's. The same instructions written in programming language is referred to as source code. To be executable by a computer, source code must be translated into object code.

technology and the master key that had been assigned to DVD CCA's licensee, Xing Technology Corporation (Xing), a manufacturer of computer DVD drives. The intended inference is that DeCSS was created, at least in part, by reverse engineering the Xing software. Since Xing licensed its software pursuant to an agreement that prohibits reverse engineering, Hoy concludes that the CSS technology contained in DeCSS was "obtained in violation of the specific provision in the Xing end-user license 'click wrap' agreement which prohibits reverse engineering." Hoy stated on information and belief that Jon Johansen, a resident of Norway, was the author of the program.

Well before DeCSS was released on the Internet, a number of people had become interested in unraveling the CSS security system. Users of the Linux computer operating system had organized a forum dedicated to finding a way to override CSS. Apparently DVD CCA had not licensed CSS to anyone making DVD drives for the Linux system, so that computers using Linux were incapable of playing DVDs. CSS was widely analyzed and discussed in the academic cryptography community. Another exchange of information took place on www.slashdot.org (Slashdot), a news Web site popular with computer programmers. As early as July 1999 comments on Slashdot revealed a worldwide interest in cracking CSS. The gist of these communications is contained in the following excerpts of a discussion that took place on July 15, 1999:

> "Yes, it is true, we have now all needed parts for software decoding of DVDs, but any software doing so will be illegal and/or non-free. [¶] ... The information about CSS was obtained by reverse engineering some DVD software decoder."
>
> "This code was released before anyone checked into the legal end of things. ... Best idea now is to download the code. Get it spread around as widely as possible. It may not be able to be used legally when all is said and done, but at least it will be out there for others to work with."
>
> "Well, it might not be the most ethical thing on earth, but if the appropriate algorithms were to be found just lying on the web, once the coders have seen them, they don't have a 'forget' button for their brains. ..."

Bunner first became aware of DeCSS on or about October 26, 1999 as a result of reading and participating in discussions on Slashdot. Bunner explained that he is a parttime user of Linux and supports its acceptance as a viable alternative to established computer operating systems such as Microsoft Windows. Bunner thought DeCSS would be useful to other Linux users. He claimed that at the time he posted the information on his Web site he had no information to suggest that the program contained any trade secrets or that it involved the misappropriation of trade secrets. There is no evidence as to the date Bunner first posted the program on his Web site.

Counsel representing the motion picture industry had become aware of the DeCSS posting on October 25, 1999. Beginning November 4, 1999, counsel sent letters to Web site operators and Internet service providers hosting Web pages that contained DeCSS or links to DeCSS and demanded the information be taken down. Sixty-six such letters were sent between November 4 and November 23, 1999. None of the letters listed in counsel's declaration were addressed to Bunner or to his Web site address. About 25 of the 66 sites were taken down. DeCSS was also removed from Johansen's Web site on or around November 8, 1999, but a link to DeCSS reappeared on the same site on or around December 11, 1999.

Meanwhile, the news that the CSS encryption system had been penetrated made headlines in Internet news magazines. *Wired News* ran several articles in the first days of November 1999 announcing the development of DeCSS. An article on November 4, 1999 said: "It shouldn't be surprising that an awful lot of people are upset at this week's Wired News reports about a utility to remove DVD security. But it's out there and people are using it." An article on *eMedia* around the same time explained that DeCSS was "available for free download from several sites on the World Wide Web."

DVD CCA filed suit on December 27, 1999, naming as defendants the operators of every infringing Web site it could identify. A hearing for a temporary restraining order was to be held the following day. In support of that application, DVD CCA informed the court that since October 25, 1999, DeCSS had been displayed on or linked to at least 118 Web pages in 11 states and 11 countries throughout the world and that approximately 93 Web pages continued to publish infringing information.

The lawsuit outraged many people in the computer programming community. A campaign of civil disobedience arose by which its proponents tried to spread the DeCSS code as widely as possible before trial. Some of the defendants simply refused to take their postings down. Some people appeared at the courthouse on December 28, 1999 to pass out diskettes and written fliers that supposedly contained the DeCSS code. They made and distributed tee shirts with parts of the code printed on the back. There were even contests encouraging people to submit ideas about how to disseminate the information as widely as possible.

C. The Trial Court's Findings

The trial court issued the preliminary injunction based upon the following findings: First, CSS is DVD CCA's trade secret and for nearly three years prior to the posting of DeCSS on defendants' Web sites, DVD CCA had exerted reasonable efforts to maintain the secrecy of CSS. The court stated that trade secret status should not be deemed destroyed merely because the information was posted on the Internet, because, "[t]o hold otherwise would do nothing less than encourage misappropriaters [*sic*] of trade secrets to post the fruits of their wrongdoing on the Internet as quickly as possible and as widely as possible thereby destroying a trade secret forever."

Second, the trial court found that the evidence was "fairly clear" that the trade secret had been obtained through a reverse engineering procedure that violated the terms of a license agreement and, based upon some defendants' boasting about their disrespect for the law, it could be inferred that all defendants knew that the trade secret had been obtained through improper means.

Third, the balancing of equities favored DVD CCA. The court determined that while the harm to defendants in being compelled to remove trade secret information from their Web sites was "truly minimal," the current and prospective harm to DVD CCA was irreparable in that DVD CCA would lose the right to protect CSS as a trade secret and to control unauthorized copying of DVD content. The court pointed out: "once this information gets into the hands of an innocent party, the Plaintiff loses their [*sic*] ability to enjoin the use of their [*sic*] trade secret. If the court does not immediately enjoin the posting of this proprietary information, the Plaintiff's right to protect this information as secret will surely be lost, given the current power of the Internet to disseminate information and the Defendants' stated determination to do so." The trial court did not expressly consider the harm to defendants' First Amendment rights.

II. DISCUSSION

A preliminary injunction is appropriate to maintain the status quo pending trial of the merits. The UTSA expressly provides for an injunction preventing the disclosure of a trade secret.[*] (§ 3426.2.) . . .

. . . .

In order to obtain an injunction prohibiting disclosure of an alleged trade secret, the plaintiff's first hurdle is to show that the information it seeks to protect is indeed a trade secret. The UTSA defines a trade secret as "information . . . that: [¶] (1) Derives independent economic value, actual or potential, from not being generally known . . . ; and [¶] (2) Is the subject of efforts that are reasonable under the circumstances to maintain its secrecy." (§ 3426.1, subd. (d).) In short, the test for a trade secret is whether the matter sought to be protected is information (1) that is valuable because it is unknown to others and (2) that the owner has attempted to keep secret. The first element is the crucial one here: in order to qualify as a trade secret, the information "must be secret, and must not be of public knowledge or of a general knowledge in the trade or business." (*Kewanee Oil Co. v. Bicron Corp.*, 416 U.S. [470,] 475 [(1974)].)

The secrecy requirement is generally treated as a relative concept and requires a fact-intensive analysis. Widespread, anonymous publication of the information over the Internet may destroy its status as a trade secret. (*Religious Technology Center v. Netcom On-Line Com.*, (N.D. Cal. 1995) 923 F. Supp. 1231, 1256.) The concern is whether the information has retained its value to the creator in spite of the publication. (See Rest. 3d Unfair Competition, § 39, com. f, p. 431.) Publication on the Internet does not necessarily destroy the secret if the publication is sufficiently obscure or transient or otherwise limited so that it does not become generally known to the relevant people, i.e., potential competitors or other persons to whom the information would have some economic value.

---

[*] [*Eds. Note*: To obtain a preliminary injunction, the plaintiff must demonstrate, among other things, a likelihood of success on the substantive merits of his or her claim. Chapter 7 addresses, in detail, the requirements for both preliminary and permanent injunctions.]

In the instant matter, the secrecy element becomes important at two points. First, if the allegedly proprietary information contained in DeCSS was already public knowledge when Bunner posted the program to his Web site, Bunner could not be liable for misappropriation by republishing it because he would not have been disclosing a trade secret. Second, even if the information was not generally known when Bunner posted it, if it had become public knowledge by the time the trial court granted the preliminary injunction, the injunction (which only prohibits *disclosure*) would have been improper because DVD CCA could not have demonstrated interim harm.

The trial court did not make an express finding that the proprietary information contained in DeCSS was not generally known at the time Bunner posted it. Indeed, there is no evidence to support such a finding. Bunner first became aware of DeCSS on or around October 26, 1999. But there is no evidence as to when he actually posted it. Indeed, neither Bunner's name nor his Web site address appears among the 66 cease and desist letters counsel sent in November. We do know, however, that by the first week in November Internet news magazines were publicizing the creation of DeCSS and informing readers that the program was available to be downloaded for free on the Internet. As early as July 1999 people in the computer programming community were openly discussing the fact that the CSS code had been reverse engineered and were brainstorming ways to be able to use it legally. That means that when DeCSS appeared in October 1999 there was a worldwide audience ready and waiting to download and repost it.

DVD CCA urges us, in effect, to ignore the fact that the allegedly proprietary information may have been distributed to a worldwide audience of millions prior to Bunner's first posting. According to DVD CCA, so long as Bunner knew or should have known that the information he was republishing was obtained by improper means, he cannot rely upon the general availability of the information to the rest of the world to avoid application of the injunction to him. In support of this position, DVD CCA contends that the denial of an injunction would offend the public policies underlying trade secret law, which are to enforce a standard of commercial ethics, to encourage research and invention, and to protect the owner's moral entitlement to the fruits of his or her labors. DVD CCA points out that these policies are advanced by making sure that those who misappropriate trade secrets do not avoid "judicial sanction" by making the secret widely available.

The first problem with this argument is that by denying a preliminary injunction the court does not per se protect a wrongdoer from judicial sanction, which in most cases would come following trial on the merits.

Second, the evidence in this case is very sparse with respect to whether the offending program was actually created by improper means. Reverse engineering alone is not improper means. Here the creator is believed to be a Norwegian resident who probably had to breach a Xing license in order to access the information he needed. We have only very thin circumstantial evidence of when, where, or how this actually happened or whether an enforceable contract prohibiting reverse engineering was ever formed.

Finally, assuming the information was originally acquired by improper means, it does not necessarily follow that once the information became *publicly* available that everyone else would be liable under the trade secret laws for re-publishing it simply because they knew about its unethical origins. In a case that receives widespread publicity, just about anyone who becomes aware of the contested information would also know that it was allegedly created by improper means. Under DVD CCA's construction of the law, in such a case the general public could theoretically be liable for misappropriation simply by disclosing it to someone else. This is not what trade secret law is designed to do.

It is important to point out that we do not assume that the alleged trade secrets contained in DeCSS became part of the public domain simply by having been published on the Internet. Rather, the evidence demonstrates that in this case, the initial publication was quickly and widely republished to an eager audience so that DeCSS and the trade secrets it contained rapidly became available to anyone interested in obtaining them. Further, the record contains no evidence as to when in the course of the initial distribution of the offending program Bunner posted it. Thus, DVD CCA has not shown a likelihood that it will prevail on the merits of its claim of misappropriation against Bunner.

. . . .

As the trial court clearly explained, the preliminary injunction prohibiting disclosure was intended to protect the trade secret. Therefore, even if Bunner was liable for misappropriation, if the information had since become generally known, a preliminary injunction prohibiting disclosure would have done nothing to protect the secret because the secret would have ceased to exist. . . .

This case is distinguishable from *Underwater Storage, Inc. v. United States Rubber Co.* (D.C. Cir. 1966) 371 F.2d 950, 955 (*Underwater Storage*), which DVD CCA cites for the proposition that "a misappropriator or his privies can[not] 'baptize' their wrongful actions by general publication of the secret." In *Underwater Storage* the defendant had misappropriated trade secrets and used them to develop a storage system for the United States Navy. After completing its work for the Navy, the defendant later published the alleged trade secrets, presumably representing them as its own technical know-how. In resolving a statute of limitations question, the appellate court rejected the contention that the subsequent publication of the secret prevented the plaintiff from seeking compensation from the original misappropriator. The court stated: "Once the secret is out, the rest of the world may well have a right to copy it at will; but this should not protect the misappropriator or his privies." (*Id.* at p. 955.) *Underwater Storage* was not concerned with the issuance of a preliminary injunction. The information was concededly public when the case was filed. The court's holding was that under the circumstances the defendant could still be liable in damages for his previous misappropriation. That holding does not alter the conclusion that a preliminary injunction cannot be used to protect a secret if there is no secret left to protect.

One of the analytical difficulties with this case is that it does not fit neatly into classic business or commercial law concepts. The typical defendant in a trade secret case is a competitor who has misappropriated the plaintiff's business secret for profit in a business venture. In that scenario, the defendant has as much interest as the plaintiff has in keeping the secret away from good faith competitors and out of the public domain. But here, according to DVD CCA it has no good faith competitors. And the alleged misappropriators not only wanted the information for themselves, they also wanted the whole world to have it.

We concur with the concerns expressed by Judge Whyte in his opinion in *Religious Technology Center, supra*, 923 F. Supp. at page 1256: "The court is troubled by the notion that any Internet user, . . . can destroy valuable intellectual property rights by posting them over the Internet, especially given the fact that there is little opportunity to screen postings before they are made. Nonetheless, one of the Internet's virtues, that it gives even the poorest individuals the power to publish to millions of readers can also be a detriment to the value of intellectual property rights. The anonymous (or judgment proof) defendant can permanently destroy valuable trade secrets, leaving no one to hold liable for the misappropriation."

There is little question that such behavior is unethical and that it probably violates other laws. But that which is in the public domain cannot be removed by action of the states under the guise of trade secret protection. (*Kewanee Oil, supra*, 416 U.S. at p. 481.)

. . . .

Notes & Questions

1. Be sure you understand the two different ways in which the public distribution of CSS affected the plaintiff's claim. Was there anything else the plaintiff could have done to protect against what happened to its trade secret? This case demonstrates one of the ways in which relying on trade secret protection for valuable information is risky: once the secret information is no longer secret, even if that revelation happens because of a misappropriation, the trade secret owner will not be able to prevail on a claim against a later distributor or user of the previously secret, but now public, information.

2. Sometimes it may be difficult to distinguish a spoiler who we want to sanction for a wrongful disclosure from a reporter engaged in scoop-driven journalism. For example, in the technology industry, many web sites and blogs exist to report on "breaking news" or new products, or even new product features, that are yet to be released. For these sites, being the first to report some piece of information results in increased readership and greater influence in the tech world, as well as higher advertising revenues. The motivations may be more laudable, but the damage from disclosure is the same. What facts should an internet reporter look for in determining whether information she is about to report qualifies for trade secret protection?

3. In 2001 Nevada amended its trade secret act, adding the following provision:

Misappropriation and posting or dissemination on Internet: Effect.
A trade secret that is misappropriated and posted, displayed or otherwise
disseminated on the Internet shall be deemed to remain a trade secret as
defined in [the Act] and not to have "ceased to exist" for the purposes of
[terminating an injunction] if:

   1. The owner, within a reasonable time after discovering that the trade
secret has been misappropriated and posted, displayed or otherwise disseminated on the Internet, obtains an injunction or order issued by a court
requiring that the trade secret be removed from the Internet; and

   2. The trade secret is removed from the Internet within a reasonable
time after the injunction or order requiring removal of the trade secret is
issued by the court.

Nev. Rev. Stat. § 600A.055. How does the Nevada approach compare to the approach taken by the court in *Bunner*? For an interesting discussion of the problems
of internet disclosure of trade secrets see, Elizabeth Rowe, *Introducing a Takedown
for Trade Secrets on the Internet,* 2007 Wis. L. Rev. 1041.

   4. In *Underwater Storage, Inc. v. United States Rubber Co.,* 371 F.2d 950 (D.C.
Cir. 1966), discussed in the *Bunner* opinion, the court recognized the defendant's
liability for misappropriation of trade secrets despite widespread publication.
What is the difference between that case and *Bunner*? Was it merely the identity
of the original person to post the trade secret on the web? Or, was it the remedy
being sought? Should Mr. Bunner still be liable to the DVD CCA for damages?

B. Reasonable Efforts to Maintain Secrecy

Rockwell Graphic Systems, Inc. v. DEV Industries, Inc.
925 F.2d 174 (7th Cir. 1991)

POSNER, JUDGE:

   This is a suit for misappropriation of trade secrets. Rockwell Graphic Systems,
a manufacturer of printing presses used by newspapers, and of parts for those
presses, brought the suit against DEV Industries, a competing manufacturer, and
against the president of DEV, who used to be employed by Rockwell. . . . The district judge granted summary judgment for the defendants upon the recommendation of a magistrate who concluded that Rockwell had no trade secrets because it
had failed to take reasonable precautions to maintain secrecy. . . .

   When we said that Rockwell manufactures both printing presses and replacement parts for its presses—"wear parts" or "piece parts," they are called—we
were speaking approximately. Rockwell does not always manufacture the parts itself. Sometimes when an owner of one of Rockwell's presses needs a particular
part, or when Rockwell anticipates demand for the part, it will subcontract the
manufacture of it to an independent machine shop, called a "vendor" by the parties. When it does this it must give the vendor a "piece part drawing" indicating

materials, dimensions, tolerances, and methods of manufacture. Without that information the vendor could not manufacture the part. Rockwell has not tried to patent the piece parts. It believes that the purchaser cannot, either by inspection or by "reverse engineering" (taking something apart in an effort to figure out how it was made), discover how to manufacture the part; to do that you need the piece part drawing, which contains much information concerning methods of manufacture, alloys, tolerances, etc. that cannot be gleaned from the part itself. So Rockwell tries—whether hard enough is the central issue in the case—to keep the piece part drawings secret, though not of course from the vendors; they could not manufacture the parts for Rockwell without the drawings. DEV points out that some of the parts are for presses that Rockwell no longer manufactures. But as long as the presses are in service—which can be a very long time—there is a demand for replacement parts.

Rockwell employed Fleck and Peloso in responsible positions that gave them access to piece part drawings. Fleck left Rockwell in 1975 and three years later joined DEV as its president. Peloso joined DEV the following year after being fired by Rockwell when a security guard caught him removing piece part drawings from Rockwell's plant. This suit was brought in 1984, and pretrial discovery by Rockwell turned up 600 piece part drawings in DEV's possession, of which 100 were Rockwell's. DEV claimed to have obtained them lawfully, either from customers of Rockwell or from Rockwell vendors, contrary to Rockwell's claim that either Fleck and Peloso stole them when they were employed by it or DEV obtained them in some other unlawful manner, perhaps from a vendor who violated his confidentiality agreement with Rockwell. Thus far in the litigation DEV has not been able to show which customers or vendors lawfully supplied it with Rockwell's piece part drawings.

The defendants persuaded the magistrate and the district judge that the piece part drawings weren't really trade secrets at all, because Rockwell made only perfunctory efforts to keep them secret. Not only were there thousands of drawings in the hands of the vendors; there were thousands more in the hands of owners of Rockwell presses, the customers for piece parts. The drawings held by customers, however, are not relevant. They are not piece part drawings, but assembly drawings. (One piece part drawing in the record is labeled "assembly," but as it contains dimensions, tolerances, and other specifications it is really a piece part drawing, despite the label.) An assembly drawing shows how the parts of a printing press fit together for installation and also how to integrate the press with the printer's other equipment. Whenever Rockwell sells a printing press it gives the buyer assembly drawings as well. These are the equivalent of instructions for assembling a piece of furniture. Rockwell does not claim that they contain trade secrets. It admits having supplied a few piece part drawings to customers, but they were piece part drawings of obsolete parts that Rockwell has no interest in manufacturing and of a safety device that was not part of the press as originally delivered but that its customers were clamoring for; more to the point, none of these drawings is among those that Rockwell claims DEV misappropriated.

    . . . .

[Defendant DEV's main argument was that Rockwell was "impermissibly sloppy in its efforts to keep the piece part drawings secret."] On this, the critical, issue, the record shows the following. (Because summary judgment was granted to DEV, we must construe the facts as favorably to Rockwell as is reasonable to do.) Rockwell keeps all its engineering drawings, including both piece part and assembly drawings, in a vault. Access not only to the vault, but also to the building in which it is located, is limited to authorized employees who display identification. These are mainly engineers, of whom Rockwell employs 200. They are required to sign agreements not to disseminate the drawings, or disclose their contents, other than as authorized by the company. An authorized employee who needs a drawing must sign it out from the vault and return it when he has finished with it. But he is permitted to make copies, which he is to destroy when he no longer needs them in his work. The only outsiders allowed to see piece part drawings are the vendors (who are given copies, not originals). They too are required to sign confidentiality agreements, and in addition each drawing is stamped with a legend stating that it contains proprietary material. Vendors, like Rockwell's own engineers, are allowed to make copies for internal working purposes, and although the confidentiality agreement that they sign requires the vendor to return the drawing when the order has been filled, Rockwell does not enforce this requirement. The rationale for not enforcing it is that the vendor will need the drawing if Rockwell reorders the part. Rockwell even permits unsuccessful bidders for a piece part contract to keep the drawings, on the theory that the high bidder this round may be the low bidder the next. But it does consider the ethical standards of a machine shop before making it a vendor, and so far as appears no shop has ever abused the confidence reposed in it.

The mere fact that Rockwell gave piece part drawings to vendors—that is, disclosed its trade secrets to "a limited number of outsiders for a particular purpose"—did not forfeit trade secret protection. *A.H. Emery Co. v. Marcan Products Corp.*, 389 F.2d 11, 16 (2d Cir. 1968). On the contrary, such disclosure, which is often necessary to the efficient exploitation of a trade secret, imposes a duty of confidentiality on the part of the person to whom the disclosure is made. . . . But with 200 engineers checking out piece part drawings and making copies of them to work from, and numerous vendors receiving copies of piece part drawings and copying them, tens of thousands of copies of these drawings are floating around outside Rockwell's vault, and many of these outside the company altogether. Although the magistrate and the district judge based their conclusion that Rockwell had not made adequate efforts to maintain secrecy in part at least on the irrelevant fact that it took no efforts at all to keep its assembly drawings secret, DEV in defending the judgment that it obtained in the district court argues that Rockwell failed to take adequate measures to keep even the piece part drawings secret. Not only did Rockwell not limit copying of those drawings or insist that copies be returned; it did not segregate the piece part drawings from the assembly drawings and institute more secure procedures for the former. So Rockwell could have done more to maintain the confidentiality of its piece part drawings than it did, and we must decide whether its failure to do more was so plain a breach of the obligation of a trade secret owner to make reasonable efforts to maintain secrecy as to justify the entry of summary judgment for the defendants.

The requirement of reasonable efforts has both evidentiary and remedial significance, and this regardless of which of the two different conceptions of trade secret protection prevails. . . . The first and more common merely gives a remedy to a firm deprived of a competitively valuable secret as the result of an independent legal wrong, which might be conversion or other trespass or the breach of an employment contract or of a confidentiality agreement. Under this approach, because the secret must be taken by improper means for the taking to give rise to liability, Ill. Rev. Stat. ch. 140, PP 352(a), (b)(1), (2)(A), (B); Restatement of Torts § 757 (1939), the only significance of trade secrecy is that it allows the victim of wrongful appropriation to obtain damages based on the competitive value of the information taken. The second conception of trade secrecy, illustrated by *E.I. duPont de Nemours & Co. v. Christopher*, 431 F.2d 1012 (5th Cir. 1970) . . . , is that "trade secret" picks out a class of socially valuable information that the law should protect even against nontrespassory or other lawful conduct—in *Christopher*, photographing a competitor's roofless plant from the air while not flying directly overhead and hence not trespassing or committing any other wrong independent of the appropriation of the trade secret itself. . . .

Since, however, the opinion in *Christopher* describes the means used by the defendant as "improper," 431 F.2d at 1015-17, which is also the key to liability under the first, more conventional conception of trade secret protection, it is unclear how distinct the two conceptions really are. It is not as if *Christopher* proscribes all efforts to unmask a trade secret. It specifically mentions reverse engineering as a proper means of doing so. *Id.* at 1015. This difference in treatment is not explained, but it may rest on the twofold idea that reverse engineering involves the use of technical skills that we want to encourage, and that anyone should have the right to take apart and to study a product that he has bought.

It should be apparent that the two different conceptions of trade secret protection are better described as different emphases. The first emphasizes the desirability of deterring efforts that have as their sole purpose and effect the redistribution of wealth from one firm to another. The second emphasizes the desirability of encouraging inventive activity by protecting its fruits from efforts at appropriation that are, indeed, sterile wealth-redistributive—not productive—activities. The approaches differ, if at all, only in that the second does not limit the class of improper means to those that fit a preexisting pigeonhole in the law of tort or contract or fiduciary duty—and it is by no means clear that the first approach assumes a closed class of wrongful acts, either.

Under the first approach, at least if narrowly interpreted so that it does not merge with the second, the plaintiff must prove that the defendant obtained the plaintiff's trade secret by a wrongful act, illustrated here by the alleged acts of Fleck and Peloso in removing piece part drawings from Rockwell's premises without authorization, in violation of their employment contracts and confidentiality agreements, and using them in competition with Rockwell. Rockwell is unable to prove directly that the 100 piece part drawings it got from DEV in discovery were stolen by Fleck and Peloso or obtained by other improper means. But if it can show that the probability that DEV could have obtained them otherwise—that is, without engaging in wrongdoing—is slight, then it will have taken a giant step toward proving what it must prove in order to recover under the first theory of trade secret protection. The greater the precautions that Rockwell took to maintain the secrecy of the piece part drawings, the lower the probability that DEV obtained them properly and the higher the probability that it obtained them through a wrongful act; the owner had taken pains to prevent them from being obtained otherwise.

Under the second theory of trade secret protection, the owner's precautions still have evidentiary significance, but now primarily as evidence that the secret has real value. For the precise means by which the defendant acquired it is less important under the second theory, though not completely unimportant; remember that even the second theory allows the unmasking of a trade secret by some means, such as reverse engineering. If Rockwell expended only paltry resources on preventing its piece part drawings from falling into the hands of competitors such as DEV, why should the law, whose machinery is far from costless, bother to provide Rockwell with a remedy? The information contained in the drawings cannot have been worth much if Rockwell did not think it worthwhile to make serious efforts to keep the information secret.

The remedial significance of such efforts lies in the fact that if the plaintiff has allowed his trade secret to fall into the public domain, he would enjoy a windfall if permitted to recover damages merely because the defendant took the secret from him, rather than from the public domain as it could have done with impunity. It would be like punishing a person for stealing property that he believes is owned by another but that actually is abandoned property. If it were true, as apparently it is not, that Rockwell had given the piece part drawings at issue to customers, and it had done so without requiring the customers to hold them in confidence, DEV could have obtained the drawings from the customers without committing any wrong. The harm to Rockwell would have been the same as if DEV had stolen the drawings from it, but it would have had no remedy, having parted with its rights to the trade secret. This is true whether the trade secret is regarded as property protected only against wrongdoers or (the logical extreme of the second conception, although no case—not even *Christopher*—has yet embraced it and the patent statute may preempt it) as property protected against the world. In the first case, a defendant is perfectly entitled to obtain the property by lawful conduct if he can, and he can if the property is in the hands of persons who themselves committed no wrong to get it. In the second case the defendant is perfectly entitled to obtain the property if the plaintiff has abandoned it by giving it away without restrictions.

It is easy to understand therefore why the law of trade secrets requires a plaintiff to show that he took reasonable precautions to keep the secret a secret. If analogies are needed, one that springs to mind is the duty of the holder of a trademark to take reasonable efforts to police infringements of his mark, failing which the mark is likely to be deemed abandoned, or to become generic or descriptive (and in either event be unprotectable). The trademark owner who fails to police his mark both shows that he doesn't really value it very much and creates a situation in which an infringer may have been unaware that he was using a proprietary mark because the mark had drifted into the public domain, much as DEV contends Rockwell's piece part drawings have done.

But only in an extreme case can what is a "reasonable" precaution be determined on a motion for summary judgment, because the answer depends on a balancing of costs and benefits that will vary from case to case and so require estimation and measurement by persons knowledgeable in the particular field of endeavor involved. On the one hand, the more the owner of the trade secret spends on preventing the secret from leaking out, the more he demonstrates that the secret has real value deserving of legal protection, that he really was hurt as a result of the misappropriation of it, and that there really was misappropriation. On the other hand, the more he spends, the higher his costs. The costs can be indirect as well as direct. The more Rockwell restricts access to its drawings, either by its engineers or by the vendors, the harder it will be for either group to do the work expected of it. Suppose Rockwell forbids any copying of its drawings. Then a team of engineers would have to share a single drawing, perhaps by passing it around or by working in the same room, huddled over the drawing. And how would a vendor be able to make a piece part—would Rockwell have to bring all that work in house? Such reconfigurations of patterns of work and production are far from costless; and therefore perfect security is not optimum security.

There are contested factual issues here, bearing in mind that what is reasonable is itself a fact for purposes of Rule 56 of the civil rules. Obviously Rockwell took some precautions, both physical (the vault security, the security guards—one of whom apprehended Peloso *in flagrante delicto*) and contractual, to maintain the confidentiality of its piece part drawings. Obviously it could have taken more precautions. But at a cost, and the question is whether the additional benefit in security would have exceeded that cost. We do not suggest that the question can be answered with the same precision with which it can be posed, but neither can we say that no reasonable jury could find that Rockwell had done enough and could then go on to infer misappropriation from a combination of the precautions Rockwell took and DEV's inability to establish the existence of a lawful source of the Rockwell piece part drawings in its possession.

This is an important case because trade secret protection is an important part of intellectual property, a form of property that is of growing importance to the competitiveness of American industry. Patent protection is at once costly and temporary, and therefore cannot be regarded as a perfect substitute. If trade secrets are protected only if their owners take extravagant, productivity-impairing measures to maintain their secrecy, the incentive to invest resources in discovering more efficient methods of production will be reduced, and with it the amount of invention. And given the importance of the case we must record our concern at the brevity of the district court's opinion granting summary judgment (one and a half printed pages). Brevity is the soul of wit, and all that, and the district judge did have the benefit of a magistrate's opinion; but it is vital that commercial litigation not appear to be treated as a stepchild in the federal courts. The future of the nation depends in no small part on the efficiency of industry, and the efficiency of industry depends in no small part on the protection of intellectual property.

The judgment is reversed and the case remanded to the district court for further proceedings consistent with this opinion. . . .

Notes & Questions

1. How does the "reasonable efforts to maintain secrecy" requirement differ from the "not readily ascertainable by proper means" requirement? Why should a state require both? Do the two different conceptions of trade secret protection discussed by Judge Posner influence whether a state should have both requirements?

2. How would you advise a client concerning how to determine whether it was taking reasonable efforts to maintain the secrecy of its trade secrets? What facts about the client's business would you need to know? Recall the *CDI Energy Services* case from earlier in this Section. What facts did the court cite to support its conclusion that the plaintiff had failed to take reasonable efforts to protect its customer lists?

3. In a portion of the *Rockwell* opinion that we omitted above, the court discussed DEV's argument that Rockwell effectively forfeited trade secret protection for its piece part drawings by stamping its admittedly non-secret assembly drawings with the same confidentiality legend it used on the piece part drawings. The court rejected the argument, concluding that such a doctrine "would make no sense." 925 F.2d at 176. (Judge Posner does not mince words.) First, even if Rockwell were, in a sense, overclaiming trade secret protection by stamping non-secret documents *Confidential*, "there are any number of innocent explanations for Rockwell's action . . . such as an excess of caution, uncertainty as to the scope of trade secret protection, concern that clerical personnel will not always be able to distinguish between assembly and piece part drawings at a glance, and the sheer economy of a uniform policy." *Id*. at 176-77. Second, adopting DEV's forfeiture doctrine "would place the owner of trade secrets on the razor's edge. If he stamped 'confidential' on every document in sight, he would" risk forfeiture; "[b]ut if he did not stamp confidential on every document he would lay himself open to an accusation that he was sloppy about maintaining secrecy." *Id*. at 177.

4. Recall the excerpt from *DVD Copy Control Ass'n v. Bunner*, 10 Cal. Rptr. 3d 185 (Cal. Ct. App. 2004) included in Section I.A. In that case, Plaintiff had argued that knowledge of CSS initially was not obtained through proper means because the only way it could have been obtained was as a result of a breach of "click-wrap" End User License Agreement (EULA). The court expressed some doubt about "whether an enforceable contract prohibiting reverse engineering was ever formed." If the elements of contract formation are shown, should a trade secret owner who distributes a product with reverse-engineerable trade secret attributes, but who prohibits the reverse engineering through contract, be able to prevail on a claim of trade secret infringement? Does such "distribution-with-contract" constitute reasonable efforts to keep the information secret? If the trade secret owner sued for breach of contract, should a court find such contractual prohibitions on

reverse engineering unenforceable on public policy grounds? Remember, a breach claim can be brought only against those in privity of contract, but a claim of trade secret misappropriate does not require privity. Review question 4 following the *Cemen Tech Inc.* case. Has your answer to that question changed?

5. Litigating a trade secret case involves some risk of disclosure of the trade secret in the proceeding itself. The UTSA requires courts to "preserve the secrecy of an alleged trade secret by reasonable means, which may include granting protective orders in connection with discovery proceedings, holding in-camera hearings, sealing the records of the action, and ordering any person involved in the litigation not to disclose an alleged trade secret without prior court approval." USTA § 5. Crafting appropriate strategies and protective orders is an important part of trade secret litigation. Nonetheless, the more people who know of a trade secret, the greater the risk of harmful disclosure. This is a consideration that must be weighed before initiating any trade secret litigation.

6. Private parties often submit business information to government bodies for regulatory compliance. When doing so, they need to ensure that they take reasonable steps to protect the secrecy of information for which they intend to preserve trade secret status. This task is made more complex by the public's right to access government information by means of, *e.g.*, the federal Freedom of Information Act ("FOIA"), 5 U.S.C. § 552. Under FOIA, a federal agency is not obliged to disclose "matters that are . . . trade secrets, and commercial or financial information obtained from a person and privileged or confidential." 5 U.S.C. § 552(b)(4). Different agencies take different approaches to identifying which information that they have received falls under the FOIA trade secret exemption. For example, the EPA provides that "[a] business which is submitting information to EPA may assert a business confidentiality claim covering the information by placing on (or attaching to) the information, at the time it is submitted to EPA, a cover sheet, stamped or typed legend, or other suitable form of notice employing language such as *trade secret, proprietary,* or *company confidential.*" 40 C.F.R. § 2.203(b). If you practice before a government agency, you will no doubt want to learn how it handles trade secret information from private parties. The government's ability to keep information secret by citing trade secret law, while helpful to industries that must submit such information to comply with various laws and regulations, raises serious policy considerations about public notice that should not be overlooked. *See* David Levine, *Secrecy and Unaccountability: Trade Secrets In Our Public Infrastructure*, 59 Fla. L. Rev. 135 (2007).

II. MISAPPROPRIATION

A. CIVIL MISAPPROPRIATION

Determining what type of information can be protected as a trade secret is only the first part of understanding legal protection for trade secrets. Not all uses or discoveries of information that qualifies as a trade secret will constitute actionable misappropriation.

Most states have passed some version of the Uniform Trade Secrets Act. This model law defines "misappropriation" in a way that embraces all the typical fact patterns constituting misappropriation. Section 1 of the UTSA provides:

> "Misappropriation" means:
>> (i) Acquisition of a trade secret of another by a person who knows or has reason to know that the trade secret was acquired by improper means; or
>> (ii) Disclosure or use of a trade secret of another without express or implied consent by a person who
>>> (A) used improper means to acquire knowledge of the trade secret; or
>>> (B) at the time of disclosure or use, knew or had reason to know that the knowledge of the trade secret was:
>>>> (I) Derived from or through a person who had utilized improper means to acquire it;
>>>> (II) Acquired under circumstances giving rise to a duty to maintain its secrecy or limit its use; or
>>>> (III) Derived from or through a person who owed a duty to the person seeking relief to maintain its secrecy or limit its use; or
>>> (C) before a material change of his [or her] position, knew or had reason to know that it was a trade secret and that knowledge of it had been acquired by accident or mistake.

The new federal statute parallels this definition. *See* 18 U.S.C. § 1839(5) (codifying the federal definition of "misappropriation"). As you can see, this definition calls out at least six different scenarios constituting misappropriation. Also, a key part of each scenario, except the last, is some reliance on "improper means." The Act provides, in turn, that "'[i]mproper means' includes theft, bribery, misrepresentation, breach or inducement of a breach of a duty to maintain secrecy or espionage through electronic or other means," UTSA § 1. The comments to section 1 indicate that "proper means" also includes both independent discovery and reverse engineering of a known product that is obtained by fair and honest means, such as purchase on the open market. *Cf.* 18 U.S.C. § 1839(6) (expressly providing that "the term 'improper means' . . . does not include reverse engineering, independent derivation, or any other lawful means of acquisition").

Imagine that an employee learns a valuable chemical formula at work and secretly sells the formula to a competitor. Under what portion of the definition of misappropriation might the employee's conduct fall? Is the competing company also a misappropriator in this scenario? If so, under what portion of the definition? Can the competing company be a misappropriator without ever using the information? Imagine a person breaks into a factory and photographs the layout of an innovative new production line. Has the person committed misappropriation?

Generally, misappropriation actions fall into two broad categories. The first, exemplified in the *duPont* case, involves competitors seeking to obtain and use trade secret information. The second involves a violation of a duty to keep the information secret. A duty to keep information secret can be created by a contract. Such contractual provisions are often included, for example, in join development agreements. These contracts are often referred to by the acronym "NDA" which stands for Non-Disclosure Agreements. A common scenario involving alleged breaches of secret-keeping duties involves an employee who is given access to trade secret information and then departs, either to start a competing business or to work for a competitor. We explore the case of departing employees in Section III.

As you read the next two cases, focus on the definitions of misappropriation and what qualifies as actionable misconduct.

E.I. duPont deNemours & Co. v. Christopher
431 F.2d 1012 (5th Cir. 1970)

GOLDBERG, JUDGE:

This is a case of industrial espionage in which an airplane is the cloak and a camera the dagger. The defendants-appellants, Rolfe and Gary Christopher, are photographers in Beaumont, Texas. The Christophers were hired by an unknown third party to take aerial photographs of new construction at the Beaumont plant of E.I. duPont deNemours & Company, Inc. Sixteen photographs of the DuPont facility were taken from the air on March 19, 1969, and these photographs were later developed and delivered to the third party.

DuPont employees apparently noticed the airplane on March 19 and immediately began an investigation to determine why the craft was circling over the plant. By that afternoon the investigation had disclosed that the craft was involved in a photographic expedition and that the Christophers were the photographers. DuPont contacted the Christophers that same afternoon and asked them to reveal the name of the person or corporation requesting the photographs. The Christophers refused to disclose this information, giving as their reason the client's desire to remain anonymous.

Having reached a dead end in the investigation, DuPont subsequently filed suit against the Christophers, alleging that the Christophers had wrongfully obtained photographs revealing DuPont's trade secrets which they then sold to the undisclosed third party. DuPont contended that it had developed a highly secret but unpatented process for producing methanol, a process which gave DuPont a competitive advantage over other producers. This process, DuPont alleged, was a trade secret developed after much expensive and time-consuming research, and a secret which the company had taken special precautions to safeguard. The area photographed by the Christophers was the plant designed to produce methanol by this secret process, and because the plant was still under construction parts of the process were exposed to view from directly above the construction area. Photographs of that area, DuPont alleged, would enable a skilled person to deduce the secret process for making methanol. DuPont thus contended that the Christophers had wrongfully appropriated DuPont trade secrets by taking the photographs and delivering them to the undisclosed third party. In its suit DuPont asked for damages to cover the loss it had already sustained as a result of the wrongful disclosure of the trade secret and sought temporary and permanent injunctions prohibiting any further circulation of the photographs already taken and prohibiting any additional photographing of the methanol plant.

. . . .

[The District Court denied the Christophers' motion to dismiss for failure to state a claim and granted DuPont's motion to compel the Christophers to disclose the name of the person to whom they had delivered the photographs. The Christophers sought immediate appeal of these interlocutory orders under 28 U.S.C. § 1291(b).]

This is a case of first impression, for the Texas courts have not faced this precise factual issue, and sitting as a diversity court we must sensitize our *Erie* antennae to divine what the Texas courts would do if such a situation were presented to them. The only question involved in this interlocutory appeal is whether DuPont has asserted a claim upon which relief can be granted. The Christophers argued both at trial and before this court that they committed no "actionable wrong" in photographing the DuPont facility and passing these photographs on to their client because they conducted all of their activities in public airspace, violated no government aviation standard, did not breach any confidential relation, and did not engage in any fraudulent or illegal conduct. In short, the Christophers argue that for an appropriation of trade secrets to be wrongful there must be a trespass, other illegal conduct, or breach of a confidential relationship. We disagree.

It is true, as the Christophers assert, that the previous trade secret cases have contained one or more of these elements. However, we do not think that the Texas courts would limit the trade secret protection exclusively to these elements. On the contrary, in *Hyde Corporation v. Huffines*, 1958, 158 Tex. 566, the Texas Supreme Court specifically adopted the rule found in the Restatement of Torts which provides:

> One who discloses or uses another's trade secret, without a privilege to do so, is liable to the other if
> (a) he discovered the secret by improper means, or
> (b) his disclosure or use constitutes a breach of confidence reposed in him by the other in disclosing the secret to him * * *.

Restatement of Torts § 757 (1939).

Thus, although the previous cases have dealt with a breach of a confidential relationship, a trespass, or other illegal conduct, the rule is much broader than the cases heretofore encountered. Not limiting itself to specific wrongs, Texas adopted subsection (a) of the Restatement which recognizes a cause of action for the discovery of a trade secret by any "improper" means.

. . . If breach of confidence were meant to encompass the entire panoply of commercial improprieties, subsection (a) of the Restatement would be either surplusage or persiflage, an interpretation abhorrent to the traditional precision of the Restatement. We therefore find meaning in subsection (a) and think that the Texas

Supreme Court clearly indicated by its adoption that there is a cause of action for the discovery of a trade secret by any "improper means." *Hyde Corporation v. Huffines, supra.*

The question remaining, therefore, is whether aerial photography of plant construction is an improper means of obtaining another's trade secret. We conclude that it is and that the Texas courts would so hold. The Supreme Court of that state has declared that "the undoubted tendency of the law has been to recognize and enforce higher standards of commercial morality in the business world." *Hyde Corporation v. Huffines, supra,* 314 S.W.2d at 773. That court has quoted with approval articles indicating that the proper means of gaining possession of a competitor's secret process is "through inspection and analysis" of the product in order to create a duplicate. *K & G Tool & Service Co. v. G & G Fishing Tool Service,* 1958, 158 Tex. 594. Later another Texas court explained:

> "The means by which the discovery is made may be obvious, and the experimentation leading from known factors to presently unknown results may be simple and lying in the public domain. But these facts do not destroy the value of the discovery and will not advantage a competitor who by unfair means obtains the knowledge *without paying the price expended by the discoverer." Brown v. Fowler,* Tex. Civ. App. 1958, 316 S.W.2d 111, 114, writ ref'd n.r.e. (emphasis added).

We think, therefore, that the Texas rule is clear. One may use his competitor's secret process if he discovers the process by reverse engineering applied to the finished product; one may use a competitor's process if he discovers it by his own independent research; but one may not avoid these labors by taking the process from the discoverer without his permission at a time when he is taking reasonable precautions to maintain its secrecy. To obtain knowledge of a process without spending the time and money to discover it independently is improper unless the holder voluntarily discloses it or fails to take reasonable precautions to ensure its secrecy.

In the instant case the Christophers deliberately flew over the DuPont plant to get pictures of a process which DuPont had attempted to keep secret. The Christophers delivered their pictures to a third party who was certainly aware of the means by which they had been acquired and who may be planning to use the information contained therein to manufacture methanol by the DuPont process. The third party has a right to use this process only if he obtains this knowledge through his own research efforts, but thus far all information indicates that the third party has gained this knowledge solely by taking it from DuPont at a time when DuPont was making reasonable efforts to preserve its secrecy. In such a situation DuPont has a valid cause of action to prohibit the Christophers from improperly discovering its trade secret and to prohibit the undisclosed third party from using the improperly obtained information.

We note that this view is in perfect accord with the position taken by the authors of the Restatement. In commenting on improper means of discovery the savants of the Restatement said:

"f. Improper means of discovery. The discovery of another's trade secret by improper means subjects the actor to liability independently of the harm to the interest in the secret. Thus, if one uses physical force to take a secret formula from another's pocket, or breaks into another's office to steal the formula, his conduct is wrongful and subjects him to liability apart from the rule stated in this Section. Such conduct is also an improper means of procuring the secret under this rule. But means may be improper under this rule even though they do not cause any other harm than that to the interest in the trade secret. Examples of such means are fraudulent misrepresentations to induce disclosure, tapping of telephone wires, eavesdropping or other espionage. A complete catalogue of improper means is not possible. In general they are means which fall below the generally accepted standards of commercial morality and reasonable conduct."

Restatement of Torts § 757, comment f at 10 (1939).

In taking this position we realize that industrial espionage of the sort here perpetrated has become a popular sport in some segments of our industrial community. However, our devotion to free wheeling industrial competition must not force us into accepting the law of the jungle as the standard of morality expected in our commercial relations. Our tolerance of the espionage game must cease when the protections required to prevent another's spying cost so much that the spirit of inventiveness is dampened. Commercial privacy must be protected from espionage which could not have been reasonably anticipated or prevented. We do not mean to imply, however, that everything not in plain view is within the protected vale, nor that all information obtained through every extra optical extension is forbidden. Indeed, for our industrial competition to remain healthy there must be breathing room for observing a competing industrialist. A competitor can and must shop his competition for pricing and examine his products for quality, components, and methods of manufacture. Perhaps ordinary fences and roofs must be built to shut out incursive eyes, but we need not require the discoverer of a trade secret to guard against the unanticipated, the undetectable, or the unpreventable methods of espionage now available.

In the instant case DuPont was in the midst of constructing a plant. Although after construction the finished plant would have protected much of the process from view, during the period of construction the trade secret was exposed to view from the air. To require DuPont to put a roof over the unfinished plant to guard its secret would impose an enormous expense to prevent nothing more than a school boy's trick. We introduce here no new or radical ethic since our ethos has never given moral sanction to piracy. The market place must not deviate far from our mores. We should not require a person or corporation to take unreasonable precautions to prevent another from doing that which he ought not do in the first place. Reasonable precautions against predatory eyes we may require, but an impenetrable fortress is an unreasonable requirement, and we are not disposed to burden industrial inventors with such a duty in order to protect the fruits of their efforts. "Improper" will always be a word of many nuances, determined by time, place, and circumstances. We therefore need not proclaim a catalogue of commercial improprieties. Clearly, however, one of its commandments does say "thou shall not appropriate a trade secret through deviousness under circumstances in which countervailing defenses are not reasonably available."

Having concluded that aerial photography, from whatever altitude, is an improper method of discovering the trade secrets exposed during construction of the DuPont plant, we need not worry about whether the flight pattern chosen by the Christophers violated any federal aviation regulations. Regardless of whether the flight was legal or illegal in that sense, the espionage was an improper means of discovering DuPont's trade secret.

The decision of the trial court is affirmed and the case remanded to that court for proceedings on the merits.

Notes & Questions

1. Would it have mattered if duPont had never planned to put a roof on the building?

2. Is the problem with the Christophers' conduct that it was too difficult to have predicted it, or too hard to have guarded against it (however likely or unlikely it was to have occurred)? Or is it a combination of both?

3. Your client wants to look in its competitor's windows using binoculars. Does *duPont* counsel against it? What about using thermal imaging equipment, aimed at the competitor's building? Or a quadcopter drone with a camera, widely available to the public through websites like Amazon? Would you answer differently if your client were in a UTSA state, rather than one using the common-law approach illustrated by *duPont*?

Before you read the next case, return to the first page of this Section and re-read the UTSA definition of misappropriation. Remember, too, that the federal statute's definition is almost identical. 18 U.S.C. 1839(5). The case below is decided under California law which includes that definition verbatim.

Silvaco Data Sys. v. Intel Corp.
109 Cal. Rptr. 3d 27 (Cal. Ct. App. 2010)

RUSHING, JUDGE:

Plaintiff Silvaco Data Systems (Silvaco) brought this action against defendant Intel Corporation (Intel) alleging that the latter had misappropriated certain trade secrets used by Silvaco in its software products. The primary gist of the claims was that Intel had used software acquired from another software concern with knowledge that Silvaco had accused that concern of incorporating source code, stolen from Silvaco, in its products. The chief question presented is whether Intel could be liable for such use if, as was effectively undisputed, it never possessed or had access to the source code but only had executable, machine-readable code compiled by its supplier from source code. We answer that question in the negative. One does not, by executing machine-readable software, "use" the underlying source code; nor does one acquire the requisite knowledge of any trade secrets embodied in that code. . . .

Background

Silvaco develops and markets computer applications for the electronic design automation (EDA) field, which covers the entire complex process of designing electronic circuits and systems. Among the various sub-categories of EDA software are circuit simulators, which permit the designer to create a virtual model of a proposed circuit in order to test its properties before incurring the expense and delay of manufacturing a working prototype. Defendant Intel, a major developer and manufacturer of integrated circuits, is a user and purchaser of EDA software, including circuit simulators. According to Silvaco's complaint, Intel has also developed some EDA software for its own use.

Among Silvaco's software products is SmartSpice, an analog circuit emulator. In December 2000, Silvaco filed a suit against Circuit Semantics, Inc. (CSI), a competing developer of EDA software, alleging that CSI, aided by two former Silvaco employees, had misappropriated trade secrets used in SmartSpice, and had incorporated them in its own product, DynaSpice. Silvaco eventually secured a judgment against CSI, including an injunction against the continued use of "technology" described in an exhibit attached to the judgment. It then brought actions against several purchasers of CSI software, including Intel. It alleged that *by using CSI's software,* these end users had misappropriated the Silvaco trade secrets assertedly incorporated in that software. Silvaco charged Intel with misappropriation of trade secrets under the California Uniform Trade Secrets Act, Civil Code sections 3426 through 3426.11 (CUTSA) . . . .

Intel . . . moved for summary judgment on the CUTSA claim, arguing that (1) CUTSA defines "misappropriation" in a way requiring the plaintiff to show that the defendant possessed "knowledge of the trade secret" (Civ. Code, § 3426.1, subd. (b)), and (2) there was no evidence that Intel ever possessed such knowledge.

In support of the motion Intel presented evidence that it never possessed the *source* code identified by Silvaco as constituting and containing its secrets, but only *executable* code supplied to it by CSI. The parties appear to agree, and we may accept for purposes of this opinion, that "source code" describes the text in which computer programs are originally written by their human authors using a high-level programming language.[4] One who possesses the source code for a program

---

[4] This is an accurate enough description where, as here, the facts conform to what one writer has called the "[s]tandard [programming] [s]cenario." (Felten, Source Code and Object Code (posted Sep. 4, 2002) <http://freedom-to-tinker.com/blog/felten/source-code-and-object-code> [as of Apr. 20, 2010] (Source Code and Object Code).) In this scenario, human programmers write a program in a high-level programming language; this source code is then processed through software known as a compiler to produce object code that may be executed on a machine. Departures from the standard scenario, however, are increasingly common. Another writer asserts that among programmers

may readily ascertain its underlying design, and may directly incorporate it, or pieces of it, into another program. In order to yield a functioning computer application, however, source code must generally be translated or "compiled" into machine-readable (executable) code. After a program is compiled, it may still be represented as text, but the text is not readily intelligible to human beings, consisting of strings of binary (base 2) or hexadecimal (base 16) numbers. For this reason, the source code for many if not most commercial software products is a secret, and may remain so despite widespread distribution of the executable program.

Intel cited discovery responses by Silvaco, and testimony by its agents and experts in this and related cases, to the effect that (1) the possession and use of DynaSpice in the form of executable object or binary code could not impart knowledge of any trade secrets embodied in the source code; and (2) Silvaco was unable to controvert Intel's evidence that it never possessed any of the source code from which DynaSpice was derived. . . .

. . . .

Discussion

. . . .

C. *Misappropriation*

1. *Disclosure; Acquisition*

Under CUTSA, misappropriation of a trade secret may be achieved through three types of conduct: "[a]cquisition," "[d]isclosure," or "[u]se." (§ 3426.1, subd. (b).) The act does not define these terms, but leaves their delineation to be adjudicated in light of the purposes and other provisions of the act. The question is whether Silvaco raised a triable issue of fact concerning Intel's commission of any of these three forms of conduct.

There is no suggestion that Intel ever *disclosed* Silvaco's source code to anyone, and it is difficult to see how it might have done so since there is no evidence that it ever had the source code to disclose. Silvaco emphasizes that wrongful *acquisition* of a trade secret may be actionable in itself. (See § 3426.1, subd. (b).) But there is no basis to suppose that Intel ever "acquired" the source code constituting the

---

the terms "source" and "object" do not really describe "well-defined classes of code" but are "actually relative terms." (Touretzky, Source vs. Object Code: A False Dichotomy (Draft Version Jul. 12, 2000) <http://www.cs.cmu.edu/~dst/DeCSS/object-code.txt> [as of Jan. 29, 2010].) He continues, "Given a device for transforming programs from one form to another, source code is what goes into the device, and object code (or 'target' code) is what comes out. The target code of one device is frequently the source code of another." The upshot seems to be that the correct use of these terms reflects variables that may not have any particular legal significance. Here, the essential variable for legal purposes is the extent to which the code reveals the underlying design, i.e., the methods and algorithms used by the developer. (See Source Code and Object Code, *supra* {suggesting distinctions based on whether code is in "'the form in which [programmers] customarily read and edit[] it'" and the extent to which it is "human-readable"}.) The distinguishing feature for our purposes is that what the parties call "source" code is written by programmers and readily understood by them, whereas object or machine code is compiled by and for machines and does not readily yield its underlying design to human understanding.

trade secrets. To acquire a thing is, most broadly, to "receive" or "come into possession of" it. (1 Oxford English Dict. (2d ed. 1989), p. 115.) But the term implies more than passive reception; it implies pointed conduct intended to secure dominion over the thing, i.e., "[t]o gain, obtain, or *get as one's own,* to gain the ownership of (by one's own exertions or qualities)." (*Id.;* see *id.* at p. 115 {"acquisition" as "[t]he action of obtaining or getting for oneself, or by one's own exertion"}.)

One does not ordinarily "acquire" a thing inadvertently; the term implies conduct directed to that objective. The choice of that term over "receive" suggests that inadvertently coming into possession of a trade secret will not constitute acquisition. Thus one who passively receives a trade secret, but neither discloses nor uses it, would not be guilty of misappropriation. We need not decide the outer limits of acquisition as contemplated by CUTSA, however, for there is no suggestion here of acquisition even in the broadest sense, i.e., that Intel ever came into possession of the source code constituting the claimed trade secrets. Indeed Silvaco does not directly argue that Intel acquired the trade secrets at issue but only that, under the terms of the statute, it *could* have done so without itself having "knowledge" of them. We doubt the soundness of this suggestion, but assuming it is correct, it remains beside the point unless Intel came into possession of the secret. Since there is no basis to find that it did, the mental state required for actionable acquisition appears to be academic.

2. *What Constitutes Use*

Silvaco has never explained how any conduct by Intel constituted "use" of its source code. Intel's only conduct in this matter, so far as the record shows, was to execute (run) the software it acquired from CSI—software which, according to Silvaco, "incorporated" or "contained" its trade secrets. We do not believe this can be viewed as a "use" of Silvaco's source code under CUTSA.

As it appears in the act, the noun "use" is surely intended in the ordinary sense, i.e., "[t]he act of *employing* a thing for any (esp. a profitable) purpose; the fact, state, or condition of being so employed; *utilization* or employment for or with some aim or purpose, *application or conversion* to some (esp. good or useful) end." (19 Oxford English Dict. (2d ed. 1989), p. 350, italics added.) The term commonly implies, if not direct physical possession, at least a certain proximity or immediacy to the thing used. Indeed a common sense of the verb "to use" is "[t]o work, employ, or manage (an implement, instrument, etc.); to manipulate, operate, or handle, esp. to some useful or desired end." (19 Oxford English Dict. (2d ed. 1989), p. 354.) Thus if one makes or receives a telephone call, he is said to "use" the telephone at his end of the call; but would not ordinarily be said to "use" the telephone at the receiving end—at least, not without qualification.

One clearly engages in the "use" of a secret, in the ordinary sense, when one directly exploits it for his own advantage, e.g., by incorporating it into his own manufacturing technique or product. But "use" in the ordinary sense is not present when the conduct consists entirely of possessing, and taking advantage of, *something that was made* using the secret. One who bakes a pie from a recipe certainly engages in the "use" of the latter; but one who eats the pie does not, by virtue of that act alone, make "use" of the recipe in any ordinary sense, and this is true even if the baker is accused of stealing the recipe from a competitor, and the diner knows of that accusation. Yet this is substantially the same situation as when one runs software that was compiled from allegedly stolen source code. The source code is the recipe from which the pie (executable program) is baked (compiled). Nor is the analogy weakened by the fact that a diner is not ordinarily said to make "use" of something he eats. His metabolism may be said to do so, or the analogy may be adjusted to replace the pie with an instrument, such as a stopwatch. A coach who employs the latter to time a race certainly makes "use" of it, but only a sophist could bring himself to say that coach "uses" trade secrets involved in the manufacture of the watch.

Strong considerations of public policy reinforce the common-sense conclusion that using a product does not constitute a "use" of trade secrets employed in its manufacture. If merely running finished software constituted a use of the source code from which it was compiled, then *every* purchaser of software would be exposed to liability if it were later alleged that the software was based in part upon purloined source code. This risk could be expected to inhibit software sales and discourage innovation to an extent far beyond the intentions and purpose of CUTSA. We therefore decline to hold that under the circumstances alleged and shown here, the mere execution of a software product constitutes "use" of the underlying source code for purposes of a misappropriation claim under CUTSA.

3. *Knowledge of the Trade Secret*

Even if Intel's use of DynaSpice could otherwise be held to constitute a misappropriation of the source code used to compile it, that conduct would only sustain liability if accompanied by the mental state prescribed by the statute—a mental state that defendant appears to have lacked. Of the five varieties of actionable "use" listed in the statute, four unmistakably require the defendant to have "knowledge of the trade secret." (§ 3426.1, subds. (b)(2)(A) {"[u]sed improper means to acquire knowledge of the trade secret"}; (b)(2)(B) {"knowledge of the trade secret was * * * (i) [d]erived * * *; [¶] (ii) [a]cquired * * *; or [¶] (iii) [d]erived * * *"}.) The fifth is arguably ambiguous on this point, in that it contemplates the defendant's actual or constructive awareness that "knowledge of [the trade secret] *had been acquired* by accident or mistake." (§ 3426.1, subd. (b)(2)(C), italics added.) The use of the passive voice here leaves open the possibility that a defendant might be liable, though himself lacking knowledge of the trade secret, if he "used" it with knowledge that *another,* on whom his use somehow depended, had acquired knowledge of it as a result of accident [or] mistake.[7]

---

[7] Nothing said here should be taken to suggest that a defendant cannot be liable for misappropriation unless he *personally* possessed knowledge of the trade secret. He can of course acquire such knowledge, and indeed can conduct the entire misappropriation, *vicariously,* e.g., through an agent. Further, *constructive* knowledge of the secret may well be sufficient, at least in some circumstances. Thus one who knowingly possesses information constituting a trade secret cannot escape liability

It is not easy to conceive how this might occur in real life. In any event it is not necessary to address the possibility at present because there is no suggestion that Silvaco's claims might be brought within this clause of the statute. Therefore, in order to prevail against Intel on its CUTSA claim, Silvaco was obliged to establish that Intel had "knowledge of the trade secret."

. . . .

"Knowledge," of course, is "the fact or condition of knowing," and in this context, "[t]he fact of knowing a thing, state, etc. * * *" (8 Oxford English Dict. (2d ed. 1989), p. 517.) To "know" a thing is to have information of that thing at one's command, in one's possession, subject to study, disclosure, and exploitation. To say that one "knows" a fact is also to say that one *possesses information* of that fact. Thus, although the *Restatement of Unfair Competition* does not identify knowledge of the trade secret as an element of a trade secrets cause of action, the accompanying comments make it clear that liability presupposes the defendant's "possession" of misappropriated information.[9] (*Rest. 3d Unfair Competition* § 40, com. d, p.457 {"To subject an actor to liability under the rules stated in Subsection (b)(2)-(4), the owner need not prove that the actor knew that *its possession of the trade secret* was wrongful; it is sufficient if the actor had reason to know. Thus, if a reasonable person in the position of the actor would have inferred that *he or she was in wrongful possession* of another's trade secret, the actor is subject to liability for any subsequent use or disclosure. A number of cases also subject an actor to liability if, based on the known facts, a reasonable person would have inquired further and learned that *possession of the information* was wrongful."}.)

The record contains no evidence that Intel ever possessed or had knowledge of any source code connected with either SmartSpice or DynaSpice. So far as the record shows, Intel never had access to that code, could not disclose any part of it to anyone else, and had no way of using it to write or improve code of its own. Intel appears to have been in substantially the same position as the customer in the pie shop who is accused of stealing the secret recipe because he bought a pie with knowledge that a rival baker had accused the seller of using the rival's stolen recipe. The customer does not, by buying or eating the pie, gain knowledge of the recipe used to make it.

. . . .

Silvaco suggests that a more expansive reading of . . . the knowledge requirement is necessary to effectuate the CUTSA's purpose of "'maintenance of standards of commercial ethics.'" (14 West's U. Laws Ann. (2005) U. Trade Secrets Act, com. to § 1, p. 538, quoting *Kewanee Oil Co. v. Bicron Corp.* (1974) 416 U.S. 470, 481.) But apart from the difficulties inherent in such an impressionistic approach to statutory construction, it is far from apparent that Intel's conduct here

---

merely because he lacks the technical expertise to understand it, or does not speak the language in which it was written.

[9] The Restatement is not, of course, a substitute for CUTSA, but it is intended to express principles reflecting the uniform act's proper application. (*Rest. 3d Unfair Competition* § 39, com. b.)

offended any sound or settled standard of commercial ethics. Intel initially purchased DynaSpice from CSI without, so far as this record shows, any inkling that
CSI might have developed that product in a questionable manner. Thereafter, having presumably undertaken to develop its own products using DynaSpice, Intel
learned of *claims* by Silvaco that CSI had misappropriated portions of its software
design—claims which CSI denied. Only when CSI entered into a stipulated judgment requiring it to stop using Silvaco code could an outsider rationally conclude
that there was substance to Silvaco's claims. But that very judgment authorized
CSI to continue marketing and supporting its products provided they were modified to excise Silvaco's trade secrets. Further litigation ensued over the extent to
which CSI had conformed to those restrictions, but that litigation appears to have
concluded favorably to CSI.

To brand Intel's conduct as unethical, we would have to conclude that any end
user of a software application must desist from its use—whatever the resulting
harm to his own business—the moment anyone *claims* that the application was
compiled from stolen source code. This would be a prescription for the stultifica-
tion of technological development and of other business activities taking place at
a considerable remove, causally and ethically, from the claimed wrong. Far from
serving the purposes of trade secrets law, such a rule would make it far too easy to
suppress competition and technological development by threatening not only
would-be competitors, but also their customers, with litigation of virtually unlimited scope. So far as this record shows, nothing Intel did offended any proper sense
of business ethics.[13]

In short, as applicable to this case the law conforms to the common-sense sup-
position that one cannot be said to have stolen something he never had. . . . So far
as appears from the record, Intel never *had* the trade secret whose theft is the subject matter of this action. That is the central defect in Silvaco's claim. Its arguments, devoted not to curing that defect but to distracting us from it, fail to suggest
any error by the trial court in granting summary judgment.

. . . .

Notes & Questions

1. Before focusing on the question of misappropriation, it is worth pausing to
consider the court's determination that source code for commercially distributed
software can be a trade secret. Footnote 4 of the court's opinion identifies the difference between computer source code and object code and describes the process
of compiling a program written in source code into object code. If the source code
for a program can be discovered through reverse engineering (called "decompil-
ing" in the software context), is the source code "readily ascertainable by proper
means"? Does it matter how difficult it is, in a given case, to engage in the process

---

[13] Nor is it easy to see why or how such a regime would be limited to software. Suppose a manufacturer hires a builder to design and construct a new manufacturing plant. Do business ethics demand
that the owner shut down the plant upon learning that a competing builder alleges the plans were
stolen from him? In other areas of law concepts such as "bona fide purchaser for value" are deployed
to constrain just such concatenations of liability.

of reverse engineering? Does it matter whether the End-User License Agreement prohibits reverse engineering?

2. Turning to the question of misapropriation, what tools does the *Silvaco* court use to interpret the statute? Which are most convincing to you? Do the metaphors of pie baking and eating and of coaches with stopwatches persuade you?

3. In interpreting the knowledge requirement the court turns to the *Restatement (Third) of Unfair Competition*, issued by the American Law Institute in January 1995. Why does the court believe this is an important source to consult in interpreting the statutory claim of trade secret misappropriation under California law? Would use of the *Restatement* be an appropriate guide for interpretation of the federal trade secret statute?

4. At the end of the opinion the court concludes with a common-sense approach indicating that "one cannot be said to have stolen something he never had." With intangibles, such as trade secret information, determining when one "has" the intangible can be a slippery task. Tangible objects—documents or products—often embody intangible information. It is far easier to determine when a defendant "has" a tangible object than it is to determine when he or she "has" the information. Is having possession of the information-bearing tangible object equivalent to possessing the information? Does possessing trade secret information amount to a misappropriation? What are the advantages and disadvantages of having a detailed statutory definition of misappropriation compared to a common law tort of misappropriation?

5. Re-read the definition of misappropriation set out at the beginning of this Section. Imagine an individual inadvertently discloses trade secret information through a misdirected email. Is the recipient of the email free to use the information they have obtained through no improper conduct on their part?

B. CRIMINAL TRADE SECRET THEFT

Trade secret misappropriation has long been, for the most part, a civil law matter between private parties—the purported secret owner and the alleged misappropriator. Beginning in the 1960s, however, legislatures began to criminalize trade secret theft, treating it as a matter of greater public concern. Today, about half the states have laws that expressly criminalize trade secret theft. Geraldine Szott Moohr, *The Problematic Role of Criminal Law in Regulating Use of Information: The Case of the Economic Espionage Act*, 80 N.C. L. Rev. 853, 875 (2004).[*] These statutes overcame what, before their enactment, had been judicial concerns about applying anti-theft statutes focused on the physical taking of tangible property to the misappropriation of intangible information. Daniel D. Fetterly, *Historical Perspectives on Criminal Laws Relating to the Theft of Trade Secrets*, 25 Bus. Law. 1535, 1537 (1970).

---

[*] To learn more about a given state's trade secret theft law, one might begin with Brian M. Malsberger, *Trade Secrets: A State-by-State Survey* (3d ed. 2006).

The California Penal Code's trade secret theft law is illustrative. Its substantive core provides as follows:

> Every person is guilty of theft who, with intent to deprive or withhold the control of a trade secret from its owner, or with an intent to appropriate a trade secret to his or her own use or to the use of another, does any of the following:
>
> (1) Steals, takes, carries away, or uses without authorization, a trade secret.
>
> (2) Fraudulently appropriates any article representing a trade secret entrusted to him or her.
>
> (3) Having unlawfully obtained access to the article, without authority makes or causes to be made a copy of any article representing a trade secret.
>
> (4) Having obtained access to the article through a relationship of trust and confidence, without authority and in breach of the obligations created by that relationship, makes or causes to be made, directly from and in the presence of the article, a copy of any article representing a trade secret.

Cal. Penal Code § 499c(b). Subsection (a) of the statute provides definitions of such key terms as "access," "article," and "trade secret."

Congress also made trade secret theft a federal crime when it enacted the Economic Espionage Act in 1996. The Act is codified at 18 U.S.C. §§ 1831-1839; the full text of the Act is set forth in the *Statutory Supplement* for these materials (available on the Semaphore Press website). Section 1831 criminalizes trade secret theft that is undertaken "intending or knowing that the offense will benefit any foreign government, foreign instrumentality, or foreign agent." 18 U.S.C. § 1831(a). Indeed, the congressional witnesses supporting the Act focused on foreign industrial espionage. *See* Moohr, *supra*, at 864-65. Section 1832, however, criminalizes theft as to any trade secret "that is related to or included in a product that is produced for or placed in interstate or foreign commerce." 18 U.S.C. § 1832(a). In other words, this act makes *domestic* trade secret theft a federal felony as well. Attempt and conspiracy are also felonies. 18 U.S.C. §§ 1832(a)(4), (5).

Many of the prosecutions in the first decade following passage of the EEA involved purely domestic trade secret theft and often involved an employee with access to trade secret information. *See, e.g.*, *United States v. Lange,* 312 F.3d 263 (7th Cir. 2002) (former employee charged with selling technical specifications for aircraft parts); *United States v. Genovese,* 409 F. Supp. 2d 253 (S.D.N.Y. 2005) (defendant charged with selling portions of Microsoft source code); *United States v. Krumrei,* 258 F.3d 535 (6th Cir. 2001) (defendant charged with selling technical information to a competitor); *United States v. Martin*, 228 F.3d 1 (1st Cir. 2000) (employee charged with passing information to competitor). The initial prosecutions helped establish the broad sweep of the federal law. *See also United States v. Hsu*, 155 F.3d 189 (3d Cir. 1998). The EEA and the use of criminal prosecutions in addition to civil liability raise many interesting legal issues that are beyond the scope of a survey course in intellectual property.

In 2016, Congress expanded the Economic Espionage Act's provisions to create a new civil cause of action under federal law, 18 U.S.C. § 1836(b), for trade secret misappropriation. *See generally* Defend Trade Secrets Act of 2016, Pub. L. 114-153, 130 Stat. 376. This new federal claim is expressly separate from any claim that state law provides. 18 U.S.C. § 1838 (providing that the DTSA "shall not be construed to preempt or displace any other remedies, whether civil or criminal, provided by United States Federal, State, commonwealth, possession, or territory law for the misappropriation of a trade secret"). As a result, trade secret plaintiffs now have their choice of a federal or state forum, and can press a state-law claim together with the federal claim using the supplemental jurisdiction statute, 28 U.S.C. § 1367(a). And, as Prof. Eric Goldman has noted, "[b]ecause of the DTSA/state law overlap, companies will need to understand and conform their practices to both the new and existing law, which increases legal costs for all companies." Eric Goldman, *The New 'Defend Trade Secrets Act' Is the Biggest IP Development in Years*, Forbes (Apr. 28, 2016), *available at* http://www.forbes.com.

III. Trade Secrets and the Departing Employee

The allocation of rights in intangible assets created by employees is a recurring issue in intellectual property. Intellectual property protections vary in how they sort out the proper balance of ownership interests. Should employers own the intangible assets generated by employees? Should the employees have any rights to the intangible asset at all, for example through co-ownership or a limited license? Or should the default rule be that the employee owns the asset but the employer obtains a limited right to use the information in its business? Generally speaking, trade secret information that employees create within the scope of their assigned duties is owned by the employer, not the employees. *See e.g. Pullman Group LLC v. Prudential Ins. Co.*, 288 A.D.2d 2 (N.Y. App. Div. 2001); *Q-Co Industries, Inc. v. Hoffman*, 625 F. Supp. 608 (S.D.N.Y. 1985).

While the rule in trade secret law is that an employer owns the trade secret information created by employees, employers do not have any claim over an employee's general skill and knowledge in a particular field. *Compare* Restatement (Third) of Unfair Competition (1995) § 42 cmt. b (trade secrets), *with id.* § 42 cmt. d (general information). An employee is free to use his or her general skill and knowledge in future employment. Defining the line between trade secret information and general skill and knowledge thus has serious ramifications for the departing employee.

A. Basic Cases

Omega Optical, Inc. v. Chroma Technology Corp.

174 Vt. 10 (2001)

Morse, Justice:

Omega Optical appeals from a judgment of the superior court in favor of defendant Chroma Technology Corporation and several other named defendants in Omega's action for trade secret misappropriation, conversion, breach of loyalty, tortious interference with business relations, unfair competition, conspiracy and breach of contract. . . .

This case arises out of events spanning several months starting in early 1991, in which a number of Omega Optical employees left the company and went into business together under the name of Chroma Technology Corporation. Chroma began making thin-film optical interference filters used in fluorescence microscopy, a product that Omega had developed and also produces. On October 1, 1996, Omega brought suit against Chroma and ten of its employees. Following a twenty-two-day bench trial, the trial court issued a 111-page decision finding in favor of defendants on all of Omega's claims.

Omega's appeal centers on its argument that, because defendants acquired substantial amounts of information that the court found was "protectible as a trade secret," it is entitled to judgment as a matter of law, notwithstanding the trial court's extensive findings that Omega failed to take reasonable steps to protect the information. The court concluded under the evidence that defendants owed no duty of confidentiality with regard to the information.

As a preliminary matter, and as Omega points out, because the operative facts of this case occurred before July 1, 1996, the effective date of the Vermont Trade Secret Act, the common law predating the act applies to this case. See 9 V.S.A. § 4609 (providing the VTSA does not apply to misappropriation occurring before its effective date). But as the trial court noted, the act and the Restatement (Third) of Unfair Competition provide guidance with regard to Omega's common law claim for misappropriation of trade secrets.

In general, liability for trade secret misappropriation in the employment context requires proof of both the existence of a trade secret as well as unauthorized disclosure or use of the secret in breach of a duty of confidence. Restatement (Third) of Unfair Competition §§40, 42 (1995). Employees, whether current or former, have a duty not to use or disclose confidential information imparted to them by their employer. Restatement (Second) Agency §396(b) (1958); see also Restatement (Third) of Unfair Competition § 42 cmt. b. The former employee's duty of confidence attaches to any information the employee knows or has reason to know is confidential. Restatement (Third) of Unfair Competition § 42 cmt. c. Whether an employee knows or should know certain information obtained from the employer is confidential can be implied from the totality of the circumstances; no explicit notice to the employee is necessarily required. . . . [S]ee also Restatement (Third) of Unfair Competition §42 cmt. c ("If an employer establishes ownership of a trade secret *and* circumstances sufficient to put the employee on notice that the information is confidential, the employment relationship will ordinarily justify the recognition of a duty of confidence.") (emphasis added). It is a fact-specific inquiry.

Omega argues, however, that employees who acquire valuable information in the course of their employment owe a duty of confidentiality to the employer merely by virtue of their status as employees, regardless of whether the employer has done anything either to protect the information or to communicate to the employees the confidential and proprietary nature of the information. This argument is simply at odds with the case law, which requires something more than the mere employer-employee relationship to establish a duty of confidentiality. . . . [O]ur decision in *McClary v. Hubbard* makes plain that the party claiming a trade secret must demonstrate that it has taken steps to ensure the information's secrecy to prevail on a claim for trade secret misappropriation. *McClary [v. Hubbard]*, 97 Vt. [222] at 232-33, 245 [(1923)].

In analyzing the facts of this case, the trial court applied the Restatement's approach to Omega's claim for trade secret misappropriation. In other words, it determined that the body of knowledge necessary to produce the thin-film optical interference filters was "sufficiently valuable, and not generally well known, that it [was] protectible as a trade secret."[1] Consistent with *McClary*, the court went on, however, to conclude that Omega had failed to take measures toward protecting the information, and the circumstances under which the employees acquired the information failed to indicate to them that the information was confidential. Therefore, the defendants owed no duty of confidentiality to Omega, and their use of the valuable information in their new venture did not constitute misappropriation of that information.

Specifically with respect to the circumstances under which the defendants acquired the information, the trial court found that while defendants were employed at Omega, the company had "no internal policies concerning confidentiality, nondisclosure or noncompetition." The court found that an open atmosphere prevailed in which employees were encouraged to share information about the development and production of filters. In contrast, during this same time, the company did have confidentiality policies with regard to proprietary information belonging to Omega's customers. The court found that Omega failed to convey during the defendants' course of employment the confidential nature of the information it now seeks to protect and, when Omega did institute a policy of confidentiality, the defendants still employed with the company refused to sign onto it and resigned. The court also found that Omega had few security measures in place prior to defendants leaving the company and the few security measures that did exist, such as signing out keys, were not enforced or even monitored. For a significant period of time, members of the public had access to work areas. The court also noted that

---

[1] Although similar, the VTSA and the Restatement take somewhat different approaches to analyzing a claim for misappropriation of trade secrets. As we noted in *Dicks v. Jensen*, 768 A.2d 1279, 1282 (Vt. 2001), the VTSA includes in its definition of "trade secret" a requirement that the information be the subject of reasonable efforts to maintain its secrecy. 9 V.S.A. § 4601(3)(B). If such efforts have not been made, the information is not a "trade secret." On the other hand, the Restatement's definition does not incorporate such a requirement. Restatement (Third) of Unfair Competition § 39. Rather, its definition merely parallels the first part of our statutory definition. Compare I., with 9 V.S.A. § 4601(3)(A). Under the Restatement, information is a "trade secret" if it is used and valued in the operation of a business and is sufficiently secret, *i.e.*, not generally well known, that it affords economic advantage to the business. Restatement (Third) of Unfair Competition § 39. Notably, the "secret" element of the Restatement definition addresses itself only to the question of whether the information bestows any competitive advantage on the employer.

many of the measures cited by Omega as security measures in the course of the trial actually evolved for different purposes and Omega's claims, such as its location as a security measure, were post hoc rationalizations.

. . . .

There was also testimony and other documentary evidence that (1) Omega had no written policy of confidentiality; (2) just prior to the time defendants were contemplating leaving Omega, the company recognized the need to establish guidelines to help employees distinguish between confidential and nonconfidential information; (3) very little awareness existed among employees as to what the company considered proprietary; and (4) record keeping at Omega was sloppy, in some cases nonexistent, and the company's technical information was not kept in an organized or centralized, controlled manner. In one instance, defendant Kebbel testified that after Omega donated one of its computers to a local daycare center, Dr. Johnson asked Kebbel if the company still had the computer because it contained information that Dr. Johnson needed. In light of this and other evidence available to the trial court, we find no error in the court's finding that Omega failed to take steps to put its employees on explicit or implicit notice that certain information conveyed to them during their employment was to be kept confidential. Hence, defendants did not receive the information in confidence and, therefore, did not breach their duty of confidence to Omega, their former employer.

Omega argues, however, that the court erred by examining the circumstances surrounding the acquisition of the information when determining whether the defendants owed a duty of confidentiality. Omega contends that it is the time of use that is determinative regarding whether former employees owe a duty of confidentiality to a former employer. Omega confuses the relevant times for determining whether a duty exists and whether the duty has been breached, that is, whether the information has been misused or misappropriated. To determine whether a duty of confidentiality arises, courts examine whether the information was acquired under circumstances that would indicate to an individual that the information is confidential, while courts look to the time of use to determine whether the individual knows that the information being used is subject to a duty of confidentiality such that the use constitutes a breach of that duty and, thus, misappropriation. . . .

The trial court's findings are supported by the record, and the trial court correctly applied the governing law. . . .

. . . .

To the degree that the additional claims are premised on Omega's claim that defendants misappropriated Omega's trade secrets, the trial court's disposition of that claim also disposes of the additional claims. Furthermore, as the trial court noted with regard to Omega's breach of loyalty claim, courts have generally held that at-will employees may plan to compete with their employer even while still employed there and may freely compete with the employer once they are no longer employed there. This behavior does not constitute a breach of a duty of loyalty.

Such at-will employees are restricted, however, from misappropriating trade secrets and soliciting customers for their new venture while still employed at the former employer. The trial court specifically found that, while defendants did formulate plans for the creation of Chroma while still working at Omega, they did not solicit Omega's customers while still employed there and continued to perform their duties at Omega in good faith. As these findings are supported by the record, we will not disturb them on appeal.

Omega argues, however, that competition for Omega customers by defendants following their departure from the company constitutes a breach of the duty of loyalty as a matter of law. But . . . when an employer does not take steps to protect information such as customer lists, competition for those customers by a former employee after that employee has left the company is legitimate.

Omega also makes a more general argument that "[a]s *former* employees of Omega, the individual defendants each . . . continued to owe Omega a duty of loyalty, which included the duty to *refrain from acting for their own benefit or the benefit of Chroma* to the detriment of Omega." (Emphasis added.) Omega cites no authority for the proposition that at-will employees continue to owe a duty of loyalty to a former employer, even after they have left that employment, that constrains them from ever acting to the detriment of that employer. Such a common law duty would prevent an employee from ever going to work for a competitor even in the absence of an agreement not to do so, an anomalous result. *Cf. Dicks*, 768 A.2d at 1285 (given restraint with which this Court enforces explicit non-compete agreements, we declined to imply one).

. . . .

Stampede Tool Warehouse, Inc. v. May

651 N.E.2d 209 (Ill. App. 1 Dist. 1995)

CERDA, JUSTICE:

. . . .

Richard Kuhn, president of Stampede, testified that Stampede Tool Warehouse is a national distributor of automotive tools and equipment. Stampede buys tools and equipment from manufacturers, then sells them to automotive jobbers, which includes service stations, tool dealers, and automotive stores. The jobbers then sell the tools to end users, who are mechanics and homeowners.

There are over 20,000 independent jobbers nationwide. Of Stampede's 11,000 customers, 35% to 40% are currently buying from them. Stampede's customer list has been developed from prospecting, which is calling end users and asking them the name of their tool jobber. Kuhn stated that there is no available public source or book where automotive jobbers are listed. Kuhn testified that the cost of getting new customers is $50,000 per month, which includes salaries, phone bills, and overhead. Stampede's salesmen spend about 60 hours per month prospecting to find about 100 new customers, half of whom actually buy from Stampede. In addition to telephone solicitation, Stampede sends out 11,000 to 12,000 catalogs each quarter. The names for the catalog list were gathered from years of prospecting and word of mouth.

. . . .

[Stampede employed defendants Mark May and Fred Moshier, from October 1990 until May 1991. Moshier and May quit and eventually went to work for a company established by Moshier. When Moshier quit, he informed Kuhn that he was establishing his own company to broker out telemarketing services to warehouses. Initially Kuhn agreed to do business with Moshier. However, after receiving an order from Moshier that came from long-standing Stampede customers, Kuhn ceased doing business with Moshier.] . . . In the two years after defendants left Stampede, the company experienced a drop of $1 million in sales.

. . . .

[Stampede brought suit against Moshier and May. The trial court entered permanent injunctions that forever restrained the defendants from doing business with any customer of Stampede to whom they had access during their employment at Stampede. The appellate court first analyzed whether Stampede's customer list was a protectable trade secret under the Illinois Trade Secret Act (ITSA) which follows the UTSA.]

Although an employee's general knowledge is not a trade secret, the general knowledge in this case is the method the defendants used in finding prospective customers, not the actual customer information. When the information learned is confidential, as here, it can be a trade secret.

In this case, the record shows the customer list is a trade secret. The customer list has been developed through the laborious method of prospecting, which requires substantial amount of time, effort, and expense by Stampede. A list of jobbers is not readily available from any one public source. Instead, the salesmen obtained a list of end users through telephone books, catalogs, and other publicly available sources; then contact those people to get the names of jobbers, who are the potential customers. The salesmen then develop a relationship with those potential customers.

In addition, Stampede protects its customer list using reasonable efforts to maintain its secrecy and confidentiality. Its offices are locked, garbage is checked daily, special computer access codes are used, customer information is limited to persons on the need to know basis, hard copies of customer lists are kept locked in the office or in Kuhn's basement at home, salesmen's call books and customer cards are kept locked up and cannot be removed from the office, and security cameras are used. Moreover, both defendants signed employee confidentiality agreements that stated that the names of Stampede's customers could not be used or disclosed because they belong to Stampede and were confidential. As a result, we conclude that Stampede's customer list is a trade secret that is protectable under ITSA.

We must now consider whether defendants misappropriated the trade secret. The ITSA provides that:

(b) 'misappropriation' means:

    * * * *

    (2) disclosure or use of a trade secret of a person without express or implied consent by another person who:

      * * * *

      (B) at the time of the disclosure or use, knew or had reason to know that knowledge of the trade secret was:

        * * * *

        (II) acquired under circumstances giving rise to a duty to maintain its secrecy or limit its use[.]

765 ILCS 1065/2 (West 1992).

Defendants assert that Stampede's customer list is not protectable because there was no physical taking of the list, but instead was committed to memory. Defendants contend that any general knowledge or information that they took with them when they left Stampede was not a misappropriation.

In response, Stampede asserts that actual misappropriation of the trade secret information was established because they intentionally copied or memorized the customer list. Stampede argues that the ITSA does not require a plaintiff to prove actual theft or conversion of physical documents embodying the trade secret information. We agree.

Although an employee may take general knowledge or information he or she has developed during their employment, he or she may not take any confidential information, including trade secrets. That taking does not have to be a physical taking by actually copying the names. A trade secret can be misappropriated by physical copying or by memorization.

There was substantial evidence that defendants misappropriated the customer list either through copying down names or through memorization. In fact, defendants admitted that they redeveloped their customer list by remembering the names and locations of at least some of their Stampede customers. Using memorization to rebuild a trade secret does not transform that trade secret from confidential information into non-confidential information. The memorization is one method of misappropriation. Since the trial court's findings were not against the manifest weight of the evidence, we affirm the injunctions.

Finally, we must consider whether the scope of the permanent injunctions are too broad. There are conflicting social and economic policy considerations in every trade secret case. On one hand, a business that has expended substantial amounts of money and time to develop secret advantages over its competitors must be protected against the misappropriation of that information by a prior employee, who was in a position of confidence and trust. On the other hand, it is a fundamental right of an individual to pursue the particular occupation for which he or she is best trained. Injunctive relief should not go beyond the need to protect the legitimate interests of the plaintiff and should not unduly burden the defendant.

Defendants assert that the permanent injunctions are too broad in scope and time to protect Stampede's interests and unnecessarily restrains the livelihood of defendants. Defendants rely on *ILG Industry, Inc.* [*v. Scott*], 49 Ill.2d [88,] 97-98 [(1971)], where the Court restricted the injunction to 18 months, which was the time that the trade secret could have been reproduced using reverse engineering. That court stated that commercial morality is preserved by preventing one from wrongfully using secret information for a period of time no longer than that required to discover or reproduce that information by lawful means.

Defendants also cite *Schulenburg* [*v. Signatrol, Inc.*], 33 Ill.2d [379,] 388 [(1965)], which held that an injunction should be limited in duration to the period of time reasonably required for defendants to legally reproduce the trade secret.

In response, Stampede asserts that the injunction was not overbroad because it expressly limited the scope of the permanent injunctions to a special list of tool jobbers that defendants were assigned while employed at Stampede. In support of that position, Stampede sites *Elmer Miller, Inc.* [*v. Landes*], 253 Ill.App.3d [123,] 135 [(1993)], where the Court found that the injunction was not overbroad.

After considering the public policy and the evidence, we hold that the permanent injunctions are overbroad in their duration, but not their scope. The injunctions are not overbroad in that they expressly limit their scope to a list of tool jobbers that defendants were assigned while employed at Stampede. However, the list of 1,500 names should be divided into two separate lists that include only the customers to which each defendant had access. The injunctions are overbroad in their duration since the average jobber remains in the industry from three to five years and purchases tools from more than one warehouse, and defendants could develop their own customer list from scratch by using the methods they learned at Stampede.

As a result, we reverse the permanent injunctions and modify them to last for four years from July 21, 1991, when the original temporary restraining order was entered. We remand the cause with directions to the circuit court to compile separate customer lists for each defendant. . . .

Notes & Questions

1. In *Omega* the court is straddling the line between the "old" law of trade secrets in Vermont, based on the common law, and the "new" law of trade secrets in Vermont, modeled after the Uniform Trade Secrets Act. Review the relevant sections of the Restatement and of the Uniform Trade Secret Act specifying what constitutes misappropriation. Does the standard for misappropriation in the employment context differ between these two sources of trade secret law?

2. To find a misappropriation in the employment context there must be both a duty and a breach of that duty. In *Omega* the defendants had not signed any agreement promising not to disclose trade secrets. Is that why defendants prevailed? Are there ways for a duty to arise other than through written agreements?

3. Earlier in this Chapter we identified three reasons why evidence of efforts to keep the information secret can be helpful: such efforts demonstrate the owner's

belief that the information is valuable in its secrecy, reflect the secret's monetary value, and reduce the likelihood that the defendant obtained the information from a legitimate or "proper" source. In *Omega* we see another reason why such efforts to maintain secrecy are important to legal protection: An employer's efforts to maintain secrecy help make employees aware of the trade secret status of the information and thus help create the duty the employees owe to their employer to maintain that secrecy.

4. The court notes that there is no restriction on an employee using "general knowledge" gained in an employment setting. Is that just another way of saying such information is "generally known" or "readily ascertainable" by people in the field?

5. A common way for an employee's duty to arise is through a written contract, such as one's employment contract. Entities can use a non-disclosure agreement, or NDA, to impose duties relating to the entity's trade secret information. NDAs can be used with not only employees, of course, but also with independent contractors (such as consultants), business partners, investors, or really anybody.

Imagine you represent a corporation that has been approached by a sole inventor who says she has some valuable information that could be beneficial to the corporation. The inventor has proposed a business relationship, but, before the inventor is willing to disclose the information, she insists the corporation sign an NDA. Because an agreement not to disclose or use information imposes obligations, an entity's willingness to sign an NDA before disclosure depends significantly on its evaluation of the net benefit to be gained by signing. Yet, how can the recipient make that evaluation without first receiving the disclosure? This problem is sometimes referred to as Arrow's Disclosure Paradox, named after economist Kenneth Arrow who described the problem. *See* Kenneth Arrow, *Economic Welfare and the Allocation of Resources for Invention,* in THE RATE AND DIRECTION OF INVENTIVE ACTIVITY: ECONOMIC AND SOCIAL FACTORS 609, 614–16 (1962).

6. One way to balance the competing interests at stake in a trade secret case that involves departing employees is to craft the injunction in the case carefully such that it both protects the former employer's valuable trade secrets and allows the former employee adequate mobility. Did the court in *Stampede* strike the right balance? If you were Stampede would you feel that the revised injunction sufficiently protected your trade secrets? If you were the employees would you be able to "pursue the particular occupation for which [you were] best trained"?

B. The Inevitable Disclosure Doctrine

Michael J. Garrison & John T. Wendt, *The Evolving Law of Employee Noncompete Agreements: Recent Trends and an Alternative Policy Approach*, 45 Am. Bus. L.J. 107, 149-52 (2008)

The inevitable disclosure doctrine preceded the widespread adoption of the UTSA and the *PepsiCo* decision.[*] As a matter of common law, the doctrine developed in a series of cases involving threatened misuse of valuable, technical trade secrets by former employees hired away by competitors seeking to gain entry into highly competitive markets. Thus, the doctrine in its infancy was designed to prevent an imminent threat of a trade secret disclosure by a former employee, a particularly damaging form of unfair competition. The leading case is *E.I. duPont de Nemours & Co. [DuPont] v. American Potash & Chemical Corp.*,[200 A.2d 428 (Del. Ch. 1964),] in which the Delaware Chancery Court granted an injunction in favor of DuPont against a former employee who had access to a valuable trade secret process. Donald Hirsch, a chemical engineer with a doctorate degree, had been involved in the long and expensive effort to develop DuPont's chloride process for manufacturing certain pigments. After American Potash was unsuccessful in securing a license from DuPont for that process, it decided to develop its own chloride process. As part of that initiative, it hired Hirsch as a technical manager for a new pigment manufacturing plant. DuPont brought suit to enforce its confidentiality agreement, and it secured a preliminary injunction stopping Hirsch from working for American Potash in the field of chloride process development.

In rejecting a summary judgment motion made by the defendants, the court relied on the inevitability of the disclosure of the trade secrets by Hirsch in finding a sufficient threat of improper disclosure. The court viewed inevitability as a factor that it was justified in considering in weighing the probabilities of a trade secret disclosure. The court reasoned that the protection of trade secrets is important to society because it encourages investment in R&D, although there is also a strong countervailing interest in protecting the employee's right to use knowledge and skills that may be "inextricably interwoven with his knowledge of the trade secrets."[*Id.* at 437.] Given other facts pointing to the probability of an unlawful disclosure, including a statement by Hirsch conceding the potential for a conflict of interest occurring in his employment with American Potash, the court left the ultimate resolution for trial, with the preliminary injunction in place to preserve the status quo.

The *DuPont* opinion was followed by the federal district court opinion in *Allis-Chalmers Manufacturing Co. v. Continental Aviation & Engineering Corp.*, [255 F. Supp. 645 (E.D. Mich. 1966),] a case that bore a striking factual similarity to *DuPont*. Allis-Chalmers marketed a specialized type of fuel injector pump. At that time, only three companies marketed this type of pump, and at least eight American and foreign companies had attempted but failed to develop comparable equipment. Allis-Chalmers negotiated with Continental Aviation regarding the sale of its distributor pumps, but Continental rejected a licensing agreement that would have granted Continental the right to use Allis-Chalmer's patent and trade secret rights in the manufacture of pumps for the military. The negotiations having failed, Continental hired George Wolff, an engineer who was instrumental in the development of Allis-Chalmers' distributor pump, to work on the design and development of Continental's fuel injection systems and pumps.

---

[*] [*Eds. Note*: The *PepsiCo* case is the next principal case in these materials.]

In granting a preliminary injunction, which prevented Wolff from doing certain work for Continental, the court found an "inevitable and imminent" risk of trade secret misappropriation. [*Id.* At 654.] This was based on the "negotiations, relating to distributor type fuel injection pumps . . . the nature of the research and development work done by Mr. Wolff . . . at Allis-Chalmers, [and] the nature of the type of work Mr. Wolff is to perform at Continental." [*Id.*] Given these facts, the court noted the "virtual impossibility of Mr. Wolff performing all of his prospective duties for Continental to the best of his ability, without in effect giving it the benefit of Allis-Chalmers' confidential information." [*Id.*]

The court granted an injunction that prevented Wolff from working in the design of distributor-type pumps, but allowed him to work for Continental on other projects, including other pumps and fuel injection systems. As in *DuPont*, the court was mindful of the competing legal principles at play and the need to protect the rights of employees to market their general knowledge and skills while at the same time protecting the intellectual property of employers. It therefore provided a limited preliminary injunction, one that was narrowly tailored to protect the threatened trade secrets.

Allis-Chalmers and *DuPont* represent the prevailing approach to the inevitable disclosure doctrine prior to *PepsiCo*. Both cases involved highly specialized and technical trade secrets that gave the businesses a substantial advantage in the market. Both also involved competitors that were apparently attempting to acquire the protected technology by hiring away high-level scientific personnel. Finally, the former employees would not have been able to perform their new responsibilities without using or disclosing their former employer's trade secrets. Because there was an imminent threat of misappropriation, and an injunction against disclosure would not have been an adequate remedy, the courts granted narrow injunctions preventing employment of the former employees in positions where the trade secrets were at risk.

PepsiCo, Inc. v. Redmond
54 F.3d 1262 (7th Cir. 1995)

FLAUM, JUDGE:

Plaintiff PepsiCo, Inc., sought a preliminary injunction against defendants William Redmond and the Quaker Oats Company to prevent Redmond, a former PepsiCo employee, from divulging PepsiCo trade secrets and confidential information in his new job with Quaker and from assuming any duties with Quaker relating to beverage pricing, marketing, and distribution. The district court agreed with PepsiCo and granted the injunction. We now affirm that decision.

I.

The facts of this case lay against a backdrop of fierce beverage-industry competition between Quaker and PepsiCo, especially in "sports drinks" and "new age drinks." Quaker's sports drink, "Gatorade," is the dominant brand in its market niche. PepsiCo introduced its Gatorade rival, "All Sport," in March and April of 1994, but sales of All Sport lag far behind those of Gatorade. Quaker also has the lead in the new-age-drink category. Although PepsiCo has entered the market through joint ventures with the Thomas J. Lipton Company and Ocean Spray Cranberries, Inc., Quaker purchased Snapple Beverage Corp., a large new-age-drink maker, in late 1994. PepsiCo's products have about half of Snapple's market share. Both companies see 1995 as an important year for their products: PepsiCo has developed extensive plans to increase its market presence, while Quaker is trying to solidify its lead by integrating Gatorade and Snapple distribution. Meanwhile, PepsiCo and Quaker each face strong competition from Coca Cola Co., which has its own sports drink, "PowerAde," and which introduced its own Snapple-rival, "Fruitopia," in 1994, as well as from independent beverage producers.

William Redmond, Jr., worked for PepsiCo in its PepsiCola North America division ("PCNA") from 1984 to 1994. Redmond became the General Manager of the Northern California Business Unit in June, 1993, and was promoted one year later to General Manager of the business unit covering all of California, a unit having annual revenues of more than 500 million dollars and representing twenty percent of PCNA's profit for all of the United States.

Redmond's relatively high-level position at PCNA gave him access to inside information and trade secrets. Redmond, like other PepsiCo management employees, had signed a confidentiality agreement with PepsiCo. That agreement stated in relevant part that he

> would not disclose at any time, to anyone other than officers or employees of [PepsiCo], or make use of, confidential information relating to the business of [PepsiCo] . . . obtained while in the employ of [PepsiCo], which shall not be generally known or available to the public or recognized as standard practices.

Donald Uzzi, who had left PepsiCo in the beginning of 1994 to become the head of Quaker's Gatorade division, began courting Redmond for Quaker in May, 1994. Redmond met in Chicago with Quaker officers in August, 1994, and on October 20, 1994, Quaker, through Uzzi, offered Redmond the position of Vice President-On Premise Sales for Gatorade. Redmond did not then accept the offer but continued to negotiate for more money. Throughout this time, Redmond kept his dealings with Quaker secret from his employers at PCNA.

On November 8, 1994, Uzzi extended Redmond a written offer for the position of Vice President-Field Operations for Gatorade and Redmond accepted. Later that same day, Redmond called William Bensyl, the Senior Vice President of Human Resources for PCNA, and told him that he had an offer from Quaker to become the Chief Operating Officer of the combined Gatorade and Snapple company but had not yet accepted it. Redmond also asked whether he should, in light of the offer, carry out his plans to make calls upon certain PCNA customers. Bensyl told Redmond to make the visits.

Redmond also misstated his situation to a number of his PCNA colleagues, including Craig Weatherup, PCNA's President and Chief Executive Officer, and Brenda Barnes, PCNA's Chief Operating Officer and Redmond's immediate superior. As with Bensyl, Redmond told them that he had been offered the position of Chief Operating Officer at Gatorade and that he was leaning "60/40" in favor of accepting the new position.

On November 10, 1994, Redmond met with Barnes and told her that he had decided to accept the Quaker offer and was resigning from PCNA. Barnes immediately took Redmond to Bensyl, who told Redmond that PepsiCo was considering legal action against him.

True to its word, PepsiCo filed this diversity suit on November 16, 1994, seeking a temporary restraining order to enjoin Redmond from assuming his duties at Quaker and to prevent him from disclosing trade secrets or confidential information to his new employer. The district court granted PepsiCo's request that same day but dissolved the order sua sponte two days later, after determining that PepsiCo had failed to meet its burden of establishing that it would suffer irreparable harm. The court found that PepsiCo's fears about Redmond were based upon a mistaken understanding of his new position at Quaker and that the likelihood that Redmond would improperly reveal any confidential information did not "rise above mere speculation."

From November 23, 1994, to December 1, 1994, the district court conducted a preliminary injunction hearing on the same matter. At the hearing, PepsiCo offered evidence of a number of trade secrets and confidential information it desired protected and to which Redmond was privy. First, it identified PCNA's "Strategic Plan," an annually revised document that contains PCNA's plans to compete, its financial goals, and its strategies for manufacturing, production, marketing, packaging, and distribution for the coming three years. Strategic Plans are developed by Weatherup and his staff with input from PCNA's general managers, including Redmond, and are considered highly confidential. The Strategic Plan derives much of its value from the fact that it is secret and competitors cannot anticipate PCNA's next moves. PCNA managers received the most recent Strategic Plan at a meeting in July, 1994, a meeting Redmond attended. PCNA also presented information at the meeting regarding its plans for Lipton ready-to-drink teas and for All Sport for 1995 and beyond, including new flavors and package sizes.

Second, PepsiCo pointed to PCNA's Annual Operating Plan ("AOP") as a trade secret. The AOP is a national plan for a given year and guides PCNA's financial goals, marketing plans, promotional event calendars, growth expectations, and operational changes in that year. The AOP, which is implemented by PCNA unit General Managers, including Redmond, contains specific information regarding all PCNA initiatives for the forthcoming year. The AOP bears a label that reads "Private and Confidential—Do Not Reproduce" and is considered highly confidential by PCNA managers.

In particular, the AOP contains important and sensitive information about "pricing architecture"—how PCNA prices its products in the marketplace. Pricing architecture covers both a national pricing approach and specific price points for given areas. Pricing architecture also encompasses PCNA's objectives for All Sport and its new age drinks with reference to trade channels, package sizes and other characteristics of both the products and the customers at which the products are aimed. Additionally, PCNA's pricing architecture outlines PCNA's customer development agreements. These agreements between PCNA and retailers provide for the retailer's participation in certain merchandising activities for PCNA products. As with other information contained in the AOP, pricing architecture is highly confidential and would be extremely valuable to a competitor. Knowing PCNA's pricing architecture would allow a competitor to anticipate PCNA's pricing moves and underbid PCNA strategically whenever and wherever the competitor so desired. PepsiCo introduced evidence that Redmond had detailed knowledge of PCNA's pricing architecture and that he was aware of and had been involved in preparing PCNA's customer development agreements with PCNA's California and California-based national customers. Indeed, PepsiCo showed that Redmond, as the General Manager for California, would have been responsible for implementing the pricing architecture guidelines for his business unit.

PepsiCo also showed that Redmond had intimate knowledge of PCNA "attack plans" for specific markets. Pursuant to these plans, PCNA dedicates extra funds to supporting its brands against other brands in selected markets. To use a hypothetical example, PCNA might budget an additional $ 500,000 to spend in Chicago at a particular time to help All Sport close its market gap with Gatorade. Testimony and documents demonstrated Redmond's awareness of these plans and his participation in drafting some of them.

Finally, PepsiCo offered evidence of PCNA trade secrets regarding innovations in its selling and delivery systems. Under this plan, PCNA is testing a new delivery system that could give PCNA an advantage over its competitors in negotiations with retailers over shelf space and merchandising. Redmond has knowledge of this secret because PCNA, which has invested over a million dollars in developing the system during the past two years, is testing the pilot program in California.

Having shown Redmond's intimate knowledge of PCNA's plans for 1995, PepsiCo argued that Redmond would inevitably disclose that information to Quaker in his new position, at which he would have substantial input as to Gatorade and Snapple pricing, costs, margins, distribution systems, products, packaging and marketing, and could give Quaker an unfair advantage in its upcoming skirmishes with PepsiCo. Redmond and Quaker countered that Redmond's primary initial duties at Quaker as Vice President– Field Operations would be to integrate Gatorade and Snapple distribution and then to manage that distribution as well as the promotion, marketing and sales of these products. Redmond asserted that the integration would be conducted according to a pre-existing plan and that his special knowledge of PCNA strategies would be irrelevant. This irrelevance would derive not only from the fact that Redmond would be implementing pre-existing plans but also from the fact that PCNA and Quaker distribute their products in entirely different ways: PCNA's distribution system is vertically integrated (i.e., PCNA owns the system) and delivers its product directly to retailers, while Quaker ships its product to wholesalers and customer warehouses and relies on independent distributors. The defendants also pointed out that Redmond had signed a confidentiality agreement with Quaker preventing him from disclosing "any confidential information belonging to others," as well as the Quaker Code of Ethics, which prohibits employees from engaging in "illegal or improper acts to acquire a competitor's trade secrets." Redmond additionally promised at the hearing that should he be faced with a situation at Quaker that might involve the use or disclosure of PCNA information, he would seek advice from Quaker's in-house counsel and would refrain from making the decision.

PepsiCo responded to the defendants' representations by pointing out that the evidence did not show that Redmond would simply be implementing a business plan already in place. On the contrary, as of November, 1994, the plan to integrate Gatorade and Snapple distribution consisted of a single distributorship agreement and a two page "contract terms summary." Such a basic plan would not lend itself to widespread application among the over 300 independent Snapple distributors. Since the integration process would likely face resistance from Snapple distributors and Quaker had no scheme to deal with this probability, Redmond, as the person in charge of the integration, would likely have a great deal of influence on the process. PepsiCo further argued that Snapple's 1995 marketing and promotion plans had not necessarily been completed prior to Redmond's joining Quaker, that Uzzi disagreed with portions of the Snapple plans, and that the plans were open to re-evaluation. Uzzi testified that the plan for integrating Gatorade and Snapple distribution is something that would happen in the future. Redmond would therefore likely have input in remaking these plans, and if he did, he would inevitably be making decisions with PCNA's strategic plans and 1995 AOP in mind. Moreover, PepsiCo continued, diverging testimony made it difficult to know exactly what Redmond would be doing at Quaker. Redmond described his job as "managing the entire sales effort of Gatorade at the field level, possibly including strategic planning," and at least at one point considered his job to be equivalent to that of a Chief Operating Officer. Uzzi, on the other hand, characterized Redmond's position as "primarily and initially to restructure and integrate our—the distribution systems for Snapple and for Gatorade, as per our distribution plan" and then to "execute marketing, promotion and sales plans in the marketplace." Uzzi also denied having given Redmond detailed information about any business plans, while Redmond described such a plan in depth in an affidavit and said that he received the information from Uzzi. Thus, PepsiCo asserted, Redmond would have a high position in the Gatorade hierarchy, and PCNA trade secrets and confidential information would necessarily influence his decisions. Even if Redmond could somehow refrain from relying on this information, as he promised he would, his actions in leaving PCNA, Uzzi's actions in hiring Redmond, and the varying testimony regarding Redmond's new responsibilities, made Redmond's assurances to PepsiCo less than comforting.

On December 15, 1994, the district court issued an order enjoining Redmond from assuming his position at Quaker through May, 1995, and permanently from using or disclosing any PCNA trade secrets or confidential information. . . . The court . . . found that Redmond's new job posed a clear threat of misappropriation of trade secrets and confidential information that could be enjoined under Illinois statutory and common law. The court also emphasized Redmond's lack of forthrightness both in his activities before accepting his job with Quaker and in his testimony as factors leading the court to believe the threat of misappropriation was real. This appeal followed.

II.

Both parties agree that the primary issue on appeal is whether the district court correctly concluded that PepsiCo had a reasonable likelihood of success on its various claims for trade secret misappropriation and breach of a confidentiality agreement. We review the district court's legal conclusions in issuing a preliminary injunction de novo and its factual determinations and balancing of the equities for abuse of discretion.

. . . .

The question of threatened or inevitable misappropriation in this case lies at the heart of a basic tension in trade secret law. Trade secret law serves to protect "standards of commercial morality" and "encourage[ ] invention and innovation" while maintaining "the public interest in having free and open competition in the manufacture and sale of unpatented goods." 2 Jager, [Trade Secrets Law] § IL.03 at IL-12. Yet that same law should not prevent workers from pursuing their livelihoods when they leave their current positions. It has been said that federal age discrimination law does not guarantee tenure for older employees. Similarly, trade secret law does not provide a reserve clause[*] for solicitous employers.

This tension is particularly exacerbated when a plaintiff sues to prevent not the actual misappropriation of trade secrets but the mere threat that it will occur. While the ITSA [Illinois Trade Secret Act] plainly permits a court to enjoin the threat of misappropriation of trade secrets, there is little law in Illinois or in this circuit establishing what constitutes threatened or inevitable misappropriation. Indeed, there are only two cases in this circuit that address the issue: *Teradyne, Inc. v. Clear Communications Corp.*, 707 F. Supp. 353 (N.D. Ill 1989), and *AMP Inc. v. Fleischhacker*, 823 F.2d 1199 (7th Cir. 1987).

In *Teradyne*, Teradyne alleged that a competitor, Clear Communications, had lured employees away from Teradyne and intended to employ them in the same field. In an insightful opinion, Judge Zagel observed that "threatened misappropriation can be enjoined under Illinois law" where there is a "high degree of probability of inevitable and immediate . . . use of . . . trade secrets." *Teradyne*, 707 F. Supp. at 356. Judge Zagel held, however, that Teradyne's complaint failed to state a claim because Teradyne did not allege "that defendants have in fact threatened to use Teradyne's secrets or that they will inevitably do so." Teradyne's claims would have passed Rule 12(b)(6) muster had they properly alleged inevitable disclosure, including a statement that Clear intended to use Teradyne's trade secrets or that the former Teradyne employees had disavowed their confidentiality agreements with Teradyne, or an allegation that Clear could not operate without Teradyne's secrets. However,

> the defendants' claimed acts, working for Teradyne, knowing its business, leaving its business, hiring employees from Teradyne and entering the same

---

* [*Eds. Note: Black's Law Dictionary* defines a "reserve clause" as "[a] clause in a professional athlete's contract restricting the athlete's right to change teams, even after the contract expires." BLACK'S LAW DICTIONARY 1334 (8th ed. 2004).]

field (though in a market not yet serviced by Teradyne) do not state a claim of threatened misappropriation. All that is alleged, at bottom, is that defendants could misuse plaintiff's secrets, and plaintiffs fear they will. This is not enough. It may be that little more is needed, but falling a little short is still falling short.

Id. at 357.

In *AMP*, . . . we emphasized that the mere fact that a person assumed a similar position at a competitor does not, without more, make it "inevitable that he will use or disclose . . . trade secret information" so as to "demonstrate irreparable injury." *Id.*

. . . .

The ITSA, *Teradyne*, and *AMP* lead to the same conclusion: a plaintiff may prove a claim of trade secret misappropriation by demonstrating that defendant's new employment will inevitably lead him to rely on the plaintiff's trade secrets. See also 1 Jager, supra, § 7.02[2][a] at 7-20 (noting claims where "the allegation is based on the fact that the disclosure of trade secrets in the new employment is inevitable, whether or not the former employee acts conciously or unconsciously"). The defendants are incorrect that Illinois law does not allow a court to enjoin the "inevitable" disclosure of trade secrets. Questions remain, however, as to what constitutes inevitable misappropriation and whether PepsiCo's submissions rise above those of the *Teradyne* and *AMP* plaintiffs and meet that standard. We hold that they do.

PepsiCo presented substantial evidence at the preliminary injunction hearing that Redmond possessed extensive and intimate knowledge about PCNA's strategic goals for 1995 in sports drinks and new age drinks. The district court concluded on the basis of that presentation that unless Redmond possessed an uncanny ability to compartmentalize information, he would necessarily be making decisions about Gatorade and Snapple by relying on his knowledge of PCNA trade secrets. It is not the "general skills and knowledge acquired during his tenure with" PepsiCo that PepsiCo seeks to keep from falling into Quaker's hands, but rather "the particularized plans or processes developed by [PCNA] and disclosed to him while the employer-employee relationship existed, which are unknown to others in the industry and which give the employer an advantage over his competitors." *AMP*, 823 F.2d at 1202. The *Teradyne* and *AMP* plaintiffs could do nothing more than assert that skilled employees were taking their skills elsewhere; PepsiCo has done much more.

Admittedly, PepsiCo has not brought a traditional trade secret case, in which a former employee has knowledge of a special manufacturing process or customer list and can give a competitor an unfair advantage by transferring the technology or customers to that competitor. PepsiCo has not contended that Quaker has stolen the All Sport formula or its list of distributors. Rather PepsiCo has asserted that Redmond cannot help but rely on PCNA trade secrets as he helps plot Gatorade and Snapple's new course, and that these secrets will enable Quaker to achieve a substantial advantage by knowing exactly how PCNA will price, distribute, and market its sports drinks and new age drinks and being able to respond strategically. This type of trade secret problem may arise less often, but it nevertheless falls within the realm of trade secret protection under the present circumstances.

Quaker and Redmond assert that they have not and do not intend to use whatever confidential information Redmond has by virtue of his former employment. They point out that Redmond has already signed an agreement with Quaker not to disclose any trade secrets or confidential information gleaned from his earlier employment. They also note with regard to distribution systems that even if Quaker wanted to steal information about PCNA's distribution plans, they would be completely useless in attempting to integrate the Gatorade and Snapple beverage lines.

The defendants' arguments fall somewhat short of the mark. Again, the danger of misappropriation in the present case is not that Quaker threatens to use PCNA's secrets to create distribution systems or coopt PCNA's advertising and marketing ideas. Rather, PepsiCo believes that Quaker, unfairly armed with knowledge of PCNA's plans, will be able to anticipate its distribution, packaging, pricing, and marketing moves. Redmond and Quaker even concede that Redmond might be faced with a decision that could be influenced by certain confidential information that he obtained while at PepsiCo. In other words, PepsiCo finds itself in the position of a coach, one of whose players has left, playbook in hand, to join the opposing team before the big game. Quaker and Redmond's protestations that their distribution systems and plans are entirely different from PCNA's are thus not really responsive.

The district court also concluded from the evidence that Uzzi's actions in hiring Redmond and Redmond's actions in pursuing and accepting his new job demonstrated a lack of candor on their part and proof of their willingness to misuse PCNA trade secrets, findings Quaker and Redmond vigorously challenge. The court expressly found that:

> Redmond's lack of forthrightness on some occasions, and out and out lies on others, in the period between the time he accepted the position with defendant Quaker and when he informed plaintiff that he had accepted that position leads the court to conclude that defendant Redmond could not be trusted to act with the necessary sensitivity and good faith under the circumstances in which the only practical verification that he was not using plaintiff's secrets would be defendant Redmond's word to that effect.

The facts of the case do not ineluctably dictate the district court's conclusion. Redmond's ambiguous behavior toward his PepsiCo superiors might have been nothing more than an attempt to gain leverage in employment negotiations. The discrepancy between Redmond's and Uzzi's comprehension of what Redmond's job would entail may well have been a simple misunderstanding. The court also pointed out that Quaker, through Uzzi, seemed to express an unnatural interest in

hiring PCNA employees: all three of the people interviewed for the position Redmond ultimately accepted worked at PCNA. Uzzi may well have focused on re-
cruiting PCNA employees because he knew they were good and not because of
their confidential knowledge. Nonetheless, the district court, after listening to the
witnesses, determined otherwise. That conclusion was not an abuse of discretion.

That conclusion also renders inapposite the defendants' reliance on *Cincinnati
Tool Steel Co. v. Breed*, 482 N.E.2d 170 (Ill. App. 2d Dist. 1985). In *Cincinnati Tool*,
the court held that the defendant's "express denial that she had disclosed or
would disclose any confidential information or that she even possessed such information" left the plaintiff without a case, one that could not be saved "merely by
offering evidence that defendant used customer and price data in her work while
employed by plaintiff." 482 N.E.2d at 180. In the instant case, the district court
simply did not believe the denials and had reason to do so.

Thus, when we couple the demonstrated inevitability that Redmond would
rely on PCNA trade secrets in his new job at Quaker with the district court's reluctance to believe that Redmond would refrain from disclosing these secrets in
his new position (or that Quaker would ensure Redmond did not disclose them),
we conclude that the district court correctly decided that PepsiCo demonstrated
a likelihood of success on its statutory claim of trade secret misappropriation.

. . . .

Bayer Corp. v. Roche Molecular Systems
72 F. Supp. 2d 1111 (N.D. Cal. 1999)

ALSUP, JUDGE:

. . . .

Last February, Pete Betzelos quit his job as HIV Marketing Manager for the
World-Wide Marketing Group at Bayer Corporation ("Bayer") to go to work for
a competitor, Roche Molecular Systems, Inc. ("Roche"), as its International Marketing Manager, HIV. There is a clear overlap in job responsibilities.

. . . Before he left to work at Roche, Mr. Betzelos worked in Bayer's Nucleic
Acid Diagnostics group, which Bayer had bought from Chiron Corporation ("Chiron") in late 1998. . . .

According to Mr. Betzelos, when he began working for Roche, he signed a contract in which he agreed not to disclose or use confidential information. On March
23, 1999, Mr. Betzelos signed a further undertaking for his employers at Roche in
which he reaffirmed his agreement that he would not use or disclose any confidential or proprietary information of Chiron Diagnostics in his performance of his
position as the International HIV Marketing Manager for Roche. The undertaking
stated as follows:

> I have had an opportunity to read the order by United States District Court
> Judge Armstrong dated March 15, 1999. I realize the extent to which Judge
> Armstrong has relied on my representations that I will not use or disclose
> the trade secrets of Chiron Diagnostics in my employment with RMS. In

addition, I have reviewed the business documents attached to Bayer's moving papers. Taking all of the foregoing into account, I now wish to reiterate and re-affirm my agreement that I will not use or disclose any confidential or proprietary information of Chiron Diagnostics in the performance of my position as the International Marketing Manager for HIV products for RMS.

The Alleged Trade Secrets

Bayer alleges that Mr. Betzelos developed or knew of many Bayer trade secrets, including marketing strategies and confidential information about Bayer's products. Bayer lists the alleged trade secrets that it fears Mr. Betzelos will use in its supplemental interrogatory responses.

According to Dr. Urdea, Bayer took steps to maintain the secrecy of its confidential information. Bayer limited access to its computers and email through use of passwords and limited access to its offices. Bayer used electronic keys to limit access to the floor where Mr. Betzelos worked. Bayer notified employees that information at Bayer was confidential, and that it should be treated as such. Bayer distributed confidentiality agreements for employees to review, sign and return. In addition, Bayer generally limited the distribution of confidential information to senior management, and to others only on a need-to-know basis. Bayer limited the distribution of strategic plans to which Mr. Betzelos had access.

Bayer alleges that at least once a Roche employee has solicited confidential Bayer information from Mr. Betzelos. Bayer points to an email from Robert Degnan, Roche's National Sales Manager for PCR, to Mr. Betzelos regarding an account named ACT-G that Bayer and Roche were allegedly competing for. Mr. Degnan wrote "[d]o you have the Chiron pricing for ACT-G? We are in the process of negotiating with them". Mr. Betzelos testified at deposition that he had not known the answer to Mr. Degnan's question and that he had not answered him. Mr. Betzelos did not report the email, nor did he tell Mr. Degnan that such a request was inappropriate.

Bayer claims that Mr. Betzelos has already used Bayer's trade secrets in three areas: reimbursement, automation, and specificity. Roche counters with evidence that the alleged trade secrets were generally known, that Roche already knew them, and/or that Roche did not use them. The details of these contentions are set forth in a separate order filed under seal to protect the details of the alleged trade secrets.

. . . .

DISCUSSION

. . . .

... The Court may enjoin "[a]ctual or threatened misappropriation" of a trade secret. Cal. Civ. Code § 3426.2(a). A corporation misappropriates a trade secret when (1) it discloses or uses the trade secret of another without express or implied consent, and (2) at the time of the disclosure or use, it knew or had reason to know that its knowledge of the trade secret was derived from a person who owed a duty to the entity seeking relief to maintain the trade secret's secrecy or limit its use. Cal. Civ. Code § 3426.1(b)(2)(B)(iii). The Court may order affirmative acts to protect a trade secret in appropriate circumstances. Cal. Civ. Code § 3426.2(c).

The Court finds it likely that Bayer will prove all the elements of its trade secrets case but one: actual or threatened use or disclosure. On that issue, Bayer raises factual circumstances of legitimate concern, but, in light of the overall record, which the Court has carefully read, Bayer has not carried its burden for obtaining preliminary relief. In brief, the Court finds that Bayer's specific evidence of actual use fails, either because the information was not really private or because it was already known by Roche. One item of contention (concerning automation) is a close question, but the evidence still tips in Roche's favor. Among other things, the counter-strategy allegedly advocated by Mr. Betzelos at Roche was already being promoted by Roche before his arrival. The Court also notes its reliance on Mr. Betzelos' undertaking, in which he affirmed that he would not use or disclose Bayer's trade secrets. The fact specifics are discussed in the sealed order.

This brings us to Bayer's theory of inevitable disclosure. Invoking this concept, Bayer argues that the disclosure of its trade secrets is unavoidable because Mr. Betzelos inevitably will use them in his new position with Roche. Citing *PepsiCo v. Redmond*, 54 F.3d 1262 (7th Cir. 1995), Bayer claims that a plaintiff may prove a claim of threatened misappropriation of a trade secret by demonstrating that a former employee's new employment will inevitably lead him to rely on plaintiff's trade secret.

. . . .

The *PepsiCo* decision has given new life to the theory of inevitable disclosure. Previously, courts had recognized the theory, but were reluctant to apply it because of strong public policies in favor of employee mobility. One commentator contends that 21 states (not including California) now allow plaintiffs to show liability for trade secrets misappropriation under the theory.[4] The theory allows plaintiff employers to demonstrate threatened misappropriation without evidence of an employee's intent to disclose trade secrets.

"In finding a likelihood of disclosure, other courts that have applied the inevitable disclosure theory have considered the degree of competition between the former and new employer, and the new employer's efforts to safeguard the former's employer's trade secrets, and the former employee's 'lack of forthrightness both in his activities before accepting his job . . . and in his testimony,' as well as the degree of similarity between the employee's former and current position." *Merck & Co. v. Lyon*, 941 F. Supp. 1443, 1460 (M.D.N.C. 1996) (holding that the inevitable disclosure theory can be applied under North Carolina law, where (1) injunction is limited to protecting specifically defined trade secrets and (2) the trade secret is clearly identified and of significant value) (internal citations omitted).

---

[4] *See* Stephen L. Sheinfeld & Jennifer M. Chow, *Protecting Employer Secrets and the "Doctrine of Inevitable Disclosure,"* 600 PLI/LIT 367 (1999).

A decision from the Southern District of Texas has identified the following factors for determining whether a trade secret will be "inevitably disclosed":

(1) Is the new employer a competitor? (2) What is the scope of the defendant's new job? (3) Has the employee been less than candid about his new position? (4) Has plaintiff clearly identified the trade secrets which are at risk? (5) Has actual trade secret misappropriation already occurred? (6) Did the employee sign a nondisclosure and/or non-competition agreement? (7) Does the new employer have a policy against use of others' trade secrets? (8) Is it possible to 'sanitize' the employee's new position?

Maxxim Medical v. Michelson, 51 F. Supp. 2d 773, 786 (S.D. Tex. 1999), *rev'd on other grounds*, 182 F.3d 915 (5th Cir. 1999). . . .

. . . .

More on point are two decisions from courts in California—decisions rejecting the theory of inevitable disclosure. *Danjaq LLC v. Sony Corp.*, 50 U.S.P.Q.2d 1638, 1999 WL 317629 (C.D. Cal. Mar. 12, 1999), overlooked by both parties, involved an allegation of actual misappropriation of trade secrets relating to a "James Bond" motion picture. There, the court stated in footnote one:

[l]acking proof of actual disclosure and actual use, the Plaintiffs fill-in the gaps in the record with the 'inevitable disclosure doctrine' articulated in *PepsiCo v. Redmond*. But the Plaintiffs' reliance on the inevitable disclosure doctrine is misplaced. *PepsiCo is not the law of the State of California or the Ninth Circuit.*

Id. at 1640, 1999 WL 317629 (citation omitted) (emphasis added).

Danjaq was followed in *Computer Sciences Corp. v. Computer Associates Int'l, Inc.*, 1999 WL 675446, *16 (C.D. Cal. Aug. 12, 1999) (reiterating that inevitable disclosure doctrine "is not the law of the State of California or the Ninth Circuit").

In determining the law of California, the Court agrees with *Danjaq* and *Computer Sciences*. California public policy favors employee mobility and freedom. California Business and Professions Code Section 16600 provides that "every contract by which anyone is restrained from engaging in a lawful profession, trade, or business of any kind is to that extent void." Reading this language broadly, California courts generally do not enforce covenants not to compete. Nor do they generally enforce covenants that seek to avoid the policy by penalizing former employees who compete with their former employers. In the words of the California Supreme Court:

Equity will to the fullest extent protect the property rights of employers in their trade secrets and otherwise, but public policy and natural justice require that equity should also be solicitous for the right inherent in all people, not fettered by negative covenants upon their part to the contrary, to follow any of the common occupations of life. Every individual possesses as a form of property, the right to pursue any calling, business or profession he may choose. A former employee has the right to engage in a competitive business for himself and to enter into competition with his former employer, even for the business of those who had formerly been the customers of his former employer, provided such competition is fairly and legally conducted.

Continental Car-Na-Var Corp. v. Moseley, 24 Cal.2d 104, 110 (1944). To the extent that the theory of inevitable disclosure creates a de facto covenant not to compete without a nontrivial showing of actual or threatened use or disclosure, it is inconsistent with California policy and case law.

In sum, the Court holds that California trade-secrets law does not recognize the theory of inevitable disclosure; indeed, such a rule would run counter to the strong public policy in California favoring employee mobility. A trade-secrets plaintiff must show an actual use or an actual threat. Once a nontrivial violation is shown, however, a court may consider all of the factors considered by the jurisdictions allowing the theory in determining the possible extent of the irreparable injury. In other words, once the employee violates the trade-secrets law in a nontrivial way, the employee forfeits the benefit of the protective policy in California. For example, that a high-level employee takes a virtually identical job at the number one competitor in a fiercely competitive industry would be a factor militating in favor of a broader injunction once sufficient evidence of a nontrivial violation is shown. In the present case, however, sufficient evidence of such a nontrivial violation has not been shown.

Balance of the Hardships

Where, as here, a moving party has shown only serious questions going to the merits, that party must also show that the balance of hardships tip sharply in its favor in order to prevail on its motion for a preliminary injunction. To determine which way the balance of the hardships tips, a court must identify the possible harm caused by the preliminary injunction against the possibility of the harm caused by not issuing it. Here, Bayer and Roche both face hardship. Should the Court decline to enjoin Mr. Betzelos from working on Roche products that compete with Bayer products, and Roche subsequently uses Bayer's trade secrets, the value of those secrets will be lost to Bayer, and damages will not repair the loss. That hardship has, however, diminished with time, for the trade secrets have lost currency. If, on the other hand, the Court enjoins Mr. Betzelos from working on Roche's HIV Amplicor products, Roche would have to start over looking for a person to fill Mr. Betzelos' position. Not only would Roche face an immediate void, but his successor, when found, would need to spend time getting up to speed. Regarding whether Roche should be enjoined from continuing to employ Mr. Betzelos, the balance of hardships does not tip sharply in favor of Bayer. Nonetheless, the balance of the hardships certainly supports the minimum relief outlined below, namely an order (1) to Mr. Betzelos directing him to honor his undertaking to the Court (quoted earlier) not to use or disclose Bayer's confidential and/or proprietary information and (2) to Mr. Betzelos and Roche to stand for the periodic examinations in discovery imposed below.

Discovery, Monitoring, and Possible Renewal of the Motion

Although Roche's counter evidence is strong enough to prevent a probability of success on the merits, Bayer has raised substantial questions about Mr. Betzelos' conduct. Although not compelling, a case has been made that he actually used his knowledge of Bayer's defensive strategy regarding automation to promote Roche's counter strategy on the same subject. Bayer demonstrates, too, a clear instance of a Roche employee trying to dig Chiron/ Bayer proprietary information from Mr. Betzelos. And, Mr. Betzelos' evasive manner at his deposition[*] likewise causes serious concern.

The Court finds the record more than ample to support periodic and targeted discovery permitting counsel for Bayer to learn of Mr. Betzelos' communications and activities most likely to put Bayer's trade secrets at risk. Accordingly, the Court will impose ongoing discovery obligations on Roche and Mr. Betzelos. Should the discovery reveal evidence of use or disclosure of Bayer's confidential or proprietary information, Bayer may renew its motion for a preliminary injunction. . . .

. . . .

Notes & Questions

1. Section 2(a) of the UTSA provides that '[a]ctual or threatened misappropriation may be enjoined." Why might a state want to empower its courts to enjoin "threatened misappropriation"?

2. Are threatened misappropriation and inevitable disclosure two names for the same theory, or are they completely different theories?

> Because of the UTSA's prohibition against threatened misappropriation, it would seem that states that have adopted the UTSA would have more favorable treatment of the inevitable disclosure doctrine than those that have not. However, in keeping with the inconsistencies in this area of the law, that is not the case. Some states that have adopted the UTSA, like California, have rejected the inevitable disclosure theory. On the other hand, New York and New Jersey have not adopted the UTSA, but they have adopted the doctrine.

Elizabeth A. Rowe, *When Trade Secrets Become Shackles: Fairness and the Inevitable Disclosure Doctrine,* 7 Tul. J. Tech. & Intell. Prop. 167 (2005).

---

* [*Eds. Note*: Earlier in the opinion the court had noted:

    Bayer took Mr. Betzelos' deposition on July 8 and 9, 1999. In his deposition, Mr. Betzelos did not always readily provide answers to Bayer's questions. His deposition transcript is replete with objections and reveals the difficulty the examining attorney had in obtaining answers to questions. The deponent, for example, asked the examining attorney what the attorney meant by "customers," by "confidential," and by "compete."]

3. Is it better to view the doctrine of inevitable disclosure as simply another tool of equity? *See* Brandy L. Treadway, *An Overview of Individual States' Application of Inevitable Disclosure: Concrete Doctrine or Equitable Tool?*, 55 SMU L. Rev. 621 (2002).

4. In the introduction to this chapter on trade secret law, we identified the balance that the law seeks to strike as one between encouraging investment in the creation of valuable information and free competition in the use of public information. The inevitable disclosure doctrine suggests a different balance—one that pits employers who have valuable information against employees who desire to work for competing companies. The more tightly trade secret law limits employee mobility, the less competition there is likely to be in the use of public information. Moreover, reduced employee mobility changes competition in labor markets: Less mobility may reduce the pressure on employers to offer salary and benefit packages calculated to retain existing high-quality employees. Pushing in the other direction, less mobility shrinks the pool of potential lateral hires, thus driving up the cost of attracting such lateral hires. Given these considerations, should courts confine the inevitable disclosure doctrine to only certain classes of employees? If so, how would you define those classes of employees? Should the classes of employees be defined by statute or should courts engage in the line drawing on a case by case basis? What are the advantages and disadvantages of each approach?

C. Noncompetition Agreements

Ronald J. Gilson, *The Legal Infrastructure of High Technology Industrial Districts: Silicon Valley, Route 128, and Covenants Not to Compete*, 74 N.Y.U. L. Rev. 575, 602-03 (1999)

A post employment covenant not to compete prevents knowledge spillover of an employer's proprietary knowledge not, as does trade secret law, by prohibiting its disclosure or use, but by blocking the mechanism by which the spillover occurs: employees leaving to take up employment with a competitor or to form a competing start-up. Such a covenant provides that, after the termination of employment for any reason, the employee will not compete with the employer in the employer's existing or contemplated business for a designated period of time—typically, one to two years—in a specified geographical region that corresponds to the market in which the employer participates.

... Given the speed of innovation and the corresponding telescoping of product life cycles, knowledge more than a year or two old likely no longer has significant competitive value. The hiatus imposed by a covenant not to compete thus assures that a departing employee will bring to a new employer only her general and industry-specific human capital. ...

William M. Landes & Richard A. Posner, *The Economic Structure of Intellectual Property Law* 365-366 (2003)

Employees' covenants not to compete with their former employer for a period of years are an important method of protecting trade secrets, because it is easier to detect and prove a violation of such a covenant than it is to discover and prove that a competitor's discovery of one's trade secret is the result of unlawful appropriation rather than of independent research. A well-known study compares high-tech firms located along Route 128 outside of Boston with similar firms in Silicon Valley.[20] Massachusetts enforces employee covenants not to compete; California does not. The study finds as one would expect that in Silicon Valley employees move around more among firms, no doubt often taking some of the previous employers' trade secrets with them. The resulting pooling of knowledge may, however, contribute more to technological progress than the greater internalization of new technological ideas that a more effective scheme for the protection of trade secrets would contribute . . . . In principle, competitors possessing "blocking" trade secrets can share them through cross-licensing, but negotiation and enforcement of licenses for the use of valuable confidential information are quite costly; . . . it is a very delicate art to negotiate for the revelation of a secret without the negotiation itself unmasking it. The informal pooling that comes from the unenforceability of employee covenants not to compete may be on balance more efficient. . . .

Office of Economic Policy, U.S. Dept. of the Treasury, *Non-compete Contracts: Economic Effects and Policy Implications* 25-26 (March 2016)

Though non-compete contracts can have important social benefits, principally related to the protection of trade secrets, a growing body of evidence suggests that they are frequently used in ways that are inimical to the interests of workers and the broader economy. Enhancing the transparency of non-competes, better aligning them with legitimate social purposes like protection of trade secrets, and instituting minimal worker protections can all help to ensure that non-compete contracts contribute to economic growth without unduly burdening workers.

Amazon.com Inc. v. Powers
2012 WL 6726538 (D. Wash. 2012)

JONES, JUDGE:

I. Introduction

This matter comes before the court on the motion of Plaintiff Amazon.com, Inc. ("Amazon") for a preliminary injunction against Defendant Daniel Powers. For the reasons stated below, the court GRANTS the motion in part and DENIES it in part. The court enters a limited preliminary injunction in Part IV of this order.

II. Background

A. Mr. Powers Worked at Amazon for Two Years.

Amazon hired Mr. Powers in mid-2010 to serve as a vice-president in its Amazon Web Services ("AWS") division. Unlike the consumer-targeting online shopping services for which Amazon initially became known, AWS caters to businesses. Mr. Powers was responsible for sales of Amazon cloud computing services. His customers were businesses who wanted to use Amazon's vast network of computing resources for their own software development, data storage, web site hosting, and the like. Mr. Powers was a Washington resident based at Amazon's Seattle headquarters, although he traveled extensively for his job.

Mr. Powers left Amazon effective July 1, 2012. No one disputes, however, that the last day on which he had access to any internal Amazon information was June

---

[20] AnnaLee Saxenian, *Regional Advantage: Culture and Competition in Silicon Valley and Route 128* (1994).

18, 2012. When he left Amazon, he took no documents in paper or electronic form. There is no evidence that, at the time of his departure, he had specific plans to work elsewhere.

B. Mr. Powers Signs Amazon's Noncompetition Agreement.

When Mr. Powers started working at Amazon, he signed a "Confidentiality, Noncompetition and Invention Assignment Agreement." ("Agreement"). The Agreement has four components that bear on this dispute.

The first component is a broad prohibition against Mr. Powers' disclosure of what the Agreement deems "Confidential Information." Confidential Information includes, but is not limited to, the identity of Amazon's customers, "data of any sort compiled by [Amazon]," including marketing data and customer data, techniques for identifying prospective customers and communicating with prospective or current customers, current or prospective marketing or pricing strategies, plans for expansion of products or services, and "any other information gained in the course of the Employee's employment . . . that could reasonably be expected to prove deleterious to [Amazon] if disclosed to third parties . . . ." Agr. ¶ 2(a)(i)-(xi). Once an employee learns "Confidential Information," he can never disclose it to anyone. Agr. ¶ 2(b)(i). The only relevant exception is for information that is in the public domain. Agr. ¶ 2(b)(ii).

The second component of the Agreement is a ban on Mr. Powers doing business with Amazon's customers or prospective customers for 18 months after his departure from Amazon. Agr. ¶ 2(c)(ii). The ban applies in any business relationship with an Amazon customer where Mr. Powers would provide a similar product or service to one that Amazon provides. *Id.* More broadly, the ban applies to any relationship with an Amazon customer where the relationship "would be competitive with or otherwise deleterious to the Company's own business relationship or anticipated business relationship" with the customer. *Id.*

The third component is an 18-month ban on Mr. Powers working in any capacity that competes with Amazon. The ban prevents him from offering a product or service in any "retail market sector, segment, or group" that Amazon did business with or planned to do business with prior to his last day at Amazon, provided that the product or service he offers is "substantially the same" as one that Amazon provides. Agr. ¶ 3(c)(iii). Taken literally, the ban has extraordinary reach: it would, for example, prevent Mr. Powers from working for a bookseller, even though he had nothing to do with Amazon's book sales while he worked there.

The fourth component is a 12-month ban, measured from Mr. Powers' last day of employment, on hiring or employing former Amazon employees. Agr. ¶ 3(b).

When Mr. Powers left Amazon, he received a substantial severance payment and signed a severance agreement. The severance agreement reiterates his obligations under the Agreement, but changes none of them. Amazon insists that it made the severance payment to reinforce the Agreement, but it offers little evidence to support that claim. It is just as likely that it paid Mr. Powers to settle disputes over his termination and Amazon stock he would have received if he had kept his job.

C. Mr. Powers Started Work at Google Three Months After He Left Amazon.

Google, Inc., hired Mr. Powers in September 2012 to work as its Director of Global Cloud Platform Sales at its Mountain View, California headquarters. As a result of this litigation, he now uses the title "Director, Google Enterprise," and he has agreed not to use a title that refers to cloud computing until the end of 2012.

Although Mr. Powers' job at Google will resemble his job at Amazon in some respects, the extent of that similarity is difficult to gauge on this record. The parties agree that part of Mr. Powers' job will be to oversee sales of Google products that do not compete with AWS offerings. In part, however, Google intends that he oversee sales of its cloud computing services. Amazon has pointed to three specific Google products (Google App Engine, Google Cloud Storage, and Google Compute Engine) that compete with various AWS products. Amazon has provided relatively little information, however, that would permit the court to assess the nature of that competition. On this record, the court can only say that Google has a variety of cloud services that it hopes to sell in approximately the same market in which AWS operates. The parties dispute, for example, whether Google App Engine competes with any AWS product. There is little evidence that would permit the court to assess the extent to which Mr. Powers' experience marketing Amazon's products would be useful in marketing Google's competing products. According to Mr. Powers, Google's cloud services are sufficiently distinct from Amazon's that his experience with Amazon services will be of little use to him at Google.

D. Google Has Temporarily Restricted Mr. Powers' Cloud Computing Work.

Before he began work on September 24, Google sent Mr. Powers a written job offer. The offer acknowledged his prior employment and imposed several restrictions. It stated that he was never to use or disclose any "confidential, proprietary, or trade secret information" of any prior employer. Powers Decl., Ex. C. It also restricted him from cloud computing work for his first six months at Google:

> [D]uring the first six months of your employment with Google, your activities will not entail participation in development of, or influencing, strategies related to product development in the areas of cloud compute, storage, database or content delivery networks products or services, other than to provide those involved in such matters with publicly available market research or customer feedback regarding Google's existing products generated after you commence work at Google.

Id. It also prevented him from working with his Amazon customers:

> During the first six (6) months from your last date of employment with your current employer, you and Google also agree that you will not participate, directly or indirectly, in sales or marketing to any customers of your prior employer that, to the best of you[r] memory, you had material direct contact or regarding whom you reviewed confidential information of your prior employer.

Id. It prohibited him from being "directly or indirectly involved in the hire of any current or former employee of your prior employer" for the twelve months following his departure from Amazon. *Id.*

When Amazon discovered that Mr. Powers had begun working at Google, it began discussions with Google about his new job. Google voluntarily disclosed Mr. Powers' new responsibilities as well as the restrictions it had already imposed on his employment.

Google's voluntary restrictions did not satisfy Amazon. It sued Mr. Powers for breach of the Agreement (and the severance agreement) and violation of Washington's version of the Uniform Trade Secrets Act. . . . Amazon first sued in King County Superior Court and sought a temporary restraining order. Mr. Powers removed the case to this court in late October, before the state court took any action.

. . . .

III. Analysis

The court may issue a preliminary injunction where a party establishes (1) a likelihood of success on the merits, (2) that it is likely to suffer irreparable harm in the absence of preliminary relief, (3) that the balance of hardships tips in its favor, and (4) that the public interest favors an injunction. *Winter v. Natural Resources Defense Council, Inc.*, 555 U.S. 7, 20 (2008). . . .

. . . .

A. Amazon is Not Likely to Succeed on the Merits of Its Claims Regarding Disclosure of Its Confidential Information or Trade Secrets.

The court's observation that there is no evidence that Mr. Powers took any documents from Amazon is a good place to begin its evaluation of the proposed injunction's restrictions on disclosure. If Mr. Powers took any confidential or trade secret information from Amazon, he took it in his memory alone.

The sole evidence Amazon offers to support its claim that Mr. Powers remembers its confidential information is the declaration of Adam Selipsky, Mr. Powers' supervisor at Amazon. Although Mr. Selipsky asserts that Mr. Powers "knows" things about Amazon, he does not acknowledge that he can only speculate about what Mr. Powers knows now, more than six months after the last day he had access to any internal Amazon information. Amazon declined the opportunity to engage in discovery that would have at least allowed it to determine what Mr. Powers still knows.

. . . .

It is likely that Mr. Powers remembers something from his time at Amazon. He no doubt remembers many of the customers with whom he dealt directly, and probably remembers significant details of the relationships between those customers and Amazon. He probably remembers more, but the court declines to speculate. Amazon had both the burden to provide evidence of what Mr. Powers knows and the opportunity to take discovery to get additional evidence. That it has not done so, even as Google and Mr. Powers have given ample time to pursue discovery by voluntarily imposing virtually every restriction Amazon seeks in its injunction, is a damaging blow to Amazon's effort to demonstrate a likelihood of success on the merits.

Not only does the court not know what Mr. Powers remembers, the court does not know whether what he remembers is useful. AWS apparently conducts "formal operations planning processes" every six months, during which AWS departments give "detailed presentations" on plans, strategy, and budget. Amazon excluded Mr. Powers from the AWS meetings that happened this past summer. Assuming that Mr. Powers attended the meetings six months prior (there is no direct evidence that he did), that means that the strategic information Mr. Powers acquired, if he remembers it, is at least a year old. Mr. Selipsky emphasizes weekly emails and reports that Mr. Powers received, which serves equally well to emphasize that Mr. Powers has had no access to this weekly material for at least 27 weeks. Perhaps Amazon's cloud computing business is structured so that even information that is as much as a year old remains competitively sensitive, but again, the court can only speculate. Putting aside Mr. Powers' relationships with Amazon customers, Amazon has provided no compelling evidence that Mr. Powers still remembers competitively sensitive information he learned at Amazon.

Relying on this hazy evidence of what Mr. Powers knows, Amazon invokes the Agreement's non-disclosure provisions and the Trade Secrets Act to prevent Mr. Powers from revealing his knowledge. For several reasons, Amazon is not likely to succeed in this effort, at least on the record before the court.

Amazon is not likely to prevail on its trade secret claim. First, with the possible exception of confidential information relating to its cloud computing customers, Amazon has not identified any trade secrets that Mr. Powers currently knows. *See Ed Nowogroski Ins., Inc. v. Rucker*, 971 P.2d 936, 942 (Wash. 1999) ("A plaintiff seeking damages for misappropriation of a trade secret . . . has the burden of proving that legally protectable secrets exist."). Amazon did not ask to file any evidence under seal, suggesting that it believes the court will divine what information is a trade secret from Mr. Selipsky's public declaration. Having scoured that declaration, the court is unable to do so. The court acknowledges that it is likely that Mr. Powers learned information that would qualify as a trade secret while he was at Amazon. But if there is trade secret information that Mr. Powers could still be expected to know, Amazon has not identified it.

The possible exception is trade secret information about Amazon's customers. Mr. Powers admits that he worked closely with 33 AWS customers. The identity of those customers is likely not a secret. Mr. Powers' unrebutted evidence shows that Amazon publicly identifies all of those entities as Amazon customers. Although a customer list can be a trade secret, Amazon has not identified a customer list or subset of a customer list that qualifies as a trade secret. It is possible, however, that Mr. Powers remembers trade secret information about Amazon's relationships with those customers. In contrast to the enormous sets of AWS data that Amazon speculates Mr. Powers still remembers, it is far more likely that he remembers information pertaining to these relatively few customers.

Even if Mr. Powers knows trade secret information about Amazon's relationship with a few customers, Google has not identified what that information is. As the court will discuss later, Washington law permits noncompetition agreements that prevent an employee from trading unfairly on customer relationships he or she built before leaving employment. An employer cannot weave a similar restriction from a nondisclosure agreement or the Trade Secret Act without identifying confidential or trade secret information with sufficient specificity. Amazon has failed to do so here. Indeed, Amazon has not identified even one of the customers about which it is so concerned, much less any specific confidential information Mr. Powers knows about that customer.

Amazon's claims based on the nondisclosure clauses of the Agreement fail for the same reasons as its trade secret claim. Amazon has not discharged its burden to identify confidential information that Mr. Powers still knows and is still competitively useful.

Even if Amazon had sustained its burden to identify confidential or trade secret information that Mr. Powers knows, it would still need to prove a threat of irreparable harm. Evidence of what Mr. Powers knows is not enough; Amazon also needs evidence that Mr. Powers is likely to disclose it. That Mr. Powers knows something is not proof that he will use that knowledge at Google. Google has already forbidden him to ever use Amazon's confidential information. Amazon's counsel conceded at oral argument that Amazon has no evidence that Mr. Powers has disclosed anything in the nearly three months since he began working at Google. Once Google lifts its self-imposed restrictions on Mr. Powers' work with its cloud computing products, Mr. Powers may have more opportunity to use what he knows about Amazon. It is that possibility that garners much of Amazon's attention. Amazon has generally failed to point to anything specific that Mr. Powers knows that he is likely to disclose at Google. It instead asserts that virtually everything Mr. Powers knows is confidential and that because of the nature of his job at Google, he must inevitably use or disclose that knowledge in his work there. Mr. Powers decries Amazon's approach as an impermissible "inevitable disclosure" argument.

Amazon has not proffered evidence from which the court can conclude that it is likely that Mr. Powers will "inevitably disclose" Amazon's confidential information. The parties debate whether Washington has ever recognized inevitable disclosure as a viable basis for a trade secret or breach of confidentiality claim. On this record, that debate is largely beside the point. The crux of an inevitable disclosure argument in this context is a showing that an employee's new job so closely resembles her old one that it would be impossible to work in that job without disclosing confidential information. Amazon has not made that showing here. It has pointed to a host of at least superficial similarities between Mr. Powers' old job and his new one, including a set of superficial similarities between Google's App Engine, Cloud Storage, and Compute Engine services and comparable AWS offerings. This effort falls short of convincing the court that Mr. Powers cannot do his new job without relying on Amazon's confidential information.

The court emphasizes the high bar for an inevitable disclosure argument for two reasons. First, if an employer cannot make a detailed showing of similarity between an employee's new job and old job, then it can hardly argue that disclosure is inevitable. Amazon's inevitable disclosure argument fails in this case for at least that reason. More importantly, however, an employer may lawfully prohibit an employee from ever disclosing its confidential information. Were inevitable disclosure as easy to establish as Amazon suggests in its motion, then a nondisclosure agreement would become a noncompetition agreement of infinite duration. As the court will now discuss in its analysis of the noncompetition clauses of the Agreement, Washington law does not permit that result.

B. With the Exception of Restrictions on Work with Former Customers, Amazon is Not Likely to Succeed on the Merits of Its Effort to Enforce the Noncompetition Clauses in the Agreement.

The Agreement contains a choice-of-law clause selecting Washington law, under which an agreement that restricts a former employee's right to compete in the marketplace is enforceable only if reasonable. The court will consider Mr. Powers' effort to avoid the choice-of-law clause in Part III.C. For now, the court applies Washington law, under which a court deciding whether a noncompetition agreement is reasonable must consider three factors:

> (1) whether restraint is necessary for the protection of the business or goodwill of the employer, (2) whether it imposes upon the employee any greater restraint than is reasonably necessary to secure the employer's business or goodwill, and (3) whether the degree of injury to the public is such loss of the service and skill of the employee as to warrant nonenforcement of the covenant.

Perry v. Moran, 748 P.2d 224, 228 (Wash. 1987). If a court finds a restraint unreasonable, it can modify the agreement by enforcing it only "to the extent reasonably possible to accomplish the contract's purpose." *Emerick v. Cardiac Study Ctr., Inc.*, 286 P.3d 689, 692 (Wash. Ct. App. 2012). Among other things, the court can reduce the duration of an unreasonably long anticompetitive restriction. *See, e.g., Perry*, 748 P.2d at 231 ("It may be that a clause forbidding service [to former clients] for a 5-year period is unreasonable as a matter of law . . . ."); *Armstrong v. Taco Time Int'l, Inc.*, 635 P.2d 1114, 1118-19 (Wash. Ct. App. 1981) (cutting five-year restriction to two and a half years). In any case, the court should protect an employer's business only "as warranted by the nature of [the] employment." *Emerick*, 286 P.3d at 692.

Applying these principles, Washington courts have typically looked favorably on restrictions against working with an employee's former clients or customers. In *Perry*, the court upheld a 20-accountant firm's noncompetition agreement preventing a departing employee from working with her former clients for about a year and a half after she left the firm.[4] 748 P.2d at 224. The court recognized the

---

[4] As written, the restrictive covenant in Perry would have lasted five years. 748 P.2d at 225. At trial, however, the employer sought to enforce it solely as to the year and a half between the employee's

employer's "legitimate interest in protecting its existing client base," and rejected the notion that lesser restrictions, like one that would only prohibit the former employee from soliciting (as opposed to working with) former clients, would be adequate to protect that interest. *Id.* at 229. Generally speaking, time-limited restrictions on business with former clients or customers survive scrutiny in Washington. *See, e.g.*, *Knight, Vale & Gregory v. McDaniel*, 680 P.2d 448, 451-52 (Wash. Ct. App. 1984) (declining to invalidate three-year restriction on accountant working with former clients); *Pac. Aerospace & Elecs., Inc. v. Taylor*, 295 F. Supp. 2d 1205, 1218 (E.D. Wash. 2003) (finding two-year restriction on solicitation of former customers to be reasonable as a matter of law); *Seabury & Smith, Inc. v. Payne Fin. Group, Inc.*, 393 F. Supp. 2d 1057, 1063 (E.D. Wash. 2005) (finding one-year restriction on working with former clients to be reasonable as a matter of law).

Washington courts have been more circumspect when considering restrictions that would prevent an employee from taking on any competitive employment. These general restrictions on competition are more suspect than mere bans on working with former clients or customers. *Perry*, 748 P.2d at 230. Courts will in some circumstances enforce general noncompetition restrictions when they apply only in a limited geographical area. *See, e.g.*, *Emerick*, 286 P.3d at 693-95 (remanding for reconsideration of necessity of five-year ban on competitive employment in a single county); *Hometask Handyman Servs., Inc. v. Cooper*, No. C07-1282RSL, 2007 U.S. Dist. LEXIS 84708, at *10-11 (W.D. Wash. Oct. 30, 2007) (granting injunction against former franchisee based on general competition restriction, but reducing area from 100-mile radius to 25-mile radius); *see also Labor Ready*, 149 F. Supp. 2d at 408 (N.D. Ill. 2001) (upholding one-year general bar on competition within 10-mile radius of former employer). Courts have also declined to enforce even geographically limited general restrictions on competition. *See A Place for Mom, Inc. v. Leonhardt*, No. C06-457P, 2006 U.S. Dist. LEXIS 58990, at *6-7, 13-14 (W.D. Wash. Aug. 4, 2006) (declining to issue injunction based on general restriction on competitive employment).

When a noncompetition agreement is targeted at a competing business, rather than an individual employee, specific circumstances can justify a general bar on competition. For example, in *Oberto Sausage Co. v. JBS S.A.*, C10-2033RSL, 2011 U.S. Dist. LEXIS 33077 (W.D. Wash. Mar. 11, 2011), the court considered a meat retailer's request for a pre-arbitration injunction against a Brazilian meat processor who had formerly supplied meat exclusively to the retailer. In that case, the retailer had worked closely with the Brazilian processor to teach its proprietary beef jerky manufacturing process. *Id.* at *3. When one of its chief competitors purchased the Brazilian processor, the court enforced a general restriction on competition within the United States, preventing its competitor from "taking a free ride on its substantial investment in training [Brazilian] employees and upgrading the [Brazilian plant] with equipment plaintiff claims it developed through

---

departure and trial. *Id.* at 231. For that reason, the court declined to decide whether a five-year restriction was too long. *Id.* at 230-31.

its confidential research and development." *Id*. at *18. The court imposed the injunction only for the length of time it took the parties to present their dispute to an arbitrator. *Id*. at *22. Similarly, in *Armstrong*, the court upheld a restriction on a former franchisee opening competing restaurants near the franchisor's restaurants, but it cut the five-year duration of the restriction in half. 635 P.2d at 1118-19.

The court distills a few general principles from these cases. First, Washington courts are relatively deferential to employers in enforcing agreements restricting a former employee's work with the employer's clients or customers. Courts are less deferential to general restrictions on competition that are not tied to specific customers. An employer can demonstrate that more general restrictions are necessary, but can do so only by pointing to specific information about the nature of its business and the nature of the employee's work. Finally, although courts are somewhat deferential about the duration or geographic extent of noncompetition agreements, they will readily shorten the duration or limit the geographic scope, especially where the employer cannot offer reasons that a longer or more expansive competitive restriction is necessary. With these principles in mind, the court considers the Agreement's 18-month restriction on working with former Amazon customers and its 18-month general noncompetition clause.

The Agreement passes muster under Washington law to the extent it seeks to prevent Mr. Powers from working with his former Amazon customers. Mr. Powers, no less than the employees in *Perry*, *Knight*, and in other Washington cases, competes unfairly with Amazon to the extent he attempts to trade at Google on customer relationships he built at Amazon. The reasonable duration of that restriction, however, is a matter of dispute. This is not a case where Mr. Powers seeks to leap from Amazon immediately to Google with his former customers in tow. He stopped working with Amazon customers more than six months ago. There is no evidence he has had contact with any of them since then. There is no direct evidence that he intends to pursue business with any of them. The only indirect evidence that he has interest in contacting his former customers is that he has chosen to fight Amazon's efforts to enforce the Agreement. Although the personal aspects of his relationships with his former customers might be expected to endure for more than six months, they might just as well extend even beyond the 18-months that the Agreement provides. Amazon has not explained why it selected an 18-month period, nor has it disputed Mr. Powers' suggestion that the Agreement he signed is a "form" agreement that Amazon requires virtually every employee to sign. Because Amazon makes no effort to tailor the duration of its competitive restrictions to individual employees, the court is not inclined to defer to its one-size-fits-all contractual choices. Amazon has not convinced the court that the aspects of Mr. Powers' relationships with customers that depend on confidential Amazon information are still viable today. On this record, the court finds it would not be reasonable to enforce the Agreement's customer-based restrictions for longer than nine months from the last date on which Mr. Powers had access to Amazon's information.

The Agreement's general noncompetition clause, in contrast to the clause targeting Amazon customers, is not reasonable. Amazon asks the court to prevent Mr. Powers from working in a competitive capacity anywhere in the world. The court is willing to assume, even though Amazon has provided no evidence, that the cloud computing business in which Google and Amazon compete is geographically far-flung. Because both companies compete globally, it is possible that Mr. Powers could inflict competitive injury on Amazon even while working a thousand miles from his Seattle-based former employer. But even if the court accepts the extraordinary geographic reach of the ban, it could not accept Amazon's implicit argument that it is impossible for Mr. Powers to compete fairly with Amazon in the cloud computing sector.

Amazon has failed to articulate how a worldwide ban on cloud computing competition is necessary to protect its business. Its ban on working with former customers serves to protect the goodwill it has built up with specific businesses. A general ban on Mr. Powers' competing against Amazon for other cloud computing customers is not a ban on unfair competition, it is a ban on competition generally.[*] Amazon cannot eliminate skilled employees from future competition by the simple expedient of hiring them. To rule otherwise would give Amazon far greater power than necessary to protect its legitimate business interest. No Washington court has enforced a restriction that would effectively eliminate a former employee from a particular business sector. This court will not be the first, particularly where Amazon has not provided enough detail about the nature of AWS's cloud computing business to convince it that an employee like Mr. Powers can only compete with AWS by competing unfairly.

Much of Amazon's argument in favor of enforcement of its general restriction on competition is cribbed from the inevitable disclosure argument it advanced in support of the Agreement's nondisclosure provisions. According to Amazon, Mr. Powers simply knows too much to compete fairly with Amazon in the cloud computing sector. The court finds these claims to be fatally nonspecific, as it explained in Part III.A. Generalized claims that a former employee cannot compete fairly are insufficient. *See Copier Specialists, Inc. v. Gillen*, 887 P.2d 919, 920 (Wash. Ct. App. 1995) (finding that the "training [an employee] acquired during his employment, without more," did not warrant enforcement of a geographically limited covenant not to compete). Before enforcing a general restriction against competition, the court would require a far more specific showing than Amazon has made here.

C. Washington Law, not California Law, Applies to Amazon's Claims Based on the Agreement.

---

* [*Eds. Note*: An agreement not to compete is, as a general matter, a violation of the antitrust laws. 15 U.S.C. § 1 ("Every contract, combination in the form of trust or otherwise, or conspiracy, in restraint of trade or commerce among the several States . . . is declared to be illegal."). Indeed, if Amazon and Google were simply to agree not to compete against each other in the cloud computing services sector, it would be not only a civil wrong, it would be a felony. *Id.* Noncompete agreements that protect trade secret information are a narrowly drawn exception to the rule.]

The court briefly addresses Mr. Powers' contention that California law, not Washington law, should apply to this dispute. . . .

. . . The threshold question in a Washington choice-of-law analysis is whether there is an actual conflict with another state's law. The court assumes without deciding that Mr. Powers correctly asserts that even the Agreement's restrictions on working with former customers would be unenforceable under California's more pro-employee approach to noncompetition agreements. Amazon does not argue otherwise.

Having identified a conflict of law, Washington choice-of-law rules require the court to consider the Agreement's choice of Washington law. Washington courts apply § 187 of the Restatement (Second) Conflict of Laws ("§ 187") when resolving "conflict of laws problems in which the parties have made an express contractual choice of law." *Erwin v. Cotter Health Ctrs.*, 167 P.3d 1112, 1120-21 (Wash. 2007). In relevant part, § 187 requires the court to enforce the parties' contractual choice of law unless "the chosen state has no substantial relationship to the parties or the transaction and there is no other reasonable basis for the parties' choice," § 187(2)(a), or "the application of the law of the chosen state would be contrary to a fundamental policy of a state which has a materially greater interest than the chosen state in the determination of the particular issue and which, under the [Restatement (Second) Conflict of Laws, § 188], would be the state of the applicable law in the absence of an effective choice of law by the parties," § 187(2)(b). No one argues that Washington lacks a substantial relationship to Mr. Powers, Amazon, and the Agreement. For that reason, Mr. Powers' plea for the application of California law requires him to, among other things, show that California has a "materially greater interest" than Washington in determining the enforcement of the Agreement.

California's interest in the enforcement of the Agreement is no greater than Washington's. Washington's willingness to enforce anticompetitive restrictions reflects a strong interest in protecting its businesses from unfair competition from former employees. California likely has a strong interest in protecting its workers from attempts by their former employers to limit their employment. Nothing, however, would permit the court to conclude that California's interest is "materially greater," especially as applied in this dispute. . . .

D. Amazon Has Made a Sufficient Showing on the Remaining Injunctive Relief Factors to Justify a Limited Injunction.

On this record, Amazon is likely to succeed on the merits only of its claim based on the Agreement's restrictions on working with former customers, although only for nine months. The court now considers whether Amazon has demonstrated a likelihood of irreparable harm, where the balance of hardships tilts, and the public interest.

Irreparable harm is a likely consequence of permitting an employee to pursue his former customers in violation of a valid restriction. The monetary damage from loss of a customer is difficult to quantify, and the damage to goodwill even more so. There is no direct evidence that Mr. Powers intends to solicit former

Amazon customers. Given his opposition to Amazon's motion, however, the court finds it likely that he would approach at least some customers (or some customers would approach him) if neither Google nor this court prevents him from doing so.

In the context of this limited injunction, the balance of hardships favors Amazon. Before Amazon even learned of Mr. Powers' work at Google, Google was willing to keep Mr. Powers from cloud computing work until six months after he began working at Google. That self-imposed restriction would have expired in late March 2013. Given that Google was willing to impose that restriction and Mr. Powers was willing to accept it, the court finds no hardship to Mr. Powers in enforcing the Agreement's more limited customer-based restrictions until March 19, 2013, nine months after Mr. Powers' last had access to Amazon information.

The public interest does not weigh heavily in favor of either party. There is no evidence that the court's decision on this injunction will impact the public.

IV. Preliminary Injunction

For the reasons stated above, the court enters the following preliminary injunction. Until March 19, 2013, unless the court orders otherwise, Defendant Daniel Powers may not directly or indirectly assist in providing cloud computing services to any current, former, or prospective customer of Amazon about whom he learned confidential information while working at Amazon. "Confidential information" has the definition the parties gave it in the Agreement.

Given the brief duration of the injunction, Google is unlikely to suffer significant financial harm. For that reason, the court will require Amazon to obtain a $100,000 bond or deposit $100,000 into the court's registry. See Fed R. Civ. P. 65(c) (requiring security "in an amount that the court considers proper to pay the costs and damages sustained by any party found to have been wrongfully enjoined"). This injunction will take effect upon Amazon's notice of a bond or cash deposit.

. . . .

Notes & Questions

1. One guarantor of reasonableness is that the former employer who seeks to enforce a noncompetition agreement actually has a legitimate interest to protect by means of such an agreement. A protectable trade secret would constitute such "legitimate interest." Assuming there is a legitimate interest to protect, the court can assess the agreement's particular terms for reasonable scope. In an excerpt at the beginning of this Subsection, Prof. Landes and Judge Posner suggest that "covenants not to compete . . . are an important method of protecting trade secrets, because it is easier to detect and prove a violation of such a covenant than it is to discover and prove that a competitor's discovery of one's trade secret is the result of unlawful appropriation rather than of independent research." Do cases like *Amazon.com Inc. v. Powers* call that "lower-cost alternative" story into question? If the reasonableness of a noncompetition agreement turns on a fact-based assessment of whether there is a valid trade secret interest to protect, does the

litigation over the covenant simply reproduce what would have been the litigation over trade secret misappropriation?

2. One context in which lawyers advise clients concerning the reasonableness of a non-competition agreement is when the departing employee or the new employer seeks counsel. In *Amazon.com Inc. v. Powers,* what steps did Google take, as the new employer of Mr. Powers, to help bolster its litigation position? Remember that even if the court refuses to enforce a non-competition agreement, an employee retains the duty not to disclose or use the trade secrets of a former employer for as long as those secrets retain the attributes that give rise to trade secret protection.

3. Lawyers drafting non-competition agreements for employees to sign need to consider the relationship between trade secret protection and the reasonableness of the scope of competition from which the employee will be prohibited. Scope can include many dimensions: duration, geography, and product or service market definition. In *Amazon.com Inc. v. Powers* the court faulted the plaintiff for not specifically tailoring the agreement for each employee. Is employee-specific tailoring an appropriate factor for courts to consider? What drafting incentives is the court creating?

If a court finds a clause to be unreasonable, should the court simply rewrite the clause to be reasonable, *e.g.,* reduce the duration from 12 months to 5 months? Alternatively, should the court refuse to enforce the clause entirely? What incentives do the alternative approaches create? Does the choice of approach turn on whether the noncompete clause is wildly unreasonable (rather than only slightly unreasonable)? Are you more or less interested in having judges "rewrite" noncompete agreements than you were in having judges establish, and refine, the misappropriation doctrine for "hot news"? Some courts consider whether the employer "overreach[ed]" or "coerc[ed]" the employee's assent to the non-competition agreement. Consider the following:

> The issue of whether a court should cure the unreasonable aspect of an overbroad employee restrictive covenant through the means of partial enforcement or severance has been the subject of some debate among courts and commentators. A legitimate consideration against the exercise of this power is the fear that employers will use their superior bargaining position to impose unreasonable anticompetitive restrictions, uninhibited by the risk that a court will void the entire agreement [thus] leaving the employee free of any restraint. The prevailing, modern view rejects a per se rule that invalidates entirely any overbroad employee agreement not to compete. Instead, when, as here, the unenforceable portion is not an essential part of the agreed exchange, a court should conduct a case specific analysis, focusing on the conduct of the employer in imposing the terms of the agreement (*see Rest. (2d) Contracts* § 184). Under this approach, if the employer demonstrates an absence of overreaching, coercive use of dominant bargaining power, or other anticompetitive misconduct, but has in good faith sought to protect a legitimate business interest, consistent with reasonable

standards of fair dealing, partial enforcement may be justified.

BDO Seidman v. Hirshberg, 712 N.E.2d 1220, 1226 (N.Y. 1999). In the *BDO Seidman* case the court concluded that partial enforcement was appropriate, noting that the covenant was not imposed as a condition of defendant's initial employment, or even his continued employment, but in connection with promotion to a position of responsibility and trust, and there was no evidence of coercion or an attempt to forestall competition. Additionally the court noted that there was no proof "that BDO imposed the covenant in bad faith, knowing full well that it was overbroad."

Based on your own work experience, what is your sense of the balance of bargaining power at key moments in the employment relationship? Is bargaining power the correct focus around which to try to balance the incentives of the employer and employee in an effort to prevent oppressive contract terms? Should non-competition agreements ever be enforced against low-wage workers such as sandwich shop employees? The national fast-food franchise chain Jimmy John's suffered serious public backlash when media put a spotlight on their non-competes. *See* Dave Jamieson, *Jimmy John's 'Oppressive' Noncompete Agreement Survives Court Challenge* (April 10, 2015), available at http://www.huffingtonpost.com/2015/04/10/jimmy-johns-noncompete-agreement_n_7042112.html. Though no Jimmy John's franchise owner sought to enforce the non-competition agreement, so far as we are aware, do you think that the agreements may have deterred at least some employees from seeking other employment, or may have needlessly complicated some departed employees' job searches? If so, should employers face any penalties for insisting that low-wage workers who learn no trade secrets of substantial value must nevertheless sign a non-compete agreement?

4. We can see the influence of California's public policy favoring employee mobility not only in trade secret cases, such as *Bayer Corp. v. Roche Molecular Systems*, but also in cases concerning the enforcement of noncompetition agreements. As employees move across state lines, different states' varied policy choices can often collide. Parties forum shop, and the states' differing policy choices play out in the context of, among other things, the civil procedure rules that you learned in your first year of law school. *See, e.g., Google, Inc. v. Microsoft Corp.*, 415 F. Supp. 2d 1018 (N.D. Cal. 2005).

In addition to the formal legal difference in California's refusal to enforce non-competition agreements, a difference in attitude may also be at play, at least in certain industries. As Professor Orly Lobel explains:

> The companies of Silicon Valley seem to have consciously understood their general interest in developing the semiconductor and high-tech industry as a whole and also understood that firms cannot learn all they need to know internally. Businesses around the Valley are not perceived as the enemy, moving between companies is not thought of as betrayal, and entrepreneurs are not labeled thieves. In fact, the firms that have been militant

in pursuing ex-employees for using their company's confidential information have damaged their reputation as responsible corporate players in the Valley, the internal morale among their best employees, and their ability to recruit new talent.

Orly Lobel, *Talent Wants to be Free: Why We Should Learn to Love Leaks, Raids, and Free Riding* 207 (2013).

5. Lawyers change jobs, just as other people do. The American Bar Association's *Model Rules of Professional Conduct*, which have been adopted in many states, provide as follows: "A lawyer shall not participate in offering or making . . . [an] employment, or other similar type of agreement that restricts the right of a lawyer to practice after termination of the relationship, except an agreement concerning benefits upon retirement[.]" Model Rule 5.6(a). Why prohibit noncompete agreements with respect to the legal profession, but adopt a reasonableness standard with respect to such agreements for other professions?

Statutory Restrictions on Non-Competition Agreements: The Oregon Example

As you have seen in the foregoing cases, state law governs the enforceability of noncompetition agreements. Each state can determine whether noncompetition agreements will be enforced and, if so, what requirements must be met to ensure enforceability. In many states that enforce noncompetition agreements, the requirements are generally found in the case law (including cases about reasonableness). In other states, statutes either prohibit enforcement, such as in California, or define requirements for enforceability.

Oregon, for example, has relied more on statutes in this area. In 2007, Oregon significantly revised its statutory requirements for enforcing noncompetition agreements. In order for a noncompetition agreement to be enforceable, several requirements have to be met. First, either the employer must give the employee written notice two weeks prior to the first day of employment that a noncompetition agreement is required as a condition of employment, or the agreement must be entered into in connection with an existing employee's "bona fide" promotion. Second, the period of noncompetition cannot be longer than two years. Third, the employer must have a protectable interest (which, in effect, justifies the need for the agreement). *See* Oregon Revised Statutes § 653.295. You can find a copy of the full text of the Oregon law in the *Statutory Supplement* for these materials (available on the Semaphore Press website <www.semaphorepress.com>).

Additionally, the legislature built protections into the statute for less-highly-compensated employees. Specifically, if an employee is an hourly employee who is subject to Oregon minimum wage and overtime laws, or if the employee makes less than the median family income for a family of four as calculated by the U.S. Census Bureau (for Oregon, about $70,000 in 2007), then the only way for an employer to have a noncompete agreement enforced is if the employer compensates the employee during the time of the noncompete. In such a case, the employer must pay the greater of either (1) 50% of the employee's annual gross salary and commissions at the time of the employee's termination, or (2) 50% of the median family income. Special rules apply for noncompetition agreements with employees who are "on-air talent" employed by broadcasting entities.

Notes & Questions

1. What do you think of the Oregon approach? What concerns might you have if you represent employers? Are all employers similarly situated? Consider a start-up company, with little capital for large salaries and a core trade secret in its business plan. What will that start-up company need to do to be able to enforce its noncompetition agreements? How might that affect competition for labor between start-ups and established companies?

2. In the Oregon statute a "protectable interest" includes an employee's access not only to trade secrets, but also "to competitively sensitive confidential business or professional information that otherwise would not qualify as a trade secret, including product development plans, product launch plans, marketing strategy or sales plans." When might information fit into the second category but *not* qualify as a trade secret?

3. One way to view the Oregon approach is as an alternative to either the "reasonableness" inquiry you saw in the *Amazon.com v. Powers* case or the balancing attempts under the inevitable disclosure doctrine for trade secrets explored earlier in this Section. Under the Oregon statute, the *legislature* has defined what constitutes a reasonable context for the enforcement of a noncompetition agreement. Does this change your view of the desirability of a judicial reasonableness inquiry? Does the Oregon statute factor in all of the appropriate considerations?

4. An open question remains under Oregon's law: Should a noncompetition agreement that meets the statutory requirements for enforceability be evaluated for "reasonableness" separately by the court? Or does compliance with the statute constitute reasonableness *per se*? This is a question of statutory interpretation and legislative intent. One the one hand, if the statute is viewed as setting out conditions under which it is appropriate to enforce the agreement, why should there be any further judicial inquiry into the agreement? (Such an approach puts a premium on the parties' freedom to enter agreements they think mutually beneficial.) On the other hand, if we view the statutory requirements as a minimum standard, further judicial scrutiny might be appropriate. (Such an approach puts a premium on people's freedom to pursue their livelihoods in a more fluid, competitive labor market.)

IV. Patent Law: A Complement to Trade Secret Law?

Trade secret law gives businesses a flexible, lower-cost way to make a return on investments in developing valuable information. But, like all things, trade secret law has some shortcomings. It is important to recognize that patent law, the next doctrinal area that we cover, is a partial substitute for trade secret law. Consider the following excerpt:

William M. Landes & Richard A. Posner, *The Economic Structure of Intellectual Property Law* 328-329 (2003)

The strongest economic arguments for the patent system (or at least *a* patent system, not necessarily the one we have and possibly one that would grant less protection to inventors) are four—and none is directly related to the traditional cost-internalization arguments for property rights in inventions. Three of the four are related to economic problems created by trade secrecy . . . .

1. In the absence of a patent option, inventors would invest many more resources in maintaining trade secre[ts] (and competitors in unmasking them) and inventive activity would be inefficiently biased toward inventions that can be kept secret. This is one point rather than two because the bias in inventive activity would be the result of inventors' trying to minimize their costs of maintaining secrecy. . . .

2. A patent option facilitates efficiency in manufacturing. The possessor of a secret process for manufacturing a product may not be the most efficient manufacturer of it. In principle, he can license the trade secret to a more efficient manufacturer. But licenses of trade secrets are even more costly than patent licenses because of the risk of inadvertent disclosure or unprovable theft of the secret is greater if the trade secret is licensed than if it is kept within a single organization. So absence of patent protection would cause inefficiencies in manufacture.

3. Suppose a firm invents a process that has value in the manufacture not only of its own products but also of products in other industries. If the process is kept secret from the world, the firm may never even learn of the other potential applications. And if it does learn of them, what is it to do? One possibility is to enter the other industries in which the process could be employed profitably and begin manufacturing their products. But apart from the delay involved in such a course of action, the firm may lack the requisite skills, knowledge, or other resources for operating effectively in those industries, even if it is a highly efficient manufacturer in its own industry, and so it either will not enter at all or, if it does, will incur needlessly high manufacturing costs. The alternative is to license the process to the firms already in those industries, but licensing of trade secrets is, as we just noted, costly because the secret is more likely to leak out the more people who are in on it. This is also a reason to doubt that merger with a firm already in the target industry is a feasible alternative to unilateral entry into the industry.

Chapter 3: Patent Law

I. The Patent System

Patent protection is exclusively a matter of federal law. 35 U.S.C. § 1 *et seq.* Congress passed the first Patent Act in 1790, made a major overhaul to the Patent Act in 1952, and—most recently—made significant changes to U.S. patent law in the America Invents Act of 2011 ("AIA"), Pub. L. 112-29, 125 Stat. 284. Congress has also made minor amendments to the patent laws from time to time, to harmonize our patent laws with those of other countries and to address transborder activity that affects the commercial value of U.S. patents. Many key concepts and statutory terms from the 1790 Act have endured down to the present day. As a result, courts in modern patent cases will often cite and discuss Supreme Court patent law decisions from the 1800s and early 1900s.

An up-to-date version of the Patent Act is available from the Patent Office:
http://www.uspto.gov/web/offices/pac/mpep/consolidated_laws.pdf
The Patent Office also maintains a separate America Invents Act Resources page:
http://www.uspto.gov/aia_implementation/resources.jsp

Only the U.S. Patent & Trademark Office, or PTO, can grant a U.S. patent, and only federal tribunals can hear patent infringement suits. Patent law protects subject matter that is different from that of the other i.p. types. We will discuss the differences in substantive scope at length in the succeeding Sections of this Chapter.

A key feature that sharply distinguishes patent law from other i.p. types is its much greater formality, with respect to both the process one must follow to obtain patent protection and the resulting stated boundary of one's right to exclude others (known as a *patent claim*). We will focus on these formal requirements before turning to patent law's substantive requirements.

A. Theoretical Underpinnings

As we explained in Chapter 1, a central rationale for intellectual property law is the need for an enforceable right to exclude that encourages creators to invest in making new inventions and other products of the mind—intangibles that creators might otherwise fail to develop for fear of free riding by others. Patent law shares this central rationale; it is the standard story for why we offer patent protection for inventions. For example, economists Scherer and Ross explain patent protection's "basic logic" this way:

The funds supporting invention and the commercial development of inventions are front-end "sunk" investments; once they have been spent, they are an irretrievable bygone. To warrant making such investments, an individual inventor or corporation must expect that once commercialization occurs, product prices can be held above postinvention production and marketing costs long enough so that the discounted present value of the profits . . . will exceed the value of the front-end investment. In other words, the investor must expect some degree of protection from competition, or some monopoly power. The patent holder's right to exclude imitating users is intended to create or strengthen that expectation.

F.M. Scherer & David Ross, *Industrial Market Structure and Economic Performance* 622 (3rd ed. 1990). This standard account of patent protection echoes, in contemporary language, the Progress Clause of the U.S. Constitution.[*] The Clause empowers Congress "[t]o promote the Progress of Science and useful Arts, by securing for limited Times to Authors and Inventors the exclusive Right to their respective Writings and Discoveries." U.S. Const. Art. I, § 8 cl. 8. We promote progress in the useful arts—our stock of tools and techniques for solving practical problems—by giving those who invent them a time-limited right to exclude others from using those new tools and techniques without permission.

Patent law also helps us accomplish two other important goals. First, by limiting protection to only those inventors who provide a written disclosure that enables skilled artisans in the pertinent field to make and use the invention, our patent system fosters the useful arts by making a public documentary record of progress in that field. It publicizes inventions. The Supreme Court has called written disclosure that every patent must have "the *quid pro quo* of the right to exclude." *Kewanee Oil Co. v. Bicron Corp.*, 416 U.S. 470, 484 (1974). In other words, patent law strikes a bargain between the inventor and the public, according to which "the public must receive meaningful disclosure in exchange for being excluded from practicing the invention[**] for a limited period of time." *Enzo Biochem, Inc. v. Gen-Probe Inc.*, 323 F.3d 956, 970 (Fed. Cir. 2002). When Congress enacted our first patent laws, this publicizing function was important; there were not lending libraries in every city and town, nor was there a thriving trade in academic and other specialized journals. Today, by contrast, there are many resources other than granted patents that engineers, scientists, and others can and do consult to learn about technological innovations, from reference works in lending libraries to the latest issue of a specialized journal in the field.

---

[*] Naming this clause presents a value choice. Some call it the Copyright & Patent Clause, though neither term appears in it. Others call it the Intellectual Property Clause, though, again, the phrase does not appear there, and the word "property," which *is* used elsewhere in the Constitution (but *not* here), is rich with connotations. Still others call it the Exclusive Rights Clause, which has the virtue of a textual connection, but focuses on the means Congress is empowered to use, rather than on the end it is empowered to pursue. We prefer to call it the Progress Clause.

[**] [*Eds. Note*: Patent practitioners use the shorthand phrase *practicing the invention*, or *practicing the claim*, to mean "making, using, selling, etc., the claimed invention." When you see this phrase, don't think "practice makes perfect," think "practicing law" or "practicing medicine."]

We no longer need the bargain—disclosure for protection—to build a thriving technology library (assuming we ever did). But the disclosures highlight that patent law functions as a complement to trade secret law, precisely by protecting inventions that *cannot* easily be kept secret. With patent law in place, "inventive activity" need not "be inefficiently biased toward inventions that can be kept secret." William M. Landes & Richard Posner, *The Economic Structure of Intellectual Property Law* 328 (2003). Where an invention *can* be kept secret, an inventor can choose trade secret law's mix of benefits and risks, or opt for patent law's different mix instead. The patent process brings rigor to the disclosure, which then serves as the public embodiment of an even trade: fair notice for fair protection.

Second, patent law helps reduce the costs of some licensing transactions, compared to trade secret licensing. It does this by reducing two of the risks endemic to licensing trade secrets—namely, inadvertent disclosure and unprovable theft. *See id.* at 328-329. Patent law, unlike trade secret law, does not recognize broad defenses for independent creation or reverse engineering; nor does patent law require proof that the alleged direct infringer copied from, or even knew about, the patent in question. As a result, if one is practicing the invention set forth in the patent without permission, and the patent is valid and enforceable, one is liable for infringement. 35 U.S.C. § 271(a). Patent licensing takes place against this backdrop of less-risky-for-the-inventor legal rules. *See generally* Robert P. Merges, *Intellectual Property and Costs of Commercial Exchange: A Review Essay*, 93 Mich. L. Rev. 1570 (1995).

B. Obtaining a U.S. Patent

As we discussed in Chapter 2, to obtain trade secret protection, one must—first and foremost—keep valuable business information a secret. One does not need a government agency's approval to generate a trade-secret asset. We will see, in Chapter 4, that copyright protection follows automatically when one fixes an original work of authorship in a tangible medium of expression; registration has benefits, but the existence of a copyright right does not wait on registration. Similarly, as we will see in Chapter 5, trademark rights follow from use of the mark in commerce; registration has benefits, but use is what creates the basic right to exclude others from using a confusingly similar mark. Patent law thus works quite differently from the other ip regimes.

To obtain patent protection, one must successfully navigate an administrative process at the PTO. This overall process, typically called "patent prosecution," involves a series of exchanges between the patent applicant and the PTO. The public, including the applicant's competitors, has no official role in this process. In a basic patent prosecution, the process begins when an inventor files an application for a patent. 35 U.S.C. §§ 111, 131. Groups of people can make an invention jointly, in which case all of them are named in the application. (We discuss this in more detail later in the chapter, in Section VII.A.) Whether the invention is made by one person or many, the vast majority of applications are prepared with the financial backing of the inventors' employers, and typically through assignment documents the employer will own any resulting patent. The application, if properly prepared, describes the invention and, through that description, enables the hypothetical *person having ordinarily skill in the art*—often called *the phosita*—to make and use the invention.[*] 35 U.S.C. § 112(a) (known as § 112, ¶ 1, before the America Invents Act). The application can also contain drawings of different embodiments of the invention. 35 U.S.C. § 113. Most importantly, the application includes a series of numbered paragraphs, called *claims*, that state the exclusionary rights to which the inventor believes she is entitled by virtue of her invention. 35 U.S.C. § 112(b) (known as § 112, ¶ 2, before the America Invents Act). These claims, which come at the very end of every issued U.S. patent, state the legal boundaries of the products (or processes) that the patent owner can exclude others from making, using, selling, offering to sell, or importing into the U.S. 35 U.S.C. §§ 154(a), 271(a).

Once the PTO receives an application, a patent examiner reviews the application and its claims to ensure that it complies with all the formal and substantive requirements of patent law. An applicant is not entitled to a given patent claim unless that claim (a) is adequately supported by the written disclosure to which it is attached, (b) constitutes patentable subject matter, (c) is useful, (d) is new, and (e) is nonobvious. (This Chapter discusses all of the formal requirements (a), as well as all of the substantive requirements (b)-(e).) An examiner communicates his or her assessment of the applicant's entitlement to a given claim in an official response, called an *Office Action*. Most often, the initial Office Action rejects one or more claims in the application for failure to comply with one or more of the requirements. The applicant can file a *Response to the Office Action*, urging the examiner to reconsider the rejection(s). Importantly, the applicant can amend the rejected claim(s) in an effort to overcome the examiner's objection(s). Indeed, the bulk of the typical patent prosecution is devoted to a series of applicant amendments to the scope of the patent claims, a process that culminates—the applicant hopes—in a set of claims that are adequately supported by the written disclosure and both new and nonobvious (when judged against the state of the art at the time the application was filed). This extended exchange between the applicant and examiner produces an official documentary history of the application's progress through the PTO; it is known as the *prosecution history*. The prosecution history can play an important role in interpreting disputed terms in a patent's claims, as well as in limiting the arguments for infringement a patentee can make—two topics we will discuss in greater detail later in this Chapter.

---

[*] Patent law's phosita, much like tort law's reasonable person, is a construct. We will learn more about the phosita's characteristics as the Chapter unfolds. For now, suffice it to say that the phosita is well-schooled in the useful art to which the patent pertains. And the patentee is expected to write the patent for a technically skilled reader like the phosita, not for an untutored generalist (like a lawyer or a judge). *See In re Nelson*, 280 F.2d 172, 181 (CCPA 1960) ("The descriptions in patents are not addressed to the public generally, to lawyers or to judges, but, as [35 U.S.C. §] 112 says, to those skilled in the art to which the invention pertains or with which it is most nearly connected."); *Phillips v. AWH Corp.*, 415 F.3d 1303, 1313 (Fed. Cir. 2005) (en banc) (quoting *In re Nelson*). Also, a word on pronunciation: some pronounce the term foz'-i-ta, and some pronounce it fo-see'-ta. Others just chuckle about the whole topic.

If an applicant cannot persuade the examiner to allow the claims to which he thinks he is entitled, he can appeal the examiner's *Final Rejection* to the appellate body within the PTO, the Patent Trial & Appeal Board or "PTAB" (known as the Board of Patent Appeals & Interferences, or "BPAI," before the America Invents Act). 35 U.S.C. § 134. If the Board affirms the examiner's rejection of the applicant's claims, the applicant can seek review of the PTO's assessment in federal court. The more common route is to appeal the PTO's rejection directly to the U.S. Court of Appeals for the Federal Circuit, located in Washington, DC.[*] 35 U.S.C. §§ 141, 144. The Federal Circuit, which in 1982 assumed appellate jurisdiction over the PTO from the U.S. Court of Customs and Patent Appeals ("CCPA"), hears all appeals arising under the U.S. patent laws, including appeals from federal district court patent infringement cases. 28 U.S.C. §§ 1292, 1295.

After a patent issues, it is subject to reconsideration by the Patent Office under statutorily prescribed conditions. One type of reconsideration is called *reexamination*. Anyone, including the patentee, a competitor, or an accused infringer, can ask the PTO to reexamine the issued claims on the basis of prior art patents or printed publications. 35 U.S.C. §§ 301, 302. The PTO Director then has the discretion to grant the requested reexamination, upon a finding that "a substantial new question of patentability affecting any claim of the patent concerned is raised by the request." 35 U.S.C. § 303(a). The patentee has an opportunity to comment on the new materials, and the requester has a right to respond to the patentee's comments. 35 U.S.C § 304. A more elaborate type of reconsideration, which allows for more participation by the patent challenger, is called *inter partes review*, or "IPR". 35 U.S.C. §§ 311-318. The AIA substantially revised this form of reconsideration. Only someone other than the patentee can request this form of review, and only a limited number of prior-art-based issues can be considered. 35 U.S.C. §§ 311(a), (b). Finally, the most elaborate and expansive type of PTO-based reconsideration, which is new with the AIA, is called *post-grant review*. 35 U.S.C. §§ 321-329. Because only patents filed after March 2013 are subject to this type of reconsideration, AIA § 6(f), there is not yet any substantial track record for this type of PTO proceeding.

The foregoing is meant to serve only as a snapshot summary of the basics of patent prosecution. As we explore various patent law doctrines, we discuss more details about patent prosecution as needed, in context.

C. Reading a U.S. Patent Document

The cardinal right in patent law is the right to exclude another's infringement. The patent embodies the grant: "Every patent shall contain . . . a grant to the patentee . . . of the right to exclude others from making, using, offering for sale, or selling the invention throughout the United States or importing the invention into

---

[*] One can instead appeal a PTO rejection to the U.S. District Court for the Eastern District of Virginia. 35 U.S.C. § 145. The District Court, unlike the Federal Circuit, can consider newly submitted evidence outside the record before the PTO. *See Kappos v. Hyatt*, 132 S. Ct. 1690 (2012) (construing the new-evidence provision of § 145). Any appeal from the District Court in a § 145 case will be heard by the Federal Circuit. 28 U.S.C. § 1295(a)(4)(C).

the United States . . . ." 35 U.S.C. § 154(a). And the Patent Act provides the cause of action: "whoever without authority makes, uses, offers to sell, or sells any patented invention, within the United States or imports into the United States any patented invention during the term of the patent therefor, infringes the patent." 35 U.S.C. § 271(a).

At the end of this Section, we provide a full copy of U.S. Patent No. 6,263,732 to Michael Hoeting & Stephen Hoeting, entitled "Measuring Cup." Patent law practitioners usually refer to a patent in short form, either by the last three digits of its official number or by the first named inventor's surname. Following that practice, one refers to this patent as "the '732 patent," or "the Hoeting patent."

The named inventor is a central figure in patent law. Indeed, under the Patent Act, an inventor's application for a patent must include a sworn oath that he or she "believes himself or herself to be the original inventor or an original joint inventor of" the invention sought to be patented. 35 U.S.C. § 115(b)(2). In other words, one cannot simply copy someone else's idea and apply for a patent on it.

We can break a patent document into three basic parts—the first page, sometimes called the cover or facing page; the patent drawings, if any; and the specification, which includes both the written description of the invention and the separately numbered claims. The *first page* of the '732 patent tells us quite a bit about it. In addition to giving us the patent's official number (in the upper right corner), title (in the upper left), and inventors' names (just below the title), we learn both the date on which the Hoetings first filed for the patent—May 18, 1999 (in the left column)—and the date on which the PTO granted the '732 patent—July 24, 2001 (upper right). It took the Hoetings just over 2 years to prosecute this patent, a typical time period for mechanical inventions like their measuring cup.

A patent's first application date and issue date are each important. Generally speaking, a patent is in force from the date it issues until the 20th anniversary of its original filing date, 35 U.S.C. § 154(a)(2), provided the patentee pays the three maintenance fees that are required to keep the patent in force. The fees are due to be paid 3.5, 7.5, and 11.5 years after grant. 35 U.S.C. § 41(b). For example, consider a patent application that was filed on September 1, 2010, and that issued as a patent on July 31, 2012. The patent will expire on September 1, 2030, so long as the maintenance fees are paid in January 2016, 2020, and 2024. Note that, because the application date sets the expiration date, an applicant has an incentive to act promptly in her dealings with the PTO; time wasted is term lost. Finally, the first page also provides us a list of the prior art patents and other publications that the PTO had before it when evaluating the Hoetings' patent application, as well as an Abstract that sets forth "the nature and gist of the technical disclosure" in the remainder of the patent, 37 C.F.R. § 1.72(b).

The *patent drawings* help us understand the invention described and claimed in the '732 patent. The six figures in this patent, which depict two different embodiments of the Hoetings' invention, contain a series of reference numbers and letters. The remainder of the patent relies on these reference numbers and letters in discussing the different embodiments of the invention.

The *specification*—the main text of the patent—includes the written description of the invention, followed by one or more numbered claim paragraphs. In the case of the '732 patent, there is one claim. The description of the '732 patent is highly conventional; this is not surprising, given the PTO regulations governing patent content. *See* 37 C.F.R. §§ 1.71-1.85. The '732 patent, like every well-written patent, also tells a story—namely, a story about a problem that existing technologies haven't solved, and the solution the applicant has invented to meet that problem. What problem(s) does the Hoeting patent identify with prior art measuring cups? You should be able to state the column and line numbers where this occurs.

The '732 patent ends with one claim, beginning at the bottom of Column 4. In a patent, a numbered claim paragraph establishes the verbal boundary for a conceptual space from which the patentee can exclude others, *i.e.*, the claimed invention. The Patent Office requires that each claim be one, and only one, sentence long.[*] As a result, the wording can look a bit stilted. The words and phrases that call out the necessary characteristics of the claim are known as claim *limitations* or claim *elements*. Any real-world item that falls within the scope of a claim is said to be *covered by* that claim; patent lawyers also say that a claim *reads on* a real-world item (or prior art patent) that falls within its scope.

Each claim in a patent provides its own separate right to exclude. *See* 35 U.S.C. § 271. As a result, one analyzes the infringement (or not) and the validity (or not) of each numbered claim individually. *See* 35 U.S.C. § 282. As you will see later in this Chapter, infringement can be *literal*—the accused item falls within the literal scope of each and every limitation of the asserted patent claim. Infringement can also be nonliteral, under a patent law doctrine known as *the doctrine of equivalents*. Nonliteral infringement occurs when the accused item falls outside the literal scope of one or more of the claim limitations, *but* for each such limitation the accused item contains a close substitute feature that is the functional equivalent of the missing literal limitation. Infringement can also be *direct*—the accused infringer is the very person who committed the infringing act—or *indirect*—the accused infringer encouraged or helped someone else to commit direct infringement. Indirect infringement most commonly applies when the claimed invention has many component parts that are assembled by an end-user. The one who makes some component(s), but does not assemble the claimed invention, is not a direct infringer. The Patent Act expands liability to reach this component-maker in circumstances where there is no innocent basis for the conduct. *See* 35 U.S.C. §§ 271(b), (c). We will not explore indirect infringement in patent law; we do, however, consider copyright law's secondary-liability doctrines in Chapter 5.

---

[*] The Manual of Patent Examining Procedure provides: "While there is no set statutory form for claims, the present Office practice is to insist that each claim must be the object of a sentence starting with 'I (or we) claim' .... Each claim begins with a capital letter and ends with a period. Periods may not be used elsewhere in the claims except for abbreviations." U.S. Dep't of Commerce, Patent & Trademark Office, *Manual of Patent Examining Procedure* § 608.01(m), at 600-128 (9th ed., Mar. 2014).

(12) **United States Patent**
Hoeting et al.

(10) **Patent No.:**   **US 6,263,732 B1**
(45) **Date of Patent:**   **Jul. 24, 2001**

(54) **MEASURING CUP**

(75) Inventors: **Michael G. Hoeting; Stephen C. Hoeting**, both of Cincinnati, OH (US)

(73) Assignee: **Bang Zoom Design**, Cincinnati, OH (US)

(*) Notice: Subject to any disclaimer, the term of this patent is extended or adjusted under 35 U.S.C. 154(b) by 0 days.

(21) Appl. No.: **09/313,686**

(22) Filed: **May 18, 1999**

(51) **Int. Cl.**[7] **G01F 19/00**
(52) **U.S. Cl.** .... **73/427**; 33/1 V; D10/46.2
(58) **Field of Search** .... 33/1 V, 1 F, 522, 33/679.1; 73/426, 427; 215/365, 366; D10/46.2; 222/23, 25, 26

(56) **References Cited**

U.S. PATENT DOCUMENTS

153,159	* 7/1874	Dinwiddie	73/427
216,530	* 6/1879	Pfitzenmeier	73/427
D. 243,500	3/1977	Cooper .	
D. 255,530	6/1980	Daenen et al. .	
D. 259,460	6/1981	Daenen et al. .	
D. 259,461	6/1981	Daenen et al. .	
D. 259,462	6/1981	Daenen et al. .	
D. 268,158	3/1983	Doyel .	
D. 272,704	2/1984	Smith .	
D. 292,381	10/1987	Kowolik et al. .	
D. 292,492	10/1987	Ross et al. .	
D. 293,770	1/1988	Ross et al. .	
D. 294,213	2/1988	Chasen .	
D. 302,920	8/1989	Ancona et al. .	
D. 303,055	8/1989	Prindle .	
D. 304,277	10/1989	Wolff et al. .	
D. 304,301	10/1989	Moss et al. .	

D. 321,328	11/1991	Duquet	D10/46.2
D. 330,863	11/1992	Green	D10/46.2
423,018	* 3/1890	Young	73/427
1,507,968	* 9/1924	Johnson	73/427
1,564,470	* 12/1925	Crimmel	73/427
1,722,101	* 7/1929	Little	73/427
2,165,045	* 7/1939	Garside	73/426
3,526,138	9/1970	Swett et al. .	
3,527,270	9/1970	Weil .	
4,073,192	2/1978	Townsend	73/429
4,283,951	8/1981	Varpio	73/426
4,566,509	1/1986	Szajna .	
4,834,251	5/1989	Yu .	
5,397,036	* 3/1995	Mainwald	73/427
5,588,747	12/1996	Blevins	73/427
5,662,249	9/1997	Grosse .	

OTHER PUBLICATIONS

"Jigger Photographs: Having four (4) semi–columns formed in relief on inside wall, of four (4) different heights. Information concerning origin and (date is unknown)."

* cited by examiner

Primary Examiner—Christopher W. Fulton
(74) *Attorney, Agent, or Firm*—Wood, Herron & Evans, L.L.P.

(57) **ABSTRACT**

A measuring cup has at least one ramp formed in relief radially inwardly on the inside surface of the measuring cup sidewall. At least one ramp rises from about the bottom edge of the sidewall to about the top edge of the sidewall. The indicia on an upwardly directed surface of the at least one ramp allows a user to look downwardly into the measuring cup to visually detect the volume level of the contents in the measuring cup, thereby eliminating the need to look horizontally at the cup at eye level. Preferably the cup has two ramps, with at least one of standard units, and another with metric units.

1 Claim, 2 Drawing Sheets

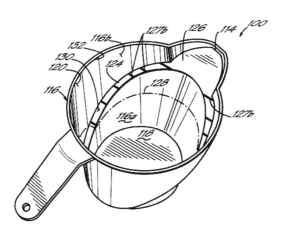

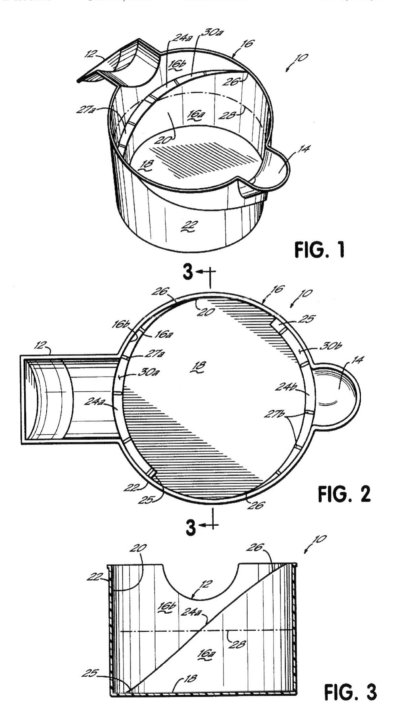

FIG. 1

FIG. 2

FIG. 3

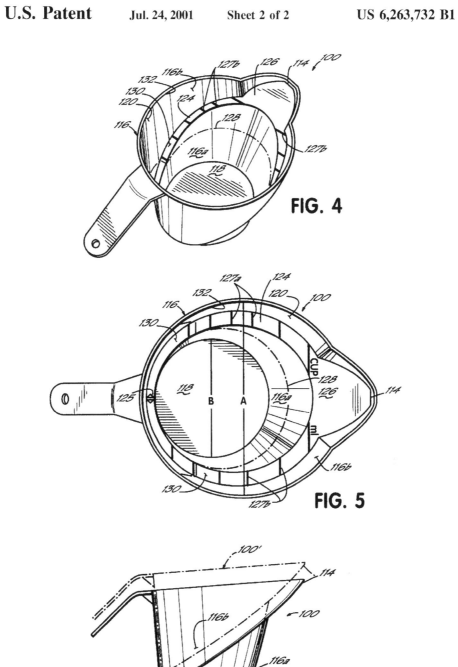

FIG. 4

FIG. 5

FIG. 6

US 6,263,732 B1

1

MEASURING CUP

FIELD OF THE INVENTIONS

This invention relates to measuring cups. More specifically, this invention relates to a measuring cup having indicia viewable from above.

BACKGROUND OF THE INVENTION

Measuring cups are known. Measuring cups can be made from a variety of materials, including plastic, metal and glass. One of the most common measuring cups found on the market today is a transparent measuring cup made of Pyrex® which is resistant to sudden changes in temperature to which it may be submitted during use.

The utility derived from a measuring cup is related to the ease with which volumetric indicia on the measuring cup's wall may be read by a user. Of course, any suitable units of measurement may be used to indicate the level to which contents have risen within a measuring cup.

Traditional measuring cups have indicia marked upon the measuring cup wall in such a manner which sometimes makes the indicia difficult to read, depending upon how precise a measurement is needed, the materials from which the measuring cup is manufactured and the physical condition of the user, for example. In the case of a measuring cup which is made from transparent or translucent material, e.g., Pyrex®, the most precise way to measure the contents contained therein is to place the measuring cup upon a level surface, pour the contents to be measured into the measuring cup and then stoop down to the vertical level of the measuring cup to attempt to visually detect the bottom of a liquid meniscus or to a level surface of solid contents. An alternative method to read the level to which contents in a transparent or translucent measuring cup have risen is to lift the measuring cup to eye level and attempt to hold the measuring cup steady while visually detecting the volume. In either use, the observer is looking in a generally horizontal direction to detect the volume.

Prior art measuring cups that are opaque are more difficult to read than transparent or translucent measuring cups. In order to read the volume of contents held within an opaque measuring cup, a user must peer over the upper margin of the measuring cup to eyeball, as closely as possible, the level to which contents have risen, either by stooping to the measuring cup's level or by lifting the measuring cup to eye level.

While the above-described methods for determining the volume of contents in a measuring cup may seem simple enough for most users, these methods can prove to be difficult for others. Users with bad knees, a bad back, or arthritis, for example, may not only have substantial difficulty in stooping over to accurately read the volume of contents in a measuring cup placed on a level surface, but may also have just as much difficulty in lifting a measuring cup to eye level and holding the cup steady to read the volume of contents held therein. When precise measurement of the volume of contents within a measuring cup is critical to a task, the simple actions of bending over or lifting a measuring cup to eye level, which seem easy to some users, may become difficult and uncomfortable for others.

Measuring the volume of cooking ingredients using prior art measuring cups can also be frustrating. As mentioned above, it can be difficult for a user to stoop over to read the level of contents when placed on a level surface or when lifted to eye level. An unsteady hand not only makes the

2

volume of contents difficult to determine when a measuring cup is lifted to eye level, but a user may spill contents or even drop the measuring cup when attempting to do so.

Measuring cups are not limited in their utility to the kitchen, of course. They may also be used for measuring proper ratios of solutions, e.g., antifreeze, the precise measurement of which is critical to its application and simplicity of determining a precise volume is necessary. Other common household solutions can be dangerous, e.g., toxic or caustic, and when a measuring cup is filled with these solutions, the possibility of spilling them within the proximity of a child or a pet greatly increases when a measuring cup must be raised to eye level to determine the volume of its contents.

It is an object of the present invention to simplify the way in which a person can accurately detect the volume of material held in a measuring cup.

It is another object of the invention to improve a measuring cup to make it more conductive to simplify an accurate volume determination.

SUMMARY OF THE INVENTION

The present invention achieves the above-stated objectives by including with a measuring cup at least one sloped ramp having an upwardly directed surface having indicia which is readily observable by an observer looking downwardly into the open end of the cup.

The structure simplifies volume determination because there is no need for the observer to move relative to the cup in order to look in a horizontal direction at the cup indicia. Thus, the possibility of spilling is reduced. Also, since the ramp preferably rises continuously and gradually from the bottom of the cup, a user who is filling the cup from above can actually see the volume indicia on the upwardly directed surface of the ramp while the cup is being filled, looking along the same line of sight generally used during filling. These advantages result from the ability to visually determine the volume of the cup by simply looking into the open upper end, and the gradual slope of the ramp.

According to a first preferred embodiment of the invention, the cup has a bottom wall and an encircling vertical sidewall, so that the cup is cylindrical in shape with an open upper end. Inside the cup, at least one ramp slopes continuously upward from the bottom wall toward the open upper end. The ramp includes an upwardly directed surface bearing printed volume indicia viewable through the open upper end to visually determine the volume of the cup contents. Preferably the cup has two ramps formed integrally along the sidewall, with one bearing standard units of measurement and the other bearing metric units. The two ramps have oppositely located bottom ends and oppositely located top ends. The cup also has a handle and a spout, with the handle located adjacent one ramp and the spout located adjacent another.

In a second embodiment, the sidewall is sloped somewhat, so that it is substantially vertical. The cup includes two integral, oppositely located ramps with adjacently located bottom ends and adjacently located top ends. The top ends feed toward the spout, and again, one ramp bears standard units and the other bears metric indicia.

With another embodiment, the cup can be formed of any suitable material and via any suitable process, although transparent and moldable material is preferred and manufactured using a molding process is also preferred.

These and other features will be more readily understood in view of the following detailed description and the drawings.

US 6,263,732 B1

3

BRIEF DESCRIPTION OF THE DRAWINGS

FIG. 1 is a perspective view of a measuring cup according to a first preferred embodiment of the invention

FIG. 2 is a top plan view of the measuring cup of FIG. 1

FIG. 3 is a cross-sectional view of the measuring cup of FIG. 2 taken along 3—3

FIG. 4 is a perspective view of a second preferred embodiment of the inventive measuring cup

FIG. 5 is a top plan view of the measuring cup of FIG. 4

FIG. 6 is side view of the measuring cup and nesting feature of a second preferred embodiment of the present inventive measuring cup.

DETAILED DESCRIPTION OF THE PREFERRED EMBODIMENT

FIGS. 1–3 show a first preferred embodiment of the present inventive measuring cup 10. Generally, the measuring cup 10 is integrally formed out of a suitable material and has a handle 12 and a spout 14 integrally attached to a substantially vertical sidewall 16. The measuring cup 10 has a base or bottom wall 18 integrally attached around its perimeter to the bottom edge of the sidewall 16. The cup 10 has an open upper end.

The wall 16 has an inside surface 20 and an outside surface 22 from which ramps 24a, 24b are formed in relief. The measuring cup 10 is molded from any suitable food grade plastic known in the art, however, it will be understood that the measuring cup 10 may be manufactured by any suitable process. It will also be understood that the measuring cup 10 may be made of any other suitable material known in the art, e.g., Pyrex®, metal.

The ramps 24a, 24b are located on opposite sides of the cup 10 but are identical in construction. Therefore, only one such ramp is described. Each ramp has a ramp base, or bottom end 25, and a ramp top or upper end 26. The ramp base 25 is located proximate the bottom edge of the sidewall 16, and the ramp top 26 is located proximate the top edge of the sidewall 16. The ramps 24a, 24b have respective ramp surfaces 30a, 30b, which are generally upwardly directed and have a substantially constant slope between the ramp base 25 and the ramp top 26. In the first preferred embodiment, the ramps 24a, 24b are oppositely disposed on the inside surface 20 of the wall 16. Also in the first preferred embodiment, the ramps 24a, 24b traverse substantially the same distance from the bottom margin of the wall 16 to the top margin of the wall 16 along the inside surface 20. It will be understood by those in the art that the ramps 24a, 24b may have a greater or lesser slope, which in turn would result in shorter or longer distances, respectively, traveled from the bottom margin to the top margin of the wall 16.

The ramps 24a, 24b have a slope great enough so that the ramps 24a, 24b do not extend more than half the circumference of the wall 16, as seen in FIG. 2. Also, the ramps 24a, 24b do not overlap each other. That is, the ramp 24a does not rise over the ramp 24b on the inside surface 20 of the wall 16. In the first preferred embodiment of the measuring cup 10, the sidewall 16 is substantially normal to the base 18, so that the cup 10 is generally cylindrical in shape. It will be understood by those in the art that the wall 16 may angle away from the perimeter of the base 18 so that the measuring cup 10 may receive a second measuring cup (not shown) therein, i.e., allow plural measuring cups 10 to stack inside each other.

Each of the ramps 24a, 24b is provided with volume indicia 27a, 27b, on the upwardly directed surface 30a, 30b,

4

so a user may easily look down into the measuring cup 10 from above and view the volume level of the contents 28 within the cup 10. In the first preferred embodiment, the ramp 24a is provided with metric indicia 27a on ramp surface 30a, and ramp 24b is provided with standard indicia 27b on ramp surface 30b. It will be understood by those in the art that the indicia 27a, 27b may be spaced differently relative to each unit of measurement on respective ramps 24a, 24b, depending on the desired slope of the ramps 24a, 24b.

FIGS. 4–6 show a second preferred embodiment of an inventive measuring cup 100. The measuring cup 100 has a sidewall 116 with an open upper end 132 having a diameter A larger than the diameter B of the bottom wall or base 118. That is, the sidewall 116 slopes outwardly away from the base 118 as the sidewall 116 rises from its bottom edge to its top edge so that at least a second measuring cup 100' can be stacked within the measuring cup 100. The measuring cup 100 has a pair of oppositely located, but identically sloped ramps 124 which are substantially continuous around the sidewall inside surface 120 from the ramp bottom 125 to the ramp top 126. That is, both ramps 124 rise symmetrically along the inside surface 120 of the sidewall 116 from about the bottom edge of the sidewall inside surface 120 generally opposite the spout 114 to the top edge of the sidewall 116 adjacent to the base of the spout 114.

Because the aperture 132 has a greater diameter A than the diameter B of the base 118, the indicia 127a, 127b along the ramps 124 are not spaced in equal intervals. That is, a given rise in level of the contents 128 near the bottom edge of the sidewall 116 requires a smaller volume than an equal rise in the level of the contents near the upper edge of the sidewall 116. As a result, the indicia 127a, 127b are spaced upon the ramps 124 closer together near the top edge of the sidewall 116 than at the bottom edge for an equivalent volume of contents 128. It will be understood by those in the art that the progressive change in the diameter of the measuring cup 100 from the base 118 to the upper edge of the sidewall 116 may also be accommodated by decreasing the slope of the ramps 124 from the lower edge of the sidewall 116 to the upper edge of the sidewall 116 while maintaining the spacing between indicia 127a, 127b along the ramps 124.

Also in this embodiment, the ramp tops 126 are coextensive with spout 114 to allow a user to more easily pour contents from the measuring cup 100 without spilling.

The sidewall 116 has a lower portion 116a below the ramps 124 which is offset inwardly by the width of the ramp surfaces 130 from an upper portion 116b of the sidewall 116. This offset allows subsequent measuring cups (not shown) to nest within the measuring cup 100 and each other when stacked.

From the above disclosure of the detailed description of the present invention and the preceding summary of the preferred embodiment, those skilled in the art will comprehend the various modifications to which the present invention is susceptible. Therefore, I desire to be limited only by the scope of the following claims and equivalents thereof.

What is claimed is:

1. A measuring device, comprising:

a bottom wall and a generally vertical and encircling side wall having a lower edge and an upper edge, said sidewall defining an upwardly opening cup with an upper end;

a spout attached integrally to said sidewall; and

a pair of continuously sloping ramps formed integrally with and radially inward in relief from said sidewall, said ramps extending from about said bottom wall

US 6,263,732 B1

5

generally opposite said spout toward said open upper end generally adjacent said spout, wherein said ramp is coextensive with said spout, said ramp having an upwardly directed surface and indicia located on said upwardly directed surface being at least one of standard

6

and metric units of measurement providing a readily observable indication of the volume of the contents contained within said cup.

*   *   *   *   *

OXO Angled Measuring Cup

OXO, a company based New York, makes cooking tools and other consumer products. You can learn more about the company at its website, www.oxo.com. One of the products OXO makes is called the Angled Measuring Cup. OXO sells

the cup design in various sizes; the 2-cup size is shown here. It is stamped with the number of the Hoeting '732 patent.

Notes & Questions

1. What does it mean in Claim 1 of the '732 patent that the sloping ramps are "coextensive with" the spout? Are the ramps of the OXO cup coextensive with its spout, as the phrase "coextensive with" is used in Claim 1? What sources should we use to arrive at or explain our answer?

2. OXO appears to have bought or licensed the '732 patent, inasmuch as it stamps the full number of the patent on each copy of the cup that it makes. It is important to understand that a patent does not provide a right to create and sell a product. Instead, it creates a right to exclude—a right to stop others from making, offering to sell, selling, or importing the claimed invention. Thus, the indication of a patent number on a product serves a different function: Manufacturers often mark commercial embodiments of patented products to provide notice to the public that the item is patented. When they do so, the Patent Act provides the patent owner with stronger remedies against infringers. 35 U.S.C. § 287(a).

II. Claim Construction & Definiteness

A. An Introduction to Claim Construction

1. Assigning the Construction Task

A patent owner can enforce her patent by filing a suit for patent infringement. In that suit, which must be brought in federal court, 28 U.S.C. § 1338, the parties are likely to disagree about whether the asserted patent claims are infringed by the accused infringer's product or process, and whether the asserted patent claims are valid (*i.e.*, whether the PTO should have granted them in the first place). In that sense, the patent system runs on claims. One cannot answer infringement or validity questions, however, without knowing the precise scope of the patent claim in question. As a result, a significant issue in virtually every patent suit is: what *is* the precise scope of the claim? Most often, the parties offer competing interpretations of one or more of the limitations in the claim. What doctrines govern how the courts decide these claim construction disputes?

In the next subsection, we present cases that discuss the methods courts must use to construe a disputed claim term. As a prelude to that discussion, however, it is useful to consider *who* should decide the meaning of a dispute claim term—a trial judge, or a jury? This *who* question has generated more than a decade of intense dispute, in both the Federal Circuit and Supreme Court.

More than 20 years ago, the Supreme Court held that a patentee has no Seventh Amendment right to have a jury decide the meaning of a disputed claim term, and that a trial judge is in the better position to construe a disputed claim term (relative to a jury). *See Markman v. Westview Instruments, Inc.*, 517 U.S. 370, 372 (1996). Thereafter, many district courts began routinely to hold pre-trial proceedings, known as "*Markman* hearings," to adjudicate the scope of the disputed terms in the asserted patent claims. These claim construction rulings are not immediately appealable unless they are combined with summary judgment in favor of one of the parties. When an appeal is timely, it will go to the Federal Circuit.

The *Markman* case appeared to leave open the question whether an appellate court should treat the trial judge's claim construction decision as a pure question of law—in which case the applicable standard of review is *de novo* review—or, instead, should treat the decision as a question of law based on underlying facts—in which case the applicable standard of review is clear error review (a more deferential standard) as to the underlying facts, and *de novo* review for the ultimate conclusion. The Federal Circuit twice held en banc, over impassioned dissents, that claim construction is treated as a pure question of law on appeal, reviewed without deference to the trial judge's decision. *See Lighting Ballast Control, LLC v. Philips Electronics North America Corp.*, 744 F.3d 1272 (Fed. Cir. 2014) (en banc); *Cybor Corp. v. FAS Technologies, Inc.*, 138 F.3d 1448 (Fed. Cir. 1998) (en banc). It is important to realize that the question generated such deep controversy because claims—like everything else in the patent document—are ostensibly drafted for the technically trained reader, while at the same time the claims have such critical legal significance. The ordinary meaning that skilled artisans assigned to a term in their field at the time the patent was sought seems like a heavily fact-laden matter. But *Markman* held that the jury has no role to play in finding those facts when the parties dispute them. The Supreme Court recently resolved the issue.

Teva Pharma. USA, Inc. v. Sandoz, Inc.

135 S. Ct. 831 (2015)

BREYER, JUSTICE:

In *Markman v. Westview Instruments, Inc.*, 517 U.S. 370 (1996), we explained that a patent claim is that "portion of the patent document that defines the scope of the patentee's rights." *Id.* at 372. We held that "the construction of a patent, including terms of art within its claim," is not for a jury but "exclusively" for "the court" to determine. *Id.* That is so even where the construction of a term of art has "evidentiary underpinnings." *Id.* at 390.

Today's case involves claim construction with "evidentiary underpinnings." And, it requires us to determine what standard the Court of Appeals should use when it reviews a trial judge's resolution of an underlying factual dispute. Should the Court of Appeals review the district court's factfinding *de novo* as it would review a question of law? Or, should it review that factfinding as it would review a trial judge's factfinding in other cases, namely by taking them as correct "unless clearly erroneous?" We hold that the appellate court must apply a "clear error," not a *de novo*, standard of review.

I

The basic dispute in this case concerns the meaning of the words "molecular weight" as those words appear in a patent claim. The petitioners, Teva Pharmaceuticals (along with related firms), own the relevant patent. The patent covers a manufacturing method for Copaxone, a drug used to treat multiple sclerosis. The drug's active ingredient, called "copolymer-1," is made up of molecules of varying sizes. And the relevant claim describes that ingredient as having "a molecular weight of 5 to 9 kilodaltons."

The respondents, Sandoz, Inc. (and several other firms), tried to market a generic version of Copaxone. Teva sued Sandoz for patent infringement. Sandoz defended the suit by arguing that the patent was invalid. The Patent Act requires that a claim "particularly poin[t] out and distinctly clai[m] the subject matter which the applicant regards as his invention." 35 U.S.C. § 112 ¶ 2 (2006 ed.); *see Nautilus, Inc. v. Biosig Instruments, Inc.*, 134 S. Ct. 2120, 2125, n. 1 (2014).[†] The phrase "molecular weight of 5 to 9 kilodaltons," said Sandoz, did not satisfy this requirement.

The reason that the phrase is fatally indefinite, Sandoz argued, is that, in the context of this patent claim, the term "molecular weight" might mean any one of three different things. The phrase might refer (1) to molecular weight as calculated by the weight of the molecule that is most prevalent in the mix that makes up copolymer-1. (The scientific term for molecular weight so calculated is, we are told, "peak average molecular weight.") The phrase might refer (2) to molecular weight as calculated by taking all the different-sized molecules in the mix that makes up copolymer-1 and calculating the average weight, i.e., adding up the weight of each molecule and dividing by the number of molecules. (The scientific term for molecular weight so calculated is, we are told, "number average molecular weight.") Or, the phrase might refer (3) to molecular weight as calculated by taking all the different-sized molecules in the mix that makes up copolymer-1 and calculating their average weight while giving heavier molecules a weight-related

---

† [*Eds. Note*: We explore the *claim definiteness* requirement in the next subsection. For purposes of this case, simply keep in mind that § 112 requires not only the numbered claim paragraphs we find in a patent, but also requires that those claims particularly point out and distinctly claim the invention. This demand for particularity and distinctness gives rise to the indefiniteness defense: A claim is invalid if its scope is indefinite. Of course, the question naturally arises, how much good-faith disagreement about the scope of a claim renders it fatally indefinite? In *Nautilus*, the 2014 case the Court cites here, the Supreme Court wrestled with that question for the first time since the 1940s. The *Nautilus* case is presented below.]

bonus when doing so. (The scientific term for molecular weight so calculated, we are told, is "weight average molecular weight.") In Sandoz's view, since Teva's patent claim does not say which method of calculation should be used, the claim's phrase "molecular weight" is indefinite, and the claim fails to satisfy the critical patent law requirement.

The District Court, after taking evidence from experts, concluded that the patent claim was sufficiently definite. Among other things, it found that in context a skilled artisan would understand that the term "molecular weight" referred to molecular weight as calculated by the first method, *i.e.*, "peak average molecular weight." In part for this reason, the District Court held the patent valid.

On appeal, the Federal Circuit held to the contrary. It found that the term "molecular weight" was indefinite. And it consequently held the patent invalid. In reaching this conclusion, the Federal Circuit reviewed de novo all aspects of the District Court's claim construction, including the District Court's determination of subsidiary facts.

Teva filed a petition for certiorari. And we granted that petition. The Federal Circuit reviews the claim construction decisions of federal district courts throughout the Nation, and we consequently believe it important to clarify the standard of review that it must apply when doing so.

II

A

Federal Rule of Civil Procedure 52(a)(6) states that a court of appeals "must not * * * set aside" a district court's "[f]indings of fact" unless they are "clearly erroneous." In our view, this rule and the standard it sets forth must apply when a court of appeals reviews a district court's resolution of subsidiary factual matters made in the course of its construction of a patent claim. We have made clear that the Rule sets forth a "clear command." *Anderson v. Bessemer City*, 470 U.S. 564, 574 (1985). "It does not make exceptions or purport to exclude certain categories of factual findings from the obligation of a court of appeals to accept a district court's findings unless clearly erroneous." *Pullman-Standard v. Swint*, 456 U.S. 273, 287 (1982). Accordingly, the Rule applies to both subsidiary and ultimate facts. And we have said that, when reviewing the findings of a "district court sitting without a jury, appellate courts must constantly have in mind that their function is not to decide factual issues de novo." *Anderson*, 470 U.S. at 573.

. . . .

Our opinion in *Markman* neither created, nor argued for, an exception to Rule 52(a). The question presented in that case was a Seventh Amendment question: Should a jury or a judge construe patent claims? 517 U.S. at 372. We pointed out that history provides no clear answer. *Id.* at 388. The task primarily involves the construction of written instruments. *Id.* at 386, 388, 389. And that task is better matched to a judge's skills. *Id.* at 388 ("The construction of written instruments is one of those things that judges often do and are likely to do better than jurors unburdened by training in exegesis"). We consequently held that claim construction falls "exclusively within the province of the court," not that of the jury. *Id.* at 372.

When describing claim construction we concluded that it was proper to treat the ultimate question of the proper construction of the patent as a question of law in the way that we treat document construction as a question of law. *Id.* at 388-391. But this does not imply an exception to Rule 52(a) for underlying factual disputes. We used the term "question of law" while pointing out that a judge, in construing a patent claim, is engaged in much the same task as the judge would be in construing other written instruments, such as deeds, contracts, or tariffs. *Id.* at 384, 386, 388, 389; *see also Motion Picture Patents Co. v. Universal Film Mfg.*, 243 U.S. 502, 510 (1917) (patent claims are "aptly likened to the description in a deed, which sets the bounds to the grant which it contains"); *Goodyear Dental Vulcanite Co. v. Davis*, 102 U.S. 222, 227 (1880) (analogizing patent construction to the construction of other written instruments like contracts). Construction of written instruments often presents a "question solely of law," at least when the words in those instruments are "used in their ordinary meaning." *Great Northern R. Co. v. Merchants Elevator Co.*, 259 U.S. 285, 291 (1922). But sometimes, say when a written instrument uses "technical words or phrases not commonly understood," *id.* at 292, those words may give rise to a factual dispute. If so, extrinsic evidence may help to "establish a usage of trade or locality." *Id.* And in that circumstance, the "determination of the matter of fact" will "preced[e]" the "function of construction." *Id.*. This factual determination, like all other factual determinations, must be reviewed for clear error.

Accordingly, when we held in *Markman* that the ultimate question of claim construction is for the judge and not the jury, we did not create an exception from the ordinary rule governing appellate review of factual matters. *Markman* no more creates an exception to Rule 52(a) than would a holding that judges, not juries, determine equitable claims, such as requests for injunctions. A conclusion that an issue is for the judge does not indicate that Rule 52(a) is inapplicable.

While we held in *Markman* that the ultimate issue of the proper construction of a claim should be treated as a question of law, we also recognized that in patent construction, subsidiary factfinding is sometimes necessary. Indeed, we referred to claim construction as a practice with "evidentiary underpinnings," a practice that "falls somewhere between a pristine legal standard and a simple historical fact." 517 U.S. at 378, 388. We added that sometimes courts may have to make "credibility judgments" about witnesses. *Id.* at 389. In other words, we recognized that courts may have to resolve subsidiary factual disputes. And, as explained above, the Rule requires appellate courts to review all such subsidiary factual findings under the "clearly erroneous" standard.

. . . .

Finally, practical considerations favor clear error review. We have previously pointed out that clear error review is "particularly" important where patent law is at issue because patent law is "a field where so much depends upon familiarity with specific scientific problems and principles not usually contained in the general storehouse of knowledge and experience." *Graver Tank & Mfg. v. Linde Air Prods.*, 339 U.S. 605, 610 (1950). A district court judge who has presided over, and listened to, the entirety of a proceeding has a comparatively greater opportunity to gain that familiarity than an appeals court judge who must read a written transcript or perhaps just those portions to which the parties have referred.

. . . .

D

Now that we have set forth why the Federal Circuit must apply clear error review when reviewing subsidiary factfinding in patent claim construction, it is necessary to explain how the rule must be applied in that context. We recognize that a district court's construction of a patent claim, like a district court's interpretation of a written instrument, often requires the judge only to examine and to construe the document's words without requiring the judge to resolve any underlying factual disputes. As all [the] parties [in this case] agree, when the district court reviews only evidence intrinsic to the patent (the patent claims and specifications, along with the patent's prosecution history), the judge's determination will amount solely to a determination of law, and the Court of Appeals will review that construction *de novo*.

In some cases, however, the district court will need to look beyond the patent's intrinsic evidence and to consult extrinsic evidence in order to understand, for example, the background science or the meaning of a term in the relevant art during the relevant time period. *See, e.g.*, *Seymour v. Osborne*, 78 U.S. (11 Wall.) 516, 546 (1871) (a patent may be "so interspersed with technical terms and terms of art that the testimony of scientific witnesses is indispensable to a correct understanding of its meaning"). In cases where those subsidiary facts are in dispute, courts will need to make subsidiary factual findings about that extrinsic evidence. These are the "evidentiary underpinnings" of claim construction that we discussed in *Markman*, and this subsidiary factfinding must be reviewed for clear error on appeal.

For example, if a district court resolves a dispute between experts and makes a factual finding that, in general, a certain term of art had a particular meaning to a person of ordinary skill in the art at the time of the invention, the district court must then conduct a legal analysis: whether a skilled artisan would ascribe that same meaning to that term *in the context of the specific patent claim under review*. That is because "[e]xperts may be examined to explain terms of art, and the state of the art, at any given time," but they cannot be used to prove "the proper or legal construction of any instrument of writing." *Winans v. New York & Erie R. Co.*, 62 U.S. (21 How.) 88, 100-101 (1859); *see also Markman*, 517 U.S. at 388 ("'Where technical terms are used, or where the qualities of substances * * * or any similar data necessary to the comprehension of the language of the patent are unknown to the judge, the testimony of witnesses may be received upon these subjects, and any other means of information be employed. *But in the actual interpretation of the patent the court proceeds upon its own responsibility, as an arbiter of the law, giving to* *the patent its true and final character and force.*'" (quoting 2 W. Robinson, *Law of Patents* § 732, pp. 482-483 (1890) (emphasis in original).

Accordingly, the question we have answered here concerns review of the district court's resolution of a subsidiary factual dispute that helps that court determine the proper interpretation of the written patent claim. The district judge, after deciding the factual dispute, will then interpret the patent claim in light of the facts as he has found them. This ultimate interpretation is a legal conclusion. The appellate court can still review the district court's ultimate construction of the claim *de novo*. But, to overturn the judge's resolution of an underlying factual dispute, the Court of Appeals must find that the judge, in respect to those factual findings, has made a clear error. Fed. Rule Civ. Proc. 52(a)(6).

In some instances, a factual finding will play only a small role in a judge's ultimate legal conclusion about the meaning of the patent term. But in some instances, a factual finding may be close to dispositive of the ultimate legal question of the proper meaning of the term in the context of the patent. Nonetheless, the ultimate question of construction will remain a legal question. Simply because a factual finding may be nearly dispositive does not render the subsidiary question a legal one. "[A]n issue does not lose its factual character merely because its resolution is dispositive of the ultimate" legal question. *Miller v. Fenton*, 474 U.S. 104, 113 (1985). It is analogous to a judge (sitting without a jury) deciding whether a defendant gave a confession voluntarily. The answer to the legal question about the voluntariness of the confession may turn upon the answer to a subsidiary factual question, say "whether in fact the police engaged in the intimidation tactics alleged by the defendant." *Id.* at 112. An appellate court will review the trial judge's factual determination about the alleged intimidation deferentially (though, after reviewing the factual findings, it will review a judge's ultimate determination of voluntariness *de novo*). *See id.* at 112-118. An appellate court similarly should review for clear error those factual findings that underlie a district court's claim construction.

III

We can illustrate our holding by considering an instance in which Teva, with the support of the Solicitor General, argues that the Federal Circuit wrongly reviewed the District Court's factual finding *de novo*. See Brief for Petitioners 54-56; Brief for United States as *Amicus Curiae* 31-32. Recall that Teva's patent claim specifies an active ingredient with a "molecular weight of about 5 to 9 kilodaltons." . . . The term might refer to the weight of the most numerous molecule, it might refer to weight as calculated by the average weight of all molecules, or it might refer to weight as calculated by an average in which heavier molecules count for more. The claim, Sandoz argues, does not tell us which way we should calculate weight.

To illustrate, imagine we have a sample of copolymer-1 (the active ingredient) made up of 10 molecules: 4 weigh 6 kilodaltons each, 3 weigh 8 kilodaltons each, and 3 weigh 9 kilodaltons each. Using the first method of calculation, the "molecular weight" would be 6 kilodaltons, the weight of the most prevalent molecule.

Using the second method, the molecular weight would be 7.5 (total weight, 75, divided by the number of molecules, 10). Using the third method, the molecular weight would be more than 8, depending upon how much extra weight we gave to the heavier molecules.

Teva argued in the District Court that the term "molecular weight" in the patent meant molecular weight calculated in the first way (the weight of the most prevalent molecule, or peak average molecular weight). Sandoz, however, argued that figure 1 of the patent showed that Teva could not be right. (We have set forth figure 1 in the Appendix, below). That figure, said Sandoz, helped to show that the patent term did not refer to the first method of calculation. Figure 1 shows how the weights of a sample's molecules were distributed in three different samples. The curves indicate the number of molecules of each weight that were present in each of the three. For example, the figure's legend says that the first sample's "molecular weight" is 7.7. According to Teva, that should mean that molecules weighing 7.7 kilodaltons were the most prevalent molecules in the sample. But,

look at the curve, said Sandoz. It shows that the most prevalent molecule weighed not 7.7 kilodaltons, but slightly less than 7.7 (about 6.8) kilodaltons. After all, the peak of the first molecular weight distribution curve (the solid curve in the figure) is not at precisely 7.7 kilodaltons, but at a point just before 7.7. Thus, argued Sandoz, the figure shows that the patent claim term "molecular weight" did not mean molecular weight calculated by the first method. It must mean something else. It is indefinite.

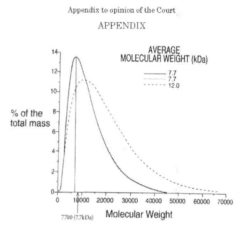

FIG. 1 (with minor additions to emphasize that the peak of the solid curve does not correspond precisely to 7.7kDa)

The District Court did not accept Sandoz's argument. Teva's expert testified that a skilled artisan would understand that converting data from a chromatogram to molecular weight distribution curves like those in figure 1 would cause the peak on each curve to shift slightly; this could explain the difference between the value indicated by the peak of the curve (about 6.8) and the value in the figure's legend (7.7). Sandoz's expert testified that no such shift would occur. The District Court credited Teva's expert's account, thereby rejecting Sandoz's expert's explanation. The District Court's finding about this matter was a factual finding—about how a skilled artisan would understand the way in which a curve created from chromatogram data reflects molecular weights. Based on that factual finding, the District Court reached the legal conclusion that figure 1 did not undermine Teva's argument that molecular weight referred to the first method of calculation (peak average molecular weight).

When the Federal Circuit reviewed the District Court's decision, it recognized that the peak of the curve did not match the 7.7 kilodaltons listed in the legend of figure 1. But the Federal Circuit did not accept Teva's expert's explanation as to how a skilled artisan would expect the peaks of the curves to shift. And it failed to accept that explanation without finding that the District Court's contrary determination was "clearly erroneous." The Federal Circuit should have accepted the District Court's finding unless it was "clearly erroneous." Our holding today makes clear that, in failing to do so, the Federal Circuit was wrong.

. . . .

We vacate the Federal Circuit's judgment, and we remand the case for further proceedings consistent with this opinion.

Notes & Questions

1. The Federal Circuit, which Congress created to hear all appeals arising under the patent laws, decides ten or more patent appeals per month, and has done so since 1982. The Supreme Court, by contrast, may hear two or three patent cases per term. The question about the proper standard of review, in *Teva*, is one to which the Federal Circuit brings a great deal of specialized experience. The Supreme Court, by contrast, brings a far more sweeping, generalist perspective to the issue. The Court also plays a key role in helping formulate the Federal Rules of Civil Procedure, one instance of which — Rule 52 — played a major role in *Teva*. What weight, if any, should one give to the Federal Circuit's greater level of experience when deciding questions about the basic structure of patent litigation? Put differently, how important is it that patent litigation be kept similar to, or instead be tailored differently from, other types of federal litigation?

2. The majority in *Teva*, analogizing patent claim construction to contract construction, posits that disputes about technical terms may require courts to consult expert witnesses and resolve fact disputes. In the dissent in *Teva*, not excerpted above, Justice Thomas (for himself and Justice Alito), analogized patent claim construction to statutory construction. In that context, trial courts consider facts, but their findings are reviewed *de novo* on appeal: "[A]lthough statutory construction may demand some inquiry into legislative 'intent,' that inquiry is analytically legal: The meaning of a statute does not turn on what an individual lawmaker intended as a matter of fact, but only on what intent has been enacted into law through the constitutionally defined channels of bicameralism and presentment. This remains so even if deciding what passed through those channels requires a court to determine a 'fact' of historical understanding through an examination of extrinsic evidence." A patent is, of course, neither a contract nor a statute, but it has some of the character of each. Which do you think is the better analogy, and why?

3. In the post-*Teva* world, how does the standard of review affect the extent to which a patentee will seek to tie her preferred construction of a disputed claim term to material that is extrinsic to the patent? How does it affect an accused infringer's strategy?

2. Claim Construction Methods

A patent is, as was noted above, quite a curious document. On one hand, the patent is a legal document that creates valuable national exclusionary rights, and we should expect a patentee to interpret claim terms in the way that best secures the patentee's personal advantage. On the other hand, the patent is a technical document drafted for knowledgeable artisans in the field, and we should expect a patentee to provide the proper technological context for the claim terms in the patent's supporting disclosure. Moreover, a trial judge who adjudicates an infringement dispute must construe disputed claim terms, but usually cannot do so without the benefit of testimony from experts in the pertinent technological field. As a result, the construing judge must take proper account of *both* the term's general meaning *and* its particular use within the context of the patent.

What is the best method for achieving a sound construction of a disputed claim term? Recall that a patent claim is, in effect, a verbal boundary line that establishes a patent owner's right to exclude others from engaging in what may otherwise be lawful, highly profitable, and socially beneficial behavior. "As courts have recognized since the requirement that one's invention be distinctly claimed became part of the patent law in 1870, the primary purpose of the requirement is 'to guard against unreasonable advantages to the patentee and disadvantages to others arising from uncertainty as to their [respective] rights.'" *Athletic Alternative, Inc. v. Prince Mfg.*, 73 F.3d 1573 (Fed. Cir. 1996) (quoting *General Electric Co. v. Wabash Appliance Corp.,* 304 U.S. 364, 369 (1938)). What clarity of notice is the public entitled to? What clarity of notice is the patentee entitled to? And what claim construction methods will help well-counseled parties reach agreements about patent scope without having to undertake the burden of litigation (which is *very* expensive!)?

Errors can happen in either direction—a court could wrongly construe a claim too narrowly, or wrongly construe it too broadly. We can frame the question of clear notice as a choice between these two types of errors. If one views the more harmful error to be a wrongful narrowing of a claim term that limits the term too strictly to the embodiments the patent describes, and thereby diminishes the patentee's right to exclude, then one will insist that a claim term has its full general meaning unless the patent document clearly signals a ground for construing the term more narrowly. If, by contrast, one views the more harmful error to be a wrongful broadening of a claim term that severs the term too sharply from the context its supporting disclosure provides, and thereby diminishes the public's right to free action, then one will insist that a claim term has its context-driven meaning unless the patent document clearly signals a ground for construing the term more broadly.

In the decade just after *Markman,* from 1995 to 2005, the Federal Circuit's case law on claim construction oscillated between these different policy views. For a brief period, a number of cases focused on the dangers of tying a claim term's meaning too closely to its supporting disclosure. Those cases, exemplified by *Texas Digital Systems, Inc. v. Telegenix, Inc.*, 308 F.3d 1193 (Fed. Cir. 2002), went so far as to enunciate a "presumption in favor of a dictionary definition" for a disputed claim term that can "be overcome [only] where the patentee, acting as his or her own lexicographer,[*] has clearly set forth an explicit definition of the term different from its ordinary meaning . . . [or] has disavowed or disclaimed scope of coverage, by using words or expressions of manifest exclusion or restriction, representing a clear disavowal of claim scope." *Id.* at 1204. At the same time, other claim construction cases stressed the importance of construing disputed claim terms in the context of the patents in which they appear. For example, in *Innova/Pure Water, Inc. v. Safari Water Filtration Systems, Inc.*, the court explained that "[t]he written description provides a context for the claims, and is appropriately resorted to for the purpose of better understanding the meaning of a claim and for showing the connection in which a device is used." 381 F.3d 1111, 1116 (Fed. Cir. 2004) (internal quotations and citations omitted).

These competing perspectives led the full court to revisit the basics of claim construction methodology in *Phillips v. AWH Corp.*, 415 F.3d 1303 (Fed. Cir. 2005) (en banc). In *Phillips*, the Federal Circuit renewed claim construction's focus on *context*. The *Phillips* court began its analysis by grounding claim construction methodology in the dictates of the Patent Act's disclosure and claiming requirements:

> The first paragraph of section 112 of the Patent Act, 35 U.S.C. § 112, states that the specification
>> shall contain a written description of the invention, and of the manner and process of making and using it, in such full, clear, concise, and exact terms as to enable any person skilled in the art to which it pertains * * * to make and use the same * * *.
> The second paragraph of section 112 provides that the specification
>> shall conclude with one or more claims particularly pointing out and distinctly claiming the subject matter which the applicant regards as his invention.
> Those two paragraphs of section 112 frame the issue of claim interpretation for us. The second paragraph requires us to look to the language of the claims to determine what "the applicant regards as his invention." On the other hand, the first paragraph requires that the specification describe the invention set forth in the claims.

415 F.3d at 1311-12. As a result, "the person of ordinary skill in the art is deemed to read the claim term not only in the context of the particular claim in which the disputed term appears, but in the context of the entire patent, including the specification." *Id.* at 1313.

The *actual invention* that the specification describes, and that the claims delineate, *animates and unifies* the claims and supporting description into a coherent whole: "The patent system is based on the proposition that claims cover only the invented subject matter." *Id.* at 1321; *see also id.* at 1324 (the goal of claim construction is "to capture the scope of the actual invention more accurately than either strictly limiting the scope of the claims to the embodiments disclosed in the

---

* [*Eds. Note*: "Lexicography" is a posh word for "dictionary writing," *i.e.*, express definitions.]

specification or divorcing the claim language from the specification"). The invention may also require the patentee to establish new meanings for familiar words, and the specification can provide those new meanings. As *Phillips* explained:

> In light of the statutory directive that the inventor provide a "full" and "exact" description of the claimed invention, the specification necessarily informs the proper construction of the claims. In *Renishaw* [*PLC v. Marposs Societa' per Azioni*], this court summarized that point succinctly:
>
>> Ultimately, the interpretation to be given a term can only be determined and confirmed with a full understanding of what the inventors actually invented and intended to envelop with the claim. The construction that stays true to the claim language and most naturally aligns with the patent's description of the invention will be, in the end, the correct construction.
>
> 158 F.3d [1243,] 1250 [(Fed. Cir. 1998)] (citations omitted).
>
> Consistent with that general principle, our cases recognize that the specification may reveal a special definition given to a claim term by the patentee that differs from the meaning it would otherwise possess. In such cases, the inventor's lexicography governs. In other cases, the specification may reveal an intentional disclaimer, or disavowal, of claim scope by the inventor. In that instance as well, the inventor has dictated the correct claim scope, and the inventor's intention, as expressed in the specification, is regarded as dispositive.

415 F.3d at 1316.

B. Contemporary Cases

Teashot.LLC v. Green Mountain Coffee Roasters, Inc.
595 Fed. Appx. 983 (Fed. Cir. 2015)

Prost, Chief Judge:

Teashot.LLC ("Teashot") appeals from a final judgment of the U.S. District Court for the District of Colorado of noninfringement of U.S. Patent No. 5,895,672 in favor of Green Mountain Coffee Roasters, Inc., Keurig, Inc., and Starbucks Corp. (collectively, "Green Mountain"). For the reasons that follow, we affirm.

Background
I. Patent

The '672 patent seeks to adapt prior art coffee pod machines to make tea without the attendant weak taste from the short brewing time. The '672 patent thus teaches the use of special tea composition in the known brewing pod or container as reflected in representative claim 1:

1. A tea extraction system for production of a serving of tea extract in a coffee brewing device, comprising:

(a) a tea extraction container for containing a tea composition, said tea extraction container comprising a sealed body having at least one internal compartment, said internal compartment containing said tea composition; wherein said sealed body is constructed of a water-permeable material which allows flow of a fluid through said sealed body to produce a tea extract from said tea composition; and,

(b) a tea composition comprising from about 2 grams to about 10 grams of tea having a particle size of from about 0.40 mm to about 0.75 mm.

II. District Court Proceedings

Teashot accuses Green Mountain's tea-brewing K-Cups of infringing the '672 patent. The accused K-Cup has a foil lid, which would be punctured by a needle to inject water during use.

The district court construed the claim element "sealed body is constructed of a water-permeable material which allows flow of a fluid through said sealed body to produce a tea extract from said tea composition" as "the portions of the sealed body into which fluid flows and out of which fluid flows are water-permeable material allowing flow of a fluid through said sealed body to produce a tea extract from said tea composition." The district court concluded that the K-Cups do not literally infringe because the K-Cups do not have a "water-permeable material" for water to flow into the sealed bodies. . . . The district court therefore entered summary judgment of non-infringement in favor of Green Mountain.

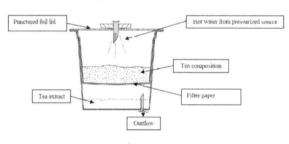

[Cross-section of a K-Cup, included in the court's opinion]

Teashot appeals the claim construction and summary judgment of no literal infringement . . . .

Discussion

I. Claim Construction

. . . .

Teashot argues that the district court's construction deviates from the claim text and improperly imports limitations from the specification by requiring fluid to flow into and out of the sealed body through water-permeable material. Green Mountain counters that the phrase "which allows," linking "water-permeable material" to the "flow of a fluid through said sealed body," requires that fluid flows through the "sealed body" via the "water-permeable material."

We agree with Green Mountain that the claim text identifies "water-permeable material" as the means through which fluid could flow through the "sealed body." The specification confirms this conclusion. Every discussion in the specification of fluid flowing through the "sealed body" refers to the "water-permeable material." The '672 patent mentions no other means through which fluid could flow through the "sealed body."

Teashot contends, however, that Figure 4 in the '672 patent teaches an embodiment in which water enters a sealed body through an opening in a material that is not otherwise permeable to water. Teashot further contends that the district court's claim construction improperly excludes this embodiment in Figure 4.

Figure 4, however, is limited to disclosing an arrangement in which multiple tea containers can be accessed individually to add different tea compositions, but used together for brewing. Figure 4 and its descriptions do not show any details of entry or exit means in the containers for water to flow through. From this silence, we cannot assume Figure 4 to depart from the consistent teachings elsewhere in the '672 patent that water can flow through a "sealed body" via a "water-permeable material." We are therefore not persuaded by Teashot that the district court erred in construing this claim element.

II. Summary Judgment of Noninfringement

. . . .

Literal infringement requires that "every limitation set forth in a claim must be found in an accused product, exactly." *Southwall Techs., Inc. v. Cardinal IG Co.*, 54 F.3d 1570, 1575 (Fed. Cir. 1995).

Teashot does not dispute that its owner and the inventor of the '672 patent, in testimonies quoted by Green Mountain, admitted that the lid of the K-Cup is not water permeable. Teashot also does not dispute the following admission that the mere puncturing of the K-Cup lid fails to transform the material into a water-permeable material:

> Q Correct me if I'm wrong, when you puncture the foil lid, the actual foil remains water impermeable, correct?
>
> A The—the foil around the hole, yes.
>
> Q Yes. The hole no longer has foil in it, correct?
>
> A Correct.
>
> Q Hence the hole.
>
> A Right.
>
> Q The water doesn't flow through the foil; it flows through an open space. Correct?
>
> A Correct.

Appellee Br. 31 (quoting Joint Appndx. A1203).

Nevertheless, Teashot contends that a factual dispute remains as to whether the K-Cup's lid, once punctured, becomes a "water-permeable material." Teashot cites no support in the record that a skilled artisan would consider "water-permeable material" to encompass material not permeable to water but having merely a puncture hole. We do not find Teashot's unsupported arguments—especially against its admissions quoted by Green Mountain—create a genuine factual dispute sufficient to survive summary judgment.

. . . .

Notes & Questions

1. Note how the claim construction dispute in *Teashot*, when joined with a summary judgment determination about the literal infringement question, resolved the case and made it suitable for appeal. This is relatively common in patent litigation.

2. Green Mountain argued, and the Federal Circuit agreed, that the text of the claim, in the way it recites the structural features and interrelates them to one another, simply rules out the possibility that fluid flows through the sealed body other than by way of a water-permeable material. The claim isn't just a list of parts; rather, it recites how the parts interact with one another to achieve the intended result. As you read the disputed patent claims throughout this chapter, consider whether thinking of them as stories about how the inventions in question actually work helps with the claim construction task.

3. Think again of the *Teva* case, and apply its lesson to *Teashot*. Does the key move here involve a factual question, or a legal question? Would a trial court be in any better position than the Federal Circuit in construing the disputed claim term in *Teashot*?

Nautilus, Inc. v. Biosig Instruments, Inc.
134 S. Ct. 2120 (2014)

GINSBURG, JUSTICE:

. . . .

Authorized by the Constitution "[t]o promote the Progress of Science and useful Arts, by securing for limited Times to * * * Inventors the exclusive Right to their * * * Discoveries," Art. I, §8, cl. 8, Congress has enacted patent laws rewarding inventors with a limited monopoly. "Th[at] monopoly is a property right," and "like any property right, its boundaries should be clear." *Festo Corp. v. Shoketsu Kinzoku Kogyo Kabushiki Co.*, 535 U.S. 722, 730 (2002). Thus, when Congress enacted the first Patent Act in 1790, it directed that patent grantees file a written specification "containing a description * * * of the thing or things * * * invented or discovered," which "shall be so particular" as to "distinguish the invention or discovery from other things before known and used." Act of Apr. 10, 1790, § 2, 1 Stat. 110.

The patent laws have retained this requirement of definiteness even as the focus of patent construction has shifted. Under early patent practice in the United States, we have recounted, it was the written specification that "represented the key to the patent." *Markman [v. Westview Instruments, Inc.*, 517 U.S. 370, 379 (1996)]. Eventually, however, patent applicants began to set out the invention's scope in a separate section known as the "claim." *See generally* 1 R. Moy, *Walker on Patents* § 4.2, pp. 4-17 to 4-20 (4th ed. 2012). The Patent Act of 1870 expressly conditioned the receipt of a patent on the inventor's inclusion of one or more such claims, described with particularity and distinctness. *See* Act of July 8, 1870, § 26,

16 Stat. 201 (to obtain a patent, the inventor must "particularly point out and distinctly claim the part, improvement, or combination which [the inventor] claims as his invention or discovery").

The 1870 Act's definiteness requirement survives today, largely unaltered. Section 112 of the Patent Act of 1952, applicable to this case, requires the patent applicant to conclude the specification with "one or more claims particularly pointing out and distinctly claiming the subject matter which the applicant regards as his invention." 35 U.S.C. § 112, ¶ 2 (2006 ed.). A lack of definiteness renders invalid "the patent or any claim in suit." § 282, ¶ 2(3).[1]

II

A

The patent in dispute, U.S. Patent No. 5,337,753 ('753 patent), issued to Dr. Gregory Lekhtman in 1994 and assigned to respondent Biosig Instruments, Inc., concerns a heart-rate monitor for use during exercise. Previous heart-rate monitors, the patent asserts, were often inaccurate in measuring the electrical signals accompanying each heartbeat (electrocardiograph or ECG signals). The inaccuracy was caused by electrical signals of a different sort, known as electromyogram or EMG signals, generated by an exerciser's skeletal muscles when, for example, she moves her arm, or grips an exercise monitor with her hand. These EMG signals can "mask" ECG signals and thereby impede their detection.

Dr. Lekhtman's invention claims to improve on prior art by eliminating that impediment. The invention focuses on a key difference between EMG and ECG waveforms: while ECG signals detected from a user's left hand have a polarity opposite to that of the signals detected from her right hand, EMG signals from each hand have the same polarity. The patented device works by measuring equalized EMG signals detected at each hand and then using circuitry to subtract the identical EMG signals from each other, thus filtering out the EMG interference.

As relevant here, the '753 patent describes a heart-rate monitor contained in a hollow cylindrical bar that a user grips with both hands, such that each hand comes into contact with two electrodes, one "live" and one "common." The device is illustrated in figure 1 of the patent . . . .

Claim 1 of the '753 patent, which contains the limitations critical to this dispute, refers to a "heart rate monitor for use by a user in association with exercise apparatus and/or exercise procedures." The claim "comprise[s]," among other elements, an "elongate member" (cylindrical bar) with a display device; "electronic circuitry including a difference amplifier"; and, on each half of the cylin-

---

[1] In the Leahy-Smith America Invents Act, Pub. L. 112-29, 125 Stat. 284, enacted in 2011, Congress amended several parts of the Patent Act. Those amendments modified §§ 112 and 282 in minor respects . . . . In any event, the amended versions of those provisions are inapplicable to patent applications filed before September 16, 2012, and proceedings commenced before September 16, 2011. Here, the application for the patent-in-suit was filed in 1992, and the relevant court proceedings were initiated in 2010. Accordingly, this opinion's citations to the Patent Act refer to the 2006 edition of the United States Code.

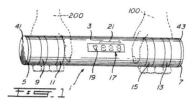

drical bar, a live electrode and a common electrode "mounted * * * in spaced relationship
with each other."[3] The claim sets forth additional elements, including that the cylindrical
bar is to be held in such a way that each of the
user's hands "contact[s]" both electrodes on
each side of the bar. Further, the EMG signals detected by the two electrode pairs
are to be "of substantially equal magnitude and phase" so that the difference amplifier will "produce a substantially zero [EMG] signal" upon subtracting the signals from one another.

B

The dispute between the parties arose in the 1990's, when Biosig allegedly disclosed the patented technology to StairMaster Sports Medical Products, Inc. According to Biosig, StairMaster, without ever obtaining a license, sold exercise machines that included Biosig's patented technology, and petitioner Nautilus, Inc.,
continued to do so after acquiring the StairMaster brand. In 2004, based on these
allegations, Biosig brought a patent infringement suit against Nautilus . . . .

With Biosig's lawsuit launched, Nautilus asked the [PTO] to reexamine the
'753 patent.[*] The reexamination proceedings centered on whether the patent
was anticipated or rendered obvious by prior art—principally, a patent issued in
1984 to an inventor named Fujisaki, which similarly disclosed a heart-rate monitor
using two pairs of electrodes and a difference amplifier. Endeavoring to distinguish the '753 patent from prior art, Biosig submitted a declaration from Dr.
Lekhtman. The declaration attested, among other things, that the '753 patent sufficiently informed a person skilled in the art how to configure the detecting electrodes so as "to produce equal EMG [signals] from the left and right hands." Although the electrodes' design variables—including spacing, shape, size, and material— cannot be standardized across all exercise machines, Dr. Lekhtman explained, a skilled artisan could undertake a "trial and error" process of equaliza-
tion. This would entail experimentation with different electrode configurations in
order to optimize EMG signal cancellation.[4] In 2010, the PTO issued a determination confirming the patentability of the '753 patent's claims.

Biosig thereafter reinstituted its infringement suit, which the parties had voluntarily dismissed without prejudice while PTO reexamination was underway. In
2011, the District Court conducted a hearing to determine the proper construction of the patent's claims, including the claim term "in spaced relationship with

---

[3] As depicted in figure 1 of the patent . . . the live electrodes are identified by numbers 9 and 13, and
the common electrodes, by 11 and 15.

[*] [*Eds. Note*: Recall that, as we discussed in Section I.B, the PTO can reexamine an issued patent to
determine whether its claims are too broad in light of newly uncovered prior art.]

[4] Dr. Lekhtman's declaration also referred to an expert report prepared by Dr. Henrietta Galiana,
Chair of the Department of Biomedical Engineering at McGill University, for use in the infringement litigation. That report described how Dr. Galiana's laboratory technician, equipped with a
wooden dowel, wire, metal foil, glue, electrical tape, and the drawings from the '753 patent, was able
in two hours to build a monitor that "worked just as described in the * * * patent."

each other." According to Biosig, that "spaced relationship" referred to the distance between the live electrode and the common electrode in each electrode pair. Nautilus, seizing on Biosig's submissions to the PTO during the reexamination, maintained that the "spaced relationship" must be a distance "greater than the width of each electrode." The District Court ultimately construed the term to mean "there is a defined relationship between the live electrode and the common electrode on one side of the cylindrical bar and the same or a different defined relationship between the live electrode and the common electrode on the other side of the cylindrical bar," without any reference to the electrodes' width.

Nautilus moved for summary judgment, arguing that the term "spaced relationship," as construed, was indefinite under § 112, ¶ 2. The District Court granted the motion. Those words, the District Court concluded, "did not tell [the court] or anyone what precisely the space should be," or even supply "any parameters" for determining the appropriate spacing.

The Federal Circuit reversed and remanded. A claim is indefinite, the majority opinion stated, "only when it is 'not amenable to construction' or 'insolubly ambiguous.'" 715 F.3d 891, 898 (2013) (quoting *Datamize, LLC v. Plumtree Software, Inc.*, 417 F.3d 1342, 1347 (Fed. Cir. 2005)). Under that standard, the majority determined, the '753 patent survived indefiniteness review. Considering first the "intrinsic evidence"—*i.e.*, the claim language, the specification, and the prosecution history—the majority discerned "certain inherent parameters of the claimed apparatus, which to a skilled artisan may be sufficient to understand the metes and bounds of 'spaced relationship.'" *Id.* at 899. These sources of meaning, the majority explained, make plain that the distance separating the live and common electrodes on each half of the bar "cannot be greater than the width of a user's hands"; that is so "because claim 1 requires the live and common electrodes to independently detect electrical signals at two distinct points of a hand." *Id.* Furthermore, the majority noted, the intrinsic evidence teaches that this distance cannot be "infinitesimally small, effectively merging the live and common electrodes into a single electrode with one detection point." *Id.* The claim's functional provisions, the majority went on to observe, shed additional light on the meaning of "spaced relationship." Surveying the record before the PTO on reexamination, the majority concluded that a skilled artisan would know that she could attain the indicated functions of equalizing and removing EMG signals by adjusting design variables, including spacing.

In a concurring opinion, Judge Schall reached the majority's result employing "a more limited analysis." *Id.* at 905. . . .

III

A

Although the parties here disagree on the dispositive question—does the '753 patent withstand definiteness scrutiny—they are in accord on several aspects of the § 112, ¶ 2 inquiry. First, definiteness is to be evaluated from the perspective of someone skilled in the relevant art. Second, in assessing definiteness, claims are to be read in light of the patent's specification and prosecution history. Third, "[d]efiniteness is measured from the viewpoint of a person skilled in [the] art *at the time the patent was filed*." Brief for Respondent 55 (emphasis added).

The parties differ, however, in their articulations of just how much imprecision § 112, ¶ 2 tolerates. In Nautilus' view, a patent is invalid when a claim is "ambiguous, such that readers could reasonably interpret the claim's scope differently." Brief for Petitioner 37. Biosig and the Solicitor General would require only that the patent provide reasonable notice of the scope of the claimed invention. See Brief for Respondent 18; Brief for United States as Amicus Curiae 9-10.

Section 112, we have said, entails a "delicate balance." *Festo*, 535 U.S. at 731. On the one hand, the definiteness requirement must take into account the inherent limitations of language. *See id.* Some modicum of uncertainty, the Court has recognized, is the "price of ensuring the appropriate incentives for innovation." *Id.* at 732. One must bear in mind, moreover, that patents are "not addressed to lawyers, or even to the public generally," but rather to those skilled in the relevant art. *Carnegie Steel Co. v. Cambria Iron Co.*, 185 U.S. 403, 437 (1902) (also stating that "any description which is sufficient to apprise [steel manufacturers] in the language of the art of the definite feature of the invention, and to serve as a warning to others of what the patent claims as a monopoly, is sufficiently definite to sustain the patent").

At the same time, a patent must be precise enough to afford clear notice of what is claimed, thereby "'appris[ing] the public of what is still open to them.'" *Markman*, 517 U.S. at 373 (quoting *McClain v. Ortmayer*, 141 U.S. 419, 424 (1891)).[6] Otherwise there would be "[a] zone of uncertainty which enterprise and experimentation may enter only at the risk of infringement claims." *United Carbon Co. v. Binney & Smith Co.*, 317 U.S. 228, 236 (1942). And absent a meaningful definiteness check, we are told, patent applicants face powerful incentives to inject ambiguity into their claims. Eliminating that temptation is in order, and "the patent drafter is in the best position to resolve the ambiguity in * * * patent claims." *Halliburton Energy Servs., Inc. v. M-I LLC*, 514 F.3d 1244, 1255 (Fed. Cir. 2008).

To determine the proper office of the definiteness command, therefore, we must reconcile concerns that tug in opposite directions. Cognizant of the competing concerns, we read § 112, ¶ 2 to require that a patent's claims, viewed in light of the specification and prosecution history, inform those skilled in the art about the scope of the invention with reasonable certainty. The definiteness requirement, so understood, mandates clarity, while recognizing that absolute precision is unattainable. The standard we adopt accords with opinions of this Court stating that "the certainty which the law requires in patents is not greater than is reasonable, having regard to their subject-matter." *Minerals Separation, Ltd. v. Hyde*, 242

---

[6] *See also United Carbon Co. v. Binney & Smith Co.*, 317 U.S. 228, 236 (1942) ("The statutory requirement of particularity and distinctness in claims is met only when they clearly distinguish what is claimed from what went before in the art and clearly circumscribe what is foreclosed from future enterprise."); *General Elec. Co. v. Wabash Appliance Corp.*, 304 U.S. 364, 369 (1938) ("The limits of a patent must be known for the protection of the patentee, the encouragement of the inventive genius of others and the assurance that the subject of the patent will be dedicated ultimately to the public.").

U.S. 261, 270 (1916). *See also United Carbon*, 317 U.S. at 236 ("claims must be reasonably clear-cut"); *Markman*, 517 U.S. at 389 (claim construction calls for "the necessarily sophisticated analysis of the whole document," and may turn on evaluations of expert testimony).

B

In resolving Nautilus' definiteness challenge, the Federal Circuit asked whether the '753 patent's claims were "amenable to construction" or "insolubly ambiguous." Those formulations can breed lower court confusion, for they lack the precision § 112, ¶ 2 demands. It cannot be sufficient that a court can ascribe some meaning to a patent's claims; the definiteness inquiry trains on the understanding of a skilled artisan at the time of the patent application, not that of a court viewing matters post hoc. To tolerate imprecision just short of that rendering a claim "insolubly ambiguous" would diminish the definiteness requirement's public-notice function and foster the innovation-discouraging "zone of uncertainty," United Carbon, 317 U.S. at 236, against which this Court has warned.

. . . [A]lthough this Court does not "micromanag[e] the Federal Circuit's particular word choice" in applying patent-law doctrines, we must ensure that the Federal Circuit's test is at least "probative of the essential inquiry." *Warner-Jenkinson Co. v. Hilton Davis Chemical Co.*, 520 U.S. 17, 40 (1997). Falling short in that regard, the expressions "insolubly ambiguous" and "amenable to construction" permeate the Federal Circuit's recent decisions concerning § 112, ¶ 2's requirement. We agree with Nautilus and its amici that such terminology can leave courts and the patent bar at sea without a reliable compass.[10]

IV

Both here and in the courts below, the parties have advanced conflicting arguments as to the definiteness of the claims in the '753 patent. Nautilus maintains that the claim term "spaced relationship" is open to multiple interpretations reflecting markedly different understandings of the patent's scope, as exemplified by the disagreement among the members of the Federal Circuit panel. Biosig responds that "spaced relationship," read in light of the specification and as illustrated in the accompanying drawings, delineates the permissible spacing with sufficient precision.

---

[10] The Federal Circuit suggests that a permissive definiteness standard "accords respect to the statutory presumption of patent validity." 715 F.3d at 902. *See also* § 282, ¶ 1 ("[a] patent shall be presumed valid," and "[t]he burden of establishing invalidity of a patent or any claim thereof shall rest on the party asserting such invalidity"); *Microsoft Corp. v. i4i Ltd.*, 131 S. Ct. 2238, 2242 (2011) (invalidity defenses must be proved by "clear and convincing evidence"). As the parties appear to agree, however, this presumption of validity does not alter the degree of clarity that § 112, ¶ 2 demands from patent applicants; to the contrary, it incorporates that definiteness requirement by reference. *See* § 282, ¶ 2(3) (defenses to infringement actions include "[i]nvalidity of the patent or any claim in suit for failure to comply with * * * any requirement of [§ 112]").

The parties nonetheless dispute whether factual findings subsidiary to the ultimate issue of definiteness trigger the clear-and-convincing evidence standard and, relatedly, whether deference is due to the PTO's resolution of disputed issues of fact. We leave these questions for another day. . . .

"[M]indful that we are a court of review, not of first view," *Cutter v. Wilkinson*, 544 U.S. 709, 718 n.7 (2005), we decline to apply the standard we have announced to the controversy between Nautilus and Biosig. As we have explained, the Federal Circuit invoked a standard more amorphous than the statutory definiteness requirement allows. We therefore follow our ordinary practice of remanding so that the Court of Appeals can reconsider, under the proper standard, whether the relevant claims in the '753 patent are sufficiently definite.

. . . .

Notes & Questions

1. Does it make sense to engage in court review of claim definiteness? Why not let the PTO be the exclusive arbiter of whether a proposed claim is, or is not, definite? An applicant can amend the claim until the PTO is satisfied with the form of the claim language. Once the patent issues, a presumption of validity attaches. As the Court indicates in footnote 10 of its opinion, however, the presumption of validity does not fit seemlessly into the definiteness inquiry.

2. On remand, the Federal Circuit—applying the new legal standard—reaffirmed its holding that the claim in *Nautilus* is not invalid for indefiniteness: "we conclude that Biosig's claims inform those skilled in the art with reasonable certainty about the scope of the invention." 783 F.3d 1374, 1382 (Fed. Cir. 2015).

3. Claim 1 of the '732 measuring cup patent requires "a pair of continuously sloping ramps formed integrally with" the measuring cup's sidewall, but then switches to the singular: "wherein said ramp is coextensive with said spout . . . ." Applying the standard in *Nautilus*, does this seeming glitch render claim 1 invalid for indefiniteness?

4. A patentee is permitted to provide express definitions for claim terms—to be her own lexicographer. Imagine a claim that recites the limitation that A is "above" B, and a written description supporting that claim that states, "In my claims, the term 'above' is defined as 'below.'" Does the *Nautilus* standard for indefiniteness provide a basis for rejecting this hypothetical patent? If not, *should* the definiteness requirement be modified to invalidate such a claim?

5. Many claims use the terms "about," "approximately," or "substantially" without raising any indefiniteness problem the Federal Circuit is prepared to remedy, at least historically. "Expressions such as 'substantially' are used in patent documents when warranted by the nature of the invention, in order to accommodate the minor variations that may be appropriate to secure the invention." *Verve, LLC v. Crane Cams, Inc.*, 311 F.3d 1116, 1120 (Fed. Cir. 2002). Given that the claim definiteness requirement's primary function is to ensure that the patent informs the public of the boundaries of the patentee's right to exclude, should we be skeptical that such claim terms serve that function?

6. Note that by learning about claim definiteness, you have just explored your first validity requirement.

Pacing Techs., LLC v. Garmin Int'l, Inc.
778 F.3d 1021 (Fed. Cir. 2015)

MOORE, JUDGE:

Pacing Technologies appeals from the district court's grant of summary judgment that Garmin International's accused products do not infringe the asserted claims of Pacing's U.S. Patent No. 8,101,843. We *affirm*.

Background

The '843 patent is directed to methods and systems for pacing users during activities that involve repeated motions, such as running, cycling, and swimming. The preferred embodiment of the '843 patent describes a method for aiding a user's pacing by providing the user with a tempo (for example, the beat of a song or flashes of light) corresponding to the user's desired pace.

Pacing alleges that Garmin GPS fitness watches and microcomputers used by runners and bikers infringe the '843 patent. The Garmin Connect website allows users to design and transfer workouts to the Garmin devices. Workouts consist of a series of intervals to which the user can assign a duration and target pace value. The devices display the intervals of a particular workout during operation, for example, by counting down the time for which the user intends to maintain a particular pace. The devices may also display the user's actual pace, e.g., 50 to 70 spm, or steps per minute. The devices do not play music or output a beat corresponding to the user's desired or actual pace.

Claim 25 of the '843 patent, the only asserted independent claim, reads as follows (emphases added):

> *A repetitive motion pacing system for pacing a user comprising*:
> a web site adapted to allowing the user to pre-select from a set of user selectable activity types an activity they wish to perform and entering one or more target tempo or target pace values corresponding to the activity;
> a data storage and *playback device*; and
> a communications device adapted to transferring data related to the pre-selected activity or the target tempo or the target pace values between the web site and the data storage and *playback device*.

The district court construed the term "playback device" as "a device capable of playing audio, video, or a visible signal." The district court also held that the preamble to claim 25 is a limitation and construed it to mean "a system for providing a sensible output for setting the pace or rate of movement of a user in performing a repetitive motion activity." This construction did not address whether the repetitive motion pacing system was required to play back the pace information using a tempo.

Garmin moved for summary judgment of noninfringement, contending that the accused devices are not "playback devices" under the district court's construction. Pacing argued that the accused devices are "playback devices" because they "play" workout information to the user, which can include the user's target and actual pace. To resolve this dispute, the district court supplemented its construction of "playback device" in the summary judgment order, holding that "[t]o be a playback device as envisioned in the patent, the device must play back the pace information." The court relied on the use of the term in the context of the specification and on its earlier decision that the preamble to claim 25 is limiting. The court granted summary judgment of noninfringement to Garmin, reasoning that while "[t]he [accused] devices repeat back or display the pace input or selections," they "do not 'play' the target tempo or pace information * * * as audio, video, or visible signals." Both parties characterize the court's construction of the term "playback device" as implicitly requiring the devices to play the pace information as a metronomic tempo, as described in the preferred embodiment of the '843 patent. . . .

Discussion

"[W]hen the district court reviews only evidence intrinsic to the patent (the patent claims and specification[], along with the patent's prosecution history), the judge's determination will amount solely to a determination of law, and the Court of Appeals will review that construction de novo." *Teva Pharm. USA Inc. v. Sandoz, Inc.*, 135 S. Ct. 831, 841 (2015). Because the only evidence at issue on appeal and presented to the district court in this claim construction was intrinsic, our review of the constructions is *de novo*. . . .

I. Claim Construction

On appeal, the parties dispute whether the asserted claims require the claimed devices to play back the pace information using a tempo, such as the beat of a song or flashes of light. This dispute turns on whether the preamble to claim 25 is limiting and on the construction of a "repetitive motion pacing system" as recited in the preamble.

We hold that the preamble to claim 25, which reads "[a] repetitive motion pacing system for pacing a user * * * ," is limiting. "Preamble language that merely states the purpose or intended use of an invention is generally not treated as limiting the scope of the claim." *Bicon, Inc. v. Straumann Co.*, 441 F.3d 945, 952 (Fed. Cir. 2006). However, "[w]hen limitations in the body of the claim rely upon and derive antecedent basis from the preamble, then the preamble may act as a necessary component of the claimed invention." *Eaton Corp. v. Rockwell Int'l Corp.*, 323 F.3d 1332, 1339 (Fed. Cir. 2003).

That is the case here. The term "user" in the preamble of claim 25 provides antecedent basis for the term "user" in the body of that claim. The body of claim 25 recites "a web site adapted to allowing *the user* to preselect from a set of user selectable activity types an activity they wish to perform and entering one or more target tempo or target pace values corresponding to the activity." (Emphasis added). The term "repetitive motion pacing system" in the preamble of claim 25 similarly provides antecedent basis for the term "repetitive motion pacing system" recited as a positive limitation in the body of claim 28, which depends from claim 25. Claim 28 of the '843 patent reads: "[t]he repetitive motion pacing system of claim 25, wherein the repetitive motion pacing system can determine a geographic location of the data storage and playback device." Because the preamble terms "user" and "repetitive motion pacing system" provide antecedent basis for and are necessary to understand positive limitations in the body of claims in the '843 patent, we hold that the preamble to claim 25 is limiting.

The plain and ordinary meaning of the phrase "repetitive motion pacing system for pacing a user" does not require the claimed system to pace the user by playing back the pace information using a tempo. However, claim terms are construed in light of the specification and prosecution history, not in isolation. *See Phillips v. AWH Corp.*, 415 F.3d 1303, 1313 (Fed. Cir. 2005) (en banc). The specification and prosecution history compel departure from the plain meaning in only two instances: lexicography and disavowal. *Thorner v. Sony Computer Entm't*, 669 F.3d 1362, 1365 (Fed. Cir. 2012). The standards for finding lexicography and disavowal are "exacting." *GE Lighting Solutions, LLC v. AgiLight, Inc.*, 750 F.3d 1304, 1309 (Fed. Cir. 2014). To act as a lexicographer, a patentee must "clearly set forth a definition of the disputed claim term" and "clearly express an intent to define the term." *Thorner*, 669 F.3d at 1365. Similarly, disavowal requires that "the specification [or prosecution history] make[] clear that the invention does not include a particular feature." *SciMed Life Sys. v. Advanced Cardiovascular Sys.*, 242 F.3d 1337, 1341 (Fed. Cir. 2001).

We have found disavowal or disclaimer based on clear and unmistakable statements by the patentee that limit the claims, such as "the present invention includes * * * " or "the present invention is * * * " or "all embodiments of the present invention are * * * ." *See, e.g., Regents of Univ. of Minn. v. AGA Med. Corp.*, 717 F.3d 929, 936 (Fed. Cir. 2013). We have found disclaimer when the specification indicated that, for "successful manufacture," a particular step was "require[d]." *Andersen Corp. v. Fiber Composites, LLC*, 474 F.3d 1361, 1367 (Fed. Cir. 2007). We have found disclaimer when the specification indicated that the invention operated by "pushing (as opposed to pulling) forces," and then characterized the "pushing forces" as "an important feature of the present invention." *SafeTCare Mfg. v. Tele-Made, Inc.*, 497 F.3d 1262, 1269-70 (Fed. Cir. 2007). We also have found disclaimer when the patent repeatedly disparaged an embodiment as "antiquated," having "inherent inadequacies," and then detailed the "deficiencies [that] make it difficult" to use. *Chi. Bd. Options Exch., Inc. v. Int'l Sec. Exch., LLC*, 677 F.3d 1361, 1372 (Fed. Cir. 2012). Likewise, we have used disclaimer to limit a claim element to a feature of the preferred embodiment when the specification described that feature as a "very important feature * * * in an aspect of the present invention," and disparaged alternatives to that feature. *Inpro II Licensing, S.A.R.L. v. T-Mobile USA Inc.*, 450 F.3d 1350, 1354-55 (Fed. Cir. 2008). When a patentee "describes the features of the 'present invention' as a whole," he alerts the reader that "this description limits the scope of the invention." *AGA Med. Corp.*, 717 F.3d at 936.

Here, the specification similarly contains a clear and unmistakable statement of disavowal or disclaimer. In a section entitled "Summary and Objects of the In-

vention," the '843 patent states that "it is a principal object of the present invention to provide a computer-implemented, network-based system having a net-
worked server, database, client computer, and input/output device for use by individuals engaged in repetitive motion activities * * * ." It then lists 18 additional
features, each time preceding the feature with the phrase "[i]t is another object of
the present invention" or "[i]t is still another object of the present invention."
This is a common practice in patent drafting. Many times, the patent drafter will
cast certain features as "an object of the present invention," and often those "objects of the present invention" correspond to features recited in the claims. That
is the case here, as many of the different "objects of the present invention" disclosed in the '843 patent are recited as features in one or more independent or
dependent claims. The characterization of a feature as "an object" or "another
object," or even as a "principal object," will not always rise to the level of disclaimer. In this case, where the patent includes a long list of different "objects of
the present invention" that correspond to features positively recited in one or
more claims, it seems unlikely that the inventor intended for each claim to be limited to all of the many objects of the invention. However, the '843 patent goes
further, and includes additional language that constitutes unmistakable disclaimer
when considered in the context of the patent as a whole. Immediately following
the enumeration of the different objects of the present invention, the '843 patent
states that "[t]hose [listed 19 objects] and other objects and features of the present
invention are accomplished, as embodied and fully described herein, by a repetitive motion pacing system that includes * * * a data storage and playback device
adapted to producing the sensible tempo." With these words, the patentee does
not describe yet another object of the invention—he alerts the reader that the invention accomplishes all of its objects and features (the enumerated 19 and all others) with a repetitive motion pacing system that includes a data storage and playback device adapted to produce a sensible tempo. In the context of this patent, this
clearly and unmistakably limits "the present invention" to a repetitive motion pacing system having a data storage and playback device that is adapted to producing
a sensible tempo.

Pacing argues that a "repetitive motion pacing system for pacing a user" cannot be limited to devices that produce a sensible tempo because the '843 patent
discloses an embodiment of a repetitive motion pacing system where the playback
device does not need to produce a sensible tempo. Pacing points to the specification's description of a repetitive motion pacing system having a playback device
that plays video landscapes to a user who is, for example, running on a treadmill,
with the video "automatically calibrated to match the speed of the user's * * *
pace," to simulate the user running through the actual landscape. Pacing argues
that if the claim is construed to limit the invention to a repetitive motion pacing
device adapted to producing a sensible tempo, this particular embodiment will not
be covered. Pacing argues that for this reason, we should reject the construction.

We disagree for two reasons. First, it is not clear that our construction excludes
this embodiment. Our construction requires the repetitive motion pacing system
to produce a sensible tempo, but it does not exclude additional features, such as

outputting video matching a user's pace. Moreover, the description of the embodiment that Pacing points to does not, as Pacing argues, exclude the production of a sensible tempo as required by the construction. Just because an embodiment does not expressly disclose a feature does not mean that embodiment excludes that feature. Second, even if Pacing is correct that this embodiment does not play a sensible tempo and therefore would be excluded under our construction, this is not a reason to ignore the specification's clear and unmistakable disavowal. It is true that constructions that exclude the preferred embodiment are disfavored. *Vitronics Corp. v. Conceptronic, Inc.*, 90 F.3d 1576, 1583 (Fed. Cir. 1996). However, in a case such as this, where the patent describes multiple embodiments, every claim does not need to cover every embodiment. *See Aug. Tech. Corp. v. Camtek, Ltd.*, 655 F.3d 1278, 1285 (Fed. Cir. 2011). This is particularly true where the plain language of a limitation of the claim does not appear to cover that embodiment. The preamble of claim 25 differs from the preambles of the other seven independent claims. Claim 25 requires a "repetitive motion pacing system for pacing a user." The plain language requires the system to pace the user. We conclude that the system of claim 25 must be capable of producing a sensible tempo for pacing the user.

II. Infringement

We hold that there is no genuine dispute of material fact as to whether the Garmin devices produce a sensible tempo. Merely displaying the rate of a user's pace—for example, displaying "100 steps per minute"—does not produce a sensible tempo. Garmin's accused devices are therefore not repetitive motion pacing devices. We affirm the district court's grant of summary judgment of noninfringement of the '843 patent.

Notes & Questions

1. The court states that "[t]he specification and prosecution history compel departure from the plain meaning in only two instances: lexicography and disavowal." Why frame lexicography and disavowal as "departure[s]" from the ordinary meaning of the claim text? After all, citing the 2005 en banc *Phillips* case, the court first emphasizes that "claim terms are construed in light of the specification and prosecution history, not in isolation." Patentee lexicography, if it takes place, is set forth in the specification or, less commonly, the prosecution history. So isn't it part of, rather than an exception to, the general rule? Similarly, disavowal of claim scope, if it takes place, occurs either in the specification or during prosecution. Isn't it, too, part of, rather than an exception to, the general rule? The court seems to say, in effect, "start with the broad ordinary meaning, and narrow it down if you must." Which error—wrongful narrowing, or wrongful overbreadth—is minimized with this approach?

2. The court insists that, to be effective, lexicography or disavowal must be *clear*, a word it uses repeatedly. Does this clarity requirement buttress, or temper, the error-minimization effect of thinking of lexicography and disavowal as exceptions to (rather than as part of) an ordinary-meaning default rule?

3. What makes the disavowal in this case so clear that it rules out the patentee's proffered construction of the disputed claim limitation? How would it benefit a patentee to include this disavowal language in its patent disclosure?

III. AN ADEQUATE WRITTEN DISCLOSURE

Recall that the Supreme Court has called the patent's required written disclosure "the *quid pro quo* of the right to exclude." *Kewanee Oil Co. v. Bicron Corp.*, 416 U.S. 470, 484 (1974). The Federal Circuit has also articulated patent law's basic bargain between the inventor and the public: "the public must receive meaningful disclosure in exchange for being excluded from practicing the invention for a limited period of time." *Enzo Biochem, Inc. v. Gen-Probe Inc.*, 323 F.3d 956, 970 (Fed. Cir. 2002). What makes a disclosure "meaningful," and thus provides adequate support for a patent claim? Whatever criteria we choose, we must bear in mind that granting too narrow a patent claim for a given disclosure errs by shorting the inventor, thus risking an inadequate incentive to invent; and that granting too broad a patent claim for a given disclosure errs by shorting the public, thus risking a windfall for the patentee that the public underwrites with inflated prices. Our criteria should also avoid deterring patenting by pushing patent documents, written for technically knowledgable readers, to morph into overlong treatises for the uninformed: "a patent need not teach, and preferably omits, what is well known in the art." *Hybritech Inc. v. Monoclonal Antibodies, Inc.*, 802 F.2d 1367, 1384 (Fed. Cir. 1986); *see also In re Gay*, 309 F.2d 769, 774 (CCPA 1962) ("Not every last detail is to be described, else patent specifications would turn into production specifications, which they were never intended to be."). Unnecessary disclosure that duplicates what the technically informed reader already knows simply drives up costs, without any corresponding benefit.

A. THE ENABLEMENT REQUIREMENT

The first subsection of 35 U.S.C. § 112 provides, in relevant part, that "[t]he specification shall contain a written description . . . of the manner and process of making and using [the claimed invention], in such full, clear, concise, and exact terms as to enable any person skilled in the art to which it pertains, or with which it is most nearly connected, to make and use the same." 35 U.S.C. § 112(a). This is the statutory source for the *enablement requirement*: a claim is invalid if the written description does not enable it. Does this mean that a phosita who follows the patent must be able to make and use the claimed invention without the need for *any* experimentation or tinkering? For example, in the '732 measuring cup patent, the specification does not describe the proper equipment for creating the requisite sloping ramps (with indicia) in the sidewall of the cup, or how to use the equipment once it's been selected. Choosing, and using, the correct machine could well involve some trial and error, couldn't it? Is that fatal to the Hoetings' claim? The answer the patent system gives will affect how much detail patents contain.

A leading Federal Circuit case on the enablement requirement, *In re Wands*, 858 F.2d 731, 735 (Fed. Cir. 1988), states that "[a] patent need not disclose what is well known in the art." Moreover, "[e]nablement is not precluded by the necessity for some experimentation such as routine screening. However, experimentation needed to practice the invention must not be undue experimentation. 'The key word is "undue," not "experimentation."'" *Id.* at 736-37 (quoting *In re Angstadt*, 537 F.2d 498, 504 (CCPA 1976)). The *Wands* court, indicating that the inquiry applies a "standard of reasonableness," *id.* at 737, highlighted a number of relevant factors: the nature of the invention and the breadth of the claims under review; the skill of those in the art and the state of that art when the patent was applied for; the relative predictability of the art, as well as any working examples or other guidance the patent specification provides; and the amount of experimentation required. *Id.* Courts continue to apply the *Wands* factors.

Importantly, the time to judge enablement is as of the *filing* date of the claim under review. But patentees may seek to enforce their exclusion rights against those who have developed alternative technologies, long after the patent application was first filed. Think about the cross-currents affecting the patentee's decision to seek broad claim scope in that later-filed infringement case. Increased breadth may make proving infringement easier, but it may also make defending the claim's validity harder. In any event, whatever the scope of the claim proves to be, § 112 requires the patentee in the patent document to enable the *full* scope of that claim: "The scope of the claims must be less than or equal to the scope of the enablement." *National Recovery Techs., Inc. v. Magnetic Separation Sys., Inc.*, 166 F.3d 1190, 1196 (Fed. Cir. 1999).

B. THE WRITTEN DESCRIPTION REQUIREMENT

The first subsection of 35 U.S.C. § 112 not only requires an enabling disclosure, it also provides that "[t]he specification shall contain a written description of the invention." This separate *written description* requirement polices another aspect of the public's bargain with the inventor: The public wants to ensure not only that it receives an enabling disclosure, but also that it is dealing with the proper bargaining partner, *i.e.*, the person who truly invented the claimed invention.

Why do we need a separate disclosure requirement to ensure that we are dealing with the right inventor? After all, if the patentee's written description enables a phosita to make and use the claimed invention, isn't it plain that the patentee is the one who invented the claimed subject matter? The answer, it turns out, is "no." This is easiest to see when we consider the fact that, during prosecution, we allow the patent applicant to amend his or her claims. We do not, however, permit the applicant to add new material to the written disclosure. 35 U.S.C. § 132(a). As an applicant amends claims over time, a gap can open—by accident, or by design—between what he described as his invention on the original filing date and what he now claims (and, if successful in obtaining a patent on the amended claims, will be able to exclude others from doing without permission).

The written description requirement's chief role is to ensure that any claims the applicant introduces after the start of patent prosecution are supported by the originally filed disclosure. The Federal Circuit applies a possession standard to analyze compliance with the requirement:

The purpose of the "written description" requirement is broader than to merely explain how to "make and use"; the applicant must also convey with reasonable clarity to those skilled in the art that, as of the filing date sought, he or she was in possession *of the invention.* The invention is, for purposes of the "written description" inquiry, *whatever is now claimed.*

Vas-Cath Inc. v. Mahurkar, 935 F.2d 1555, 1563-64 (Fed. Cir. 1991).

The written description requirement can also invalidate even an originally filed claim, if the supporting disclosure fails to show that the inventor possessed the claimed invention at the time the application was filed. *See Ariad Pharms., Inc. v. Eli Lilly & Co.,* 598 F.3d 1336 (Fed. Cir. 2010) (en banc). Consider, for example, the so-called "super aspirin" pain relievers Vioxx, Celebrex, and Bextra, that achieved great fame and notoriety in the late 1990s. The active ingredient in this family of drugs is a compound that selectively inhibits an inflammation-causing enzyme (COX-2) without inhibiting a similar enzyme (COX-1) that protects the stomach lining. Regular aspirin inhibits both COX-2 and COX-1, indiscriminately. *University of Rochester v. G.D. Searle & Co.,* 358 F.3d 916, 917-18 (Fed. Cir. 2004). In one of the earliest-filed patent applications in this area, the University of Rochester included the following originally filed claim in 1992:

1. A method for selectively inhibiting [COX]-2 activity in a human host, comprising administering a non-steroidal compound that selectively inhibits activity of the [COX]-2 gene product to a human host in need of such treatment.

Id. at 918. It was undisputed, however, that in the University's supporting disclosure, no compounds that would perform the claimed method were disclosed, nor was there any evidence that such a compound was known at the time the Rochester scientists filed the application. The Federal Circuit struck down these method claims for failure to satisfy the written description requirement: "generalized language may not suffice if it does not convey the detailed identity of an invention. In this case, there is no language here, generalized or otherwise, that describes compounds that achieve the claimed effect." *Id.* at 923. To appreciate the strictly functional (and, frankly, uninformative) nature of the University's claim, compare the following hypothetical claim: "A method for curing cancer in a human host, comprising administering a compound that cures cancer to a human host in need of such treatment." Without a supporting disclosure that identifies specific compounds, this looks more like a hopeful letter to Santa than a well-earned right to exclude.

C. THE BEST MODE REQUIREMENT

The first subsection of 35 U.S.C. § 112 concludes with the requirement that the specification "shall set forth the best mode contemplated by the inventor or joint inventor of carrying out the invention." 35 U.S.C. § 112(a). This is a separate requirement. In other words, a claim can be invalid, even though it is both enabled and described by the patent, if the specification fails to "set forth the best mode." However, as a result of the America Invents Act of 2011, "the failure to disclose the best mode shall not be a basis on which any claim of a patent may be canceled or held invalid or otherwise unenforceable." 35 U.S.C. § 282(b)(3)(A). In short, there is no longer "best-mode failure" defense in litigation. The Patent Office, however, purports to apply the requirement during prosecution. U.S. Dep't of Commerce, Patent & Trademark Office, *Manual of Patent Examining Procedure* § 2165 (9th ed., Mar. 2014).

IV. PATENTABLE SUBJECT MATTER AND UTILITY

We have considered key *formal* aspects of patent law—the nature of a patent claim and how to construe it, and the claim definiteness, enablement, written description, and best mode requirements. Patent law also imposes four *substantive* requirements: patentable subject matter, utility, novelty, and nonobviousness. In this Section, we consider the first two of these substantive requirements.

A. PATENTABLE SUBJECT MATTER

What is patentable? What is not? The operative statutory provision has, with the exception of one word, been the same since 1793:

> Whoever invents or discovers any new and useful process, machine, manufacture, or composition of matter, or any new and useful improvement thereof, may obtain a patent therefor, subject to the conditions and requirements of this title.

35 U.S.C. § 101. The word "process" replaced the word "art" in 1952.

What common thread runs through these statutory categories of patentable subject matter? Human intervention appears to be key to at least three of the four groups: "machines" and "manufactures" are not found in nature, and naturally occurring materials are not "compos[ed]." And the *patentable subject matter* requirement has, in fact, been easy to apply for most technologies. From fluoxetine hydrochloride, the active ingredient in Prozac (U.S. Patent No. 4,314,081) to bubble wrap (U.S. Patent No. 3,142,599) to the airplane (U.S. Patent No. 821,393), practical solutions to concrete problems fall comfortably within the scope of § 101.

Two technologies—genome modifications and computer-implemented processes—have presented challenging line-drawing problems. In two cases, decided a year apart over 30 years ago, the Supreme Court considered the patent-eligibility of inventions from each of these fields. The first case, *Diamond v. Chakrabarty*, 447 U.S. 303 (1980), pitted the PTO—which had rejected the claims—against microbiologist Ananda Chakrabarty. Chakrabarty had developed a "human-made, genetically engineered bacterium [that was] capable of breaking down multiple components of crude oil." *Id.* at 305. The PTO "concluded that § 101 was not intended to cover living things such as these laboratory created micro-organisms." *Id.* at 306. The Supreme Court, in a 5-4 decision, disagreed. As a general matter, the Court concluded that the text of § 101 sweeps broadly: "In choosing such expansive terms as 'manufacture' and 'composition of matter,' modified by the comprehensive 'any,' Congress plainly contemplated that the patent laws would be given wide scope." *Id.* at 308. Using legislative history, the Court put a gloss on the text: "The Committee Reports accompanying the 1952 Act inform us that Congress intended statutory subject matter to 'include anything under the sun that is made by man.'" *Id.* at 309 (quoting Senate and House reports). This "made by man" gloss effectively resolved the case in favor of inventor Chakrabarty: "His claim is not to a hitherto unknown natural phenomenon, but to a nonnaturally occurring manufacture or composition of matter—a product of human ingenuity . . . ." *Id.* The fact that Chakrabarty's manufacture was also a living organism proved irrelevant to the Supreme Court.*

The following year, the Court took up the patentability of a process that included, among other limitations, the use of a digital computer to run a complex computation. The Court had, in the then-recent past, rejected the patentability of two such inventions on the ground that they were tantamount to abstract mathematical algorithms. *See Parker v. Flook*, 437 U.S. 584 (1978); *Gottschalk v. Benson*, 409 U. S. 63 (1972). In *Diamond v. Diehr*, 450 U.S. 175 (1981), by contrast, the Court followed *Chakrabarty*'s more expansive approach. And just as in *Chakrabarty*, the vote was a closely divided 5-4. The technology in *Diehr* was a computer-controlled, heated pressure mold for curing synthetic rubber into finished products. The patent claim recited not only the mold, but also the computer controller for the mold and the long-known algorithm the controller used to calculate the best time to stop the cure and open the mold. Claim 1 from the application is set forth in the margin.† The inventors in *Diehr*

---

* Since 1987, the PTO has taken the position that, although organisms more complex than the bacterium in *Chakrabarty*—such as transgenic cows, goats, and pigs—are patentable, the PTO cannot issue a patent claim that covers a human being. *See* JANICE M. MUELLER, PATENT LAW 382-83 (4th ed. 2013) (discussing post-*Chakrabarty* developments). In 2011, in Section 33 of the America Invents Act, Congress provided that "[n]otwithstanding any other provision of law, no patent may issue on a claim directed to or encompassing a human organism." 125 Stat. 284, 340.

† Claim 1 in the *Diehr* application was as follows, with bracketed numbers and letters that we have added to make the limitations/elements of the claim more clear:

   1. A method of operating a rubber-molding press for precision molded compounds with the aid of a digital computer, comprising:
      [a] providing said computer with a database for said press, including at least,
         [1] natural logarithm conversion data (ln),
         [2] the activation energy constant (C) unique to each batch of said compound being molded, and
         [3] a constant (x) dependent upon the geometry of the particular mold of the press,
      [b] initiating an interval timer in said computer upon the closure of the press for monitoring the elapsed time of said closure,
      [c] constantly determining the temperature (Z) of the mold at a location closely adjacent to the mold cavity in the press during molding,
      [d] constantly providing the computer with the temperature (Z),
      [e] repetitively calculating in the computer, at frequent intervals during each cure, the Arrhenius equation for reaction time during the cure, which is
         [1] ln v = CZ + x
         [2] where *v* is the total required cure time,
      [f] repetitively comparing in the computer at said frequent intervals during the cure each said calculation of the total required cure time calculated with the Arrhenius equation and said elapsed time, and
      [g] opening the press automatically when a said comparison indicates equivalence.

characterize[d] their contribution to the art to reside in the process of constantly measuring the actual temperature inside the mold. These temperature measurements are then automatically fed into a computer which repeatedly recalculates the cure time by use of the Arrhenius equation. When the recalculated time equals the actual time that has elapsed since the press was closed, the computer signals a device to open the press. According to the respondents, the continuous measuring of the temperature inside the mold cavity, the feeding of this information to a digital computer which constantly recalculates the cure time, and the signaling by the computer to open the press are all new in the art.

450 U.S. at 178-79. The Court rejected the PTO's decision to deny patent protection, concluding that

> a physical and chemical process for molding precision synthetic rubber products falls within the § 101 categories of possibly patentable subject matter. That respondents' claims involve the transformation of an article, in this case raw, uncured synthetic rubber, into a different state or thing cannot be disputed. The respondents' claims describe in detail a step-by-step method for accomplishing such, beginning with the loading of a mold with raw, uncured rubber and ending with the eventual opening of the press at the conclusion of the cure. Industrial processes such as this are the types which have historically been eligible to receive the protection of our patent laws.

Id. at 184. Moreover, its conclusion was "not altered by the fact that, in several steps of the process, a mathematical equation and a programmed digital computer are used." In the Court's view, the *Diehr* inventors did "not seek to patent a mathematical formula. Instead, they s[ought] patent protection for a process of curing synthetic rubber." *Id.* at 187. And although "[t]heir process admittedly employ[ed] a well-known mathematical equation, . . . they d[id] not seek to preempt the use of that equation." *Id.*; *see also id.* at 188 ("Arrhenius' equation is not patentable in isolation, but when a process for curing rubber is devised which incorporates in it a more efficient solution of the equation, that process is, at the very least, not barred at the threshold by § 101."); *id.* at 192 ("[W]hen a claim containing a mathematical formula implements or applies that formula in a structure or process which, when considered as a whole, is performing a function which the patent laws were designed to protect (*e.g.,* transforming or reducing an article to a different state or thing), then the claim satisfies the requirements of § 101.").

Notes & Questions

1. Why *not* allow the patenting of a naked mathematical algorithm (note that "algorithm" is a fancy word for "formula")? A formula is created by one or more people, and thus seemingly satisfies the "made by man" criterion; and it is a process (or can be written as one, to put it in the form of a patent claim). One objection might be that such a claim, if allowed, would be infringed by anyone who "used" the algorithm, including merely by thinking through the formula. Should it make

a difference whether the patent claim is drafted in a way that literally covers a particular human thought?

2. Assuming the claim is limited to a nonhuman computing environment, such as a silicon chip, would the claimed invention be so far upstream from practical applications that transaction costs for end-users might explode as they try to assemble permissions from a vast number of patent owners, thereby blocking access to those practical applications? Prof. Michael Heller has called such an assembly problem a "tragedy of the anticommons," *i.e.*, a problem of property underuse resulting from too many property owners with veto power. *See* Michael A. Heller, *The Tragedy of the Anticommons: Property in the Transition from Marx to Markets*, 111 Harv. L. Rev. 621 (1998); Michael A. Heller & Rebecca Eisenberg, *Can Patents Deter Innovation? The Anticommons in Biomedical Research*, 280 Science 698 (1998). A tragedy of the commons, by contrast, is a problem of property overuse resulting from too few property owners with an incentive to manage a resource sustainably.

After *Diehr*, patent applicants sought to locate the dividing line between a patentable practical application of a formula, algorithm, or idea on the one hand, and an unpatentable abstract idea on the other hand, by submitting multiple claims to progressively less and less contextualized algorithms in the patent application. The Supreme Court stayed out of the patentable subject matter area for almost 30 years, allowing the caselaw to develop in the newly created Federal Circuit. The Federal Circuit proved quite receptive to the patentability of computer implemented processes. For a decade, the Federal Circuit tested patent-eligibility by inquiring whether the claimed invention yields "a useful, concrete, and tangible result." *State Street Bank & Trust Co. v. Signature Financial Group, Inc.*, 149 F.3d 1368, 1373 (Fed. Cir. 1998). A number that represented a real-world item qualified as such a result in the *State Street* case. *Id.* at 1373 (holding "that the transformation of data, representing discrete dollar amounts, by a machine through a series of mathematical calculations into a final share price, constitutes a practical application of a mathematical algorithm").

In 2006, three members of the Supreme Court expressed concern that the Federal Circuit's approach to the issue had become too permissive. Justice Breyer, for himself and Justices Stevens and Souter, dissented from the Court's dismissal of review in a case concerning the patentability of a two-step process for diagnosing a particular vitamin deficiency. Quoting the Federal Circuit's "useful, concrete, and tangible result" test for patentable subject matter from *State Street*, Justice Breyer noted that the Supreme Court "has never made such a statement and, if taken literally, the statement would cover instances where this Court has held the contrary." *Laboratory Corp. v. Metabolite Labs.*, 548 U.S. 124, 136 (2006) (Breyer, J., dissenting from dismissal of certiorari as improvidently granted).

In 2010, the Supreme Court returned to this doctrine. Specifically, in *Bilski v. Kappos*, 561 U.S. 593 (2010), the Court struck down claims to a risk-hedging method as too abstract for patenting, likening the claims to those invalidated in the pre-*Diehr* cases of *Flook* and *Benson*. The day after it decided *Bilski*, the Supreme Court vacated the Federal Circuit's decision in a case called *Prometheus Labs. v. Mayo Collaborative Servs.*, 581 F.3d 1336 (Fed. Cir. 2009), sending the case back for reconsideration in light of *Bilski*. In *Mayo*, the Federal Circuit had upheld the medical-diagnostic-method claims at issue, on the ground that practicing the method entailed the physical transformation of matter. Claim 1 of one of the key patents at issue is set forth below:

> A method of optimizing therapeutic efficacy for treatment of an immune-mediated gastrointestinal disorder, comprising:
>
> (a) administering a drug providing 6-thioguanine to a subject having said immune-mediated gastrointestinal disorder; and
>
> (b) determining the level of 6-thioguanine in said subject having said immune-mediated gastrointestinal disorder,
>
> wherein the level of 6-thioguanine less than about 230 pmol per 8 x 10⁸ red blood cells indicates a need to increase the amount of said drug subsequently administered to said subject and
>
> wherein the level of 6-thioguanine greater than about 400 pmol per 8 x 10⁸ red blood cells indicates a need to decrease the amount of said drug subsequently administered to said subject.

U.S. Patent No. 6,355,623.

The Federal Circuit, in the post-*Bilski* remand decision, once again held that the claims at issue constituted patentable subject matter. 628 F.3d 1347 (Fed. Cir. 2010). As you will see in a moment, the Supreme Court struck down these patents when it took up the case again in its October 2011 Term.

Mayo Collaborative Servs. v. Prometheus Labs., Inc.

566 U.S. 66 (2012)

BREYER, JUSTICE:

. . . .

The Court has long held that [§ 101] contains an important implicit exception. "[L]aws of nature, natural phenomena, and abstract ideas" are not patentable. *Diamond v. Diehr*, 450 U.S. 175, 185 (1981); *see also Bilski v. Kappos*, 130 S.Ct. 3218, 3225 (2010); *Diamond v. Chakrabarty*, 447 U.S. 303, 309 (1980); *Le Roy v. Tatham*, 14 How. 156, 175 (1853); *O'Reilly v. Morse*, 15 How. 62, 112-120 (1854). Thus, the Court has written that "a new mineral discovered in the earth or a new plant found in the wild is not patentable subject matter. Likewise, Einstein could not patent his celebrated law that $E=mc^2$; nor could Newton have patented the law of gravity. Such discoveries are 'manifestations of * * * nature, free to all men and reserved exclusively to none.'" *Chakrabarty*, 447 U.S. at 309 (quoting *Funk Brothers Seed Co. v. Kalo Inoculant Co.*, 333 U.S. 127, 130 (1948)).

"Phenomena of nature, though just discovered, mental processes, and abstract intellectual concepts are not patentable, as they are the basic tools of scientific and technological work." *Gottschalk v. Benson*, 409 U.S. 63, 67 (1972). And monopolization of those tools through the grant of a patent might tend to impede innovation more than it would tend to promote it.

The Court has recognized, however, that too broad an interpretation of this exclusionary principle could eviscerate patent law. For all inventions at some level embody, use, reflect, rest upon, or apply laws of nature, natural phenomena, or abstract ideas. Thus, in *Diehr* the Court pointed out that "'a process is not unpatentable simply because it contains a law of nature or a mathematical algorithm.'" 450 U.S. at 187 (quoting *Parker v. Flook*, 437 U.S. 584, 590 (1978)). It added that "an application of a law of nature or mathematical formula to a known structure or process may well be deserving of patent protection." *Diehr*, 450 U.S. at 187. And it emphasized Justice Stone's similar observation in *Mackay Radio & Telegraph Co. v. Radio Corp. of America*, 306 U.S. 86 (1939):

> "While a scientific truth, or the mathematical expression of it, is not a patentable invention, a novel and useful structure created with the aid of knowledge of scientific truth may be."

450 U.S. at 188 (quoting *Mackay Radio*, 306 U.S. at 94). *See also Funk Brothers*, 333 U.S. at 130 ("If there is to be invention from [a discovery of a law of nature], it must come from the application of the law of nature to a new and useful end").

Still, as the Court has also made clear, to transform an unpatentable law of nature into a patent-eligible application of such a law, one must do more than simply state the law of nature while adding the words "apply it." *See, e.g., Benson*, 409 U.S. at 71-72.

The case before us lies at the intersection of these basic principles. It concerns patent claims covering processes that help doctors who use thiopurine drugs to treat patients with autoimmune diseases determine whether a given dosage level is too low or too high. The claims purport to apply natural laws describing the relationships between the concentration in the blood of certain thiopurine metabolites and the likelihood that the drug dosage will be ineffective or induce harmful side-effects. We must determine whether the claimed processes have transformed these unpatentable natural laws into patent-eligible applications of those laws. We conclude that they have not done so and that therefore the processes are not patentable.

Our conclusion rests upon an examination of the particular claims before us in light of the Court's precedents. Those cases warn us against interpreting patent statutes in ways that make patent eligibility "depend simply on the draftsman's art" without reference to the "principles underlying the prohibition against patents for [natural laws]." *Flook*, 437 U.S. at 593. They warn us against upholding patents that claim processes that too broadly preempt the use of a natural law. *Morse*, 15 How. at 112-120; *Benson*, 409 U.S. at 71-72. And they insist that a process that focuses upon the use of a natural law also contain other elements or a combination of elements, sometimes referred to as an "inventive concept," sufficient to ensure that the patent in practice amounts to significantly more than a patent upon the natural law itself. *Flook*, 437 U.S. at 594; *see also Bilski*, 130 S. Ct. at 3230

("[T]he prohibition against patenting abstract ideas 'cannot be circumvented by attempting to limit the use of the formula to a particular technological environment' or adding 'insignificant postsolution activity'" (quoting *Diehr*, 450 U.S. at 191-92)).

We find that the process claims at issue here do not satisfy these conditions. In particular, the steps in the claimed processes (apart from the natural laws themselves) involve well-understood, routine, conventional activity previously engaged in by researchers in the field. At the same time, upholding the patents would risk disproportionately tying up the use of the underlying natural laws, inhibiting their use in the making of further discoveries.

I

A

The patents before us concern the use of thiopurine drugs in the treatment of autoimmune diseases, such as Crohn's disease and ulcerative colitis. When a patient ingests a thiopurine compound, his body metabolizes the drug, causing metabolites to form in his bloodstream. Because the way in which people metabolize thiopurine compounds varies, the same dose of a thiopurine drug affects different people differently, and it has been difficult for doctors to determine whether for a particular patient a given dose is too high, risking harmful side effects, or too low, and so likely ineffective.

At the time the discoveries embodied in the patents were made, scientists already understood that the levels in a patient's blood of certain metabolites, including, in particular, 6-thioguanine and its nucleotides (6-TG) and 6-methyl-mercaptopurine (6-MMP), were correlated with the likelihood that a particular dosage of a thiopurine drug could cause harm or prove ineffective. *See* U.S. Patent No. 6,355,623, col. 8, ll. 37-40. ("Previous studies suggested that measurement of 6-MP metabolite levels can be used to predict clinical efficacy and tolerance to azathioprine or 6-MP" (citing Cuffari, Théorêt, Latour, & Seidman, *6-Mercaptopurine Metabolism in Crohn's Disease: Correlation with Efficacy and Toxicity*, 39 Gut 401 (1996))). But those in the field did not know the precise correlations between metabolite levels and likely harm or ineffectiveness. The patent claims at issue here set forth processes embodying researchers' findings that identified these correlations with some precision.

More specifically, the patents [at issue here]—U.S. Patent No. 6,355,623 (the '623 patent) and U.S. Patent No. 6,680,302 (the '302 patent)—embody findings that concentrations in a patient's blood of 6-TG or of 6-MMP metabolite beyond a certain level (400 and 7000 picomoles per 8×10^8 red blood cells, respectively) indicate that the dosage is likely too high for the patient, while concentrations in the blood of 6-TG metabolite lower than a certain level (about 230 picomoles per 8×10^8 red blood cells) indicate that the dosage is likely too low to be effective.

The patent claims seek to embody this research in a set of processes. Like the Federal Circuit we take as typical claim 1 of the '623 patent, which describes one of the claimed processes . . . . [The text of this claim is set forth in the Notes & Questions above.]

For present purposes we may assume that the other claims in the patents do not differ significantly from claim 1.

B

Respondent, Prometheus Laboratories, Inc. (Prometheus), is the sole and exclusive licensee of the '623 and '302 patents. It sells diagnostic tests that embody the processes the patents describe. For some time petitioners, Mayo Clinic Rochester and Mayo Collaborative Services (collectively Mayo), bought and used those tests. But in 2004 Mayo announced that it intended to begin using and selling its own test—a test using somewhat higher metabolite levels to determine toxicity (450 pmol per 8×10^8 for 6-TG, and 5700 pmol per 8×10^8 for 6-MMP). Prometheus then brought this action claiming patent infringement.

The District Court found that Mayo's test infringed claim 7 of the '623 patent. In interpreting the claim, the court accepted Prometheus' view that the toxicity-risk level numbers in Mayo's test and the claim were too similar to render the tests significantly different. . . . The District Court also accepted Prometheus' view that a doctor using Mayo's test could violate the patent even if he did not actually alter his treatment decision in the light of the test. . . .

Nonetheless the District Court ultimately granted summary judgment in Mayo's favor. The court reasoned that the patents effectively claim natural laws or natural phenomena—namely the correlations between thiopurine metabolite levels and the toxicity and efficacy of thiopurine drug dosages—and so are not patentable.

On appeal, the Federal Circuit reversed. It pointed out that in addition to these natural correlations, the claimed processes specify the steps of (1) "administering a [thiopurine] drug" to a patient and (2) "determining the [resulting metabolite] level." These steps, it explained, involve the transformation of the human body or of blood taken from the body. Thus, the patents satisfied the Circuit's "machine or transformation test," which the court thought sufficient to "confine the patent monopoly within rather definite bounds," thereby bringing the claims into compliance with §101. 581 F.3d 1336, 1345, 1346-1347 (2009) (internal quotation marks omitted).

Mayo filed a petition for certiorari. We granted the petition, vacated the judgment, and remanded the case for reconsideration in light of *Bilski*, which clarified that the "machine or transformation test" is not a definitive test of patent eligibility, but only an important and useful clue. On remand the Federal Circuit reaffirmed its earlier conclusion. It thought that the "machine-or-transformation test," understood merely as an important and useful clue, nonetheless led to the "clear and compelling conclusion * * * that the * * * claims * * * do not encompass laws of nature or preempt natural correlations." 628 F.3d 1347, 1355 (2010). Mayo again filed a petition for certiorari, which we granted.

II

Prometheus' patents set forth laws of nature—namely, relationships between concentrations of certain metabolites in the blood and the likelihood that a dosage of a thiopurine drug will prove ineffective or cause harm. Claim 1, for example, states that if the levels of 6-TG in the blood (of a patient who has taken a dose of a thiopurine drug) exceed about 400 pmol per 8×10^8 red blood cells, then the administered dose is likely to produce toxic side effects. While it takes a human action (the administration of a thiopurine drug) to trigger a manifestation of this relation in a particular person, the relation itself exists in principle apart from any human action. The relation is a consequence of the ways in which thiopurine compounds are metabolized by the body—entirely natural processes. And so a patent that simply describes that relation sets forth a natural law.

The question before us is whether the claims do significantly more than simply describe these natural relations. To put the matter more precisely, do the patent claims add enough to their statements of the correlations to allow the processes they describe to qualify as patent-eligible processes that apply natural laws? We believe that the answer to this question is no.

A

If a law of nature is not patentable, then neither is a process reciting a law of nature, unless that process has additional features that provide practical assurance that the process is more than a drafting effort designed to monopolize the law of nature itself. A patent, for example, could not simply recite a law of nature and then add the instruction "apply the law." Einstein, we assume, could not have patented his famous law by claiming a process consisting of simply telling linear accelerator operators to refer to the law to determine how much energy an amount of mass has produced (or vice versa). Nor could Archimedes have secured a patent for his famous principle of flotation by claiming a process consisting of simply telling boat builders to refer to that principle in order to determine whether an object will float.

What else is there in the claims before us? The process that each claim recites tells doctors interested in the subject about the correlations that the researchers discovered. In doing so, it recites an "administering" step, a "determining" step, and a "wherein" step. These additional steps are not themselves natural laws but neither are they sufficient to transform the nature of the claim.

First, the "administering" step simply refers to the relevant audience, namely doctors who treat patients with certain diseases with thiopurine drugs. That audience is a pre-existing audience; doctors used thiopurine drugs to treat patients suffering from autoimmune disorders long before anyone asserted these claims. In any event, the "prohibition against patenting abstract ideas 'cannot be circumvented by attempting to limit the use of the formula to a particular technological environment.'" *Bilski*, 130 S. Ct. at 3230 (quoting *Diehr*, 450 U.S. at 191-192).

Second, the "wherein" clauses simply tell a doctor about the relevant natural laws, at most adding a suggestion that he should take those laws into account when treating his patient. That is to say, these clauses tell the relevant audience about the laws while trusting them to use those laws appropriately where they are relevant to their decision-making (rather like Einstein telling linear accelerator operators about his basic law and then trusting them to use it where relevant).

Third, the "determining" step tells the doctor to determine the level of the relevant metabolites in the blood, through whatever process the doctor or the laboratory wishes to use. As the patents state, methods for determining metabolite levels were well known in the art. '623 patent, col. 9, ll. 12-65. Indeed, scientists routinely measured metabolites as part of their investigations into the relationships between metabolite levels and efficacy and toxicity of thiopurine compounds. '623 patent, col. 8, ll. 37-40. Thus, this step tells doctors to engage in well-understood, routine, conventional activity previously engaged in by scientists who work in the field. Purely "conventional or obvious" "[pre]-solution activity" is normally not sufficient to transform an unpatentable law of nature into a patent-eligible application of such a law. *Flook*, 437 U.S. at 590; *see also Bilski*, 130 S. Ct. at 3230 ("[T]he prohibition against patenting abstract ideas 'cannot be circumvented by' * * * adding 'insignificant post-solution activity'" (quoting *Diehr*, 450 U.S. at 191-92)).

Fourth, to consider the three steps as an ordered combination adds nothing to the laws of nature that is not already present when the steps are considered separately. *See Diehr*, 450 U.S. at 188. Anyone who wants to make use of these laws must first administer a thiopurine drug and measure the resulting metabolite concentrations, and so the combination amounts to nothing significantly more than an instruction to doctors to apply the applicable laws when treating their patients.

The upshot is that the three steps simply tell doctors to gather data from which they may draw an inference in light of the correlations. To put the matter more succinctly, the claims inform a relevant audience about certain laws of nature; any additional steps consist of well-understood, routine, conventional activity already engaged in by the scientific community; and those steps, when viewed as a whole, add nothing significant beyond the sum of their parts taken separately. For these reasons we believe that the steps are not sufficient to transform unpatentable natural correlations into patentable applications of those regularities.

B

1

A more detailed consideration of the controlling precedents reinforces our conclusion. The cases most directly on point are *Diehr* and *Flook*, two cases in which the Court reached opposite conclusions about the patent eligibility of processes that embodied the equivalent of natural laws. . . .

[In *Diehr* the] Court pointed out that the basic mathematical equation, like a law of nature, was not patentable . . . [but that] the overall process [was] patent eligible because of the way the additional steps of the process integrated the equation into the [rubber curing] process as a whole. . . . And so the patentees did not "seek to pre-empt the use of [the] equation," but sought "only to foreclose from others the use of that equation in conjunction with all of the other steps in their claimed process." *Diehr*, 450 U.S. at 187. These other steps apparently added to the formula something that in terms of patent law's objectives had significance — they transformed the process into an inventive application of the formula.

. . . .

. . . [In *Flook* the Court] characterized the claimed process as doing nothing other than "provid[ing] a[n unpatentable] formula for computing an updated alarm limit." *Flook*, 437 U.S. at 586. . . . [T]he other steps in the process did not limit the claim to a particular application. . . . "[P]ostsolution activity" that is purely "conventional or obvious" the Court wrote, "can[not] transform an unpatentable principle into a patentable process." *Id.* at 589, 590.

The claim before us presents a case for patentability that is weaker than the (patent-eligible) claim in *Diehr* and no stronger than the (unpatentable) claim in *Flook*. Beyond picking out the relevant audience, namely those who administer doses of thiopurine drugs, the claim simply tells doctors to: (1) measure (somehow) the current level of the relevant metabolite, (2) use particular (unpatentable) laws of nature (which the claim sets forth) to calculate the current toxicity/inefficacy limits, and (3) reconsider the drug dosage in light of the law. These instructions add nothing specific to the laws of nature other than what is well-understood, routine, conventional activity, previously engaged in by those in the field. And since they are steps that must be taken in order to apply the laws in question, the effect is simply to tell doctors to apply the law somehow when treating their patients. The process in *Diehr* was not so characterized; that in *Flook* was characterized in roughly this way.

. . . .

3

The Court has repeatedly emphasized . . . a concern that patent law not inhibit further discovery by improperly tying up the future use of laws of nature. Thus, in *Morse* the Court set aside as unpatentable Samuel Morse's general claim for "'the use of the motive power of the electric or galvanic current * * * however developed, for making or printing intelligible characters, letters, or signs, at any distances,'" 15 How. at 86. The Court explained:

> For aught that we now know some future inventor, in the onward march of science, may discover a mode of writing or printing at a distance by means of the electric or galvanic current, without using any part of the process or combination set forth in the plaintiff's specification. His invention may be less complicated—less liable to get out of order—less expensive in construction, and in its operation. But yet if it is covered by this patent the inventor could not use it, nor the public have the benefit of it without the permission of this patentee.

Id. at 113.

Similarly, in *Benson* the Court said that the claims before it were "so abstract and sweeping as to cover both known and unknown uses of the [mathematical formula]." 409 U.S. at 67, 68. In *Bilski* the Court pointed out that to allow "petitioners to patent risk hedging would pre-empt use of this approach in all fields." 130 S. Ct. at 3231. And in *Flook* the Court expressed concern that the claimed process was simply "a formula for computing an updated alarm limit," which might "cover a broad range of potential uses." 437 U.S. at 586.

These statements reflect the fact that, even though rewarding with patents those who discover new laws of nature and the like might well encourage their discovery, those laws and principles, considered generally, are "the basic tools of scientific and technological work." *Benson*, 409 U.S. at 67. And so there is a danger that the grant of patents that tie up their use will inhibit future innovation premised upon them, a danger that becomes acute when a patented process amounts to no more than an instruction to "apply the natural law," or otherwise forecloses more future invention than the underlying discovery could reasonably justify. *See generally* Lemley, Risch, Sichelman, & Wagner, *Life After* Bilski, 63 Stan. L. Rev. 1315 (2011) (arguing that §101 reflects this kind of concern); see also C. Bohannan & H. Hovenkamp, *Creation without Restraint: Promoting Liberty and Rivalry in Innovation* 112 (2012) ("One problem with [process] patents is that the more abstractly their claims are stated, the more difficult it is to determine precisely what they cover. They risk being applied to a wide range of situations that were not anticipated by the patentee"); W. Landes & R. Posner, *The Economic Structure of Intellectual Property Law* 305-306 (2003) (The exclusion from patent law of basic truths reflects "both * * * the enormous potential for rent seeking that would be created if property rights could be obtained in them and * * * the enormous transaction costs that would be imposed on would-be users [of those truths]").

The laws of nature at issue here are narrow laws that may have limited applications, but the patent claims that embody them nonetheless implicate this concern. They tell a treating doctor to measure metabolite levels and to consider the resulting measurements in light of the statistical relationships they describe. In doing so, they tie up the doctor's subsequent treatment decision whether that treatment does, or does not, change in light of the inference he has drawn using the correlations. And they threaten to inhibit the development of more refined treatment recommendations (like that embodied in Mayo's test), that combine Prometheus' correlations with later discovered features of metabolites, human physiology or individual patient characteristics. The "determining" step too is set forth in highly general language covering all processes that make use of the correlations after measuring metabolites, including later discovered processes that measure metabolite levels in new ways.

We need not, and do not, now decide whether were the steps at issue here less conventional, these features of the claims would prove sufficient to invalidate them. For here, as we have said, the steps add nothing of significance to the natural laws themselves. Unlike, say, a typical patent on a new drug or a new way of using an existing drug, the patent claims do not confine their reach to particular applications of those laws. The presence here of the basic underlying concern that these patents tie up too much future use of laws of nature simply reinforces our conclusion that the processes described in the patents are not patent eligible, while eliminating any temptation to depart from case law precedent.

III

We have considered several further arguments in support of Prometheus' position. But they do not lead us to adopt a different conclusion. . . .

. . . .

[Finally], Prometheus, supported by several amici, argues that a principle of law denying patent coverage here will interfere significantly with the ability of medical researchers to make valuable discoveries, particularly in the area of diagnostic research. That research, which includes research leading to the discovery of laws of nature, is expensive; it "ha[s] made the United States the world leader in this field"; and it requires protection. Brief for Respondent 52.

Other medical experts, however, argue strongly against a legal rule that would make the present claims patent eligible, invoking policy considerations that point in the opposite direction. The American Medical Association, the American College of Medical Genetics, the American Hospital Association, the American Society of Human Genetics, the Association of American Medical Colleges, the Association for Molecular Pathology, and other medical organizations tell us that if "claims to exclusive rights over the body's natural responses to illness and medical treatment are permitted to stand, the result will be a vast thicket of exclusive rights over the use of critical scientific data that must remain widely available if physicians are to provide sound medical care." Brief for American College of Medical Genetics *et al.* as Amici Curiae 7; see also App. to Brief for Association Internationale pour la Protection de la Propriété Intellectuelle *et al.* as Amici Curiae A6, A16 (methods of medical treatment are not patentable in most of Western Europe).

We do not find this kind of difference of opinion surprising. Patent protection is, after all, a two-edged sword. On the one hand, the promise of exclusive rights provides monetary incentives that lead to creation, invention, and discovery. On the other hand, that very exclusivity can impede the flow of information that might permit, indeed spur, invention, by, for example, raising the price of using the patented ideas once created, requiring potential users to conduct costly and time-consuming searches of existing patents and pending patent applications, and requiring the negotiation of complex licensing arrangements. At the same time, patent law's general rules must govern inventive activity in many different fields of human endeavor, with the result that the practical effects of rules that reflect a general effort to balance these considerations may differ from one field to another. *See* Bohannan & Hovenkamp, *Creation without Restraint* at 98-100.

In consequence, we must hesitate before departing from established general legal rules lest a new protective rule that seems to suit the needs of one field produce unforeseen results in another. And we must recognize the role of Congress in crafting more finely tailored rules where necessary. *Cf.* 35 U.S. C. §§ 161-164 (special rules for plant patents). We need not determine here whether, from a policy perspective, increased protection for discoveries of diagnostic laws of nature is desirable.

*   *   *

For these reasons, we conclude that the patent claims at issue here effectively claim the underlying laws of nature themselves. The claims are consequently invalid. . . .

Notes & Questions

1. The *Diehr* and *Bilski* cases focused on the exclusion for abstract ideas. The *Mayo* case focuses on the exclusion of laws of of nature, and borrows freely from *Diehr* and *Bilski*. Should these exceptions have similar contours? Do they raise the same concerns?

2. The Court emphasizes, in almost all its decisions about what is and is not § 101 patentable subject matter, that patentable subject matter is only one of several criteria necessary for a claim to be patentable. In Sections V and VI of this Chapter we explore the novelty and non-obviousness requirements, which together require that, to be patentable, the invention must be both new in the art and a significant technological advance over the prior art. Is the Court in *Mayo* concerned about the "newness" of the claimed process? Note how many times it refers to "conventional" or "obvious" steps in the process. Would a court be better off skipping the inquiry into the § 101 requirement and instead using the novelty and non-obviousness requirement? *Compare MySpace v. GraphOn*, 672 F.3d 1250, 1251 (Fed. Cir. 2012) (noting that courts can "avoid the swamp of verbiage that is § 101" and the "murky morass that is § 101 jurisprudence" by instead addressing other statutory requirements for validity) *and id.* at 1264 (Mayer, J., dissenting) ("This court must first resolve the issue of whether the GraphOn patents are directed to an unpatentable 'abstract idea' before proceeding to consider subordinate issues related to obviousness and anticipation. *See Bilski v. Kappos*, 130 S. Ct. 3218, 3225 (2010) (noting that whether claims are directed to statutory subject matter is a 'threshold test') . . . ."). Revisit this question after you have learned more about those requirements.

3. The preceding questions present the patentable-subject-matter issue in a context that is familiar for patent prosecutors and patent litigators—namely, "is this specific claim patentable under § 101, or not?" A general business lawyer or intellectual property lawyer may encounter the patentable-subject-matter issue in a different context, before specific claim language—or, indeed, any portion of a patent document—has been drafted. For example, a designer or engineer employed at a company may ask in-house counsel, "I've come up with a new business structure for selling and servicing the product . . . is it patentable?" After *Bilski* and *Mayo*, what's the best way to approach such a question?

After *Mayo*, the Supreme Court vacated and remanded a case to the Federal Circuit for reconsideration. That case, *Myriad Genetics*, related to whether isolated, purified DNA molecules are patentable subject matter under § 101. On remand, the Federal Circuit divided on the question and, in the October 2012 Term, the Supreme Court took up the *Myriad Genetics* case again. *Association for Molecular Pathology v. Myriad Genetics, Inc.*, 569 U.S. 576 (2013). Patentee Myriad had "discovered the precise location and sequence of two human genes, mutations of which can substantially increase the risks of breast and ovarian cancer." *Id.* at 579. These two genes are referred to as BRCA1 and BRCA2. To protect its commercial position in selling diagnostic tests for those mutations, Myriad obtained a family of patents. Some of the patent claims were to isolated, purified DNA, while others were to complementary DNA ("cDNA"), a synthetically created molecule that contains the same protein-coding information found in natural DNA but that omits the noncoding portions also found in natural DNA. *Id.* at 582-85. The Court, harking back to the *Chakrabarty* case, held that Myriad's claims to isolated natural DNA were *not* patentable, but that its claims to synthetic cDNA *were* patentable:

> It is undisputed that Myriad did not create or alter any of the genetic information encoded in the BRCA1 and BRCA2 genes. The location and order of the nucleotides existed in nature before Myriad found them. Nor did Myriad create or alter the genetic structure of DNA. Instead, Myriad's principal contribution was uncovering the precise location and genetic sequence of the BRCA1 and BRCA2 genes within chromosomes 17 and 13. The question is whether this renders the genes patentable.

> Myriad recognizes that our decision in *Chakrabarty* is central to this inquiry. In *Chakrabarty*, scientists added four plasmids to a bacterium, which enabled it to break down various components of crude oil. 447 U.S. at 305 & n.1. The Court held that the modified bacterium was patentable. It explained that the patent claim was "not to a hitherto unknown natural phenomenon, but to a nonnaturally occurring manufacture or composition of matter—a product of human ingenuity 'having a distinctive name, character [and] use.'" *Id.* at 309-310 (quoting *Hartranft v. Wiegmann*, 121 U.S. 609, 615 (1887); alteration in original). The *Chakrabarty* bacterium was new "with markedly different characteristics from any found in nature," 447 U.S. at 310, due to the additional plasmids and resultant "capacity for degrading oil." *Id.* at 305 n.1. In this case, by contrast, Myriad did not create anything. To be sure, it found an important and useful gene, but separating that gene from its surrounding genetic material is not an act of invention.

> . . . .

> Nor are Myriad's claims saved by the fact that isolating DNA from the human genome severs chemical bonds and thereby creates a nonnaturally occurring molecule. Myriad's claims are simply not expressed in terms of chemical composition, nor do they rely in any way on the chemical changes that result from the isolation of a particular section of DNA. Instead, the claims understandably focus on the genetic information encoded in the BRCA1 and BRCA2 genes. If the patents depended upon the creation of a unique molecule, then a would-be infringer could arguably avoid at least Myriad's patent claims on entire genes (such as claims 1 and 2 of the '282 patent) by isolating a DNA sequence that included both the BRCA1 or BRCA2 gene and one additional nucleotide pair. Such a molecule would not be chemically identical to the molecule "invented" by Myriad. But Myriad obviously would resist that outcome because its claim is concerned primarily with the information contained in the genetic sequence, not with the specific chemical composition of a particular molecule.

. . . .

cDNA does not present the same obstacles to patentability as naturally occurring, isolated DNA segments. As already explained, creation of a cDNA sequence from mRNA results in a [coding-portions]-only molecule that is not naturally occurring.[8] Petitioners concede that cDNA differs from natural DNA in that "the non-coding regions have been removed." They nevertheless argue that cDNA is not patent eligible because "[t]he nucleotide sequence of cDNA is dictated by nature, not by the lab technician." That may be so, but the lab technician unquestionably creates something new when cDNA is made. cDNA retains the naturally occurring [coding regions] of DNA, but it is distinct from the DNA from which it was derived. As a result, cDNA is not a "product of nature" and is patent eligible under § 101, except insofar as very short series of DNA may have no intervening [non-coding regions] to remove when creating cDNA. In that situation, a short strand of cDNA may be indistinguishable from natural DNA.[9]

Id. at 590-95.

Notes & Questions

1. An isolated natural DNA sequence is not patentable subject matter, but a human-made cDNA sequence *is* patentable subject matter. The difference is the human intervention: "the lab technician unquestionably creates something new when cDNA is made." Of course, chemically isolating a naturally occurring DNA fragment is also the work of a "lab technician." Does the Court explain why the latter lab-tech work leads to categorically different treatment than the former lab-tech work? In any event, if generating cDNA from naturally occurring mRNA—the biotech equivalent of applying Silly Putty to the Sunday comics page—is sufficient, note that the amount of human intervention required to earn patentable-subject-matter status is actually quite small.

2. Consider a more humble, and perhaps more familiar, fact pattern: Picking a newly found variety of apple off a tree does not seem sufficient to render the apple patentable subject matter under § 101. Coring and slicing the newly found apple, however, *does* seem sufficient: the fruit technician unquestionably creates something new. Or does she? Perhaps this is still too much like merely picking the apple

---

[8] Some viruses rely on an enzyme called reverse transcriptase to reproduce by copying RNA into cDNA. In rare instances, a side effect of a viral infection of a cell can be the random incorporation of fragments of the resulting cDNA, known as a pseudogene, into the genome. Such pseudogenes serve no purpose; they are not expressed in protein creation because they lack genetic sequences to direct protein expression. *See* J. Watson *et al.*, *Molecular Biology of the Gene* 142, 144 fig. 7-5 (6th ed. 2008). Perhaps not surprisingly, given pseudogenes' apparently random origins, petitioners "have failed to demonstrate that the pseudogene consists of the same sequence as the BRCA1 cDNA." *Association for Molecular Pathology v. United States Patent and Trademark Office*, 689 F.3d 1303, 1356 n.5 (Fed. Cir. 2012). The possibility that an unusual and rare phenomenon *might* randomly create a molecule similar to one created synthetically through human ingenuity does not render a composition of matter nonpatentable.

[9] We express no opinion whether cDNA satisfies the other statutory requirements of patentability. *See, e.g.*, 35 U.S.C. §§ 102, 103, and 112.

off the tree? What's the criterion for deciding how much human intervention is enough?

3. Everything turns, in *Myriad*, on which DNA sequences occur naturally and which do not. Footnote 8 of the Court's opinion demonstrates this quite graphically, as the Court wrestles with an argument about a naturally occurring species of cDNA known as a pseudogene. If cDNA weren't newly created, as a general matter, it would be the same as isolated DNA. Why doesn't the fact that pseudogenes exist in nature doom the patentability of cDNA claimed in the patent at issue in *Myriad*?

4. *Myriad* does not take the same two separate analytical steps that *Mayo* takes (in section II of that opinion)—namely, first determining if the claim recites improper subject matter (such as a law of nature), and then determining if the claim "do[es] significantly more than simply" recite the improper subject matter. But one can explain *Myriad*'s holding in terms of those two steps in *Mayo*. In effect, *Myriad* gets resolved at the first step, with two contrasting answers. The claim to isolated DNA is a natural phenomenon (step 1), and there is nothing else at all to the claim (step 2). The claim to cDNA is *not* a natural phenomenon (step 1), so the *Mayo*-style inquiry ends there.

In 1980 and 1981, the Supreme Court decided the *Chakrabarty* and *Diehr* cases concerning the patentability of inventions in the biosciences and computer industries. Thirty years later, the Court returned to the topic of patentable subject matter, deciding the *Bilski*, *Mayo*, and *Myriad Genetics* cases. In the most recent Supreme Court decision on this doctrine, decided in June 2014, the Court synthesized its approaches in *Bilski* and *Mayo* into a unified two-step analysis. As you read the case, consider whether it improves the analysis to synthesize the prior cases in this way.

Alice Corp. v. CLS Bank Int'l

134 S. Ct. 2347 (2014)

THOMAS, JUSTICE:

The patents at issue in this case disclose a computer-implemented scheme for mitigating "settlement risk" (*i.e.*, the risk that only one party to a financial transaction will pay what it owes) by using a third-party intermediary. The question presented is whether these claims are patent eligible under 35 U.S.C. § 101, or are instead drawn to a patent-ineligible abstract idea. We hold that the claims at issue are drawn to the abstract idea of intermediated settlement, and that merely requiring generic computer implementation fails to transform that abstract idea into a patent-eligible invention. We therefore affirm the judgment of the United States Court of Appeals for the Federal Circuit.

I

A

Petitioner Alice Corporation is the assignee of several patents that disclose schemes to manage certain forms of financial risk.[1] According to the specification largely shared by the patents, the invention "enabl[es] the management of risk relating to specified, yet unknown, future events." The specification further explains that the "invention relates to methods and apparatus, including electrical computers and data processing systems applied to financial matters and risk management."

The claims at issue relate to a computerized scheme for mitigating "settlement risk"—*i.e.*, the risk that only one party to an agreed-upon financial exchange will satisfy its obligation. In particular, the claims are designed to facilitate the exchange of financial obligations between two parties by using a computer system as a third-party intermediary.[2] The intermediary creates "shadow" credit and debit records (*i.e.*, account ledgers) that mirror the balances in the parties' real-world accounts at "exchange institutions" (*e.g.*, banks). The intermediary updates the shadow records in real time as transactions are entered, allowing "only those transactions for which the parties' updated shadow records indicate sufficient resources to satisfy their mutual obligations." 717 F.3d 1269, 1285 (Fed. Cir. 2013) (Lourie, J., concurring). At the end of the day, the intermediary instructs the relevant financial institutions to carry out the "permitted" transactions in accordance with the updated shadow records, thus mitigating the risk that only one party will perform the agreed-upon exchange.

In sum, the patents in suit claim (1) the foregoing method for exchanging obligations (the method claims), (2) a computer system configured to carry out the method for exchanging obligations (the system claims), and (3) a computer-readable medium containing program code for performing the method of exchanging

---

[1] The patents at issue are United States Patent Nos. 5,970,479 (the '479 patent), 6,912,510, 7,149,720, and 7,725,375.

[2] The parties agree that claim 33 of the '479 patent is representative . . . . Claim 33 recites:

> A method of exchanging obligations as between parties, each party holding a credit record and a debit record with an exchange institution, the credit records and debit records for exchange of predetermined obligations, the method comprising the steps of:
>
> (a) creating a shadow credit record and a shadow debit record for each stakeholder party to be held independently by a supervisory institution from the exchange institutions;
>
> (b) obtaining from each exchange institution a start-of-day balance for each shadow credit record and shadow debit record;
>
> (c) for every transaction resulting in an exchange obligation, the supervisory institution adjusting each respective party's shadow credit record or shadow debit record, allowing only these transactions that do not result in the value of the shadow debit record being less than the value of the shadow credit record at any time, each said adjustment taking place in chronological order, and
>
> (d) at the end-of-day, the supervisory institution instructing on[e] of the exchange institutions to exchange credits or debits to the credit record and debit record of the respective parties in accordance with the adjustments of the said permitted transactions, the credits and debits being irrevocable, time invariant obligations placed on the exchange institutions.

obligations (the media claims). All of the claims are implemented using a computer; the system and media claims expressly recite a computer, and the parties have stipulated that the method claims require a computer as well.

B

Respondents CLS Bank International and CLS Services Ltd. (together, CLS Bank) operate a global network that facilitates currency transactions. In 2007, CLS Bank filed suit against petitioner, seeking a declaratory judgment that the claims at issue are invalid, unenforceable, or not infringed. Petitioner counterclaimed, alleging infringement. Following this Court's decision in *Bilski v. Kappos*, 561 U.S. 593 (2010), the parties filed cross-motions for summary judgment on whether the asserted claims are eligible for patent protection under 35 U.S.C. § 101. The District Court held that all of the claims are patent ineligible because they are directed to the [same] abstract idea . . . .

A divided panel of the United States Court of Appeals for the Federal Circuit reversed, holding that it was not "manifestly evident" that petitioner's claims are directed to an abstract idea. 685 F.3d 1341, 1352, 1356 (Fed. Cir. 2012). The Federal Circuit granted rehearing en banc, vacated the panel opinion, and affirmed the judgment of the District Court in a one-paragraph per curiam opinion. 717 F.3d at 1273. Seven of the ten participating judges agreed that petitioner's method and media claims are patent ineligible. With respect to petitioner's system claims, the en banc Federal Circuit affirmed the District Court's judgment by an equally divided vote.

Writing for a five-member plurality, Judge Lourie concluded that all of the claims at issue are patent ineligible. In the plurality's view, under this Court's decision in *Mayo Collaborative Services v. Prometheus Laboratories, Inc.*, 132 S. Ct. 1289 (2012), a court must first "identif[y] the abstract idea represented in the claim," and then determine "whether the balance of the claim adds 'significantly more.'" 717 F.3d at 1286. The plurality concluded that petitioner's claims "draw on the abstract idea of reducing settlement risk by effecting trades through a third-party intermediary," and that the use of a computer to maintain, adjust, and reconcile shadow accounts added nothing of substance to that abstract idea.

. . . .

II

Section 101 of the Patent Act defines the subject matter eligible for patent protection. . . .

"We have long held that this provision contains an important implicit exception: Laws of nature, natural phenomena, and abstract ideas are not patentable." *Association for Molecular Pathology v. Myriad Genetics, Inc.*, 133 S. Ct. 2107, 2116 (2013). We have interpreted § 101 and its predecessors in light of this exception for more than 150 years. *O'Reilly v. Morse*, 15 How. 62, 112-120 (1854); *Le Roy v. Tatham*, 14 How. 156, 174-175 (1853).

We have described the concern that drives this exclusionary principle as one of pre-emption. *See, e.g., Bilski,* 561 U.S. at 611-612 (upholding the patent "would preempt use of this approach in all fields, and would effectively grant a monopoly over an abstract idea"). Laws of nature, natural phenomena, and abstract ideas are "the basic tools of scientific and technological work." *Myriad*, 133 S. Ct. at 2116. "[M]onopolization of those tools through the grant of a patent might tend to impede innovation more than it would tend to promote it," thereby thwarting the primary object of the patent laws. *Mayo*, 132 S. Ct. at 1293; see U.S. Const., Art. I, § 8, cl. 8 (Congress "shall have Power * * * To promote the Progress of Science and useful Arts"). We have "repeatedly emphasized this * * * concern that patent law not inhibit further discovery by improperly tying up the future use" of these building blocks of human ingenuity. *Mayo*, 132 S. Ct. at 1301 (citing *Morse*, 15 How. at 113).

At the same time, we tread carefully in construing this exclusionary principle lest it swallow all of patent law. *Mayo*, 132 S. Ct. at 1293. At some level, "all inventions * * * embody, use, reflect, rest upon, or apply laws of nature, natural phenomena, or abstract ideas." *Id.* Thus, an invention is not rendered ineligible for patent simply because it involves an abstract concept. *See Diamond v. Diehr*, 450 U.S. 175, 187 (1981). "[A]pplication[s]" of such concepts "'to a new and useful end,'" we have said, remain eligible for patent protection. *Gottschalk v. Benson*, 409 U.S. 63, 67 (1972).

Accordingly, in applying the § 101 exception, we must distinguish between patents that claim the "buildin[g] block[s]" of human ingenuity and those that integrate the building blocks into something more, *Mayo*, 132 S. Ct. at 1303, thereby "transform[ing]" them into a patent-eligible invention, *id.* at 1294. The former "would risk disproportionately tying up the use of the underlying" ideas, *id.*, and are therefore ineligible for patent protection. The latter pose no comparable risk of preemption, and therefore remain eligible for the monopoly granted under our patent laws.

III

In *Mayo Collaborative Services v. Prometheus Laboratories, Inc.*, 132 S. Ct. 1289, we set forth a framework for distinguishing patents that claim laws of nature, natural phenomena, and abstract ideas from those that claim patent-eligible applications of those concepts. First, we determine whether the claims at issue are directed to one of those patent-ineligible concepts. *Id.* at 1296-97. If so, we then ask, "[w]hat else is there in the claims before us?" *Id.* at 1297. To answer that question, we consider the elements of each claim both individually and "as an ordered combination" to determine whether the additional elements "transform the nature of the claim" into a patent-eligible application. *Id.* 1297-98. We have described step two of this analysis as a search for an "'inventive concept'"—*i.e.*, an element or combination of elements that is "sufficient to ensure that the patent in practice amounts to significantly more than a patent upon the [ineligible concept] itself." *Id.* at 1294.[3]

---

[3] Because the approach we made explicit in *Mayo* considers all claim elements, both individually and in combination, it is consistent with the general rule that patent claims "must be considered as a whole." *Diamond v. Diehr*, 450 U.S. 175, 188 (1981); *see Parker v. Flook*, 437 U.S. 584, 594 (1978)

A

We must first determine whether the claims at issue are directed to a patent-ineligible concept. We conclude that they are: These claims are drawn to the abstract idea of intermediated settlement.

The "abstract ideas" category embodies "the longstanding rule that '[a]n idea of itself is not patentable.'" *Benson*, 409 U.S. at 67. In *Benson*, for example, this Court rejected as ineligible patent claims involving an algorithm for converting binary-coded decimal numerals into pure binary form, holding that the claimed patent was "in practical effect * * * a patent on the algorithm itself." 409 U.S. at 71-72. And in *Parker v. Flook*, 437 U.S. 584, 594-595 (1978), we held that a mathematical formula for computing "alarm limits" in a catalytic conversion process was also a patent-ineligible abstract idea.

We most recently addressed the category of abstract ideas in *Bilski v. Kappos*, 561 U.S. 593 (2010). . . .

"[A]ll members of the Court agree[d]" that the patent at issue in *Bilski* claimed an "abstract idea." *Id.* at 609; *see also id.* at 619 (Stevens, J., concurring in the judgment). Specifically, the claims described "the basic concept of hedging, or protecting against risk." *Id.* at 611. The Court explained that "'[h]edging is a fundamental economic practice long prevalent in our system of commerce and taught in any introductory finance class.'" *Id.* "The concept of hedging" as recited by the claims in suit was therefore a patent-ineligible "abstract idea, just like the algorithms at issue in *Benson* and *Flook*." *Id.*

It follows from our prior cases, and *Bilski* in particular, that the claims at issue here are directed to an abstract idea. Petitioner's claims involve a method of exchanging financial obligations between two parties using a third-party intermediary to mitigate settlement risk. The intermediary creates and updates "shadow" records to reflect the value of each party's actual accounts held at "exchange institutions," thereby permitting only those transactions for which the parties have sufficient resources. At the end of each day, the intermediary issues irrevocable instructions to the exchange institutions to carry out the permitted transactions.

On their face, the claims before us are drawn to the concept of intermediated settlement, *i.e.*, the use of a third party to mitigate settlement risk. Like the risk hedging in *Bilski*, the concept of intermediated settlement is "'a fundamental economic practice long prevalent in our system of commerce.'" *Id.*; *see, e.g.*, Emery, *Speculation on the Stock and Produce Exchanges of the United States*, in 7 Studies in History, Economics and Public Law 283, 346-356 (1896) (discussing the use of a "clearing-house" as an intermediary to reduce settlement risk). The use of a third-party intermediary (or "clearing house") is also a building block of the modern economy. *See, e.g.*, Yadav, *The Problematic Case of Clearinghouses in Complex*

("Our approach * * * is * * * not at all inconsistent with the view that a patent claim must be considered as a whole").

Markets, 101 Geo. L.J. 387, 406-412 (2013); J. Hull, *Risk Management and Financial Institutions* 103-104 (3d ed. 2012). Thus, intermediated settlement, like hedging, is an "abstract idea" beyond the scope of § 101.

Petitioner acknowledges that its claims describe intermediated settlement, see Brief for Petitioner 4, but rejects the conclusion that its claims recite an "abstract idea." Drawing on the presence of mathematical formulas in some of our abstract-ideas precedents, petitioner contends that the abstract-ideas category is confined to "preexisting, fundamental truth[s]" that "'exis[t] in principle apart from any human action.'" *Id.* at 23, 26 (quoting *Mayo*, 132 S. Ct. at 1297).

Bilski belies petitioner's assertion. The concept of risk hedging we identified as an abstract idea in that case cannot be described as a "preexisting, fundamental truth." The patent in *Bilski* simply involved a "series of steps instructing how to hedge risk." 561 U.S. at 599. Although hedging is a longstanding commercial practice, *id.* at 599, it is a method of organizing human activity, not a "truth" about the natural world "that has always existed," Brief for Petitioner 22. One of the claims in *Bilski* reduced hedging to a mathematical formula, but the Court did not assign any special significance to that fact, much less the sort of talismanic significance petitioner claims. Instead, the Court grounded its conclusion that all of the claims at issue were abstract ideas in the understanding that risk hedging was a "fundamental economic practice." 561 U.S. at 611.

In any event, we need not labor to delimit the precise contours of the "abstract ideas" category in this case. It is enough to recognize that there is no meaningful distinction between the concept of risk hedging in *Bilski* and the concept of intermediated settlement at issue here. Both are squarely within the realm of "abstract ideas" as we have used that term.

B

Because the claims at issue are directed to the abstract idea of intermediated settlement, we turn to the second step in *Mayo*'s framework. We conclude that the method claims, which merely require generic computer implementation, fail to transform that abstract idea into a patent-eligible invention.

1

At *Mayo* step two, we must examine the elements of the claim to determine whether it contains an "'inventive concept'" sufficient to "transform" the claimed abstract idea into a patent-eligible application. 132 S. Ct. 1294, 1298. A claim that recites an abstract idea must include "additional features" to ensure "that the [claim] is more than a drafting effort designed to monopolize the [abstract idea]." *Id.* 1297. *Mayo* made clear that transformation into a patent-eligible application requires "more than simply stat[ing] the [abstract idea] while adding the words 'apply it.'" *Id.* 1294.

Mayo itself is instructive. The patents at issue in *Mayo* claimed a method for measuring metabolites in the bloodstream in order to calibrate the appropriate dosage of thiopurine drugs in the treatment of autoimmune diseases. *Id.* at 1295. The respondent in that case contended that the claimed method was a patent-eligible application of natural laws that describe the relationship between the concentration of certain metabolites and the likelihood that the drug dosage will be harmful or ineffective. But methods for determining metabolite levels were already "well known in the art," and the process at issue amounted to "nothing significantly more than an instruction to doctors to apply the applicable laws when treating their patients." *Id.* at 1297-98. "Simply appending conventional steps, specified at a high level of generality," was not "enough" to supply an "inventive concept." *Id.* at 1300, 1297, 1294.

The introduction of a computer into the claims does not alter the analysis at *Mayo* step two. In *Benson*, for example, we . . . "held that simply implementing a mathematical principle on a physical machine, namely a computer, [i]s not a patentable application of that principle." *Mayo*, 132 S. Ct. at 1301 (citing *Benson*, 409 U.S. at 64).

Flook is to the same effect. . . . In holding that the process was patent ineligible, we rejected the argument that "implement[ing] a principle in some specific fashion" will "automatically fal[l] within the patentable subject matter of § 101." *Id.* at 593. Thus, "*Flook* stands for the proposition that the prohibition against patenting abstract ideas cannot be circumvented by attempting to limit the use of [the idea] to a particular technological environment." *Bilski*, 561 U.S. at 610-611.

In *Diehr*, 450 U.S. 175, by contrast, we held that a computer-implemented process for curing rubber was patent eligible, but not because it involved a computer. The claim employed a "well-known" mathematical equation, but it used that equation in a process designed to solve a technological problem in "conventional industry practice." *Id.* at 177, 178. The invention in Diehr used a "thermocouple" to record constant temperature measurements inside the rubber mold—something "the industry ha[d] not been able to obtain." *Id.* at 178 & n.3. The temperature measurements were then fed into a computer, which repeatedly recalculated the remaining cure time by using the mathematical equation. *Id.* at 178-79. These additional steps, we recently explained, "transformed the process into an inventive application of the formula." *Mayo*, 132 S. Ct. at 1299. In other words, the claims in *Diehr* were patent eligible because they improved an existing technological process, not because they were implemented on a computer.

These cases demonstrate that the mere recitation of a generic computer cannot transform a patent-ineligible abstract idea into a patent-eligible invention. Stating an abstract idea "while adding the words 'apply it'" is not enough for patent eligibility. *Mayo*, 132 S. Ct. at 1294. Nor is limiting the use of an abstract idea "'to a particular technological environment.'" *Bilski*, 561 U.S. at 610-11. Stating an abstract idea while adding the words "apply it with a computer" simply combines those two steps, with the same deficient result. Thus, if a patent's recitation of a computer amounts to a mere instruction to "implemen[t]" an abstract idea "on * * * a computer," *Mayo*, 132 S. Ct. at 1301, that addition cannot impart patent eligibility. This conclusion accords with the preemption concern that undergirds our § 101 jurisprudence. Given the ubiquity of computers, wholly generic computer implementation is not generally the sort of "additional featur[e]" that provides any "practical assurance that the process is more than a drafting effort designed to monopolize the [abstract idea] itself." *Mayo*, 132 S. Ct. at 1297.

The fact that a computer "necessarily exist[s] in the physical, rather than purely conceptual, realm," Brief for Petitioner 39, is beside the point. There is no dispute that a computer is a tangible system (in § 101 terms, a "machine"), or that many computer-implemented claims are formally addressed to patent-eligible subject matter. But if that were the end of the § 101 inquiry, an applicant could claim any principle of the physical or social sciences by reciting a computer system configured to implement the relevant concept. Such a result would make the determination of patent eligibility "depend simply on the draftsman's art," *Flook*, 437 U.S. at 593, thereby eviscerating the rule that "'[l]aws of nature, natural phenomena, and abstract ideas are not patentable,'" *Myriad*, 133 S. Ct. at 2116.

2

The representative method claim in this case recites the following steps: (1) "creating" shadow records for each counterparty to a transaction; (2) "obtaining" start-of-day balances based on the parties' real-world accounts at exchange institutions; (3) "adjusting" the shadow records as transactions are entered, allowing only those transactions for which the parties have sufficient resources; and (4) issuing irrevocable end-of-day instructions to the exchange institutions to carry out the permitted transactions. Petitioner principally contends that the claims are patent eligible because these steps "require a substantial and meaningful role for the computer." Brief for Pet'r 48. As stipulated, the claimed method requires the use of a computer to create electronic records, track multiple transactions, and issue simultaneous instructions; in other words, "[t]he computer is itself the intermediary." *Id.*

In light of the foregoing, the relevant question is whether the claims here do more than simply instruct the practitioner to implement the abstract idea of intermediated settlement on a generic computer. They do not.

Taking the claim elements separately, the function performed by the computer at each step of the process is "[p]urely conventional." *Mayo*, 132 S. Ct. at 1298. Using a computer to create and maintain "shadow" accounts amounts to electronic recordkeeping—one of the most basic functions of a computer. *See, e.g.*, *Benson*, 409 U.S. at 65 (noting that a computer "operates * * * upon both new and previously stored data"). The same is true with respect to the use of a computer to obtain data, adjust account balances, and issue automated instructions; all of these computer functions are "well-understood, routine, conventional activit[ies]" previously known to the industry. *Mayo*, 132 S. Ct. 1298. In short, each step does no more than require a generic computer to perform generic computer functions.

Considered "as an ordered combination," the computer components of petitioner's method "ad[d] nothing * * * that is not already present when the steps are considered separately." *Id.* Viewed as a whole, petitioner's method claims simply recite the concept of intermediated settlement as performed by a generic computer. *See* 717 F.3d at 1286 (Lourie, J., concurring) (noting that the representative method claim "lacks any express language to define the computer's participation"). The method claims do not, for example, purport to improve the functioning of the computer itself. *See id.* ("There is no specific or limiting recitation of * * * improved computer technology * * * ."); Brief for United States as *Amicus Curiae* 28-30. Nor do they effect an improvement in any other technology or technical field. *See, e.g., Diehr,* 450 U.S. at 177-178. Instead, the claims at issue amount to "nothing significantly more" than an instruction to apply the abstract idea of intermediated settlement using some unspecified, generic computer. *Mayo,* 132 S. Ct. 1298. Under our precedents, that is not "enough" to transform an abstract idea into a patent-eligible invention. *Id.* at 1297.

C

Petitioner's claims to a computer system and a computer-readable medium fail for substantially the same reasons. Petitioner conceded below that its media claims rise or fall with its method claims. As to its system claims, petitioner emphasizes that those claims recite "specific hardware" configured to perform "specific computerized functions." Brief for Petitioner 53. But what petitioner characterizes as specific hardware—a "data processing system" with a "communications controller" and "data storage unit," for example—is purely functional and generic. Nearly every computer will include a "communications controller" and "data storage unit" capable of performing the basic calculation, storage, and transmission functions required by the method claims. See 717 F.3d at 1290 (Lourie, J., concurring). As a result, none of the hardware recited by the system claims "offers a meaningful limitation beyond generally linking 'the use of the [method] to a particular technological environment,' that is, implementation via computers." *Id.* at 1291 (quoting *Bilski,* 561 U.S. at 610-611).

Put another way, the system claims are no different from the method claims in substance. The method claims recite the abstract idea implemented on a generic computer; the system claims recite a handful of generic computer components configured to implement the same idea. This Court has long "warn[ed] * * * against" interpreting § 101 "in ways that make patent eligibility 'depend simply on the draftsman's art.'" *Mayo,* 132 S. Ct. 1294 (quoting *Flook,* 437 U.S. at 593); *see Flook,* 437 U.S. at 590 ("The concept of patentable subject matter under § 101 is not 'like a nose of wax which may be turned and twisted in any direction * * * .'") (quoting *White v. Dunbar,* 119 U.S. 47, 51 (1886)). Holding that the system claims are patent eligible would have exactly that result.

Because petitioner's system and media claims add nothing of substance to the underlying abstract idea, we hold that they too are patent ineligible under § 101.

. . . .

SOTOMAYOR, JUSTICE, concurring (with Justices Ginsburg & Breyer):

I adhere to the view that any "claim that merely describes a method of doing business does not qualify as a 'process' under § 101." *Bilski v. Kappos,* 561 U.S. 593, 614 (2010) (Stevens, J., concurring in the judgment). As in *Bilski,* however, I further believe that the method claims at issue are drawn to an abstract idea. *Cf.* 561 U.S. at 619 (opinion of Stevens, J.). I therefore join the opinion of the Court.

Notes & Questions

1. How much work does the first step of the Court's two-step test do? What guidance does the Court provide for determining whether a claim is "directed to" or "drawn to" an ineligible "abstract idea"? Are all process claims going to meet this test? And how about the second step of the test? When will a business method patent contain an "inventive concept" sufficient to make the claim patent-eligible subject matter? Would a bright line rule barring patents on business methods be a better solution?

2. Is the "inventive concept" requirement constitutionally mandated? What word in Art. I, § 8 cl. 8, might require such a standard? Patentable subject matter is just the first of the substantive requirements. In addition to the utility requirement, which we address next, a claim must also be novel and not obvious, requirements we explore in the next Section. The Court's reliance on an "inventive concept" in the patentable subject matter inquiry, in some ways, addresses concerns similar to those that underlie the novelty and nonobviousness requirements. Keep this in mind as you study those requirements.

3. Recall the Court's statement in *Diehr*, in 1981: "This Court has undoubtedly recognized limits to § 101, and every discovery is not embraced within the statutory terms. Excluded from such patent protection are laws of nature, natural phenomena, and abstract ideas." 450 U.S. at 185. In four years, the Supreme Court examined the contours of all three exclusions: laws of nature in *Mayo* (2012), natural phenomena in *Myriad* (2013), and abstract ideas in *Bilski* (2010) and *Alice* (2014). What unified theory of the scope of § 101, if any, emerges from this quartet of cases?

B. Utility

The word "useful"—which appears twice in 35 U.S.C. § 101, as you saw at the start of this Section—is the germ of another substantive patent law requirement. According to the *utility* requirement, a claimed invention is unpatentable unless it is useful.

Perhaps the first thing one thinks, when confronted with this rule, is, why would the applicant try to patent something if it *wasn't* useful? Obtaining a patent isn't cheap,[*] after all. Won't this issue police itself? Perhaps the next thing one thinks is, if the claimed invention isn't useful, what's the harm in letting someone patent it? No one else will want to use the invention (it's not useful), and thus the right to exclude others from using it won't affect anyone. One answer to these questions is that they take too static a view of how we learn about the varied uses

---

[*] The filing fees for a utility patent begin at $330, but upon issuance of the patent, an issuance fee of $1,510 is required. 35 U.S.C. § 41(a)(1)(A), (a)(4)(A). And there are the legal fees involved, for anyone who hires a patent lawyer: The American Intellectual Property Law Association 2017 Economic Survey reports that the median price for legal services for the preparation and filing of a basic utility patent in 2016 was $8,523.

of things over time. In the chemical arts, in particular, one may be able to synthe-size a chemical long before one recognizes that it is useful as, for example, a drug. A patent claim, once granted, lasts for twenty years from the date the application for it was first filed (if one pays the required maintenance fees). What if a chemi-cal's value as a powerful drug is discovered during that time? Should we care whether we've granted patent protection to someone other than the one who dis-covered the chemical's medical value? If we grant a patent on the chemical as soon as it's been synthesized, will the patentee even have the right incentive to explore the chemical's utility as a drug (or for other purposes)? Difficult utility doctrine questions rarely arise, and when they do the case usually involves the chemical or biotechnological arts.

For example, in the Supreme Court's most recent utility doctrine case, issued over 50 years ago, the invention in question was a method for making a chemical compound. *Brenner v. Manson*, 383 U.S. 519 (1966). Researchers were interested in the compound because a structurally similar compound had shown some anti-cancer efficacy; but the compound itself—at the time the applicant, Manson, filed his application—had not been shown to be effective against cancer. The com-pound and its production method were, in other words, part of an industry-wide search for cancer cures. The Supreme Court concluded that Manson had not demonstrated the requisite utility for the method, as of his filing date, and thus had not complied with § 101. Usefulness, for purposes of § 101, requires a cur-rently available benefit that is both specific and substantial:

> The basic *quid pro quo* contemplated by the Constitution and the Congress for granting a patent monopoly is the benefit derived by the public from an invention with substantial utility. Unless and until a process is refined and developed to this point—where specific benefit exists in currently available form—there is insufficient justification for permitting an applicant to en-gross what may prove to be a broad field.

Id. at 534-35. In other words, as the Court colorfully put it, "a patent is not a hunt-ing license. It is not a reward for the search, but compensation for its successful conclusion." *Id.* at 536.

"For the common law, possession or 'occupancy' is the origin of property." Carol M. Rose, *Possession as the Origin of Property*, 52 U. CHI. L. REV. 73, 74 (1985). The utility requirement, along with the written description requirement (dis-cussed in Section II), comprises patent law's possession standard. In the 1L course on real property law, you probably encountered cases such as *Pierson v. Post*, 3 Cai. R. 175 (N.Y. Sup. Ct. 1805), the famous dispute over which of two hunters owned a fox carcass—Post, who had given chase, and Pierson, who darted in at the last moment to kill the fox. (The court held for Pierson.) Perhaps you considered re-lated cases about finding buried treasure, drilling for oil, digging a water well, or even staking out a frequency on the broadcast radio dial. In all these cases, courts have struggled with how best to define possession—what counts as possession, and why?—in light of the resource's characteristics and our need for or reliance on it. For patent law, concrete solutions to practical problems are the resource.

Many people, especially in the pharmaceutical art, race to be the first to identify an effective drug for a disease. *Brenner* involved such a race. What counts as adequate identification to earn patent rights? The question is less whether someone will earn the patent right, than who among all the racing parties will win. (Someone almost certainly will.) The answer: Until one can state a specific and substantial use for a compound, one does not possess the compound, or a method for making it, as an invention. Put another way, we want to keep multiple parties in the race at least until one of them identifies a specific, substantial use. Or so *Brenner* tells us. (Similarly, *Post* tells us we want to keep multiple parties in the race until we get a dead—rather than merely a frightened—fox.) What would the consequence be if we held that less information constituted utility? What would the consequence be if we required even more information to show utility? Remember that patent protection not only requires disclosure, it facilitates disclosure of useful information that might otherwise be kept secret.

In part, the *Brenner* Court based its holding on a hunch about how disclosure and research in pharmaceuticals will develop in the absence of patent protection on processes that produce substances for which there is, as yet, no known use. In some ways, all of intellectual property is based on a hunch—a hunch that providing a right to exclude will induce the creation of more of the things that qualify for the exclusionary right. In dynamic markets empirical evidence demonstrating how inventors would behave with and without patent protection is incredibly tough to obtain. In the absence of empirical data, how should the decision be made? Should Congress or the Court be making the determination? In the years since *Brenner* there have not been any efforts to amend the patent statute to permit patents for processes that produce substances for which there is no known use. Does this mean the Court decided the question correctly?

In any event, *Brenner* remains the controlling standard for assessing patentable utility. *See In re Fisher*, 421 F.3d 1365 (Fed. Cir. 2005) (rejecting claims on gene fragments where the inventor could not state a specific, substantial utility for any protein for which those genes code). And be careful not to make too much out of *Brenner*, as a matter of routine patent practice. Outside the biomedical and chemical fields, most inventions easily meet the utility requirement.

Should we use the utility requirement to address concerns about the ethical or moral dimensions of technologies? Assuming the answer is "yes," is it clear whether one who objects to the technology should *support* patent protection (to make the technology more costly, and thus less widespread), or *oppose* patent protection (to reduce the incentive to invent in the area to begin with)? Does the answer depend on whether there are patent-independent incentives to make those inventions?

Juicy Whip, Inc. v. Orange Bang, Inc.
185 F.3d 1364 (Fed. Cir. 1999)

BRYSON, JUDGE:

The district court in this case held a patent invalid for lack of utility on the ground that the patented invention was designed to deceive customers by imitating another product and thereby increasing sales of a particular good. We reverse and remand.

I

Juicy Whip, Inc., is the assignee of United States Patent No. 5,575,405, which is entitled "Post-Mix Beverage Dispenser With an Associated Simulated Display of Beverage." A "post-mix" beverage dispenser stores beverage syrup concentrate and water in separate locations until the beverage is ready to be dispensed. The syrup and water are mixed together immediately before the beverage is dispensed, which is usually after the consumer requests the beverage. In contrast, in a "pre-mix" beverage dispenser, the syrup concentrate and water are pre-mixed and the beverage is stored in a display reservoir bowl until it is ready to be dispensed. The display bowl is said to stimulate impulse buying by providing the consumer with a visual beverage display. A pre-mix display bowl, however, has a limited capacity and is subject to contamination by bacteria. It therefore must be refilled and cleaned frequently.

The invention claimed in the '405 patent is a post-mix beverage dispenser that is designed to look like a pre-mix beverage dispenser. The claims require the post-mix dispenser to have a transparent bowl that is filled with a fluid that simulates the appearance of the dispensed beverage and is resistant to bacterial growth. The claims also require that the dispenser create the visual impression that the bowl is the principal source of the dispensed beverage, although in fact the beverage is mixed immediately before it is dispensed, as in conventional post-mix dispensers.

Claim 1 is representative of the claims at issue. It reads as follows:

In a post-mix beverage dispenser of the type having an outlet for discharging beverage components in predetermined proportions to provide a serving of dispensed beverage, the improvement which comprises:

a transparent bowl having no fluid connection with the outlet and visibly containing a quantity of fluid;

said fluid being resistant to organic growth and simulating the appearance of the dispensed beverage;

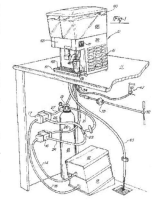

said bowl being positioned relative to the outlet to create the visual impression that said bowl is the reservoir and principal source of the dispensed beverage from the outlet; and

said bowl and said quantity of fluid visible within said bowl cooperating to create the visual impression that multiple servings of the dispensed beverage are stored within said bowl.

Juicy Whip sued defendants Orange Bang, Inc., and Unique Beverage Dispensers, Inc., (collectively, "Orange Bang") in the United States District Court for the Central District of California, alleging that they were infringing the claims of the '405 patent. Orange Bang moved for summary judgment of invalidity, and the district court granted Orange Bang's motion on the ground that the invention lacked utility and thus was unpatentable under 35 U.S.C. § 101.

The court concluded that the invention lacked utility because its purpose was to increase sales by deception, i.e., through imitation of another product. The court explained that the purpose of the invention "is to create an illusion, whereby customers believe that the fluid contained in the bowl is the actual beverage that they are receiving, when of course it is not." Although the court acknowledged Juicy Whip's argument that the invention provides an accurate representation of the dispensed beverage for the consumer's benefit while eliminating the need for retailers to clean their display bowls, the court concluded that those claimed reasons for the patent's utility "are not independent of its deceptive purpose, and are thus insufficient to raise a disputed factual issue to present to a jury." The court further held that the invention lacked utility because it "improves the prior art only to the extent that it increases the salability of beverages dispensed from post-mix dispensers"; an invention lacks utility, the court stated, if it confers no benefit to the public other than the opportunity for making a product more salable. Finally, the court ruled that the invention lacked utility because it "is merely an imitation of the pre-mix dispenser," and thus does not constitute a new and useful machine.

II

Section 101 of the Patent Act of 1952, 35 U.S.C. § 101, provides that "[w]hoever invents or discovers any new and useful process, machine, manufacture, or composition of matter, or any new and useful improvement thereof," may obtain a patent on the invention or discovery. The threshold of utility is not high: An invention is "useful" under section 101 if it is capable of providing some identifiable benefit. *See Brenner v. Manson*, 383 U.S. 519, 534 (1966); *Brooktree Corp. v. Advanced Micro Devices, Inc.*, 977 F.2d 1555, 1571 (Fed. Cir. 1992) ("To violate § 101 the claimed device must be totally incapable of achieving a useful result"); *Fuller v. Berger*, 120 F. 274, 275 (7th Cir. 1903) (test for utility is whether invention "is incapable of serving any beneficial end").

To be sure, since Justice Story's opinion in *Lowell v. Lewis*, 15 F. Cas. 1018 (C.C.D. Mass. 1817), it has been stated that inventions that are "injurious to the well-being, good policy, or sound morals of society" are unpatentable. As examples of such inventions, Justice Story listed "a new invention to poison people, or to promote debauchery, or to facilitate private assassination." *Id.* at 1019. Courts have continued to recite Justice Story's formulation, but the principle that inventions are invalid if they are principally designed to serve immoral or illegal purposes has not been applied broadly in recent years. For example, years ago courts invalidated patents on gambling devices on the ground that they were immoral, *see e.g., Brewer v. Lichtenstein*, 278 F. 512 (7th Cir. 1922); *Schultze v. Holtz*, 82 F. 448 (N.D. Cal. 1897); *National Automatic Device Co. v. Lloyd*, 40 F. 89 (N.D. Ill. 1889), but that is no longer the law, *see In re Murphy*, 200 USPQ 801 (PTO Bd. App. 1977).

In holding the patent in this case invalid for lack of utility, the district court relied on two Second Circuit cases dating from the early years of this century, *Rickard v. Du Bon*, 103 F. 868 (2d Cir. 1900), and *Scott & Williams, Inc. v. Aristo Hosiery Co.*, 7 F.2d 1003 (2d Cir. 1925). In the *Rickard* case, the court held invalid a patent on a process for treating tobacco plants to make their leaves appear spotted. At the time of the invention, according to the court, cigar smokers considered cigars with spotted wrappers to be of superior quality, and the invention was designed to make unspotted tobacco leaves appear to be of the spotted—and thus more desirable—type. The court noted that the invention did not promote the burning quality of the leaf or improve its quality in any way; "the only effect, if not the only object, of such treatment, is to spot the tobacco, and counterfeit the leaf spotted by natural causes." *Id.* at 869.

The *Aristo Hosiery* case concerned a patent claiming a seamless stocking with a structure on the back of the stocking that imitated a seamed stocking. The imitation was commercially useful because at the time of the invention many consumers regarded seams in stockings as an indication of higher quality. The court noted that the imitation seam did not "change or improve the structure or the utility of the article," and that the record in the case justified the conclusion that true seamed stockings were superior to the seamless stockings that were the subject of the patent. *See Aristo Hosiery*, 7 F.2d at 1004. "At best," the court stated, "the seamless stocking has imitation marks for the purposes of deception, and the idea prevails that with such imitation the article is more salable." *Id.* That was not enough, the court concluded, to render the invention patentable.

We decline to follow *Rickard* and *Aristo Hosiery*, as we do not regard them as representing the correct view of the doctrine of utility under the Patent Act of 1952. The fact that one product can be altered to make it look like another is in itself a specific benefit sufficient to satisfy the statutory requirement of utility.

It is not at all unusual for a product to be designed to appear to viewers to be something it is not. For example, cubic zirconium is designed to simulate a diamond, imitation gold leaf is designed to imitate real gold leaf, synthetic fabrics are designed to simulate expensive natural fabrics, and imitation leather is designed to look like real leather. In each case, the invention of the product or process that makes such imitation possible has "utility" within the meaning of the patent statute, and indeed there are numerous patents directed toward making one product imitate another. *See, e.g.*, U.S. Pat. No. 5,762,968 (method for producing imitation grill marks on food without using heat); U.S. Pat. No. 5,899,038 (laminated flooring imitating wood); U.S. Pat. No. 5,571,545 (imitation hamburger). Much of the value of such products resides in the fact that they appear to be something they are not. Thus, in this case the claimed post-mix dispenser meets the statutory requirement of utility by embodying the features of a post-mix dispenser while imitating the visual appearance of a pre-mix dispenser.

The fact that customers may believe they are receiving fluid directly from the display tank does not deprive the invention of utility. Orange Bang has not argued that it is unlawful to display a representation of the beverage in the manner that fluid is displayed in the reservoir of the invention, even though the fluid is not what the customer will actually receive. Moreover, even if the use of a reservoir containing fluid that is not dispensed is considered deceptive, that is not by itself sufficient to render the invention unpatentable. The requirement of "utility" in patent law is not a directive to the Patent and Trademark Office or the courts to serve as arbiters of deceptive trade practices. Other agencies, such as the Federal Trade Commission and the Food and Drug Administration, are assigned the task of protecting consumers from fraud and deception in the sale of food products. *Cf. In re Watson*, 517 F.2d 465, 474-76 (CCPA 1975) (stating that it is not the province of the Patent Office to determine, under section 101, whether drugs are safe). As the Supreme Court put the point more generally, "Congress never intended that the patent laws should displace the police powers of the States, meaning by that term those powers by which the health, good order, peace and general welfare of the community are promoted." *Webber v. Virginia*, 103 U.S. 344, 347-48 (1880).

Of course, Congress is free to declare particular types of inventions unpatentable for a variety of reasons, including deceptiveness. *Cf.* 42 U.S.C. § 2181(a) (exempting from patent protection inventions useful solely in connection with special nuclear material or atomic weapons). Until such time as Congress does so, however, we find no basis in section 101 to hold that inventions can be ruled unpatentable for lack of utility simply because they have the capacity to fool some members of the public. The district court therefore erred in holding that the invention of the '405 patent lacks utility because it deceives the public through imitation in a manner that is designed to increase product sales.

Notes & Questions

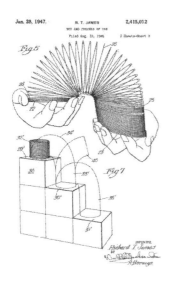

1. Toys are patentable; consider, for example, the humble Slinky, U.S. Patent No. 2,415,012. Should they be?

2. The European Patent Convention, in Article 53(a), provides that "European patents shall not be granted in respect of . . . inventions the publication or exploitation of which would be contrary to 'ordre public' or morality . . . ." Should Congress amend the Patent Act to contain a similar provision, as a sort of moral utility doctrine? If so, how would we determine whether something fell within the prohibited area? For example, should the transgenic bacterium from the *Chakrabarty* case be excluded from patentability under such a provision? How about more complex organisms, including mammals? *See generally* Margo A. Bagley, *Patent First, Ask Questions Later: Morality & Biotechnology in Patent Law*, 45 Wm. & Mary L. Rev. 469 (2003). Although Congress recently ratified the Patent Office

practice of refusing patents on a human organism, the Patent Office has granted patents on a variety of non-naturally occurring non-human mammals. See, e.g. U.S. Patent No. 4,736,866 (issued in 1988) (claiming a transgenic non-human mammal with an oncogene), U.S. Patent No. 7,550,649 (issued in 2009) (claiming a transgenic mouse that is a model for Parkinson's disease).

V. Novelty & Statutory Bars

Patent protection is reserved for inventions that are *new* to the art (and not merely new to the individual inventor). Only new inventions promote the constitutional goal of "progress"—forward movement—in the useful arts to which they pertain. Indeed, the Supreme Court has made clear that, under the Progress Clause, "Congress may not authorize the issuance of patents whose effects are to remove existent knowledge from the public domain, or to restrict free access to materials already available." *Graham v. John Deere Co.*, 383 U.S. 1, 6 (1966).

The easy part is stating the axiom that only new inventions are patentable. The hard part is crafting a statutory framework that puts the principle into practice, sorting the old from the new in a productive, reliable way. And the challenge of learning and applying the novelty rules is especially significant in the U.S. at this time because, as a result of recent changes to the Patent Act, different novelty rules apply depending on when the U.S. patent was sought and issued.

The Patent Act framework is in 35 U.S.C. § 102. Section 102 now exists, simultaneously, in two quite different forms. Both remain relevant, because each one applies to a different set of U.S. patents. This will remain true for many years. First, the 1952 Act version of this provision, with its seven subsections, has produced a rich, challenging body of case law. This Survey book focuses on that case law. Second, the America Invents Act of 2011 makes profound changes to some, but not all, aspects of § 102. The new § 102 (AIA) governs patents the applications for which were filed on or after March 16, 2013. The Patent Office is therefore applying the new § 102 now, to applications filed after that date. The old § 102 (1952 Act version), however, *will continue to govern* patents that issued from applications filed before March 16, 2013. In addition, given the 20-years-from-filing term that patents enjoy, it is no exaggeration to say old § 102 (1952 Act) will be applicable in patent litigation for *decades*. In this Section, we focus on the two most important subsections of the *old* (1952 Act) version of the statute, § 102(a) and § 102(b). At the end of the Section, we highlight the changes that the America Invents Act brought about in March 2013. Litigation in which the new § 102 (AIA) governs is only in its infancy. Unless otherwise noted, a reference to § 102 is to the old version based on 1952 Act.

A. An Introduction to the Old § 102 (1952 Act)

New is a relative term—new, compared to what? Section 102's main task is to create a baseline for the state of the art; we then assess an invention's novelty against this baseline. In designing our patent system, we also want to encourage inventors to disclose their inventions earlier rather than later. Earlier disclosure keeps the publicly known state of the art closer to the leading edge of technology, and thus helps others in the field avoid wasting research dollars chasing a solution that someone else has already achieved. Section 102's secondary task, then, is to encourage inventors to file for patent protection sooner than they otherwise might.

Section 102's basic approach is to state conditions that, if proved, negate patentability: "A person *shall* be entitled to a patent *unless* . . . ." 35 U.S.C. § 102 (emphasis added). The section provides seven independent negating conditions. If any one is met, the claim is not patentable. We focus here on the first two conditions. As you read them, identify (1) the key date on which the subsection focuses attention, and (2) the actors whose activities can negate patentability. The first two negating conditions are as follows:

> (a) the invention was known or used by others in this country, or patented or described in a printed publication in this or a foreign country, before the invention thereof by the applicant for patent, or
> (b) the invention was patented or described in a printed publication in this or a foreign country or in public use or on sale in this country, more than one year prior to the date of the application for patent in the United States
> . . .

35 U.S.C. §§ 102(a), (b). Again, this code, although superseded by the AIA for patents filed on or after March 16, 2013, still governs patents from before that date.

Subsection (a) is our fundamental *novelty* provision. One looks for public signs of the invention from a time "before the invention thereof by the applicant." Put another way, the inventor's own activities, taking place after she has made the invention, *cannot* negate patentability under § 102(a).

What public signs of the invention are material under § 102(a)? The statute distinguishes between public information in the form of documents—things "patented or described in a printed publication in this or a foreign country"—and public information in a form other than a document—things "known or used by others in this country." It makes this distinction to enable a geographical differentiation. An eligible pre-invention *document* negates novelty no matter where in the world the document comes from, and no matter what language it appears in. So long as it qualifies as a "printed publication"—an analysis we take up in Subsection V.C—a 50-year old article written in a Russian science journal, for example, is just as eligible to negate patentability as a 5-year old article from *USA Today*. Public knowledge or use by others, by contrast, qualifies for consideration only if it took place in the U.S. For example, if the claimed invention was used by others publicly in Brownsville, Texas, before the applicant (re)invented it, it's unpatentable. If, however, it was in public use next door in Matamoros, Mexico, it's patentable (at least as far as § 102(a) is concerned).[*]

---

[*] One of the many ways the America Invents Act changes § 102 is that it ends this geographic disparity, putting public acts and items on the same worldwide footing as documents.

Note an important premise of how § 102(a) works: what matters is the *actual date of invention*, not merely the date on which the applicant filed her application with the PTO. An applicant can remove a printed publication from consideration as prior art under § 102(a) by showing that, even though the publication in question predates her filing at the PTO, the publication came after the actual date of her invention.

The focus on the date of invention in § 102(a) is in keeping with the 1952 Act's choice to employ a first-to-invent standard, rather than a first-to-file standard. When two or more different applicants applied for the same patent claim, the PTO—using a procedure known as an *interference*—had to adjudicate which of them was the first inventor. 35 U.S.C. §§ 102(g), 135. (The test for determining which invention is "first" is beyond the scope of this Chapter.) For many decades, the U.S. was the only country to use this first-to-invent approach; other countries rely on a first-to-file system. That difference has disappeared for applications filed after March 15, 2013: The most significant change made by the America Invents Act (AIA), and the way it redefines what does and does not constitute prior art under § 102, is the move to a type of first-to-file system. We discuss this change in more detail below, in Subsection V.D.

Even before the AIA, our first-to-invent system was qualified by the one-year period established in § 102(b). Thus, *when* you show up at the PTO *does* affect how we assess an invention's novelty. Specifically, one's filing date helps determine the content of the prior art for purposes of § 102(b). Subsection (b) is our fundamental *statutory bar* provision. Unlike § 102(a), § 102(b) embraces information created by the inventor *after her invention date*, along with information created by others. The key term is the one-year grace period: an item is eligible for consideration if it was published, or occurred, "more than one year prior to the date of the [patent] application" in question. *It does not matter, under § 102(b), whether the item originates from people other than the inventor, or from the inventor himself.* If the inventor fully described the invention in a journal article published more than a year before the application's filing date, the claim is unpatentable. Similarly, if someone else fully described the invention in a journal article published more than a year before the application's filing date, the claim is unpatentable. This remains true even if the inventor-applicant could prove an invention date that pre-dates the publication of the other inventor's article. In this respect, § 102(b) is also an independent novelty provision, creating a separate ground of exclusion. Like § 102(a), § 102(b) incorporates a geographic disparity, paired with a distinction between a document category ("patented or described in a printed publication in this or a foreign country") and an other-than-document category ("in public use or on sale in this country"). Patent practitioners call subsection (b)'s exclusions the "statutory bars" because they include inventor activities that can bar an inventor from obtaining a patent on an invention that, but for the inventor's activity, would be new.

An important word of caution is in order: Although the U.S. allows inventors one year of commercial sales or printed publication exposure prior to submitting their patent applications, nearly all other countries have an absolute novelty rule with no grace period at all (much less a one-year grace period). In such a country, the patent application must be on file prior to even the first sale or public use. If a U.S. inventor plans to seek patent protection in other countries, it is best to file an application for protection before any sales activity or public demonstration of the invention. Whether filed in the U.S. or abroad, that first application can, through a series of steps established by multinational patent treaties, provide the foundation for a transnational family of patents covering the invention.

To make this discussion concrete, let's consider a series of hypotheticals. Imagine a patent application that claims "a widget comprising an A, a B, and a C," **filed on September 1, 2008**. The applicant can prove, with documents and a prototype, that she **finished inventing the widget on June 1, 2008**. Investigation yields the following items, each one of which fully and expressly describes, or embodies, a widget comprising an A, a B, and a C (arranged as they are in the claim) and how to make the widget. Each of these items, if made part of the prior art by §§ 102(a) or (b), proves that the invention is old and thus unpatentable. We must therefore determine whether each item qualifies as prior art under §§ 102(a) or (b).

1. A widget that was developed (start to finish) and put on sale to the public on July 15, 2008. *Analysis*—This widget was not in public use or on sale more than a year before the applicant's September 1, 2008, filing date *i.e.*, on or before August 31, 2007. It is thus not a prior art reference under § 102(b). It was also not used or known by others, *i.e.*, publicly known or used, before the applicant's invention date of June 1, 2008. It is thus not a prior art reference under § 102(a).

2. A journal article from Finland, in Finnish, published and indexed in many public libraries no later than February 1, 2008. *Analysis*—This article was not published and indexed more than a year before September 1, 2008. It is thus not a prior art reference under § 102(b). It was, however, published and indexed before the applicant made her invention on June 1, 2008. It is thus an invalidating prior art reference under § 102(a).

3. A presentation the applicant made at a trade show, with a paper handout describing the claimed invention, on August 1, 2008. *Analysis*—This presentation was not published to the trade more than a year before September 1, 2008. It is thus not a prior art reference under § 102(b). Because it is the applicant's own activity, *after* having made the invention, it cannot—of course—be a prior art reference under § 102(a).

4. A trade catalog from Canada, in English, published and sent to members of the trade no later than August 1, 2007. *Analysis*—This catalog was published and sent more than a year before September 1, 2008. It is thus an invalidating prior art reference under § 102(b). The catalog was also published and sent before the applicant made her invention on June 1, 2008. It is thus also, separately, an invalidating prior art reference under § 102(a).

Now let's change a key fact: The applicant delayed—for whatever reason—and filed her patent application on **August 2, 2009** (not September 1, 2008).

5. The July 15, 2008 widget (# 1 above). *Analysis*—Same result for § 102(a); it was not publicly known or used by others until after the applicant's invention date. For § 102(b), however, the result has changed. The widget was in public use and on sale more than a year before the applicant's August 2, 2009, filing date, *i.e.*, on or before August 1, 2008. It is thus an invalidating prior art reference under § 102(b).

6. The article from Finland (#2 above). *Analysis*—Same result for § 102(a); it pre-dates the applicant's invention date. For § 102(b), however, the result has changed. The article was published and indexed more than a year before August 2, 2009. It is thus an invalidating prior art reference under § 102(b).

7. The inventor's August 1, 2008, trade show presentation (#3 above). *Analysis*—Same result for § 102(a); the applicant's own post-invention activity cannot be a § 102(a) prior art reference. For § 102(b), however, the result has changed. The presentation was published to the trade more than a year before August 2, 2009 (on August 1, 2008). The applicant has, through her own delay, thrown up a bar to her claim.

8. The catalog from Canada (#4 above). *Analysis*—Same results.

The pattern from the above examples highlights a difference between subsections (a) and (b). An item's prior art status under § 102(a) doesn't change with the passage of time; the item either predates the actual invention date or it doesn't. An item's prior art status under § 102(b), by contrast, *does* change over time. Specifically, after one makes one's invention, the longer one waits to file an application, the more items become eligible for consideration as prior art under § 102(b). With each passing day, an additional day's worth of material becomes eligible as prior art—namely, whatever appeared in a printed publication more than a year before that newly passed day. Put differently, an article published on January 1, 2012, isn't § 102(b) prior art against anything until January 2, 2013, at which time it becomes prior art against newly filed applications, *no matter when* the inventions claimed in those applications were *actually* invented. This is the feature of § 102(b) that created an incentive for applicants to file sooner rather than later.

Note, too, that the fact that an item is (or is not) eligible to be considered as a reference under one of these two subsections (quite apart from its substantive content) says little about whether or not it is also eligible to be considered as a reference under the other subsection. This is an important practice tip: When analyzing novelty, you must assess a potential prior art reference's eligibility under each subsection separately.

B. Anticipation's Identity & Enablement Requirements

Thus far, we have focused on one aspect of a § 102 analysis—namely, whether the item in question is eligible for consideration as part of the prior art against which we assess a given claim's novelty. Next we must consider what the prior art reference must contain in order to negate patentability. There are at least two crucial steps in that analysis. First, does the prior art reference describe or show each and every limitation of the claimed invention? Second, does the prior art reference enable the phosita to make the claimed invention?[*] If the answer to both these questions is "yes," the reference defeats the claim's novelty. In patent law parlance, we say that the prior art reference *anticipates* the claim, or the claim *reads on* the prior art reference. Consider the following celebrated anticipation case.

Titanium Metals Corp. of America v. Banner
778 F.2d 775 (Fed. Cir. 1985)

RICH, JUDGE:

This appeal is from an Order of the United States District Court for the District of Columbia in a civil action brought pursuant to 35 U.S.C. § 145 against Donald W. Banner as Commissioner of Patents and Trademarks authorizing the Commissioner to issue to appellee a patent containing claims 1, 2, and 3 of patent application serial No. 598,935 for "TITANIUM ALLOY." The Commissioner has appealed. We reverse.

Background

The inventors, Loren C. Covington and Howard R. Palmer, employees of appellee to whom they have assigned their invention and the application thereon, filed an application on March 29, 1974, serial No. 455,964, to patent an alloy they developed. The application involved in this appeal . . . [contains] the three claims on appeal. The alloy is made primarily of titanium (Ti) and contains small amounts of nickel (Ni) and molybdenum (Mo) as alloying ingredients to give the alloy certain desirable properties, particularly corrosion resistance in hot brine solutions, while retaining workability so that articles such as tubing can be fabricated from it by rolling, welding and other techniques. The inventors apparently also found that iron content should be limited, iron being an undesired impurity rather than an alloying ingredient. They determined the permissible ranges of the components, above and below which the desired properties were not obtained. A precise definition of the invention sought to be patented is found in the claims, set forth below, claim 3 representing the preferred composition, it being understood, however, that no iron at all would be even more preferred.

> 1. A titanium base alloy consisting essentially by weight of about 0.6% to 0.9% nickel, 0.2% to 0.4% molybdenum, up to 0.2% maximum iron, balance titanium, said alloy being characterized by good corrosion resistance in hot brine environments.
> 2. A titanium base alloy as set forth in Claim 1 having up to 0.1% iron, balance titanium.
> 3. A titanium base alloy as set forth in Claim 1 having 0.8% nickel, 0.3%

---

[*] Note well, this is *not* the same question as the enablement inquiry under § 112, ¶ 1, explored in Section III above, which requires that a claim's supporting written description enable the phosita *both* to make *and to use* the claimed invention. Anticipation's enablement standard is lower, requiring only that the prior art reference enable the phosita to make the claimed invention; it need not enable one to use the invention. *See, e.g., In re Schoenwald*, 964 F.2d 1122, 1124 (Fed. Cir. 1992); *In re Hafner*, 410 F.2d 1403, 1405 (CCPA 1969).

molybdenum, up to 0.1% maximum iron, balance titanium.

The examiner's final rejection, repeated in his Answer on appeal to the Patent and Trademark Office (PTO) Board of Appeals (board), was on the grounds that claims 1 and 2 are anticipated (fully met) by, and claim 3 would have been obvious from, an article by Kalabukhova and Mikheyew, *Investigation of the Mechanical Properties of Ti-Mo-Ni Alloys*, Russian Metallurgy (Metally) No. 3, pages 130-133 (1970) (in the court below and hereinafter called "the Russian article") under 35 U.S.C. §§ 102 and 103, respectively. The board affirmed the examiner's rejection.

. . .

. . . The Russian article is short (3 pages), highly technical, and contains 10 graphs as part of the discussion. As its title indicates, it relates to ternary Ti-Mo-Ni alloys, the subject of the application at bar. The examiner and the board both found that it would disclose to one skilled in the art an alloy on which at least claims 1 and 2 read, so that those claims would not be allowable under the statute because of lack of novelty of their subject matter. Since the article does not specifically disclose such an alloy *in words*, a little thinking is required about what it would disclose to one knowledgeable about Ti-Ni-Mo alloys. The PTO did that thinking as follows:

> Figure 1c [a graph] shows data for the ternary titanium alloy which contains Mo and Ni in the ratio of 1:3. Amongst the actual points on the graph is one at 1% Mo + Ni. At this point, the amounts of Mo and Ni would be 0.25% and 0.75% respectively. A similar point appears on the graph shown in Figure 2 of the article.
>
> * * * *
>
> Appellants do not deny that the data points are disclosed in the reference. In fact, the Hall affidavit indicates at least two specific points (at 1% and 1.25% Mo + Ni) which would represent a description of alloys falling within the scope of the instant claims.

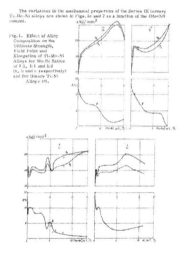

On that basis, the board found that the claimed alloys were not new, because they were disclosed in the prior art. It having been argued that the Russian article contains no disclosure of corrosion-resistant *properties* of any of the alloys, the board held: "The fact that a particular property or the end use for this alloy as contemplated by appellants was not recognized in the article is of no consequence." It therefore held the Russian article to be an anticipation, noting that although the article does not discuss corrosion resistance, it does disclose other properties such as strength and ductility. The PTO further points out that the authors of the reference must have made the alloys to obtain the data points.

Being dissatisfied with the decision of the board, Titanium Metals Corporation of America, as assignee of the Covington and Palmer application, then brought an action in the District Court for the District of Columbia against the Commissioner pursuant to 35 U.S.C. § 145 . . . .

The case came on for trial on January 24, 1980, before the Honorable John G. Penn and was concluded in two and a half hours. The testimony of one witness was heard by the court, Dr. James C. Williams, professor at Carnegie-Mellon University in Pittsburgh and an expert in titanium metallurgy. His testimony was about equally divided between direct and cross examination.

. . . .

. . . The court then concluded that claims 1-3 were not anticipated and that claim 3 was wrongly rejected as directed to obvious subject matter. In the court's view, Dr. Williams' testimony tipped the scales in favor of issuing a patent.

. . . .

A. Anticipation, § 102

. . . .

We are left in no doubt that the court was impressed by the totality of the evidence that the applicants for patent had discovered or invented and disclosed knowledge which is not to be found in the reference, nor do we have any doubt about that ourselves. But those facts are beside the point. The patent law imposes certain fundamental conditions for patentability, paramount among them being the condition that what is sought to be patented, as determined by the claims, be new. . . . The title of the application here involved is "Titanium Alloy," a composition of matter. Surprisingly, in all of the evidence, nobody discussed the key issue of whether the alloy was new, which is the essence of the anticipation issue, including the expert Dr. Williams. Plaintiff's counsel, bringing Dr. Williams' testimony to its climax, after he had explained the nature of the ingredients, the alloys made therefrom, and their superior corrosion resistance in hot brine, etc., repetitively asked him such questions as "Does the [Russian] article *direct you* as one skilled in the art to a titanium alloy having nickel present in an amount between .6 and .9 percent molybdenum in an amount between .2 and .4 percent?" (emphasis ours) followed by "Is there anything mentioned in the article about corrosion resistance?" Of course, the answers were emphatically negative. But this and like testimony does not deal with the critical question: do claims 1 and 2, to which the questions obviously relate, *read on or encompass* an alloy which was already known by reason of the disclosure of the Russian article?

Section 102, the usual basis for rejection for lack of novelty or anticipation, lays down certain principles for determining the novelty [of an invention], among which are the provisions in § 102(a) and (b) that the claimed invention has not been "described in a printed publication in this or a foreign country," either (a) before the invention by the applicant or (b) more than one year before the application date to which he is entitled (strictly a "loss of right" provision similar to novelty). Either provision applies in this case, the Russian article having a date some 5 years prior to the filing date and its status as "prior art" not being questioned. The PTO was never specific as to what part of § 102 applies, merely rejecting on

§ 102. The question, therefore, is whether claims 1 and 2 encompass and, if allowed, would enable plaintiff-appellee to exclude others from making, using, or selling an alloy described in the Russian article.

To answer the question we need only turn to the affidavit of James A. Hall, a metallurgist employed by appellee's TIMET Division, who undertook to analyze the Russian article disclosure by calculating the ingredient percentages shown in the graph data points, which he presented in tabular form. There are 15 items in his table. The second item shows a titanium base alloy containing 0.25% by weight Mo and 0.75% Ni and this is squarely within the ranges of 0.2-0.4% Mo and 0.6-0.9% Ni of claims 1 and 2. As to that disclosed alloy of the prior art, there can be no question that claims 1 and 2 read on it and would be infringed by anyone making, using, or selling it. Therefore, *the statute prohibits* a patent containing them. This seems to be a case either of not adequately considering the novelty requirement of the statute, the true meaning of the correlative term "anticipation," or the meaning of the claims.

By reason of the court's quotations from cases holding that a reference is not an anticipation which does not enable one skilled in the art to practice the claimed invention, it appears that the trial court thought there was some deficiency in the Russian article on that score. Enablement in this case involves only being able to make the alloy, given the ingredients and their proportions without more. The evidence here, however, clearly answers that question in two ways. Appellee's own patent application does not undertake to tell anyone how to make the alloy it describes and seeks to patent. It assumes that those skilled in the art would know how. Secondly, appellee's expert, Dr. Williams, testified on cross examination that given the alloy information in the Russian article, he would know how to prepare the alloys "by at least three techniques." Enablement is not a problem in this case.

As we read the situation, the court was misled by the arguments and evidence to the effect that the inventors here found out and disclosed in their application many things that one cannot learn from reading the Russian article and that this was sufficient in law to justify granting them a patent for their contributions—such things as what good corrosion resistance the claimed alloys have against hot brine, which possibly was not known, and the range limits of the Ni and Mo content, outside of which that resistance diminishes, which are teachings of very useful information. These things the applicants teach the art and the Russian article does not. Indeed, appellee's counsel argued in his opening statement to the trial court that the PTO's refusal of a patent was "directly contrary to the requirement of Article I, Section 8, of the Constitution," which authorizes Congress to create a patent law. But throughout the trial counsel never came to grips with the real issues: (1) what do the claims cover and (2) is what they cover new? Under the laws Congress wrote, they must be considered. Congress has not seen fit to permit the patenting of an old alloy, known to others through a printed publication, by one who has discovered its corrosion resistance or other useful properties, or has found out to what extent one can modify the composition of the alloy without losing such properties.

It is also possible that the trial court did not properly interpret the claims and took them to be directed only to the applicants' discoveries about the properties of the alloys instead of to the alloys themselves, as they are, possibly because of the phrase at the end of claim 1, "characterized by good corrosion resistance in hot brine environments," which applies to the other two dependent claims also. No light is shed by its opinion on what the court thought the claims mean as the opinion does not construe the claims. . . . It is the correct and necessary construction of all three claims that they simply define titanium base alloys. Claims 1 and 2 state certain narrow limits within which the alloying ingredients, Mo and Ni, are present and necessarily cover a number of alloys. Claim 3 is specific to a single alloy. This said, it is immaterial, on the issue of their novelty, what inherent properties the alloys have or whether these applicants discovered certain inherent properties.

The trial court and appellee have relied on *In re Wilder* [429 F.2d 447 (CCPA 1970)], but they have both failed to note those portions of that opinion most relevant to the present case. The issue there, as here, was anticipation of certain claims. Wilder argued "that even though there may be a technical anticipation, the discovery of the new property and the recitation of this property in the claims 'lends patentable novelty' to the claims." The court answered:

> However, recitation, in a claim to a composition, of a particular property said to be possessed by the recited composition, be that property newly-discovered or not, does not necessarily change the scope of the subject matter otherwise defined by that claim.

429 F.2d at 450. The court in that case also said:

> [W]e start with the proposition that claims cannot be obtained to that which is not new. This was the basis of the holding in *In re Thuau*, 135 F.2d 344 (CCPA 1943). It was the law then, is now and will be until Congress decrees otherwise.

Id. It is also an elementary principle of patent law that when, as by a recitation of ranges or otherwise, a claim covers several compositions, the claim is "anticipated" if *one* of them is in the prior art. *In re Petering*, 301 F.2d 676, 682 (CCPA 1962).

For all of the foregoing reasons, the court below committed clear error and legal error in authorizing the issuance of a patent on claims 1 and 2 since, properly construed, they are anticipated under § 102 by the Russian article which admittedly discloses an alloy on which these claims read.

B. Obviousness, § 103

Little more need be said in support of the examiner's rejection of claim 3, affirmed by the board, on the ground that its more specific subject matter would have been obvious at the time the invention was made from the knowledge disclosed in the reference.

As admitted by appellee's affidavit evidence from James A. Hall, the Russian article discloses two alloys having compositions very close to that of claim 3, which is 0.3% Mo and 0.8% Ni, balance titanium. The two alloys in the prior art have 0.25% Mo-0.75% Ni and 0.31% Mo-0.94% Ni, respectively. The proportions are so close that prima facie one skilled in the art would have expected them to have the same properties. Appellee produced no evidence to rebut that prima facie case. The specific alloy of claim 3 must therefore be considered to have been obvious from known alloys.

. . . .

Notes & Questions

1. The Russian article does not contain a sentence that describes, verbatim, the same alloy formulation as inventors Covington and Palmer sought to claim. The court concludes, however, that the explicit disclosure in the article's figures anticipates, and thus invalidates, claims 1 and 2. Note also that the single prior art alloy anticipates claims that recite a range of values for the ingredients. If claims 1 and 2 were permitted, the patentee would be able to exclude others from making the alloy with ingredients proportioned as shown in the article's figures. This conclusion exemplifies the patent law adage: that which would literally infringe if later in time anticipates if earlier in time. (Because both anticipation and literal infringement are about mapping claim limitations onto external items, this symmetry isn't surprising.) The requirement of novelty focuses on what the public already possesses. If the public already possess the invention, it would be inappropriate to provide an exclusive right to that invention.

2. The court notes that, on the facts of this case, it doesn't matter whether the claim for invalidity is asserted under § 102(a) or § 102(b), as they stood in the 1952 Patent Act. Be sure you understand why that is the case.

3. Everyone agrees that the Russian article does not disclose, in *any* way, the alloy's corrosion resistance in hot brine. But such corrosion resistance is an inherent property of the alloy. Covington and Palmer cannot claim the alloy itself, as if it were new, simply by reciting a newly discovered property of the old alloy. Can you draft a claim that captures the corrosion resistance invention and that the Russian article doesn't anticipate? With that claim, what prior art search would you want to conduct to ensure that the claimed subject matter is new?

4. Why doesn't the alloy in the Russian article anticipate claim 3? What disclosure is lacking? Note that the court shifts to a different analysis, from the question of novelty to the question of nonobviousness. The nonobviousness inquiry, which we will take up in detail in Section VI, is also based on a comparison of the claimed invention with the state of the art at the time the invention was made. Unlike novelty, which requires that the claimed invention be set forth, expressly or inherently, in a single prior art reference, obviousness—which also invalidates the claim—can be shown by combining multiple prior art references, or by demonstrating that the claim is an obvious modification of a single prior art reference.

C. Categories of "Prior Art"

1. What Is a "Printed Publication"?

Both §§ 102(a) and (b) embrace, as potentially anticipatory prior art, anything "described in a printed publication in this or a foreign country" in the relevant time frame.* What, then, counts as a *printed publication*?

In re Klopfenstein
380 F.3d 1345 (Fed. Cir. 2004)

Prost, Judge:

Carol Klopfenstein and John Brent appeal a decision from the Patent and Trademark Office's Board of Patent Appeals and Interferences ("Board") upholding the denial of their patent application. The Board upheld the Patent and Trademark Office's ("PTO's") initial denial of their application on the ground that the invention described in the patent application was not novel under 35 U.S.C. § 102(b) because it had already been described in a printed publication more than one year before the date of the patent application. We affirm.

BACKGROUND
A.

The appellants applied for a patent on October 30, 2000. Their patent application, Patent Application Serial No. 09/699,950 ("the '950 application"), discloses methods of preparing foods comprising extruded soy cotyledon fiber ("SCF"). The '950 application asserts that feeding mammals foods containing extruded SCF may help lower their serum cholesterol levels while raising HDL cholesterol levels. The fact that extrusion reduces cholesterol levels was already known by those of ordinary skill in the art that worked with SCF. What was not known at the time was that double extrusion increases this effect and yielded even stronger results.

In October 1998, the appellants, along with colleague M. Liu, presented a printed slide presentation ("Liu" or "the Liu reference") entitled "Enhancement of Cholesterol-Lowering Activity of Dietary Fibers By Extrusion Processing" at a meeting of the American Association of Cereal Chemists ("AACC"). The fourteen-slide presentation was printed and pasted onto poster boards. The printed slide presentation was displayed continuously for two and a half days at the AACC meeting.

In November [1998], the same slide presentation was put on display for less than a day at an Agriculture Experiment Station ("AES") at Kansas State University.

---

* There is also a category of things "patented . . . in this or a foreign country," in both § 102(a) and § 102(b). Given the way patent documents are now publicized and indexed, this separate "patented" category is of little practical significance: Any government grant of a right-to-exclude-others that we would be prepared to label a "patent" is almost certainly also a "printed publication."

Both parties agree that the Liu reference presented to the AACC and at the AES in 1998 disclosed every limitation of the invention disclosed in the '950 patent application. Furthermore, at neither presentation was there a disclaimer or notice to the intended audience prohibiting note-taking or copying of the presentation. Finally, no copies of the presentation were disseminated either at the AACC meeting or at the AES, and the presentation was never catalogued or indexed in any library or database.

. . . .

DISCUSSION

A.

Where no facts are in dispute, the question of whether a reference represents a "printed publication" is a question of law. *In re Cronyn,* 890 F.2d 1158, 1159 (Fed. Cir. 1989). Questions of law appealed from a Board decision are reviewed de novo. *In re Bass,* 314 F.3d 575, 576 (Fed. Cir. 2002).

The only question in this appeal is whether the Liu reference constitutes a "printed publication" for the purposes of 35 U.S.C. § 102(b). As there are no factual disputes between the parties in this appeal, the legal issue of whether the Liu reference is a "printed publication" will be reviewed de novo.

B.

The appellants argue on appeal that the key to establishing whether or not a reference constitutes a "printed publication" lies in determining whether or not it had been disseminated by the distribution of reproductions or copies and/or indexed in a library or database. They assert that because the Liu reference was not distributed and indexed, it cannot count as a "printed publication" for the purposes of 35 U.S.C. § 102(b). To support their argument, they rely on several precedents from this court and our predecessor court on "printed publications."[1] They argue that *In re Cronyn, In re Hall,* 781 F.2d 897 (Fed. Cir. 1986), *Massachusetts Institute of Technology v. AB Fortia,* 774 F.2d 1104 (Fed. Cir. 1985) ("*MIT*"), and *In re Wyer,* 655 F.2d 221 (CCPA 1981), among other cases, all support the view that distribution and/or indexing is required for something to be considered a "printed publication."[2]

We find the appellants' argument unconvincing and disagree with their characterization of our controlling precedent. Even if the cases cited by the appellants relied on inquiries into distribution and indexing to reach their holdings, they do not limit this court to finding something to be a "printed publication" *only* when

---

[1] In their brief, the appellants note that there is scant legislative history to guide us in determining the meaning of the term "printed publication." Accordingly, and rightfully, they have based the bulk of their argument on the controlling precedent of this court and its predecessor court.

[2] Appellants acknowledge that our precedent considers the term "printed publication" to be a unitary concept that may not correspond exactly to what the term "printed publication" meant when it was introduced into the patent statutes in 1836. *In re Wyer,* 655 F.2d at 226. Indeed, the question to be resolved in a "printed publication" inquiry is the extent of the reference's "accessibility to at least the pertinent part of the public, of a perceptible description of the invention, in whatever form it may have been recorded." *Id.*

there is distribution and/or indexing. Indeed, the key inquiry is whether or not a reference has been made "publicly accessible." As we have previously stated,

> The statutory phrase "printed publication" has been interpreted to mean that before the critical date the reference must have been sufficiently accessible to the public interested in the art; dissemination and public accessibility are the keys to the legal determination whether a prior art reference was "published."

In re Cronyn, 890 F.2d at 1160 (quoting *Constant v. Advanced Micro-Devices, Inc.,* 848 F.2d 1560, 1568 (Fed. Cir. 1988)).[3] For example, a public billboard targeted to those of ordinary skill in the art that describes all of the limitations of an invention and that is on display for the public for months may be neither "distributed" nor "indexed"—but it most surely is "sufficiently accessible to the public interested in the art" and therefore, under controlling precedent, a "printed publication." Thus, the appellants' argument that "distribution and/or indexing" are the key components to a "printed publication" inquiry fails to properly reflect what our precedent stands for.

Furthermore, the cases that the appellants rely on can be clearly distinguished from this case. *Cronyn* involved college students' presentations of their undergraduate theses to a defense committee made up of four faculty members. Their theses were later catalogued in an index in the college's main library. The index was made up of thousands of individual cards that contained only a student's name and the title of his or her thesis. The index was searchable by student name and the actual theses themselves were neither included in the index nor made publicly accessible. We held that because the theses were only presented to a handful of faculty members and "had not been cataloged [sic] or indexed in a meaningful way," they were not sufficiently publicly accessible for the purposes of 35 U.S.C. § 102(b). *In re Cronyn,* 890 F.2d at 1161.

In *Hall,* this court determined that a thesis filed and indexed in a university library did count as a "printed publication." The *Hall* court arrived at its holding after taking into account that copies of the indexed thesis itself were made freely available to the general public by the university more than one year before the filing of the relevant patent application in that case. But the court in *Hall* did not rest its holding merely on the indexing of the thesis in question. Instead, it used indexing as a factor in determining "public accessibility." As the court asserted:

> The ["printed publication"] bar is grounded on the principle that once an invention is in the public domain, it is no longer patentable by anyone. . . . Because there are many ways in which a reference may be disseminated to the interested public, "public accessibility" has been called the touchstone

---

[3] While the *Cronyn* court held "dissemination" to be necessary to finding something to be a "printed publication," the court there used the word "disseminate" in its literal sense, *i.e.,* "make widespread" or "to foster general knowledge of." *Webster's Third New International Dictionary* 656 (1993). The court did not use the word in the narrower sense the appellants have employed it, which requires distribution of reproductions or photocopies.

in determining whether a reference constitutes a "printed publication" bar under 35 U.S.C. § 102(b).

In re Hall, 781 F.2d at 898-99.

In *MIT,* a paper delivered orally to the First International Cell Culture Congress was considered a "printed publication." In that case, as many as 500 persons having ordinary skill in the art heard the presentation, and at least six copies of the paper were distributed. The key to the court's finding was that actual copies of the presentation were distributed. The court did not consider the issue of indexing. The *MIT* court determined the paper in question to be a "printed publication" but did not limit future determinations of the applicability of the "printed publication" bar to instances in which copies of a reference were actually offered for distribution. *MIT,* 774 F.2d at 1108-10.[4]

Finally, the *Wyer* court determined that an Australian patent application kept on microfilm at the Australian Patent Office was "sufficiently accessible to the public and to persons skilled in the pertinent art to qualify as a 'printed publication.'" *In re Wyer,* 655 F.2d at 226. The court so found even though it did not determine whether or not there was "actual viewing or dissemination" of the patent application. *Id.* It was sufficient for the court's purposes that the records of the application were kept so that they could be accessible to the public.[5] *Id.* According to the *Wyer* court, the entire purpose of the "printed publication" bar was to "prevent withdrawal" of disclosures "already in the possession of the public" by the issuance of a patent. *Id.*

Thus, throughout our case law, public accessibility has been the criterion by which a prior art reference will be judged for the purposes of § 102(b). Oftentimes courts have found it helpful to rely on distribution and indexing as proxies for public accessibility. But when they have done so, it has not been to the exclusion of all other measures of public accessibility. In other words, distribution and indexing are not the only factors to be considered in a § 102(b) "printed publication" inquiry.

C.

The determination of whether a reference is a "printed publication" under 35 U.S.C. § 102(b) involves a case-by-case inquiry into the facts and circumstances surrounding the reference's disclosure to members of the public. . . .

---

[4] With regard to scientific presentations, it is important to note that an entirely oral presentation at a scientific conference that includes neither slides nor copies of the presentation is without question not a "printed publication" for the purposes of 35 U.S.C. § 102(b). Furthermore, a presentation that includes a transient display of slides is likewise not necessarily a "printed publication." *See, e.g., Regents of the Univ. of Cal. v. Howmedica, Inc.,* 530 F. Supp. 846, 860 (D.N.J. 1981) (holding that "the projection of slides at the lecture [that] was limited in duration and could not disclose the invention to the extent necessary to enable a person of ordinary skill in the art to make or use the invention" was not a "printed publication"). While *Howmedica* is not binding on this court, it stands for the important proposition that the mere presentation of slides accompanying an oral presentation at a professional conference is not per se a "printed publication" for the purposes of § 102(b).

[5] Unlike in *Cronyn,* it was the actual patent application — and not just an index card searchable by author name only — that was made publicly accessible.

. . . .

Given that the Liu reference was never distributed to the public and was never indexed, we must consider several factors relevant to the facts of this case before determining whether or not it was sufficiently publicly accessible in order to be considered a "printed publication" under § 102(b). These factors aid in resolving whether or not a temporarily displayed reference that was neither distributed nor indexed was nonetheless made sufficiently publicly accessible to count as a "printed publication" under § 102(b). The factors relevant to the facts of this case are: the length of time the display was exhibited, the expertise of the target audience, the existence (or lack thereof) of reasonable expectations that the material displayed would not be copied, and the simplicity or ease with which the material displayed could have been copied. Only after considering and balancing these factors can we determine whether or not the Liu reference was sufficiently publicly accessible to be a "printed publication" under § 102(b).

The duration of the display is important in determining the opportunity of the public in capturing, processing and retaining the information conveyed by the reference. The more transient the display, the less likely it is to be considered a "printed publication." *See, e.g., Howmedica,* 530 F.Supp. at 860 (holding that a presentation of lecture slides that was of limited duration was insufficient to make the slides "printed publications" under § 102(b)). Conversely, the longer a reference is displayed, the more likely it is to be considered a "printed publication." In this case, the Liu reference was displayed for a total of approximately three days. It was shown at the AACC meeting for approximately two and a half days and at the AES at Kansas State University for less than one day.

The expertise of the intended audience can help determine how easily those who viewed it could retain the displayed material. As Judge Learned Hand explained in *Jockmus v. Leviton,* 28 F.2d 812, 813-14 (2d Cir. 1928), a reference, "however ephemeral its existence," may be a "printed publication" if it "goes direct to those whose interests make them likely to observe and remember whatever it may contain that is new and useful." In this case, the intended target audience at the AACC meeting was comprised of cereal chemists and others having ordinary skill in the art of the '950 patent application. The intended viewers at the AES most likely also possessed ordinary skill in the art.

Whether a party has a reasonable expectation that the information it displays to the public will not be copied aids our § 102(b) inquiry. Where professional and behavioral norms entitle a party to a reasonable expectation that the information displayed will not be copied, we are more reluctant to find something a "printed publication." This reluctance helps preserve the incentive for inventors to participate in academic presentations or discussions. Where parties have taken steps to prevent the public from copying temporarily posted information, the opportunity for others to appropriate that information and assure its widespread public accessibility is reduced. These protective measures could include license agreements, nondisclosure agreements, anticopying software or even a simple disclaimer informing members of the viewing public that no copying of the information will be allowed or countenanced. Protective measures are to be considered insofar as they create a reasonable expectation on the part of the inventor that the displayed information will not be copied. In this case, the appellants took no measures to protect the information they displayed—nor did the professional norms under which they were displaying their information entitle them to a reasonable expectation that their display would not be copied. There was no disclaimer discouraging copying, and any viewer was free to take notes from the Liu reference or even to photograph it outright.

Finally, the ease or simplicity with which a display could be copied gives further guidance to our § 102(b) inquiry. The more complex a display, the more difficult it will be for members of the public to effectively capture its information. The simpler a display is, the more likely members of the public could learn it by rote or take notes adequate enough for later reproduction. The Liu reference was made up of 14 separate slides. One slide was a title slide; one was an acknowledgement slide; and four others represented graphs and charts of experiment results. The other eight slides contained information presented in bullet point format, with no more than three bullet points to a slide. Further, no bullet point was longer than two concise sentences. Finally, as noted earlier, the fact that extrusion lowers cholesterol levels was already known by those who worked with SCF. The discovery disclosed in the Liu reference was that double extrusion increases this effect. As a result, most of the eight substantive slides only recited what had already been known in the field, and only a few slides presented would have needed to have been copied by an observer to capture the novel information presented by the slides.

Upon reviewing the above factors, it becomes clear that the Liu reference was sufficiently publicly accessible to count as a "printed publication" for the purposes of 35 U.S.C. § 102(b). The reference itself was shown for an extended period of time to members of the public having ordinary skill in the art of the invention behind the '950 patent application. Those members of the public were not precluded from taking notes or even photographs of the reference. And the reference itself was presented in such a way that copying of the information it contained would have been a relatively simple undertaking for those to whom it was exposed—particularly given the amount of time they had to copy the information and the lack of any restrictions on their copying of the information. For these reasons, we conclude that the Liu reference was made sufficiently publicly accessible to count as a "printed publication" under § 102(b).

. . . .

Notes & Questions

1. Is a publicly available, search-engine-indexed web page a "printed publication" for purposes of § 102? If yes, what evidentiary challenge remains for using it effectively as a prior art reference against a patent claim?

2. In footnote 4, the court indicates that "an entirely oral presentation at a scientific conference that includes neither slides nor copies of the presentation is without question not a 'printed publication.'" What if a digital audio recording of

the presentation were posted on a public web page as an mp3 file, under a descriptive title of the presentation? Would it matter whether the audio file's content were word-searchable, the way text is?

3. What is the downside, if any, to using a multi-factor "printed publication" test tailored to the facts of a given case, some of which may point in contrary directions? (For example, should *Klopfenstein* have been decided the same way if the slides on the poster had said "CONFIDENTIAL," and there had been only one conference?) How would you develop a list of "best practices" for the inventors and patent attorneys at a business to avoid creating printed publication bars to the company's future patent claims?

4. A patent applicant and her agents, including her lawyer(s), owe the Patent Office a duty of candor and good faith during patent prosecution. This duty, which is set forth in both Patent Office rules, 37 C.F.R. § 1.56, and many judicial opinions, obliges an applicant to tell the Office any "material" information he knows. In *Klopfenstein*, for example, the Office learned about the inventors' poster presentations from the applicants themselves. Violating the duty of candor, called "inequitable conduct" and treated as a type of fraud, renders the entire patent unenforceable in court. A full exploration of these matters is beyond the scope of this Chapter; we do not intend, however, to minimize either the gravity of the lawyer's duty of candor or the severe sanctions one's state bar disciplinary system may impose for violations. *See* Edwin S. Flores & Sanford E. Warren, Jr., *Inequitable Conduct, Fraud, and Your License to Practice Before the United States Patent and Trademark Office*, 8 Tex. Intell. Prop. L.J. 299 (2000). Today, the leading case on inequitable conduct is *Therasense, Inc. v. Becton, Dickinson & Co.*, 649 F.3d 1276 (Fed. Cir. 2011) (en banc).

2. What Is "In Public Use"?

Section 102(a) provides that if the invention is "used by others" prior to the invention date, that will invalidate a patent. Similarly, § 102(b) provides that if the technology is "in public use" more than one year prior to the application, the patent will be barred. Despite the difference in language, courts generally interpret what qualifies as a public use similarly. It is easy enough to imagine a straightforward case of public use: An inventor takes his new ring-toss game to the public beach every day during the summer months, playing the game with friends and passers-by alike. The ring-toss game is plainly "in public use." But must a use be this longstanding or widespread to be qualify as invalidating prior art?

Beachcombers v. Wildewood Creative Prods.
31 F.3d 1154 (Fed. Cir. 1994)

PLAGER, JUDGE:

This is a patent infringement case involving an improved kaleidoscope as described and claimed in the patent-in-suit, U.S. Patent No. 4,740,046 (the '046 patent). A jury . . . found in favor of the accused infringer, WildeWood Creative Products, Inc. (WildeWood), finding that the asserted claims of the '046 patent were invalid . . . .

. . . .

The jury found that the ODYLIC [a prototype of the allegedly infringing kaleidoscope] was in public use by virtue of its display to guests at a party hosted by Carolyn Bennett, the designer and developer of the device, on April 12, 1985.[*] The jury's factual findings in support of its conclusion are entitled to a deferential standard of review. . . .

Bennett testified as to the specifics of the party in question. She explained that the April 12, 1985 date was corroborated by an entry in her date book and discussions with friends; that between 20-30 guests were present; that she personally demonstrated the device to some of the guests for the purpose of getting feedback on the device; and that she made no efforts to conceal the device or keep anything about it secret.

Mary Ann Regan, Bennett's friend and a guest at the party, also testified at trial. She said that she observed the ODYLIC at Bennett's party, and remembered noticing its distinctive characteristic — the liquid-filled tube extending through the end of the barrel. She testified that a lot of other guests were present at the time she observed the device; that other guests were looking at the device and picking it up; and that Bennett never asked her to maintain in secrecy any information about the device.

[Beachcombers] asks us to discount this evidence on the basis it is insufficiently corroborated by documentary evidence, that Bennett was not a disinterested witness, that it is inconsistent with Bennett's testimony that she did not begin actively commercializing the device until 1986, that it is inconsistent with her deposition testimony, and that Bennett has no independent recollection of the date of the party.

This we may not do. Bennett's oral testimony was adequately corroborated by physical evidence, *i.e.*, the entry in her date book and the ODYLIC itself, and, if accepted at face value, meets the clear and convincing standard required. The jury chose to believe the evidence in spite of these purported deficiencies, which it was entitled to do as these factors all relate to Bennett's and Regan's credibility. We have no warrant to override the jury's decision on the basis of these factors. That would usurp the prescribed function of the jury.

[Beachcombers] also questions the substantiality of the evidence in light of *Moleculon Research Corp. v. CBS, Inc.,* 793 F.2d 1261 (Fed. Cir. 1986), a case in which the court found that a display of a device to friends and colleagues of the inventor was subject to an implied restriction of confidentiality, and thus did not constitute a "public use." But in *Moleculon* the court found that the inventor at all times retained control over the use of the device as well as over the distribution of information concerning it. *Id.* at 1266. Here, there was evidence upon which the jury could have reasonably concluded that Bennett did not retain control over the

---

* [*Eds. Note*: The '046 patent's filing date was May 27, 1986, making May 27, 1985, its critical date under § 102(b). Anything "in public use" before that date was prior art, and thus was potentially invalidating.]

use of the device and the future dissemination of information about it — Bennett's testimony that her purpose in demonstrating the device at her party was to generate discussion and garner feedback, that she never imposed any secrecy or confidentiality obligations, and Regan's testimony to the effect that she did not believe she was subject to any secrecy or confidentiality restrictions — notwithstanding the closeness and ongoing nature of Bennett's relationship with her guests. We have no warrant or reason to disturb that finding.

Thus, there is sufficient evidence to sustain the conclusion that the ODYLIC was in public use prior to May 27, 1985. . . .

Notes & Questions

1. Is it the number of people who interact with the claimed invention, or the fact that none of them were obliged to keep the invention a secret, that makes the difference in determining whether the use was "public"?

2. The focus of § 102(a)'s "used by others" provision, like § 102(b)'s "in public use" provision, is any use that generates *publicly* available information. As the Federal Circuit explained the principle many years ago:

> [N]o patent should be granted which withdraws from the public domain technology already available to the public. It is available, in legal theory at least, when it is described in the world's accessible literature, including patents, or has been publicly known or in the public use or on sale 'in this country.' 35 U.S.C. § 102(a) and (b). That is the real meaning of 'prior art' in legal theory—it is knowledge that is available . . . at a given time, to a person of ordinary skill in an art. Society, speaking through Congress and the courts, has said 'thou shalt not take it away.'

Kimberly-Clark Corp. v. Johnson & Johnson, 745 F.2d 1437, 1453-54 (Fed. Cir. 1984)

3. The Problem of Secretly Exploited Methods & Forfeited Rights

Beachcombers tells us how to analyze a *third party*'s activities that may have put a technology in public use for purposes of §§ 102(a) and (b). What if, however, the activity was the would-be patentee's activity? And what if that activity commercially exploited, for example, a method to make items for sale? Or a method to service items on a paid basis? The would-be patentee could use the method in secret, in a way that would not inform the public of the method's details, while at the same time reaping commercial rewards from the general public. Imagine, for example, that Athos has invented a new method for sharpening kitchen knives. He keeps the method a secret, opens a knife-sharpening service to the public, and accepts payment to sharpen knives (in secret, using his method). After the shop has been open for more than a year, can Athos apply for a patent on his method? Or has he forfeited the right, under § 102(b)? What if Porthos independently develops the same method—can Porthos obtain a patent on it, or is Athos' year-plus sales activity prior art that prevents Porthos from patenting, again per § 102(b)?

Consider Porthos first. And recall what the *text* of § 102(b) says: one can obtain a patent unless "the invention was . . . in public use or on sale in this country, more than one year prior to the date of the application for patent . . . ." Was the Athos method in public use or on sale? It does not seems so, as to Porthos. The public didn't learn anything about the Athos method from Athos' secret sharpening activity. Notice what we focus on—not *who* was engaged in the secret activity, but *what* the public learned from that activity. That focus closely tracks the statutory text, which tells us to ask about the publicity of the *invention* (what) not the *inventor* (who). And, in promptly applying for a patent, Porthos has done something we value: Rather than exploit the method in secret, Porthos has taught the invention to the public, who will be able to freely use the method after the patent expires.

Now consider Athos. The text of § 102(b) is the same. The secret activity's failure to teach the invention to the public is also the same. Is the legal outcome the same? Or does § 102(b) bar Athos from patenting a method that he himself has been commercially exploiting—albeit in a way that does not teach the public the invention—for more than a year before he files his patent application? Ever since the Supreme Court's decision in *Pennock v. Dialogue*, 27 U.S. 1 (1829), which invalidated a method patent on the ground that the inventor had commercially exploited his method in secret for seven years,[6] the federal courts have held that a patentee's commercial activity outside the one-year period in § 102(b) forfeits the patent claim on the method secretly used for that commercial activity. *See W.L. Gore & Assocs., Inc. v. Garlock, Inc.*, 721 F.2d 1540, 1550 (Fed. Cir. 1983) (stating the principle, in *dicta*); *D.L. Auld Co. v. Chroma Graphics Corp.*, 714 F.2d 1144, 1147-48 (Fed. Cir. 1983) (invalidating a claim). Put differently, the activity bars a patent to the exploiter, who knows the secret, but does not bar third parties, who do not learn anything from the exploiter's activity. The modern classic among these forfeiture cases is Judge Learned Hand's decision in *Metallizing Engineering Co. v. Kenyon Bearing & Auto Parts Co.*, 153 F.2d 516 (2d Cir. 1946). It also has its critics. *See, e.g.*, Dmitry Karshtedt, *Did Learned Hand Get It Wrong? The Questionable Patent Forfeiture Rule of* Metallizing Engineering, 57 Vill. L. Rev. 261 (2012).

4. When, Exactly, Is an Invention "On Sale"?

Section 102(b) bars one from obtaining a patent on an invention that has been "on sale" in the U.S. "more than one year prior to the date of the [patent] application." This statutory bar, along with the "in public use" bar also contained in § 102(b) and introduced above, encourages earlier patent filing by invalidating delayed claims. The bars also stop inventors from, in effect, improperly extending their patent terms backward in time (during an earlier start-up period). When, exactly, will an invention be determined to be "on sale," and thus begin the one-year grace-period clock? Note that § 102(a) does not identify "on sale" as a category of invalidating prior art. However, sales activity may well prove that an invention was "known or used by others" for purposes of § 102(a).

---

[6] He had invented a method for joining lengths of water hose that could withstand higher pressures. He sold thousands of feet of the hose over a number of years, though he had kept the production method a secret. The Supreme Court struck down the patent claim.

Pfaff v. Wells Electronics, Inc.
525 U.S. 55 (1998)

STEVENS, JUSTICE:

Section 102(b) of the Patent Act of 1952 provides that no person is entitled to patent an "invention" that has been "on sale" more than one year before filing a patent application. We granted certiorari to determine whether the commercial marketing of a newly invented product may mark the beginning of the 1-year period even though the invention has not yet been reduced to practice.[2]

I

On April 19, 1982, petitioner, Wayne Pfaff, filed an application for a patent on a computer chip socket. Therefore, April 19, 1981, constitutes the critical date for purposes of the on-sale bar of 35 U. S. C. § 102(b); if the 1-year period began to run before that date, Pfaff lost his right to patent his invention.

Pfaff commenced work on the socket in November 1980, when representatives of Texas Instruments asked him to develop a new device for mounting and removing semiconductor chip carriers. In response to this request, he prepared detailed engineering drawings that described the design, the dimensions, and the materials to be used in making the socket. Pfaff sent those drawings to a manufacturer in February or March 1981.

Prior to March 17, 1981, Pfaff showed a sketch of his concept to representatives of Texas Instruments. On April 8, 1981, they provided Pfaff with a written confirmation of a previously placed oral purchase order for 30,100 of his new sockets for a total price of $91,155. In accord with his normal practice, Pfaff did not make and test a prototype of the new device before offering to sell it in commercial quantities.[3]

The manufacturer took several months to develop the customized tooling necessary to produce the device, and Pfaff did not fill the order until July 1981. The evidence therefore indicates that Pfaff first reduced his invention to practice in the

---

[2] "A process is reduced to practice when it is successfully performed. A machine is reduced to practice when it is assembled, adjusted and used. A manufacture is reduced to practice when it is completely manufactured. A composition of matter is reduced to practice when it is completely composed." *Corona Cord Tire Co. v. Dovan Chemical Corp.,* 276 U. S. 358, 383 (1928).

[3] At his deposition, respondent's counsel engaged in the following colloquy with Pfaff:

Q. Now, at this time [late 1980 or early 1981] did we [sic] have any prototypes developed or anything of that nature, working embodiment?
A. No.
Q. It was in a drawing. Is that correct?
A. Strictly in a drawing. Went from the drawing to the hard tooling. That's the way I do my business.
Q. 'Boom-boom'?
A. You got it.
Q. You are satisfied, obviously, when you come up with some drawings that it is going to go-'it works'?
A. I know what I'm doing, yes, most of the time.

summer of 1981. The socket achieved substantial commercial success before Patent No. 4,491,377 ('377 patent) issued to Pfaff on January 1, 1985.

After the patent issued, petitioner brought an infringement action against respondent, Wells Electronics, Inc., the manufacturer of a competing socket. . . .

. . . [T]he District Court rejected respondent's § 102(b) defense because Pfaff had filed the application for the '377 patent less than a year after reducing the invention to practice.

The Court of Appeals reversed . . . . 124 F.3d 1429 (Fed. Cir. 1997). Four of the claims (1, 6, 7, and 10) described the socket that Pfaff had sold to Texas Instruments prior to April 8, 1981. Because that device had been offered for sale on a commercial basis more than one year before the patent application was filed on April 19, 1982, the court concluded that those claims were invalid under § 102(b). That conclusion rested on the court's view that as long as the invention was "substantially complete at the time of sale," the 1-year period began to run, even though the invention had not yet been reduced to practice. *Id.* at 1434. . . .

. . . .

II

The primary meaning of the word "invention" in the Patent Act unquestionably refers to the inventor's conception rather than to a physical embodiment of that idea. The statute does not contain any express requirement that an invention must be reduced to practice before it can be patented. . . .

It is well settled that an invention may be patented before it is reduced to practice. In 1888, this Court upheld a patent issued to Alexander Graham Bell even though he had filed his application before constructing a working telephone. Chief Justice Waite's reasoning in that case merits quoting at length:

> It is quite true that when Bell applied for his patent he had never actually transmitted telegraphically spoken words so that they could be distinctly heard and understood at the receiving end of his line, but in his specification he did describe accurately and with admirable clearness his process, that is to say, the exact electrical condition that must be created to accomplish his purpose, and he also described, with sufficient precision to enable one of ordinary skill in such matters to make it, a form of apparatus which, if used in the way pointed out, would produce the required effect, receive the words, and carry them to and deliver them at the appointed place. * * * A good mechanic of proper skill in matters of the kind can take the patent and, by following the specification strictly, can, without more, construct an apparatus which, when used in the way pointed out, will do all that it is claimed the method or process will do * * * .

The law does not require that a discoverer or inventor, in order to get a patent for a process, must have succeeded in bringing his art to the highest degree of perfection. It is enough if he describes his method with sufficient clearness and precision to enable those skilled in the matter to understand what the process is, and if he points out some practicable way of putting it into operation.

The Telephone Cases, 126 U. S. 1, 535-536 (1888).

When we apply the reasoning of *The Telephone Cases* . . . it is evident that Pfaff could have obtained a patent on his novel socket when he accepted the purchase order from Texas Instruments for 30,100 units. At that time he provided the manufacturer with a description and drawings that had "sufficient clearness and precision to enable those skilled in the matter" to produce the device. *Id.* at 536. The parties agree that the sockets manufactured to fill that order embody Pfaff's conception as set forth in claims 1, 6, 7, and 10 of the '377 patent. We can find no basis . . . for concluding that Pfaff's invention was not "on sale" within the meaning of the statute until after it had been reduced to practice.

III

Pfaff nevertheless argues that longstanding precedent, buttressed by the strong interest in providing inventors with a clear standard identifying the onset of the 1-year period, justifies a special interpretation of the word "invention" as used in § 102(b). We are persuaded that this nontextual argument should be rejected.

As we have often explained, most recently in *Bonito Boats, Inc. v. Thunder Craft Boats, Inc.,* 489 U.S. 141, 151 (1989), the patent system represents a carefully crafted bargain that encourages both the creation and the public disclosure of new and useful advances in technology, in return for an exclusive monopoly for a limited period of time. The balance between the interest in motivating innovation and enlightenment by rewarding invention with patent protection on the one hand, and the interest in avoiding monopolies that unnecessarily stifle competition on the other, has been a feature of the federal patent laws since their inception. . . .

Consistent with these ends, § 102 of the Patent Act serves as a limiting provision, both excluding ideas that are in the public domain from patent protection and confining the duration of the monopoly to the statutory term.

We originally held that an inventor loses his right to a patent if he puts his invention into public use before filing a patent application. "His voluntary act or acquiescence in the public sale and use is an abandonment of his right." *Pennock v. Dialogue,* 2 Pet. 1, 24 (1829) (Story, J.). A similar reluctance to allow an inventor to remove existing knowledge from public use undergirds the on-sale bar.

Nevertheless, an inventor who seeks to perfect his discovery may conduct extensive testing without losing his right to obtain a patent for his invention — even if such testing occurs in the public eye. The law has long recognized the distinction between inventions put to experimental use and products sold commercially. In 1878, we explained why patentability may turn on an inventor's use of his product.

It is sometimes said that an inventor acquires an undue advantage over the public by delaying to take out a patent, inasmuch as he thereby preserves the monopoly to himself for a longer period than is allowed by the policy of the law; but this cannot be said with justice when the delay is occasioned by a *bona fide* effort to bring his invention to perfection, or to ascertain whether it will answer the purpose intended. * * * [I]t is the interest of the public, as well as himself, that the invention should be perfect and properly tested, before a patent is granted for it. *Any attempt to use it for a profit, and not by way of experiment, for a longer period than two years before the application, would deprive the inventor of his right to a patent.*

Elizabeth v. Pavement Co., 97 U. S. 126, 137 (1878) (emphasis added).

The patent laws therefore seek both to protect the public's right to retain knowledge already in the public domain and the inventor's right to control whether and when he may patent his invention. The Patent Act of 1836, 5 Stat. 117, was the first statute that expressly included an on-sale bar to the issuance of a patent. . . . [T]hat provision precluded patentability if the invention had been placed on sale at any time before the patent application was filed. In 1839, Congress ameliorated that requirement by enacting a 2-year grace period in which the inventor could file an application. 5 Stat. 353.

In *Andrews v. Hovey,* 123 U. S. 267, 274 (1887), we noted that the purpose of that amendment was "to fix a period of limitation which should be certain"; it required the inventor to make sure that a patent application was filed "within two years from the completion of his invention," *id.* In 1939, Congress reduced the grace period from two years to one year. 53 Stat. 1212.

Petitioner correctly argues that these provisions identify an interest in providing inventors with a definite standard for determining when a patent application must be filed. A rule that makes the timeliness of an application depend on the date when an invention is "substantially complete" seriously undermines the interest in certainty. Moreover, such a rule finds no support in the text of the statute. Thus, petitioner's argument calls into question the standard applied by the Court of Appeals, but it does not persuade us that it is necessary to engraft a reduction to practice element into the meaning of the term "invention" as used in § 102(b).

The word "invention" must refer to a concept that is complete, rather than merely one that is "substantially complete." It is true that reduction to practice ordinarily provides the best evidence that an invention is complete. But just because reduction to practice is sufficient evidence of completion, it does not follow that proof of reduction to practice is necessary in every case. Indeed, both the facts of *The Telephone Cases* and the facts of this case demonstrate that one can prove that an invention is complete and ready for patenting before it has actually been reduced to practice.

We conclude, therefore, that the on-sale bar applies when two conditions are satisfied before the critical date.

First, the product must be the subject of a commercial offer for sale. An inventor can both understand and control the timing of the first commercial marketing of his invention. The experimental use doctrine, for example, has not generated concerns about indefiniteness, and we perceive no reason why unmanageable uncertainty should attend a rule that measures the application of the on-sale bar of

§ 102(b) against the date when an invention that is ready for patenting is first marketed commercially. In this case the acceptance of the purchase order prior to April 8, 1981, makes it clear that such an offer had been made, and there is no question that the sale was commercial rather than experimental in character.

Second, the invention must be ready for patenting. That condition may be satisfied in at least two ways: by proof of reduction to practice before the critical date; or by proof that prior to the critical date the inventor had prepared drawings or other descriptions of the invention that were sufficiently specific to enable a person skilled in the art to practice the invention. In this case the second condition of the on-sale bar is satisfied because the drawings Pfaff sent to the manufacturer before the critical date fully disclosed the invention.

The evidence in this case thus fulfills the two essential conditions of the on-sale bar. As succinctly stated by Learned Hand:

> [I]t is a condition upon an inventor's right to a patent that he shall not exploit his discovery competitively after it is ready for patenting; he must content himself with either secrecy, or legal monopoly.

Metallizing Engineering Co. v. Kenyon Bearing & Auto Parts Co., 153 F.2d 516, 520 (2d Cir. 1946).

. . . When Pfaff accepted the purchase order for his new sockets prior to April 8, 1981, his invention was ready for patenting. The fact that the manufacturer was able to produce the socket using his detailed drawings and specifications demonstrates this fact. Furthermore, those sockets contained all the elements of the invention claimed in the '377 patent. Therefore, Pfaff's '377 patent is invalid because the invention had been on sale for more than one year in this country before he filed his patent application. Accordingly, the judgment of the Court of Appeals is affirmed.

Notes & Questions

1. Be careful to distinguish the sale of the invention from the sale of the intangible asset of the patent itself. Only the former creates a potential statutory bar. Granting a license to a patent, or pending patent application, in return for a lump sum payment or a royalty does not amount to a sale for purposes of § 102(b). *See In re Kollar*, 286 F.3d 1326, 1333 (Fed. Cir. 2002) (finding on-sale bar not triggered by an agreement that "did not involve the sale of a product of the claimed process, but rather provided . . . a license to practice the claimed process and 'information defining an embodiment' of that process.").

2. The incentive mechanism for early filing that § 102(b) embodies can work well only if would-be patentees can determine whether they have, in fact, triggered the one-year countdown with a particular use or sale. We should fashion our statutory bar tests with this need for predictability in view. How well does the Court's *Pfaff* opinion accomplish this goal?

5. The Exception for "Experimental Use"

The policy preference for earlier filing expressed in the statutory bar provisions is in some tension with another policy preference—namely, our desire to give an inventor adequate time, and a proper context, for testing whether her invention truly works for its intended purpose. As you saw in the *Pfaff* opinion above, this is called "making an actual reduction to practice," or "actually reducing the invention to practice." Since the Supreme Court's seminal decision in *City of Elizabeth v. American Nicholson Pavement Co.*, 97 U.S. 126 (1877), an inventor can negate an otherwise invalidating public use or sale of the invention outside the grace period by showing that "the delay is occasioned by a *bona fide* effort to bring his invention to perfection." *Id.* at 137. The variability of appropriate testing methods and circumstances tends to make the overall inquiry less, not more, predictable.

Clock Spring, L.P. v. Wrapmaster, Inc.
560 F.3d 1317 (Fed. Cir. 2009)

DYK, JUDGE:

Clock Spring, L.P. brought suit alleging that Wrapmaster, Inc. infringed the claims of U.S. Patent No. 5,632,307 . . . . The '307 patent claims methods for repairing damaged high-pressure gas pipes. On summary judgment the [district court] held that the claims of the '307 patent were invalid due to obviousness . . . . We affirm the summary judgment of invalidity because we conclude that the claims of the '307 patent are invalid as a matter of law, due to prior public use. We do not reach the issue of invalidity due to obviousness. . . .

Background

Both Clock Spring and Wrapmaster are high-pressure gas pipeline repair companies. Clock Spring is the exclusive licensee of the '307 patent. The '307 patent has five independent claims and thirty-eight dependent claims. All are method claims. Claim 1 of the '307 patent reads as follows:

> A method for repairing a pipe adapted to carry an internal load directed radially outward therefrom, *said pipe having a defective region* defined by at least one cavity extending from an outer surface of said pipe toward the center of said pipe but not extending completely through the wall of said pipe, said method comprising the steps of:
> providing a filler material *having a workable uncured state* and a rigid cured state,
> *filling said cavity to at least said outer surface of said pipe with said filler material in said workable state*,
> providing at least one band having a plurality of elastic convolutions of high tensile strength material,
> *while said filler material is in said workable state*, wrapping said plurality of convolutions of said high tensile strength material about said pipe to form a coil overlying stud filler material[,]
> tightening said coil about said pipe so that said filler material completely fills that portion of said cavity underlying said coil[,] securing at least one

of said convolutions to an adjacent one of said convolutions, and
  permitting said filler material to cure to said rigid state, whereby a load carried by said pipe is transferred substantially instantaneously from said pipe
to said coil.

(Emphases added). The parties appeared to agree, or at least not contest, that the
main distinctive feature over the prior art is wrapping the pipe while the filler is in
an uncured state so as to ensure smooth and continuous contact between the wrap
and the pipe. The other independent claims (claims 38, 39, 42, and 43) also require
wrapping in an uncured state, but address different types of defects and repair
methods. The various dependent claims add further limitations for the properties
of the materials used in the individual steps of the method . . . .

In 2005 Clock Spring filed an infringement suit against Wrapmaster alleging
infringement of all the claims of the '307 patent. . . .

After discovery, Wrapmaster filed a summary judgment motion of invalidity of
all the claims of the '307 patent . . . .

The invalidity summary judgment motion argued that the claims were invalid
due to a prior public use under 35 U.S.C. § 102(b) in October 1989, in Cuero,
Texas, more than one year before the patent application was filed in 1992. The
motion was supported by a 1994 Gas Research Institute report (hereinafter "1994
GRI report") regarding the demonstration made by named inventor Norman C.
Fawley.[1] GRI, since renamed the Gas Technology Institute, is a non-profit research and development organization which was entitled to receive royalty payments from Clock Spring on the '307 patent. The motion also urged that the
claims were invalid on grounds of obviousness based on a number of prior art patents.

Clock Spring opposed the motion. Clock Spring did not dispute that the 1989
demonstration was public, or that it involved the limitations of the patent with one
exception. Clock Spring apparently urged that the 1989 demonstration had not
involved the application of the wrap with an uncured filler, and that the use had
been experimental. Clock Spring also urged that the patent claims were not obvious.

The district court referred the motion to a magistrate judge for recommendations. The magistrate judge recommended that the district court grant summary
judgment of invalidity with respect to the claims of the '307 patent.

The magistrate judge first addressed Wrapmaster's contention that the '307
patent is invalid due to prior public use. The magistrate judge concluded that the
1994 GRI report proved that there was no genuine issue of material fact regarding
whether the filler compound was uncured when the wrap was applied to the pipe.
The magistrate judge also rejected Clock Spring's argument that the use was experimental. Based on this, the magistrate judge recommended finding that the
1989 demonstration triggered the public use bar under 35 U.S.C. § 102(b).

---

[1] Fawley also prepared this 1994 GRI report.

The magistrate judge then addressed Wrapmaster's contention that the claims of the '307 patent are invalid due to obviousness. . . .

On review in the district court Clock Spring objected to the magistrate judge's recommendations . . . . In support of its argument on experimental use to the district court, Clock Spring submitted new evidence including additional GRI reports (some of which mentioned the 1989 demonstration) and a 28-page report by NCF Industries, Inc.[3] concerning the 1989 demonstration. Though characterizing the late submission of these documents as "clearly improper," the district court considered them and concluded that Clock Spring had "raise[d] a fact question about whether the 1989 installation was experimental," relying on the NCF report, a 1993 GRI report, and a 1998 GRI report. The district court did not explain why these reports raised a genuine issue of material fact. The district court thus rejected the magistrate's recommendation concerning the public use bar. However, the district judge agreed with the magistrate judge as to obviousness and granted summary judgment of invalidity due to obviousness.

    . . . .

Discussion

I

Although the district court granted summary judgment of invalidity on obviousness, Wrapmaster contends that the invalidity decision can be sustained on the separate ground of prior public use. Relying on the 1994 GRI report and the NCF report, Wrapmaster contends that the 1989 demonstration was a public use of the method of claim 1 because the method was demonstrated to the public almost three years before the priority date of the '307 patent application, September 9, 1992. We agree.

We may affirm a grant of summary judgment on a ground supported in the record but not adopted by the district court if we conclude that "there [wa]s no genuine issue as to any material fact and * * * the movant [wa]s entitled to a judgment as a matter of law." Fed. R. Civ. P. 56(c).

    . . . .

For a challenger to prove a patent claim invalid under § 102(b), the record must show by clear and convincing evidence that the claimed invention was in public use before the patent's critical date. See Adenta GmbH v. OrthoArm, Inc., 501 F.3d 1364, 1371 (Fed. Cir. 2007). The critical date is "one year prior to the date of the application for patent in the United States," here September 9, 1991. See 35 U.S.C. § 102(b). "[A] public use includes any public use of the claimed invention by a person other than the inventor who is under no limitation, restriction or obligation of secrecy to the inventor." Adenta, 501 F.3d at 1371 (quotation and alteration marks omitted). In order for a use to be public within the meaning of § 102(b), there must be a public use with all of the claim limitations.

---

[3] Fawley, one of the named inventors in the '307 patent, was the president of NCF Industries, Inc.

There is no dispute that the 1989 demonstration was public. In fact, representatives of several other domestic gas transmission companies were present at the demonstration, and there was no suggestion that they were under an obligation of confidentiality. This demonstration was accessible to the public. *See Am. Seating Co. v. USSC Group, Inc.*, 514 F.3d 1262, 1267 (Fed. Cir. 2008) ("An invention is in public use if it is shown to or used by an individual other than the inventor under no limitation, restriction, or obligation of confidentiality.").

. . . .

. . . Clock Spring claims that the 1989 demonstration was an experimental use and not a prior public use.

The experimental use exception is not a doctrine separate or apart from the public use bar. *EZ Dock, Inc. v. Schafer Sys. Inc.*, 276 F.3d 1347, 1351-52 (Fed. Cir. 2002). Rather, something that would otherwise be a public use may not be invalidating if it qualifies as an experimental use. *Electromotive Div. of Gen. Motors Corp. v. Transp. Sys. Div. of Gen. Elec. Co.*, 417 F.3d 1203, 1211 (Fed. Cir. 2005) (limiting "experimentation sufficient to negate a pre-critical date public use or commercial sale to cases where the testing was performed to perfect claimed features, or * * * to perfect features inherent to the claimed invention"). In *Allen Engineering Corp. v. Bartell Industries, Inc.*, we catalogued a set of factors that in previous cases had been found instructive, and in some cases dispositive, for determining commercial versus experimental uses. 299 F.3d 1336, 1353 (Fed. Cir. 2002). These factors include:

> (1) the necessity for public testing, (2) the amount of control over the experiment retained by the inventor, (3) the nature of the invention, (4) the length of the test period, (5) whether payment was made, (6) whether there was a secrecy obligation, (7) whether records of the experiment were kept, (8) who conducted the experiment, (9) the degree of commercial exploitation during testing, (10) whether the invention reasonably requires evaluation under actual conditions of use, (11) whether testing was systematically performed, (12) whether the inventor continually monitored the invention during testing, and (13) the nature of contacts made with potential customers.

Id. (quotation and alteration marks omitted). Though a prior commercial sale and not a prior public use was at issue in *Allen Engineering*, the factors explicated are equally relevant to an analysis of experimental use.

We have said that lack of control over the invention during the alleged experiment, while not always dispositive, may be so. *Atlanta Attachment Co. v. Leggett & Platt, Inc.*, 516 F.3d 1361, 1366 (Fed. Cir. 2008). In that case, we held that a public use had occurred, finding "dispositive" the fact that the patentee "did not have control over the alleged testing," which was performed by its customer. *Id.* Clock Spring argues that Fawley, a named inventor, exercised tight control over the demonstration, as shown through the detailed reports made of the demonstration. But, the detailed reports do not provide evidence that Fawley controlled the demonstration. An independent observer "analyzed and recorded" the 1989 demonstration. Three of the eleven Clock Spring installations were done by the pipeline's personnel. None of these individuals was under Fawley's control or surveillance. We need not, however, rely on lack of control as establishing public use because we conclude that the use cannot qualify as experimental for other reasons.

A use may be experimental only if it is designed to (1) test claimed features of the invention or (2) to determine whether an invention will work for its intended purpose—itself a requirement of patentability. *See In re Omeprazole patent Litig.*, 536 F.3d 1361, 1373-75 (Fed. Cir. 2008). In other words, an invention may not be ready for patenting if claimed features or overall workability are being tested. But, there is no experimental use unless claimed features or overall workability are being tested for purposes of the filing of a patent application.[8] *See EZ Dock*, 276 F.3d at 1352, 1354; *Weatherchem Corp. v. J.L. Clark, Inc.*, 163 F.3d 1326, 1333 (Fed. Cir. 1998) (stating that the public use provision strives to provide "inventors with a definite standard for determining when a patent application must be filed") (quotation marks omitted). Indeed, the experimental use negation of the § 102(b) bar only exists to allow an inventor to perfect his discovery through testing without losing his right to obtain a patent for his invention. *See EZ Dock*, 276 F.3d at 1352.

Clock Spring does not urge that refining the claim limitations was the subject of the 1989 demonstration. Rather, Clock Spring argues that the demonstration was experimental because the 1989 demonstration was designed to determine durability of the method, *i.e.*, its suitability for the intended purpose. *See City of Elizabeth v. Am. Nicholson Pavement Co.*, 97 U.S. 126, 136 (1877). The reports make no such explicit statement. The NCF report states that "[t]he purpose of this demonstration * * * was to demonstrate to Panhandle Eastern attendants and guests the steps of application and the ability of minimally-trained crews to make Clock Spring installations." The 1994 GRI report states that "[t]his demonstration was designed to familiarize pipeline personnel with the Clock Spring technology, and to begin training of maintenance personnel in the use of the coil pass installation *method*." (Emphasis added). The demonstration was similarly described to the [PTO] during prosecution, where the applicant stated that the purpose of the demonstration was to seek "input from people in the industry on the performance of the bands and the practicality of their installation techniques."

To be sure, the 1994 GRI report can be read as suggesting that the 1989 demonstration was for durability testing because it states that "recovery and analysis of installed composite after several years of exposure in pipeline settings was the only means of verifying the long-term performance of [the clock spring's] composites in moist soils." Clock Spring's problem, however, is that no report in the record states, or in any way suggests, that the 1989 demonstration was designed to test

---

[8] To be sure, an applicant may in some circumstances elect to delay filing an application and continue testing until an actual reduction to practice has occurred; such testing may nonetheless be for the purpose of filing an application. Of course, it is clear from this court's case law that experimental use cannot negate a public use when it is shown that the invention was reduced to practice before the experimental use, even if an application has not yet been filed. *See Omeprazole patent Litig.*, 536 F.3d at 1372.

durability for the purposes of the patent application to the PTO. In fact, the reports make clear that the durability testing was for "acceptance by regulators and the pipeline industry," and that the 1989 installation was not dug up and examined until almost a year after the 1992 patent application. Thus, even if durability were being tested, it was not for purposes of the patent application, and cannot bring the experimental use exception into play. By filing the 1992 application, the inventors represented that the invention was then ready for patenting, and studies done thereafter cannot justify an earlier delay in filing the application under the rubric of experimental use.

Finally, Clock Spring asserts that because the Department of Transportation did not grant any installation waivers until 1993, the 1989 demonstration must have been experimental. This terse argument is unsupported by any citation to law. That the inventors were not legally allowed to perform the method on a pipeline in commercial operation, does not mean that a public use did not occur. The former fact has absolutely nothing to do with the latter question.

In summary, during the 1989 demonstration, all elements of the repair method in claim 1 of the '307 patent were performed. There was no evidence that the overall suitability of the '307 patent's method nor any of the claim elements was being tested as would be required for experimental use. Accordingly, claim 1 of the '307 patent is invalid due to prior public use.

Clock Spring has not contended that the remaining independent claims of the '307 patent (claims 38, 39, 42, and 43) could be valid if claim 1 was invalid, under § 102(b). However, Clock Spring does argue that the dependent claims are not invalid and should have each been addressed separately. This is the first time that Clock Spring has made this argument. Wrapmaster had filed a motion for summary judgment, contending that all of the claims of the '307 patent are invalid due to prior public use. In its opposition to Wrapmaster's motion for summary judgment, Clock Spring did not assert that the dependent claims needed to be separately addressed but, instead, essentially conceded that if claim 1 was invalid the other claims were also invalid. Clock Spring did not even address the dependent claims to the district court on review of the magistrate judge's recommendation. Clock Spring has waived its current argument that the invalidity of each of the dependent claims needs to be addressed separately.

In light of our finding of invalidity due to prior public use, we do not reach the obviousness question.

. . . .

Notes & Questions

1. As the court notes, the inventor's own activities can result in a public use if the invention is shown to an individual other than the inventor under no limitation, restriction, or obligation of confidentiality. Thus, non-disclosure agreements that you learned about in the trade secret chapter can play an important role in preserving an inventor's ability to obtain a patent, while allowing for some sharing of the technology more than one year prior to filing for a patent. While a written

NDA can be extremely helpful in demonstrating an obligation of confidentiality, the courts have been willing to imply a duty of confidentiality based on the circumstances.

2. Experimental use can also negate an otherwise barring sale. However, in the context of a sale, in addition to the inventor's control of the testing, "customer awareness ordinarily must be proven if experimentation is to be found." *General Motors Corp. v. General Elec. Co.*, 417 F.3d 1203, 1214-15 (Fed. Cir. 2005).

3. In *Pfaff*, the Supreme Court articulates the two requirements necessary for an invention to be "on-sale" for purposes of § 102(b). In addition to a "commercial offer of sale" the invention must be "ready for patenting." As the Federal Circuit emphasizes in *Clock Spring*, the "ready for patenting" requirement applies in a "public use" analysis as well. Consider the kaleidoscope in *Beachcombers v. Wildewood Creative Prods.*, 31 F.3d 1154 (Fed. Cir. 1994). What if the ODYLIC prototype that was demonstrated at the party was still being tested to see if it would actually work under the full range of ordinary conditions? Would that use have constituted "public use" or would it be appropriate to say the invention was not yet "ready for patenting?" Alternatively, in that situation, is it better to rely on the test for "experimental use" to determine whether the admittedly "public use" is sufficient to bar the patent? What if, after testing, no modifications are made, *i.e.*, the prototype was, in fact, identical to the patent that was eventually filed? Is the result under a "ready for patenting" standard different than an inquiry based on experimental use?

4. In *Pfaff*, the Supreme Court criticizes the Federal Circuit's use of a "substantially complete" standard to determine whether a claimed invention was on sale, concluding that such a standard "seriously undermines the interest in certainty." 525 U.S. at 65-66. At the same time, the Supreme Court expresses confidence that "[t]he experimental use doctrine," which can negate the invalidating force of a sale or use, has not resulted in unmanageable uncertainty. *Id.* at 67. After reading the Federal Circuit's analysis in *Clock Spring v. Wrapmaster*, do you share the Supreme Court's confidence?

5. Imagine you are in-house counsel at a firm with an active product testing practice. What protocols would you request to avoid accidentally crossing the statutory-bar boundaries that *Clock Spring v. Wrapmaster* discusses? Does the complexity of the statutory bar analysis, including experimental use negation of the bars, tilt the field unfairly against small businesses (who don't have as much money to pay lawyers to counsel them through that complexity)?

D. Novelty Under the New § 102 (AIA)

The America Invents Act of 2011 made major changes to some of the architecture of § 102, while at the same time retaining many of the basic building blocks of the novelty and statutory-bar standards. The new § 102 applies to patent applications filed on or after March 16, 2013. Applications filed before that date continue to be evaluated under the pre-AIA version, both in the PTO and in any subsequent litigation involving the patent.

The new version of § 102 eliminates the distinction between actual invention date and filing date that separated the 1952 Act version of § 102, in favor of a single operative date—namely, the filing date. The new version also eliminates the geographic disparity between documents from anywhere in the world and public activities in the U.S. But it retains the concepts related to the prior art categories we explored above: printed publication, public use, and on sale. The new version of § 102 also provides a 1-year grace period, although it implements that grace period in a different way from the pre-AIA statutory bar provision.

The core of the new § 102 is as follows: "A person shall be entitled to a patent unless . . . the claimed invention was patented, described in a printed publication, or in public use, on sale, or otherwise available to the public before the effective filing date of the claimed invention . . . ." 35 U.S.C. § 102(a)(1) (effective March 16, 2013). Note the new category of potential prior art: material that was "otherwise available to the public." The AIA does not provide a formal definition of public availability. The principal committee report on the law, from the House Judiciary Committee, explains that "the phrase 'available to the public' is added to clarify the broad scope of relevant prior art, as well as to emphasize the fact that it must be publicly accessible." H. Rep. 112-98 at 43. It will no doubt take some time for the courts to map the contours of this new category of prior art.

The revised version of the grace period is an exemption from the force of subsection (a)(1), just quoted. It provides as follows:

> A disclosure made 1 year or less before the effective filing date of a claimed invention shall not be prior art to the claimed invention under subsection (a)(1) if–
>   (A) the disclosure was made by the inventor or joint inventor or by another who obtained the subject matter disclosed directly or indirectly from the inventor or a joint inventor; or
>   (B) the subject matter disclosed had, before such disclosure, been publicly disclosed by the inventor or a joint inventor or another who obtained the subject matter disclosed directly or indirectly from the inventor or a joint inventor.

35 U.S.C. § 102(b)(1) (effective March 16, 2013). This grace period protects the inventor for disclosures she makes, or that originate from her, but it does *not* protect her from independent third-party disclosures that occur between the time she invents and the time she files her application (as did the combination of the old §§ 102(a) and (b)). The incentive to file promptly has thus become even larger. This new grace period also creates an incentive for prompt disclosure followed by prompt filing with the Patent Office, if one is not interested in foreign patent protection. This is so because, under (b)(1)(B), an inventor's public disclosure negates any subsequent public disclosure by another. So long as the inventor applies for a patent within one year of her public disclosure, she will be able to obtain her patent without concern for the intervening disclosures by others.

To get a flavor for how this new version of § 102 will play out, we can reconsider some of the hypotheticals we analyzed under the 1952 Act version of the statute, in Section V.A, with new dates that trigger the America Invents Act amendments to § 102. Imagine a patent application that claims "a widget comprising an A, a B, and a C," **filed on September 1, 2013**. The applicant can prove, with documents and a prototype, that she **finished inventing the widget on June 1, 2011**. Investigation yields the following items, each one of which fully and expressly describes, or embodies, a widget comprising an A, a B, and a C (arranged as they are in the claim) and how to make the widget. Each of these items, if made part of the prior art by the combined operation of §§ 102(a)(1) and (b)(1), proves that the invention is old and thus unpatentable. We must therefore determine whether each item qualifies as prior art under the new § 102.

1. A widget that was developed (start to finish) and put on sale to the public on July 15, 2013 by a third party. *Analysis*—This widget was on sale to the public before the applicant's September 1, 2013, filing date. The third-party sale before her filing date makes the widget prior art under § 102(a)(1). Finally, her earlier invention date is completely irrelevant. § 102(a)(1) bars the patent.

2. A journal article by a third party from Finland, in Finnish, published and indexed in many public libraries no later than February 1, 2013. *Analysis*—This article was published and indexed before the applicant's September 1, 2013, filing date. It is prior art under § 102(a)(1). Again, her earlier invention date is completely irrelevant. § 102(a)(1) bars the patent.

3. A presentation the applicant made at a trade show, with a paper handout describing the claimed invention, on November 1, 2012. *Analysis*—This presentation by the applicant was not published to the trade more than a year before the September 1, 2013, filing date. It is thus not a prior art reference, by reason of the exemption that § 102(b)(1)(A) provides an inventor for her own disclosures. § 102(a), therefor, does not bar the patent.

Recall the list of "best practices" that you made after reading the *Klopfenstein* decision concerning what constitutes a "printed publication" for purposes of the pre-AIA statutory bars. If one is interested in avoiding invalidity, one needs to be concerned with "disclosures" to the public, which presumably will include any item that previously qualified as a "printed publication." If one makes such a disclosure, a grace period of one year remains viable under the AIA, with the added benefit that any disclosures of the invention by others that take place *after* the inventor-applicant's first disclosure cannot be used as invalidating prior art. 35 U.S.C. § 102(b)(1)(B). For example, consider the fact pattern of hypothetical #1 above: Even though the third party's sale came after the applicant's actual invention date, and even though the sale was not more than a year before the applicant's filing date, it was an invalidating disclosure. (It would not have been invalidating under the old § 102.) Now combine hypo #1 with hypo #3: The applicant had made a public disclosure of the invention prior to the July 15, 2013 third-party sales, in the form of the November 1, 2012 trade show presentation. Now, the third-party sales are not invalidating. Note, however, that if the trade show had taken place before September 1, 2012, *i.e.* more than one year prior to the filing of the patent application, the trade-show disclosure would be invalidating prior art.

In the context of a survey course such as this, we can't do more than scratch the surface of the new regime the American Invents Act creates for patent applications filed in March 2013 and thereafter. Indeed, there is a great deal about the new § 102 that we have omitted, just as we omitted five of the seven subsections of the old § 102. And as we noted at the beginning of this Section, these two novelty regimes will exist side-by-side for many years to come. The PTO and the courts will work through the ambiguities and complexities of the America Invents Act, just as they worked through the ambiguities and complexities of the 1952 Patent Act. If you practice in the patent law field, you will play your part in fleshing out the full meaning of the new regime and ushering out the old.

VI. THE NONOBVIOUSNESS REQUIREMENT

There is nothing rare about invention. Each and every one of us makes inventions all the time. Little problems in our daily lives give rise to little inventions to solve them.[7]—Judge Giles Sutherland Rich (1978)

A. AN INTRODUCTION TO § 103

We face new practical problems, large and small, every day. How can we create millions of doses of flu vaccine safely and quickly? Provide cleaner, cheaper energy for homes and businesses? Increase a lightbulb's life? Decrease a photocopier's paper jams? Or guard our hands from too-hot disposable coffee cups? These practical problems drive us to find new solutions and to improve on existing ones. We rely on people's natural, irrepressible creativity and insight in the search for solutions. And, "add[ing] the fuel of interest to the fire of genius,"[8] we maintain a special reward system with our patent laws, to induce people to invest in creating solutions to especially tough practical problems.

How creative must an invention be, then, to merit patent protection? Every modern general patent system has restricted patents to inventions that are new. "This rule is easily justified, because it prevents already existing matter from falling under a new set of" reward-based rights to exclude others.[9] But novelty is not enough. To merit patent protection, an invention must also be *nonobvious* at the time it is made, *i.e.*, it must be an invention that the phosita would not readily make, given what ordinary artisans know and the typical problems in the art they readily solve at that time.[10] "This requirement asks whether an invention is a big

---

[7] Giles S. Rich, *Escaping the Tyranny of Words—Is Evolution in Legal Thinking Impossible?*, 14 Fed. Cir. B.J. 193, 197 (2004) (reprinted from *Nonobviousness—The Ultimate Condition of Patentability* 3:301 (John F. Witherspoon ed. 1980)).

[8] Abraham Lincoln, *Second Lecture on Discoveries and Inventions*, Delivered to the Phi Alpha Society of Illinois College at Jacksonville, Illinois (Feb. 11, 1859), 3 *Collected Works of Abraham Lincoln* 357 (R. Basler ed., 1953).

[9] John F. Duffy & Robert P. Merges, *The Story of* Graham v. John Deere Company: *Patent Law's Evolving Standard of Creativity, in* Intellectual Property Stories 108, 112 (Jane C. Ginsburg & Rochelle Cooper Dreyfuss eds., 2006).

[10] 35 U.S.C. § 103. In other patent systems, this requirement is called the *inventive step* requirement, rather than the nonobviousness requirement.

enough technical advance; even if an invention is new and useful, it will still not merit a patent if it represents merely a trivial step forward in the art."[11] Nonobviousness is, in this sense, "the final gatekeeper of the patent system."[12]

The nonobviousness requirement entered U.S. law with the Supreme Court's 1851 decision in *Hotchkiss v. Greenwood*, 52 U.S. 248, and Congress has since codified it in what is now 35 U.S.C. § 103. In *Hotchkiss*, the Court struck down a patent claim on a clay doorknob on the ground that the new doorknob configuration was too small an improvement to merit protection. The new configuration included a clay knob around a dovetail-based metal rod; the prior art included clay knobs with straight rods and metal or wood knobs with dovetail rods. The Court assumed, for purposes of argument, "that, by connecting the clay or porcelain knob with the metallic shank in this well-known [dovetail] mode, an article is produced better and cheaper than in the case of the metallic or wood knob." *Id*. at 266. Nevertheless, it held the new configuration to be unpatentable. According to the Court, an invention is not patentable unless its achievement is marked by "more ingenuity and skill . . . than were possessed by an ordinary mechanic acquainted with the business." *Id*. at 267. The Court thus contrasted "the work of the skillful mechanic" from "that of the inventor." *Id*.

Similarly, in *Reckendorfer v. Faber*, 92 U.S. 347 (1875), the Court struck down a patent claiming the combination of a pencil with an attached rubber eraser on the ground that "[a]n instrument or manufacture which is the result of mechanical skill merely is not patentable." *Id*. at 356. The Court continued as follows: "Mechanical skill is one thing; invention is a different thing. Perfection of workmanship, however much it may increase the convenience, extend the use, or diminish expense, is not patentable. The distinction between mechanical skill, with its conveniences and advantages and inventive genius, is recognized in all the cases." *Id*. at 356-57. And in *Atlantic Works v. Brady*, 107 U.S. 192 (1883), the Court struck down a patent claiming a particular dredge boat improvement, explaining that "[t]he process of development in manufactures creates a constant demand for new appliances, which the skill of ordinary head-workmen and engineers is generally adequate to devise, and which, indeed, are the natural and proper outgrowth of such development"; and concluding that "[t]o grant to a single party a monopoly of every slight advance made, except where the exercise of invention, somewhat above ordinary mechanical or engineering skill, is distinctly shown, is unjust in principle and injurious in its consequences." *Id*. at 199-200. These concepts are with us still.

Why demand that an applicant's creative efforts reach this higher standard before rewarding her with a patent? It is socially wasteful for us to pay a patent-backed premium for an innovation that we are almost certain to receive for free

---

[11] Robert P. Merges, *Commercial Success and Patent Standards: Economic Perspectives on Innovation*, 76 Cal. L. Rev. 803, 812 (1988).

[12] *Id*.

and in about the same amount of time.[13] When an ordinary artisan encounters a new problem, she will create a new ordinary invention—an obvious invention— as a matter of course. We do not need to provide a reward to draw into existence these obvious inventions that fall within the ordinary artisan's skill: The need to solve practical problems is sufficient to spark the development of obvious inventions, and their suitability for the needs they satisfy is itself a sufficient reward.

The common sense proposition that we should not pay a premium for that which naturally arises for free, and quickly, is strengthened when we consider the social costs of an imaginary alternative regime in which novelty alone is sufficient achievement to earn a patent reward.[14] It would complicate matters considerably, and for no good reason, *both* to grant patents on obvious inventions *and* to allow patentees to enforce their rights to exclude against the many others who will doubtless come up with the very same obvious inventions independently. (Of course, the more obvious the invention is, the more such other people there will be.) The reward of a patent would not be needed to induce the obvious invention's creation in the first place, yet users would be burdened with liability, licensing, or inventing around the patent. These added costs would be passed along to consumers, with no offsetting benefit (*i.e.*, an invention that would not otherwise have been made). To prevent conferring this hyper-reward on the one who happened to do first what many others would soon do in the ordinary course, we would need to narrow patent liability from a use-based standard with no defense for independent creation (as it exists now) to a copying-based standard with a full defense for independent creation (much as copyright law uses now). Such an approach to patent law has not been our tradition. Instead, patent law remains a tort for which no showing of knowledge or copying need be made.

Because nonobviousness divides the patentably new from the merely new, "the precise level of creativity needed to support a patent . . . is one of the most important policy issues in all of patent law."[15] From 1851 onward, the courts continued to elaborate on the nonobviousness standard in an ad hoc way, applying what was then called *the invention requirement*. These decades of case law are a rich field for analysis and reflection, and more than a little frustration. Indeed, Judge Learned Hand observed in 1950, with the benefit of forty years' service as a federal judge, that the question "whether there is a patentable invention . . . is as fugitive, impalpable, wayward, and vague a phantom as exits in the whole paraphernalia of legal concepts." *Harries v. Air King Prods.*, 183 F.2d 158, 162 (2d Cir. 1950). We cannot even summarize this vast body of law, much less analyze it in detail. It is sufficient for our purposes to note two aspects of the case law as it stood in 1952,

---

[13] Giles S. Rich, *Principles of Patentability*, 14 Fed. Cir. B.J. 135, 140 (2004) (explaining that inventions "produced . . . by the expected skill of ordinary workers in the arts . . . are never patentable," "[b]ecause they will be made anyway, without the 'fuel of interest' which the patent system supplies") (reprinted from 28 Geo. Wash. L. Rev. 393 (1960)).

[14] *See generally* Duffy & Merges, *supra* note 3, at 111-14.

[15] *Id.* at 110.

when Congress added an explicit nonobviousness requirement to the Patent Act for the first time.

First, as it built upon *Hotchkiss* against the backdrop of more vigorous, far-reaching use of the antitrust laws to combat monopolies, the Supreme Court appeared to put increasing stress on the magnitude of the difference between an invention marked by unpatentable mechanical skill and one marked by patentable ingenuity. As the Court stressed the difference between ordinary skill and ingenuity, the distance between them seemed to grow, with patentable achievement growing ever more elusive. Two cases, in particular, have come to symbolize the perceived anti-patent exesses of that time—namely, *Cuno Engineering Corp. v. Automatic Devices Corp.*, 314 U.S. 84 (1941), and *Great Atlantic & Pacific Tea Co. v. Supermarket Equipment Co.*, 340 U.S. 147 (1950). In both cases, the Court strove to find positive proof that the invention under review embodied a sufficiently high level of inventive accomplishment.

In *Cuno*, the Court struck down a patent on a self-timed, cordless cigarette lighter for a car dashboard. The new car lighter brought together two pre-existing technologies, (a) a thermostatically controlled timing mechanism with an autorelease and (b) a cordless lighter. The Court was prepared to "concede that the functions performed by the [claimed] combination were new and useful." 314 U.S. at 90. More, of course, was required. "Since *Hotchkiss v. Greenwood*," the Court observed, "it has been recognized that if an improvement is to obtain the privileged position of a patent more ingenuity must be involved than the work of a mechanic skilled in the art." *Id.* So far, so good; this merely restates *Hotchkiss*. But the Court continued: "That is to say, the new device, however useful it may be, must reveal *the flash of creative genius*, not merely the skill of the calling." *Id.* at 91 (emphasis added). It also cautioned that "[s]trict application of that test is necessary lest" patents be granted "on each slight technological advance in the art." *Id* at 92. The Court's operative phrase, "the flash of creative genius," created a furor in the established patent bar. As the renowned patent expert Judge Giles Sutherland Rich once observed, the *Cuno* opinion "drove patent lawyers up the wall."[16] It seemed to many to raise the nonobviousness hurdle far higher than it previously had been set.

Nine years later, in *Great Atlantic*, the Supreme Court struck down a patent on an improved grocery store checkout counter. According to the Court, where "invention, if it exists at all, is only in bringing old elements together," it is necessary to subject the claim to a "rather severe test." 340 U.S. at 150, 152. Indeed, the Court warned, "[c]ourts should scrutinize combination patent claims with a care proportioned to the difficulty and improbability of finding invention in an assembly of elements." *Id. Great Atlantic* offers multiple verbal formulations to describe this scrutiny. The one that gained notoriety at the time, however, was to examine whether "the whole [of the claimed invention] in some way exceeds the sum of its parts." *Id.* This seeming demand for synergy, when combined with *Cuno*'s "flash

---

[16] Giles S. Rich, *Laying the Ghost of the "Invention" Requirement*, 14 Fed. Cir. B.J. 163, 167 (2004) (reprinted from 1 APLA Q.J. 26 (1972)).

of creative genius" standard, arguably threatened to make the nonobviousness hurdle all but insuperable.

The second notable aspect of the pre-1952 case law is the lower courts' increased reliance on nonobviousness tests that entailed not an imaginative construction of the ordinary artisan's insights at an earlier moment in time, but rather a focus on the inferences one could draw from the course of technological development in the pertinent field. Chief among these inferences is that an invention would not have been obvious where, once made, it satisfied a recognized, long felt need that other artisans had tried and failed to meet. The reasoning is straightforward: "The driving force behind innovation is the need for improvement of existing technology. A defect in a product or process spurs the businessman to deploy resources for discovering a solution. . . . Existence of the defect creates a demand for its correction, and it is reasonable to infer that the defect would not persist were the solution 'obvious.'"[17] Or, as Judge Learned Hand once pithily observed, "in judging what requires uncommon ingenuity, the best standard is what common ingenuity has failed for long to contrive under the same incentive."[18] This inference, and others like it, would later come to be grouped under the heading *secondary considerations.*

Congress, amid these cross-cutting currents of heightened Supreme Court scepticism and varied analyses in the lower courts, recalibrated the nonobviousness standard as part of its wholesale recodification of the patent laws in 1952. Responding to concerted reform efforts from the patent bar and the Patent Office, Congress for the first time included an explicit nonobviousness standard in the Patent Act:

> A patent may not be obtained though the invention is not identically disclosed or described as set forth in section 102 of this title, if the differences between the subject matter sought to be patented and the prior art are such that the subject matter as a whole would have been obvious at the time the invention was made to a person having ordinary skill in the art to which said

---

[17] Richard L. Robins, Note, *Subtests of "Nonobviousness": A Nontechnical Approach to Patent Validity,* 112 U. Pa. L. Rev. 1169, 1172 (1964).

[18] *W. States Mach. Co. v. S.S. Hepworth Co.,* 147 F.2d 345, 347 (2d Cir. 1945). *See also Lyon v. Bausch & Lomb Optical Co.,* 224 F.2d 530, 535 (2d Cir. 1955) (Hand, L.) ("The most competent workers in the field had at least ten years been seeking a hardy, tenacious coating to prevent reflection; there had been a number of attempts, none satisfactory; meanwhile nothing in the implementary arts had been lacking to put the advance into operation; when it appeared, it supplanted the existing practice and occupied substantially the whole field. We do not see how any combination of evidence could more completely demonstrate that, simple as it was, the change had not been 'obvious * * * to a person having ordinary skill in the art'—§ 103."); *Ruben Condenser Co. v. Aerovox Corp.,* 77 F.2d 266, 268 (2d Cir. 1935) (Hand, L.) ("If the machine or composition appears shortly after some obstacle to its creation, technical or economic, has been removed, we should scrutinize its success jealously; if at about the same time others begin the same experiments in the same or nearby fields, or if these come to fruition soon after the patentee's, the same is true. Such a race does not indicate invention. We should ask how old was the need; for how long could known materials and processes have filled it; how long others had unsuccessfully tried for an answer.").

subject matter pertains. Patentability shall not be negatived by the manner in which the invention was made.

35 U.S.C. § 103(a).[19]

Five things about the statute are noteworthy here. First, it adopts as its central perspective "a person having ordinary skill in the art," thus evoking the "ordinary mechanic" of *Hotchkiss*. Second, it states only a negative test, *i.e.*, it denies patentability to that which would have been obvious, without stating a positive standard of ingenuity or inventiveness. Third, it expressly forbids the use of hindsight by instructing that the obviousness of the invention is to be judged as of "the time the invention was made."[20] Fourth, it adds an extra guard against hindsight by restricting the body of prior art with which to assess patentability to "the art to which [the invention] pertains," acknowledging that many groundbreaking inventions have come when artisans made creative use of teachings from fields far removed from the pertinent art. (Ordinary artisans, one imagines, do not look to the teachings of far-flung fields.) Fifth, it underscores that the test is objective, not subjective, by ruling out of bounds any consideration of the specific "manner in which the invention was made." Combined with the framing of the test in the negative, this last point effectively eliminates *Cuno*'s "flash of creative genius" standard (as was intended). The lower courts quickly recognized that, with § 103, Congress had reset the nonobviousness standard back to the *Hotchkiss* level.[21]

The Supreme Court first confronted § 103 in a pair of cases involving obviousness attacks on three different patents—*Graham v. John Deere Co.*, 383 U.S. 1 (1966), and *United States v. Adams*, 383 U.S. 39 (1966). The Court heard argument in both cases on the same day (October 14, 1965), and rendered decisions in both cases on the same day (February 21, 1966). Both patents at issue in *Graham*—one for a third-generation plow shank clamp, and another for a spray bottle shipping cap—fell to a successful obviousness attack, while the patent at issue in *Adams*—for a water-activated field battery—survived. Far more important than the results, however, is the Court's reasoning in these landmark cases.

Graham established the fundamental framework U.S. courts still use today to analyze whether an invention would have been obvious to the ordinary artisan at the time of the invention. First, the Court "concluded that the 1952 Act was intended to codify judicial precedents embracing the principle long ago announced by this Court in *Hotchkiss v. Greenwood*." 383 U.S. at 3-4. It also emphasized, canvassing the history of the early U.S. patent system, "[t]he inherent problem was to develop some means of weeding out [for patent protection] those inventions which would not be disclosed or devised but for the inducement of a patent." *Id.*

---

[19] The America Invents Act amended § 103 to track changes made elsewhere in the statute.

[20] The courts have long recognized that hindsight bias—the tendency, in hindsight, to overestimate what could have been predicted with foresight—can distort the nonobviousness inquiry. *See, e.g.*, *Loom Co. v. Higgins*, 105 U.S. 580, 591 (1881) ("Now that it has succeeded, it may seem very plain to any one that he could have done it as well. This is often the case with inventions of the greatest merit.").

[21] *See, e.g.*, *Lyon v. Bausch & Lomb Optical Co.*, 224 F.2d 530, 534-37 & n.6 (2d Cir. 1955).

at 11. Second, it dismissed *Cuno*'s "flash of creative genius" notion as nothing "but a rhetorical embellishment" and a "rhetorical restate[ment]." *Id.* at 15 n.7. Third, and most important, the Court concluded that "the §103 condition . . . lends itself to several basic factual inquiries." *Id.* at 17. The inquiries the Court described are as follows:

> Under § 103, the scope and content of the prior art are to be determined; differences between the prior art and the claims at issue are to be ascertained; and the level of ordinary skill in the pertinent art resolved. Against this background, the obviousness or nonobviousness of the subject matter is determined. Such secondary considerations as commercial success, long felt but unsolved needs, failure of others, etc., might be utilized to give light to the circumstances surrounding the origin of the subject matter sought to be patented.

Id. at 17-18.

The *Graham* Court made no mention whatever of requiring that the prior art provide the ordinary artisan a suggestion or motivation to combine or modify the prior art's teachings in a way that yields the invention under review. The Court did, however, highlight the helpful role that secondary-considerations evidence can play in preventing hindsight bias, thus adding yet a third layer of protection to the two that § 103 already contained: "Such inquiries . . . may also serve to guard against slipping into use of hindsight and to resist the temptation to read into the prior art the teachings of the invention in issue." *Id.* at 36 (internal quotation and citation omitted). It is interesting, too, that the Court did *not* state in general terms *how* one was to go about determining the presence or absence of obviousness. Instead, it simply analyzed (over an additional 20 pages) the plow clamp and spray cap inventions, concluding that each would have been obvious to the ordinary artisan in light of the pertinent prior art.

Adams, unlike *Graham*, provides a powerful example of a nonobvious invention. Adams, the patentee, had invented "the first practical, water-activated, constant potential battery which could be fabricated and stored indefinitely without any fluid in its cells." 383 U.S. at 43. What was most compelling about the evidence in *Adams* was the degree to which conventional wisdom in the battery art deterred the ordinary artisan from bringing together the electrode and electrolyte components that Adams combined. When Adams demonstrated his new battery "before the experts of the United States Army Signal Corps" in early 1942, they "did not believe the battery was workable." *Id.* at 44. By late 1943, however, the Signal Corps—unbeknownst to Adams—had contracted with various suppliers for mass production of this highly effective battery, to aid the war effort. The Court summarized the degree to which Adams' invention bucked conventional wisdom this way:

> Despite the fact that each of the elements of the Adams battery was well known in the prior art, to combine them as did Adams required that a person reasonably skilled in the prior art must ignore that (1) batteries which continued to operate on an open circuit and which heated in normal use

[like the Adams battery] were not practical; and (2) water-activated batteries were successful only when combined with electrolytes detrimental to the use of magnesium. These long-accepted factors, when taken together, would, we believe, deter any investigation into such a combination [of water and magnesium] as is used by Adams.

Id. at 51-52. The way the prior art would have deterred the ordinary artisan from making the Adams battery clearly impressed the Court.

The lower courts have, for 50 years, applied the *Graham* framework when analyzing the question of nonobviousness. When, in 1982, the Federal Circuit took up the task of hearing all appeals arising under the U.S. patent laws, it applied the *Graham* framework, but it also supplemented *Graham* with an inquiry that came to be known as the *teaching/suggestion/motivation test*, or TSM test. *See, e.g., ACS Hospital Sys. v. Montefiore Hospital*, 732 F.2d 1572, 1577-78 (Fed. Cir. 1984). This TSM test added to § 103's existing hindsight-prevention mechanisms by requiring that, to prove a claimed invention would have been obvious, one must prove that "the prior art as a whole would have suggested [the invention] to one skilled in the art." *Envt'l Designs, Ltd. v. Union Oil Co.*, 713 F.2d 693, 698 (Fed. Cir. 1983).

For example, in a 1999 case overturning the PTO's determination that a decorative jack-o-lantern plastic lawn bag for leaves would have been obvious, the Federal Circuit succinctly explained that

the best defense against the subtle but powerful attraction of a hindsight-based obviousness analysis is rigorous application of the requirement for a showing of the teaching or motivation to combine prior art references. Combining prior art references without evidence of such a suggestion, teaching, or motivation simply takes the inventor's disclosure as a blueprint for piecing together the prior art to defeat patentability—the essence of hindsight.

In re Dembiczak, 175 F.3d 994, 999 (Fed. Cir. 1999) (citations omitted). In this case, the court reversed the PTO's obviousness rejection due to the lack of sufficient record evidence of a suggestion or motivation to combine the prior art jack-o-lantern paper bags with the prior art orange plastic trash bags. *Id.* at 999-1001. A PTO rejection absent such evidence indicates improper reliance on hindsight, however, only to the extent that an ordinary artisan is incapable of creatively, spontaneously remixing known prior art elements when confronted with a new problem. The Federal Circuit did not consider the ordinary artisan's ordinary creativity.

Thus matters stood, until April 2007.

B. Contemporary Nonobviousness Analysis

One prior art patent describes an adjustable gas pedal for a car. Another prior art patent describes a pedal-mounted sensor to link the car pedal to a computer-controlled throttle. Would it have been obvious to a pedal engineer of ordinary skill, in February 1998, to combine the two prior art items into a sensor-bearing adjustable gas pedal? If so, the sensor-bearing pedal was not properly patentable, even if it was new at that time.

On roughly these facts, the Supreme Court granted review in *KSR International Co. v. Teleflex Inc.* to decide "[w]hether the Federal Circuit ha[d] erred in holding that a claimed invention cannot be held 'obvious,' and thus unpatentable under 35 U.S.C. § 103, in the absence of some proven 'teaching, suggestion, or motivation' that would have led a person of ordinary skill in the art to combine the relevant prior art teachings in the manner claimed."[22] The Court's decision in *KSR* follows.

KSR International Co. v. Teleflex Inc.

550 U.S. 398 (2007)

Kennedy, Justice:

Teleflex Incorporated and its subsidiary Technology Holding Company—both referred to here as Teleflex—sued KSR International Company for patent infringement. The patent at issue, U.S. Patent No. 6,237,565, is entitled "Adjustable Pedal Assembly With Electronic Throttle Control." The patentee is Steven J. Engelgau, and the patent is referred to as "the Engelgau patent." Teleflex holds the exclusive license to the patent.

Claim 4 of the Engelgau patent describes a mechanism for combining an electronic sensor with an adjustable automobile pedal so the pedal's position can be transmitted to a computer that controls the throttle in the vehicle's engine. When Teleflex accused KSR of infringing the Engelgau patent by adding an electronic sensor to one of KSR's previously designed pedals, KSR countered that claim 4 was invalid under the Patent Act, 35 U.S.C. §103, because its subject matter was obvious.

. . . .

I
A

In car engines without computer-controlled throttles, the accelerator pedal interacts with the throttle via cable or other mechanical link. The pedal arm acts as a lever rotating around a pivot point. In a cable-actuated throttle control the rotation caused by pushing down the pedal pulls a cable, which in turn pulls open valves in the carburetor or fuel injection unit. The wider the valves open, the more fuel and air are released, causing combustion to increase and the car to accelerate. When the driver takes his foot off the pedal, the opposite occurs as the cable is released and the valves slide closed.

In the 1990's it became more common to install computers in cars to control engine operation. Computer-controlled throttles open and close valves in response to electronic signals, not through force transferred from the pedal by a mechanical link. Constant, delicate adjustments of air and fuel mixture are possible.

---

[22] Petition for Writ of Certiorari at i, *KSR Int'l Co. v. Teleflex Inc.*, 127 S. Ct. 1727 (2007) (No. 04-1530), 2005 WL 835463 (internal quotations omitted).

The computer's rapid processing of factors beyond the pedal's position improves fuel efficiency and engine performance.

For a computer-controlled throttle to respond to a driver's operation of the car, the computer must know what is happening with the pedal. A cable or mechanical link does not suffice for this purpose; at some point, an electronic sensor is necessary to translate the mechanical operation into digital data the computer can understand.

Before discussing sensors further we turn to the mechanical design of the pedal itself. In the traditional design a pedal can be pushed down or released but cannot have its position in the footwell adjusted by sliding the pedal forward or back. As a result, a driver who wishes to be closer or farther from the pedal must either reposition himself in the driver's seat or move the seat in some way. In cars with deep footwells these are imperfect solutions for drivers of smaller stature. To solve the problem, inventors, beginning in the 1970's, designed pedals that could be adjusted to change their location in the footwell. Important for this case are two adjustable pedals disclosed in U.S. Patent Nos. 5,010,782 (filed July 28, 1989) (Asano) and 5,460,061 (filed Sept. 17, 1993) (Redding). The Asano patent reveals a support structure that houses the pedal so that even when the pedal location is adjusted relative to the driver, one of the pedal's pivot points stays fixed. The pedal is also designed so that the force necessary to push the pedal down is the same regardless of adjustments to its location. The Redding patent reveals a different, sliding mechanism where both the pedal and the pivot point are adjusted.

We return to sensors. Well before Engelgau applied for his challenged patent, some inventors had obtained patents involving electronic pedal sensors for computer-controlled throttles. These inventions, such as the device disclosed in U.S. Patent No. 5,241,936 (filed Sept. 9, 1991) ('936), taught that it was preferable to detect the pedal's position in the pedal assembly, not in the engine. The '936 patent disclosed a pedal with an electronic sensor on a pivot point in the pedal assembly. U.S. Patent No. 5,063,811 (filed July 9, 1990) (Smith) taught that to prevent the wires connecting the sensor to the computer from chafing and wearing out, and to avoid grime and damage from the driver's foot, the sensor should be put on a fixed part of the pedal assembly rather than in or on the pedal's footpad.

In addition to patents for pedals with integrated sensors inventors obtained patents for self-contained modular sensors. A modular sensor is designed independently of a given pedal so that it can be taken off the shelf and attached to mechanical pedals of various sorts, enabling the pedals to be used in automobiles with computer-controlled throttles. One such sensor was disclosed in U.S. Patent No. 5,385,068 (filed Dec. 18, 1992) ('068). In 1994, Chevrolet manufactured a line of trucks using modular sensors "attached to the pedal support bracket, adjacent to the pedal and engaged with the pivot shaft about which the pedal rotates in operation." 298 F. Supp. 2d 581, 589 (E.D. Mich. 2003).

The prior art contained patents involving the placement of sensors on adjustable pedals as well. For example, U.S. Patent No. 5,819,593 (filed Aug. 17, 1995) (Rixon) discloses an adjustable pedal assembly with an electronic sensor for detecting the pedal's position. In the Rixon pedal the sensor is located in the pedal footpad. The Rixon pedal was known to suffer from wire chafing when the pedal was depressed and released.

This short account of pedal and sensor technology leads to the instant case.

B

KSR, a Canadian company, manufactures and supplies auto parts, including pedal systems. Ford Motor Company hired KSR in 1998 to supply an adjustable pedal system for various lines of automobiles with cable-actuated throttle controls. KSR developed an adjustable mechanical pedal for Ford and obtained U.S. Patent No. 6,151,976 (filed July 16, 1999) ('976) for the design. In 2000, KSR was chosen by General Motors Corporation (GMC or GM) to supply adjustable pedal systems for Chevrolet and GMC light trucks that used engines with computer-controlled throttles. To make the '976 pedal compatible with the trucks, KSR merely took that design and added a modular sensor.

Teleflex is a rival to KSR in the design and manufacture of adjustable pedals. As noted, it is the exclusive licensee of the Engelgau patent. Engelgau filed the patent application on August 22, 2000 as a continuation of a previous application for U.S. Patent No. 6,109,241, which was filed on January 26, 1999.[*] He has sworn he invented the patent's subject matter on February 14, 1998. The Engelgau patent discloses an adjustable electronic pedal described in the specification as a "simplified vehicle control pedal assembly that is less expensive, and which uses fewer parts and is easier to package within the vehicle." Engelgau, col. 2, lines 2–5. Claim 4 of the patent, at issue here, describes:

A vehicle control pedal apparatus comprising:
> a support adapted to be mounted to a vehicle structure;
> an adjustable pedal assembly having a pedal arm moveable in for[e] and aft directions with respect to said support;
> a pivot for pivotally supporting said adjustable pedal assembly with respect to said support and defining a pivot axis; and
> an electronic control attached to said support for controlling a vehicle system;
> said apparatus characterized by said electronic control being responsive to said pivot for providing a signal that corresponds to pedal arm position as said pedal arm pivots

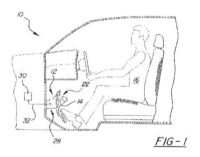

FIG-1

---

* [*Eds. Note*: As you know from your study of novelty earlier in this Chapter, the original filing date of a patent application is a crucial benchmark for what constitutes prior art and what does not. The Patent Act allows a patent applicant to use a series of applications, grouped together as a family of a "parent" application and "child" continuation applications that depend from the parent, to pursue the asserted entitlement to a given claim based on the supporting disclosure in the specification. The "child" continuation applications are given the filing date of the "parent" application. The key procedural constraint is that a child application cannot claim the benefit of its parent's original filing date unless the child application is filed before the parent application is formally issued by the Patent Office or, alternatively, formally abandoned by the applicant. 35 U.S.C. § 120 (setting forth the copendency requirement).]

about said pivot axis between rest and applied positions wherein the position of said pivot remains constant while said pedal arm moves in fore and aft directions with respect to said pivot.

Id. col. 6, lines 17–36 (diagram numbers omitted). We agree with the District Court that the claim discloses "a position-adjustable pedal assembly with an electronic pedal position sensor attached to the support member of the pedal assembly. Attaching the sensor to the support member allows the sensor to remain in a fixed position while the driver adjusts the pedal." 298 F. Supp. 2d at 586-587.

Before issuing the Engelgau patent the U.S. Patent & Trademark Office (PTO) rejected one of the patent claims that was similar to, but broader than, the present claim 4. The claim did not include the requirement that the sensor be placed on a fixed pivot point. The PTO concluded the claim was an obvious combination of the prior art disclosed in Redding and Smith, explaining:

> "Since the prior ar[t] references are from the field of endeavor, the purpose disclosed * * * would have been recognized in the pertinent art of Redding. Therefore it would have been obvious * * * to provide the device of Redding with the * * * means attached to a support member as taught by Smith."

Id. at 595 [quoting the PTO]. In other words Redding provided an example of an adjustable pedal and Smith explained how to mount a sensor on a pedal's support structure, and the rejected patent claim merely put these two teachings together.

Although the broader claim was rejected, claim 4 was later allowed because it included the limitation of a fixed pivot point, which distinguished the design from Redding's. *Id.* Engelgau had not included Asano among the prior art references, and Asano was not mentioned in the patent's prosecution. Thus, the PTO did not have before it an adjustable pedal with a fixed pivot point. The patent issued on May 29, 2001 and was assigned to Teleflex.

Upon learning of KSR's design for GM, Teleflex sent a warning letter informing KSR that its proposal would violate the Engelgau patent. "'Teleflex believes that any supplier of a product that combines an adjustable pedal with an electronic throttle control necessarily employs technology covered by one or more'" of Teleflex's patents. *Id.* at 585. KSR refused to enter a royalty arrangement with Teleflex; so Teleflex sued for infringement, asserting KSR's pedal infringed the Engelgau patent and two other patents. *Id.* Teleflex later abandoned its claims regarding the other patents and dedicated the patents to the public. The remaining contention was that KSR's pedal system for GM infringed claim 4 of the Engelgau patent. Teleflex has not argued that the other three claims of the patent are infringed by KSR's pedal, nor has Teleflex argued that the mechanical adjustable pedal designed by KSR for Ford infringed any of its patents.

C

The District Court granted summary judgment in KSR's favor. After reviewing the pertinent history of pedal design, the scope of the Engelgau patent, and the relevant prior art, the court considered the validity of the contested claim. By direction of 35 U.S.C. § 282, an issued patent is presumed valid. The District

Court applied *Graham*'s framework to determine whether under summary judgment standards KSR had overcome the presumption and demonstrated that claim 4 was obvious in light of the prior art in existence when the claimed subject matter was invented.

The District Court determined, in light of the expert testimony and the parties' stipulations, that the level of ordinary skill in pedal design was "an undergraduate degree in mechanical engineering (or an equivalent amount of industry experience) [and] familiarity with pedal control systems for vehicles." 298 F. Supp. 2d at 590. The court then set forth the relevant prior art, including the patents and pedal designs described above.

Following *Graham*'s direction, the court compared the teachings of the prior art to the claims of Engelgau. It found "little difference." *Id.* Asano taught everything contained in claim 4 except the use of a sensor to detect the pedal's position and transmit it to the computer controlling the throttle. That additional aspect was revealed in sources such as the '068 patent and the sensors used by Chevrolet.

Under the controlling cases from the Court of Appeals for the Federal Circuit, however, the District Court was not permitted to stop there. The court was required also to apply the TSM test. The District Court held KSR had satisfied the test. It reasoned (1) the state of the industry would lead inevitably to combinations of electronic sensors and adjustable pedals, (2) Rixon provided the basis for these developments, and (3) Smith taught a solution to the wire chafing problems in Rixon, namely locating the sensor on the fixed structure of the pedal. This could lead to the combination of Asano, or a pedal like it, with a pedal position sensor.

The conclusion that the Engelgau design was obvious was supported, in the District Court's view, by the PTO's rejection of the broader version of claim 4. Had Engelgau included Asano in his patent application, it reasoned, the PTO would have found claim 4 to be an obvious combination of Asano and Smith, as it had found the broader version an obvious combination of Redding and Smith. As a final matter, the District Court held that the secondary factor of Teleflex's commercial success with pedals based on Engelgau's design did not alter its conclusion. The District Court granted summary judgment for KSR.

With principal reliance on the TSM test, the Court of Appeals reversed. It ruled the District Court had not been strict enough in applying the test, having failed to make "'finding[s] as to the specific understanding or principle within the knowledge of a skilled artisan that would have motivated one with no knowledge of [the] invention' * * * to attach an electronic control to the support bracket of the Asano assembly." 119 Fed. Appx. at 288 (brackets in original) (quoting *In re Kotzab*, 217 F.3d 1365, 1371 (Fed. Cir. 2000)). The Court of Appeals held that the District Court was incorrect that the nature of the problem to be solved satisfied this requirement because unless the "prior art references address[ed] the precise problem that the patentee was trying to solve," the problem would not motivate an inventor to look at those references. 119 Fed. Appx. at 288.

Here, the Court of Appeals found, the Asano pedal was designed to solve the "'constant ratio problem'"—that is, to ensure that the force required to depress the pedal is the same no matter how the pedal is adjusted—whereas Engelgau sought to provide a simpler, smaller, cheaper adjustable electronic pedal. *Id.* As for Rixon, the court explained, that pedal suffered from the problem of wire chafing but was not designed to solve it. In the court's view Rixon did not teach anything helpful to Engelgau's purpose. Smith, in turn, did not relate to adjustable pedals and did not "necessarily go to the issue of motivation to attach the electronic control on the support bracket of the pedal assembly." *Id.* When the patents were interpreted in this way, the Court of Appeals held, they would not have led a person of ordinary skill to put a sensor on the sort of pedal described in Asano.

That it might have been obvious to try the combination of Asano and a sensor was likewise irrelevant, in the court's view, because "'"[o]bvious to try" has long been held not to constitute obviousness.'" *Id.* at 289 (quoting *In re Deuel*, 51 F.3d 1552, 1559 (Fed. Cir. 1995)).

The Court of Appeals also faulted the District Court's consideration of the PTO's rejection of the broader version of claim 4. The District Court's role, the Court of Appeals explained, was not to speculate regarding what the PTO might have done had the Engelgau patent mentioned Asano. Rather, the court held, the District Court was obliged first to presume that the issued patent was valid and then to render its own independent judgment of obviousness based on a review of the prior art. The fact that the PTO had rejected the broader version of claim 4, the Court of Appeals said, had no place in that analysis.

The Court of Appeals further held that genuine issues of material fact precluded summary judgment. Teleflex had proffered statements from one expert that claim 4 "'was a simple, elegant, and novel combination of features,'" 119 Fed. Appx. at 290, compared to Rixon, and from another expert that claim 4 was nonobvious because, unlike in Rixon, the sensor was mounted on the support bracket rather than the pedal itself. This evidence, the court concluded, sufficed to require a trial.

II

A

We begin by rejecting the rigid approach of the Court of Appeals. Throughout this Court's engagement with the question of obviousness, our cases have set forth an expansive and flexible approach inconsistent with the way the Court of Appeals applied its TSM test here. To be sure, *Graham* recognized the need for "uniformity and definiteness." 383 U.S. at 18. Yet the principles laid down in *Graham* reaffirmed the "functional approach" of *Hotchkiss*. *See* 383 U.S. at 12. To this end, *Graham* set forth a broad inquiry and invited courts, where appropriate, to look at any secondary considerations that would prove instructive. *Id.* at 17.

Neither the enactment of § 103 nor the analysis in *Graham* disturbed this Court's earlier instructions concerning the need for caution in granting a patent based on the combination of elements found in the prior art. For over a half century, the Court has held that a "patent for a combination which only unites old elements with no change in their respective functions . . . obviously withdraws what is already known into the field of its monopoly and diminishes the resources available to skillful men." *Great Atlantic & Pacific Tea Co.* v. *Supermarket Equipment Corp.*, 340 U.S. 147, 152 (1950). This is a principal reason for declining to allow patents for what is obvious. The combination of familiar elements according to known methods is likely to be obvious when it does no more than yield predictable results. Three cases decided after *Graham* illustrate the application of this doctrine.

In *United States* v. *Adams*, 383 U.S. 39 (1966), a companion case to *Graham*, the Court considered the obviousness of a "wet battery" that varied from prior designs in two ways: It contained water, rather than the acids conventionally employed in storage batteries; and its electrodes were magnesium and cuprous chloride, rather than zinc and silver chloride. The Court recognized that when a patent claims a structure already known in the prior art that is altered by the mere substitution of one element for another known in the field, the combination must do more than yield a predictable result. It nevertheless rejected the Government's claim that Adams's battery was obvious. The Court relied upon the corollary principle that when the prior art teaches away from combining certain known elements, discovery of a successful means of combining them is more likely to be nonobvious. When Adams designed his battery, the prior art warned that risks were involved in using the types of electrodes he employed. The fact that the elements worked together in an unexpected and fruitful manner supported the conclusion that Adams's design was not obvious to those skilled in the art.

In *Anderson's-Black Rock, Inc.* v. *Pavement Salvage Co.*, 396 U.S. 57 (1969), the Court elaborated on this approach. The subject matter of the patent before the Court was a device combining two pre-existing elements: a radiant-heat burner and a paving machine. The device, the Court concluded, did not create some new synergy: The radiant-heat burner functioned just as a burner was expected to function; and the paving machine did the same. The two in combination did no more than they would in separate, sequential operation. In those circumstances, "while the combination of old elements performed a useful function, it added nothing to the nature and quality of the radiant-heat burner already patented," and the patent failed under § 103.

Finally, in *Sakraida* v. *AG Pro, Inc.*, 425 U.S. 273 (1976), the Court derived from the precedents the conclusion that when a patent "simply arranges old elements with each performing the same function it had been known to perform" and yields no more than one would expect from such an arrangement, the combination is obvious.

The principles underlying these cases are instructive when the question is whether a patent claiming the combination of elements of prior art is obvious. When a work is available in one field of endeavor, design incentives and other market forces can prompt variations of it, either in the same field or a different one. If a person of ordinary skill can implement a predictable variation, § 103 likely bars its patentability. For the same reason, if a technique has been used to improve one device, and a person of ordinary skill in the art would recognize that it would improve similar devices in the same way, using the technique is obvious unless its actual application is beyond his or her skill. *Sakraida* and *Anderson's-Black Rock* are illustrative—a court must ask whether the improvement is more than the predictable use of prior art elements according to their established functions.

Following these principles may be more difficult in other cases than it is here because the claimed subject matter may involve more than the simple substitution of one known element for another or the mere application of a known technique to a piece of prior art ready for the improvement. Often, it will be necessary for a court to look to interrelated teachings of multiple patents; the effects of demands known to the design community or present in the marketplace; and the background knowledge possessed by a person having ordinary skill in the art, all in order to determine whether there was an apparent reason to combine the known elements in the fashion claimed by the patent at issue. To facilitate review, this analysis should be made explicit. As our precedents make clear, however, the analysis need not seek out precise teachings directed to the specific subject matter of the challenged claim, for a court can take account of the inferences and creative steps that a person of ordinary skill in the art would employ.

B

When it first established the requirement of demonstrating a teaching, suggestion, or motivation to combine known elements in order to show that the combination is obvious, the Court of Customs and Patent Appeals captured a helpful insight. As is clear from cases such as *Adams*, a patent composed of several elements is not proved obvious merely by demonstrating that each of its elements was, independently, known in the prior art. Although common sense directs one to look with care at a patent application that claims as innovation the combination of two known devices according to their established functions, it can be important to identify a reason that would have prompted a person of ordinary skill in the relevant field to combine the elements in the way the claimed new invention does. This is so because inventions in most, if not all, instances rely upon building blocks long since uncovered, and claimed discoveries almost of necessity will be combinations of what, in some sense, is already known.

Helpful insights, however, need not become rigid and mandatory formulas; and when it is so applied, the TSM test is incompatible with our precedents. The obviousness analysis cannot be confined by a formalistic conception of the words teaching, suggestion, and motivation, or by overemphasis on the importance of published articles and the explicit content of issued patents. The diversity of inventive pursuits and of modern technology counsels against limiting the analysis in this way. In many fields it may be that there is little discussion of obvious techniques or combinations, and it often may be the case that market demand, rather than scientific literature, will drive design trends. Granting patent protection to advances that would occur in the ordinary course without real innovation retards progress and may, in the case of patents combining previously known elements, deprive prior inventions of their value or utility.

In the years since the Court of Customs and Patent Appeals set forth the essence of the TSM test, the Court of Appeals no doubt has applied the test in accord with these principles in many cases. There is no necessary inconsistency between the idea underlying the TSM test and the *Graham* analysis. But when a court transforms the general principle into a rigid rule that limits the obviousness inquiry, as the Court of Appeals did here, it errs.

C

The flaws in the analysis of the Court of Appeals relate for the most part to the court's narrow conception of the obviousness inquiry reflected in its application of the TSM test. In determining whether the subject matter of a patent claim is obvious, neither the particular motivation nor the avowed purpose of the patentee controls. What matters is the objective reach of the claim. If the claim extends to what is obvious, it is invalid under § 103. One of the ways in which a patent's subject matter can be proved obvious is by noting that there existed at the time of invention a known problem for which there was an obvious solution encompassed by the patent's claims.

The first error of the Court of Appeals in this case was to foreclose this reasoning by holding that courts and patent examiners should look only to the problem the patentee was trying to solve. 119 Fed. Appx. at 288. The Court of Appeals failed to recognize that the problem motivating the patentee may be only one of many addressed by the patent's subject matter. The question is not whether the combination was obvious to the patentee but whether the combination was obvious to a person with ordinary skill in the art. Under the correct analysis, any need or problem known in the field of endeavor at the time of invention and addressed by the patent can provide a reason for combining the elements in the manner claimed.

The second error of the Court of Appeals lay in its assumption that a person of ordinary skill attempting to solve a problem will be led only to those elements of prior art designed to solve the same problem. *Id.* The primary purpose of Asano was solving the constant ratio problem; so, the court concluded, an inventor considering how to put a sensor on an adjustable pedal would have no reason to consider putting it on the Asano pedal. *Id.* Common sense teaches, however, that familiar items may have obvious uses beyond their primary purposes, and in many cases a person of ordinary skill will be able to fit the teachings of multiple patents together like pieces of a puzzle. Regardless of Asano's primary purpose, the design provided an obvious example of an adjustable pedal with a fixed pivot point; and the prior art was replete with patents indicating that a fixed pivot point was an ideal mount for a sensor. The idea that a designer hoping to make an adjustable electronic pedal would ignore Asano because Asano was designed to solve the constant ratio problem makes little sense. A person of ordinary skill is also a person of ordinary creativity, not an automaton.

The same constricted analysis led the Court of Appeals to conclude, in error, that a patent claim cannot be proved obvious merely by showing that the combination of elements was "obvious to try." *Id.* at 289 (internal quotation marks omitted). When there is a design need or market pressure to solve a problem and there are a finite number of identified, predictable solutions, a person of ordinary skill has good reason to pursue the known options within his or her technical grasp. If this leads to the anticipated success, it is likely the product not of innovation but of ordinary skill and common sense. In that instance the fact that a combination was obvious to try might show that it was obvious under § 103.

The Court of Appeals, finally, drew the wrong conclusion from the risk of courts and patent examiners falling prey to hindsight bias. A factfinder should be aware, of course, of the distortion caused by hindsight bias and must be cautious of arguments reliant upon *ex post* reasoning. *See Graham*, 383 U.S. at 36 (warning against a "temptation to read into the prior art the teachings of the invention in issue" and instructing courts to "'guard against slipping into the use of hindsight'" (quoting *Monroe Auto Equipment Co.* v. *Heckethorn Mfg. & Supply Co.*, 332 F.2d 406, 412 (6th Cir. 1964))). Rigid preventative rules that deny factfinders recourse to common sense, however, are neither necessary under our case law nor consistent with it.

. . . .

III

When we apply the standards we have explained to the instant facts, claim 4 must be found obvious. We agree with and adopt the District Court's recitation of the relevant prior art and its determination of the level of ordinary skill in the field. As did the District Court, we see little difference between the teachings of Asano and Smith and the adjustable electronic pedal disclosed in claim 4 of the Engelgau patent. A person having ordinary skill in the art could have combined Asano with a pedal position sensor in a fashion encompassed by claim 4, and would have seen the benefits of doing so.

. . . .

B

The District Court was correct to conclude that, as of the time Engelgau designed the subject matter in claim 4, it was obvious to a person of ordinary skill to combine Asano with a pivot-mounted pedal position sensor. There then existed a marketplace that created a strong incentive to convert mechanical pedals to electronic pedals, and the prior art taught a number of methods for achieving this advance. The Court of Appeals considered the issue too narrowly by, in effect, asking whether a pedal designer writing on a blank slate would have chosen both Asano and a modular sensor similar to the ones used in the Chevrolet truckline and disclosed in the '068 patent. The District Court employed this narrow inquiry as well, though it reached the correct result nevertheless. The proper question to have asked was whether a pedal designer of ordinary skill, facing the wide range of needs created by developments in the field of endeavor, would have seen a benefit to upgrading Asano with a sensor.

In automotive design, as in many other fields, the interaction of multiple components means that changing one component often requires the others to be modified as well. Technological developments made it clear that engines using computer-controlled throttles would become standard. As a result, designers might have decided to design new pedals from scratch; but they also would have had reason to make pre-existing pedals work with the new engines. Indeed, upgrading its own pre-existing model led KSR to design the pedal now accused of infringing the Engelgau patent.

For a designer starting with Asano, the question was where to attach the sensor. The consequent legal question, then, is whether a pedal designer of ordinary skill starting with Asano would have found it obvious to put the sensor on a fixed pivot point. The prior art discussed above leads us to the conclusion that attaching the sensor where both KSR and Engelgau put it would have been obvious to a person of ordinary skill.

The '936 patent taught the utility of putting the sensor on the pedal device, not in the engine. Smith, in turn, explained to put the sensor not on the pedal's footpad but instead on its support structure. And from the known wire-chafing problems of Rixon, and Smith's teaching that "the pedal assemblies must not precipitate any motion in the connecting wires," Smith, col. 1, ll. 35–37, the designer would know to place the sensor on a nonmoving part of the pedal structure. The most obvious nonmoving point on the structure from which a sensor can easily detect the pedal's position is a pivot point. The designer, accordingly, would follow Smith in mounting the sensor on a pivot, thereby designing an adjustable electronic pedal covered by claim 4.

Just as it was possible to begin with the objective to upgrade Asano to work with a computer-controlled throttle, so too was it possible to take an adjustable electronic pedal like Rixon and seek an improvement that would avoid the wire-chafing problem. Following similar steps to those just explained, a designer would learn from Smith to avoid sensor movement and would come, thereby, to Asano because Asano disclosed an adjustable pedal with a fixed pivot.

. . . .

Like the District Court, finally, we conclude Teleflex has shown no secondary factors to dislodge the determination that claim 4 is obvious. Proper application of *Graham* and our other precedents to these facts therefore leads to the conclusion that claim 4 encompassed obvious subject matter. As a result, the claim fails to meet the requirement of § 103.

. . . .

IV

A separate ground the Court of Appeals gave for reversing the order for summary judgment was the existence of a dispute over an issue of material fact. We disagree with the Court of Appeals on this point as well. To the extent the court understood the *Graham* approach to exclude the possibility of summary judgment when an expert provides a conclusory affidavit addressing the question of obviousness, it misunderstood the role expert testimony plays in the analysis. In considering summary judgment on that question the district court can and should take into account expert testimony, which may resolve or keep open certain questions of fact. That is not the end of the issue, however. The ultimate judgment of obviousness is a legal determination. *Graham*, 383 U.S. at 17. Where, as here, the content of the prior art, the scope of the patent claim, and the level of ordinary skill in the art are not in material dispute, and the obviousness of the claim is apparent in light of these factors, summary judgment is appropriate. Nothing in the declarations proffered by Teleflex prevented the District Court from reaching the careful conclusions underlying its order for summary judgment in this case.

*   *   *

We build and create by bringing to the tangible and palpable reality around us new works based on instinct, simple logic, ordinary inferences, extraordinary ideas, and sometimes even genius. These advances, once part of our shared knowledge, define a new threshold from which innovation starts once more. And as progress beginning from higher levels of achievement is expected in the normal course, the results of ordinary innovation are not the subject of exclusive rights under the patent laws. Were it otherwise patents might stifle, rather than promote, the progress of useful arts. *See* U.S. Const., Art. I, §8, cl. 8. These premises led to the bar on patents claiming obvious subject matter established in *Hotchkiss* and codified in § 103. Application of the bar must not be confined within a test or formulation too constrained to serve its purpose.

. . . .

Notes & Questions

1. The Court acknowledges that hindsight bias can lead one to incorrectly deny patent protection to an invention that would not have been obvious. At the same time, the Court expresses concern about the harm to innovation and competition from an incorrect grant of patent protection to an obvious invention. Recall the discussion about error costs at the start of the Trade Secret Chapter. Can the Court's disagreement with the Federal Circuit be explained as a disagreement concerning which is worse, erroneous grants of patent rights or erroneous denials?

2. Perhaps the most jarring sentence in the Court's opinion, for those patent law practitioners who had grown accustomed to the Federal Circuit's TSM test, is the following: "A person of ordinary skill is also a person of ordinary creativity, not an automaton." 550 U.S. at 421. ("Automaton" is a posh word for robot.) How does the phosita's ordinary creativity affect how we estimate "whether the improvement [of a claimed invention] is more than the predictable use of prior art elements according to their established functions"? *Id.* at 417.

3. In October 2007, the PTO published new guidelines to help examiners recognize "[e]xemplary rationales that may support a conclusion of obviousness" under § 103, in light of the Supreme Court's analysis in *KSR*. These guidelines, now codified in §§ 2141 and 2143 of the *Manual of Patent Examining Procedure*, detail seven rationales that one can properly use to reject a claim for obviousness:

> (A) Combining prior art elements according to known methods to yield predictable results;
> (B) Simple substitution of one known element for another to obtain predictable results;
> (C) Use of known technique to improve similar devices (methods, or products) in the same way;

(D) Applying a known technique to a known device (method, or product) ready for improvement to yield predictable results;

(E) "Obvious to try" – choosing from a finite number of identified, predictable solutions, with a reasonable expectation of success;

(F) Known work in one field of endeavor may prompt variations of it for use in either the same field or a different one based on design incentives or other market forces if the variations are predictable to one of ordinary skill in the art;

(G) Some teaching, suggestion, or motivation in the prior art that would have led one of ordinary skill to modify the prior art reference or to combine prior art reference teachings to arrive at the claimed invention.

MPEP § 2141, at 2100-143 – 2100-144 (9th ed. Mar. 2014). Note that only one of the seven—letter (G)—is the Federal Circuit's pre-*KSR* suggestion test; the other six are rooted in the text of the *KSR* decision.

4. The Court noted that Teleflex dedicated certain of its patents to the public. The Patent Act provides a mechanism for doing so. 35 U.S.C. § 253. What might motivate a patent holder to make such a public dedication?

5. Another surprising aspect of the Supreme Court's analysis, compared to settled Federal Circuit law, is that one can prove that a claimed invention would have been obvious "by showing that the combination of elements [in the claim] was 'obvious to try.'" *KSR*, 550 U.S. at 421. *Cf. Gillette Co. v. S.C. Johnson & Son, Inc.*, 919 F.2d 720, 725 (Fed. Cir. 1990) ("[W]e have consistently held that "obvious to try" is not to be equated with obviousness under 35 U.S.C. § 103.").

St. Jude Medical, Inc. v. Access Closure, Inc.
729 F.3d 1369 (Fed. Cir. 2013)

PLAGER, JUDGE:

This is a patent case. Access Closure, Inc. (ACI), the defendant at trial, appeals from several rulings . . . in favor of St. Jude Medical, Inc. and St. Jude Medical Puerto Rico, LLC (collectively "St. Jude"), plaintiffs patentees. The rulings relate to three patents that St. Jude asserted against ACI . . . [including] U.S. Patent No. 5,275,616 to Fowler (the '616 patent) and U.S. Patent No. 5,716,375 to Fowler (the '375 patent) . . . .

ACI appeals . . . the district court's ruling[] . . . that ACI was not entitled to JMOL that the Fowler patents are invalid for obviousness. For the reasons that follow, we . . . affirm the district court's ruling that the Fowler patents are nonobvious and not shown to be invalid.

I. Background

The patents in this appeal relate to methods and devices for sealing a 'vascular puncture.' A vascular puncture occurs when a medical procedure requires a medical professional to puncture through the skin and into a vein or artery to insert a medical device, such as a catheter, into a patient's vasculature. After such a procedure concludes, the medical professional typically removes the medical device from the vasculature.

Prior to the development of the technology at issue in this case, the medical professional was then required to apply external pressure to the puncture site until clotting occurred. Due to a variety of factors, the medical professional often had to apply pressure to the puncture site for an extended period of time. This caused discomfort to the patient and increased the recovery time. The [asserted] patents disclose a variety of methods and devices for sealing a vascular puncture with the objective of improving patient recovery.

. . . .

Like [some additional St. Jude] patent[s], the Fowler patents disclose devices and methods for closing a vascular puncture with a plug, but the Fowler patents also disclose a balloon catheter with a balloon configured to position the plug. '616 patent, col. 4, l. 46 – col. 5, l.6. A user inserts the balloon catheter in the puncture tract until the balloon is positioned in the vessel, and the user inflates the balloon. The user then inserts a plug such that the plug contacts the inflated balloon; the inflated balloon prevents the plug from extending into the vessel. After the plug is positioned, the user removes the balloon catheter, leaving the plug in the puncture to promote healing. This arrangement is seen in Figure 3 of the Fowler patents

. . . .

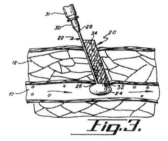

Two claims from the Fowler patents, dependent claim 14 of the '616 patent and independent claim 21 of the '375 patent, are involved in this appeal. Claim 21 is generally representative and recites, "a vessel plug" and "at least a portion of [a] positioning member [that] is expandable * * * to position said vessel plug in the incision proximally of the

Fig.3.

blood vessel such that said vessel plug obstructs the flow of blood through the incision without extending into the blood vessel." '375 patent, col. 10, ll. 12-22.

C. Procedural History

St. Jude filed its complaint in the United States District Court for the Western District of Arkansas on October 22, 2008, alleging that ACI infringed several of its patents, including . . . the Fowler patents. . . . The parties proceeded to trial before a jury.

. . . .

With regard to the Fowler patents, the jury found that ACI had infringed claim 14 of the Fowler '616 patent, and claim 21 of the Fowler '375 patent. The jury also found that ACI willfully infringed the Fowler claims. Regarding validity, the jury found that the Fowler claims were not obvious and thus valid.

After the jury rendered its verdict, ACI filed a renewed JMOL motion on various issues. The district court denied ACI's motion.

. . . .

II. Discussion

. . . .

. . . To argue obviousness, ACI relies on a 1988 article by Kenichi Takayasu, M.D., *et al.*, that describes a method and device for sealing a puncture in a liver vein with a compressed "gelfoam stick." Takayasu *et al.*, *A New Hemostatic Procedure for Percutaneous Transhepatic Portal Vein Catheterization*, Jpn. J. Clin. Oncol. 18:227-30 (1988) ("Takayasu"). ACI also argues obviousness based on a 1971 article by Karl Smiley that describes a technique for blocking bleeding from a vessel using a balloon catheter. Smiley and Perry, *Balloon Catheter Tamponade of Major Vascular Wounds*, Am. J. of Surgery, 326-27 (1971) ("Smiley").

According to ACI, Takayasu and Smiley establish that there was a known problem in the art—achieving hemostasis in punctured or damaged blood vessels—and the prior art references both disclosed methods of solving that problem. ACI argues that a person of ordinary skill in the art would have reasonably expected to permanently stop bleeding in blood vessels by combining the techniques described in Takayasu and Smiley in the manner claimed by the Fowler patents. For these reasons, ACI argues that claim 14 of the Fowler '616 patent, and claim 21 of the Fowler '375 patent are invalid for obviousness under 35 U.S.C. § 103.

In response, St. Jude contends that both prior art references fail to disclose the Fowler patent's claimed feature of a balloon configured to operate as a positioning device to prevent a plug from entering a blood vessel. In particular, St. Jude argues that the references lack "an elongate positioning member * * * to position said vessel plug * * * without extending into the blood vessel" from claim 21 of the '375 patent and "inflating a member on an insertion member to identify the location of the blood vessel adjacent to the incision" and "positioning the vessel plug in the incision such that the distal end of the vessel plug is located proximally of the blood vessel" from . . . claim 14 of the '616 patent.

We review the jury's conclusions on obviousness without deference and the underlying findings of fact for substantial evidence. *Johns Hopkins Univ. v. Datascope Corp.*, 543 F.3d 1342, 1345 (Fed. Cir. 2008). We see no error in the district court's legal conclusion of nonobviousness. Neither Takayasu nor Smiley discloses a balloon configured to operate as a positioning device to prevent a plug from entering a blood vessel as claimed in the Fowler patents. Takayasu discloses no balloon at all, and Smiley discloses a balloon that is used to control the bleeding of a gunshot wound or an abdominal aneurism.

Moreover, we are not persuaded by ACI's arguments to combine the teachings of Takayasu and Smiley. ACI points to Takayasu and Smiley, arguing that achieving hemostasis in blood vessels was a known problem in the art. But we note that Takayasu and Smiley both teach the same thing to overcome this problem: inserting an object into a wound to help achieve hemostasis. Takayasu teaches inserting a gelfoam stick, and Smiley teaches inserting a balloon. One of ordinary skill in the art at the time of the invention would have viewed the balloon and the gelfoam stick as substitutes to achieve the same hemostasis objective, not as complementary devices to achieve the positioning benefit of St. Jude's claimed invention.

ACI also notes that Takayasu discloses visualizing the placement of the gelfoam stick with x-ray and ultrasound imaging. Presumably, ACI is suggesting that since x-ray and ultrasound were used to place a gelfoam stick, it would have been obvious to one of skill in the art at the time of the invention to use a balloon to prevent a plug from entering a blood vessel. But ACI fails to explain why it would have been obvious to do so. Certainly the balloon in Smiley is not used to position a plug as claimed, and ACI provides no evidence that a balloon had ever been so used. Nor does ACI explain why common sense would lead one of ordinary skill in the art to abandon the noninvasive x-ray and ultrasound technology in favor of an invasive, inflatable balloon.

Therefore, while Takayasu does disclose using x-ray and ultrasound to help place a gelfoam stick, between this disclosure and the claimed invention lies a logical chasm—a chasm not bridged by the prior art, common sense, or ACI's statements that the claimed invention was obvious. Even under our "expansive and flexible" obviousness analysis (*see KSR Int'l Co. v. Teleflex Inc.*, 550 U.S. 398, 415 (2007)), we must guard against "hindsight bias" and "*ex post* reasoning" (*id.* at 421). Doing so here compels us to reject ACI's argument.

ACI further contends that, because St. Jude's expert Dr. Kovacs allegedly provided only conclusory testimony, the district court erred by rejecting ACI's proposed combination. Dr. Kovacs testified that the proposed combination of Takayasu and Smiley was "very, very far out and it makes no sense to me whatsoever" and "[c]ombining two references which are independently farfetched is farfetched times farfetched, it's farfetched squared."

Even if Dr. Kovacs provided conclusory testimony, ACI did not carry its burden of proving invalidity. We have held that "[s]ince we must presume a patent valid, the patent challenger bears the burden of proving the factual elements of invalidity by clear and convincing evidence." *Pfizer, Inc. v. Apotex, Inc.*, 480 F.3d 1348, 1359 (Fed. Cir. 2007). This burden of proof never shifts to the patentee. *Id.* ACI failed to carry its burden of proof. Therefore, ACI's arguments about Dr. Kovacs are immaterial.

We find each of ACI's obviousness arguments without merit and affirm the district court's refusal to grant ACI's renewed motion for JMOL.

. . . .

Notes & Questions

1. Does the *St. Jude* court do an effective job identifying a gap too big for the phosita's ordinary creativity to fill? Or does *St. Jude* suggest that the Federal Circuit's overemphasis on guarding against hindsight bias, criticized in *KSR*, may continue?

C. What Prior Art is Pertinent Under § 103?

The nonobviousness requirement for patentability is clearly a higher hurdle than the novelty requirement. If one must clear the highest hurdle in order to obtain a patent, why retain the lower hurdle of novelty? In part, the answer lies in understanding the relevant prior art that can be consulted under these two requirements.

Under § 102, a prior art reference is eligible for consideration in a novelty inquiry if it comes from the right time period. If that prior art reference anticipates, it does not matter that the reference comes from a technological field far removed from that of the claimed invention.

Under § 103, by contrast, a reference is not eligible for consideration in a non-obviousness inquiry unless it is *both* properly prior art under § 102 *and* topically pertinent to the claimed invention. This is sometimes called the *analogous art* requirement. *See, e.g., In re Bigio*, 381 F.3d 1320, 1325 (Fed. Cir. 2004) ("References within the statutory terms of 35 U.S.C. § 102 qualify as prior art for an obviousness determination only when analogous to the claimed invention."). All the prior art in *KSR* was pertinent; as a result, the Supreme Court had no need to discuss, and did not disturb, the existing analogous-art jurisprudence. The case below remains a leading case on the issue.

In re Clay
966 F.2d 656 (Fed. Cir. 1992)

LOURIE, JUDGE:

Carl D. Clay appeals the decision of the [PTO Board] affirming the rejection of claims 1-11 and 13 as being unpatentable under 35 U.S.C. § 103. These are all the remaining claims in application Serial No. 245,083, filed April 28, 1987, entitled "Storage of a Refined Liquid Hydrocarbon Product." We reverse.

BACKGROUND

Clay's invention, assigned to Marathon Oil Company, is a process for storing refined liquid hydrocarbon product in a storage tank having a dead volume between the tank bottom and its outlet port. The process involves preparing a gelation solution which gels after it is placed in the tank's dead volume; the gel can easily be removed by adding to the tank a gel-degrading agent such as hydrogen peroxide. Claims 1, 8, and 11 are illustrative of the claims on appeal:

1. A process for storing a refined liquid hydrocarbon product in a storage tank having a dead volume between the bottom of said tank and an outlet port in said tank, said process comprising:

preparing a gelation solution comprising an aqueous liquid solvent, an acrylamide polymer and a crosslinking agent containing a polyvalent metal cation selected from the group consisting of aluminum, chromium and mixtures thereof, said gelation solution capable of forming a rigid crosslinked polymer gel which is substantially insoluble and inert in said refined liquid hydrocarbon product;

placing said solution in said dead volume;

gelling said solution substantially to completion in said dead volume to produce said rigid gel which substantially fills said dead volume; and

storing said refined liquid hydrocarbon product in said storage tank in contact with said gel without substantially contaminating said product with said gel and without substantially degrading said gel.

8. The process of claim 1 further comprising removing said rigid gel from said dead volume by contacting said gel with a chemical agent which substantially degrades said gel to a flowing solution.

11. The process of claim 1 wherein said gelation solution further comprises an aqueous liquid contaminant present in said dead volume which dissolves in said solution when said solution is placed in said dead volume.

Two prior art references were applied against the claims on appeal. They were U.S. Patent 4,664,294 (Hetherington), which discloses an apparatus for displacing dead space liquid using impervious bladders, or large bags, formed with flexible membranes; and U.S. Patent 4,683,949 (Sydansk), also assigned to Clay's assignee, Marathon Oil Company, which discloses a process for reducing the permeability of hydrocarbon-bearing formations and thus improving oil production, using a gel similar to that in Clay's invention.

The Board agreed with the examiner that, although neither reference alone describes Clay's invention, Hetherington and Sydansk combined support a conclusion of obviousness. It held that one skilled in the art would glean from Hetherington that Clay's invention "was appreciated in the prior art and solutions to that problem generally involved filling the dead space with *something*." Opinion at 3 (emphasis in original).

The Board also held that Sydansk would have provided one skilled in the art with information that a gelation system would have been impervious to hydrocarbons once the system gelled. The Board combined the references, finding that the "cavities" filled by Sydansk are sufficiently similar to the "volume or void space" being filled by Hetherington for one of ordinary skill to have recognized the applicability of the gel to Hetherington.

DISCUSSION

The issue presented in this appeal is whether the Board's conclusion was correct that Clay's invention would have been obvious from the combined teachings of Hetherington and Sydansk. Although this conclusion is one of law, such determinations are made against a background of several factual inquiries, one of which is the scope and content of the prior art.

A prerequisite to making this finding is determining what is "prior art," in order to consider whether "the differences between the subject matter sought to be patented and the prior art are such that the subject matter as a whole would have been obvious at the time the invention was made to a person having ordinary skill in the art." 35 U.S.C. § 103. Although § 103 does not, by its terms, define the "art to which [the] subject matter [sought to be patented] pertains," this determination is frequently couched in terms of whether the art is analogous or not, *i.e.*, whether the art is "too remote to be treated as prior art." *In re Sovish*, 769 F.2d 738 (Fed. Cir. 1985).

Clay argues that the claims at issue were improperly rejected over Hetherington and Sydansk, because Sydansk is nonanalogous art. Whether a reference in the prior art is "analogous" is a fact question. . . .

Two criteria have evolved for determining whether prior art is analogous: (1) whether the art is from the same field of endeavor, regardless of the problem addressed, and (2) if the reference is not within the field of the inventor's endeavor, whether the reference still is reasonably pertinent to the particular problem with which the inventor is involved. *In re Deminski*, 796 F.2d 436, 442 (Fed. Cir. 1986); *In re Wood*, 599 F.2d 1032, 1036 (CCPA 1979).

The Board found Sydansk to be within the field of Clay's endeavor because, as the Examiner stated, "one of ordinary skill in the art would certainly glean from [Sydansk] that the rigid gel as taught therein would have a number of applications within the manipulation of the storage and processing of hydrocarbon liquids * * * [and that] the gel as taught in Sydansk would be expected to function in a similar manner as the bladders in the Hetherington patent." These findings are clearly erroneous.

The PTO argues that Sydansk and Clay's inventions are part of a common endeavor—"maximizing withdrawal of petroleum stored in petroleum reservoirs." However, Sydansk cannot be considered to be within Clay's field of endeavor merely because both relate to the petroleum industry. Sydansk teaches the use of a gel in unconfined and irregular volumes within generally underground natural oil-bearing formations to channel flow in a desired direction; Clay teaches the introduction of gel to the confined dead volume of a man-made storage tank. The Sydansk process operates in extreme conditions, with petroleum formation temperatures as high as 115°C and at significant well bore pressures; Clay's process apparently operates at ambient temperature and atmospheric pressure. Clay's field of endeavor is the storage of refined liquid hydrocarbons. The field of endeavor of Sydansk's invention, on the other hand, is the extraction of crude petroleum. The Board clearly erred in considering Sydansk to be within the same field of endeavor as Clay's.

Even though the art disclosed in Sydansk is not within Clay's field of endeavor, the reference may still properly be combined with Hetherington if it is reasonably pertinent to the problem Clay attempts to solve. A reference is reasonably pertinent if, even though it may be in a different field from that of the inventor's endeavor, it is one which, because of the matter with which it deals, logically would have commended itself to an inventor's attention in considering his problem. Thus, the purposes of both the invention and the prior art are important in determining whether the reference is reasonably pertinent to the problem the invention attempts to solve. If a reference disclosure has the same purpose as the claimed invention, the reference relates to the same problem, and that fact supports use of that reference in an obviousness rejection. An inventor may well have been motivated to consider the reference when making his invention. If it is directed to a different purpose, the inventor would accordingly have had less motivation or occasion to consider it.

Sydansk's gel treatment of underground formations functions to fill anomalies[1] so as to improve flow profiles and sweep efficiencies of injection and production fluids through a formation, while Clay's gel functions to displace liquid product from the dead volume of a storage tank. Sydansk is concerned with plugging formation anomalies so that fluid is subsequently diverted by the gel into the formation matrix, thereby forcing bypassed oil contained in the matrix toward a production well. Sydansk is faced with the problem of recovering oil from rock, i.e., from a matrix which is porous, permeable sedimentary rock of a subterranean formation where water has channeled through formation anomalies and bypassed oil present in the matrix. Such a problem is not reasonably pertinent to the particular problem with which Clay was involved—preventing loss of stored product to tank dead volume while preventing contamination of such product. Moreover, the subterranean formation of Sydansk is not structurally similar to, does not operate under the same temperature and pressure as, and does not function like Clay's storage tanks. *See In re Ellis*, 476 F.2d 1370, 1372 (CCPA 1973) ("the similarities and differences in structure and function of the invention disclosed in the references * * * carry far greater weight [in determining analogy]").

A person having ordinary skill in the art would not reasonably have expected to solve the problem of dead volume in tanks for storing refined petroleum by considering a reference dealing with plugging underground formation anomalies. The Board's finding to the contrary is clearly erroneous. Since Sydansk is non-analogous art, the rejection over Hetherington in view of Sydansk cannot be sustained.

. . . .

Notes & Questions

1. The court concludes that the Sydansk reference is not pertinent to the question whether Clay's invention would have been obvious. But, as the court notes, the Sydansk patent is assigned on its face to Marathon Oil, just as Clay's application is. (In other words, Sydansk, like Clay, worked for Marathon Oil at the time he made his invention.) Why doesn't this common ownership make Sydansk at least presumptively pertinent when analyzing the patentability of Clay?

2. What evidence should the PTO, or a court, use to determine the claimed invention's "field of endeavor"? A prior art reference's "field of endeavor"? And what is the right level of generality at which to state the "field of endeavor"? Would it be a good idea to require an applicant to make a binding statement of his or her field of endeavor in the patent itself?

3. Under *KSR*, would Clay's invention have been obvious from the Hetherington reference alone? The PTO does not appear to have pursued that line of reasoning at the time. If the Clay application were before it today, should it pursue this analysis?

---

[1] Sydansk refers to an anomaly, one of two general region types in an oil-bearing geological formation, as "a volume or void space [e.g., 'streaks, fractures, fracture networks, vugs, solution channels, caverns, washouts, cavities, etc.'] in the formation having very high permeability relative to the matrix [the other region type, consisting of homogeneous porous rock]."

VII. Infringement

A valid patent claim provides the patentee with a right to exclude. The Patent Act defines the nature of this right by stating the activities that constitute infringement:

> [W]hoever without authority makes, uses, offers to sell, or sells any patented invention, within the United States or imports into the United States any patented invention during the term of the patent therefor, infringes the patent

35 U.S.C. § 271(a). This is not the only type of infringement the Act specifies, but it is the most basic type of infringement, and we can describe other types using this one as our baseline. The Act further provides that "[a] patentee shall have remedy by civil action for infringement of his patent." 35 U.S.C. § 281. To understand who can sue for infringement, then, we need to know the scope of the term "patentee." We begin our exploration of infringement doctrines there.

A. Ownership & Standing

"Ownership springs from invention." *Teets v. Chromalloy Gas Turbine Corp.*, 83 F.3d 403, 407 (Fed. Cir. 1996). The statute effectively dictates this: "An application for patent shall be made, or authorized to be made, by the inventor" of the claimed subject matter. 35 U.S.C. § 111(a)(1). And recall, from the *Pfaff* case (included in Section V of this Chapter), that "[t]he primary meaning of the word 'invention' in the Patent Act unquestionably refers to the inventor's conception rather than to a physical embodiment of that idea." *Pfaff v. Wells Electronics, Inc.*, 525 U.S. 55, 60 (1998). "Conception," in short, "is the touchstone of inventorship, the completion of the mental part of invention."[*] *Burroughs Wellcome Co. v. Barr Labs., Inc.*, 40 F.3d 1223, 1227-28 (Fed. Cir. 1994). In the simplest case, a solo inventor conceives of an invention and applies for patent protection. He owns the patent, if any, that results.

But solo invention and application are atypical today. In the typical case, an application's named inventor makes the invention in the workplace and applies for a patent with the aid of her employer. If the employer obtained a written assignment from the employee-inventor, such as a blanket assignment contained in an employment agreement, the employer owns the application and the patent that results from it. Although the Patent Act requires that assignments of patents, as well as assignments of applications for patents, be in writing, 35 U.S.C. § 261, ¶ 2, the employer can apply for the patent even if the assignor-employee refuses to help. The Act provides that "[a] person to whom the inventor has assigned or is under an obligation to assign the invention may make an application for patent." 35 U.S.C. § 118; *see also* 1 U.S.C. § 1 ("In determining the meaning of any Act of

---

[*] What constitutes "conception"? According to the Federal Circuit, "[c]onception is the formation in the mind of the inventor, of a definite and permanent idea of the complete and operative invention, as it is hereafter to be applied in practice." *Hybritech, Inc. v. Monoclonal Antibodies, Inc.*, 802 F.2d 1367, 1376 (Fed. Cir. 1986) (internal quotation omitted).

Congress, unless the context indicates otherwise . . . the words 'person' and 'who-ever' include corporations, companies, associations, firms, partnerships, societies, and joint stock companies, as well as individuals[.]"). The employer, as the inventor's successor by assignment, is now the "patentee" for Patent Act purposes, *see* 35 U.S.C. § 100(d) ("The word 'patentee' includes not only the patentee to whom the patent was issued but also the successors in title to the patentee."), and thus can sue others to recover for infringement. Absent the written assignment, the most an employer could have is a type of implied license to the patent, sometimes referred to as a "shop right." As a mere licensee, the employer will not have standing to sue for infringement.

The Patent Act also allows for *joint* inventorship, where two or more people working together contribute to the conception of the invention. Importantly, one can be a joint inventor without contributing as large a share as the others, and one can co-own a patent as a joint inventor without contributing to the subject matter of each and every claim in the patent. "A contribution to one claim is enough." *See Ethicon, Inc. v. U.S. Surgical Corp.*, 135 F.3d 1456, 1460 (Fed. Cir. 1998). These conclusions flow naturally from the statute, which expressly provides that "[i]nventors may apply for a patent jointly even though (1) they did not physically work together or at the same time, (2) each did not make the same type or amount of contribution, or (3) each did not make a contribution to the subject matter of every claim of the patent." 35 U.S.C. § 116(a). The tacit theory of joint invention seems to be a large corporate R&D department, like the Dow Chemical or Bell Labs of yesteryear, where managers carefully, systematically orchestrate the work of multimember teams according to clear plans. What makes this "big lab" notion of joint inventorship especially significant is that, absent a writing to the contrary, every joint owner is free to license others to practice any claim in the patent. *See Ethicon*, 135 F.3d at 1465-66 (holding that one joint inventor validly licensed the defendant, thus ending the case). This is so because, under the Act, "each of the joint owners of a patent may make, use, offer to sell, or sell the patented invention within the United States, or import the patented invention into the United States, without the consent of and without accounting to the other owners." 35 U.S.C. § 262. All of this is straightforward to arrange and document in corporate R&D departments at 3M, DuPont, Apple, or Microsoft, but can present challenges for smaller, less-sophisticated entities.

A patentee can, of course, license another party to practice the invention and retain ownership of the patent herself. This possible division between ownership and authority to practice the invention raises a question about standing to sue for infringement. Specifically, who has it? The patentee, as we noted above, has standing to sue for infringement. 35 U.S.C. § 281. What about a licensee? The basic division the cases draw is between exclusive licensees, who *can* sue for infringement, and nonexclusive licensees, who *cannot*. *See Intellectual Property Development, Inc. v. TCI Cablevision*, 248 F.3d 1333, 1342-48 (Fed. Cir. 2001) (explaining and applying these principles). Where an exclusive licensee sues, he needs to bring the patent owner into the case, and the courts will compel the patentee's joinder if need be. *See id.*

B. Direct and Indirect Infringement

Section 271(a), the core infringement provision, actually prohibits not one act of infringement, but five: making, using, offering to sell, selling, or importing. One can infringe, for example, by making a claimed invention even if one doesn't sell it. Similarly, one can infringe by using a claimed invention even if one didn't make it. What all these infringing acts have in common, however, is that they are carried out by the accused infringer (or one acting under that accused's direction and control) and embrace each and every limitation in the claim (either literally or equivalently—more of which in a moment). In patent law parlance, these are acts of *direct infringement*.

We have been talking about direct infringement from the beginning of this Chapter, to a greater or lesser degree. Recall the *Teashot v. Green Mountain* tea-brewing "K-cup" case, which we considered earlier. Teashot, the patentee, sued Green Mountain, alleging that Green Mountain directly infringed Teashot's patent by making and selling foil-covered tea pods. The key issue in the case was claim construction—as it so often is—but the setting for that issue was the patentee's allegation of direct infringement by the defendant. One proves direct infringement by a mapping process of sorts, demonstrating that the defendant's accused product or process meets every limitation of the claim considered individually. As the Federal Circuit recently summarized:

> Infringement is determined, even at summary judgment, through a two-step inquiry. First, the claims are properly construed and then those construed terms are compared to the accused product.

University of Pittsburgh v. Varian Medical Sys., 561 Fed. Appx. 934, 942 (Fed. Cir. 2014).

For example, take U.S. Patent No. 5,443,036 to Kevin Amiss, entitled "Method of Exercising a Cat." Claim 1 of the '036 patent is as follows:

> A method of inducing aerobic exercise in an unrestrained cat comprising the steps of:
> (a) directing an intense coherent beam of invisible light produced by a hand-held laser apparatus to produce a bright highly-focused pattern of light at the intersection of the beam and an opaque surface, said pattern being of visual interest to a cat; and
> (b) selectively redirecting said beam out of the cat's immediate reach to induce said cat to run and chase said beam and pattern of light around an exercise area.

'036 Patent, col. 2, ln. 60—col. 3, ln. 2. (To see the method in action, look at search results for "laser cat toy" at YouTube.) Figure 1 depicts use of the method.

To prove that someone has directly infringed the exercising
method in claim 1, Amiss must
prove that the accused infringer
carried out the recited method
steps. Given that most of this activity takes place in people's private
homes, and is thus hard to learn

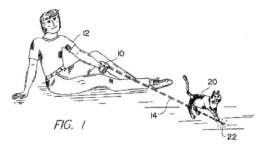

FIG. 1

about, how is this claimed method of value to Amiss? Would the patent be more
valuable to Amiss if it empowers him to exclude others from marketing hand-held
lasers for cat play? Someone who markets a laser for cat play is *not* carrying out
the method steps in Claim 1 of the '036 patent, but that seller *is* helping someone
else—a cat owner—practice that method by providing both the means and the
know-how.

Fortunately for a patentee such as Amiss, the law has long recognized *indirect
infringement* liability as well. This type of infringement exists where the accused
party has not itself directly carried out each and every limitation of a claimed invention, but it *has* brought someone else to the brink of infringement in a way that
the law deems blameworthy. Consider, for example, *Wallace v. Holmes*, 29 F. Cas.
74 (C.C.D. Conn. 1871) (Case No. 17,100), the first U.S. patent case finding indirect infringement liability. The patentee owned a claim to a combination of a metal
burner and a glass chimney. The accused infringer made and sold only the burner,
leaving its customers to provide their own glass chimneys (which were readily
available as a separate item). The accused burner maker was not a direct infringer,
because it wasn't making the claimed combination. The court, however, rejected
the notion that the burner maker was blameless. To the contrary, the court held
that the accused burner maker had sold the burners "for the express purpose of
assisting, and making profit by assisting, in a gross infringement of the complain-
ants' patent." *Id.* at 80. Rather than leave the patentee to sue the individual customers as direct infringers (a legally available but quite impractical option), the
court let the patentee recover against its blameworthy competitor. The Patent Act
codifies our contemporary indirect infringement doctrines, known as *active inducement* and *contributory infringement*, in subsections (b) and (c) of § 271. In addition, unless the third party to whom the defendant was contributing or inducing
actually infringed the patent, one cannot be held liable as an indirect infringer. *See
Everpure, Inc. v. Cuno, Inc.*, 875 F.2d 300, 304 (Fed. Cir. 1989). Although the full
details of the doctrines of indirect liability are beyond the scope of our survey, it
is important to recognize that one cannot escape patent infringement liability by
avoiding practicing the claim oneself while at the same time aiding and abetting
some other party's direct infringement.

C. Literal Infringement

Literal infringement occurs where each and every limitation of the claimed invention is present in another's product or process. *Cybor*, 138 F.3d at 1467. For
example, if the claim is for "a widget comprising an A, a B, and a C," one would

literally infringe the claim if, without authority, one made (or used, or sold, etc.) a widget comprising an A, a B, and a C. In patent jargon we would say that the claimed invention "reads on" the "accused" product. Importantly, because the courts have long construed the claim term "comprising" as a term of art that signals an open set of limitations, one would literally infringe the claim if, without authority, one made a widget comprising an A, a B, a C, and a D; or an A, a B, a C, and an E; etc. When the claim says "comprising," an accused item with the stated limitations infringes, whatever else it has. (The term of art to close one's claim is "consisting of.") A process claim is analyzed in a similar way. If the claim is for "a process for gadgeting comprising the steps A, B, and C," one would literally infringe the claim if, without authority, one carried out the steps A, B, and C.

From what you have studied thus far, it should be clear that, generally speaking, one can defend against a charge of infringement by showing that (a) the claim, when properly construed, does not read on the accused product or process, (b) the accused infringer has a license to practice the invention, or (c) the asserted patent claim is invalid. If the defense to infringement is patent invalidity, the accused infringer bears the burden of proving invalidity by clear and convincing evidence, *see* 35 U.S.C. § 282, and the material we covered in Sections III–VI is relevant. If the defense involves interpreting the scope of the claims, the material we explored in Section II of this Chapter is in play. Generally, if the claim is for literal infringement and the defense is that the patent does not read on the accused product or process, the issue is one of claim construction. For example, in the Amiss patent, does the claim term "cat" include big game cats, such as lions or tigers? Often, once the court construes the claim, the issue of infringement is resolved straightaway because the facts about the accused product or process are clear. Consider a recent Federal Circuit case, which illustrates the relationship between claim construction and the analysis of a literal-infringement question.

Nassau Precision Casting Co. v. Acushnet Co.
566 Fed. Appx. 933 (Fed. Cir. 2014)

TARANTO, JUDGE:

Nassau Precision Casting Co. owns United States Patent No. 5,486,000, entitled "Weighted Golf Iron Club Head." In September 2010, Nassau sued Acushnet Company [and others] for infringement of the '000 patent. Nassau appeals from a decision . . . that granted Acushnet's motion for summary judgment of noninfringement. For the reasons set out below, we affirm in part, vacate in part, and remand.

Background

The '000 patent describes what it says is an improvement in the distribution of weight within the head of a golf club. The purpose of the invention is to achieve "sweet spot-enhancement, i.e., significant improvement in the ball-striking efficacy of the club head, while maintaining the same starting overall weight of the club head." '000 patent, col. 1. The '000 patent states that "in the typical use of a golf club iron the ball is never intentionally struck near or at the top edge of the club face, but always at the 'sweet spot' or below," and therefore calls for removing material (generally metal or graphite) from the "top edge central portion" of the golf club head and relocating it to a lower position, preferably near the bottom edge of the golf club head. '000 patent, cols. 2, 3. Figures 5 and 10 of the '000 patent depict a prior art golf club head and a golf club head contemplated by the '000 patent, respectively[.]

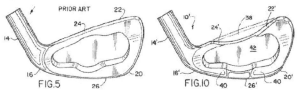

Area 38 of figure 10 shows where material has been removed. The '000 patent calls for relocating the removed material (for instance, to areas 40), rather than eliminating it entirely, so as not to alter "the overall swing weight of the club head, which typically is selected according to the size and handicap of the golfer and should remain unchanged." '000 patent, Abstract.

The two asserted claims recite "methods" of improving weight distribution in a club head by removing construction material from the central portion of the top surface of the club head and relocating it to the bottom areas of the toe and heel of the club head. Claims 1 and 2 read:

> 1. In a golf iron club head of a type having a ball-striking body of weight-imparting construction material inclined at a selected angle for driving a struck golf ball a corresponding selected height during its trajectory, said body having spaced-apart top and bottom surfaces bounding a ball-striking surface therebetween, the method of improving weight distribution comprising removing construction material from said top surface, relocating said removed construction material from said top surface to clearance positions below said top surface located adjacent opposite ends of said bottom[] surface[,] whereby said removed construction material from a location not used during ball-striking service of said golf iron[] is of no adverse consequence thereto[,] and said removed construction material in said relocated positions contributes to increasing said height attained by a struck golf ball.
>
> 2. A method of improving the weight distribution of a selected construction material constituting a golf iron club head with a ball striking surface bounded in a vertical perspective by top and bottom surfaces and in a horizontal perspective by toe and heel portions said method comprising the steps of removing construction material from a central portion of said top surface, determining the weight of said removed construction material, and embodying as part of selected bottom areas of said toe and heel of said club head said removed construction material having said determined weight, whereby the weight is distributed to said selected bottom area without any increase in the overall weight of the club head.

'000 patent, col. 4.

On September 16, 2010, Nassau brought this action, accusing Acushnet of infringing claims 1 and 2 of the '000 patent by making, offering to sell, and selling its Cobra S9, Cobra S9 Second Generation, King Cobra UFi, and Cobra S2 clubs. The following characteristics of the named clubs—not of the processes by which they were designed or manufactured—are not in dispute. None of the accused clubs has a concave-from-above topline like the one depicted in figure 10 of the '000 patent; the face of each club, in the upper portion, has a profile like that of the prior-art club head shown in figure 5. But the top surface (not face) of each club contains a channel in the metal/graphite construction material, and a lightweight polymer insert fills that channel. To ensure that the total weight of the club head is the same as if the metal/graphite construction material filled the channel, the bottom heel and toe of the club head contain more metal/graphite construction material than if no channel existed at the top—the extra amount equal in weight to the construction material "missing" from the top *minus* the weight of the polymer insert.

On April 17, 2013, the district court granted summary judgment of non-infringement in favor of Acushnet. The district court concluded that "the ordinary meaning of terms in Claims 1 and 2 such as 'remove,' 'relocate,' and 'construction material' is the shifting or redistributing of mass or weight in a golf club head design in order to achieve an optimal weight distribution." Treating that language as not referring to any actual process of manufacture or design, but to where weight is in a club compared to where it would be in some other model club, the court nevertheless held both claims not to be infringed as a matter of law.

As to claim 2, the district court reasoned that the claim requires that "all of the construction material or weight removed from the top of the club be embodied in the selected bottom areas," while not changing the net weight of the club. In the accused clubs, however, an amount of construction material equal to only the difference in weight between the polymer insert and the removed construction material is embodied in the bottom areas of the accused clubs—not the entire "determined weight" of the removed material. For that reason, the district court found that the accused clubs could not infringe claim 2.

As to claim 1, the district court construed the phrase "removed construction material from a location not used during ball-striking service" as "prohibit[ing] removal of construction material from the club head *face*," because "any area of the club head face may be used to strike a golf ball." The district court based its conclusion on testimony stating that the upper portion of the *face* near the "topline"—the line where the face meets the top surface of the club—is sometimes, though rarely, used to strike a ball. Having so construed the claim element, the district court then concluded that the accused clubs did not meet this element, even though it is undisputed that in the accused clubs material is "removed" only from the top surface of the club head, not the face. In moving from claim construction to application, the court thus seems to have broadened its view to extend beyond the face of the club head to encompass the top surface, although none of the material cited from the patent or the evidence says that the top surface is ever used for striking the ball.

On May 28, 2013, the district court entered an amended final judgment in favor of Achushnet on Nassau's infringement claims, dismissing as moot Acushnet's counterclaim for invalidity of the '000 patent. . . .

Discussion

. . . .

A

We begin with claim 2. We hold that the district court correctly ruled that claim 2 is not infringed. For this purpose, it suffices to focus, as Nassau does and the district court did regarding claim 2, on a comparison of two clubs—one before and one after the claimed process of "removing" material from the top surface and "embodying * * * said removed construction material" on bottom areas, whether in making any of the clubs or in designing them. '000 patent, col. 4.

Claim 2 requires that the weight of the club remain unchanged after relocating to the bottom of the club the construction material removed from the top, so all the removed material (in weight) must be relocated to the bottom. It is undisputed that, in the accused clubs, the "before" and "after" clubs have exactly the same weight after the combined steps of relocating construction material from the top to the bottom and inserting the polymer material at the top. Oral Arg. at 26:41-27:19. As a result, as the district court correctly concluded, not all the construction material allegedly removed from the top (or its equivalent in weight) could be placed on the bottom of the accused clubs. The polymer insert weighs something. For the weight to remain the same in the resulting club, as is undisputedly true of the accused clubs, only an amount of the removed construction material with the polymer insert weight subtracted could have been relocated to the bottom areas of the accused clubs. Claim 2 is not satisfied.

. . . .

B

With respect to claim 1, we vacate the district court's summary judgment of non-infringement. For this purpose, it is again sufficient to focus, as Nassau does and the district court did, on a comparison of two clubs—one before and one after the process of "removing" material from the top surface and "relocating said removed construction material" to certain bottom-area positions, either in making any of the clubs or in designing them. '000 patent, col. 4. The sole claim element that the district court found missing in the accused clubs is, on the record before us, undisputedly present. We remand for further proceedings on claim 1, which presents further claim construction and invalidity issues that we flag but do not definitively resolve.

1

The district court granted summary judgment of noninfringement of claim 1 based on the claim language requiring that the "removed construction material" be "from a location not used during ball-striking service," and Nassau's apparent concession that the entire club head face may be used (if unintentionally) during ballstriking service. Focusing just on that element alone, we conclude that the district court's holding that the limitation was not met is incorrect. Indeed, on the current record, the undisputed facts show that the limitation is met.

Nassau accused Acushnet of "remov[ing] material from behind the top face of the club head," and Acushnet asserts that the "channel in the center topline area of the club" containing the polymer insert (which displaced the allegedly removed construction material) "undisputedly sits behind a face at the center topline." Oral Arg. at 25:00-25:30. Thus, it is apparently undisputed that in none of the accused clubs was material removed from the face of the club head. Under the district court's construction of the pertinent claim element, the removed construction material must be from an area other than the club head face. Under Nassau's still broader construction of the claim element, the removed construction material must be either from club head areas other than the club head face or from the upper portion of the club head face (which is not normally and intentionally used by golfers for striking the ball). Under either construction, the accused clubs meet the requirement: it is undisputed that all of the removed material is from areas other than the club head face.

We need not choose between the district court's construction and Nassau's construction in order to vacate summary judgment of non-infringement. Both constructions readily encompass the accused clubs on the record before us. Resolving this aspect of the infringement inquiry therefore does not require us to decide which of the competing constructions is correct. *See, e.g., EMI Grp. N. Am., Inc. v. Intel Corp.*, 157 F.3d 887, 895 (Fed. Cir. 1998) (finding it "irrelevant whether the district court achieved a technologically perfect definition, because there [was] no dispute that the corresponding step of the [accused] process [was] within the literal scope of [the term], however * * * defined").

Acushnet's validity defenses to be litigated on remand could, of course, be affected by which construction is chosen. Keeping in mind that "'[a] patent may not, like a "nose of wax," be twisted one way to avoid anticipation and another to find infringement,'" *Amazon.com, Inc. v. Barnesandnoble.com, Inc.*, 239 F.3d 1343, 1351 (Fed. Cir. 2001) (quoting *Sterner Lighting, Inc. v. Allied Elec. Supply, Inc.*, 431 F.2d 539, 544 (5th Cir. 1970)), Nassau may be held to its insistence on its broader construction when it comes to assessing invalidity. We do not so decide, but leave the matter for remand if further construction of this claim element is needed.

2

Before entering summary judgment of noninfringement based on the "location not used" claim element, the district court rejected Acushnet's claim construction argument that the "method" language of claim 1 (and of claim 2, for that matter) refers to a physical manufacturing process, in which a club head is initially formed one way, material is then physically removed from certain (top) areas, and that material is then moved to certain (bottom) areas. Acushnet renews its argument here as an alternative ground to affirm the summary judgment of noninfringement. Like the district court, we reject Acushnet's manufacturing-process construction.

Acushnet's argument starts with the important fact that the claim language uses the fundamental patent-law language of "method," which refers to a process of taking specified actions over time. What the language does not plainly require, however, is that the actions are steps in a manufacturing process. And as the district court explained, such a construction is unreasonable here. Nothing in the patent's specification refers to the steps of physically constructing a golf club; and in the absence of a description of the physical construct-remove-relocate club-head-making process that Acushnet's interpretation would require, it is evidently undisputed that a skilled artisan would not think of making a golf club that way (instead of making and using a mold in the ultimately desired shape). Acushnet's manufacturing-process construction is so surprising in context that it must be rejected where the language does not make it unavoidable.

Importantly, an alternative construction is readily available that gives the unmistakable language of "method" its due. . . . [T]he [c]ourt construes Claims 1 and 2 as describing steps taken in designing a golf club head with the physical characteristics described in the '000 Patent. . . . The claims thus refer to a process of designing a golf club, not only a golf club having certain structural features that might be called its "design."

A "method" in a claim, one of the most common and basic terms of patent drafting, is a "process," and "method" and "process" have a clear, settled meaning: a set of actions, necessarily taken over time. *Limelight Networks, Inc. v. Akamai Networks, Inc.*, 134 S. Ct. 2111, 2117 (2014) ("[a] method patent claims a number of steps" to be "carried out"); *Bilski v. Kappos*, 130 S. Ct. 3218, 3228 (2010) (relying on definition of "method" as "way or manner of doing anything"). "Method" is not a technical term; its meaning is therefore not a matter of skilled artisans' understandings. It is a patent-law term with a stable, unambiguous meaning that distinguishes the subject matter from "product" subject matter. It would do unacceptable violence to that meaning to treat the claim here, reciting a "method," as one to a product defined simply by structural features.

This conclusion is reinforced by the fact that claim 1, in stating the steps of the "method," uses action words: "removing," "relocating." Moreover, while it calls for a comparison of a resulting club to some template, . . . it provides no identification of the template in terms of structural properties. The only template indicated by the language is a temporal one: the club with which a designing process began. That temporal comparison, built into the claim language, reinforces the steps-over-time meaning of "method" and the action words of the claim.

Finally, the "method" language of claim 1 contrasts with the unmistakable structural product language that Acushnet used when, in some of its own patents, it claimed golf club heads that embodied designs with weight distributions different from prior-art clubs. See U.S. Patent No. 7,481,718, cols. 5-6; U.S. Patent No. 7,524,250, cols. 9-10; U.S. Patent No. 7,819,757, cols. 17-20. The specifications of those patents contain language loosely describing the design with language akin to the removal/relocation language Nassau placed in its claim. But Acushnet knew that, when it came to the claims, precision was needed in using the standard language of patent claiming for the fundamental choice about what category of subject matter it was claiming. Nassau could have claimed a product in structural terms, but it did not.

Performance of the claimed steps by a designer of the accused clubs is therefore required by claim 1. . . . .

Acushnet argues on appeal that there is no evidence from which a reasonable jury could find that Acushnet performed the designing process required by claim 1. But the district court did not examine the evidence on whether Acushnet performed that process. We think it premature to rule on Acushnet's contention. Rather, we leave initial examination of the evidence on this question—whether direct or indirect evidence—to the district court.

. . . .

Notes & Questions

1. Note the relationship between claim construction and infringement, on the one hand, and claim construction and validity, on the other. Once the court construes the claim, that construction is examined in the determining the claim's validity and in determining infringement. Why might a broad construction be both helpful and detrimental? Why might a narrow construction be both helpful and detrimental?

D. The Doctrine of Equivalents

Suppose that an unauthorized party's product or process does not literally meet one or more of the limitations in a patent claim. For example, recalling the Amiss cat-exercise patent from the beginning of this Section, if Joe uses a handheld laser to exercise his dog Harlan—in steps that would otherwise fall within the scope of Claim 1 of the '036 patent—he is plainly *not* literally infringing Claim 1. Whatever the claim term "cat" covers, it does not *literally* cover a dog. Does that difference from the claim's literal scope entirely defeat infringement liability? According to the *doctrine of equivalents*, a doctrine that allows for *nonliteral* infringement, the answer is "no." As the Federal Circuit has stated, "[a] claim of infringement under the doctrine of equivalents modifies th[e] second step [of the infringement analysis, *i.e.*, the comparison step,] by requiring that the fact finder determine whether differences between particular elements of the accused device and the asserted claims are insubstantial." *Cybor Corp. v. FAS Technologies, Inc.*, 138 F.3d 1448, 1467 (Fed. Cir. 1998) (en banc) (Mayer, J., concurring in judgment). In the context of Claim 1 of the '036 patent, practicing the exercising method with a dog may be insubstantially different from practicing it with a cat. Put another way, in the Amiss method, a dog may be the *equivalent* of a "cat."

1. The Basics

The U.S. Supreme Court introduced the *doctrine of equivalents* into our patent jurisprudence (although it did not use that name for it) in *Winans v. Denmead*, 56 U.S. 330 (1854). One should appreciate that the *Winans* case arose before the modern patent claim took its settled form and central role in determining infringement

liability. In the case, patentee Ross Winans sued the Denmeads
for infringing his patent on a rail car. The asserted claim recited
that the rail car was "in the form of a frustum of a cone." *Id.* at
342. A cone is curved, of course, and a frustum is a portion of a
solid that lies between two parallel planes cutting the solid (such
as is illustrated at right). The Denmeads' car was not conical,
however; it was octagonal. They argued—and the lower court
agreed—that their straight-edged rail car could not infringe, because it was not conical or curved. Winans maintained that the

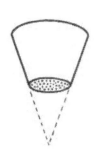

Denmeads should be held liable because their octagonal car achieved the same
practical result, and on the same practical design basis, as Winans' conical car.
Indeed, even the Denmeads' expert witness agreed "that the octagon car, practically, was as good as the conical ones; and that, substantially, [he] saw no difference between the two." *Id.* at 340.

Splitting five to four, the Supreme Court agreed with Winans that the Denmeads should be held liable as infringers. Concerned that, under the opposite approach, "the property of inventors would be valueless," *id.* at 342, the Court reasoned as follows:

> Where form and substance are inseparable, it is enough to look at the
> form only. Where they are separable; where the whole substance of the in-
> vention may be copied in a different form, it is the duty of courts and juries
> to look through the form for the substance of the invention—for that which
> entitled the inventor to his patent, and which the patent was designed to
> secure; where it is found, there is an infringement; and it is not a defen[s]e,
> that it is embodied in a form not described, and in terms [not] claimed by
> the patentee.

Id. at 343. The dissent in *Winans*, pointing to the then-current Patent Act language
akin to our § 112(a), predicted that the majority's theory of nonliteral infringement
would work great unfairness: "Nothing, in the administration of this [patent] law,
will be more mischievous, more productive of oppressive and costly litigation, of
exorbitant and unjust pretensions and vexatious demands, more injurious to labor,
than" abandoning claim language as the touchstone for infringement liability. *Id.*
at 347 (Campbell, J., dissenting).

Nearly a century later, after modern claiming practice had taken full hold, the
Supreme Court revisited the question whether one could infringe a patent equivalently, rather than literally. Splitting six to two, the Court reaffirmed *Winans*, and
on the same basic policy rationale, in *Graver Tank & Mfg. Co. v. Linde Air Products
Co.*, 339 U.S. 605 (1949). According to Justice Jackson,

> courts have . . . recognized that to permit imitation of a patented invention
> which does not copy every literal detail would be to convert the protection
> of the patent grant into a hollow and useless thing. Such a limitation would
> leave room for—indeed, encourage—the unscrupulous copyist to make un-
> important and insubstantial changes and substitutions in the patent which,
> though adding nothing, would be enough to take the copied matter outside

the claim, and hence outside the reach of law. One who seeks to pirate an invention, like one who seeks to pirate a copyrighted book or play, may be expected to introduce minor variations to conceal and shelter the piracy. Outright and forthright duplication is a dull and very rare type of infringement. To prohibit no other would place the inventor at the mercy of verbalism, and would be subordinating substance to form. It would deprive him of the benefit of his invention, and would foster concealment, rather than disclosure, of inventions, which is one of the primary purposes of the patent system.

Id. at 607. Justice Black's dissent largely echoed Justice Campbell's dissent in *Winans*: "Hereafter, a manufacturer cannot rely on what the language of a patent claims. He must be able, at the peril of heavy infringement damages, to forecast how far a court relatively unversed in a particular technological field will expand the claim's language after considering the testimony of technical experts in that field." *Id.* at 617 (Black, J., dissenting).

The basic tensions that *Winans* laid bare and that *Graver* rehearsed—between the patent claim as a source of fair notice to the public and nonliteral infringement as a guarantor of fair value for the patentee—remain with us today. Since *Graver*, however, the story of the doctrine of equivalents has largely been one of contraction. The paramount importance of claims has carried the day, in that the courts have developed a number of legal restrictions that, when triggered by the patentee, put the doctrine of equivalents off limits as a theory of recovery.

2. Contemporary Equivalents Analysis

Liability under the doctrine of equivalents presents two overarching questions. One is factual: Is the alleged equivalent an insubstantial change from the recited claim limitation? *Cybor*, 138 F.3d at 1467. The most common way to assess this is to determine whether the alleged equivalent performs substantially the same function, in substantially the same way, to achieve substantially the same result, as the claim limitation. In patent parlance, this analysis is called *the function/way/result test*.

The other overarching question is legal: Do any of the legal restrictions hemmed around the doctrine of equivalents put this theory of recovery off limits for the patentee? We look at three of these legal limitations in the materials that follow. The first legal restriction against using the doctrine of equivalents is *prosecution history estoppel*, the second is the *all limitations rule* (also known, in more recent cases, as the *claim vitiation rule*), and the third is the *disclosure-dedication rule*. These legal limitations present questions for the court, and can thus be used to resolve a case on summary judgment (or on motion for judgment as a matter of law).

a. Prosecution History Estoppel

Festo Corp. v. Shoketsu Kinzoku Kogyo Kabushiki Co.
535 U.S. 722 (2002)

KENNEDY, JUSTICE:

This case requires us to address once again the relation between two patent law concepts, the doctrine of equivalents and the rule of prosecution history estoppel. The Court considered the same concepts in *Warner-Jenkinson Co.* v. *Hilton Davis Chemical Co.,* 520 U. S. 17 (1997), and reaffirmed that a patent protects its holder against efforts of copyists to evade liability for infringement by making only insubstantial changes to a patented invention. At the same time, we appreciated that by extending protection beyond the literal terms in a patent the doctrine of equivalents can create substantial uncertainty about where the patent monopoly ends. If the range of equivalents is unclear, competitors may be unable to determine what is a permitted alternative to a patented invention and what is an infringing equivalent.

To reduce the uncertainty, *Warner-Jenkinson* acknowledged that competitors may rely on the prosecution history, the public record of the patent proceedings. In some cases the Patent and Trademark Office (PTO) may have rejected an earlier version of the patent application on the ground that a claim does not meet a statutory requirement for patentability. When the patentee responds to the rejection by narrowing his claims, this prosecution history estops him from later arguing that the subject matter covered by the original, broader claim was nothing more than an equivalent. Competitors may rely on the estoppel to ensure that their own devices will not be found to infringe by equivalence.

In the decision now under review the Court of Appeals for the Federal Circuit held that by narrowing a claim to obtain a patent, the patentee surrenders all equivalents to the amended claim element. Petitioner asserts this holding departs from past precedent in two respects. First, it applies estoppel to every amendment made to satisfy the requirements of the Patent Act and not just to amendments made to avoid pre-emption by an earlier invention, *i.e.,* the prior art. Second, it holds that when estoppel arises, it bars suit against every equivalent to the amended claim element. The Court of Appeals acknowledged that this holding departed from its own cases, which applied a flexible bar when considering what claims of equivalence were estopped by the prosecution history. . . .

We granted certiorari to consider these questions.

I

Petitioner Festo Corporation owns two patents for an improved magnetic rodless cylinder, a piston-driven device that relies on magnets to move objects in a conveying system. The device has many industrial uses and has been employed in machinery as diverse as sewing equipment and the Thunder Mountain ride at Disney World. Although the precise details of the cylinder's operation are not essential here, the prosecution history must be considered.

Petitioner's patent applications, as often occurs, were amended during the prosecution proceedings. The application for the first patent, the Stoll Patent (U.S. Patent No. 4,354,125), was amended after the patent examiner rejected the initial application because the exact method of operation was unclear and some claims were [drafted] in an impermissible way [under § 112]. The inventor, Dr. Stoll, submitted a new application designed to meet the examiner's objections . . . . The second patent, the Carroll Patent (U.S. Patent No. 3,779,401), was also amended during a reexamination proceeding. . . . Both amended patents added a new limitation—that the inventions contain a pair of sealing rings, each having a lip on one side, which would prevent impurities from getting on the piston assembly. The amended Stoll Patent added the further limitation that the outer shell of the device, the sleeve, be made of a magnetizable material.

After Festo began selling its rodless cylinder, respondents (whom we refer to as SMC) entered the market with a device similar, but not identical, to the ones disclosed by Festo's patents. SMC's cylinder, rather than using two one-way sealing rings, employs a single sealing ring with a two-way lip. Furthermore, SMC's sleeve is made of a nonmagnetizable alloy. SMC's device does not fall within the literal claims of either patent, but petitioner contends that it is so similar that it infringes under the doctrine of equivalents.

SMC contends that Festo is estopped from making this argument because of the prosecution history of its patents. The sealing rings and the magnetized alloy in the Festo product were both [listed as claim limitations] for the first time in the amended applications. In SMC's view, these amendments narrowed the earlier [claims], surrendering alternatives that are the very points of difference in the competing devices—the sealing rings and the type of alloy used to make the sleeve. As Festo narrowed its claims in these ways in order to obtain the patents, says SMC, Festo is now estopped from saying that these features are immaterial and that SMC's device is an equivalent of its own.

The United States District Court for the District of Massachusetts disagreed. It held that Festo's amendments were not made to avoid prior art, and therefore the amendments were not the kind that give rise to estoppel. A panel of the Court of Appeals for the Federal Circuit affirmed. We granted certiorari, vacated, and remanded in light of our intervening decision in *Warner-Jenkinson Co.* v. *Hilton Davis Chemical Co.* After a decision by the original panel on remand, the Court of Appeals ordered rehearing en banc to address questions that had divided its judges since our decision in *Warner-Jenkinson*.

The en banc court reversed, holding that prosecution history estoppel barred Festo from asserting that the accused device infringed its patents under the doctrine of equivalents. The court held, with only one judge dissenting, that estoppel arises from any amendment that narrows a claim to comply with the Patent Act, not only from amendments made to avoid prior art. More controversial in the Court of Appeals was its further holding: When estoppel applies, it stands as a complete bar against any claim of equivalence for the element that was amended. . . . In the court's view a complete-bar rule, under which estoppel bars all claims of equivalence to the narrowed element, would promote certainty in the determination of infringement cases.

.....

II

The patent laws "promote the Progress of Science and useful Arts" by rewarding innovation with a temporary monopoly. U.S. Const., Art. I, § 8, cl. 8. The monopoly is a property right; and like any property right, its boundaries should be clear. This clarity is essential to promote progress, because it enables efficient investment in innovation. A patent holder should know what he owns, and the public should know what he does not. For this reason, the patent laws require inventors to describe their work in "full, clear, concise, and exact terms," 35 U.S.C. § 112, as part of the delicate balance the law attempts to maintain between inventors, who rely on the promise of the law to bring the invention forth, and the public, which should be encouraged to pursue innovations, creations, and new ideas beyond the inventor's exclusive rights.

Unfortunately, the nature of language makes it impossible to capture the essence of a thing in a patent application. The inventor who chooses to patent an invention and disclose it to the public, rather than exploit it in secret, bears the risk that others will devote their efforts toward exploiting the limits of the patent's language:

> An invention exists most importantly as a tangible structure or a series of drawings. A verbal portrayal is usually an afterthought written to satisfy the requirements of patent law. This conversion of machine to words allows for unintended idea gaps which cannot be satisfactorily filled. Often the invention is novel and words do not exist to describe it. The dictionary does not always keep abreast of the inventor. It cannot. Things are not made for the sake of words, but words for things.

Autogiro Co. of America v. *United States,* 384 F. 2d 391, 397 (Ct. Cl. 1967).

The language in the patent claims may not capture every nuance of the invention or describe with complete precision the range of its novelty. If patents were always interpreted by their literal terms, their value would be greatly diminished. Unimportant and insubstantial substitutes for certain elements could defeat the patent, and its value to inventors could be destroyed by simple acts of copying. For this reason, the clearest rule of patent interpretation, literalism, may conserve judicial resources but is not necessarily the most efficient rule. The scope of a patent is not limited to its literal terms but instead embraces all equivalents to the claims described. See *Winans* v. *Denmead,* 15 How. 330, 347 (1854).

It is true that the doctrine of equivalents renders the scope of patents less certain. It may be difficult to determine what is, or is not, an equivalent to a particular element of an invention. If competitors cannot be certain about a patent's extent, they may be deterred from engaging in legitimate manufactures outside its limits, or they may invest by mistake in competing products that the patent secures. In addition the uncertainty may lead to wasteful litigation between competitors, suits that a rule of literalism might avoid. These concerns with the doctrine of equivalents, however, are not new. Each time the Court has considered the doctrine, it has acknowledged this uncertainty as the price of ensuring the appropriate incentives for innovation, and it has affirmed the doctrine over dissents that urged a more certain rule. When the Court in *Winans* v. *Denmead* first adopted what has become the doctrine of equivalents, it stated that "[t]he exclusive right to the thing patented is not secured, if the public are at liberty to make substantial copies of it, varying its form or proportions." *Id.* at 343. The dissent argued that the Court had sacrificed the objective of "[f]ul[l]ness, clearness, exactness, preciseness, and particularity, in the description of the invention." *Id.* at 347.

The debate continued in *Graver Tank & Mfg. Co.* v. *Linde Air Products Co.*, 339 U.S. 605 (1950), where the Court reaffirmed the doctrine. . . .

Most recently, in *Warner-Jenkinson*, the Court reaffirmed that equivalents remain a firmly entrenched part of the settled rights protected by the patent. A unanimous opinion concluded that if the doctrine is to be discarded, it is Congress and not the Court that should do so:

> [T]he lengthy history of the doctrine of equivalents strongly supports adherence to our refusal in *Graver Tank* to find that the Patent Act conflicts with that doctrine. Congress can legislate the doctrine of equivalents out of existence any time it chooses. The various policy arguments now made by both sides are thus best addressed to Congress, not this Court.

520 U.S. at 28.

III

Prosecution history estoppel requires that the claims of a patent be interpreted in light of the proceedings in the PTO during the application process. Estoppel is a "rule of patent construction" that ensures that claims are interpreted by reference to those "that have been cancelled or rejected." *Schriber-Schroth Co.* v. *Cleveland Trust Co.*, 311 U.S. 211, 220-221 (1940). The doctrine of equivalents allows the patentee to claim those insubstantial alterations that were not captured in drafting the original patent claim but which could be created through trivial changes. When, however, the patentee originally claimed the subject matter alleged to infringe but then narrowed the claim in response to a rejection, he may not argue that the surrendered territory comprised unforeseen subject matter that should be deemed equivalent to the literal claims of the issued patent. On the contrary, "[b]y the amendment [the patentee] recognized and emphasized the difference between the two phrases[,] * * * and [t]he difference which [the patentee] thus disclaimed must be regarded as material." *Exhibit Supply Co.* v. *Ace Patents Corp.*, 315 U.S. 126, 136-137 (1942).

A rejection indicates that the patent examiner does not believe the original claim could be patented. While the patentee has the right to appeal, his decision to forgo an appeal and submit an amended claim is taken as a concession that the invention as patented does not reach as far as the original claim. Were it otherwise, the inventor might avoid the PTO's gatekeeping role and seek to recapture in an infringement action the very subject matter surrendered as a condition of receiving the patent.

Prosecution history estoppel ensures that the doctrine of equivalents remains tied to its underlying purpose. Where the original application once embraced the purported equivalent but the patentee narrowed his claims to obtain the patent or to protect its validity, the patentee cannot assert that he lacked the words to describe the subject matter in question. The doctrine of equivalents is premised on language's inability to capture the essence of innovation, but a prior application describing the precise element at issue undercuts that premise. In that instance the prosecution history has established that the inventor turned his attention to the subject matter in question, knew the words for both the broader and narrower claim, and affirmatively chose the latter.

A

The first question in this case concerns the kinds of amendments that may give rise to estoppel. Petitioner argues that estoppel should arise when amendments are intended to narrow the subject matter of the patented invention, for instance, amendments to avoid prior art, but not when the amendments are made to comply with requirements concerning the form of the patent application. . . .

Petitioner is correct that estoppel has been discussed most often in the context of amendments made to avoid the prior art. See *Exhibit Supply Co., supra,* at 137; *Keystone Driller Co.* v. *Northwest Engineering Corp.,* 294 U.S. 42, 48 (1935). Amendment to accommodate prior art was the emphasis, too, of our decision in *Warner-Jenkinson, supra,* at 30. It does not follow, however, that amendments for other purposes will not give rise to estoppel. Prosecution history may rebut the inference that a thing not described was indescribable. That rationale does not cease simply because the narrowing amendment, submitted to secure a patent, was for some purpose other than avoiding prior art.

We agree with the Court of Appeals that a narrowing amendment made to satisfy any requirement of the Patent Act may give rise to an estoppel. As that court explained, a number of statutory requirements must be satisfied before a patent can issue. The claimed subject matter must be useful, novel, and not obvious. 35 U.S.C. §§ 101-103. In addition, the patent application must describe, enable, and set forth the best mode of carrying out the invention. 35 U.S.C. § 112. These latter requirements must be satisfied before issuance of the patent, for exclusive patent rights are given in exchange for disclosing the invention to the public. What is claimed by the patent application must be the same as what is disclosed in the specification; otherwise the patent should not issue. The patent also should not issue if the other requirements of § 112 are not satisfied, and an applicant's failure to meet these requirements could lead to the issued patent being held invalid in later litigation.

. . . .

Estoppel arises when an amendment is made to secure the patent and the amendment narrows the patent's scope. If a § 112 amendment is truly cosmetic, then it would not narrow the patent's scope or raise an estoppel. On the other hand, if a § 112 amendment is necessary and narrows the patent's scope—even if only for the purpose of better description—estoppel may apply. A patentee who narrows a claim as a condition for obtaining a patent disavows his claim to the broader subject matter, whether the amendment was made to avoid the prior art or to comply with § 112. We must regard the patentee as having conceded an inability to claim the broader subject matter or at least as having abandoned his right to appeal a rejection. In either case estoppel may apply.

B

Petitioner concedes that the limitations at issue—the sealing rings and the composition of the sleeve—were made for reasons related to § 112, if not also to avoid the prior art. Our conclusion that prosecution history estoppel arises when a claim is narrowed to comply with § 112 gives rise to the second question presented: Does the estoppel bar the inventor from asserting infringement against any equivalent to the narrowed element or might some equivalents still infringe? The Court of Appeals held that prosecution history estoppel is a complete bar, and so the narrowed element must be limited to its strict literal terms. Based upon its experience the Court of Appeals decided that the flexible-bar rule is unworkable because it leads to excessive uncertainty and burdens legitimate innovation. For the reasons that follow, we disagree with the decision to adopt the complete bar.

Though prosecution history estoppel can bar a patentee from challenging a wide range of alleged equivalents made or distributed by competitors, its reach requires an examination of the subject matter surrendered by the narrowing amendment. The complete bar avoids this inquiry by establishing a *per se* rule; but that approach is inconsistent with the purpose of applying the estoppel in the first place — to hold the inventor to the representations made during the application process and to the inferences that may reasonably be drawn from the amendment. By amending the application, the inventor is deemed to concede that the patent does not extend as far as the original claim. It does not follow, however, that the amended claim becomes so perfect in its description that no one could devise an equivalent. After amendment, as before, language remains an imperfect fit for invention. The narrowing amendment may demonstrate what the claim is not; but it may still fail to capture precisely what the claim is. There is no reason why a narrowing amendment should be deemed to relinquish equivalents unforeseeable at the time of the amendment and beyond a fair interpretation of what was surrendered. Nor is there any call to foreclose claims of equivalence for aspects of the invention that have only a peripheral relation to the reason the amendment was submitted. The amendment does not show that the inventor suddenly had more foresight in the drafting of claims than an inventor whose application was granted without amendments having been submitted. It shows only that he was familiar with the broader text and with the difference between the two. As a result, there is no more reason for holding the patentee to the literal terms of an amended claim than there is for abolishing the doctrine of equivalents altogether and holding every patentee to the literal terms of the patent.

This view of prosecution history estoppel is consistent with our precedents and respectful of the real practice before the PTO. . . . [W]e have consistently applied the doctrine in a flexible way, not a rigid one. We have considered what equivalents were surrendered during the prosecution of the patent, rather than imposing a complete bar that resorts to the very literalism the equivalents rule is designed to overcome.

The Court of Appeals ignored the guidance of *Warner-Jenkinson*, which instructed that courts must be cautious before adopting changes that disrupt the settled expectations of the inventing community. See 520 U.S. at 28. In that case we made it clear that the doctrine of equivalents and the rule of prosecution history estoppel are settled law. The responsibility for changing them rests with Congress. . . .

In *Warner-Jenkinson* we struck the appropriate balance by placing the burden on the patentee to show that an amendment was not for purposes of patentability:

> Where no explanation is established, however, the court should presume that the patent application had a substantial reason related to patentability for including the limiting element added by amendment. In those circumstances, prosecution history estoppel would bar the application of the doctrine of equivalents as to that element.

Id. at 33.

When the patentee is unable to explain the reason for amendment, estoppel not only applies but also "bar[s] the application of the doctrine of equivalents as to that element." *Id.* These words do not mandate a complete bar; they are limited to the circumstance where "no explanation is established." They do provide, however, that when the court is unable to determine the purpose underlying a narrowing amendment — and hence a rationale for limiting the estoppel to the surrender of particular equivalents — the court should presume that the patentee surrendered all subject matter between the broader and the narrower language.

Just as *Warner-Jenkinson* held that the patentee bears the burden of proving that an amendment was not made for a reason that would give rise to estoppel, we hold here that the patentee should bear the burden of showing that the amendment does not surrender the particular equivalent in question. . . . The patentee, as the author of the claim language, may be expected to draft claims encompassing readily known equivalents. A patentee's decision to narrow his claims through amendment may be presumed to be a general disclaimer of the territory between the original claim and the amended claim. *Exhibit Supply,* 315 U.S. at 136-137 ("By the amendment [the patentee] recognized and emphasized the difference between the two phrases and proclaimed his abandonment of all that is embraced in that difference"). There are some cases, however, where the amendment cannot reasonably be viewed as surrendering a particular equivalent. The equivalent may have been unforeseeable at the time of the application; the rationale underlying the amendment may bear no more than a tangential relation to the equivalent in question; or there may be some other reason suggesting that the patentee could not reasonably be expected to have described the insubstantial substitute in question. In those cases the patentee can overcome the presumption that prosecution history estoppel bars a finding of equivalence.

This presumption is not, then, just the complete bar by another name. Rather, it reflects the fact that the interpretation of the patent must begin with its literal claims, and the prosecution history is relevant to construing those claims. When the patentee has chosen to narrow a claim, courts may presume the amended text was composed with awareness of this rule and that the territory surrendered is not an equivalent of the territory claimed. In those instances, however, the patentee still might rebut the presumption that estoppel bars a claim of equivalence. The patentee must show that at the time of the amendment one skilled in the art could not reasonably be expected to have drafted a claim that would have literally encompassed the alleged equivalent.

IV

On the record before us, we cannot say petitioner has rebutted the presumptions that estoppel applies and that the equivalents at issue have been surrendered. Petitioner concedes that the limitations at issue — the sealing rings and the composition of the sleeve — were made in response to a rejection for reasons under § 112, if not also because of the prior art references. As the amendments were made for a reason relating to patentability, the question is not whether estoppel applies but what territory the amendments surrendered. While estoppel does not effect a complete bar, the question remains whether petitioner can demonstrate that the narrowing amendments did not surrender the particular equivalents at issue. On these questions, SMC may well prevail, for the sealing rings and the composition of the sleeve both were noted expressly in the prosecution history. These matters, however, should be determined in the first instance by further proceedings in the Court of Appeals or the District Court.

. . . .

Notes & Questions

1. If the Court is right about language's inability to fully capture an invention in words, why have claims at all? Would it be better to simply require an enabling disclosure and let the fact finder determine the scope of protection at trial?

2. The Court states that Congress can, any time it chooses, "legislate the doctrine of equivalents out of existence." Given what we know about basic interest group politics, is the Court's statement realistic? Assume, for the sake of argument, that you think the doctrine of equivalents strikes the wrong balance between public freedom and patentee value. Does the Court, as the doctrine's creator, bear any responsibility for ending it?

3. One consequence of prosecution history estoppel is that the same final claim language produces different rights to exclude, depending on whether the operative limitation in a case was part of the originally filed claim or, in an alternative reality, was introduced as a narrowing amendment. In the first world, the patentee does not face a presumptive surrender of claim scope as to the limitation. In the second world, the patentee *does* face the presumption. And it is hard to rebut this presumption! Think of the matter in incentive terms. What are we encouraging with this different treatment? Is it good policy to encourage that (whatever it is)?

4. On remand, 344 F.3d 1359 (Fed. Cir. 2003), the Federal Circuit fleshed out the showings needed for each of the three rebuttals to the new *Festo* presumptive surrender:

(a) *The unforeseeability rebuttal* "presents an objective inquiry, asking whether the alleged equivalent would have been unforeseeable to one of ordinary skill in the art at the time of the amendment. Usually, if the alleged equivalent represents later-developed technology (*e.g.,* transistors in relation to vacuum tubes, or Velcro® in relation to fasteners) or technology that was not known in the relevant art, then it would not have been foreseeable. In contrast, old technology, while not always foreseeable, would more likely have been foreseeable. Indeed, if the alleged equivalent were known in the prior art in the field of the invention, it certainly should have been foreseeable at the time of the amendment. By its very nature, objective unforeseeability depends on underlying factual issues relating to, for example, the state of the art and the understanding of a hypothetical person of ordinary skill in the art at the time of the amendment. Therefore, in determining whether an alleged equivalent would have been unforeseeable, a district court may hear expert testimony and consider other extrinsic evidence relating to the relevant factual inquiries." *Id.* at 1369.

(b) *The tangentiality rebuttal* "asks whether the reason for the narrowing amendment was peripheral, or not directly relevant, to the alleged equivalent. Although we cannot anticipate the instances of mere tangentialness that may arise, we can say that an amendment made to avoid prior art that contains the equivalent in question is not tangential; it is central to allowance of the claim. Moreover, much like the inquiry into whether a patentee can rebut the *Warner-Jenkinson* presumption that a narrowing amendment was made for a reason of patentability, the inquiry into whether a patentee can rebut the *Festo* presumption under the 'tangential' criterion focuses on the patentee's objectively apparent reason for the narrowing amendment. As we have held in the *Warner-Jenkinson* context, that reason should be discernible from the prosecution history record, if the public notice function of a patent and its prosecution history is to have significance. Moreover, whether an amendment was merely tangential to an alleged equivalent necessarily requires focus on the context in which the amendment was made; hence the resort to the prosecution history. Thus, whether the patentee has established a merely tangential reason for a narrowing amendment is for the court to determine from the prosecution history record without the introduction of additional evidence, except, when necessary, testimony from those skilled in the art as to the interpretation of that record." *Id.* at 1369-70.

(c) *The "some other reason" rebuttal,* "while vague, must be a narrow one; it is available in order not to totally foreclose a patentee from relying on reasons, other than unforeseeability and tangentialness, to show that it did not surrender the alleged equivalent. Thus, the third criterion may be satisfied when there was some reason, such as the shortcomings of language, why the patentee was prevented from describing the alleged equivalent when it narrowed the claim. When at all possible, determination of the third rebuttal criterion should also be limited to the prosecution history record." *Id.* at 1370.

5. Festo filed its infringement suit against SMC in August 1988. The Federal Circuit issued its final decision in the case, rejecting infringement under the doctrine of equivalents, on July 5, 2007, 493 F.3d 1368, and the Supreme Court denied review on June 9, 2008, 553 U.S. 1093.

6. Congress established the Federal Circuit in 1982 to bring uniformity to the field of patent law, by (a) providing appellate jurisdiction for patent cases regardless of the location of the district court in which the infringement suit was filed, and (b) combining it with appellate review of PTO rejections of patent applications. In addition to disparate patent law standards and the resultant forum shopping, the rate of patent invalidity findings was thought to weaken the patent system's incentives to innovate and to disclose; Congress sought to ameliorate these problems. The Federal Circuit has certainly made the law more uniform, given that circuit splits are ruled out by design, and has invalidated patents less frequently than did the regional courts of appeals. On the other hand, having only one appellate court hear a particular type of case can lead some to perceive that appellate court as a quasi-supreme court on that topic. For about 12 years after the establishment of the Federal Circuit, the Supreme Court granted review in only a small number of cases, cases that focused on procedural matters rather than substantive patent law. This pattern reinforced the notion that the Federal Circuit was becoming, effectively, the last judicial word on U.S. patent law. Since 1995, however, the Supreme Court has become more and more deeply involved with substantive patent law. In *KSR*, the Supreme Court overruled longstanding Federal Circuit jurisprudence on the nonobviousness requirement. In *Festo*, the Supreme Court rejected the Federal Circuit's conclusion that a flexible prosecution history bar to equivalent infringement had become unworkable. In *Pfaff*, the Supreme Court rejected the Federal Circuit's framework for determining when a claimed invention was complete enough to be "on sale" under § 102(b). In *Mayo* and *Myriad*, the Supreme Court reversed Federal Circuit decisions on the scope of patentable subject matter under § 101. Do these decisions say something about the wisdom of having a single intermediate appellate court for patent law? For an intriguing proposal that we return to a multi-circuit-court model, see Craig Allen Nard & John F. Duffy, *Rethinking Patent Law's Uniformity Principle*, 101 Nw. U. L. Rev. 1619 (2007).

b. The All Limitations Rule

As the Supreme Court noted at the outset in *Festo*, that case was the second time in five years that the Court had considered the fundamentals of the doctrine of equivalents. In the first of those two cases, *Warner-Jenkinson Co. v. Hilton Davis Chemical Co.,* 520 U. S. 17 (1997), the Court conceded—even as it decided that it was for Congress, not the Court, to eliminate the doctrine of equivalents—that "the doctrine of equivalents, as it has come to be applied since *Graver Tank* [(1950)], has taken on a life of its own, unbounded by the patent claims," and that "[t]here can be no denying that the doctrine of equivalents, when applied broadly, conflicts with the definitional and public-notice functions of the statutory claiming requirement." *Id.* at 28-29. To remedy the threat posed by a freewheeling doctrine of equivalents entirely cut loose from the rigors of claim language, the Court held that "[e]ach element contained in a patent claim is deemed material to defining the scope of the patented invention, and thus the doctrine of equivalents must be applied to individual elements of the claim, not to the invention as a whole." *Id.* Moreover, "[i]t is important to ensure that the application of the doctrine, even as to an individual element, is not allowed such broad play as to effectively eliminate that element in its entirety." *Id.*; *see also id.* at 39 n.8 ("if a theory of equivalence would entirely vitiate a particular claim element, partial or complete judgment [of noninfringement] should be rendered by the court"), *id.* at 40 (requiring "[a] focus on individual elements and a special vigilance against allowing the concept of equivalence to eliminate completely any such elements").

This rule, that "the doctrine of equivalents must be applied to individual elements of the claim, not to the invention as a whole," is known as the All Limitations Rule. Sometimes the rule is also referred to as "the claim vitiation doctrine." The Federal Circuit has relied on this rule to conclude, as a matter of law, that the doctrine of equivalents is unavailable in a variety of fact settings. A common thread running through at least some of these cases is that the accused infringer could frame the purported equivalent as the natural-language-opposite of the claim limitation in question:

- *Freedman Seating Co. v. American Seating Co.*, 420 F.3d 1350 (Fed. Cir. 2005)—a rotatably mounted seat cannot be the equivalent of a slidably mounted seat

- *Asyst Techs., Inc. v. Emtrak, Inc.*, 402 F.3d 1188 (Fed. Cir. 2005)—an unmounted part cannot be the equivalent of a mounted part

- *Novartis Pharm. Corp. v. Abbott Labs.*, 375 F.3d 1328 (Fed. Cir. 2004)—a surfactant cannot be the equivalent of a nonsurfactant

- *Moore U.S.A., Inc. v. Standard Register Co.*, 229 F.3d 1091, 1106 (Fed. Cir. 2000)—a minority length cannot be the equivalent of a majority length

It is important to recognize, however, that the Federal Circuit has resisted stating any general formula to guide its application of the All Limitations Rule:

> There is no set formula for determining whether a finding of equivalence would vitiate a claim limitation, and thereby violate the all limitations rule. Rather, courts must consider the totality of the circumstances of each case and determine whether the alleged equivalent can be fairly characterized as an insubstantial change from the claimed subject matter without rendering the pertinent limitation meaningless.

Freedman Seating Co., 420 F.3d at 1359. As the next case reflects, the Federal Circuit has more recently described the limiting force of the All Limitations Rule merely as an application of the traditional summary judgment standard.

Cadence Pharm. Inc. v. Exela PharmSci Inc.
780 F.3d 1364 (Fed. Cir. 2015)

Linn, Judge:

In this Hatch-Waxman Act litigation [against Exela PharmSci Inc., a generic drug maker seeking to market a generic drug before a patent on the drug expires, Exela], . . . appeal[s] the district court's construction of certain claim terms of U.S. Patent[] No. . . . 6,992,218 and its rulings that Exela infringed . . . the '218 patent. For the reasons set forth *infra*, we affirm.

I. Background

. . . .

SCR Pharmatop and Cadence Pharmaceuticals, Inc. (collectively "Cadence") are the owner and exclusive licensee, respectively, of the . . . '218 patent[]. [It is] directed to aqueous phenol formulations—particularly acetaminophen (sometimes referred to as "paracetamol").

. . . .

The '218 patent claims priority to a French application filed on June 6, 2000. The '218 patent discloses a method for obtaining stable acetaminophen formulations by deoxygenating solutions with an inert gas to achieve oxygen concentrations below 2 parts-per-million ("ppm"). Claim 1 of the '218 patent is the only independent claim, and recites (with the edits from the certificate of correction in brackets and the disputed terms highlighted):

> 1. A method for preparing an *aqueous solution* with an active [principle of phenolic] nature susceptible to oxidation, which is paracetamol, while preserving for a prolonged period, comprising deoxygenation of the *solution* by bubbling with at least one inert gas and/or placing under vacuum, until the oxygen content is below 2 ppm, and optionally the aforementioned *aqueous solution* with an active principle is topped with an inert gas atmosphere heavier than air and placed in a closed container in which the prevailing pressure is 65,000 Pa maximum, and the oxygen content of the *aqueous solution* is below 2 ppm, and optionally the deoxygenation of the *solution* is completed by addition of an antioxidant.

. . . .

Cadence Pharmaceuticals Inc. markets an injectable acetaminophen product, which is approved by the Food & Drug Administration and is distributed under the name Ofirmev. The FDA's Approved Drug Products with Therapeutic Equivalence Evaluations (better known as the "Orange Book") lists the . . . '218 patent[] in connection with Ofirmev.

Exela filed an Abbreviated New Drug Application ("ANDA") with the FDA, seeking approval of a generic equivalent of Ofirmev. The ANDA included [the typical] certification [that a generic drug maker provides] . . . stating that the . . . '218 patent [was] invalid and not infringed. In response, Cadence sued Exela for infringing . . . claims 1, 3, 4 and 19 of the '218 patent pursuant to 35 U.S.C. § 271(e)(2).

The district court found . . . the '218 patent not invalid and infringed under the doctrine of equivalents. Exela appeals . . . .

II. Discussion

. . . .

"Infringement, either literal or under the doctrine of equivalents, is a question of fact that we review for clear error when tried without a jury." *Ultra-Tex Surfaces, Inc. v. Hill Bros. Chem. Co.*, 204 F.3d 1360, 1363 (Fed. Cir. 2000). "A factual finding is clearly erroneous if, despite some supporting evidence, we are left with the definite and firm conviction that a mistake has been made." *Ferring B.V. v. Watson Labs.*, 764 F.3d 1401, 1406 (Fed. Cir. 2014). Whether the doctrine of equivalents would vitiate a claimed element is a question of law that we review *de novo*. *Cordis Corp. v. Bos. Scientific Corp.*, 561 F.3d 1319, 1330 (Fed. Cir. 2009).

. . . .

Exela argues . . . that the district court erred in holding that Exela's process infringed the asserted claims of the '218 patent under the doctrine of equivalents, contending that reducing the amount of dissolved oxygen to below 2 ppm before acetaminophen is added is substantially different from reducing the dissolved oxygen content after acetaminophen is added. . . .

. . . .

The district court construed the terms "aqueous solution" and "solution" in claim 1 of the '218 patent as "[a] composition containing water as a solvent and an active ingredient susceptible to oxidation." The district court thus concluded that the claimed step of "deoxygenation of the solution" required that an active ingredient already be dissolved. In other words, the district court interpreted the claim to directly cover only the method of first dissolving an active ingredient to form a solution and then deoxygenating the solution. Exela's accused process, by contrast, first deoxygenates a solvent and only then adds an active ingredient. Accordingly, the district court found that Exela did not literally infringe claim 1.

Nevertheless, the district court found that Exela's ANDA formulation infringed claim 1 under the doctrine of equivalents. It found that the timing of the addition of the active ingredient did not matter and ruled that the differences between the claimed steps and Exela's method were insubstantial.

Exela argues that the district court clearly erred in finding that there was no substantial difference between deoxygenating before or after forming the solution. Exela contends that Cadence's expert's testimony on this point was conclusory and improperly compared Exela's process, which did not involve stoppering under vacuum, with a process that did. Cadence disputes that there was clear error and contends that its expert's testimony supports the district court's decision as does the fact that Exela's formulation achieves similar stability to the formulation described in the '218 patent.

We agree with Cadence and find no clear error in the district court's finding of infringement under the doctrine of equivalents. The district court relied on the testimony of Cadence's expert, Dr. Orr, "that adding acetaminophen before or after the deoxygenation step would have no impact on the stability of the final product." Dr. Orr explained that this was so because "in both cases you're trying to deoxygenate your solution. In both cases, you're employing bubbling to do that. And the results that you achieve under this prolonged period of—of bubbling is still a solution of less than two parts per million." This testimony supports the district court's finding that changing the timing of the deoxygenation step was an insubstantial difference. The correctness of this conclusion is confirmed by the district court's finding and Exela's accession that its formulation is, in fact, stable. Exela's speculation that other differences between its formulation and the claimed formulation may be responsible for stability is insufficient to create a definite and firm conviction that the district court made a mistake.

The district court also did not accept Exela's argument that this scope of equivalents would vitiate a limitation of the claim. Exela challenges that determination and contends that deoxygenating after adding the active ingredient is the "antithesis" of deoxygenating before adding the active ingredient and that because such a substitution would "vitiate" the claimed limitation, there can be no finding of equivalence. It maintains that the facts here are analogous to *Planet Bingo, LLC v. GameTech International, Inc.*, where we held that determining a winning combination after a game started could not be equivalent to a claim that recited "a predetermined winning combination." 472 F.3d 1338, 1345 (Fed. Cir. 2006). Cadence responds that deoxygenating prior to adding the active ingredient is insubstantially different from deoxygenating after and that reference to "vitiation" is inappropriate. According to Cadence, the finding of vitiation in *Planet Bingo* was premised on the fact that the difference in timing was substantial.

Exela's reliance on *Planet Bingo* is misplaced. *Planet Bingo*'s holding was based on a finding that a combination determined before a game was substantially different, factually, from a combination determined after the game started. *See Brilliant Instruments, Inc. v. GuideTech, LLC*, 707 F.3d 1342, 1347 (Fed. Cir. 2013) (explaining the rationale for *Planet Bingo* as "two elements likely are not insubstantially different when they are polar opposites"). Exela's understanding of *Planet Bingo* (Dec. 13, 2006) is also expressly at odds with this court's holding in *DePuy Spine, Inc. v. Medtronic Sofamor Danek, Inc.*, 469 F.3d 1005 (Fed. Cir. Nov. 20, 2006), decided just a few weeks before *Planet Bingo*. *DePuy Spine* explained:

> A holding that the doctrine of equivalents cannot be applied to an accused device because it "vitiates" a claim limitation is nothing more than a conclusion that the evidence is such that no reasonable jury could conclude that an element of an accused device is equivalent to an element called for in the claim, or that the theory of equivalence to support the conclusion of infringement otherwise lacks legal sufficiency.

469 F.3d at 1018-19, *cited with approval* in *Voda v. Cordis Corp.*, 536 F.3d 1311, 1325 n.5 (Fed. Cir. 2008).

Exela fundamentally misunderstands the doctrine of claim vitiation. "Vitiation" is not an exception or threshold determination that forecloses resort to the doctrine of equivalents, but is instead a legal conclusion of a lack of equivalence based on the evidence presented and the theory of equivalence asserted. We have repeatedly reaffirmed this proposition. *See VirnetX, Inc. v. Cisco Sys., Inc.*, 767 F.3d 1308, 1323 (Fed. Cir. 2014); *Ring & Pinion Serv. Inc. v. ARB Corp. Ltd.*, 743 F.3d 831, 836 (Fed. Cir. 2014); *Charles Mach. Works, Inc. v. Vermeer Mfg.*, 723 F.3d 1376, 1380 (Fed. Cir. 2013). Characterizing an element of an accused product as the "antithesis" of a claimed element is also a conclusion that should not be used to overlook the factual analysis required to establish whether the differences between a claimed limitation and an accused structure or step are substantial *vel non*. The determination of equivalence depends not on labels like "vitiation" and "antithesis" but on the proper assessment of the language of the claimed limitation and the substantiality of whatever relevant differences may exist in the accused structure. *See Graver Tank & Mfg. Co. v. Linde Air Prods. Co.*, 339 U.S. 605, 610-12 (1950) (finding that a welding process that used manganese, a non-alkaline metal, could be equivalent to the claimed "alkaline earth metal," even though an alkaline metal can formally be described as the "antithesis" of a non-alkaline metal). *But see Moore U.S.A., Inc. v. Standard Register Co.*, 229 F.3d 1091, 1106 (Fed. Cir. 2000) ("it would defy logic to conclude that a minority—the very antithesis of a majority— could be insubstantially different from a claim limitation requiring a majority, and no reasonable juror could find otherwise").

Since a reasonable trier of fact could (and, in fact, did) conclude that Exela's process is insubstantially different from that recited in the claims, the argument that a claim limitation is vitiated by the district court's application of the doctrine of equivalents is both incorrect and inapt. Therefore, we affirm the district court's determination of infringement of claim 1. Because Exela does not assert any independent bases for not infringing dependent claims 3, 4 and 19, the district court's finding of infringement of these claims is also affirmed.

. . . .

Notes & Questions

1. How would you have drafted this claim if you wanted to ensure that a process that deoxygenates would constitute literal infringement, whether deoxygenation happened before or after the dissolving of acetaminophen into the final mixture? In light of your answer, do you agree that the court should be indifferent to the way that the claim was *actually* drafted?

2. Recall the excerpt from *Teashot.LLC v. Green Mountain Coffee Roasters, Inc.*, 595 Fed. Appx. 983 (Fed. Cir. 2015), included in Section II.B of this chapter. After construing the phrase "water-permeable material" contained in the claim language, the court concluded that the foil lid of the K-cups did not literally infringe. The plaintiff had also asserted a claim of equivalent infringement but failed to preserve that argument. If the argument had been preserved, how should the court have ruled on the question of equivalent infringement?

c. The Disclosure-Dedication Rule

The constraint that prosecution history estoppel imposes on claim scope, by cutting off the doctrine of equivalents, creates a potentially perverse incentive for a would-be patentee at the application drafting stage. Specifically, a patent drafter may reason that, to avoid having to narrow a claim by amendment during prosecution (and thereby create an estoppel), the best strategy is to claim an invention narrowly but to disclose it broadly. The narrow claim is less likely to lead to a narrowing amendment. The broad disclosure is evidence the patentee can use later to show that an accused infringer's conduct is equivalent: the infringer is using the very technology that the patentee explained—in the patent itself!—is interchangeable.

This strategy would make sense for a patentee. But it is terrible patent policy. First, the strategy undermines the patent document's public notice function. The general public could fairly conclude that a patentee who discloses material but makes no effort to claim exclusionary rights over it either (a) doesn't believe such rights can be obtained, or (b) doesn't want to claim them (for whatever reason). The foregoing strategy, if the court were to permit it, would be a form of unfair surprise on the public.

Second, it expressly contemplates a patentee obtaining exclusionary rights with claim coverage that the PTO never had the chance to examine. The strategy subverts patent examination. Suppose, for example, that an examiner has identified a prior art patent that is germane to an application's broad disclosure, but the applicant has proffered narrow claims that need not rely on that broad disclosure. In such a case, the examiner would quite properly refrain from rejecting the claims on novelty or nonobviousness grounds, even though, had the patentee sought broader claims, they would surely have elicited those prior-art-based rejections. Without a legal barrier to curb the reach of equivalent infringement on these facts, a court would, in effect, give a patentee a right to exclude that he could not have obtained from the PTO in the first instance.

Happily, the disclosure-dedication rule prevents patentees from carrying out the "claim narrowly, disclose broadly" strategy just outlined. The Federal Circuit reaffirmed this rule, which is rooted in two Supreme Court cases from the 1880s, in *Johnson & Johnston Assocs. v. R.E. Service Co.*, 285 F.3d 1046 (Fed. Cir. 2002) (en banc). In that case, a patentee claimed a metal-layer component for making circuit boards. The claim recited that the metal layer comprised "a sheet of aluminum," but the supporting disclosure stated that "[w]hile aluminum is currently the preferred material for the substrate, other metals, such as stainless steel or nickel alloys, may be used." R.E. Service Co., the accused infringer, used steel sheets. The Federal Circuit rejected J&J's reliance on the doctrine of equivalents, holding that "when a patent drafter discloses but declines to claim subject matter, as in this case, this action dedicates that unclaimed subject matter to the public." *Id.* at 1054. By enforcing the disclosure-dedication rule, "the courts avoid the problem of extending the coverage of an exclusive right to encompass more than that properly examined by the PTO." *Id.* at 1055; *see also Keystone Bridge Co. v. Phoenix Iron Co.,* 95 U.S. 274, 278 (1877) ("[T]he courts have no right to enlarge a patent beyond the scope of its claim as allowed by the Patent Office, or the appellate tribunal to which contested applications are referred.").

E. Additional Defenses

As we noted at the start of this part, generally speaking, the primary defenses to accusations of patent infringement are noninfringment (*i.e.*, arguments about patent scope) and invalidity. There are additional defenses, rooted in both case law and Patent Act provisions. In a survey course such as this, we do not explore these defenses in any detail. We do, however, want to alert you to the fact that they exist. Some of the more prominent of these additional defenses are as follows:

Inequitable Conduct—Under this doctrine, a patent is unenforceable in court if the patentee procured it from the Patent Office through inequitable conduct, *i.e.*, by making material misrepresentations, by omission or commission, with the intent to mislead. The taint of inequitable conduct, if found, cannot be removed. The entire patent is rendered unenforceable. Today, the leading case on inequitable conduct is *Therasense, Inc. v. Becton, Dickinson & Co.*, 649 F.3d 1276 (Fed. Cir. 2011) (en banc). *See also American Calcar, Inc. v. American Honda Motor Co.*, 768 F.3d 1185 (Fed. Cir. 2014) (affirming inequitable conduct, applying *Therasense*).

Exhaustion—This doctrine, also known as "first sale," prevents a patentee from suing downstream buyers and users for infringement once the patentee (or its licensee) has sold a patented item. The first authorized sale of the patented item is said to "exhaust" the patentee's right to exclude others. Thus, for example, one who buys an item from a licensed producer is not liable for infringement if she resells it to someone else. In patent and trademark law, the first sale doctrine is decisional; in copyright law it is statutory. We explore the first sale doctrine in Chapter 6.

Repair—Under the doctrine of repair, it is not a liability-triggering "mak[ing]" of a claimed invention to repair an item one has purchased from the patentee. For example, if a patentee sells a milling machine that lasts for years with sharp blades that wear out in weeks or months, replacing those blades is lawful repair. The cases contrast lawful repair with unlawful "reconstruction," *i.e.*, making whole a new embodiment of the claimed invention. Not surprisingly, the combined complexity of claim language and technological facts makes for a rich set of cases that parse permissible repair from impermissible reconstruction.

Experimental Use—This noninfringement doctrine, which the Federal Circuit has *sharply* curtailed in recent cases, should not be confused with experimental use negation of the on-sale or public-use bars of § 102(b). Under the experimental use defense, it is not infringement to make or use the claimed invention solely "for amusement, to satisfy idle curiosity, or for strictly philosophical inquiry." *Embrex, Inc. v. Service Eng'g Corp.*, 216 F.3d 1343, 1349 (Fed. Cir. 2000).

Drug Data Preparation Safe Harbor—Since 1984, the Patent Act has included a safe harbor against infringement for those whose otherwise unlawful activities are part of an effort to apply for FDA permission to sell a generic version of a preexisting patented drug. 35 U.S.C. § 271(e)(1). This safe harbor is but one part of a quite complex framework that Congress created in 1984 to better harmonize the interaction of the patent and drug approval systems, including the prompt development and sale of generic versions of drugs as the patent rights covering those drugs expire (or are held to be invalid).

Prior User Rights—Since 1999, the Patent Act has included an infringement defense for one who, at least a year before the patentee filed for its patent, had actually reduced the same invention to practice itself and used it commercially. 35 U.S.C. § 273. Initially this defense was limited to business method patents. In the America Invents Act, Congress expanded the scope of this prior user right significantly. Prior to the AIA, if someone was exploiting an invention as a trade secret, there was a risk that a third party might invent the same invention and obtain a patent on it. (Recall our exploration of the judicial construction of § 102(b) in part V.C.3 above, according to which secret exploitation may bar you from obtaining a patent for yourself, even though it does *not* create prior art that prevents another from obtaining a patent if they separately invent it.) That third party could then sue others for infringement, including those who were using the invention as a trade secret. Thus, electing to forgo patent protection in favor of trade secret involved some risk. That risk has been significantly reduced with the expansion of the prior commercial use defense.

VIII. DESIGN PATENTS

When someone refers to "a patent on something," or to patents in general, she is almost certainly referring to a *utility* patent—the sort of patent we have been discussing thus far in this Chapter. A utility patent protects a new and *useful* solution to a practical problem. *See* 35 U.S.C. § 101 (permitting patents on "new and useful" inventions). But think of a consumer product that you have enjoyed. Perhaps it was a smartphone, or an athletic shoe, or a lamp, or a kitchen gadget. Did part of your enjoyment come from the beauty—the aesthetic appeal—of the product, quite apart from its usefulness? The patent system provides protection for a product's aesthetic design, by means of a *design* patent. 35 U.S.C. §§ 171-173. In this way, the patent system encourages firms to innovate in the ornamental design of products, as well as in the functionality of objects.

What does a design patent look like? The most important thing it does is depict the design for which exclusive rights are claimed. In fact, a design patent can have one, and only one, claim, and it's always worded the same: "The ornamental design" for an item, "as shown and described." 37 C.F.R. § 1.153(a). Consider U.S. Patent No. D427,484 to James Ethridge, for the design of a BBQ grill, the first two pages of which are shown below.

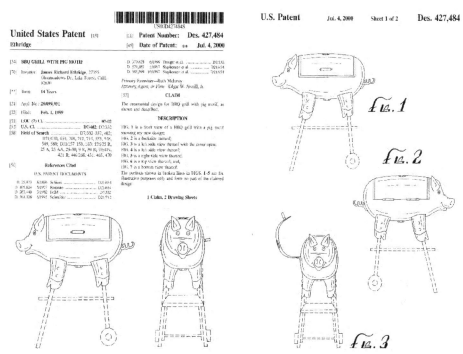

The D'484 patent protects the ornamental pig-like shape of the grill, not the functionality of this, or any other, barbecue grill. Note, too, that the claimed design is to the grill body, and not to the supporting legs. The D'484 denotes this by using broken, rather than solid, lines to show the supporting legs: "Structure that is not part of the claimed design, but is considered necessary to show the environment in which the design is associated, may be represented in the drawing by broken lines. This includes any portion of an article in which the design is embodied or applied to that is not considered part of the claimed design." *Manual of Patent Examining Procedure* § 1503.02, ¶ III.

A. OVERVIEW OF DESIGN PATENTS

The provisions in the Patent Act relating to design patents are few in number and spare in words. They rely on the other parts of the Patent Act to do the heavy lifting: "The provisions of this title relating to patents for inventions shall apply to patents for designs, except as otherwise provided." 35 U.S.C. § 171, ¶ 2. There are a total of three substantive provisions, and only two are of interest here. One sets the design patent term at "15 years from the date of grant." 35 U.S.C. § 173.[*] There are no maintenance fees required to keep a design patent in force. 35 U.S.C. § 41(b)(3). The other provision defines design-patent-eligible subject matter: "Whoever invents any new, original and ornamental design for an article of manufacture may obtain a patent therefor." 35 U.S.C. § 171, ¶ 1.

---

[*] Congress changed the design-patent term in 2012, from 14 years to 15 years, to bring the U.S. in line with a 1999 international agreement on the protection of industrial designs. Patent Law Treaties Implementation Act of 2012, Pub. L. 112-211, § 102(7), 126 Stat. 1527, 1532. The longer term applies to design patents filed on or after May 13, 2015. *Id.* at § 103(a)(2).

Three differences stand out when one compares the subject matter eligible for design patent protection with the subject matter eligible for utility patents. First, design patents are limited to manufactures, whereas utility patents can cover both products and processes. Second, a design patent must be "original" as well as new. Third, a design patent must be "ornamental," whereas a utility patent must be "useful." As for the seeming originality requirement in § 171, ¶ 1, it has played little to no role in the development of design patent law. As the Federal Circuit recently explained,

> The originality requirement in § 171 dates back to 1842 when Congress enacted the first design patent law. The purpose of incorporating an originality requirement is unclear; it likely was designed to incorporate the copyright concept of originality—requiring that the work be original with the author, although this concept did not find its way into the language of the Copyright Act until 1909. In any event, the courts have not construed the word "original" as requiring that design patents be treated differently than utility patents.

International Seaway Trading Corp. v. Walgreens Corp., 589 F.3d 1233, 1238 (Fed. Cir. 2009) (footnote omitted).

The ornamentality requirement, by contrast, plays a significant role in design patent law. It is what separates design-patent protection, which focuses on the appearance of a manufacture, from utility-patent protection, which focuses on the functionality of manufacture (or process). Courts police the ornamentality requirement by screening *out* designs that are primarily functional: "if the design claimed in a design patent is dictated solely by the function of the article of manufacture, the patent is invalid because the design is not ornamental." *Best Lock Corp. v. Ilco Unican Corp.*, 94 F.3d 1563, 1566 (Fed. Cir. 1996). But, of course, given that design patents are confined to articles of manufacture, and "[a]n article of manufacture necessarily serves a utilitarian purpose" to some degree, *L.A. Gear, Inc. v. Thom McAn Shoe Co.*, 988 F.2d 1117, 1123 (Fed. Cir. 1993), we need a standard for sorting the primarily functional from the primarily ornamental. The standard is one of necessity: "the design of a useful article is deemed to be functional when the appearance of the claimed design is 'dictated by' the use or purpose of the article. If the particular design is essential to the use of the article, it can not be the subject of a design patent." *Id.* An important piece of evidence that a design is ornamental, rather than functional, is the existence of alternative design options: "When there are several ways to achieve the function of an article of manufacture, the design of the article is more likely to serve a primarily ornamental purpose." *Id.* Additionally, like utility patents, design patents protect only innovations that are new under § 102 and nonobvious under § 103. *International Seaway*, 589 F.3d at 1238.

The rights that one obtains with a design patent are identical to those under utility patents: the right to exclude others from making, using, selling, offering to sell, or importing the claimed design. The infringement analysis, however, differs. Whereas in utility patent infringement analysis we examine the accused product or process to determine if all limitations of the claim are present, either literally or equivalently, with design patents we look to see if the "accused article embodies the patented design or any colorable imitation thereof." *Egyptian Goddess, Inc. v. Swisa, Inc.,* 543 F.3d 665, 678 (Fed. Cir. 2008) (en banc). Specifically, there "'must be sameness of appearance'" *Id.* at 670 (quoting *Gorham Co. v. White,* 81 U.S. 511, 526-28 (1871)). The articulation of the test in *Gorham,* which focuses on an ordinary observer's response to the design, remains in effect today:

> [I]f, in the eye of an ordinary observer, giving such attention as a purchaser usually gives, two designs are substantially the same, if the resemblance is such as to deceive such an observer, inducing him to purchase one supposing it to be the other, the first one patented is infringed by the other.

81 U.S. at 528. Finally, if infringement is proven, there is an additional damages option for a design patent owner, who can elect to recover "the extent of [the infringer's] total profit." 35 U.S.C. § 289; *see Samsung Elecs. Co. v. Apple Inc.*, 137 S. Ct. 429 (2016) (construing § 289 as to multi-component manufactures).

B. Ornamentality—Out of Sight, Out of Protection?

Many items of manufacture are visible when they are made and while they are waiting to be sold, but not routinely seen in regular operation. For example, an ink cartridge for an inkjet printer is usually concealed in the printer housing when the printer is in ordinary use. Does it make sense to think of the design of such an item as "ornamental"? For most of its useful life, it makes no appeal to a consumer's aesthetic sensibility. On the other hand, the design's visual appeal might play some role in attracting consumers to it in the first place, and thus we might wish to enable producers to better compete against one another by protecting their investments in providing new visual designs for the item. The way we resolve this question is far from a trivial matter. If design patents are available only as to those manufactures that are visible throughout their useful lives, this form of intellectual property will play no role in, for example, the market for internal replacement parts. The two cases below show the expansion of design patent law.

In re Stevens
173 F.2d 1015 (CCPA 1949)

HATFIELD, JUDGE:

This is an appeal from the decision of the Board of Appeals of the United States Patent Office affirming the decision of the Primary Examiner rejecting the single claim in appellant's application for a patent for a design for a rotary brush.
. . .

. . . .

Appellant's application discloses a rotary brush of the type used in vacuum cleaners. The brush consists of an elongated cylinder having two rows of brush tufts projecting from the cylindrical surface. Each row of tufts extends in a generally spiral form from one end of the cylinder to the other, there being a gap in each row at the central portion of the cylinder. The brush tufts of one row are thicker than those of the other.

. . . .

The appealed claim was . . . rejected on the ground that since the article to which appellant's design is applied is hidden in normal use, its appearance is immaterial and, if any invention is involved, it is obviously utilitarian in character. Although appellant's application does not specify the use for which his brush is intended, it was stated by counsel for appellant in his remarks accompanying an amendment which appears in the record, that the brush "is particularly designed as a rotary brush for a vacuum cleaner," and it is evident from the construction of the brush that that is its intended use. In such use it would normally be concealed, although it would, of course, be visible when purchased as a replacement and, as suggested by counsel for appellant, it might be used in a glass demonstration model.

It has been held repeatedly that articles which are concealed or obscure in normal use are not proper subjects for design patents, since their appearance cannot be a matter of concern. Articles of the type referred to are: horseshoe calks, *Rowe v. Blodgett & Clapp Co.*, 112 F. 61 (2d Cir. [1901]); a fastener for machinery belts, *Eaton v. Lewis*, 115 F. 635 [(1902)]; a spool for a typewriter ribbon, *Wagner Typewriter Co. v. F.S. Webster Co.*, 144 F. 405 [(1906)]; and an automobile tire, *North British Rubber Co. v. Racine Rubber Tire Co.*, 271 F. 936 (2d Cir. [1921]). The doctrine announced in those cases was approved by this court in *In re Koehring*, 37 F.2d 421 [(2d Cir. 1930)].

It is true, as pointed out by counsel for appellant, that appellant's brush is or may be visible at some times and under some circumstances, but that is equally true of the articles referred to in the decisions hereinbefore cited. Almost every article is visible when it is made and while it is being applied to the position in which it is to be used. Those special circumstances, however, do not justify the granting of a design patent on an article such as that here under consideration which is always concealed in its normal and intended use. The ornamental appearance of such an article is a matter of such little concern that it cannot be said to possess patentability as a design. We are of opinion, therefore, that the rejection of appellant's claim was proper.

For the reasons stated, the decision of the Board of Appeals is affirmed.

In re Webb
916 F.2d 1553 (Fed. Cir. 1990)

CLEVENGER, JUDGE:

This is an appeal from a decision of the [PTO] Board of Patent Appeals & Interferences ("Board") affirming the final rejection of the sole claim of appellants' ("Webb") U.S. Design Patent Application Serial No. 833,470. The claim for "[t]he ornamental design for a grooved femoral hip stem prosthesis as shown and described," was "rejected as being unpatentable under 35 U.S.C. § 171 as being directed to non-statutory subject matter." The design can be appreciated from Figure 2 of the application reproduced below.

The Board affirmed the Examiner's holding that the design, "clearly not intended to be visible in actual use," "is not proper subject matter under 35 U.S.C. § 171." The Board's decision creates a per se rule that a design for an article which will not be visible in the final use for which the article was created is non-statutory subject matter even if the design is observed at some stage of the article's commercial life. We reverse and remand.

I

Hip stem prostheses of the design invented by Webb are metallic implants that are generally used by orthopedic surgeons to supplant the functioning of a diseased or broken femur, near the hip, where the femur is joined to the pelvis. According to Webb, and not disputed by the [PTO], surgeons are made aware of differing brands and types of prostheses through advertisements in professional journals and through trade shows, where the prostheses themselves are displayed. Advertisements that were put in the record prominently and visually display the features of the prostheses. Furthermore, the applicant's agent submitted that "an implant's appearance is observed by potential and actual purchasers, surgeons, nurses, operating room staff, and other hospital personnel." After purchase, the prosthesis is surgically implanted into a patient's body where the implant is to remain indefinitely. Neither party disputes that, after implantation, the prosthesis is no longer visible to the naked eye.

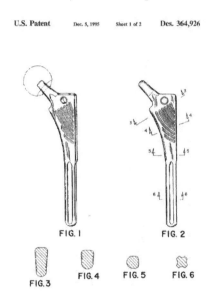

U.S. Patent   Dec. 5, 1995   Sheet 1 of 2   Des. 364,926

FIG. 1   FIG. 2

FIG. 3   FIG. 4   FIG. 5   FIG. 6

II

In the Initial Office Action, the Examiner rejected the claim "under 35 U.S.C. § 171 for the reason that the instant article is believed to be devoid of ornamentality, as comprehended by the statute. Articles of this type are not only completely hidden in use, but are devised to satisfy purely structural and mechanical requirements as well." The Examiner thus found the article to be unpatentable subject matter for two reasons: because it was purely functional and because it was concealed in normal use. In reply, Webb argued that the design was not purely functional since a "prosthetic implant could utilize the mechanical/utilitarian features/concepts . . . and have a totally different visual appearance." Webb also argued that the "visual appearance can certainly draw attention to a particular implant at a trade show or in advertising" and, therefore, the design was visible during normal use.

In the Final Office Action, the Examiner stated:

Applicant argues that, while the design is functional in nature, it is still ornamental. While this may be true, it has been held that articles which are hidden in use are not proper subject matter for design patents.

* * * *

* * * There is not sound reason or logic for "normal use" to include the repair, service, replacement, sale or display of the article which incorporates the claimed design. While such occasions are of course "normal" in the sense of commonplace or routine occasions of an item's use, for patent purposes "normal use" should be limited to the ordinary functioning for which it was designed, not incidents in the article's life which are not integral to its function or purpose. Items are not designed for sale, display, replacement or repair.

The Board did not address the issue of functionality of the claimed design that had been raised in the Examiner's Initial Action. It affirmed the Examiner's final rejection of the claim as unpatentable subject matter because the article was not visible in what the Board considered to be its normal or intended use.

. . . .

IV

The issuance of design patents is limited by statute to designs that are ornamental. 35 U.S.C. § 171. Our predecessor court has affirmed the rejection of design applications that cannot be perceived in their normal and intended uses. For instance, the Court of Customs and Patent Appeals affirmed the rejection of a design claim for a vent tube placed in the wall of a frame house, stating that "[i]t is well-settled that patentability of a design cannot be based on elements which are concealed in the normal use of the device to which the design is applied." *In re Cornwall*, 230 F.2d 457, 459 (CCPA 1956). Even earlier, that court affirmed the rejection of a design claim for a vacuum cleaner brush. *In re Stevens*, 173 F.2d 1015 (CCPA 1949). . . .

We read those cases to establish a reasonable general rule that presumes the absence of ornamentality when an article may not be observed. This is a sound rule of thumb, but it is not dispositive. In each case, the inquiry must extend to whether at some point in the life of the article an occasion (or occasions) arises when the appearance of the article becomes a "matter of concern." [*Stevens*, 173 F.2d at 1019.]

Here, we read the Board's decision to have established a per se rule under § 171 that if an article is hidden from the human eye when it arrives at the final use of its functional life, a design upon that article cannot be ornamental. The rule in *Stevens* does not compel the Board's decision. Instead, *Stevens* instructs us to decide whether the "article such as here under consideration"—a hip stem implant—"is always concealed in its normal and intended use." The issue before us, then, is whether "normal and intended use" of these prosthetic devices is confined to their final use.

V

Although we agree that "normal and intended use" excludes the time during which the article is manufactured or assembled, it does not follow that evidence that an article is visible at other times is legally irrelevant to ascertaining whether the article is ornamental for purposes of § 171. Contrary to the reasoning of the Examiner in this case, articles are designed for sale and display, and such occasions are normal uses of an article for purposes of § 171. The likelihood that articles would be observed during occasions of display or sale could have a substantial influence on the design or ornamentality of the article. "The law manifestly contemplates that giving certain new and original appearances to a manufactured article may enhance its salable value . . . ." *Gorham Co. v. White*, 81 U.S. (14 Wall.) 511, 525 (1871).

In short, we construe the "normal and intended use" of an article to be a period in the article's life, beginning after completion of manufacture or assembly and ending with the ultimate destruction, loss, or disappearance of the article. Although the period includes all commercial uses of the article prior to its ultimate destination, only the facts of specific cases will establish whether during that period the article's design can be observed in such a manner as to demonstrate its ornamentality.

It is possible, as in *Stevens*, that although an article may be sold as a replacement item, its appearance might not be of any concern to the purchaser during the process of sale. Indeed, many replacement items, including vacuum cleaner brushes, are sold by replacement or order number, or they are noticed during sale only to assess functionality. In such circumstances, the PTO may properly conclude that an application provides no evidence that there is a period in the commercial life of a particular design when its ornamentality may be a matter of concern. However, in other cases, the applicant may be able to prove to the PTO that the article's design is a "matter of concern" because of the nature of its visibility at some point between its manufacture or assembly and its ultimate use. Many commercial items, such as colorful and representational vitamin tablets, or caskets, have designs clearly intended to be noticed during the process of sale and equally clearly intended to be completely hidden from view in the final use. Here, for example, there was ample evidence that the features of the device were displayed in advertisements and in displays at trade shows. That evidence was disregarded by the Board because, in its view, doctors should select implants solely for their functional characteristics, not their design. It is not the task of the Board to make such presumptions.

. . . .

Notes & Questions

1. The basic structure of patent prosecution puts the burden on the PTO to demonstrate a basis for denying a patent, once a patent application in the proper form makes the prima facie case for an applicant's entitlement. How will the PTO know, in a given design patent case, whether the item in question is or is not, in the court's words, "designed for sale and display"? Should the PTO amend its

regulations to require verification, in a design patent application, that the article's design will be relied upon in the marketing of the article? What if the applicant doesn't actually make or sell the article in question (as is often the case with utility patents)? Should the PTO require the applicant to establish that items of this type are generally marketed in a way that relies on their visual appeal?

2. The court seems to place great weight on the fact that the patented prosthetic's design was displayed in advertisements and at trade shows, to promote the article. Is the best way for a would-be patentee to avoid a *Stevens/Webb* problem to make sure to feature the visual design in advertisements and at one or more trade shows, even if the article will be sold thereafter by replacement or order number *without* reliance on the visual design?

C. FUNCTIONALITY—SEPARATING ORNAMENTALITY FROM UTILITY

Rosco, Inc. v. Mirror Lite Co.
304 F.3d 1373 (Fed. Cir. 2002)

DYK, JUDGE:

Rosco, Inc. ("Rosco") appeals the decision of the United States District Court for the Eastern District of New York finding Rosco's design patent, United States Design Patent No. 346,357 ("the '357 patent"), invalid as functional . . .

BACKGROUND

Rosco and Mirror Lite are competitors in the school bus mirror market. This dispute involves "cross-view" mirrors, which are convex, three-dimensional, curved surface mirrors mounted on the front fender of a school bus, enabling the bus driver to view the front and passenger side of a school bus. . . .

Each party owns a patent that it alleged was infringed by the other. . . .

1. Rosco's '357 Design Patent

Rosco's '357 design patent relates to an oval, highly convex cross-view mirror with a black, flat metal backing. Rosco applied for the patent on April 14, 1992, and the patent issued on April 26, 1994. Rosco alleged that Mirror Lite infringed the '357 design patent. Mirror Lite argued that the '357 design patent was invalid as functional and therefore was not infringed.

. . . .

DISCUSSION

. . . .

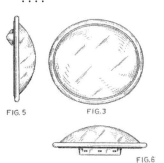

FIG. 5   FIG. 3   FIG. 4

FIG.6

Rosco's '357 design patent shows a highly convex, curved-surface, three-dimensional oval mirror with a black, flat metal backing. In May 1992, Rosco began manufacturing the mirror of the '357 patent under the name "Eagle Eye."

Rosco alleged that Mirror Lite infringed the '357 patent by manufacturing and selling a duplicate of Rosco's mirror under the name "Hawk Eye." . . . .

We apply a stringent standard for invalidating a design patent on grounds of functionality: the design of a useful article is deemed functional where "the appearance of the claimed design is 'dictated by' the use or purpose of the article." *L.A. Gear, Inc. v. Thom McAn Shoe Co.*, 988 F.2d 1117, 1123 (Fed. Cir. 1993). "[T]he design must not be governed solely by function, *i.e.*, that this is not the only possible form of the article that could perform its function." *Seiko Epson Corp. v. Nu-Kote Int'l, Inc.*, 190 F.3d 1360, 1368 (Fed. Cir. 1999). "When there are several ways to achieve the function of an article of manufacture, the design of the article is more likely to serve a primarily ornamental purpose." *L.A. Gear*, 988 F.2d at 1123. That is, if other designs could produce the same or similar functional capabilities, the design of the article in question is likely ornamental, not functional. Invalidity of a design patent claim must be established by clear and convincing evidence.

The district court found that because the mirror's oval shape, the asserted point of novelty of the '357 patent, "of necessity dictates its function," the '357 patent was invalid as functional. The court based its determination of functionality on its findings that the mirror of the '357 patent offered a unique field of view (when compared to Mirror Lite's Bus Boy mirror); that Rosco represented to the Patent & Trademark Office that its mirror provided a superb field of view; and that Rosco marketed the mirror of the '357 patent as more "aerodynamic" than other cross-view mirrors.

The mere fact that the invention claimed in the design patent exhibited a superior field of view over a single predecessor mirror (here, the Bus Boy) does not establish that the design was "dictated by" functional considerations, as required by *L.A. Gear*. The record indeed reflects that other mirrors that have non-oval shapes also offer that particular field of view. Similarly, nothing in the record connects the oval shape of the patented design with aerodynamics, and the record shows that other non-oval shaped mirrors have the same aerodynamic effect.

Mirror Lite has not shown by clear and convincing evidence that there are no designs, other than the one shown in Rosco's '357 patent, that have the same functional capabilities as Rosco's oval mirror. Under these circumstances it cannot be said that the claimed design of the '357 patent was dictated by functional considerations. We reverse the district court and hold that the '357 patent claim was not shown to be invalid on functionality grounds.

. . . .

PHG Techs. v. St. John Cos.
469 F.3d 1361 (Fed. Cir. 2006)

PROST, JUDGE:

Defendant-Appellant, St. John Companies, Inc. ("St. John"), appeals the decision of the United States District Court for the Middle District of Tennessee granting a preliminary injunction in favor of Plaintiff-Appellee, PHG Technologies, L.L.C. ("PHG"). Because we find that St. John has raised a substantial question of the validity of the two patents at issue, the district court abused its discretion by granting PHG's motion for a preliminary injunction. Therefore, we vacate the preliminary injunction.

I. BACKGROUND

PHG and its predecessors have been in the business of selling certain medical patient identification labels as well as identification labeling software in the United States since 1995. PHG owns the two design patents at issue in this case: United States Patent Nos. D496,405 (the '405 patent) and D503,197 (the '197 patent). The '405 patent claims "[t]he ornamental design for the medical label sheet, as shown." The '197 patent claims "[t]he ornamental design for a label pattern for a medical label sheet, as shown." Figure 1 from the '405 patent and figure 1 from the '197 patent appear below, respectively[.]

As can be seen, both designs include eleven rows of labels, with each row containing three labels. The first nine rows are depicted to contain three labels of equal size, the size being consistent with a standard medical chart label. The tenth and eleventh row each contain differently-sized labels which apparently correspond to the size of a pediatric and adult patient wristband respectively. The difference between the two patents is that the border is part of the design claimed in the '405 patent but not part of the design claimed in the '197 patent. The '405 and '197 patents depend from a utility patent application, No. 09/952,425 (the '425 utility application), which is still pending at the United States Patent & Trademark Office.[*]

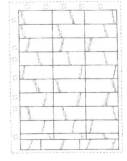

St. John also sells medical patient identification labels in the United States in competition with PHG. On May 13, 2004, before PHG's design patents issued, PHG informed St. John by letter that the design of St. John's medical label sheet infringed the intellectual property rights of PHG and that PHG anticipated that patents covering the accused design would be issued in the future. St. John did not respond to the May 13th letter and continued to sell its medical label sheet. After the two patents issued, PHG filed suit on August 11, 2005 alleging, *inter alia*, that St. John's medical label sheet infringed the '405 and '197 patents.

On August 26, 2005, two weeks after filing suit, PHG moved for a preliminary injunction against St. John's continued sale of its accused medical label sheet. The district court held an evidentiary hearing on November 22, 2005. St. John argued

---

* [*Eds. Note*: A design patent can rely on an earlier-filed parent application for a utility patent, so long as the parent application contains figures that adequately show the ornamental design for which a design patent is later sought. See p. 242, n.*, for further information concerning the way that a family of continuation applications can share an original filing date.]

that the patented medical label sheets are primarily functional and presented evidence from the prosecution history of the '425 utility application and from an affidavit submitted by Adam Press, St. John's Chief Executive Officer, in support of its argument. PHG presented the testimony of Mr. Moyer, one of the inventors of the patents at issue. Mr. Moyer testified that he and Mr. Stewart, his co-inventor, experimented with different configurations of the medical labels and chose the claimed designs because they were the "most aesthetically pleasing to us."

On December 5, 2005, the district court granted PHG's motion for a preliminary injunction . . .

. . . .

II. DISCUSSION

. . . .

St. John asserts that the district court erred in finding that the patented designs are primarily ornamental rather than merely a byproduct of functional considerations. In support of its assertion of functionality, St. John points to various statements made by PHG in the prosecution of the '425 utility application and to statements made by Mr. Press in an affidavit submitted to the court. St. John argues that the statements made during prosecution and those submitted by Mr. Press constitute a clear and convincing showing of functionality. Further, St. John asserts that because PHG presented no evidence to rebut St. John's showing of invalidity, the district court clearly erred in finding that the patented designs are primarily ornamental.

PHG responds that the district court correctly determined that St. John failed to raise a substantial question regarding the functionality of the designs because the patented designs were not dictated by the use or purpose of the article of manufacture—a medical label sheet. PHG concedes that the design has functional features but argues that the arrangement of the different sizes of labels on the sheet is primarily ornamental because, as found by the district court, "there are a multitude of ways to arrange different sizes of labels on an 8½ x 11 inch sheet." Further, PHG accuses St. John of focusing solely on the individual features of the claimed designs rather than analyzing the overall appearance to determine if the designs were dictated by functional considerations.

. . . The district court's sole finding with regard to St. John's assertion of invalidity was that the design was not dictated by its function because "[t]he testimony revealed [that] there are a multitude of ways to arrange different sizes of labels on an 8½ x 11 inch sheet." In support, the district court noted that "Brian Moyer testified that PHG considered various arrangements for medical label sheets and settled on the design ultimately patented because it had 'the best flow and look.'"
. . .

. . . As the statute indicates, a design patent is directed to the appearance of an article of manufacture. *L.A. Gear, Inc. v. Thom McAn Shoe Co.*, 988 F.2d 1117, 1123 (Fed. Cir. 1993). "If the patented design is primarily functional rather than ornamental, the patent is invalid." *Power Controls Corp. v. Hybrinetics, Inc.*, 806 F.2d 234, 238 (Fed. Cir. 1986). The design of a useful article is deemed to be functional when "the appearance of the claimed design is 'dictated by' the use or purpose of the article." *L.A. Gear*, 988 F.2d at 1123.

"[T]he determination of whether the patented design is dictated by the function of the article of manufacture must ultimately rest on an analysis of its overall appearance." *Berry Sterling Corp. v. Pescor Plastics, Inc.*, 122 F.3d 1452, 1455 (Fed. Cir. 1997). Our cases reveal a "list of . . . considerations for assessing whether the patented design as a whole—its overall appearance—was dictated by functional considerations," including:

> whether the protected design represents the best design; *whether alternative designs would adversely affect the utility of the specified article*; whether there are any concomitant utility patents; whether the advertising touts particular features of the design as having specific utility; and whether there are any elements in the design or an overall appearance clearly not dictated by function.

Id. at 1456 (emphasis added). In particular, we have noted that "[t]he presence of alternative designs may or may not assist in determining whether the challenged design can overcome a functionality challenge. Consideration of alternative designs, if present, is a useful tool that may allow a court to conclude that a challenged design is not invalid for functionality." *Id.* . . .

Our case law makes clear that a full inquiry with respect to alleged alternative designs includes a determination as to whether the alleged "alternative designs would adversely affect the utility of the specified article," such that they are not truly "alternatives" within the meaning of our case law. In this case, while the district court relied exclusively on its finding that there were a multitude of alternative designs, the court did not make any findings with respect to whether any of the alternatives would adversely affect the utility of the medical label sheet. One might presume that the district court's findings with respect to alternatives implicitly include the additional finding that the alternatives did not adversely affect the utility of the medical label sheet. The difficulty in doing so in this case, however, is that the district court makes no reference to St. John's evidence that the overall arrangement of the labels on the medical label sheet was dictated by the use and purpose of the medical label sheet and that alternative designs lacking that arrangement would adversely affect the utility of the sheet. Specifically, St. John presented Mr. Press's affidavit, in which he stated:

> The labels for use on the wristbands themselves are located on the bottom two rows of the sheet as these are usually the first labels used when a patient is admitted to a medical facility. The lower right hand corner is the easiest location for a right-handed user to remove the label as it is flush to an edge and unencumbered by a file or binder clip along the top or left hand margins. By placing the labels for the wristbands at the bottom of the page, the subsequent removal of additional labels adjacent to the removed label is facilitated.

Press Aff. ¶ 5.d.

Mr. Press's affidavit constitutes evidence that alternative designs, which do not include the "novel feature" of PHG's design—the placement of various sizes of medical labels at the bottom of the sheet—would adversely affect the utility of the medical label sheet. It articulates a clear functional reason why the use and purpose of the article of manufacture dictated that the "wristband" labels be located at the bottom of the sheet. Additionally, PHG's statements during prosecution of the '425 utility application indicate that there were functional reasons for each of the other features of the medical label sheet, including: for creating one sheet containing labels of different sizes; for the particular sizes of each differently-sized label; for the size of the sheet itself; and for including holes along the side and top of the sheet.

While a district court's determination as to whether a design is primarily ornamental is reviewed for clear error, in this case there is no explicit finding by the court on whether the alleged alternatives are in fact functionally equivalent (*i.e.*, that the alternatives do not adversely affect the utility of the medical label sheet), or any mention or finding whatsoever with respect to the evidence presented in Mr. Press's affidavit. The evidence presented by St. John, in our view, was sufficient to raise a substantial question of invalidity. The only evidence presented by PHG and relied upon by the district court was Mr. Moyer's testimony that he and his co-inventor chose the patented designs because they had "the best flow and look." PHG did not offer testimony refuting the assertions made in Mr. Press's affidavit—that functional considerations dictated the medical label design, specifically the "novel feature" of the differently-sized labels being placed at the bottom of the sheet. In fact, on cross-examination Mr. Moyer testified that the original intent in designing a medical label sheet with differently-sized labels was "functional." Therefore, this case is clearly distinguishable from *L.A. Gear*, in which the patentee introduced evidence indicating that "a myriad of athletic shoe designs" could achieve the same functions that were achieved by the patented designs and "[i]t was not disputed that there were other ways of designing athletic shoes to perform the functions of the elements of the [patented] design." 988 F.2d at 1123.

Further, we reject PHG's assertion that St. John's analysis focuses solely on the individual features of the designs rather than their overall appearance. The evidence presented by St. John not only addresses the individual features of the designs, but also their overall appearance. Mr. Press's statements directly pertain to the overall arrangement of the designs as a whole and indicate that the use and purpose of the medical label sheet dictate that the wristband-sized labels be located at the bottom of the sheet. His statements reasonably indicate that once the location of the wristband-sized labels has been dictated by the use and purpose of the medical label sheet, the location of the remaining labels is necessarily dictated as well. This is because the remaining labels, as well as the medical label sheet itself, are of standard size. Therefore, in order to maximize the efficient use of space on the sheet, the location and number of the medical chart and record labels is dictated by the placement of the wristband-sized labels at the bottom of the sheet. St. John's evidence thus directly pertains to, and is sufficient to raise a substantial question with respect to, whether the overall appearance of the patented designs is "dictated by" the medical label sheet's use and purpose. Because St. John has satisfied its burden of raising a substantial question of invalidity, the district court's finding that PHG was likely to show that the patented designs were primarily ornamental is clearly erroneous.

. . . .

Accordingly, we vacate the district court's grant of a preliminary injunction.

Notes & Questions

1. In *Rosco*, the court calls its functionality test a "stringent" standard, but is that accurate? The cases show that, if there are multiple designs that all effectively serve the function in question—serving as a driving mirror, providing hospital intake labels—then the claimed design is almost certainly *not* functional for design patent purposes. Won't this usually be true for most manufactures? When would the presence of alternative designs be a poor proxy for determining that a claimed design lacks functionality?

2. Blocking a design from design patent protection on the ground that it is functional channels would-be applicants toward the utility patent system when what they want is a strong exclusion right for a given utility. This principle—a strong exclusion right for a utility comes from a utility patent, and not from anything else—is a foundational principle in intellectual property law. Trade secret protection is a weak exclusion right because independent development, such as by reverse engineering, is a complete defense. As you will see in later chapters, copyright's *useful article* doctrine and trademark's *functionality* doctrine also channel the protection of utility toward the utility patent system and away from other forms. Only the rigors of patent examination—including ensuring a good fit between disclosure and claim scope, and testing for novelty and nonobviousness— are sufficient to justify giving an innovator the strong exclusion right. Without the rigors of patent examination, no patent-style protection should be permitted for a given functionality.

3. Think about this channeling role in terms of the possible errors. If we test for a design's functionality, we might wrongly bar a nonfunctional (*i.e.*, a purely ornamental) design from design patent protection. Alternatively, we might wrongly permit a functional design to garner design patent protection, even though it should have been forced off into the utility patent regime. How bad is that mistake? After all, design patents, like utility patents, are examined for novelty and nonobviousness, as you will see in more detail below. And the term of a design patent is only 15 years from the issue date, even shorter than that of a utility patent. On the other hand, the disclosure requirement is not as burdensome for designs; the drawings show the design, and the claims are coextensive with the drawings. The public is thus getting less disclosure in the bargain. If this categorization mistake isn't terribly serious, perhaps it makes sense for the functionality exclusion to be hard to trigger in design patent law (as it appears to be). When you

learn about copyright and trademark's functionality exclusions, be sure to consider (a) whether the exclusion is easier or harder to trigger in those domains, (b) and how that answer relates to the distance between the rigors of obtaining a utility patent and the ease of obtaining a copyright or trademark.

D. INFRINGEMENT—THE ORDINARY OBSERVER TEST

Recall that, in the case of a utility patent, we have a claim that recites the limitations we look for in the accused product or process to determine if it infringes. In the case of a design patent, we don't have a verbal boundary with carefully enumerated limitations. We have, instead, a design "as shown and described," with a series of drawings. How does one determine whether an accused design infringes?

As we noted above, in the introductory overview of design patents, the Supreme Court set forth an "ordinary observer" approach to analyzing infringement in *Gorham Co. v. White*, 81 U.S. 511 (1871). The Court used the adjective "ordinary" to contrast the proper inquiry from the one the lower court had used in the case, which focused on expert evaluation:

> An engraving which has many lines may present to the eye the same picture, and to the mind the same idea or conception as another with much fewer lines. The design, however, would be the same. So a pattern for a carpet, or a print may be made up of wreaths of flowers arranged in a particular manner. Another carpet may have similar wreaths, arranged in a like manner, so that none but very acute observers could detect a difference. Yet in the wreaths upon one there may be fewer flowers, and the wreaths may be placed at wider distances from each other. Surely in such a case the designs are alike. . . . The court below was of opinion that the test of a patent for a design is not the eye of an ordinary observer. . . . There must, he thought, be a [side-by-side] comparison of the features which make up the two designs. With this we cannot concur. Such a test would destroy all the protection which the act of Congress intended to give. There never could be piracy of a patented design, for human ingenuity has never yet produced a design, in all its details, exactly like another, so like, that an expert could not distinguish them. No counterfeit bank note is so identical in appearance with the true that an experienced artist cannot discern a difference. . . . The purpose of the law must be effected if possible; but, plainly, it cannot be if, while the general appearance of the design is preserved, minor differences of detail in the manner in which the appearance is produced, observable by experts, but not noticed by ordinary observers, by those who buy and use, are sufficient to relieve an imitating design from condemnation as an infringement.
>
> We hold, therefore, that if, in the eye of an ordinary observer, giving such attention as a purchaser usually gives, two designs are substantially the same, if the resemblance is such as to deceive such an observer, inducing him to purchase one supposing it to be the other, the first one patented is infringed by the other.

Id. at 526-28. In a more recent case, the Federal Circuit reaffirmed the *Gorham Co. v. White* "ordinary observer" test. *Egyptian Goddess, Inc. v. Swisa, Inc.*, 543 F.3d 665, 678 (Fed. Cir. 2008) (en banc). The *Egyptian Goddess* court also explained that, under the Supreme Court's decision in *Smith v. Whitman Saddle Co.*, 148 U.S. 674 (1893), and its progeny, "the ordinary observer is deemed to view the differences between the patented design and the accused product *in the context of the prior art.*" 543 F.3d at 676. The backdrop of the prior art can strongly affect one's perception of similarity in design: "When the differences between the claimed and accused design are viewed in light of the prior art, the attention of the hypothetical ordinary observer will be drawn to those aspects of the claimed design that differ from the prior art. And when the claimed design is close to the prior art designs, small differences between the accused design and the claimed design are likely to be important to the eye of the hypothetical ordinary observer. . . . If the accused design has copied a particular feature of the claimed design that departs conspicuously from the prior art, the accused design is naturally more likely to be regarded as deceptively similar to the claimed design, and thus infringing." *Id.* at 676-77. Consider how the Federal Circuit uses the ordinary observer analysis, with the prior-art backdrop, in the following case.

Wallace v. Ideavillage Products Corp.
640 Fed. Appx. 970 (Fed. Cir. 2016)

NEWMAN, JUDGE:

Ms. Allyson Wallace owns U.S. Design Patent No. D485,990 for an "ornamental design for a body washing brush, as shown and described" in six drawings. Ms. Wallace, proceeding pro se, sued Ideavillage Products Corporation for patent infringement . . . based on the Ideavillage product called the "Spin Spa." Ideavillage moved for summary judgment of noninfringement. . . . The district court, reviewing the record and receiving argument, granted summary judgment to Ideavillage. For the reasons set forth below, we affirm the district court's judgment.

. . . .

. . . The standard for design patent infringement is summarized in *Egyptian Goddess, Inc. v. Swisa, Inc.*, 543 F.3d 665, 678 (Fed. Cir. 2008) (en banc), applying the "ordinary observer" test of *Gorham* . . . . The ordinary observer test proceeds in two stages. "In some instances, the claimed design and the accused design will be sufficiently distinct that it will be clear without more that the patentee has not met its burden of proving the two designs would appear 'substantially the same' to the ordinary observer . . . ." *Id.* "In other instances, when the claimed and accused designs are not plainly dissimilar, resolution of the question whether the ordinary observer would consider the two designs to be substantially the same will benefit from a comparison of the claimed and accused designs with the prior art . . . ." *Id.*

The district court applied both stages and determined under both tests that an ordinary observer would not consider Ms. Wallace's patented design and the Ideavillage design to be substantially the same.

A

With respect to the first stage, the district court highlighted six differences between the Ideavillage design and the claimed design:

(1) '990 patent has a straight handle, while the Ideavillage product has a bent or curved handle;

(2) '990 patent has a finger grip with a hill and valley design, while the Ideavillage product has no such finger grip;

(3) '990 patent has a flat threaded opening at the base of the handle, while the Ideavillage product has a closed pointed end with an aperture where a rope can be attached;

(4) '990 patent has a round head with a two tiered brush, while the Ideavillage product has an oblong head without a two tier brush;

(5) '990 patent has a protrusion at the back of the head, while Ideavillage product has a smooth back without any protrusion; and finally,

(6) '990 patent has no decoration on the back of the handle, while Ideavillage product has an oval at the neck of the handle and an oval group in the back of the handle.

In granting summary judgment, the district court

> acknowledge[d] manifest differences in the overall appearance of the '990 patent and the [Ideavillage] product. Indeed, a comparison supports a finding that these two designs are sufficiently distinct and Ms. Wallace cannot, as a matter of law, prove that the designs appear substantially the same. To the ordinary observer, in other words, the two designs do not look substantially the same.

We have compared the D'990 patent's drawings to the photographs of the accused product in Ms. Wallace's expert report and agree with the district court that no reasonable fact-finder would find them to be substantially the same under the first stage of the ordinary observer test. The district court compared the following designs:

D'990 Patent   **Ideavillage's Accused Product**

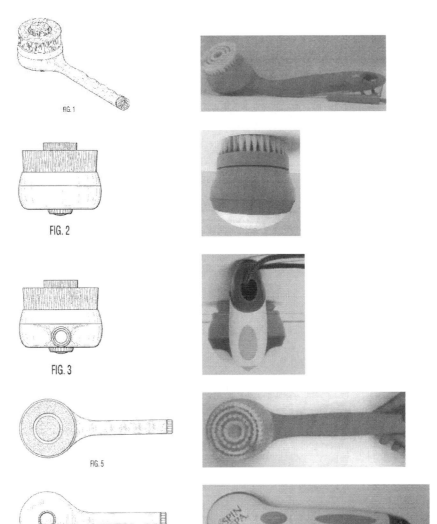

FIG. 1

FIG. 2

FIG. 3

FIG. 5

FIG. 6

The district court properly found that the D'990 patent's design and the accused product are plainly dissimilar. Bath brushes with a generally rounded head and roughly cylindrical handle were shown in the prior art, and the curved shape of the handle, the angled connection between the handle and brush, the ovoid design of the head, and the surface details are such that an ordinary observer viewing both designs would not confuse one product for the other.

B

"In an effort to assure a fair and complete decision on this record," the district court proceeded to the second stage of the ordinary observer test. Comparing the claimed design with figures from the prior art, U.S. Patent No. 4,417,826 (the '826 Patent), reinforces the district court's findings under the first stage of the test:

D'990 Patent

FIG. 1

Ideavillage's Accused Product

Prior Art '826 Patent

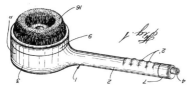

Other prior art views also show the known characteristics of the design.

D'990 Patent

FIG. 5

FIG. 6

Prior Art '826 Patent

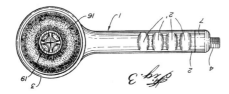

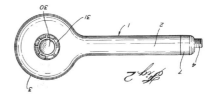

Ideavillage's Accused Product

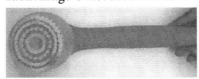

The district court found that:

[P]laintiff's '990 patent displays significant similarities with the design of the '826 patent. Amoung other things, each design has a rounded head and a straight handle with "hill and valley" finger grip. Each design has a round protrusion on the backside of the rounded head.

The district court properly concluded that "in light of the similarites between the '826 patent and the '990 patent . . . no reasonable ordinary observer, familiar with the prior art, would be deceived into believing the [Ideavillage] [p]roduct is the same as the design depicted in the '990 patent."

The district court correctly applied the law, that "differences between the claimed and accused designs that might not be noticeable in the abstract can become significant to the hypothetical ordinary observer who is conversant with the prior art." *Egyptian Goddess*, 543 F.3d at 678.

The judgment of noninfringement is affirmed. . . .

Notes & Questions

1. The articulation of the test for infringement in the context of design patents, which are pictorial rather than delineated in verbal claim language, is quite different from the utility patent context. In the case of a design patent, we don't distinguish between literal and equivalent infringement. Instead, we embrace both per-

fect duplication of and mere approximation to a design with the same liability standard—namely, whether an ordinary observer, familiar with the prior art, would be deceived into believing the accused design is the same as the patented design. How close can one come to the depicted design before infringing? For example, consider whether the design patent at the beginning of this section is infringed by the product here, at left.

Does your answer depend on how many prior-art pig-shaped grills there are?

2. The plaintiff-patentee, Wallace, did not appear to have enormous resources to spend on her case against Ideavillage. As a result, the question of consumer survey evidence simply did not arise. What if it had? Should we require, or presumptively disfavor, a patentee who fails to introduce, as part of its prima facie case, survey evidence that shows how ordinary consumers familiar with the relevant prior art perceive the accused product? Even if "the ordinary observer" is a hypothetical construct, wouldn't it be helpful to know how often consumers to whom competing products are directed would mistake one for the other or rate them as highly similar? As you will see in later Chapters, the "ordinary observer" and the "ordinary customer" are also highly relevant figures in copyright and trademark law.

3. Recall that, in *Egyptian Goddess*, the Federal Circuit made a specific empirical assertion about observer psychology—namely, that "[w]hen the claimed design is close to the prior art designs, small differences between the accused design and the claimed design are likely to be important to the eye of the hypothetical ordinary observer." The assertion seems plausible. But what if it is false, or true only some of the time under certain conditions? For example, what if the assertion is true only when the items in question are all presented side by side on a merchandise counter (rather than in isolation, in a single-brand store or single-brand website)? The *Egyptian Goddess* court's apparent indifference to actual evidence (or the lack of it) about ordinary-observer psychology is quite common in intellectual property law, as it is in law more generally.

E. Anticipation & Obviousness

In Section V.B, in the notes following the *Titanium Metals* case, we noted an adage in patent law that highlights the mirror-image identity requirements for anticipation and literal infringement—that which would literally infringe if later in time anticipates if earlier in time. Given the test for design-patent infringement, what do you predict is the proper way to analyze design-patent anticipation?

OraLabs, Inc. v. The Kind Group LLC
No. 13-cv-00170, 2015 WL 4538444 (D. Colo. July 28, 2015)

Brimmer, Judge:

  . . . .

Kind Group is the assignee of U.S. Patent No. D644,939, a design patent that covers the "ornamental design for the spherically-shaped lip balm," as shown in the following five figures:

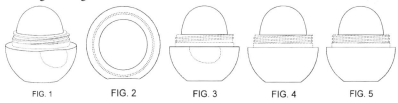

FIG. 1   FIG. 2   FIG. 3   FIG. 4   FIG. 5

In 2009, the Kind Group began selling the eos Smooth Sphere (the "Smooth Sphere"), a lip balm container based on the design of the '939 Patent. . . .

OraLabs is a manufacturer and distributor of cosmetic products. In 2011, due to OraLabs' customer interest, OraLabs developed a product similar to the Smooth Sphere, which it called the Chap-Ice Lip Revolution (the "Lip Revo") . . . . Both the Lip Revo and Smooth Sphere are relatively inexpensive products, typically priced between $3 and $4.

OraLabs filed this case on January 24, 2013, seeking a declaration that [among other things] the '939 Patent is invalid . . .

The parties filed cross-motions for summary judgment. . . . Kind Group moves for summary judgment that . . . the '939 Patent is valid.

. . . .

. . . OraLabs bears the burden to "call to the court's attention the prior art that an ordinary observer is most likely to regard as highlighting the differences between the claimed and accused design." *Egyptian Goddess*, 543 F.3d at 679. OraLabs identifies three prior art references that it claims would magnify the differences between the Lip Revo and the '939 Patent in the eyes of an ordinary observer . . . [including] U.S. Patent No[]. D554, 529 ("Green I") . . . .

Green I describes a design for a spherically shaped perfume applicator. . . .

. . . .

Kind Group seeks summary judgment that the '939 Patent is not invalid as anticipated . . . . "A patent is presumed valid, and the burden of establishing invalidity as to any claim of a patent rests upon the party asserting such invalidity. Clear and convincing evidence is required to invalidate a patent." *Aero Prods. Int'l, Inc. v. Intex Recreation Corp.*, 466 F.3d 1000, 1015 (Fed. Cir. 2006).

. . . .

A patent may be found invalid as a matter of law if it is anticipated. 35 U.S.C. § 102. To determine whether a design patent is invalid as anticipated by prior art, courts apply the same "ordinary observer" test used for infringement. *Int'l Seaway Trading Corp. v. Walgreens Corp.*, 589 F.3d 1233, 1240 (Fed. Cir. 2009). Thus, anticipation in the design context involves comparing the patent-in-suit with the relevant prior art and determining whether a hypothetical ordinary observer would find that the designs are substantially the same. *See id.* OraLabs notes that its expert, Michael Thuma, found that the '939 Patent was anticipated by Green I, with the caveat that Mr. Thuma assumed that the flat edge in the '939 Patent was functional and thus not part of the claimed design. Kind Group argues that Mr. Thuma's conclusion was based on an improper reduction of the '939 Patent and Green I to two verbal elements: a hemispherical base and a hemispherical applicator. The Court need not decide at this juncture whether the flat edge on the side wall in the '939 Patent is functional. The Court finds that, even accepting OraLabs' construction, which omits the flat edge, no jury could conclude that the ordinary observer would believe that the '939 Patent and Green I are the same design. Figures 1-3 of Green I are depicted below:

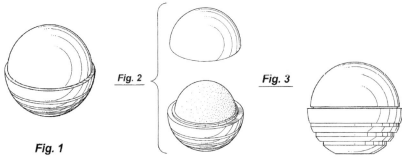

Fig. 1   Fig. 2   Fig. 3

Comparing the overall design of the Green I figures with the figures that comprise the scope of the '939 Patent, the Court finds that an ordinary observer would not believe that the '939 Patent and Green I are substantially the same. First, as noted above, the '939 Patent is egg-shaped while the Green I design appears to be a perfect sphere. Second, the product mound of Green I has the same shape as the base, a perfect sphere, while the product mound of the '939 Patent is smaller proportionately and has an ovoid shape. These differences in proportion and overall shape result in a different overall visual impression. *See Voltstar Techs., Inc. v. Amazon.com, Inc.*, 2014 WL 3725860, at *4 (N.D. Ill. July 28, 2014) (granting summary judgment where the ordinary observer would note that the patent-in-suit had a "squatter proportion" than the accused design and where the accused design created an "oval effect" as opposed to the patent-in-suit, which "create[d] an effect of a flat top and bottom with rounded corners and slightly arced sides").

. . . .

Notes & Questions

1. Legal decisions require courts to articulate the reasons for their decisions. Notice how many words the courts use to describe the similarity and differences between the relevant designs. The ability to describe both images and ornamental characteristics of tangible objects is a skill that a well-trained intellectual property lawyer must have. Be sure to pay close attention to the descriptions and ask yourself if there are alternatives. For example, how might an attorney for the BBQ Pig design patent owner describe the BBQ pig depicted in the Notes & Questions following the *Wallace* case?

When we studied the nonobviousness requirement in Section VI, we saw that the Supreme Court's 2007 decision in *KSR v. Teleflex* marked a departure from the way the Federal Circuit had analyzed nonobviousness since that court began work in 1982. Ornamental designs, like useful inventions, are patentable only if they are nonobvious under § 103. Below is a recent Federal Circuit case analyzing design-patent nonobviousness. As you read it, consider the ways in which the nonobviousness analysis for design patents differs from the analysis for utility patents, and whether the differences comport with the Supreme Court's decision in *KSR*.

MRC Innovations, Inc. v. Hunter Mfg.
747 F.3d 1326 (Fed. Cir. 2014)

PROST, JUDGE:

MRC Innovations, Inc. appeals from a final judgment of . . . invalidity with respect to U.S. Design Patent Nos. D634,488 and D634,487. For the reasons stated below, we affirm.

Background

MRC is the owner by assignment of both patents-in-suit. The '488 patent claims an ornamental design for a football jersey for a dog, while the '487 patent does the same for a baseball jersey, as shown below:

'488 Patent	'487 Patent
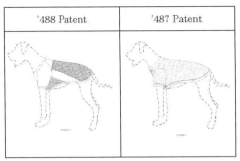	

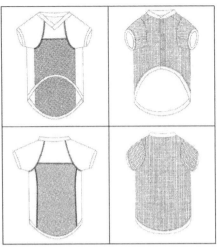

Mark Cohen is the named inventor of both patents; he is the principal shareholder of MRC and assigned his rights in both patents to that company. Appellee Hunter Manufacturing, LLP, is a retailer of licensed sports consumer products, including pet jerseys. Prior to September 9, 2009, Hunter purchased pet jerseys for dogs from Mark Cohen through companies with which he was affiliated. For example, Cohen supplied Hunter with a "V2" football jersey through the Stephen Gould Corporation and through Fun-in-Games, Inc. ("FiG"). Similarly, Cohen supplied Hunter, through FiG, with a green pet jersey bearing a Philadelphia Eagles logo, which Hunter then sold through third-party retailers such as Wal-Mart and PetSmart prior to July 30, 2009. The V2 and Eagles jerseys are depicted below:

V2 Jersey	Eagles Jersey

Cohen asserts that in 2009 he designed another pet jersey, known as the "V3" jersey, which would later become the subject of the '488 patent. Hunter began purchasing the V3 jersey from FiG sometime after September 8, 2009. On September 8, 2010, Cohen filed a patent application for both the V3 jersey and the baseball equivalent that would later become the subject of the '487 patent.

In December 2010, Cohen informed Hunter that he no longer intended to do business with Hunter because Hunter was having difficulty making payments. Hunter then sought proposals from other companies to manufacture and supply it with pet jerseys like the V3. Ultimately, Hunter contracted with another supplier, appellee CDI International, Inc., to supply Hunter with pet jerseys.

Both patents-in-suit eventually issued on March 15, 2011.

MRC filed suit against both Hunter and CDI for willful infringement of both patents. The district court granted summary judgment in favor of Hunter and CDI

on the grounds that both patents are invalid as obvious under 35 U.S.C. § 103(a).
. . .

. . . .

Discussion

. . . .

I. The '488 Patent

The district court concluded that the '488 patent would have been obvious in view of several prior art pet jerseys. MRC now appeals that determination.

Obviousness is a question of law that is reviewed de novo, based on underlying factual questions that are reviewed for clear error following a bench trial. *Honeywell Int'l, Inc. v. United States*, 609 F.3d 1292, 1297 (Fed. Cir. 2010). The underlying factual inquiries include: (1) the scope and content of the prior art; (2) the level of ordinary skill in the art; (3) the differences between the claimed invention and the prior art; and (4) objective evidence of non-obviousness. *Graham v. John Deere Co.*, 383 U.S. 1, 17 (1966). Summary judgment of obviousness is appropriate if "the content of the prior art, the scope of the patent claim, and the level of ordinary skill in the art are not in material dispute, and the obviousness of the claim is apparent in light of these factors." *TriMed, Inc. v. Stryker Corp.*, 608 F.3d 1333, 1341 (Fed. Cir. 2010) (citing *KSR Int'l Co. v. Teleflex Inc.*, 550 U.S. 398, 427 (2007)).

In the context of design patents, "'the ultimate inquiry under section 103 is whether the claimed design would have been obvious to a designer of ordinary skill who designs articles of the type involved.'" *Titan Tire Corp. v. Case New Holland, Inc.*, 566 F.3d 1372, 1380-81 (Fed. Cir. 2009) (quoting *Durling v. Spectrum Furniture Co.*, 101 F.3d 100, 103 (Fed. Cir. 1996)). To answer this question, a court must first determine "whether one of ordinary skill would have combined teachings of the prior art to create the same overall visual appearance as the claimed design." *Durling*, 101 F.3d at 103. That inquiry involves a two-step process. First, the court must identify "a single reference, 'a something in existence, the design characteristics of which are basically the same as the claimed design.'" *Id.* (quoting *In re Rosen*, 673 F.2d 388, 391 (CCPA 1982)). The "basically the same" test requires consideration of the "visual impression created by the patented design as a whole." *Id.* We have noted that "the trial court judge may determine almost instinctively whether the two designs create basically the same visual impression," but "must communicate the reasoning behind that decision." *Id.*

Once the primary reference is found, other "secondary" references "may be used to modify it to create a design that has the same overall visual appearance as the claimed design." *Id.* These secondary references must be "'so related [to the primary reference] that the appearance of certain ornamental features in one would suggest the application of those features to the other.'" *Id.* (quoting *In re Borden*, 90 F.3d 1570, 1575 (Fed. Cir. 1996) (alteration in original)).

A. Primary Reference

The district court used the "Eagles" pet jersey as the "primary reference" under step one of the *Durling* analysis. MRC argues that this was legally erroneous

because there are significant differences between the Eagles jersey and the patented design of the '488 patent. Specifically, there are three differences: (1) the
patented design has a V-neck collar where the Eagles jersey has a round neck;
(2) the patented design contains an interlock fabric panel on the side portion of
the design rather than mesh; and (3) the patented design contains additional ornamental surge stitching on the rear portion of the jersey. MRC argues that the
district court overlooked these differences by focusing on the claimed design at
"too high a level of abstraction." *High Point Design LLC v. Buyers Direct, Inc.*, 730
F.3d 1301, 1314 (Fed. Cir. 2013) (citing *Apple, Inc. v. Samsung Elecs. Co.*, Ltd., 678
F.3d 1314, 1331 (Fed. Cir. 2012)). If the district court had translated the claimed
design into a verbal description as required by *High Point*, MRC insists, it would
have concluded that neither the Eagles jersey nor any other prior art reference
contained design characteristics that were "basically the same" as the claimed design.

As an initial matter, it is true that the district court did not expressly undertake
to translate the claimed design into a verbal description. However, *High Point*
makes clear that the purpose of requiring district courts to describe the claimed
design in words is so that the parties and appellate courts can discern the trial
court's reasoning in identifying a primary reference. *See id.* (citing *Durling*, 101
F.3d at 103). It is entirely clear from the district court's opinion what it considered
to be the relevant design characteristics of the '488 patented design.

First, the district court pointed out three key similarities between the claimed
design and the Eagles jersey: an opening at the collar portion for the head, two
openings and sleeves stitched to the body of the jersey for limbs, and a body portion on which a football logo is applied. If the district court's analysis had ended
there, it might indeed have failed to meet the *High Point* verbal description requirement. However, the district court went on to point out two additional similarities between the two designs: first, the Eagles jersey is made "primarily of a
mesh and interlock fabric"; and second, it contains at least some ornamental surge
stitching—both features found in the '488 claimed design. The district court also
went on to acknowledge the three major differences between the two designs that
are enumerated above. Taking all of those things together (the at least five design
characteristics that the claimed design shares with the Eagles jersey and three design characteristics that differ from it), the district court painted a clear picture of
the claimed design. The district court did far more than merely ask whether the
Eagles jersey disclosed the "general concept" of a pet jersey; it thoroughly considered the "distinctive 'visual appearances' of the reference and the claimed design." *Apple*, 678 F.3d at 1332 (quoting *Durling*, 101 F.3d at 104). Thus, the district
court did not err by failing to provide an express verbal description of the claimed
design; rather, it described the claimed design in the context of comparing it to
the prior art.

Nor did the district court err in finding that the design characteristics of the
'488 design created "basically the same" overall visual impression as the Eagles
jersey prior art reference. As the district court noted, both designs contain the
same overall shape, similar fabric, and ornamental surge stitching. That there are

slight differences in the precise placement of the interlock fabric and the ornamental stitching does not defeat a claim of obviousness; if the designs were identical, no obviousness analysis would be required.[1] Indeed, we have permitted prior art designs to serve as "primary references" when their differences are as great or greater than the differences in this case. *See Jore Corp. v. Kouvato, Inc.*, 117 Fed. Appx. 761, 763 (Fed. Cir. 2005) (finding prior art drill bit to be a primary reference despite containing a smooth cylindrical shaft rather than the grooved hexagonal shaft of the claimed design); *In re Nalbandian*, 661 F.2d 1214, 1217-18 (CCPA 1981) (finding tweezer design obvious in light of prior art reference that contained vertical rather than horizontal fluting and straight rather than curved pincers).

. . . .

B. Secondary References

After concluding that the Eagles jersey could be a "primary reference," the district court determined that the V2 jersey and another reference known as the "Sporty K9" jersey were "so related to the primary reference" that they could serve as "secondary references" that would motivate the skilled artisan to make the claimed design.

The district court found that both jerseys suggested the use of a V-neck pattern and non-mesh fabric on the side panels—the first two differences described above. MRC argues that the district court erred by failing to explain why a skilled artisan would have chosen to incorporate those features of the V2 and Sporty K9 jerseys with the Eagles jersey.

We disagree. It is true that "[i]n order for secondary references to be considered * * * there must be some suggestion in the prior art to modify the basic design with features from the secondary references." *In re Borden*, 90 F.3d at 1574. However, we have explained this requirement to mean that "the teachings of prior art designs may be combined only when the designs are 'so related that the appearance of certain ornamental features in one would suggest the application of those features to the other.'" *Id.* at 1575 (quoting *In re Glavas*, 230 F.2d 447, 450 (CCPA 1956)). In other words, it is the mere similarity in appearance that itself provides the suggestion that one should apply certain features to another design.

In re Borden also discussed what is required for a reference to be considered sufficiently "related" for that test to apply. There, we noted that the secondary

---

[1] This conclusion is not inconsistent with the law of this circuit on design patent infringement. In that context, we have often noted that design patents have "almost no scope" beyond the precise images shown in the drawings. *In re Mann*, 861 F.2d 1581, 1582 (Fed. Cir. 1988). However, in practice, our focus on the "overall visual appearance" of a claimed design rather than on individual features has led us to find products infringing despite differences in specific ornamental features. For example, in *Crocs, Inc. v. Int'l Trade Comm'n*, 598 F.3d 1294 (Fed. Cir. 2010), we concluded that all of the accused products infringed the asserted design patents despite the fact that two of the infringing products (the Groovy DAWGS shoes and Big DAWGS shoes) contained a wider shoe front with an additional row of holes, and another infringing product (the Effervescent Waldies AT shoe) contained square holes on the top of the shoe rather than round ones. *Id.* at 1303-06.

references were "closely akin" to the claimed design, and relied heavily on the fact that "the two missing design elements [were] not taken from unrelated references, but [were] found in other dual-chamber containers." *Id.* Thus, those references could be used "to bridge the small gap between the [primary] container and Borden's claimed design." *Id.* So too, here, the secondary references that the district court relied on were not furniture, or drapes, or dresses, or even human football jerseys; they were football jerseys designed to be worn by dogs. Moreover, . . . the V2 could easily have served as a primary reference itself, so similar is its overall visual appearance to that of the claimed design and the Eagles jersey. We therefore agree that those references were "so related" to the Eagles jersey that the striking similarity in appearance across all three jerseys would have motivated a skilled designer to combine features from one with features of another.

With respect to the only remaining difference between the Eagles jersey and the '488 claimed design—the presence of additional ornamental surge stitching running down the rear of the jersey—the district court acknowledged that no prior art reference contained exactly that same stitching on the rear of the jersey, but nevertheless concluded that this was not a "substantial" difference that created a patentably distinct design, but rather was a "*de minimis* change[] which would be well within the skill of an ordinary designer in the art[,]" citing *In re Carter*, 673 F.2d 1378, 1380 (CCPA 1982).

MRC argues that adding *any* ornamental feature to a primary reference that is not suggested by the prior art is, by definition, more than *de minimis*. But our case law plainly contradicts that position; on numerous occasions we have invalidated design patents despite the inclusion of ornamental features that were entirely absent from prior art designs. *See, e.g., In re Nalbandian*, 661 F.2d at 1217 (different shape of fluting on finger grips and different shape of pincers were de minimis differences in design for tweezers); *In re Carter*, 673 F.2d at 1380 (modifications to the waistband of an infant garment were "*de minimis* changes which would be well within the skill of an ordinary designer in the art"); *In re Cooper*, 480 F.2d 900, 901-02 (CCPA 1973) (affirming Board's conclusion that numerous changes to the design of a prior art building—including a single rather than double door and the addition of windows—were *de minimis* because the overall impression was still a building that looked like a barrel).

Here, the Eagles jersey had already disclosed the use of ornamental surge stitching. The only additional step needed was to extend the stitching down the sides of the rear of the jersey. Moreover, the V2 jersey plainly suggested the addition of vertical lines down the rear of the jersey through the use of the seams between the two types of fabric. We agree with the district court that adding ornamental surge stitching on top of a preexisting seam was an insubstantial change that would have been obvious to a skilled designer.[6]

---

[6] To be clear, we do not intend to suggest that merely because one prior art reference used ornamental surge stitching, *any* use of such stitching would have been a *de minimis* change. Rather, the addition of the surge stitching in this case was *de minimis* because it merely

C. Secondary Considerations

In support of the non-obviousness of its patents, MRC submitted evidence relating to commercial success, copying, and acceptance by others. First, with respect to commercial success, MRC argued that the sales of the V3 jersey (which embodied the patented design) were more successful than sales of the V2 design it replaced. Second, MRC alleged that Hunter and CDI chose to copy the patented V3 design rather than other available non-patented designs. And third, MRC points out that it has granted a license on the '488 patent.

The district court noted that the only evidence in support of these secondary considerations was the testimony of the inventor himself, which was "unpersuasive to demonstrate a genuine dispute" of fact sufficient to defeat summary judgment of obviousness. MRC argues that in so doing, the district court effectively dismissed the uncontroverted secondary considerations evidence as if it did not exist, rather than construing the evidence in the favor of the non-moving party—in this case, MRC. MRC also points out that Hunter and CDI failed to provide any evidence of another explanation for those secondary considerations besides a nexus with the claimed design.

Here, however, MRC has the standard backwards. As the patentee, it was MRC's burden of production to demonstrate a nexus between the claimed design and the secondary considerations. *See Crocs*, 598 F.3d at 1311. MRC presented no evidence whatsoever that the commercial success and copying were related to the merits of the claimed invention. Merely stating—with no supporting figures or data—that the V3 was more successful than the V2 is insufficient on its own to establish that the V3 has been a "commercial success" and that its success was attributable to the claimed design features. Moreover, the only license MRC produced was between MRC and FiG, a company that is at least partially owned by Mark Cohen, who also owns MRC.

Thus, although the district court's analysis of secondary considerations was admittedly somewhat cursory, we do not believe that the evidence of record before the district court created a genuine dispute of material fact; to the contrary, even construing the evidence in the light most favorable to MRC, MRC had not established a nexus between the secondary considerations and the claimed design that was sufficient to overcome the other evidence of obviousness.

. . . .

II. The '487 Patent

A similar analysis applies to the '487 patent. The district court also found the '487 patent obvious, relying on the baseball version of the Sporty K9 jersey as the "primary reference," and the V2 and Eagles football jerseys as the secondary references. . . .

[The Federal Circuit affirmed the invalidity of the '487 patent as well.]

---

followed the visual lines created by the seams of the V2 jersey; in other words, it served only to highlight a design feature that had already existed in the V2 prior art jersey.

Notes & Questions

1. How does the test for obviousness of a design patent differ from the test used for utility patents? Does requiring a "primary reference" make more sense in the context of a design patent? In what way are the requirements for the allowable "secondary references" similar to the requirements for qualifying prior art that are explored in the *In re Clay* decision excerpted in Section VI?

2. The strictures of the nonobviousness analysis for design patents—especially the need for a suitably similar "primary reference" upon which to ground the whole analysis—arguably makes it harder to reject a design patent for obviousness than to reject a mechanical utility patent, such as the adjustable pedal at issue in *KSR*. Is that justifiable, as a matter of innovation policy—which innovations are socially more valuable or beneficial, and thus which innovations do we want to more strongly encourage with larger rewards?

CHAPTER 4: COPYRIGHT LAW

I. THEORY AND COPYRIGHTABLE SUBJECT MATTER

Copyright protection, like patent protection, is governed exclusively by federal statute. 17 U.S.C. § 101 et seq. Congress passed the current Copyright Act in 1976 and it took effect on January 1, 1978. Fueled in large part by ongoing technological change, Congress has amended the 1976 Copyright Act many times in the three decades since its passage. The breadth of creations that are subject to copyright protection, the scope of the rights granted to copyright owners, and the ease of reproducing many works in the digital age make copyright law a particularly challenging and important area of intellectual property today. The Copyright Act grants a copyright owner limited rights to exclude others from reproducing the copyrighted expression, publicly performing or displaying the copyrighted expression, distributing copies of the copyrighted expression, and from creating derivative works based on the copyrighted expression.

A copy of the Copyright Act is available from CALI's:
http://www.cali.org/books/united-states-copyright-law
A copy may also be obtained from the Copyright Office:
http://www.copyright.gov/title17/

A. THEORETICAL UNDERPINNINGS

As we explored at the beginning of Chapter 1, intangible assets have a risk of underproduction because of their characteristics of non-rivalrous consumption and non-excludability. In the United States, copyright law, like patent law, is expressly utilitarian: granting exclusive rights in expressive works is meant to address the underproduction risk. The Constitutional provision that grants Congress the power to adopt the Copyright Act acknowledges that progress in knowledge and learning is the purpose that the exclusive rights are meant to achieve:

> The Congress shall have power . . . To promote the Progress of Science and useful Arts, by securing for limited Times to Authors and Inventors the exclusive Right to their respective Writings and Discoveries[.]

U.S. CONST. Art. I, § 8, cl. 8. At the time of the founding, "science" meant broadly, knowledge and learning. The framers believed that Congress needed the power to grant authors exclusive rights in their writings in order to achieve progress as a society. After all, we were a fledgling nation, without the resources to engage in significant government patronage as a way to guard against the risk of underproduction. Thus, giving Congress the authority to harness the power of the market as a way to provide the incentive to create and distribute expressive works was a smart choice. At the same time, the framers were aware of the history of copyright protection in Europe as a tool of government censorship and press control and thus sought to guard against this dark side of copyright protection by enumerating the purpose of the exclusive rights. In fact, some evidence exists that granting copyright protection was meant to ensure a free press separate from the government influence that comes with government funding. *See* Neil Weinstock Netanel, *Copyright and a Democratic Civil Society*, 106 Yale L.J. 283 (1996).

The basic utilitarian rationale that you learned about for patents applies as well for copyrights. While human beings are naturally creative and expressive creatures, they also need to support themselves and their families. Creators need a certain amount of remuneration in order to invest their time and labor in creative pursuits instead of other activities that might result in higher rewards. Additionally, the works that are created need to be communicated or distributed to the public if we are to achieve the progress in knowledge and learning that the Constitution specifies as the purpose behind granting exclusive rights. Thus copyright law is meant to provide an incentive not just for creation but for dissemination as well.

Until recently, widespread dissemination entailed significant costs: printing presses, typesetting, paper, ink, and shipping costs. If a competing book publisher could wait until sales indicated which book was a blockbuster hit and then begin making copies of that book, the price would be driven down through competition. This freewheeling competition, unconstrained by copyright law, would result in the initial publisher earning less and perhaps not even recouping its investment in obtaining the work from the author. Without copyright protection, the author also would likely earn far less in her publishing agreement with the initial publisher. Thus, copyright protection helps provide the incentives necessary for both creation and dissemination.

The internet has changed the economics of dissemination of expressive works, making the investment required to achieve widespread dissemination far smaller. This change in market dynamics for some methods of distribution presents challenges to the legitimacy of the utilitarian theory of copyright: disseminators may not need the same level of monetary reward in order to create the incentive necessary to engage in the dissemination activity. Congress has amended the Copyright Act in response to the technological changes brought about by the internet, but those changes have been to increase the protections afforded to copyright owners, not to decrease them. While such a change might seem counterintuitive, the content owning industries (publishing, motion pictures, and sound recording) lobbied vigorously for the changes, arguing that the ease of digital reproduction meant infringement would be rampant and would significantly reduce the value of the copyright, undermining the very incentive copyright was meant to provide.

While the utilitarian justification for copyright protection is dominant in the United States, it is not the only possible rationale for granting exclusive rights to the creators of expressive works. One can also rely on Lockean labor theory to justify granting creators rights in the products of their mind. Locke argued that property rights come into being as one mixed one's labor with things found in nature. His examples involved men who gathered acorns and apples from the commons and thus owned them. John Locke, *Two Treatises on Government*, Book II, ch. V (1690). If physical labor and intellectual labor should be treated equally, then those who create expressive works are as deserving of a "property" right as those who engage in physical labor. *See* Alfred Yen, *Restoring the Natural Law: Copyright as Labor and Possession*, 51 Ohio St. L.J. 517 (1990).

It is also possible to consider granting a right in expressive content as a way to protect an extension of oneself. Under this justification, the connection between creator and creation is a unique bond that should be legally protected by granting the creator control over the expression. The economic consequences of granting such rights are not important under this rationale; instead, the creator's ability to protect the artistic integrity of the work is paramount. Several European countries base their copyright law on this justification. *See* Jane C. Ginsburg, *A Tale of Two Copyrights: Literary Property in Revolutionary France and America*, 64 Tul. L. Rev. 991, 992 (1990). These countries grant copyright protection for what we refer to as "moral rights," *i.e.*, the right to protect the integrity of the expression from modification, as well as a right to be recognized as the creator of a work, called a right of attribution. In the United States, these types of rights are only granted to "works of visual art," defined quite narrowly by the statute, 17 U.S.C. § 101, and the protections granted are more limited than in European copyright law. *See* 17 U.S.C. § 106A.

B. COPYRIGHTABLE SUBJECT MATTER

Copyrightable works range from traditional categories of works that many think of when they think of copyrighted material, such as literary works and music, to new categories brought about by technological change, such as computer software and the audiovisual output of video games. Such works even include creations that few realize are subject to copyright protection, such as architectural works, toys and games. The Copyright Act specifies what can be protected:

> (a) Copyright protection subsists, in accordance with this title, in original works of authorship fixed in any tangible medium of expression, now known or later developed, from which they can be perceived, reproduced, or otherwise communicated, either directly or with the aid of a machine or device.

17 U.S.C. § 102. Generally, this provision is read to require two elements in order to obtain copyright protection: originality and fixation.

1. Fixation

The statute provides guidance on what it means for a work to be "fixed." First, § 102 itself indicates that fixation can be in "any tangible medium of expression" and expresses a technologically forward-looking view, encompassing future mediums of expression without need for statutory revision. This broad scope for fixation is also contained in the definition of fixation:

A work is "fixed" in a tangible medium of expression when its embodiment in a copy or phonorecord, by or under the authority of the author, is sufficiently permanent or stable to permit it to be perceived, reproduced, or otherwise communicated for a period of more than transitory duration.

17 U.S.C. § 101. These linguistic choices by Congress were born of experience. Past copyright acts had used technologically specific language that had created problems when new methods of expression developed, such as photography. Additionally, the language indicating that direct perception is not necessary, but that perception, reproduction or communication can occur "with the aid of machine or device" also was meant to address problematic experiences with new technology, such as video tape that required a machine in order to view the images on the tape. The broad approach in the current statute means that a copyrighted work can be "fixed" in such mediums as digital files, fabric, wood, paper, and everything and anything in between. Only expressive works that are fleeting will not meet the requirement of a lasting "for a period of more than transitory duration."

Courts have been quite liberal in their interpretation of even this provision, concluding, for example, that a work contained in temporary computer RAM memory is sufficiently fixed for purposes of the statute. *MAI Systems Corp. v. Peak Computer, Inc.*, 991 F.2d 511 (9th Cir. 1993). However, the Second Circuit recently put some limits on that holding:

> [W]e construe *MAI Systems* and its progeny as holding that loading a program into a computer's RAM can result in copying that program. We do not read *MAI Systems* as holding that, as a matter of law, loading a program into a form of RAM always results in copying. Such a holding would read the 'transitory duration' language out of the definition . . . .

Cartoon Network LP v. CSC Holdings, Inc., 536 F.3d 121, 128 (2d Cir. 2008). In that case the Second Circuit was considering the design of Cablevision's remote DVR, in which the buffer memory constantly overwrote programming data as it was broadcast.

> No bit of data remains in any buffer for more than a fleeting 1.2 seconds. And unlike the data in cases like *MAI Systems*, which remained embodied in the computer's RAM memory until the user turned the computer off, each bit of data here is rapidly and automatically overwritten as soon as it is processed. While our inquiry is necessarily fact-specific, and other factors not present here may alter the duration analysis significantly, these facts strongly suggest that the works in this case are embodied in the buffer for only a ''transitory'' period, thus failing the duration requirement.

Id. at 129-30. In the age of digital connectivity, determining what constitutes a sufficiently "fixed" copy for purposes of the Copyright Act requires attention to the technical details.

2. Originality

Unlike the way it does with fixation, the statute does not define the second requirement for protection: originality. Instead, Supreme Court precedent provides guidance on what the originality requirement entails.

Bleistein v. Donaldson Lithographing Co.
188 U.S. 239 (1903)

HOLMES, JUSTICE:

. . . The alleged infringements consisted in the copying in reduced form of three chromolithographs prepared by employees of the plaintiffs for advertisements of a circus owned by one Wallace. Each of the three contained a portrait of Wallace in the corner, and lettering bearing some slight relation to the scheme of decoration, indicating the subject of the design and the fact that the reality was to be seen at the circus. One of the designs was of an ordinary ballet, one of a number of men and women, described as the Stirk family, performing on bicycles, and one of groups of men and women whitened to represent statues. The circuit court directed a verdict for the defendant on the ground that the chromolithographs were not within the protection of the copyright law, and this ruling was sustained by the circuit court of appeals.

. . . .

. . . It is obvious . . . that the plaintiff's case is not affected by the fact, if it be one, that the pictures represent actual groups—visible things. They seem from the testimony to have been composed from hints or description, not from sight of a performance. But even if they had been drawn from the life, that fact would not deprive them of protection. The opposite proposition would mean that a portrait by Velasquez or Whistler was common property because others might try their hand on the same face. Others are free to copy the original. They are not free to copy the copy. The copy is the personal reaction of an individual upon nature. Personality always contains something unique. It expresses its singularity even in handwriting, and a very modest grade of art has in it something irreducible which is one man's alone. That something he may copyright unless there is a restriction in the words of the act.

If there is a restriction, it is not to be found in the limited pretensions of these particular works. The least pretentious picture has more originality in it than directories and the like, which may be copyrighted. . . .

There is no reason to doubt that these prints, in their ensemble and in all their details, in their design and particular combinations of figures, lines, and colors, are the original work of the plaintiffs' designer. If it be necessary, there is express testimony to that effect.

We assume that the construction of Rev. Stat. § 4952, allowing a copyright to the "author, designer, or proprietor . . . of any engraving, cut, print . . . [or]

chromo," is affected by the act of 1874, § 3. That section provides that, "in the construction of this act, the words 'engraving,' 'cut,' and 'print' shall be applied only to pictorial illustrations or works connected with the fine arts."

We see no reason for taking the words "connected with the fine arts" as qualifying anything except the word "works," but it would not change our decision if we should assume further that they also qualified "pictorial illustrations," as the defendant contends.

These chromolithographs are "pictorial illustrations." The word "illustrations" does not mean that they must illustrate the text of a book, and that the etchings of Rembrandt or Muller's engraving of the Madonna di San Sisto could not be protected today if any man were able to produce them. Again, the act, however construed, does not mean that ordinary posters are not good enough to be considered within its scope. The antithesis to "illustrations or works connected with the fine arts" is not works of little merit or of humble degree, or illustrations addressed to the less educated classes; it is "prints or labels designed to be used for any other articles of manufacture." Certainly works are not the less connected with the fine arts because their pictorial quality attracts the crowd, and therefore gives them a real use—if use means to increase trade and to help to make money. A picture is nonetheless a picture, and nonetheless a subject of copyright, that it is used for an advertisement. And if pictures may be used to advertise soap, or the theater, or monthly magazines, as they are, they may be used to advertise a circus. Of course, the ballet is as legitimate a subject for illustration as any other. A rule cannot be laid down that would excommunicate the paintings of Degas.

. . . .

It would be a dangerous undertaking for persons trained only to the law to constitute themselves final judges of the worth of pictorial illustrations, outside of the narrowest and most obvious limits. At the one extreme, some works of genius would be sure to miss appreciation. Their very novelty would make them repulsive until the public had learned the new language in which their author spoke. It may be more than doubted, for instance, whether the etchings of Goya or the paintings of Manet would have been sure of protection when seen for the first time. At the other end, copyright would be denied to pictures which appealed to a public less educated than the judge. Yet if they command the interest of any public, they have a commercial value—it would be bold to say that they have not an aesthetic and educational value—and the taste of any public is not to be treated with contempt. It is an ultimate fact for the moment, whatever may be our hopes for a change. That these pictures had their worth and their success is sufficiently shown by the desire to reproduce them without regard to the plaintiffs' rights. We are of opinion that there was evidence that the plaintiffs have rights entitled to the protection of the law. . . .

Notes & Questions

1. *Bleistein* is most famous for Holmes' pronouncement on the judiciary's inability to judge the artistic merit of a work. Often cited in First Amendment cases,

this decision is the genesis of the aesthetic nondiscrimination doctrine in copyright law. The aesthetic nondiscrimination doctrine means that when determining whether a work contains sufficient originality one is not supposed to inquire as to the artistic merit of the work. If artistic merit is not part of originality, what is?

2. Under the utilitarian justification for copyright, do advertisements really need the protection of copyright as an incentive to induce their creation? In *Bleistein* itself, Wallace had commissioned Bleistein to create the posters, and, running low on copies, Wallace ordered additional copies made by the Donaldson Lithographing Co. *See* Diane Leenheer Zimmerman, *The Story of* Bleistein v. Donaldson Lithographing Company: *Originality as a Vehicle for Copyright Inclusivity,* in INTELLECTUAL PROPERTY STORIES (Jane C. Ginsburg & Rochelle Cooper Drefuss eds., 2006). Might there have been some merit to the idea that we don't need to guard against the underprotection of posters advertising a circus by providing the lithographing company the incentive of copyright protection? Is copyright in this scenario—an independent creator, and a business customer— helping the creator guard against a customer's change of mind about paying the creator's full price? If the original lithographing company could have bargained to be the exclusive poster supplier for the Wallace Shows, and it simply failed to do so, why use copyright law to fill that contractual gap?

3. The Court states: "That these pictures had their worth and their success is sufficiently shown by the desire to reproduce them without regard to the plaintiffs' rights . . . . We are of the opinion that there was evidence that the plaintiffs have rights entitled to the protection of the law." Is this "if value, then right" approach a good idea? Is it a good way to judge whether a work contains the "originality" required to obtain a copyright?

The most recent Supreme Court precedent on originality involves a claim of copyright protection in a telephone directory. To understand the context of the claim, we must first explore some additional basics in the statute.

In addition to providing a general statement of the types of works protected by copyright, § 102(a) provides a list of "categories" of works eligible for protection:

Works of authorship include the following categories:
(1) literary works;
(2) musical works, including any accompanying words;
(3) dramatic works, including any accompanying music;
(4) pantomimes and choreographic works;
(5) pictorial, graphic, and sculptural works;
(6) motion pictures and other audiovisual works;
(7) sound recordings; and
(8) architectural works.

Some of these categories of works, although not all, are defined in § 101. Additionally, § 103 identifies two special types of works that are subject to copyright protection: derivative works and compilations. It is this last type of work that is at issue in the *Feist* decision.

Feist Publications, Inc. v. Rural Telephone Service Co.
499 U.S. 340 (1991)

O'CONNOR, JUSTICE:

. . . .

I

Rural Telephone Service Company is a certified public utility that provides telephone service to several communities in northwest Kansas. It is subject to a state regulation that requires all telephone companies operating in Kansas to issue annually an updated telephone directory. Accordingly, as a condition of its monopoly franchise, Rural publishes a typical telephone directory, consisting of white pages and yellow pages. The white pages list in alphabetical order the names of Rural's subscribers, together with their towns and telephone numbers. The yellow pages list Rural's business subscribers alphabetically by category, and feature classified advertisements of various sizes. Rural distributes its directory free of charge to its subscribers, but earns revenue by selling yellow pages advertisements.

Feist Publications, Inc., is a publishing company that specializes in area-wide telephone directories. Unlike a typical directory, which covers only a particular calling area, Feist's area-wide directories cover a much larger geographical range, reducing the need to call directory assistance or consult multiple directories. The Feist directory that is the subject of this litigation covers 11 different telephone service areas in 15 counties and contains 46,878 white pages listings—compared to Rural's approximately 7,700 listings. Like Rural's directory, Feist's is distributed free of charge and includes both white pages and yellow pages. Feist and Rural compete vigorously for yellow pages advertising.

As the sole provider of telephone service in its service area, Rural obtains subscriber information quite easily. Persons desiring telephone service must apply to Rural and provide their names and addresses; Rural then assigns them a telephone number. Feist is not a telephone company, let alone one with monopoly status, and therefore lacks independent access to any subscriber information. To obtain white pages listings for its area-wide directory, Feist approached each of the 11 telephone companies operating in northwest Kansas and offered to pay for the right to use its white pages listings.

Of the 11 telephone companies, only Rural refused to license its listings to Feist. Rural's refusal created a problem for Feist, as omitting these listings would have left a gaping hole in its area-wide directory, rendering it less attractive to potential yellow pages advertisers. In a decision subsequent to that which we review here, the District Court determined that this was precisely the reason Rural refused to license its listings. The refusal was motivated by an unlawful purpose "to extend its monopoly in telephone service to a monopoly in yellow pages advertising." *Rural Telephone Service Co. v. Feist Publications, Inc.*, 737 F. Supp. 610, 622 (D. Kan. 1990).

Unable to license Rural's white pages listings, Feist used them without Rural's consent. Feist began by removing several thousand listings that fell outside the geographic range of its area-wide directory, then hired personnel to investigate the 4,935 that remained. These employees verified the data reported by Rural and sought to obtain additional information. As a result, a typical Feist listing includes the individual's street address; most of Rural's listings do not. Notwithstanding these additions, however, 1,309 of the 46,878 listings in Feist's 1983 directory were identical to listings in Rural's 1982-1983 white pages. Four of these were fictitious listings that Rural had inserted into its directory to detect copying.

Rural sued for copyright infringement in the District Court for the District of Kansas, taking the position that Feist, in compiling its own directory, could not use the information contained in Rural's white pages. Rural asserted that Feist's employees were obliged to travel door-to-door or conduct a telephone survey to discover the same information for themselves. Feist responded that such efforts were economically impractical and, in any event, unnecessary, because the information copied was beyond the scope of copyright protection. The District Court granted summary judgment to Rural, explaining that "[c]ourts have consistently held that telephone directories are copyrightable" and citing a string of lower court decisions. 663 F. Supp. 214, 218 (1987). In an unpublished opinion, the Court of Appeals for the Tenth Circuit affirmed "for substantially the reasons given by the district court." 916 F.2d 718 (1990). We granted certiorari, 498 U.S. 808 (1990), to determine whether the copyright in Rural's directory protects the names, towns, and telephone numbers copied by Feist.

II

A

This case concerns the interaction of two well-established propositions. The first is that facts are not copyrightable; the other, that compilations of facts generally are. Each of these propositions possesses an impeccable pedigree. That there can be no valid copyright in facts is universally understood. The most fundamental axiom of copyright law is that "[n]o author may copyright his ideas or the facts he narrates." *Harper & Row, Publishers, Inc. v. Nation Enterprises*, 471 U.S. 539, 556 (1985). Rural wisely concedes this point, noting in its brief that "[f]acts and discoveries, of course, are not themselves subject to copyright protection." At the same time, however, it is beyond dispute that compilations of facts are within the subject matter of copyright. Compilations were expressly mentioned in the Copyright Act of 1909, and again in the Copyright Act of 1976.

There is an undeniable tension between these two propositions. Many compilations consist of nothing but raw data—i.e., wholly factual information not accompanied by any original written expression. On what basis may one claim a copyright in such a work? Common sense tells us that 100 uncopyrightable facts do not magically change their status when gathered together in one place. Yet copyright law seems to contemplate that compilations that consist exclusively of facts are potentially within its scope.

The key to resolving the tension lies in understanding why facts are not copyrightable. The *sine qua non* of copyright is originality. To qualify for copyright protection, a work must be original to the author. Original, as the term is used in copyright, means only that the work was independently created by the author (as opposed to copied from other works), and that it possesses at least some minimal degree of creativity. 1 M. Nimmer & D. Nimmer, *Copyright* §§ 2.01[A], [B] (1990) (hereinafter Nimmer). To be sure, the requisite level of creativity is extremely low; even a slight amount will suffice. The vast majority of works make the grade quite easily, as they possess some creative spark, "no matter how crude, humble or obvious" it might be. *Id.* § 1.08[C][1]. Originality does not signify novelty; a work may be original even though it closely resembles other works, so long as the similarity is fortuitous, not the result of copying. To illustrate, assume that two poets, each ignorant of the other, compose identical poems. Neither work is novel, yet both are original and, hence, copyrightable.

Originality is a constitutional requirement. The source of Congress' power to enact copyright laws is Article I, § 8, cl. 8, of the Constitution, which authorizes Congress to "secur[e] for limited Times to Authors . . . the exclusive Right to their respective Writings." In two decisions from the late 19th Century—*The Trade-Mark Cases*, 100 U.S. 82 (1879); and *Burrow-Giles Lithographic Co. v. Sarony*, 111 U.S. 53 (1884)—this Court defined the crucial terms "authors" and "writings." In so doing, the Court made it unmistakably clear that these terms presuppose a degree of originality.

In *The Trade-Mark Cases*, the Court addressed the constitutional scope of "writings." For a particular work to be classified "under the head of writings of authors," the Court determined, "originality is required." 100 U.S. at 94. The Court explained that originality requires independent creation plus a modicum[*] of creativity:

> [W]hile the word *writings* may be liberally construed, as it has been, to include original designs for engraving, prints, &c., it is only such as are *original*, and are founded in the creative powers of the mind. The writings which are to be protected are *the fruits of intellectual labor*, embodied in the form of books, prints, engravings, and the like.

Id. (emphasis in original).

In *Burrow-Giles*, the Court distilled the same requirement from the Constitution's use of the word "authors." The Court defined "author," in a constitutional sense, to mean "he to whom anything owes its origin; originator; maker." 111 U.S. at 58 (internal quotations omitted). As in *The Trade-Mark Cases*, the Court emphasized the creative component of originality. It described copyright as being limited to "original intellectual conceptions of the author," *id.*, and stressed the

---

[*] [*Eds. Note*: A "modicum" is simply a small portion—a limited quantity or amount.]

importance of requiring an author who accuses another of infringement to prove "the existence of those facts of originality, of intellectual production, of thought, and conception." *Id.* 111 U.S. 590.

The originality requirement articulated in *The Trade-Mark Cases* and *Burrow-Giles* remains the touchstone of copyright protection today. It is the very "premise of copyright law." *Miller v. Universal City Studios, Inc.*, 650 F.2d 1365, 1368 (5th Cir. 1981). Leading scholars agree on this point. As one pair of commentators succinctly puts it: "The originality requirement is *constitutionally mandated* for all works." Patterson & Joyce, *Monopolizing the Law: The Scope of Copyright Protection for Law Reports and Statutory Compilations*, 36 UCLA L. Rev. 719, 763, n. 155 (1989) (emphasis in original) (hereinafter Patterson & Joyce).

It is this bedrock principle of copyright that mandates the law's seemingly disparate treatment of facts and factual compilations. "No one may claim originality as to facts." [Nimmer] § 2.11[A], p. 2-157. This is because facts do not owe their origin to an act of authorship. The distinction is one between creation and discovery: the first person to find and report a particular fact has not created the fact; he or she has merely discovered its existence. To borrow from *Burrow-Giles*, one who discovers a fact is not its "maker" or "originator." 111 U.S. at 58. "The discoverer merely finds and records." Nimmer § 2.03[E]. Census-takers, for example, do not "create" the population figures that emerge from their efforts; in a sense, they copy these figures from the world around them. Denicola, *Copyright in Collections of Facts: A Theory for the Protection of Nonfiction Literary Works*, 81 Colum. L. Rev. 516, 525 (1981) (hereinafter Denicola). Census data therefore do not trigger copyright, because these data are not "original" in the constitutional sense. Nimmer § 2.03[E]. The same is true of all facts — scientific, historical, biographical, and news of the day. "[T]hey may not be copyrighted, and are part of the public domain available to every person." *Miller, supra*, at 1369.

Factual compilations, on the other hand, may possess the requisite originality. The compilation author typically chooses which facts to include, in what order to place them, and how to arrange the collected data so that they may be used effectively by readers. These choices as to selection and arrangement, so long as they are made independently by the compiler and entail a minimal degree of creativity, are sufficiently original that Congress may protect such compilations through the copyright laws. Thus, even a directory that contains absolutely no protectible written expression, only facts, meets the constitutional minimum for copyright protection if it features an original selection or arrangement.

This protection is subject to an important limitation. The mere fact that a work is copyrighted does not mean that every element of the work may be protected. Originality remains the *sine qua non* of copyright; accordingly, copyright protection may extend only to those components of a work that are original to the author. Patterson & Joyce 800-802; Ginsburg, *Creation and Commercial Value: Copyright Protection of Works of Information*, 90 Colum. L. Rev. 1865, 1868, and n. 12 (1990) (hereinafter Ginsburg). Thus, if the compilation author clothes facts with an original collocation of words, he or she may be able to claim a copyright in this written expression. Others may copy the underlying facts from the publication, but not the precise words used to present them. In *Harper & Row*, for example, we explained that President Ford could not prevent others from copying bare historical facts from his autobiography, see 471 U.S. at 556-557, but that he could prevent others from copying his "subjective descriptions and portraits of public figures." *Id.* at 563. Where the compilation author adds no written expression, but rather lets the facts speak for themselves, the expressive element is more elusive. The only conceivable expression is the manner in which the compiler has selected and arranged the facts. Thus, if the selection and arrangement are original, these elements of the work are eligible for copyright protection. See Patry, *Copyright in Compilations of Facts (or Why the "White Pages" Are Not Copyrightable)*, 12 Com. & Law 37, 64 (Dec. 1990) (hereinafter Patry). No matter how original the format, however, the facts themselves do not become original through association. See Patterson & Joyce 776.

This inevitably means that the copyright in a factual compilation is thin. Notwithstanding a valid copyright, a subsequent compiler remains free to use the facts contained in another's publication to aid in preparing a competing work, so long as the competing work does not feature the same selection and arrangement. As one commentator explains it:

> [N]o matter how much original authorship the work displays, the facts and ideas it exposes are free for the taking. . . . [T]he very same facts and ideas may be divorced from the context imposed by the author, and restated or reshuffled by second comers, even if the author was the first to discover the facts or to propose the ideas.

Ginsburg 1868.

It may seem unfair that much of the fruit of the compiler's labor may be used by others without compensation. As Justice Brennan has correctly observed, however, this is not "some unforeseen byproduct of a statutory scheme." *Harper & Row*, 471 U.S. at 589 (dissenting opinion). It is, rather, "the essence of copyright," *id.*, and a constitutional requirement. The primary objective of copyright is not to reward the labor of authors, but "[t]o promote the Progress of Science and useful Arts." Art. I, § 8, cl. 8. To this end, copyright assures authors the right to their original expression, but encourages others to build freely upon the ideas and information conveyed by a work. *Harper & Row, supra*, 471 U.S. at 556-557. This principle, known as the idea/expression or fact/expression dichotomy, applies to all works of authorship. As applied to a factual compilation, assuming the absence of original written expression, only the compiler's selection and arrangement may be protected; the raw facts may be copied at will. This result is neither unfair nor unfortunate. It is the means by which copyright advances the progress of science and art.

. . . .

This, then, resolves the doctrinal tension: Copyright treats facts and factual compilations in a wholly consistent manner. Facts, whether alone or as part of a compilation, are not original, and therefore may not be copyrighted. A factual compilation is eligible for copyright if it features an original selection or arrangement of facts, but the copyright is limited to the particular selection or arrangement. In no event may copyright extend to the facts themselves.

B

. . . .

[The Court discussed how lower courts had misunderstood the previous copyright statute, the Copyright Act of 1909, Section 5 of which had contained the word "compilation."] [T]he fact that factual compilations were mentioned specifically in § 5 led some courts to infer erroneously that directories and the like were copyrightable per se, "without any further or precise showing of original—personal—authorship." Ginsburg 1895.

Making matters worse, these courts developed a new theory to justify the protection of factual compilations. Known alternatively as "sweat of the brow" or "industrious collection," the underlying notion was that copyright was a reward for the hard work that went into compiling facts. . . .

The "sweat of the brow" doctrine had numerous flaws, the most glaring being that it extended copyright protection in a compilation beyond selection and arrangement—the compiler's original contributions—to the facts themselves. Under the doctrine, the only defense to infringement was independent creation. . . . "Sweat of the brow" courts thereby eschewed the most fundamental axiom of copyright law—that no one may copyright facts or ideas.

Decisions of this Court applying the 1909 Act make clear that the statute did not permit the "sweat of the brow" approach. The best example is *International News Service v. Associated Press*, 248 U.S. 215 (1918). In that decision, the Court flatly rejected, the notion that the copyright in an article extended to the factual information it contained:

> [T]he news element—the information respecting current events contained in the literary production—is not the creation of the writer, but is a report of matters that ordinarily are publici juris; it is the history of the day.

Id.

. . . "Sweat of the brow" courts took a contrary view; they handed out proprietary interests in facts and declared that authors are absolutely precluded from saving time and effort by relying upon the facts contained in prior works. In truth, "[i]t is just such wasted effort that the proscription against the copyright of ideas and facts . . . [is] designed to prevent." *Rosemont Enterprises, Inc. v. Random House, Inc.*, 366 F.2d 303, 310 (2d Cir. 1966).

> [P]rotection for the fruits of such research . . . may, in certain circumstances, be available under a theory of unfair competition. But to accord copyright protection on this basis alone distorts basic copyright principles in that it creates a monopoly in public domain materials without the necessary justification of protecting and encouraging the creation of 'writings' by 'authors.'

Nimmer § 3.04, p. 3-23 (footnote omitted).

C

. . . .

§ 102(b) . . . identifies specifically those elements of a work for which copyright is not available:

> [I]n no case does copyright protection for an original work of authorship extend to any idea, procedure, process, system, method of operation, concept, principle, or discovery, regardless of the form in which it is described, explained, illustrated, or embodied in such work.

§ 102(b) is universally understood to prohibit any copyright in facts. *Harper Row, supra*, at 556. Accord, Nimmer § 2.03[E] (equating facts with "discoveries"). As with § 102(a), Congress emphasized that § 102(b) did not change the law, but merely clarified it: "Section 102(b) in no way enlarges or contracts the scope of copyright protection under the present law. Its purpose is to restate . . . that the basic dichotomy between expression and idea remains unchanged." H.R. Rep. at 57; S. Rep. at 54, U.S. Code Cong. & Admin. News 1976, p. 5670.

Congress took another step to minimize confusion by deleting the specific mention of "directories . . . and other compilations" in § 5 of the 1909 Act. As mentioned, this section had led some courts to conclude that directories were copyrightable per se, and that every element of a directory was protected. In its place, Congress enacted two new provisions. First, to make clear that compilations were not copyrightable per se, Congress provided a definition of the term "compilation." Second, to make clear that the copyright in a compilation did not extend to the facts themselves, Congress enacted 17 U.S.C. § 103.

The definition of "compilation" is found in § 101 of the 1976 Act. It defines a "compilation" in the copyright sense as "a work formed by the collection and assembly of preexisting materials or of data *that* are selected, coordinated, or arranged *in such a way that* the resulting work, as a whole, constitutes an original work of authorship." (Emphasis added.)

The purpose of the statutory definition is to emphasize that collections of facts are not copyrightable per se. . . .

. . . .

The key to the statutory definition is the second requirement. It instructs courts that, in determining whether a fact-based work is an original work of authorship, they should focus on the manner in which the collected facts have been selected, coordinated, and arranged. This is a straightforward application of the originality requirement. Facts are never original, so the compilation author can claim originality, if at all, only in the way the facts are presented. To that end, the statute dictates that the principal focus should be on whether the selection, coordination, and arrangement are sufficiently original to merit protection.

Not every selection, coordination, or arrangement will pass muster. This is plain from the statute. It states that, to merit protection, the facts must be selected, coordinated, or arranged "in such a way" as to render the work as a whole original. This implies that some "ways" will trigger copyright, but that others will not. See Patry 57, and n. 76. . . .

As discussed earlier, however, the originality requirement is not particularly stringent. A compiler may settle upon a selection or arrangement that others have used; novelty is not required. Originality requires only that the author make the selection or arrangement independently (i.e., without copying that selection or arrangement from another work), and that it display some minimal level of creativity. Presumably, the vast majority of compilations will pass this test, but not all will. There remains a narrow category of works in which the creative spark is utterly lacking or so trivial as to be virtually nonexistent. Such works are incapable of sustaining a valid copyright.

Even if a work qualifies as a copyrightable compilation, it receives only limited protection. This is the point of § 103 of the Act. Section 103 explains that "[t]he subject matter of copyright . . . includes compilations," § 103(a), but that copyright protects only the author's original contributions—not the facts or information conveyed:

> [T]he copyright in a compilation . . . extends only to the material contributed by the author of such work, as distinguished from the preexisting material employed in the work, and does not imply any exclusive right in the preexisting material.

§ 103(b).

As § 103 makes clear, copyright is not a tool by which a compilation author may keep others from using the facts or data he or she has collected.

> [T]he most important point here is one that is commonly misunderstood today: copyright . . . has no effect one way or the other on the copyright or public domain status of the preexisting material.

H.R. Rep. at 57; S. Rep. at 55, U.S. Code Cong. & Admin. News 1976, p. 5670. The 1909 Act did not require, as "sweat of the brow" courts mistakenly assumed, that each subsequent compiler must start from scratch, and is precluded from relying on research undertaken by another. Rather, the facts contained in existing works may be freely copied, because copyright protects only the elements that owe their origin to the compiler—the selection, coordination, and arrangement of facts.

III

There is no doubt that Feist took from the white pages of Rural's directory a substantial amount of factual information. At a minimum, Feist copied the names, towns, and telephone numbers of 1,309 of Rural's subscribers. Not all copying, however, is copyright infringement. To establish infringement, two elements must be proven: (1) ownership of a valid copyright, and (2) copying of constituent elements of the work that are original. The first element is not at issue here; Feist appears to concede that Rural's directory, considered as a whole, is subject to a valid copyright because it contains some foreword text, as well as original material in its yellow pages advertisements.

The question is whether Rural has proved the second element. In other words, did Feist, by taking 1,309 names, towns, and telephone numbers from Rural's white pages, copy anything that was "original" to Rural? Certainly, the raw data does not satisfy the originality requirement. Rural may have been the first to discover and report the names, towns, and telephone numbers of its subscribers, but this data does not "ow[e] its origin'" to Rural. *Burrow-Giles*, 111 U.S. at 58. Rather, these bits of information are uncopyrightable facts; they existed before Rural reported them, and would have continued to exist if Rural had never published a telephone directory. The originality requirement "rule[s] out protecting . . . names, addresses, and telephone numbers of which the plaintiff, by no stretch of the imagination, could be called the author." Patterson & Joyce 776.

Rural essentially concedes the point by referring to the names, towns, and telephone numbers as "preexisting material." Section 103(b) states explicitly that the copyright in a compilation does not extend to "the preexisting material employed in the work."

The question that remains is whether Rural selected, coordinated, or arranged these uncopyrightable facts in an original way. As mentioned, originality is not a stringent standard; it does not require that facts be presented in an innovative or surprising way. It is equally true, however, that the selection and arrangement of facts cannot be so mechanical or routine as to require no creativity whatsoever. The standard of originality is low, but it does exist. As this Court has explained, the Constitution mandates some minimal degree of creativity, and an author who claims infringement must prove "the existence of . . . intellectual production, of thought, and conception." *Burrow-Giles, supra*, 111 U.S. at 59-60.

The selection, coordination, and arrangement of Rural's white pages do not satisfy the minimum constitutional standards for copyright protection. As mentioned at the outset, Rural's white pages are entirely typical. Persons desiring telephone service in Rural's service area fill out an application, and Rural issues them a telephone number. In preparing its white pages, Rural simply takes the data provided by its subscribers and lists it alphabetically by surname. The end product is a garden-variety white pages directory, devoid of even the slightest trace of creativity.

Rural's selection of listings could not be more obvious: it publishes the most basic information—name, town, and telephone number—about each person who applies to it for telephone service. This is "selection" of a sort, but it lacks the modicum of creativity necessary to transform mere selection into copyrightable expression. Rural expended sufficient effort to make the white pages directory useful, but insufficient creativity to make it original.

We note in passing that the selection featured in Rural's white pages may also fail the originality requirement for another reason. Feist points out that Rural did not truly "select" to publish the names and telephone numbers of its subscribers; rather, it was required to do so by the Kansas Corporation Commission as part of its monopoly franchise. See 737 F.Supp. at 612. Accordingly, one could plausibly conclude that this selection was dictated by state law, not by Rural.

Nor can Rural claim originality in its coordination and arrangement of facts. The white pages do nothing more than list Rural's subscribers in alphabetical order. This arrangement may, technically speaking, owe its origin to Rural; no one disputes that Rural undertook the task of alphabetizing the names itself. But there is nothing remotely creative about arranging names alphabetically in a white pages directory. It is an age-old practice, firmly rooted in tradition and so commonplace that it has come to be expected as a matter of course. See *Brief for Information Industry Association et al. as Amici Curiae* 10 (alphabetical arrangement "is universally observed in directories published by local exchange telephone companies"). It is not only unoriginal, it is practically inevitable. This time-honored tradition does not possess the minimal creative spark required by the Copyright Act and the Constitution.

We conclude that the names, towns, and telephone numbers copied by Feist were not original to Rural, and therefore were not protected by the copyright in Rural's combined white and yellow pages directory. As a constitutional matter, copyright protects only those constituent elements of a work that possess more than a de minimis quantum of creativity. Rural's white pages, limited to basic subscriber information and arranged alphabetically, fall short of the mark. As a statutory matter, 17 U.S.C. § 101 does not afford protection from copying to a collection of facts that are selected, coordinated, and arranged in a way that utterly lacks originality. Given that some works must fail, we cannot imagine a more likely candidate. Indeed, were we to hold that Rural's white pages pass muster, it is hard to believe that any collection of facts could fail.

Because Rural's white pages lack the requisite originality, Feist's use of the listings cannot constitute infringement. This decision should not be construed as demeaning Rural's efforts in compiling its directory, but rather as making clear that copyright rewards originality, not effort. As this Court noted more than a century ago, "'great praise may be due to the plaintiffs for their industry and enterprise in publishing this paper, yet the law does not contemplate their being rewarded in this way.'" *Baker v. Selden*, 101 U.S. at 105.

Notes & Questions

1. What can you tell about the requirement of "originality" from the *Feist* decision? Focus on the standards articulated by the Court. Would the following works satisfy the originality requirement:

    a. an email to a family member describing what you did yesterday;
    b. a doodle jotted in the margin of your book while listening to a lecture on copyright law;
    c. a snapshot of a child playing with a puppy;
    d. a tape of birds, crickets, and other forest sounds you record while walking through the woods.

2. In *Feist*, the Court refers to two earlier Supreme Court cases: *The Trade-Mark Cases* and *Burrow-Giles*. *The Trade-Mark Cases* involved a constitutional challenge to Congress' authority to protect trademarks, an issue we address in Chapter 5. *Burrow-Giles* involved a claim of copyright in the photograph of the famous author Oscar Wilde (shown on the left). The defendant argued that protecting photographs was beyond the scope of Congressional authority granted by the Constitution: "a photograph being a reproduction, on paper, of the exact features of some natural object, or of some person, is not a writing of which the producer is the author." The Court concluded that "the constitution is broad enough to cover an act authorizing copyright of photographs, so far as they are representatives of original intellectual conceptions of the author." In addressing whether the photograph at issue qualified for protection, the Court quoted from the trial court's opinion:

The third finding of facts says, in regard to the photograph in question, that it is a 'useful, new, harmonious, characteristic, and graceful picture, and that plaintiff made the same * * * entirely from his own original mental conception, to which he gave visible form by posing the said Oscar Wilde in front of the camera, selecting and arranging the costume, draperies, and other various accessories in said photograph, arranging the subject so as to present graceful outlines, arranging and disposing the light and shade, suggesting and evoking the desired expression, and from such disposition, arrangement, or representation, made entirely by plaintiff, he produced the picture in suit.' These findings, we think, show this photograph to be an original work of art, the product of plaintiff's intellectual invention, of which plaintiff is the author, and of a class of inventions for which the constitution intended that congress should secure to him the exclusive right to use, publish, and sell . . . .

Burrow-Giles Lithographic Company v. Sarony, 111 U.S. 53, 60 (1884). The Court has used various phrases to describe originality: "original intellectual conception of the author," the "fruits of intellectual labor," a "modicum of creativity." Do any of these linguistic formulations help you in judging the originality of the works identified in question 1?

3. The Court uses the example of a census as a work that would not be protected because the population figures would not be "original" to the creators of the census. The federal census is not protected by copyright for another reason: 17 U.S.C. §105 specifies that works of the federal government are not eligible for copyright protection. Thus, even if a work created by the federal government (this includes all works created by employees of the federal government that are created within the scope of their employment) contains sufficient originality, it is statutorily ineligible for protection. Why might Congress have adopted this rule? This

statutory prohibition on protection for government works does *not* apply to state or local governments, only the federal government. Public policy, not statute, also dictates that laws themselves, whether statutory or decisional, are not copyrightable. *See Wheaton v. Peters*, 33 U.S. (8 Pet.) 591 (1834); *Veeck v. Southern Building Code Congress Int'l*, 293 F.3d 791 (5th Cir. 2002) (en banc).

4. In *Feist* the Court emphasizes how the underlying goal of copyright protection guides our understanding of the scope of the rights granted to copyright owners. Progress is achieved not solely by having more works created and distributed, but also by allowing others to build upon the ideas and facts contained in a work, so long as the copyrighted expression of those ideas or facts is not reproduced. The unprotected elements of a copyrighted work are said to be in the "public domain"—free for anyone to use and to copy. We explore the idea-expression distinction in the next Section of this Chapter. The public domain also contains the full expression of copyrighted works, once the copyright has expired. We address the duration of copyright protection in Section III of this Chapter.

5. The Court refers to the ability of others to copy what is in the public domain as the "essence of copyright," not some unfair and unforeseen by-product of a statutory scheme. In other words, it is a feature, not a bug. However, the collection of public domain facts may have cost the compiler significant sums. The effort or expense expended does not equal originality. Not only is the court rejecting the sweat of the brow doctrine in *Feist*, it is also rejecting the "if value, then right" premise for copyright protection. Does this mean that compilers of data lack the incentive necessary to create important and valuable databases?

Many databases of public domain information rely upon limiting physical access and contractual restrictions to obtain a level of protection that will make their enterprise profitable. Westlaw and Lexis are two familiar examples. Law schools must pay a subscription charge for Westlaw and Lexis access for their students (contributing to the cost of legal education), and, when obtaining a password, students enter into a contract with Westlaw and Lexis. Did you read that contract when you accepted the password, or clicked "I agree"?

6. The Court references the idea that for a work that has a low degree of originality the copyright is "thin," meaning that to infringe that work a second author would need to reproduce more of the copyrighted expression. Some courts have even begun using a standard of "virtual identity" in order to find infringement certain types of works. We will turn to the tests for infringement in Section IV. Varying levels of protection based on the level of originality the work contains is also an important dynamic for achieving the goal of copyright. However, it makes it more difficult to determine whether infringement is present in any particular case.

7. Did Rural need the incentive that copyright offers in order to create and distribute the white pages of the phone book? Was it monetary remuneration that Rural sought in the publication of its phone book?

8. The Supreme Court is clear that "originality" is a constitutional requirement. As a constitutional requirement, Congress would be acting outside of its

authority if it adopted a copyright law that protected works that lack originality. The Constitution, however, grants Congress other legislative powers, for example in the Commerce Clause. When do the limitations of Art. I, sec. 8, cl. 8 apply to acts of Congress? The Second Circuit has addressed this issue in the context of a criminal statute that prohibits the distribution of bootleg recordings of live musical performances. *United States v. Martignon*, 492 F.3d 140 (2d Cir. 2007). The court held that "Congress exceeds its power under the Commerce Clause by transgressing limitations of the Copyright Clause only when (1) the law it enacts is an exercise of the power granted Congress by the Copyright Clause and (2) the resulting law violates one or more specific limits of the Copyright Clause." *Id.* 149. The court held that to be an exercise of Copyright Clause authority the law must create, bestow, or allocate property rights in expression. Because the provision at issue in the case was a criminal provision creating "a power in the government to protect the interest of performers from commercial predations," *id.* at 151, the Second Circuit concluded that it was not an exercise of the Copyright Clause power and therefore was not subject to the limitations in that clause. Under the Second Circuit's test, how easy will it be for Congress to avoid the limits of the Copyright Clause?

9. Section 102(b) is the statutory embodiment of the idea/expression dichotomy:

> [I]n no case does copyright protection for an original work of authorship extend to any idea, procedure, process, system, method of operation, concept, principle, or discovery, regardless of the form in which it is described, explained, illustrated, or embodied in such work.

"Facts" are not included in the list of items not protected. The Court, instead, focused on "facts" not owing their origin to the creator but rather being "found" or "discovered." Is it possible that some "facts" are created by authors? *See Castle Rock Entertainment, Inc. v. Carol Publishing Group*, 150 F.3d 132, 139 (2nd Cir. 1998) (citing *Feist* and distinguishing between "true facts" and "created facts," which constitute original and thus protectable expression").

II. THE BOUNDARIES OF COPYRIGHTABLE SUBJECT MATTER

A. THE IDEA-EXPRESSION DISTINCTION

In the *Feist* opinion included in Section I of this Chapter, the Supreme Court discusses fundamental limitations on the scope of copyrightable subject matter. One is that copyright protects the particular expression, not the facts or general ideas expressed. This is often identified as the idea-expression distinction. An important role for this distinction is clearly to distinguish between the subject matter of copyright and the subject matter of patent.

Baker v. Selden
101 U.S. 99 (1879)

BRADLEY, JUSTICE:

. . . .

The book or series of books of which the complainant claims the copyright consists of an introductory essay explaining [a peculiar] system of book-keeping . . . , to which are annexed certain forms or blanks, consisting of ruled lines, and headings, illustrating the system and showing how it is to be used and carried out in practice. This system effects the same results as book-keeping by double entry; but, by a peculiar arrangement of columns and headings, presents the entire operation, of a day, a week, or a month, on a single page, or on two pages facing each other, in an account-book. The defendant uses a similar plan so far as results are concerned; but makes a different arrangement of the columns, and uses different headings. If the complainant[] . . . had the exclusive right to the use of the system explained in his book, it would be difficult to contend that the defendant does not infringe it, notwithstanding the difference in his form of arrangement; but if it be assumed that the system is open to public use, it seems to be equally difficult to contend that the books made and sold by the defendant are a violation of the copyright of the complainant's book considered merely as a book explanatory of the system. Where the truths of a science or the methods of an art are the common property of the whole world, any author has the right to express the one, or explain and use the other, in his own way. As an author, Selden explained the system in a particular way. It may be conceded that Baker makes and uses account-books arranged on substantially the same system; but the proof fails to show that he has violated the copyright of Selden's book, regarding the latter merely as an explanatory work; or that he has infringed Selden's right in any way, unless the latter became entitled to an exclusive right in the system.

The evidence of the complainant is principally directed to the object of showing that Baker uses the same system as that which is explained and illustrated in Selden's books. It becomes important, therefore, to determine whether, in obtaining the copyright of his books, he secured the exclusive right to the use of the system or method of book-keeping which the said books are intended to illustrate and explain. It is contended that he has secured such exclusive right, because no one can use the system without using substantially the same ruled lines and headings which he was appended to his books in illustration of it. In other words, it is contended that the ruled lines and headings, given to illustrate the system, are a part of the book, and, as such, are secured by the copyright; and that no one can make or use similar ruled lines and headings, or ruled lines and headings made and arranged on substantially the same system, without violating the copyright. And this is really the question to be decided in this case. Stated in another form, the question is, whether the exclusive property in a system of book-keeping can be claimed, under the law or copyright, by means of a book in which that system is explained? The complainant's bill, and the case made under it, are based on the hypothesis that it can be.

. . . .

There is no doubt that a work on the subject of book-keeping, though only explanatory of well-known systems, may be the subject of a copyright; but, then, it
is claimed only as a book. Such a book may be explanatory either of old systems,
or of an entirely new system; and, considered as a book, as the work of an author,
conveying information on the subject of book-keeping, and containing detailed explanations of the art, it may be a very valuable acquisition to the practical knowledge of the community. But there is a clear distinction between the book, as such,
and the art which it is intended to illustrate. The mere statement of the proposition is so evident, that it requires hardly any argument to support it. The same
distinction may be predicated of every other art as well as that of book-keeping. A
treatise on the composition and use of medicines, be they old or new; on the construction and use of ploughs, or watches, or churns; or on the mixture and application of colors for painting or dyeing; or on the mode of drawing lines to produce
the effect of perspective, would be the subject of copyright; but no one would contend that the copyright of the treatise would give the exclusive right to the art or
manufacture described therein. The copyright of the book, if not pirated from
other works, would be valid without regard to the novelty, or want of novelty, of
its subject-matter. The novelty of the art or thing described or explained has nothing to do with the validity of the copyright. To give to the author of the book an
exclusive property in the art described therein, when no examination of its novelty
has ever been officially made, would be a surprise and a fraud upon the public.
That is the province of letters-patent, not of copyright. The claim to an invention
or discovery of an art or manufacture must be subjected to the examination of the
Patent Office before an exclusive right therein can be obtained; and it can only be
secured by a patent from the government.

The difference between the two things, letters-patent and copyright, may be
illustrated by reference to the subjects just enumerated. Take the case of medicines. Certain mixtures are found to be of great value in the healing art. If the
discoverer writes and publishes a book on the subject (as regular physicians generally do), he gains no exclusive right to the manufacture and sale of the medicine;
he gives that to the public. If he desires to acquire such exclusive right, he must
obtain a patent for the mixture as a new art, manufacture, or composition of matter. He may copyright his book, if he pleases; but that only secures to him the
exclusive right of printing and publishing his book. So of all other inventions or
discoveries.

The copyright of a book on perspective, no matter how many drawings and
illustrations it may contain, gives no exclusive right to the modes of drawing described, though they may never have been known or used before. By publishing
the book, without getting a patent for the art, the latter is given to the public. The
fact that the art [is] described in the book by illustrations of lines and figures which
are reproduced in practice in the application of the art, makes no difference. Those
illustrations are the mere language employed by the author to convey his ideas
more clearly. Had he used words of description instead of diagrams (which merely
stand in the place of words), there could not be the slightest doubt that others,

applying the art to practical use, might lawfully draw the lines and diagrams which were in the author's mind, and which he thus described by words in his book.

The copyright of a work on mathematical science cannot give to the author an exclusive right to the methods of operation which he propounds, or to the diagrams which he employs to explain them, so as to prevent an engineer from using them whenever occasion requires. The very object of publishing a book on science or the useful arts is to communicate to the world the useful knowledge which it contains. But this object would be frustrated if the knowledge could not be used without incurring the guilt of piracy of the book. And where the art it teaches cannot be used without employing the methods and diagrams used to illustrate the book, or such as are similar to them, such methods and diagrams are to be considered as necessary incidents to the art, and given therewith to the public; not given for the purpose of publication in other works explanatory of the art, but for the purpose of practical application.

Of course, these observations are not intended to apply to ornamental designs, or pictorial illustrations addressed to the taste. Of these it may be said, that their form is their essence, and their object, the production of pleasure in their contemplation. This is their final end. They are as much the product of genius and the result of composition, as are the lines of the poet or the historian's period. On the other hand, the teachings of science and the rules and methods of useful art have their final end in application and use; and this application and use are what the public derive from the publication of a book which teaches them. But as embodied and taught in a literary composition or book, their essence consists only in their statement. This alone is what is secured by the copyright. The use by another of the same methods of statement, whether in words or illustrations, in a book published for teaching the art, would undoubtedly be an infringement of the copyright.

Recurring to the case before us, we observe that Charles Selden, by his books, explained and described a peculiar system of book-keeping, and illustrated his method by means of ruled lines and blank columns, with proper headings on a page, or on successive pages. Now, whilst no one has a right to print or publish his book, or any material part thereof, as a book intended to convey instruction in the art, any person may practise and use the art itself which he has described and illustrated therein. The use of the art is a totally different thing from a publication of the book explaining it. The copyright of a book on book-keeping cannot secure the exclusive right to make, sell, and use account-books prepared upon the plan set forth in such book. Whether the art might or might not have been patented, is a question which is not before us. It was not patented, and is open and free to the use of the public. And, of course, in using the art, the ruled lines and headings of accounts must necessarily be used as incident to it.

The plausibility of the claim put forward by the complainant in this case arises from a confusion of ideas produced by the peculiar nature of the art described in the books which have been made the subject of copyright. In describing the art, the illustrations and diagrams employed happen to correspond more closely than usual with the actual work performed by the operator who uses the art. Those

illustrations and diagrams consist of ruled lines and headings of accounts; and it
is similar ruled lines and headings of accounts which, in the application of the art,
the book-keeper makes with his pen, or the stationer with his press; whilst in most
other cases the diagrams and illustrations can only be represented in concrete
forms of wood, metal, stone, or some other physical embodiment. But the principle is the same in all. The description of the art in a book, though entitled to the
benefit of copyright, lays no foundation for an exclusive claim to the art itself. The
object of the one is explanation; the object of the other is use. The former may be
secured by copyright. The latter can only be secured, if it can be secured at all, by
letters-patent.

. . . .

The conclusion to which we have come is, that blank account-books are not
the subject of copyright; and that the mere copyright of Selden's book did not
confer upon him the exclusive right to make and use account-books, ruled and
arranged as designated by him and described and illustrated in said book.

Notes & Questions

1. Some say that *Baker* created a blank form doctrine. Indeed, Copyright Office
regulations prohibit copyright in blank forms that lack the requisite originality:

§ 202.1 Material not subject to copyright
The following are examples of works not subject to copyright and applications for registration of such works cannot be entertained:
    (a) Words and short phrases such as names, titles, and slogans; familiar
symbols or designs; mere variations of typographic ornamentation, lettering or coloring; mere listing of ingredients or contents;
    (b) Ideas, plans, methods, systems, or devices, as distinguished from
the particular manner in which they are expressed or described in a writing;
    (c) Blank forms, such as time cards, graph paper, account books, dia-
ries, bank checks, scorecards, address books, report forms, order forms and
the like, which are designed for recording information and do not in themselves convey information;
    (d) Works consisting entirely of information that is common property
containing no original authorship, such as, for example: Standard calendars, height and weight charts, tape measures and rulers, schedules of
sporting events, and lists or tables taken from public documents or other
common sources.
    (e) Typeface as typeface.

37 C.F.R. § 202.1. Notice that this regulation prohibits copyright protection for far
more than just blank forms. With respect to the last exclusion—"[t]ypeface as
typeface"—how else might one protect typeface with copyright? Blake Fry, *Why
Typefaces Proliferate Without Copyright Protection*, 8 J. Telecomm. & High Tech. L.
425 (2010).

2. *Baker* is about far more than just the copyrightability of blank forms. As it
was in *Feist*, the Court in *Baker* was concerned with achieving the ultimate goal of

copyright: progress in knowledge and learning. Leaving certain aspects of a work free for others to use is important to that underlying goal because it lowers the next creator's expression costs and allows the public to learn and implement what they learn from works of authorship. Congress included this important limitation on what copyright protects in the statute:

> (b) In no case does copyright protection for an original work of authorship extend to any idea, procedure, process, system, method of operation, concept, principle, or discovery, regardless of the form in which it is described, explained, illustrated, or embodied in such work.

17 U.S.C. § 102(b). Consider Selden's system of bookkeeping. While a relatively simple system, it is now widely used as a standard for accountants. Once Selden explained the system in his book, the principal way for him to profit from this new accounting method was to sell the blank forms. Copyright law, however, does not provide a person in Selden's shoes the exclusive right to profit in that way. Does this seem unfair? Recall the Supreme Court's discussion in *Feist* concerning the fairness in profiting off of someone else's labor in collecting facts by copying those facts (without going through the expense of collecting the facts independently). Is it any more, or less, unfair than allowing a second-comer to set up another gas station across the street from the person who first realized the location's desirability?

3. Leaving ideas, facts, processes and methods of operation free for the public to use is consistent not just with the utilitarian justification for copyright protection, but also with Lockean labor theory. John Locke acknowledged that in giving a property right to one who has mixed his labor with the things found in the "common," there still needed to be "enough, and as good left in common for others." Locke's statement concerning leaving enough for others is sometimes referred to as Locke's proviso. The requirement that the ideas expressed in a copyrighted work are free for others to use is one of the items that is left "in common" for others to use and build upon.

4. As articulated in *Baker,* the distinction between protected expression and unprotected ideas, processes, and methods of operation, also serves an important boundary policing mechanism. Some things are to be protected, if at all, by utility patents. To interpret copyright to protect those things would be, as the Court puts, it "a fraud upon the public." The examination process for registering a copyright is not nearly as extensive as for patents. And, as we discuss in more detail in Section III of this Chapter, registration is not even required in order to obtain copyright protection. Policing the line between patents and copyrights can be difficult in other areas of copyright and is explored in the remainder of this Section.

5. The idea-expression distinction creates a related concept in copyright law known as "merger." The merger doctrine prohibits protecting even the expression if it is merged with the idea expressed. In such a situation, to protect the expression would be to grant protection for the idea, and copyright protection should never extend to the idea. Courts often determine whether there is merger by examining if there are alternative ways of expressing the same idea. If there are many

different means of expression, then there is no "merger"—one can copy the idea without copying the expression used by the other. For example, the rules of a sweepstake or a game may not be copyrightable because of merger. *See Morrissey v. Procter & Gamble Co.*, 379 F.2d 675, 678-79 (1st Cir. 1967) (sweepstake rules); *Allen v. Academic Games League of America, Inc.*, 89 F.3d 614 (9th Cir. 1996) (game rules). Consider the instructions on operating a toy helicopter. Should the instructions be copyrighted? *See Lanard Toys, Ltd. v. Novelty Inc.*, 511 F. Supp. 2d 1020 (C.D. Cal. 2007) (expressing doubt). Defining what the "idea" is can present real challenges.

B. Useful Articles

When most people think about what copyright law protects, their first thoughts are usually about books, movies, and music. In addition, the current Copyright Act also protects "pictorial, graphic, and sculptural works," 17 U.S.C. § 102(a)(5). The Act defines this trio of work types to include "applied art," and emphasizes that these "works shall include works of artistic craftsmanship insofar as their form but not their mechanical or utilitarian aspects are concerned." 17 U.S.C. § 101. *Useful articles*, a specially defined term in the Act, can have designs that merit copyright protection—namely, where they "incorporate[] pictorial, graphic, or sculptural features that can be identified separately from, and are capable of existing independently of, the utilitarian aspects of the article."

The Supreme Court recently returned to the question of the scope of copyright for the design of a useful article. To set the stage for that return, we begin our exploration of the topic with the Supreme Court's last foray on the topic, which continues to have relevance today. At the time of this earlier case, the controlling part of the Copyright Act provided protection for "works of art" and "reproductions of a work of art."

Mazer v. Stein

347 U.S. 201 (1954)

REED, J.:

. . . .

. . . Respondents are partners in the manufacture and sale of electric lamps. One of the respondents created original works of sculpture in the form of human figures by traditional clay-model technique. From this model, a production mold for casting copies was made. . . . [T]he statuettes were sold in quantity throughout the country both as lamp bases and as statuettes. The sales in lamp form accounted for all but an insignificant portion of respondents' sales.

. . . .

. . . Petitioners in their petition for certiorari present a single question:

"Can statuettes be protected in the United States by copyright when the copyright applicant intended primarily to use the statuettes in the form of lamp bases to be made and sold in quantity and carried the intentions into effect?

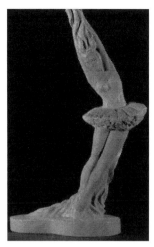

[*Eds. Note*: Image from Plaintiff's copyright registration]

"Stripped down to its essentials, the question presented is: Can a lamp manufacturer copyright his lamp bases?"

The first paragraph accurately summarizes the issue. The last gives it a quirk that unjustifiably, we think, broadens the controversy. The case requires an answer, not as to a manufacturer's right to register a lamp base but as to an artist's right to copyright a work of art intended to be reproduced for lamp bases. . . . Petitioners question the validity of a copyright of a work of art for "mass" production. "Reproduction of a work of art" does not mean to them unlimited reproduction. Their position is that a copyright does not cover industrial reproduction of the protected article. . . .

. . . .

. . . In 1790 the First Congress conferred a copyright on "authors of any map, chart, book or books already printed." Later, designing, engraving and etching were included; in 1831 musical compositions; dramatic compositions in 1856; and photographs and negatives thereof in 1865. The Act of 1870 defined copyrightable subject matter as:

"* * * any book, map, chart, dramatic or musical composition, engraving, cut, print, or photograph or negative thereof, or of a painting, drawing, chromo, *statue, statuary, and of models or designs intended to be perfected as works of the fine arts.*"

(Emphasis supplied.) The italicized part added three-dimensional work[s] of art to what had been protected previously. . . .

. . . .

The successive acts, the legislative history of the 1909 Act and the practice of the Copyright Office unite to show that "works of art" and "reproductions of works of art" are terms that were intended by Congress to include the authority to copyright these statuettes. Individual perception of the beautiful is too varied a power to permit a narrow or rigid concept of art. . . .

The conclusion that the statues here in issue may be copyrighted goes far to solve the question whether their intended reproduction as lamp stands bars or invalidates their registration. This depends solely on statutory interpretation. . . .

But petitioners assert that congressional enactment of the design patent laws should be interpreted as denying protection to artistic articles embodied or reproduced in manufactured articles. . . . Their argument is that design patents require the critical examination given patents to protect the public against monopoly. . . . Petitioner urges that overlapping of patent and copyright legislation so as to give an author or inventor a choice between patents and copyrights should not be permitted. We assume petitioner takes the position that protection for a statuette for industrial use can only be obtained by patent, if any protection can be given.

As we have held the statuettes here involved copyrightable, we need not decide the question of their patentability. Though other courts have passed upon the issue as to whether allowance by the election of the author or patentee of one bars a grant of the other, we do not. We do hold that the patentability of the statuettes, fitted as lamps or unfitted, does not bar copyright as works of art. Neither the Copyright Statute nor any other says that because a thing is patentable it may not be copyrighted. We should not so hold.

Unlike a patent, a copyright gives no exclusive right to the art disclosed; protection is given only to the expression of the idea—not the idea itself. . . . Regulation § 202.8 [37 C.F.R. § 202.8 (1949)], makes clear that artistic articles are protected in "form but not their mechanical or utilitarian aspects." The dichotomy of protection for the aesthetic is not beauty and utility but art for the copyright and the invention of original and ornamental design for design patents. We find nothing in the copyright statute to support the argument that the intended use or use in industry of an article eligible for copyright bars or invalidates its registration. We do not read such a limitation into the copyright law.

   . . . .

Douglas, Justice, concurring (with Justice Black):

   . . . .

. . . The Copyright Office has supplied us with a long list of such articles which have been copyrighted—statuettes, book ends, clocks, lamps, door knockers, candlesticks, inkstands, chandeliers, piggy banks, sundials, salt and pepper shakers, fish bowls, casseroles, and ash trays. Perhaps these are all "writings" in the constitutional sense. But to me, at least, they are not obviously so. It is time that we came to the problem full face. I would accordingly put the case down for reargument.

Despite Justice Douglas' reluctance to treat "clocks, lamps, door knockers" and similar items as "writings" for Progress Clause purposes, permitting Congress to provide them copyright protection, many such products rely on copyright to prevent competitors from copying. Today's Copyright Act uses different language to identify the products of industrial design that blend utility with creative aesthetic expression. The challenge the Act poses is that we must parse the utility, which is not copyrightable, from the expressive features, which are.

Star Athletica, L.L.C. v. Varsity Brands, Inc.

137 S. Ct. 1002 (2017)

Thomas, J.:

Congress has provided copyright protection for original works of art, but not for industrial designs. The line between art and industrial design, however, is

Can design be copyrighted? Work of art?

often difficult to draw. This is particularly true when an industrial design incorporates artistic elements. . . .

    . . . .

                                                    I

Respondents Varsity Brands, Inc., [and others] design, make, and sell cheerleading uniforms. Respondents have obtained or acquired more than 200 U.S. copyright registrations for two-dimensional designs appearing on the surface of their uniforms and other garments. These designs are primarily "combinations, positionings, and arrangements of elements" that include "chevrons * * *, lines, curves, stripes, angles, diagonals, inverted [chevrons], coloring, and shapes." . . .

Images from Appendix of the Court's decision

Petitioner Star Athletica, L.L.C., also markets and sells cheerleading uniforms. Respondents sued petitioner for infringing their copyrights in the five designs. The District Court entered summary judgment for petitioner on respondents' copyright claims on the ground that the designs did not qualify as protectable pictorial, graphic, or sculptural works. . . .

The Court of Appeals for the Sixth Circuit reversed. . . .

    . . . .

                                                    II

    . . . .

The Copyright Act . . . establishes a special rule for copyrighting a pictorial, graphic, or sculptural work incorporated into a "useful article," which is defined as "an article having an intrinsic utilitarian function that is not merely to portray the appearance of the article or to convey information." [17 U.S.C. § 101.] The statute does not protect useful articles as such. Rather, "the design of a useful article" is "considered a pictorial, graphical, or sculptural work only if, and only to the extent that, such design incorporates pictorial, graphic, or sculptural features that can be identified separately from, and are capable of existing independently of, the utilitarian aspects of the article." [17 U.S.C. § 101 (defining "pictorial, graphic, sculptural works").]

Courts, the Copyright Office, and commentators have described the analysis undertaken to determine whether a feature can be separately identified from, and exist independently of, a useful article as "separability." In this case, our task is to determine whether the arrangements of lines, chevrons, and colorful shapes appearing on the surface of respondents' cheerleading uniforms are eligible for copyright protection as separable features of the design of those cheerleading uniforms.

A

As an initial matter, we must address whether separability analysis is necessary in this case.

1

Respondents argue that "[s]eparability is only implicated when a [pictorial, graphic, or sculptural] work is the 'design of a useful article.'" They contend that the surface decorations in this case are "two-dimensional graphic designs that appear *on* useful articles," but are not themselves designs *of* useful articles. Consequently, the surface decorations are protected two-dimensional works of graphic art without regard to any separability analysis under §101. Under this theory, two-dimensional artistic features on the surface of useful articles are "inherently separable." Brief for Respondents 26.

This argument is inconsistent with the text of § 101. The statute requires separability analysis for any "pictorial, graphic, or sculptural features" incorporated into the "design of a useful article." "Design" refers here to "the combination" of "details" or "features" that "go to make up" the useful article. 3 *Oxford English Dictionary* 244 (def. 7, first listing) (1933) (OED). Furthermore, the words "pictorial" and "graphic" include, in this context, two-dimensional features such as pictures, paintings, or drawings. See 4 *id.* at 359 (defining "[g]raphic" to mean "[o]f or pertaining to drawing or painting"); 7 *id.* at 830 (defining "[p]ictorial" to mean "of or pertaining to painting or drawing"). And the statute expressly defines "[p]ictorial, graphical, and sculptural works" to include "two-dimensional * * * works of * * * art." § 101. The statute thus provides that the "design of a useful article" can include two-dimensional "pictorial" and "graphic" features, and separability analysis applies to those features just as it does to three-dimensional "sculptural" features.

       . . . .

B

We must now decide when a feature incorporated into a useful article "can be identified separately from" and is "capable of existing independently of" "the utilitarian aspects" of the article. This is not a free-ranging search for the best copyright policy, but rather "depends solely on statutory interpretation." *Mazer v. Stein*, 347 U.S. 201, 214 (1954). "The controlling principle in this case is the basic and unexceptional rule that courts must give effect to the clear meaning of statutes as written." *Estate of Cowart v. Nicklos Drilling Co.*, 505 U.S. 469, 476 (1992). We thus begin and end our inquiry with the text, giving each word its "ordinary, contemporary, common meaning." *Walters v. Metropolitan Ed. Enters.* 519 U.S. 202, 207 (1997) (internal quotation marks omitted). We do not, however, limit this inquiry to the text of § 101 in isolation. "[I]nterpretation of a phrase of uncertain reach is not confined to a single sentence when the text of the whole statute gives instruction as to its meaning." *Maracich v. Spears*, 133 S. Ct. 2191, 2203 (2013). We thus "look to the provisions of the whole law" to determine §101's meaning. *United States v. Heirs of Boisdoré*, 8 How. 113, 122 (1849).

1

The statute provides that a "pictorial, graphic, or sculptural featur[e]" incorporated into the "design of a useful article" is eligible for copyright protection if it (1) "can be identified separately from," and (2) is "capable of existing independently of, the utilitarian aspects of the article." § 101. The first requirement—separate identification—is not onerous. The decisionmaker need only be able to look at the useful article and spot some two- or three- dimensional element that appears to have pictorial, graphic, or sculptural qualities.

The independent-existence requirement is ordinarily more difficult to satisfy. The decisionmaker must determine that the separately identified feature has the capacity to exist apart from the utilitarian aspects of the article. See 2 OED 88 (def. 5) (defining "[c]apable" of as "[h]aving the needful capacity, power, or fitness for"). In other words, the feature must be able to exist as its own pictorial, graphic, or sculptural work as defined in § 101 once it is imagined apart from the useful article. If the feature is not capable of existing as a pictorial, graphic, or sculptural work once separated from the useful article, then it was not a pictorial, graphic, or sculptural feature of that article, but rather one of its utilitarian aspects.

Of course, to qualify as a pictorial, graphic, or sculptural work on its own, the feature cannot itself be a useful article or "[a]n article that is normally a part of a useful article" (which is itself considered a useful article). § 101 [(defining "useful article")]. Nor could someone claim a copyright in a useful article merely by creating a replica of that article in some other medium—for example, a cardboard model of a car. Although the replica could itself be copyrightable, it would not give rise to any rights in the useful article that inspired it.

2

The statute as a whole confirms our interpretation. The Copyright Act provides "the owner of [a] copyright" with the "exclusive righ[t] * * * to reproduce the copyrighted work in copies." § 106(1). The statute clarifies that this right "includes the right to reproduce the [copyrighted] work in or on any kind of article, whether useful or otherwise." § 113(a). Section 101 is, in essence, the mirror image of § 113(a). Whereas § 113(a) protects a work of authorship first fixed in some tangible medium other than a useful article and subsequently applied to a useful article, § 101 protects art first fixed in the medium of a useful article. The two provisions make clear that copyright protection extends to pictorial, graphic, and sculptural works regardless of whether they were created as free-standing art or as features of useful articles. The ultimate separability question, then, is whether the feature for which copyright protection is claimed would have been eligible for copyright protection as a pictorial, graphic, or sculptural work had it originally been fixed in some tangible medium other than a useful article before being applied to a useful article.

3

This interpretation is also consistent with the history of the Copyright Act. . . .

Two of *Mazer*'s holdings are relevant here. First, the Court held that the respondents owned a copyright in the statuette even though it was intended for use as a lamp base. . . .

Second, the Court held that it was irrelevant to the copyright inquiry whether the statuette was initially created as a freestanding sculpture or as a lamp base. . . .

Shortly thereafter, the Copyright Office enacted a regulation implementing the holdings of *Mazer*. . . .

Congress essentially lifted the language governing protection for the design of a useful article directly from the post-*Mazer* regulations and placed it into § 101 of the 1976 Act. Consistent with *Mazer*, the approach we outline today interprets §§ 101 and 113 in a way that would afford copyright protection to the statuette in *Mazer* regardless of whether it was first created as a standalone sculptural work or as the base of the lamp.

C

In sum, a feature of the design of a useful article is eligible for copyright if, when identified and imagined apart from the useful article, it would qualify as a pictorial, graphic, or sculptural work either on its own or when fixed in some other tangible medium.

Applying this test to the surface decorations on the cheerleading uniforms is straightforward. First, one can identify the decorations as features having pictorial, graphic, or sculptural qualities. Second, if the arrangement of colors, shapes, stripes, and chevrons on the surface of the cheerleading uniforms were separated from the uniform and applied in another medium—for example, on a painter's canvas—they would qualify as "two-dimensional * * * works of * * * art," § 101. And imaginatively removing the surface decorations from the uniforms and applying them in another medium would not replicate the uniform itself. Indeed, respondents have applied the designs in this case to other media of expression—different types of clothing—without replicating the uniform. The decorations are therefore separable from the uniforms and eligible for copyright protection.[1]

The dissent argues that the designs are not separable because imaginatively removing them from the uniforms and placing them in some other medium of

---

[1] We do not today hold that the surface decorations are copyrightable. We express no opinion on whether these works are sufficiently original to qualify for copyright protection or on whether any other prerequisite of a valid copyright has been satisfied.

expression—a canvas, for example—would create "pictures of cheerleader uniforms." Petitioner similarly argues that the decorations cannot be copyrighted because, even when extracted from the useful article, they retain the outline of a cheerleading uniform.

This is not a bar to copyright. Just as two-dimensional fine art corresponds to the shape of the canvas on which it is painted, two-dimensional applied art correlates to the contours of the article on which it is applied. A fresco painted on a wall, ceiling panel, or dome would not lose copyright protection, for example, simply because it was designed to track the dimensions of the surface on which it was painted. Or consider, for example, a design etched or painted on the surface of a guitar. If that entire design is imaginatively removed from the guitar's surface and placed on an album cover, it would still resemble the shape of a guitar. But the image on the cover does not "replicate" the guitar as a useful article. Rather, the design is a two-dimensional work of art that corresponds to the shape of the useful article to which it was applied. The statute protects that work of art whether it is first drawn on the album cover and then applied to the guitar's surface, or vice versa. . . .

To be clear, the only feature of the cheerleading uniform eligible for a copyright in this case is the two-dimensional work of art fixed in the tangible medium of the uniform fabric. Even if respondents ultimately succeed in establishing a valid copyright in the surface decorations at issue here, respondents have no right to prohibit any person from manufacturing a cheerleading uniform of identical shape, cut, and dimensions to the ones on which the decorations in this case appear. They may prohibit only the reproduction of the surface designs in any tangible medium of expression—a uniform or otherwise.[2]

D

Petitioner . . . raise[s] several objections to the approach we announce today. None is meritorious.

1

Petitioner first argues that our reading of the statute is missing an important step. It contends that a feature may exist independently only if it can stand alone as a copyrightable work *and* if the useful article from which it was extracted would remain equally useful. . . .

The debate over the relative utility of a plain white cheerleading uniform is unnecessary. The focus of the separability inquiry is on the extracted feature and not on any aspects of the useful article that remain after the imaginary

---

[2] The dissent suggests that our test would lead to the copyrighting of shovels. But a shovel, like a cheerleading uniform, even if displayed in an art gallery, is "an article having an intrinsic utilitarian function that is not merely to portray the appearance of the article or to convey information." 17 U.S.C. § 101. It therefore cannot be copyrighted. A drawing of a shovel could, of course, be copyrighted. And, if the shovel included any artistic features that could be perceived as art apart from the shovel, and which would qualify as protectable pictorial, graphic, or sculptural works on their own or in another medium, they too could be copyrighted. But a shovel as a shovel cannot.

extraction. The statute does not require the decisionmaker to imagine a fully functioning useful article without the artistic feature. . . .

. . . Indeed, such a requirement would deprive the *Mazer* statuette of protection had it been created first as a lamp base rather than as a statuette. Without the base, the "lamp" would be just a shade, bulb, and wires. The statute does not require that we imagine a nonartistic replacement for the removed feature to determine whether that *feature* is capable of an independent existence.

. . . .

. . . [T]his has been the rule since *Mazer*. In holding that the statuette was protected, the Court emphasized that the 1909 Act abandoned any "distinctions between purely aesthetic articles and useful works of art." Congress did not enact such a distinction in the 1976 Act. Were we to accept petitioner's argument that the only protectable features are those that play absolutely no role in an article's function, we would effectively abrogate the rule of *Mazer* and read "applied art"out of the statute.

. . . .

2

Petitioner next argues that we should incorporate two "objective" components, into our test to provide guidance to the lower courts: (1) "whether the design elements can be identified as reflecting the designer's artistic judgment exercised independently of functional influence," and (2) whether "there is [a] substantial likelihood that the pictorial, graphic, or sculptural feature would still be marketable to some significant segment of the community without its utilitarian function."

We reject this argument because neither consideration is grounded in the text of the statute. The first would require the decisionmaker to consider evidence of the creator's design methods, purposes, and reasons. The statute's text makes clear, however, that our inquiry is limited to how the article and feature are perceived, not how or why they were designed.

The same is true of marketability. Nothing in the statute suggests that copyrightability depends on market surveys. Moreover, asking whether some segment of the market would be interested in a given work threatens to prize popular art over other forms, or to substitute judicial aesthetic preferences for the policy choices embodied in the Copyright Act. *See Bleistein v. Donaldson Lithographing Co.*, 188 U.S. 239, 251 (1903).

. . . .

3

Finally, petitioner argues that allowing the surface decorations to qualify as a "work of authorship" is inconsistent with Congress' intent to entirely exclude industrial design from copyright. Petitioner notes that Congress refused to pass a provision that would have provided limited copyright protection for industrial designs, including clothing, when it enacted the 1976 Act and that it has enacted laws protecting designs for specific useful articles—semiconductor chips and boat hulls, see 17 U. S. C. §§901-914, 1301-1332—while declining to enact other industrial design statutes. From this history of failed legislation petitioner reasons that Congress intends to channel intellectual property claims for industrial design into design patents. . . .

We do not share petitioner's concern. As an initial matter, "[c]ongressional inaction lacks persuasive significance" in most circumstances. *Pension Benefit Guaranty Corporation* v. *LTV Corp.*, 496 U.S. 633, 650 (1990) (internal quotation marks omitted). Moreover, we have long held that design patent and copyright are not mutually exclusive. See *Mazer*, 347 U.S., at 217. Congress has provided for limited copyright protection for certain features of industrial design, and approaching the statute with presumptive hostility toward protection for industrial design would undermine Congress' choice. In any event, as explained above, our test does not render the shape, cut, and physical dimensions of the cheerleading uniforms eligible for copyright protection.

III

We hold that an artistic feature of the design of a useful article is eligible for copyright protection if the feature (1) can be perceived as a two- or three-dimensional work of art separate from the useful article and (2) would qualify as a protectable pictorial, graphic, or sculptural work either on its own or in some other medium if imagined separately from the useful article. Because the designs on the surface of respondents' cheerleading uniforms in this case satisfy these requirements, the judgment of the Court of Appeals is affirmed.

[JUSTICE GINSBURG's opinion concurring in the judgment is omitted.]

BREYER, JUSTICE, dissenting (with Justice Kennedy):

I agree with much in the Court's opinion. But I do not agree that the designs that Varsity Brands, Inc., submitted to the Copyright Office are eligible for copyright protection. Even applying the majority's test, the designs *cannot* "be perceived as * * * two- or three-dimensional work[s] of art separate from the useful article."

Look at the designs that Varsity submitted to the Copyright Office. See Appendix to opinion of the Court. You will see only pictures of cheerleader uniforms. And cheerleader uniforms are useful articles. A picture of the relevant design features, whether separately "perceived" on paper or in the imagination, is a picture of, and thereby "replicate[s]," the underlying useful article of which they are a part. Hence the design features that Varsity seeks to protect are not "capable of existing independently o[f] the utilitarian aspects of the article." 17 U.S.C. § 101.

I

The relevant statutory provision says that the "design of a useful article" is copyrightable "only if, and only to the extent that, such design incorporates pictorial, graphic, or sculptural features that can be identified separately from, and are capable of existing independently of, the utilitarian aspects of the article." But what, we must ask, do the words "identified separately" mean? Just when is a design separate from the "utilitarian aspect of the [useful] article?" The most direct, helpful aspect of the Court's opinion answers this question by stating:

"Nor could someone claim a copyright in a useful article merely by creating a replica of that article in some other medium—for example, a cardboard model of a car. Although the replica could itself be copyrightable, it would not give rise to any rights in the useful article that inspired it."

Exactly so. These words help explain the Court's statement that a copyrightable work of art must be "perceived as a two- or three-dimensional work of art separate from the useful article." They help clarify the concept of separateness. And they reflect long held views of the Copyright Office. See *Compendium of U.S. Copyright Office Practices* § 924.2(B) (3d ed. 2014) (*Compendium*).

Consider, for example, the explanation that the House Report for the Copyright Act of 1976 provides. It says:

"Unless the shape of an automobile, airplane, ladies' dress, food processor, television set, or any other industrial product contains some element that, *physically or conceptually*, can be identified as separable from the utilitarian aspects of that article, the design would not be copyrighted * * * ." H.R. Rep. No. 94-1476, pp. 55 (emphasis added).

These words suggest two exercises, one physical, one mental. Can the design features (the picture, the graphic, the sculpture) be physically removed from the article (and considered separately), all the while leaving the fully functioning utilitarian object in place? If not, can one nonetheless conceive of the design features separately without replicating a picture of the utilitarian object? If the answer to either of these questions is "yes," then the design is eligible for copyright protection. Otherwise, it is not. The abstract nature of these questions makes them sound difficult to apply. But with the Court's words in mind, the difficulty tends to disappear.

An example will help. Imagine a lamp with a circular marble base, a vertical 10-inch tall brass rod (containing wires) inserted off center on the base, a light bulb fixture emerging from the top of the brass rod, and a lampshade sitting on top. In front of the brass rod a porcelain Siamese cat sits on the base facing

APPENDIX TO OPINION OF BREYER, J.

Fig. 1   Fig. 2

outward. Obviously, the Siamese cat is *physically separate* from the lamp, as it could be easily removed while leaving both cat and lamp intact. And, assuming it otherwise qualifies, the designed cat is eligible for copyright protection.

Now suppose there is no long brass rod; instead the cat sits in the middle of the base and the wires run up through the cat to the bulbs. The cat is not physically separate from the lamp, as the reality of the lamp's construction is such that an effort to physically separate the cat and lamp will destroy both cat and lamp. The two are integrated into a single functional object, like the similar configuration of the ballet dancer statuettes that formed the lamp bases at issue in *Mazer v. Stein*, 347 U.S. 201 (1954). But we can easily imagine the cat on its own, as did Congress when conceptualizing the ballet dancer. See H.R. Rep. at 55 (the statuette in *Mazer* was "incorporated into a product without losing its ability to exist independently as a work of art"). In doing so, we do not create a mental picture of a lamp (or, in the Court's words, a "replica" of the lamp), which is a useful article. We simply perceive the cat separately, as a small cat figurine that could be a copyrightable design work standing alone that does not replicate the lamp. Hence the cat is *conceptually separate* from the utilitarian article that is the lamp.

Case law, particularly case law that Congress and the Copyright Office have considered, reflects the same approach. Congress cited examples of copyrightable design works, including "a carving on the back of a chair" and "a floral relief design on silver flatware." H.R. Rep. at 55. Copyright Office guidance on copyrightable designs in useful articles include "an engraving on a vase," "[a]rtwork printed on a t-shirt," "[a] colorful pattern decorating the surface of a shopping bag," "[a] drawing on the surface of wallpaper," and "[a] floral relief decorating the handle of a spoon." *Compendium* § 924.2(B). . . .

By way of contrast, Van Gogh's painting of a pair of old shoes, though beautifully executed and copyrightable as a painting, would not qualify for a shoe design copyright. 17 U.S.C. §§ 113(a)–(b). Courts have similarly denied copyright protection to objects that begin as three-dimensional designs, such as measuring spoons shaped like heart-tipped arrows, *Bonazoli v. R.S.V.P. Int'l, Inc.*, 353 F. Supp. 2d 218, 226–227 (D.R.I. 2005); candleholders shaped like sailboats, *Design Ideas, Ltd. v. Yankee Candle Co.*, 889 F. Supp. 2d 1119, 1128 (C.D. Ill. 2012); and wire spokes on a wheel cover, *Norris Industries, Inc. v. International Tel. & Tel. Corp.*, 696 F.2d 918, 922–924 (11th Cir. 1983). None of these designs could qualify for copyright protection that would prevent others from selling spoons, candleholders, or wheel covers with the same design. Why not? Because in each case the design is not separable from the utilitarian aspects of the object to which it relates. The designs cannot be physically separated because they themselves

APPENDIX TO OPINION OF BREYER, J.

Fig. 3: Vincent Van Gogh, "Shoes"

make up the shape of the spoon, candleholders, or wheelcovers of which they are a part. And spoons, candleholders, and wheel covers are useful objects, as are the old shoes depicted in Van Gogh's painting. More importantly, one cannot easily imagine or otherwise conceptualize the design of the spoons or the candleholders or the shoes *without that picture, or image, or replica being a picture of spoons, or candleholders, or wheel covers, or shoes.* The designs necessarily bring along the underlying utilitarian object. Hence each design is not conceptually separable from the physical useful object.

The upshot is that one could copyright the floral design on a soupspoon but one could not copyright the shape of the spoon itself, no matter how beautiful, artistic, or esthetically pleasing that shape might be: A picture of the shape of the spoon is also a picture of a spoon; the picture of a floral design is not. See *Compendium* § 924.2(B).

To repeat: A separable design feature must be "capable of existing independently" of the useful article as a separate artistic work that is not itself the useful article. If the claimed feature could be extracted without replicating the useful article of which it is a part, and the result would be a copyrightable artistic work standing alone, then there is a separable design. But if extracting the claimed features would necessarily bring along the underlying useful article, the design is not separable from the useful article. In many or most cases, to decide whether a design or artistic feature of a useful article is conceptually separate from the article itself, it is enough to imagine the feature on its own and ask, "Have I created a picture of a (useful part of a) useful article?" If so, the design is not separable from the useful article. If not, it is.

In referring to imagined pictures and the like, I am not speaking technically. I am simply trying to explain an intuitive idea of what separation is about, as well as how I understand the majority's opinion. So understood, the opinion puts design copyrights in their rightful place. The law has long recognized that drawings or photographs of real world objects are copyrightable as drawings or photographs, but the copyright does not give protection against others making the underlying useful objects. That is why a copyright on Van Gogh's painting would prevent others from reproducing that painting, but it would not prevent others from reproducing and selling the comfortable old shoes that the painting depicts. Indeed, the purpose of § 113(b) was to ensure that " 'copyright in a pictorial, graphic, or sculptural work, portraying a useful article as such, does not extend to the manufacture of the useful article itself.' " H.R. Rep. at 105.

II

To ask this kind of simple question—does the design picture the useful article?—will not provide an answer in every case, for there will be cases where it is difficult to say whether a picture of the design is, or is not, also a picture of the useful article. But the question will avoid courts focusing primarily upon what I believe is an unhelpful feature of the inquiry, namely, whether the design can be imagined as a "two- or three-dimensional work of art." That is because virtually any industrial design can be thought of separately as a "work of art": Just imagine a frame surrounding the design, or its being placed in a gallery. Consider Marcel Duchamp's "ready-mades" series, the functional mass-produced objects he designated as art. What is there in the world that, viewed through an esthetic lens, cannot be seen as a good, bad, or indifferent work of

art? What design features could not be imaginatively reproduced on a painter's canvas?
Indeed, great industrial design may well include design that is inseparable from the useful article—where, as Frank Lloyd Wright
put it, "form and function are one." F.
Wright, *An Autobiography* 146 (1943). Where
they are one, the designer may be able to obtain 15 years of protection through a design patent. But, if they are one, Congress did not intend a century or more of copyright protection.

Fig. 4: Marcel Duchamp,
"In Advance of the Broken Arm"

III

. . . .

The Constitution grants Congress primary responsibility for assessing comparative costs and benefits and drawing copyright's statutory lines. Courts must
respect those lines and not grant copyright protection where Congress has
decided not to do so. And it is clear that Congress has not extended broad copyright protection to the fashion design industry.

Congress' decision not to grant full copyright protection to the fashion industry has not left the industry without protection. Patent design protection is
available. A maker of clothing can obtain trademark protection under the Lanham
Act for signature features of the clothing. And a designer who creates an original
textile design can receive copyright protection for that pattern as placed, for
example, on a bolt of cloth, or anything made with that cloth. . . .

The fashion industry has thrived against this backdrop, and designers have
contributed immeasurably to artistic and personal self-expression through clothing. But a decision by this Court to grant protection to the design of a garment
would grant the designer protection that Congress refused to provide. It would
risk increased prices and unforeseeable disruption in the clothing industry, which
in the United States alone encompasses nearly $370 billion in annual spending
and 1.8 million jobs. Brief for Council of Fashion Designers of America, Inc.,
as *Amicus Curiae* 3–4 (citing U.S. Congress, Joint Economic Committee, *The
New Economy of Fashion* 1 (2016)). That is why I believe it important to emphasize those parts of the Court's opinion that limit the scope of its interpretation.
That language, as I have said, makes clear that one may not "claim a copyright
in a useful article merely by creating a replica of that article in some other medium," which "would not give rise to any rights in the useful article that inspired
it."

IV

If we ask the "separateness" question correctly, the answer here is not difficult
to find. . . . Can the design features in Varsity's pictures exist separately from the
utilitarian aspects of a dress? Can we extract those features as copyrightable

design works standing alone, without bringing along, via picture or design, the dresses of which they constitute a part?

Consider designs 074, 078, and 0815. They certainly look like cheerleader uniforms. That is to say, they look like pictures of cheerleader uniforms, just like Van Gogh's old shoes look like shoes. I do not see how one could see them otherwise. Designs 299A and 2999B present slightly closer questions. They omit some of the dresslike context that the other designs possess. But the necklines, the sleeves, and the cut of the skirt suggest that they too are pictures of dresses. Looking at all five of Varsity's pictures, I do not see how one could conceptualize the design features in a way that does not picture, not just artistic designs, but dresses as well.

Were I to accept the majority's invitation to "imaginatively remov[e]" the chevrons and stripes *as they are arranged* on the neckline, waistline, sleeves, and skirt of each uniform, and apply them on a "painter's canvas," that painting would be of a cheerleader's dress. The esthetic elements on which Varsity seeks protection exist only as part of the uniform design—there is nothing to separate out but for dress-shaped lines that replicate the cut and style of the uniforms. Hence, each design is not physically separate, nor is it conceptually separate, from the useful article it depicts, namely, a cheerleader's dress. They cannot be copyrighted.

Varsity, of course, could have sought a design patent for its designs. Or, it could have sought a copyright on a textile design, even one with a similar theme of chevrons and lines.

But that is not the nature of Varsity's copyright claim. It has instead claimed ownership of the particular "'treatment and arrangement'" of the chevrons and lines of the design as they appear at the neckline, waist, skirt, sleeves, and overall cut of each uniform. Brief for Respondents 50. The majority imagines that Varsity submitted something different—that is, only the surface decorations of chevrons and stripes, as in a textile design. As the majority sees it, Varsity's copyright claim would be the same had it submitted a plain rectangular space depicting chevrons and stripes, like swaths from a bolt of fabric. But considered on their own, the simple stripes are plainly unoriginal. Varsity, then, seeks to do indirectly what it cannot do directly: bring along the design and cut of the dresses by seeking to protect surface decorations whose "treatment and arrangement" are *coextensive with that design and cut.* As Varsity would have it, it would prevent its competitors from making useful three-dimensional cheerleader uniforms by submitting plainly unoriginal chevrons and stripes as cut and arranged on a useful article. But with that cut and arrangement, the resulting pictures on which Varsity seeks protection do not simply depict designs. They depict clothing. They depict the useful articles of which the designs are inextricable parts. And Varsity cannot obtain copyright protection that would give them the power to prevent others from making those useful uniforms, any more than Van Gogh can copyright comfortable old shoes by painting their likeness.

I fear that, in looking past the three-dimensional design inherent in Varsity's claim by treating it as if it were no more than a design for a bolt of cloth, the majority has lost sight of its own important limiting principle. One may not "claim a copyright in a useful article merely by creating a replica of that article in some other medium," such as in a picture. That is to say, one cannot obtain a copyright that would give its holder "any rights in the useful article that inspired it."

With respect, I dissent.

Notes & Questions

1. The Court identifies the "ultimate separability question" as "whether the feature for which copyright protection is claimed would have been eligible for copyright protection as a pictorial, graphic, or sculptural work had it originally been fixed in some tangible medium other than a useful article before being applied to a useful article." Try applying that test to the following objects:

How would you advise a client who had designed the above products and was interested in preventing a competitor from copying them? Is copyright protection a viable option? How about a design patent?

2. Features that one can readily physically separate from a larger useful article, such as the lunging jaguar hood ornament on a sedan, are the "easy" cases under the Court's framework. The elements that cannot be so easily physically separated are the ones that cause the problems. This is what Justice Breyer is describing with his cat lamps.

3. The Court rejected a test that would have asked, essentially, whether the separable pictorial, graphic, or sculptural elements were marketable as art because it thought such a test ran afoul of the *Bleistein* aesthetic non-discrimination doctrine. Is the test the Court adopts consistent with a policy of aesthetic non-discrimination? In his dissent Justice Breyer thinks his analysis of the separability question avoids asking whether something is "art." Does it? Is it possible to analyze conceptual separability without making some determination concerning what is, and what is not, art? For a discussion of this issue, *see* Alfred C. Yen, *Copyright Opinions and Aesthetic Theory*, 71 S. Cal. L. Rev. 247 (1998). Does the hesitancy of the courts to pronounce what is "art" cause too much to meet the subject matter eligibility requirement of copyright?

4. Recall that copyright protection, unlike patent protection, is conferred without any expert agency examination for novelty or nonobviousness, and without any mandatory enabling disclosure of how to make and use the useful article in

question. How does the separability test for the design of a useful article compare to the design patent test for whether an element is primarily functional rather than ornamental?

5. In footnote 1 of *Star Athletica*, the Court clarifies that it is not ruling on whether the separable graphic elements of the cheerleader uniforms are sufficiently original to qualify for protection. Justic Breyer also refers to the "plainly unoriginal chevrons and stripes." Do plaintiff's separable pictorial, graphic, or sculptural features meet the originality requirement, stated in *Feist*?

6. Separability is only necessary when the pictorial, graphic or sculptural elements are incorporated into a useful article, defined as "an article having an intrinsic utilitarian function that is not merely to portray the appearance of the article or to convey information." 17 U.S.C. § 101. While an automobile or an airplane would qualify as a useful article, as the Court notes, a toy airplane is not a useful article. *Gay Toys, Inc. v. Buddy L. Corp.*, 703 F.2d 970 (6th Cir. 1983) (finding that "other than the portrayal of a real airplane, a toy airplane, like a painting, has no intrinsic utilitarian function"). Clothing is generally considered a useful article — it is meant to provide protection and warmth, as well as conformity with social norms. Therefore, the protectability of clothing is subject to the separability analysis, and only those pictorial, graphic or sculptural elements that are separable are eligible for protection. As the Court noted, "[e]ven if respondents ultimately succeed in establishing a valid copyright in the surface decorations at issue here, respondents have no right to prohibit any person from manufacturing a cheerleading uniform of identical shape, cut, and dimensions to the ones on which the decorations in this case appear." Congress has repeatedly declined to adopt proposed bills that would provide a separate three-year term of copyright-style protection for fashion designs. If you were in Congress would you support such legislation? Does the lack of protection for clothing design make the *Star Athletica* decision more, or less, understandable?

C. Computer Software

Computer software is eligible for copyright protection. The Copyright Act contains the following definition:

> "Literary works" are works, other than audiovisual works, expressed in words, numbers, or other verbal or numerical symbols or indicia, regardless of the nature of the material objects, such as books, periodicals, manuscripts, phonorecords, film, tapes, disks, or cards, in which they are embodied.

17 U.S.C. § 101. Pursuant to this definition, copyright law protects computer software as a literary work. Be careful with this classification, however, as it applies only to the software code itself. The output of some programs may also have protection as a different type of copyrighted work. For example, a video game consists of images and sounds that qualify the output for protection as an audiovisual work.

The application of copyright to software has some interesting twists and tangles as well. Consider the copyright doctrines that are addressed in Section I of this chapter: fixation and originality. Computer software is most often fixed in computer memory. The originality of software, following *Feist*, comes from being independently created and having a modicum of creativity. Like other forms of authorial expression, computer programs build on what already exists, but the additional new expression, whether it be combining bits of code in creative ways or in writing all new code, gives rise to protection for the resulting software program.

But computer software fits uneasily into the copyright scheme. On the one hand, computer software can be written in a complicated language and thus involves a certain amount of expressive activity on the programmer's part. On the other hand, computer software is, at bottom, functional—it operates a machine. The Act itself acknowledges this when it defines a computer program as "a set of statements or instructions to be used directly or indirectly in a computer in order to bring about a certain result." 17 U.S.C. § 101.

The functional aspect of computer software presents challenges for determining what in software, exactly, copyright law protects. As you saw in *Baker v. Selden*, leaving certain aspects of a work free for others to use is important to the underlying goal that copyright law seeks to serve. Further, the limitation on what copyright protects contained in § 102(b) includes aspects that are highly relevant to software: "In no case does copyright protection for an original work of authorship extend to any idea, procedure, process, system, method of operation, concept, principle, or discovery . . . ." 17 U.S.C. § 102(b).

In *Baker* the Supreme Court emphasized the importance of ensuring that copyright does not function as a back-door route to patent protection. If someone wants protection for the functional aspects of a literary work, a system of bookkeeping for example, they must make the disclosures and meet the high thresholds required for patent protection. Similarly, the statutory prohibition on copyright protection for functional aspects of pictorial, graphic, or sculptural works that qualify as useful articles seeks to keep a boundary line between copyrights and patents. Because software is a literary work, not a pictorial, graphic, or sculptural work, the useful article doctrine's built-in statutory prohibition on protecting functional aspects does not apply. Importantly, § 102(b) applies to all works. Recall, as we discussed in Chapter 3, that in the U.S. patent system, protection is now accepted for computer-implemented methods so long as those methods meet the requirements of patentability. *See Alice Corp. v. CLS Bank Int'l*, 134 S. Ct. 2347 (2014). Recall, as well, that an aspect of the idea/expression distinction is the concept of "merger" introduced above in Note 5 following the *Baker* case. The merger doctrine prohibits protecting even the expression if it is merged with the unprotectable ideas, processes or method of operations expressed. In such a situation, to protect the expression would be to grant protection for the idea, and copyright protection should never extend to the idea. Courts often determine whether there is merger by examining if there are alternative ways of expressing the same idea. Some computer programs are very simple, consisting of only a few lines of code. Such programs may not be copyrightable because there is only one or very few ways of expressing the idea of the program. *See Lexmark Intern., Inc. v. Static Control Components, Inc.*, 387 F.3d 522 (6th Cir. 2004) (finding the Toner Loading Program contained on ink cartridges used in computer printers was ineligible for copyright protection).

In a survey course it is difficult to delve too deeply into the doctrines that have developed concerning copyright protection for computer programs. The most recent high-profile case to address the topic, *Oracle America, Inc. v. Google Inc.*, 750 F.3d 1339, 1366 (Fed. Cir. 2014), *cert. denied*, 135 S. Ct. 2887 (2015), has generated a wealth of court opinions and commentary. The case concerns the copyrightability of Application Programing Interfaces (or APIs) in the Java programming language, and Google's copying of 37 different Oracle-created APIs in creating a virtual machine for its Android operating system. Google asserted that the APIs were methods of operation and thus were not copyrightable under § 102(b) and the merger doctrine. On appeal, the Federal Circuit held that the APIs were eligible for copyright protection but remanded the case for a determination of whether Google's copying of them was fair use. A jury found in favor of Google, but on appeal the Federal Circuit reversed, holding that Google's copying was not fair use, as a matter of law. *Oracle Am. Inc. v. Google LLC*, 886 F.3d 1179 (2018).

III. Owning and Maintaining a Copyright

A. Authors as Initial Owners of Copyright

In the United States, when an original work of authorship is fixed in a tangible medium of expression, a copyright exists: no registration, notice, or even distribution of copies of the work is required to obtain the rights granted by the federal Copyright Act. The Copyright Act specifies that ownership of the copyright initially vests in the "author" or "authors" of the work. 17 U.S.C. § 201. The Act does not, however, define what it means to be an "author." Typically, identifying the author of a work is easy. There are two situations, however, in which the issue becomes more complicated. The first is when more than one person is involved in the creation of a work and the second is when the work is created as a "Work Made for Hire," a category the Copyright Act specifically defines. We begin this Section with these two wrinkles in the fabric of authorship under the Copyright Act.

1. Authors and Joint Authors

Imagine three friends, Larry, Moe and Curly, out for a day of sightseeing with their cameras. Larry says to Moe and Curly "stand here, put your arms over each others' shoulders and face me" and then steps back and takes a photograph. Assuming the resulting photograph contains sufficient originality, who is the "author" of this work? The Supreme Court addressed this question in the case involving the picture of Oscar Wilde, described in Section I of this Chapter. In that case, the Court decided that the photographer was the author: "the person who has superintended the arrangement, who has actually formed the picture by putting the persons in position, and arranging the place where the people are to be — the man who is the effective cause of that." *Burrow-Giles v. Sarony*, 111 U.S. 53 (1884). The Court said that an "author," in the sense that the framers used the term in the Constitution, was "he to whom anything owes its origin; originator; maker; one who completes a work of science or literature." *Id.*

What if Larry, Moe and Curly work in a much more collaborative way, discussing who should stand where, what should be in the background, and even what facial expressions those in the picture should make. Who would be the "author" of the resulting work? If they are joint authors they are each entitled to equal undivided interests in the whole work—in other words, each joint author has the right to use or to license the work as he or she wishes. Each joint author must also, however, account to the other joint author(s) for any profits from that exploitation of the work. *See, e.g.*, *Greene v. Ablon*, 794 F.3d 133, 154-56 (1st Cir. 2015) (reviewing issues in an accounting between two joint authors).

The Copyright Act defines a "joint work" as "a work prepared by two or more authors with the intention that their contributions be merged into inseparable or interdependent parts of a unitary whole." 17 U.S.C. § 101 (1994). The legislative history indicates that the key element "is the intention at the time the writing is done that the parts be absorbed or combined into an integrated unit." H.R. Rep. No. 1476, 94th Cong. 120, 121 (1976), reprinted in 1976 U.S. Code Cong. & Admin. News 5659, 5735. The statutory definition could be read to encompass anyone who contributes to a work of authorship in some way. Consider an editor who makes useful revisions to a writer's submitted drafts. The writer and the editor intend for their contributions to be merged into inseparable parts of a unitary whole, yet very few editors and even fewer writers would expect the editor to be accorded the status of joint author, enjoying an undivided half interest in the copyright in the published work. *See Childress v. Taylor,* 945 F.2d 500 (2d Cir. 1991).

To address this danger of an overbroad construction of the statute, the Second Circuit adopted a two-pronged test for determining joint authorship. A party seeking to establish joint authorship bears the burden of establishing that each of the putative co-authors (1) made independently copyrightable contributions to the work; and (2) fully intended to be co-authors. *See id.* at 507-08. The facts of *Childress* exemplify the kinds of claims of joint authorship that fail the two-prong test. Actress Clarice Taylor wrote a script based on the life of legendary comedienne Jackie "Moms" Mabley, but Taylor was unable to get it produced as a play. Taylor convinced playwright Alice Childress to rescue the project by writing a new script. After Childress' completion of the script, Taylor took a copy of Childress' copyrighted play and produced it at another theater without permission. *See id.* at 503. Childress sued Taylor for copyright infringement, and Taylor asserted a defense of co-authorship. *See id.* at 504. The court concluded that there was "no evidence that [Taylor's contribution] ever evolved into more than the helpful advice that might come from the cast, the directors, or the producers of any play." *Id.* at 509. On that basis, the court rejected Taylor's claim of joint authorship.

In a later case involving a dramatrug's claim of co-authorship in the Broadway smash success *Rent*, the Second Circuit further explained the second prong, which it called a requirement of "mutual intent." *Thomson v. Larson*, 147 F.3d 195, 201 (2d Cir. 1998):

Childress and its progeny . . . do not explicitly define the nature of the necessary intent to be co-authors.[16] The court stated that "[i]n many instances, a useful test will be whether, in the absence of contractual arrangements concerning listed authorship, each participant intended that all would be identified as co-authors." *Childress*, 945 F.2d at 508. But it is also clear that the intention standard is not strictly subjective. In other words, co-authorship intent does not turn solely on the parties' own words or professed state of mind. *See id.* ("[J]oint authorship can exist without any explicit discussion of this topic by the parties."). Rather, the *Childress* court suggested a more nuanced inquiry into factual indicia of ownership and authorship, such as how a collaborator regarded herself in relation to the work in terms of billing and credit, decisionmaking, and the right to enter into contracts. *See id.* at 508-09. . . .

. . . [T]he *Childress* court emphasized that the requirement of intent is particularly important where "one person * * * is indisputably the dominant author of the work and the only issue is whether that person is the sole author or she and another * * * are joint authors." *Id.* "Care must be taken * * * to guard against the risk that a sole author is denied exclusive authorship status simply because another person render[s] some form of assistance." *Id.* at 504; see also *Erickson*, 13 F.3d at 1069 ("Those seeking copyrights would not seek further refinement that colleagues may offer if they risked losing their sole authorship.").

Id. at 201-202.

In *Thomson* the court looked at the objective evidence of the playwright Larson's intent. The court examined whether the contributor had decision making authority over what changes were made and what was included in a work. *Id.* at 202. *See also Erickson*, 13 F.3d at 1071-72 (an actor's suggestion of text does not support a claim of co-authorship where the sole author determined whether and where such contributions were included in the work). The way in which the parties bill or credit themselves also can be significant objective indicia of mutual intent. In *Thomson* the court noted, "'billing or credit is * * * a window on the mind of the party who is responsible for giving the billing or the credit.' And a writer's attribution of the work to herself alone is 'persuasive proof . . . that she intended this particular piece to represent her own individual authorship' and is 'prima facie proof that [the] work was not intended to be joint.'" *Thomson*, 203 (quoting *Weissmann v. Freeman*, 868 F.2d 1313, 1320 (2d Cir.1989)). Finally, the court looked to written agreements not only between the parties, but also agreements

---

[16] The court in fact suggests that the test of co-authorship intent will vary depending on the specific factual circumstances. *See* 945 F.2d at 508 (stating that the issue of co-authorship intent "requires less exacting consideration in the context of traditional forms of collaboration, such as between the creators of the words and music of a song"); *see, e.g., Edward B. Marks Music Corp. v. Jerry Vogel Music Co.,* 140 F.2d 266 (2d Cir. 1944), *modified by*, 140 F.2d 268 (2d Cir. 1944) (finding that a lyricist and composer were co-authors where the lyricist wrote the words for a song, intending that someone else would eventually write the music).

with outsiders. If the dominant author consistently insists in agreements with others that he be listed as the sole author, and no agreement with the putative joint author indicates an intent to be listed as co-authors, that can be persuasive evidence of a lack of intent to be co-authors. Previous experience with co-authorship also can prove an author's awareness of the significance of such designation.

The first prong of the *Childress* test, contributing something that is copyrightable, may help to provide ex ante clarity for parties that are collaborating. But what if one person contributes the ideas and the other person acts as the scribe in translating those ideas into fixed form? *Childress* would exclude the idea contributor from claiming joint authorship. In *Gaiman v. McFarlane*, 360 F.3d 644 (7th Cir. 2004), however, the Seventh Circuit provided this illustration:

> One professor has brilliant ideas but can't write; another is an excellent writer, but his ideas are commonplace. So they collaborate on an academic article, one contributing the ideas, which are not copyrightable, and the other the prose envelope, and . . . they sign as coauthors.

Id. at 659. Following the *Childress* test the professor with the brilliant idea would not be a joint author. Judge Posner, a former law professor, wrote the opinion in *Gaiman*. He concluded that the two professors' "intent to be the joint owners of the copyright in the article would be plain, and that should be enough to constitute them joint authors within the meaning of 17 U.S.C. § 201(a)." *Id.*

Notes & Questions

1. Consider the hypothetical of Moe, Larry, and Curly posed at the beginning of this Section. Are they joint authors? What would each need to show in order to be a joint author in any copyrights in the photographs? Is the two-prong *Childress* test biased towards sole authorship? Consider the possibility that the analysis could, and perhaps should, be tailored to the kind of work in question, *e.g.*, photographs, stage plays, or tattoos (coauthored by the tattoo artist and the recipient). Does the statutory definition provide sufficient leeway for consideration of expectations and customs in different creative communities?

2. Re-read the definition of joint authorship. Is the Second Circuit right to impose a requirement of contributing separately copyrightable elements? Is the Second Circuit correct that mutual intent to be co-authors is required?

3. When parties working together on a work are not joint authors, they are left to protect their interests through contract. What is the likelihood of a contract protecting contributors when the typical claim for joint authorship occurs when the parties did not even discuss copyright ownership, let alone any royalty arrangements? Are the standards for determining joint authorship clear enough that the parties will know when they need to use contracts to protect their interests?

2. Works Made for Hire

Community for Creative Nonviolence v. Reid
490 U.S. 730 (1989)

MARSHALL, JUSTICE:

In this case, an artist and the organization that hired him to produce a sculpture contest the ownership of the copyright in that work. To resolve this dispute, we must construe the "work made for hire" provisions of the Copyright Act of 1976 (Act or 1976 Act), 17 U.S.C. §§ 101 and 201(b), and in particular, the provision in § 101, which defines as a "work made for hire" a "work prepared by an employee within the scope of his or her employment" (hereinafter § 101(1)).

Petitioners are the Community for Creative Non-Violence (CCNV), a non-profit unincorporated association dedicated to eliminating homelessness in America, and Mitch Snyder, a member and trustee of CCNV. In the fall of 1985, CCNV decided to participate in the annual Christmastime Pageant of Peace in Washington, D.C., by sponsoring a display to dramatize the plight of the homeless. As the District Court recounted:

> Snyder and fellow CCNV members conceived the idea for the nature of the display: a sculpture of a modern Nativity scene in which, in lieu of the traditional Holy Family, the two adult figures and the infant would appear as contemporary homeless people huddled on a streetside steam grate. The family was to be black (most of the homeless in Washington being black); the figures were to be life-sized, and the steam grate would be positioned atop a platform 'pedestal,' or base, within which special-effects equipment would be enclosed to emit simulated 'steam' through the grid to swirl about the figures. They also settled upon a title for the work—'Third World America'—and a legend for the pedestal: 'and still there is no room at the inn.'

Snyder made inquiries to locate an artist to produce the sculpture. He was referred to respondent James Earl Reid, a Baltimore, Maryland, sculptor. In the course of two telephone calls, Reid agreed to sculpt the three human figures. CCNV agreed to make the steam grate and pedestal for the statue. Reid proposed that the work be cast in bronze, at a total cost of approximately $100,000 and taking six to eight months to complete. Snyder rejected that proposal because CCNV did not have sufficient funds, and because the statue had to be completed by December 12 to be included in the pageant. Reid then suggested, and Snyder agreed, that the sculpture would be made of a material known as "Design Cast 62," a synthetic substance that could meet CCNV's monetary and time constraints, could be tinted to resemble bronze, and could withstand the elements. The parties agreed that the project would cost no more than $15,000, not including Reid's services, which he offered to donate. The parties did not sign a written agreement. Neither party mentioned copyright.

After Reid received an advance of $3,000, he made several sketches of figures in various poses. At Snyder's request, Reid sent CCNV a sketch of a proposed sculpture showing the family in a crèche like setting: the mother seated, cradling a baby in her lap; the father standing behind her, bending over her shoulder to touch the baby's foot. Reid testified that Snyder asked for the sketch to use in raising funds for the sculpture. Snyder testified that it was also for his approval. Reid sought a black family to serve as a model for the sculpture. Upon Snyder's suggestion, Reid visited a family living at CCNV's Washington shelter but decided that only their newly born child was a suitable model. While Reid was in Washington, Snyder took him to see homeless people living on the streets. Snyder pointed out that they tended to recline on steam grates, rather than sit or stand, in order to warm their bodies. From that time on, Reid's sketches contained only reclining figures.

Throughout November and the first two weeks of December 1985, Reid worked exclusively on the statue, assisted at various times by a dozen different people who were paid with funds provided in installments by CCNV. On a number of occasions, CCNV members visited Reid to check on his progress and to coordinate CCNV's construction of the base. CCNV rejected Reid's proposal to use suitcases or shopping bags to hold the family's personal belongings, insisting instead on a shopping cart. Reid and CCNV members did not discuss copyright ownership on any of these visits.

On December 24, 1985, 12 days after the agreed-upon date, Reid delivered the completed statue to Washington. There it was joined to the steam grate and pedestal prepared by CCNV and placed on display near the site of the pageant. Snyder paid Reid the final installment of the $15,000. The statue remained on display for a month. In late January 1986, CCNV members returned it to Reid's studio in Baltimore for minor repairs. Several weeks later, Snyder began making plans to take the statue on a tour of several cities to raise money for the homeless. Reid objected, contending that the Design Cast 62 material was not strong enough to withstand the ambitious itinerary. He urged CCNV to cast the statue in bronze at a cost of $35,000, or to create a master mold at a cost of $5,000. Snyder declined to spend more of CCNV's money on the project.

In March 1986, Snyder asked Reid to return the sculpture. Reid refused. He then filed a certificate of copyright registration for "Third World America" in his name and announced plans to take the sculpture on a more modest tour than the one CCNV had proposed. Snyder, acting in his capacity as CCNV's trustee, immediately filed a competing certificate of copyright registration.

Snyder and CCNV then commenced this action . . . seeking return of the sculpture and a determination of copyright ownership. The District Court granted a preliminary injunction, ordering the sculpture's return. After a 2-day bench trial, the District Court declared that "Third World America" was a "work made for hire" under § 101 of the Copyright Act and that Snyder, as trustee for CCNV, was the exclusive owner of the copyright in the sculpture. 652 F. Supp. at 1457. The court reasoned that Reid had been an "employee" of CCNV within the meaning of § 101(1) because CCNV was the motivating force in the statue's production. Snyder and other CCNV members, the court explained, "conceived the idea of a contemporary Nativity scene to contrast with the national celebration of the season," and "directed enough of [Reid's] effort to assure that, in the end, he had produced what they, not he, wanted." *Id.* at 1456.

The Court of Appeals for the District of Columbia Circuit reversed and remanded, holding that Reid owned the copyright because "Third World America" was not a work for hire. . . .

We granted certiorari to resolve a conflict among the Courts of Appeals over the proper construction of the "work made for hire" provisions of the Act. . . .

II

A.

The Copyright Act of 1976 provides that copyright ownership "vests initially in the author or authors of the work." 17 U.S.C. § 201(a). As a general rule, the author is the party who actually creates the work, that is, the person who translates an idea into a fixed, tangible expression entitled to copyright protection. § 102. The Act carves out an important exception, however, for "works made for hire." If the work is for hire, "the employer or other person for whom the work was prepared is considered the author" and owns the copyright, unless there is a written agreement to the contrary. § 201(b). Classifying a work as "made for hire" determines not only the initial ownership of its copyright, but also the copyright's duration, § 302(c), and the owners' renewal rights, § 304(a), termination rights, § 203(a), and right to import certain goods bearing the copyright, § 601(b)(1). See 1 M. Nimmer & D. Nimmer, *Nimmer on Copyright* § 5.03[A], pp. 5-10 (1988). The contours of the work for hire doctrine therefore carry profound significance for freelance creators—including artists, writers, photographers, designers, composers, and computer programmers—and for the publishing, advertising, music, and other industries which commission their works.

Section 101 of the 1976 Act provides that a work is "for hire" under two sets of circumstances:

> (1) a work prepared by an employee within the scope of his or her employment; or
> (2) a work specially ordered or commissioned for use as a contribution to a collective work, as a part of a motion picture or other audiovisual work, as a translation, as a supplementary work, as a compilation, as an instructional text, as a test, as answer material for a test, or as an atlas, if the parties expressly agree in a written instrument signed by them that the work shall be considered a work made for hire.

Petitioners do not claim that the statue satisfies the terms of § 101(2). Quite clearly, it does not. Sculpture does not fit within any of the nine categories of "specially ordered or commissioned" works enumerated in that subsection, and no written agreement between the parties establishes "Third World America" as a work for hire.

The dispositive inquiry in this case therefore is whether "Third World America" is "a work prepared by an employee within the scope of his or her employment" under § 101(1). The Act does not define these terms. In the absence of such guidance, four interpretations have emerged. The first holds that a work is prepared by an employee whenever the hiring party[6] retains the right to control the product. Petitioners take this view. A second, and closely related, view is that a work is prepared by an employee under § 101(1) when the hiring party has actually wielded control with respect to the creation of a particular work. This approach was formulated by the Court of Appeals for the Second Circuit, *Aldon Accessories Ltd. v. Spiegel, Inc.*, 738 F.2d 548 (2d. Cir. 1984), and adopted by the Fourth Circuit, *Brunswick Beacon, Inc. v. Schock-Hopchas Publishing Co.*, 810 F.2d 410 (4th Cir. 1987), the Seventh Circuit, *Evans Newton, Inc. v. Chicago Systems Software*, 793 F.2d 889 (7th Cir. 1986), and, at times, by petitioners. A third view is that the term "employee" within § 101(1) carries its common-law agency law meaning. This view was endorsed by the Fifth Circuit in *Easter Seal Society for Crippled Children & Adults of Louisiana, Inc. v. Playboy Enterprises*, 815 F.2d 323 (5th Cir. 1987), and by the Court of Appeals below. Finally, respondent and numerous *amici curiae* contend that the term "employee" only refers to "formal, salaried" employees. The Court of Appeals for the Ninth Circuit recently adopted this view. See *Dumas v. Gommerman*, 865 F.2d 1093 (9th Cir. 1989).

The starting point for our interpretation of a statute is always its language. The Act nowhere defines the terms "employee" or "scope of employment." It is, however, well established that "[w]here Congress uses terms that have accumulated settled meaning under . . . the common law, a court must infer, unless the statute otherwise dictates, that Congress means to incorporate the established meaning of these terms." *NLRB v. Amax Coal Co.*, 453 U.S. 322, 329 (1981). In the past, when Congress has used the term "employee" without defining it, we have concluded that Congress intended to describe the conventional master-servant relationship as understood by common-law agency doctrine. See, *e.g.*, *Kelley v. Southern Pacific Co.*, 419 U.S. 318, 322-323 (1974). Nothing in the text of the work for hire provisions indicates that Congress used the words "employee" and "employment" to describe anything other than "'the conventional relation of employer and employee.'" *Kelley, supra*, 419 U.S. at 323. On the contrary, Congress' intent to incorporate the agency law definition is suggested by § 101(1)'s use of the term, "scope of employment," a widely used term of art in agency law. *See Restatement (Second) of Agency* § 228 (1958) (hereinafter *Restatement*).

In past cases of statutory interpretation, when we have concluded that Congress intended terms such as "employee," "employer," and "scope of employment" to be understood in light of agency law, we have relied on the general common law of agency, rather than on the law of any particular State, to give meaning to these terms. See, *e.g.*, *Kelley*, 419 U.S. at 323-324, and n. 5. This practice reflects the fact that "federal statutes are generally intended to have uniform nationwide

---

[6] By "hiring party," we mean to refer to the party who claims ownership of the copyright by virtue of the work for hire doctrine.

application." *Mississippi Band of Choctaw Indians v. Holyfield,* 490 U.S. 30, 43 (1989). Establishment of a federal rule of agency, rather than reliance on state agency law, is particularly appropriate here given the Act's express objective of creating national, uniform copyright law by broadly pre-empting state statutory and common-law copyright regulation. See 17 U.S.C. § 301(a). We thus agree with the Court of Appeals that the term "employee" should be understood in light of the general common law of agency.

In contrast, neither test proposed by petitioners is consistent with the text of the Act. The exclusive focus of the right to control the product test on the relationship between the hiring party and the product clashes with the language of § 101(1), which focuses on the relationship between the hired and hiring parties. The right to control the product test also would distort the meaning of the ensuing subsection, § 101(2). Section 101 plainly creates two distinct ways in which a work can be deemed for hire: one for works prepared by employees, the other for those specially ordered or commissioned works which fall within one of the nine enumerated categories and are the subject of a written agreement. The right to control the product test ignores this dichotomy by transforming into a work for hire under § 101(1) any "specially ordered or commissioned" work that is subject to the supervision and control of the hiring party. Because a party who hires a "specially ordered or commissioned" work by definition has a right to specify the characteristics of the product desired, at the time the commission is accepted, and frequently until it is completed, the right to control the product test would mean that many works that could satisfy § 101(2) would already have been deemed works for hire under § 101(1). Petitioners' interpretation is particularly hard to square with § 101(2)'s enumeration of the nine specific categories of specially ordered or commissioned works eligible to be works for hire, *e.g.,* "a contribution to a collective work," "a part of a motion picture," and "answer material for a test." The unifying feature of these works is that they are usually prepared at the instance, direction, and risk of a publisher or producer. By their very nature, therefore, these types of works would be works by an employee under petitioners' right to control the product test.

The actual control test, articulated by the Second Circuit in *Aldon Accessories,* fares only marginally better when measured against the language and structure of § 101. Under this test, independent contractors who are so controlled and supervised in the creation of a particular work are deemed "employees" under § 101(1). Thus work for hire status under § 101(1) depends on a hiring party's *actual* control of, rather than *right* to control, the product. *Aldon Accessories,* 738 F.2d at 552. Under the actual control test, a work for hire could arise under § 101(2), but not under § 101(1), where a party commissions, but does not actually control, a product which falls into one of the nine enumerated categories. Nonetheless, we agree with the Court of Appeals for the Fifth Circuit that "[t]here is simply no way to milk the 'actual control' test of *Aldon Accessories* from the language of the statute." *Easter Seal Society,* 815 F.2d at 334. Section 101 clearly delineates between works prepared by an employee and commissioned works. Sound though other distinctions might be as a matter of copyright policy, there is no statutory support for an additional dichotomy between commissioned works that are actually controlled and supervised by the hiring party and those that are not.

We therefore conclude that the language and structure of § 101 of the Act do not support either the right to control the product or the actual control approaches.[8] The structure of § 101 indicates that a work for hire can arise through one of two mutually exclusive means, one for employees and one for independent contractors, and ordinary canons of statutory interpretation indicate that the classification of a particular hired party should be made with reference to agency law.

This reading of the undefined statutory terms finds considerable support in the Act's legislative history. The Act, which almost completely revised existing copyright law, was the product of two decades of negotiation by representatives of creators and copyright-using industries, supervised by the Copyright Office and, to a lesser extent, by Congress. . . .

. . . .

In response to objections by book publishers that the preliminary draft bill limited the work for hire doctrine to "employees," the 1964 revision bill expanded the scope of the work for hire classification to reach, for the first time, commissioned works. The bill's language, proposed initially by representatives of the publishing industry, retained the definition of work for hire insofar as it referred to "employees," but added a separate clause covering commissioned works, without regard to the subject matter, "if the parties so agree in writing." Those representing authors objected that the added provision would allow publishers to use their superior bargaining position to force authors to sign work for hire agreements, thereby relinquishing all copyright rights as a condition of getting their books published.

In 1965, the competing interests reached a historic compromise, which was embodied in a joint memorandum submitted to Congress and the Copyright Office, incorporated into the 1965 revision bill, and ultimately enacted in the same form and nearly the same terms 11 years later, as § 101 of the 1976 Act. The compromise retained as subsection (1) the language referring to "a work prepared by an employee within the scope of his employment." However, in exchange for concessions from publishers on provisions relating to the termination of transfer rights, the authors consented to a second subsection which classified four categories of commissioned works as works for hire if the parties expressly so agreed in writing: works for use "as a contribution to a collective work, as a part of a motion picture, as a translation, or as supplementary work." S. 1006, H.R. 4347, H.R. 5680, H.R. 6835, 89th Cong., 1st Sess., § 101 (1965). The interested parties selected these categories because they concluded that these commissioned works,

---

[8] We also reject the suggestion of respondent and *amici* that the § 101(1) term "employee" refers only to formal, salaried employees. . . . The Act does not say "formal" or "salaried" employee, but simply "employee." Moreover, respondent and those *amici* who endorse a formal, salaried employee test do not agree upon the content of this test. Compare, *e.g.,* Brief for Respondent 37 (hired party who is on payroll is an employee within § 101(1)) with Tr. of Oral Arg. 31 (hired party who receives a salary or commissions regularly is an employee within § 101(1)); and Brief for Volunteer Lawyers for the Arts, Inc., et al. as *Amici Curiae* 4 (hired party who receives *a* salary *and* is treated as an employee for Social Security and tax purposes is an employee within § 101(1)). . . .

although not prepared by employees and thus not covered by the first subsection, nevertheless should be treated as works for hire because they were ordinarily prepared "at the instance, direction, and risk of a publisher or producer." [Supplementary Report of the Register of Copyrights on the General Revision of the U.S. Copyright Law: 1965 Revision Bill, 89th Cong., 1st Sess., Copyright Law Revision, pt. 6, p 67 (H.R. Judiciary Comm. Print 1965)]. The Supplementary Report emphasized that only the "four special cases specifically mentioned" could qualify as works made for hire; "[o]ther works made on special order or commission would not come within the definition." *Id.* at 67-68.

. . . .

Thus, the legislative history of the Act is significant for several reasons. First, the enactment of the 1965 compromise with only minor modifications demonstrates that Congress intended to provide two mutually exclusive ways for works to acquire work for hire status: one for employees and the other for independent contractors. Second, the legislative history underscores the clear import of the statutory language: only enumerated categories of commissioned works may be accorded work for hire status. . . . [I]mporting a test based on a hiring party's right to control, or actual control of, a product would unravel the "'carefully worked out compromise aimed at balancing legitimate interests on both sides.'" H.R. Rep. No. 2237, [89th Cong., 2nd Sess., 114 (1966)].

. . . .

Finally, petitioners' construction of the work for hire provisions would impede Congress' paramount goal in revising the 1976 Act of enhancing predictability and certainty of copyright ownership. In a "copyright marketplace," the parties negotiate with an expectation that one of them will own the copyright in the completed work. *Dumas,* 865 F.2d at 1104-1105, n. 18. With that expectation, the parties at the outset can settle on relevant contractual terms, such as the price for the work and the ownership of reproduction rights.

To the extent that petitioners endorse an actual control test, CCNV's construction of the work for hire provisions prevents such planning. Because that test turns on whether the hiring party has closely monitored the production process, the parties would not know until late in the process, if not until the work is completed, whether a work will ultimately fall within § 101(1). Under petitioners' approach, therefore, parties would have to predict in advance whether the hiring party will sufficiently control a given work to make it the author. . . . Moreover, petitioners' interpretation "leaves the door open for hiring parties, who have failed to get a full assignment of copyright rights from independent contractors falling outside the subdivision (2) guidelines, to unilaterally obtain work-made-for-hire rights years after the work has been completed as long as they directed or supervised the work, a standard that is hard not to meet when one is a hiring party." Hamilton, *Commissioned Works as Works Made for Hire Under the 1976 Copyright Act: Misinterpretation and Injustice,* 135 U. Pa. L. Rev. 1281, 1304 (1987).

In sum, we must reject petitioners' argument. Transforming a commissioned work into a work by an employee on the basis of the hiring party's right to control, or actual control of, the work is inconsistent with the language, structure, and legislative history of the work for hire provisions. To determine whether a work is for hire under the Act, a court first should ascertain, using principles of general common law of agency, whether the work was prepared by an employee or an independent contractor. After making this determination, the court can apply the appropriate subsection of § 101.

B

We turn, finally, to an application of § 101 to Reid's production of "Third World America." In determining whether a hired party is an employee under the general common law of agency, we consider the hiring party's right to control the manner and means by which the product is accomplished. Among the other factors relevant to this inquiry are the skill required; the source of the instrumentalities and tools; the location of the work; the duration of the relationship between the parties; whether the hiring party has the right to assign additional projects to the hired party; the extent of the hired party's discretion over when and how long to work; the method of payment; the hired party's role in hiring and paying assistants; whether the work is part of the regular business of the hiring party; whether the hiring party is in business; the provision of employee benefits; and the tax treatment of the hired party. *See Restatement* § 220(2) (setting forth a nonexhaustive list of factors relevant to determining whether a hired party is an employee).[31] No one of these factors is determinative.

Examining the circumstances of this case in light of these factors, we agree with the Court of Appeals that Reid was not an employee of CCNV but an independent contractor. True, CCNV members directed enough of Reid's work to ensure that he produced a sculpture that met their specifications. But the extent of control the hiring party exercises over the details of the product is not dispositive. Indeed, all the other circumstances weigh heavily against finding an employment relationship. Reid is a sculptor, a skilled occupation. Reid supplied his own tools. He worked in his own studio in Baltimore, making daily supervision of his activities from Washington practicably impossible. Reid was retained for less than two months, a relatively short period of time. During and after this time, CCNV had no right to assign additional projects to Reid. Apart from the deadline for completing the sculpture, Reid had absolute freedom to decide when and how long to work. CCNV paid Reid $15,000, a sum dependent on "completion of a specific job, a method by which independent contractors are often compensated." *Holt v. Winpisinger*, 811 F.2d 1532, 1540 (1987). Reid had total discretion in hiring and paying assistants. "Creating sculptures was hardly 'regular business' for CCNV." Indeed, CCNV is not a business at all. Finally, CCNV did not pay payroll or Social Security taxes, provide any employee benefits, or contribute to unemployment insurance or workers' compensation funds.

Because Reid was an independent contractor, whether "Third World America" is a work for hire depends on whether it satisfies the terms of § 101(2). This

---

[31] In determining whether a hired party is an employee under the general common law of agency, we have traditionally looked for guidance to the *Restatement of Agency*.

petitioners concede it cannot do. Thus, CCNV is not the author of "Third World America" by virtue of the work for hire provisions of the Act. However, as the Court of Appeals made clear, CCNV nevertheless may be a joint author of the sculpture if, on remand, the District Court determines that CCNV and Reid prepared the work "with the intention that their contributions be merged into inseparable or interdependent parts of a unitary whole." 17 U.S.C. § 101. . . .

. . . .

Notes & Questions

1. *CCNV* represents an excellent reminder that copyright law is a creature of statute. Many, although not all, of the issues that arise can be resolved through ordinary principles of statutory construction. As the Copyright Act specifies, the employer is the author of any work created by "an employee within the scope of his or her employment." *CCNV* is clear in the order in which to analyze the issues. First one determines whether the creator of a work is an employee or an independent contractor. If the individual is an employee, pursuant to the first statutory definition of a work made for hire, the next inquiry will be to examine whether the work was created "within the scope of his or her employment." 17 U.S.C. § 101. The Supreme Court holds that determining whether someone is an "employee" should be guided by the general principles of the common law of agency. Lower courts have also looked to principles of common law agency to determine whether the work was prepared within the scope of employment. Determining a particular employee's "scope of employment" can be more difficult when employees shift their hours, telecommute, and engage in a variety of practices that blur the lines between "work" and "non-work" activities.

2. If, after applying the general principles of agency, the creator is determined to be an independent contractor, the only way the resulting work can be a work made for hire is if the requirements of the second definition are met: (1) the work is within one of the nine categories specified in the statute, *and* (2) there is a signed writing specifying the work will be a work made for hire. Examine those nine types of works. What do they have in common? Does the justification contained in legislative history that the Supreme Court cites make sense? Are there other types of works that should be eligible for work made for hire status? How about sound recordings or computer programs? Would it make sense to provide that any work that is "prepared at the instance, direction, and risk of the hiring party" can be a work made for hire under the second definition?

In 1999 Congress added sound recordings to the list in §101(2) as a "technical amendment" contained in an omnibus bill amending the patent, trademark, and telecommunications laws. Intellectual Property and Communications Omnibus Reform Act of 1999, §1011(d), *as enacted by* Pub. L. 106-113, §1000(a)(9), 113 Stat. 1501A-521, -544 (1999). Following expressions of outrage by recording artists who accused the recording industry of staging a copyright grab, Congress repealed the amendment less than a year later. Work Made for Hire and Copyright Corrections Act of 2000, Pub. L. 106-379, §2, 114 Stat. 1444, 1444 (2000). Consider the context in which many commercial sound recordings are created. Should those works be works made for hire? How should copyright law treat the contributions of the sound engineers, studio musicians and back-up singers? Are those individuals joint authors? If they are not employees, the absence of "sound recordings" from the list in subsection (2) of the definition of work made for hire creates pressure for the transfer, or at least license, of any intellectual property rights those individuals might have.

3. As the Supreme Court identifies, establishing rules that provide for certainty is one of the goals in this area of copyright law. Certainty allows parties to contract around the default rules provided by the statute. But, even without certainty, contracts can provide for alternatives. For example, a hiring party can require a creator to sign a contract that states that the works created by the creator are works made for hire, and, in the alternative, if the works are determined not to be works made for hire, that the artist assigns the copyrights in the works to the hiring party. This type of agreement is common in the recording industry for sound recordings. What difference does it make whether the hiring party obtains the copyright through the work made for hire designation or through assignment? In *CCNV* the Court identifies three consequences of a work being a work made for hire: initial vesting of copyright ownership, duration of copyright protection, and the ability to terminate transfers. It is this last difference that may have the most consequence. As subsection D, below, explores, the Copyright Act provides certain authors with an ability to terminate transfers of copyright interests 35 years after the date of the transfer agreement. The statute expressly provides that individuals who create works made for hire have no such termination rights.

4. The 1976 Act was created in large part through negotiation and compromise between the various interests groups most affected by copyright law. At the time that meant lobby groups for creators and large distributors such as publishers, the recording industry, and the motion picture companies. See Jessica D. Litman, *Copyright, Compromise, and Legislative History*, 72 Cornell L. Rev. 857 (1987) ("As often as not, the reason a specific provision was deemed to be good was that industry representatives had agreed on it."). With the exception of the authors' lobbies, the interests of individuals were not well represented in these negotiations because, in large measure, the rules of copyright did not affect the everyday practices of individuals. Today, however, with individuals possessing the power to be creators and world-wide distributors of copies of works, a copyright law written to address a different era may not fit well.

5. The Court did not address whether CCNV and Mr. Reid were joint authors. Applying the test from *Thomson*, are they joint authors? On remand the parties settled their dispute with an agreement to credit Mr. Reid as the sole author of the work and to give CCNV sole ownership of the sculpture. The settlement agreement also gave both parties the right to reproduce the work in two dimensions, while Mr. Reid obtained the sole right to make three-dimensional reproductions. *See CCNV v. Reid,* 1991 Copr. L. Dec. (CCH) ¶26,753 (D.D.C. Jan. 7, 1991).

B. Formalities

1. Registration & Deposit

Copyright protection today does not require any registration with the Copyright Office. There are, however, benefits to obtaining a registration. First, registration is required in order to commence a lawsuit.[*] 17 U.S.C. § 411. Second, timely registration permits a copyright owner to seek statutory damages, 17 U.S.C. § 412, which can be helpful if actual damages are small or hard to prove, and can be a powerful factor in obtaining a prompt settlement from an alleged infringer. Third, timely registration makes a copyright owner eligible for an award of attorney's fees in a successful infringement lawsuit. *Id.* For these benefits, "timely" means registering the work before the infringement commences.[**] Fourth, registration within five years of publication carries with it a presumption of "the validity of the copyright and of the facts stated in the certificate." 17 U.S.C. § 410(c). Fifth, registration permits the copyright owner to seek the assistance of the U.S. Customs and Border Protection in stopping the importation of infringing goods. 19 C.F.R. § 133.31. Sixth, registration creates a publicly accessible record of copyright ownership providing contact information for those interested in obtaining a license to use the work.

Registration is a straightforward process, and can even be accomplished online, with several different application forms depending on the nature of the work. The current filing fee is $35 per application if filed electronically. A registrant is required to also deposit copies of the work with the Copyright Office. The Copyright Office provides many informative circulars to assist in completing the different forms and determining the deposit requirements. For more information see http://www.copyright.gov.

The Copyright Office reviews the application to determine copyrightability. This review is far less scrutinizing than the examination process for patent applications. Even if the Copyright Office denies registration, the applicant can still bring a claim for copyright infringement, but she must notify the Register of Copyrights, who may then become a party to the action regarding the issue of registrability. 17 U.S.C. § 411(a).

2. Notice

A proper notice of copyright contains three elements: (1) the symbol ©, the word "Copyright," or the abbreviation "Copr."; (2) the year of first publication of the work; and (3) the name of the owner of copyright in the work. 17 U.S.C. § 401.

---

[*] The only exceptions to this rule are for (1) claims of moral-rights violations under §106A made by authors of "works of visual art," and (2) infringement claims for works that are not "United States works." 17 U.S.C. § 411. Both of these terms of art are specifically defined in the statute. *See* 17 U.S.C. § 101.

[**] In the case of a published work, if the infringement commences soon after publication, the copyright owner has three months following publication to file for registration. 17 U.S.C. § 412.

Today, notice of copyright is not required in order for a work to be protected by the federal Copyright Act. Prior to the 1976 Act, publication of a work with proper notice was the point at which federal protection began. Unpublished works were previously protected by state copyright law, with publication extinguishing the rights created by state law. Thus, *publication* of a work *without notice* divested an author his state law copyright and, at the same time, failed to achieve federal copyright protection. A work published without proper copyright notice was, in short, in the public domain.

The 1976 Act changed the point in time at which federal copyright protection begins: federal copyright protection begins upon fixation. The 1976 Act also eliminated all state law protection for fixed works. 17 U.S.C. § 301. At the same time, publication without notice of the work continued to constitute a divesture of the copyright. However, the 1976 Act softened this potential loss of protection with some limited provisions to permit "cure" of the omission of notice. *See* 17 U.S.C. § 405. In 1989, when the United States joined the major international copyright treaty, the Berne Convention, it eliminated the notice requirement completely, as the treaty required. *See* Berne Convention Implementation Act of 1988, Pub. L. 100-567, 102 Stat. 2854. Thus, beginning March 1, 1989, published copies of copyrighted works no longer require copyright notices to maintain federal copyright protection.

Even though notice is not required to obtain or maintain copyright protection, there are benefits to using a proper copyright notice. First, such notice provides information to those who encounter a copy of the work. The notice not only indicates that the copyright owner is aware of his or her rights, but also provides the name of the copyright owner. This information can deter infringement and facilitate licensing. Second, the Copyright Act provides that if notice of copyright appears on published copies "to which a defendant in a copyright infringement suit had access, then no weight shall be given to . . . a defendant's interposition of a defense based on innocent infringement in mitigation of actual or statutory damages . . . ." 17 U.S.C. § 401(d).[*]

3. Transfers of Copyright Ownership and the Potential for Recording Transfers

To be valid, a transfer of copyright ownership must be in writing. 17 U.S.C. § 204. This statute of frauds provision applies not only to full transfers of all ownership rights, but also to exclusive licenses and "hypothecation"—essentially the pledge of a copyright as collateral for a loan. 17 U.S.C. § 101 (definition of "transfer"). Given the writing requirement found in § 204, if an agreement is oral, the most that can be conveyed is a non-exclusive license. *See Effects Associates, Inc. v. Cohen*, 908 F.2d 55 (9th Cir. 1990).

---

[*] As you will learn later in this Chapter, copyright infringement is a strict liability offense. A defendant's assertion of "innocent infringement" can only serve, potentially, to reduce a damages award, it cannot absolve the defendant of liability for infringement.

Transfers of copyright may be recorded with the Copyright Office. While the transfer itself must be in writing to be valid, recordation of the transfer is not required. Recording the transfer with the Copyright Office, however, can be a wise decision. In addition to facilitating identification of the current copyright owners by potential licensees, the statute provides that a proper recordation of the document puts "all persons on constructive notice of the facts stated in the recorded document." 17 U.S.C. § 205. This constructive notice thus also serves to establish priority of rights if an unscrupulous copyright owner attempts to transfer rights to more than one buyer.

C. DURATION

The basic rule for the duration of copyright is simple to state: copyright protection lasts for the life of the author plus 70 years. 17 U.S.C. § 302. Unfortunately, the real duration of any particular work is complicated by a several different factors. First, if the work is a work made for hire, a duration based on the "life" of the "author" could potentially last forever; for example a corporation under the work for hire doctrine is the "author" of works created by employees within the scope of their employment, yet corporations can exist for centuries. The statute therefore has a separate duration rule for works made for hire: copyright protection lasts for 95 years from publication of the work, or 120 years from creation, whichever expires first. *Id.* That duration also applies to anonymous works and works published under a pseudonym. *Id.*

The duration rules stated in the preceding paragraph apply to works created after January 1, 1978, the effective date of the 1976 Copyright Act. For works created earlier, the rules are far more complicated, relying on a basic structure of a 28 year initial term following publication of the work and a 67 year renewal term, for a total of 95 years of possible protection. But even that is a simplified statement of those duration rules.

The first Copyright Act in the United States granted an initial 14 year term with a 14 year renewal term, for a possible total of 28 years. Over the years Congress has seen fit to lengthen the term, first by extending the length of the renewal term and then the initial term, and then in the 1976 Copyright Act moving to what is referred to as a "unitary term"—a single term with no renewal required. Copyright duration today is thus extremely different from the duration with which the Framers were familiar.

The most recent lengthening of the copyright term occurred in 1998 with the passage of the Sony Bono Copyright Term Extension Act (CTEA), which added 20 years to all of the terms of copyright resulting in the durations described in the first paragraph of this subsection. Eric Eldred mounted an ultimately unsuccessful challenge to the constitutionality of the CTEA. Mr. Eldred ran a website devoted to literary works that were in the public domain. Prior to the CTEA, on December 31 of each year, the copyright would expire in a year's worth of works[*] and Mr.

---

[*] By statute, copyright terms run to the end of the calendar year in which they would expire regardless of the actual date of publication or the actual date of death of the author. 17 U.S.C. § 305.

Eldred would make those works available through his website. Passage of the CTEA meant that he would have to wait until 2018, 20 years after passage of the CTEA, for the copyright in more works to expire. Assisted by Professor Lawrence Lessig and the Harvard Law School's Berkman Center for Internet and Society, Mr. Eldred challenged the retroactive extension of the duration of copyright for works that were already in existence at the time of passage of the CTEA.

Applying the deferential standard of rational-basis review, the Supreme Court refused to find that Congress had exceeded it constitutional authority in enacting the CTEA. *Eldred v. Ashcroft*, 537 U.S. 186 (2003). Justifications for the CTEA's adoption included a 1993 European Union directive that established a copyright term in Europe of life plus 70 and, importantly, denied anything beyond life plus 50 to authors from countries whose copyright terms did not match the life-plus-70 EU choice. *Id.* at 205. The Court also noted that "Congress passed the CTEA in light of demographic, economic, and technological changes, and rationally credited projections that longer terms would encourage copyright holders to invest in the restoration and public distribution of their works." The Court then turned to a more sweeping argument about the policy goal copyright is intended to serve:

> . . . [P]etitioners contend that the CTEA's extension of existing copyrights does not "promote the Progress of Science" as contemplated by the pre-ambular language of the Copyright Clause. Art. I, § 8, cl. 8. To sustain this objection, petitioners do not argue that the Clause's preamble is an independently enforceable limit on Congress' power. Rather, they maintain that the preambular language identifies the sole end to which Congress may legislate; accordingly, they conclude, the meaning of "limited Times" must be "determined in light of that specified end." The CTEA's extension of existing copyrights categorically fails to "promote the Progress of Science," petitioners argue, because it does not stimulate the creation of new works but merely adds value to works already created.
>
> As petitioners point out, we have described the Copyright Clause as "both a grant of power and a limitation," *Graham v. John Deere Co. of Kansas City,* 383 U.S. 1, 5 (1966), and have said that "[t]he primary objective of copyright" is "[t]o promote the Progress of Science," *Feist,* 499 U.S. at 349. The "constitutional command," we have recognized, is that Congress, to the extent it enacts copyright laws at all, create a "system" that "promote[s] the Progress of Science." *Graham,* 383 U.S. at 6.
>
> We have also stressed, however, that it is generally for Congress, not the courts, to decide how best to pursue the Copyright Clause's objectives. The justifications we earlier set out for Congress' enactment of the CTEA, provide a rational basis for the conclusion that the CTEA "promote[s] the Progress of Science."

Id. at 211-213. The use of "rational-basis" review made the CTEA's constitutionality a foregone conclusion. The lower court had rejected Eldred's argument that the application of the CTEA to existing copyrights should be subjected to heightened scrutiny under the First Amendment because of the burdens on speech, stating that copyright laws were "categorically immune from challenges under the First Amendment." 239 F.3d at 375. The Supreme Court, for its part, concluded that the lower court's statement was too broad, but also held that when "Congress has not altered the traditional contours of copyright protection, further First Amendment scrutiny is unnecessary." *Eldred v. Ashcroft*, 537 U.S. at 221. The Court did not specify in *Eldred* what "the traditional contours of copyright protection" entail. More recently, however, in *Golan v. Holder*, the Court—in an opinion by Justice Ginsburg, who also authored *Eldred*—clarified the meaning of the phrase the Court used in *Eldred*:

> We then described the "traditional contours" of copyright protection, *i.e.*, the "idea/expression dichotomy" and the "fair use" defense. Both are recognized in our jurisprudence as "built-in First Amendment accommodations." *Eldred*, 537 U.S. at 219; *see Harper & Row*, 471 U.S. at 560 (First Amendment protections are "embodied in the Copyright Act's distinction between copyrightable expression and uncopyrightable facts and ideas," and in the "latitude for scholarship and comment" safeguarded by the fair use defense).

132 S. Ct. 873, 890 (2012) (rejecting a challenge to the constitutionality of § 514, which in 1994 granted copyright protection to certain preexisting works of Berne Convention member countries that were protected in their country of origin but had lacked protection in the U.S.).

Notes & Questions

1. The Supreme Court also rejected plaintiff's argument that extending the duration of existing copyrights violated the command in the Constitution that copyrights be granted for "limited times." So long as the extended term had an end, the Court concluded the term was limited. The full title of the Act challenged in *Eldred* is the Sony Bono Copyright Term Extension Act. Sonny Bono was a congressman from California who had been a part of the singing duo "Sonny and Cher" in the 1970's. The Act was named for him when he met an untimely death in a skiing accident shortly before the passage of law. The CTEA's legislative history indicates that he "wanted the term of copyright protection to last forever." 144 Cong. Rec. H9952 (daily ed. Oct. 7, 1998) (Statement of Rep. Mary Bono). At what point would a copyright term violate the Constitution's "limited times" requirement? What if, for example, Congress amended the Act to provide for a 300-year copyright term?

2. The internet has brought about the possibility for capturing revenue on the waning side of the demand curve, sometimes referred to as the "long tail." Internet publication allows publishers, for example, to keep all of their titles "in print" without actually having to print and store copies in inventory until a customer orders (and pays for) a copy. Should the increased efficiency in capturing long tail revenues affect the duration of copyright protection? If so, in which direction does this point—longer or shorter terms?

3. The elimination of formalities described earlier in this Section also had a dramatic effect on the quantity of creative works entering the public domain. When renewal was required, Copyright Office records indicate that only a small fraction of eligible copyrights were renewed. Christopher Sprigman, *Reform(al-iz)ing Copyright,* 57 Stan. L. Rev. 485 (2003). Those works whose copyrights were not renewed entered the public domain after only 28 years. The elimination of the notice requirement also has had a significant effect on the quantity of creative works that remain subject to copyright protection. Adding those effects to an increased duration provides a fuller picture of the lack of growth of works in the public domain.

4. The ease with which copyright protection attaches, the long duration of copyright protection, and the lack of any requirement for registration, or updating ownership information for works previously registered with the Copyright Office, creates many works that have been referred to as "orphan works," *i.e.,* works whose copyright owners cannot be located. The Copyright Office proposed legislation to permit certain uses of orphan works after a "diligent search" yielded no contactable copyright owner, but that legislation did not become law. Google's project to scan millions of books resulted in a class-action lawsuit, the proposed settlement of which would have provided some ability for Google to use orphan works with a lump sum set aside for copyright owners that resurfaced. That settlement was eventually rejected by the court. *Authors Guild v. Google, Inc.*, 770 F. Supp. 2d 666 (S.D.N.Y. 2011). Should we be concerned about the problem of orphan works? What might be the appropriate way to address the problem? The Patent Act requires the payment of maintenance fees 3.5, 7.5, and 11.5 years after grant to obtain the full term of protection. 35 U.S.C. § 41(b). Would periodic maintenance fees make sense for copyrights?

5. The extension of copyright term the CTEA confers results in works taking longer to enter the public domain. But, can Congress take works that have entered the public domain and grant a copyright in those works? That was the question raised by a challenge to a provision that "restored" copyright protection for foreign authors whose copyrights had entered the public domain—so far as U.S. law was concerned—due to a failure to comply with the U.S. formality requirements (for example, renewal registration and the inclusion of a copyright notice on published copies of a work). Once again the Supreme Court rejected the plaintiffs' challenge: "Neither the Copyright and Patent Clause nor the First Amendment, we hold, makes the public domain, in any and all cases, a territory that works may never exit." *Golan v. Holder*, 132 S. Ct. 873, 885 (2012). Given the Court's recent rejections of constitutional challenges to copyright legislation, what formal limits are there to what Congress may do with copyright term under the Progress Clause? What practical limits are there?

D. RENEWALS AND TERMINATIONS OF TRANSFERS

Prior to the 1976 Act, the copyright term was divided into two parts: an initial 28 year term and a renewal term. In order to obtain protection beyond the initial term, copyright owners had to file for renewal with the Copyright Office. Because the renewal term was considered to be a "new estate" that was contingent on the proper filing of a renewal, an assignment of the "copyright" did not include an assignment of the renewal term. Thus, if an author assigned his "copyright" to a publisher, at the end of the initial term the author could file for renewal and regain ownership of the copyright for the period of the renewal term despite the previous assignment. Publishers did not like this outcome and turned to contract to avoid it: author assignments were drafted to include both the initial term and the renewal term as well. The Supreme Court upheld these assignments. *See Fred Fisher Music Co. v. M. Witmark & Sons*, 318 U.S. 643, 651 (1943). But the Copyright Act has always provided a set of forced inheritance rules for the renewal term of a copyright. If the author was not living at the end of the initial term, the copyright in the renewal term vested in the widow or widower of the author, children and lineal descendants, or, if none of those existed, in the executor of the author's estate. Thus, even if the author assigned the renewal term, the author possessed only a future interest, contingent upon the vesting of the renewal in him, the author. For the renewal term to vest in him, he would need to be alive at the time of vesting. Death of the author before vesting resulted in the renewal term vesting in someone else. Some publishers required spouses of authors to sign agreements as well, and these were also upheld by the courts.

Congress created the renewal term out of concern for the valuation problem inherent in creative works, a valuation problem that is particularly acute prior to the commercial exploitation of a work. If it turned out that the work was far more successful than the initial price paid to the author would indicate, the vesting rules of the renewal term created an opportunity for authors to renegotiate their bargains. Judicial enforcement of renewal-term assignments lessened the benefits of that part of copyright law for authors.

When Congress decided to change the duration rules to a unitary term in the 1976 Copyright Act, it sought to provide a more robust way to permit authors and their heirs to revisit the valuation question by creating a provision for terminating transfers.

The Copyright Act provides that transfers by authors of a copyright interest that are executed on or after January 1, 1978, can be terminated 35 years after the date of the transfer agreement. 17 U.S.C. § 203. Unlike the prior Act's renewal vesting rules, this termination of transfer right cannot be circumvented by contract. The Copyright Act specifically nullifies such agreements: "Termination of the grant may be effected notwithstanding any agreement to the contrary, including an agreement to make a will or to make any future grant." 17 U.S.C. § 203(a)(5). As a result, even if an author's copyright transfer agreement states that the transfer lasts "in perpetuity," the author can still terminate it 35 years later. The statute specifies that this termination right applies only to agreements made by authors, not subsequent transfers by assignees. Therefore, if an author assigns her copyright to a publisher, and the publisher grants an exclusive license to a movie producer, neither the publisher nor the author can terminate the movie maker's license. The statute also expressly excludes transfer agreements concerning works made for hire. 17 U.S.C. § 203(a).

In order to effect a termination, an author or her widower, children, etc., must comply with specific notice requirements contained in the statute and further delineated in Copyright Office regulations. 37 C.F.R. § 201.10 (2009). Once such a termination occurs, all rights that had been transferred revert back to the author, or to the statutorily defined class of successors. If an author or her heirs fails to take action to terminate the transfer in the proscribed timeframe, then the transfer continues pursuant to its terms.

Given that the § 203 termination applies to agreements entered into in 1978 or later, and the author must wait 35 years before terminating (1978 + 35 = 2013), litigation concerning § 203 terminations is just beginning.[*] For example, Victor Willis, a member of the band *The Village People*, successfully terminated the assignment of his copyright interest in 33 musical works, including the hit *YMCA*. *Scorpio Music S.A. v. Willis*, 2012 WL 1598043, at *2 (S.D. Cal. May 7, 2012) (holding that a joint author who separately transfers his copyright interest may unilaterally terminate that grant).

IV. THE RIGHTS OF A COPYRIGHT OWNER

Once a work that is eligible for copyright protection is fixed in a tangible medium of expression and the owner of the copyright is identified, the next logical question to ask is: what rights does a copyright owner have? The Copyright Act, § 106 grants to copyright owners the following exclusive rights:

(1) to reproduce the copyrighted work in copies or phonorecords [*the reproduction right*];
(2) to prepare derivative works based upon the copyrighted work [*the derivative work right*];
(3) to distribute copies or phonorecords of the copyrighted work to the public by sale or other transfer of ownership, or by rental, lease, or lending [*the public distribution right*];
(4) in the case of literary, musical, dramatic, and choreographic works, pantomimes, and motion pictures and other audiovisual works, to perform the copyrighted work publicly [*the public performance right*];
(5) in the case of literary, musical, dramatic, and choreographic works, pantomimes, and pictorial, graphic, or sculptural works, including the individual images of a motion picture or other audiovisual work, to display the copyrighted work publicly [*the public display right*];
(6) in the case of sound recordings, to perform the copyrighted work publicly by means of a digital audio transmission [*the digital public performance right for sound recordings*].

The Act specifies that all of these rights are "[s]ubject to sections 107 through 121." 17 U.S.C. § 106. Those sections provide limitations on these rights. Thus to

---

[*] There is another termination provision, under §304, that involves pre-1978 agreements and concerns the lengthening of copyright duration. That provision has been litigated for quite some time. *See, e.g.*, *Mills Music, Inc. v. Snyder*, 469 U.S. 153 (1985).

truly understand the rights of the copyright owner, one must read both the exclusive rights provision, § 106, and all of the limitations, §§ 107 to 121. This Section focuses on the exclusive rights and explores the basic elements a copyright owner must prove to demonstrate infringement. Some of the limitations on a copyright owner's rights are mentioned in this Section. Section V focuses on the important limitation of fair use, codified in § 107 of the Act, and the relationship between the rights granted to a copyright owner and the fair use limitation. However, a full exploration of the rights and limitations is left for an advanced class in copyright law.

An important aspect of copyright infringement to note at the outset is the strict liability nature of the offense. The statute does not require any level of scienter for civil liability for infringement. The Copyright Act does provide for criminal penalties for certain types of infringement. To be criminally liable, the defendant must engage in his or her conduct "willfully." 17 U.S.C. § 506.

A. Establishing a Prima Facie Case of Infringement

A prima facie case of copyright infringement is typically said to consist of proof of two basic elements: (1) ownership of a valid copyright, and (2) an unauthorized exercise of a § 106 right. The first element is really two separate items, introduced in Sections I through III of this Chapter: ownership and validity. Proof of the second element takes different shapes depending on the right allegedly violated by the defendant. For example, if the right allegedly violated is the public performance right, typically the prima facie case will revolve around whether the performance engaged in was sufficiently public, whereas if the right allegedly violated is the distribution right, proof that the defendant transferred of ownership of tangible objects, i.e., "copies," will be the primary issue.

Peters v. West

692 F.3d 629 (7th Cir. 2012)

WOOD, JUDGE:

    . . . .

I

Vince P describes himself in the complaint as an up-and-coming hip-hop artist and songwriter. In 2006, as he was beginning his career in music, he wrote and recorded a song entitled *Stronger,* which is about the competitive—indeed cutthroat—nature of the hip-hop and rap world. For clarity, we refer to this as *Stronger (VP).* Vince P's music apparently captured the attention of someone at Interscope Records; that person told him that the company would devote "substantial resources" to producing Vince P's inaugural album, but only if he could procure the services of a good executive producer.

His search led him to John Monopoly, a well-known producer and—importantly for our purposes—a close friend and business manager to Kanye West. Vince P sent several of his songs to Monopoly, who liked what he heard enough to schedule a meeting. On November 12, 2006, Vince P and Monopoly met at the latter's home in Chicago, where Vince P played several of his recordings, including *Stronger (VP)*. At the conclusion of their meeting, Vince P left a CD of some of his songs—including *Stronger (VP)*—with Monopoly. Eventually, Monopoly agreed to be Vince P's executive producer, so long as Interscope Records was willing to fund the recording project. That funding, however, fell through, and so the project stalled.

In July 2007, less than a year after the November 2006 meeting between Vince P and Monopoly, West released his own single titled *Stronger*. (We call this *Stronger (KW)*.) It was a huge hit. The song earned the #1 spot in several Billboard charts, the single sold over three million copies, and it eventually earned West a Grammy for Best Rap Solo Performance. Vince P, however, was not among its fans. He noticed what he thought were several infringing similarities between his 2006 song and West's more recent release. Vince P also saw that Monopoly was listed as a manager on the notes to West's album GRADUATION, on which *Stronger (KW)* appears. Vince P attempted to contact West, but he was rebuffed by West's representatives, and so he turned to the federal courts. . . . Vince P sued West in the U.S. District Court for the Northern District of Illinois. That court dismissed Vince P's complaint under Federal Rule of Civil Procedure 12(b)(6), and he now appeals.

II

We review the district court's order granting West's motion to dismiss *de novo*. *Justice v. Town of Cicero,* 577 F.3d 768, 771 (7th Cir. 2009). We "construe the complaint in the light most favorable to the plaintiff," and we therefore draw all plausible inferences in Vince P's favor. *Tamayo v. Blagojevich,* 526 F.3d 1074, 1081 (7th Cir. 2008). As a practical matter for the present case, this means that we assume as true all of Vince P's allegations regarding Monopoly's early access to Vince P's song and his claims about the close relationship between Monopoly and Kanye West. We review *de novo* the district court's determinations regarding the similarity between the two songs as well as its ultimate conclusion of noninfringement. *Intervest Constr. Inc. v. Canterbury Estate Homes, Inc.,* 554 F.3d 914, 919–20 (11th Cir. 2008).

Vince P's complaint contains only one claim: his allegation that *Stronger (KW)* infringes his valid copyright in *Stronger (VP)*. Proving infringement of a copyright owner's exclusive right under 17 U.S.C. § 106(1) (the reproduction right) requires proof of "(1) ownership of a valid copyright, and (2) copying of constituent elements of the work that are original." *Feist Publ'ns, Inc. v. Rural Tel. Serv. Co.,* 499 U.S. 340, 361 (1991).

A

Copyright "registration made before or within five years after the first publication of the work shall constitute prima facie evidence of the validity of the copyright." 17 U.S.C. § 410(c). Vince P applied for copyright registration in *Stronger (VP)* on March 28, 2010, which is well within the statutory five-year window beginning in 2006. West appropriately does not challenge Vince P's copyright registration, nor does he otherwise question the validity of Vince P's copyright owner-ship in *Stronger (VP)*. Vince P has thus made a *prima facie* showing of his owner-ship in the whole of the lyrics to his song.

Nevertheless, whether the parts of that song that West allegedly copied are, on their own, entitled to copyright protection is a separate question. If the copied parts are not, on their own, protectable expression, then there can be no claim for infringement of the reproduction right.

B

Satisfied that Vince P has shown valid copyright ownership, we turn our atten-tion to the question of copying. The standard for copying is surprisingly muddled. Where direct evidence, such as an admission of copying, is not available (as is typ-ically the case) a plaintiff may prove copying by showing that the defendant had the opportunity to copy the original (often called "access") and that the two works are "substantially similar," thus permitting an inference that the defendant actu-ally did copy the original. The various efforts to define these two key concepts, however, have unfortunately had the unintended effect of obscuring rather than clarifying the issues. This court has said that substantial similarity can be shown by evidence of "actual copying" and "improper appropriation." *Incredible Techs., Inc. v. Virtual Techs., Inc.,* 400 F.3d 1007, 1011 (7th Cir. 2005). Thus, we permit copying to be proven by evidence of access, actual copying, and improper appro-priation. . . .

Other circuits have also had trouble expressing the test with any clarity. The First Circuit, for example, finds copying where the plaintiff has shown substantial similarity, access, and probative similarity. *T-Peg, Inc. v. Vermont Timber Works, Inc.,* 459 F.3d 97, 111-12 (1st Cir. 2006). The formulation found in the Second Circuit requires proof of improper appropriation and actual copying; the latter is shown by proving access and probative similarity. *Jorgensen v. Epic/Sony Records,* 351 F.3d 46, 51 (2d Cir. 2003). The Eleventh Circuit takes still a different ap-proach, requiring either "striking similarity" or access and merely probative sim-ilarity. *Peter Letterese & Assocs. v. World Institute of Scientology Enterprises,* 533 F.3d 1287, 1300-01 (11th Cir. 2008); see also *La Resolana Architects, PA v. Reno, Inc.,* 555 F.3d 1171, 1178-79 (10th Cir. 2009) (applying same test). See also *Universal Furniture Int'l, Inc. v. Collezione Europa USA, Inc.,* 618 F.3d 417, 435 (4th Cir. 2010) (access, intrinsic similarity, and extrinsic similarity); *Frye v. YMCA Camp Kitaki,* 617 F.3d 1005, 1008 (8th Cir. 2010) (same); *Armour v. Knowles,* 512 F.3d 147, 152 (5th Cir. 2007) (factual copying and substantial similarity, where factual copying is shown either by striking similarity, or access and probative similarity); *Bridgeport Music, Inc. v. UMG Recordings, Inc.,* 585 F.3d 267, 274 (6th Cir. 2009) (access and substantial similarity, or "a high degree of similarity"); *Kay Berry, Inc. v. Taylor Gifts, Inc.,* 421 F.3d 199, 207-08 (3d Cir. 2005) (access, copying, and im-proper appropriation).

Despite all of this confusing nomenclature, this strikes us as a "pseudo-con-flict": despite the conflicting and confusing verbiage, the outcomes do not appear to differ. Fundamentally, proving the basic tort of infringement simply requires the plaintiff to show that the defendant had an actual opportunity to copy the original (this is because independent creation is a defense to copyright infringement), and that the two works share enough unique features to give rise to a breach of the duty not to copy another's work. Our analysis will follow this structure.

i

We begin with the question of opportunity. We already know (for purposes of this Rule 12(b)(6) inquiry) that Monopoly had access to Vince P's song and that Monopoly has a close relationship with West. These allegations are more than enough to support an inference that West had an opportunity to copy *Stronger (VP)*. Not only did Monopoly actually hear Vince P's song: he also twice received copies of it, once before their November 2006 meeting and again on a CD during that meeting. Furthermore, Monopoly is credited with acting as West's manager on the GRADUATION album. This evidence of close collaboration between West and Monopoly suggests that Monopoly may have passed Vince P's song on to West during the production of the album, and that West could have used that song in crafting his own hit single. Viewed together, these allegations, taken as true, suggest that Monopoly and West had ample access to *Stronger (VP),* and that this access gave West an opportunity to copy the song.

ii

But even assuming that West had the opportunity to copy the lyrics to *Stronger (VP),* the question remains whether the complaint plausibly alleges that he actually did so. Before we can answer this question, we must confront the differences among the circuits about the relation between proof of access and evidence of similarity. Some circuits follow an "inverse ratio" rule, under which the strength of proof of similarity varies inversely with the proof of access (*i.e.,* strong proof of access allows for only weak proof of similarity, and vice versa). *Three Boys Music Corp. v. Bolton,* 212 F.3d 477, 485 (9th Cir. 2000). Other courts have rejected the inverse-ratio rule. . . . *Arc Music Corp. v. Lee,* 296 F.2d 186 (2d Cir. 1961).

This court's rule has not been so explicit, although we have occasionally endorsed something that comes close to this inverse approach. In *Selle v. Gibb,* 741 F.2d 896, 903 n. 4 (7th Cir. 1984), we held that "degree of similarity required to establish an inference of access [should be] in an inverse ratio to the quantum of direct evidence adduced to establish access." More recently, we noted that "similarity that is so close as to be highly unlikely to have been an accident of independent creation *is* evidence of access." *Ty, Inc. v. GMA Accessories, Inc.,* 132 F.3d 1167, 1170 (7th Cir. 1997) (emphasis in original); but see *id.* (noting that such similarity cannot be evidence of access when both are copies of something in the public domain). Thus, in both *Selle* and *GMA Accessories,* we noted that evidence that two works are very similar can suggest that the alleged infringer had access to the original.

Notably, however, we have never endorsed the other side of the inverse relation: the idea that a "high degree of access" justifies a "lower standard of proof" for similarity. *Three Boys Music,* 212 F.3d at 485. As we explained above, evidence of access is required because independent creation is a defense to copyright infringement, and so a plaintiff must show that the defendant had an opportunity to copy her original work. This issue is independent of the question whether an alleged infringer breached his duty not to copy another's work. See *GMA Accessories,* 132 F.3d at 1170. Once a plaintiff establishes that a defendant could have copied her work, she must separately prove—regardless of how good or restricted the opportunity was—that the allegedly infringing work is indeed a copy of her original. In this case, Vince P has adequately pleaded that West had an opportunity to copy his song, but that does not help him prove similarity. Vince P must show that West actually copied his song by pointing to similarities between the two works. We are not persuaded that the similarities alleged by Vince P rise to the level of copyright infringement.

. . . [W]e give the two "hooks," which provide the backdrop to the discussion that follows:

Stronger (VP) [Hook]
What don't kill me make me stronger
The more I blow up the more you wronger
You copied my CD you can feel my hunger
The wait is over couldn't wait no longer

Stronger (KW) [Hook]
N-N-N-now th-th-that don't kill me
Can only make me stronger
I need you to hurry up now
Cause I can't wait much longer
I know I got to be right now
Cause I can't get much wronger
Man I've been waitin' all night now
That's how long I've been on ya.

Three features in particular of *Stronger (KW)* form the basis of Vince P's argument that West's song infringes his. First, he notes that the hooks of both songs derive from the same common maxim and that they implement similar rhyme schemes (stronger, wronger, etc.). Second, he points to the songs' shared title, which . . . derives from Nietzsche. Finally, he notes that both songs contain "incongruous" references to the British model Kate Moss, who is not usually featured in rap or hip-hop lyrics.

Nietzsche's phrase "what does not kill me, makes me stronger" comes from *Twilight of the Idols* (1888). Although the fact that both songs quote from a 19th century German philosopher might, at first blush, seem to be an unusual coincidence, West correctly notes that the aphorism has been repeatedly invoked in song lyrics over the past century. Notably, an even more recent popular song—one that held the top spot in the Billboard Hot 100 chart at about the same time as oral argument in this case—also shares this key feature with both West's and Vince P's songs. See Gary Trust, *Kelly Clarkson Returns to Hot 100 Peak, The Wanted Hit Top 10,* BILLBOARD, *available at* http://www.billboard.com/#/news/kelly-clarkson-returns-to-hot-100-peak-the-1006316152.story (last visited July 13, 2012) (discussing Stronger (What Doesn't Kill You), performed by Kelly Clarkson). The ubiquity of this common saying, together with its repeated use in other songs, suggests that West's title and lyric do not infringe on Vince P's song.

Next, Vince P claims that West's song infringes on the rhyme pattern he uses in the hook. But this argument misapprehends the nature of Vince P's rights. Copyright protects actual expression, not methods of expression. 17 U.S.C. § 102(b); *Baker v. Selden,* 101 U.S. 99, 104 (1879). Just as a photographer cannot claim copyright in the use of a particular aperture and exposure setting on a given lens, no poet can claim copyright protection in the form of a sonnet or a limerick. Similarly, Vince P cannot claim copyright over a tercet. See *Steele v. Turner Broad. Sys. Inc.,* 646 F. Supp. 2d 185, 192 (D. Mass. 2009) ("A common rhyme scheme or structure does not qualify as original expression protectable under federal copyright law."). (We note for the sake of precision that, although Vince P seems to be claiming protection over a "triple rhyme," a closer examination of his lyrics reveals that he actually uses a soft quadruple monorhyme (stronger, wronger, hunger, longer). West, by contrast, uses two soft four-line schemes (stronger and longer, and wronger and "on ya.").) Nor are we persuaded that the particular rhymes of stronger, longer, and wronger qualify for copyright protection. See *Prunte v. Universal Music Grp.,* 699 F. Supp. 2d. 15, 29 (D.D.C. 2010) (no protection for rhyming "-ill" sound).

We turn then to the songs' references to Kate Moss, a well-known supermodel. In Vince P's song, the line is "Trying to get a model chick like Kate Moss"; in West's it is "You could be my black Kate Moss tonight." Vince P argues that his lyrical reference to Kate Moss "as a paragon of female beauty" is so unique as to "undermine[ ] the possibility of coincidental similarity." We cannot go that far. In the first place, the lines are entirely different. In the second, analogizing to models as a shorthand for beauty is, for better or for worse, commonplace in our society. The particular selection of Kate Moss, who is very famous in her own right, adds little to the creative choice. And finally, the name alone cannot constitute protectable expression. *Feist,* 499 U.S. at 347.

Even viewing all of these elements in combination, we conclude that Vince P has not plausibly alleged that *Stronger (KW)* infringes on *Stronger (VP)*. Vince P's theory is that the combination of the songs' similar hooks, their shared title, and their references to Kate Moss would permit a finding of infringement. But, as we have discussed, in the end we see only two songs that rhyme similar words, draw from a commonplace maxim, and analogize feminine beauty to a specific successful model. These songs are separated by much more than "small cosmetic differences," *JCW,* 482 F.3d at 916; rather, they share only small cosmetic similarities. This means that Vince P's claim for copyright infringement fails as a matter of law. The judgement of the district court is *Affirmed.*

Kay Berry, Inc. v. Taylor Gifts, Inc.
421 F.3d 199 (3d Cir. 2005)

Van Antwerpen, Judge:

Before us is an appeal from an order granting summary judgment in favor of the defendant-Appellees on a copyright infringement claim. Appellant Kay Berry, Inc. ("Kay Berry") claims that Appellees Taylor Gifts, Inc. ("Taylor") and Bandwagon, Inc. ("Bandwagon") infringed its copyright on its sculptural work—a garden rock cast with a poem found in the public domain. The United States District Court for the Western District of Pennsylvania granted summary judgment . . . . We will reverse.

I. Factual Background and Procedural History

Kay Berry designs, manufactures, markets and sells "Garden Accent Rocks," which it describes as decorative, cement-cast, outdoor sculptures typically resembling rocks or stones, inscribed with writings. . . .

One of Kay Berry's best-selling Garden Accent Rocks is Sculpture No. 646, a rectangular object having a stone-like appearance and a verse inscribed on the face. The verse appears in five lines, inscribed in a right-leaning font with the first letter of each word capitalized . . . .

If Tears Could Build A Stairway, And Memories A Lane, I'd Walk Right Up To Heaven And Bring You Home Again.

[*Eds. Note*: image from Kay Berry's website]

During 2003, Bandwagon began supplying to Taylor, and Taylor began marketing and selling, a "Memory Stone," which was similar to Kay Berry's Sculpture No. 646. Like Sculpture No. 646, the Memory Stone was a rectangular object with a stone-like appearance featuring the exact same verse that appears on Sculpture No. 646. The Memory Stone's verse was also laid out in the same five-line format, each word also began with a capital letter, and the entire verse also appeared in a right-leaning font.

Kay Berry sued Taylor and Bandwagon for copyright infringement . . . .

. . . .

B. *Sculpture No. 646 is entitled to Copyright Protection*

. . . [T]he District Court concluded that Sculpture No. 646 lacked any protectible configuration or design. We disagree.

"To qualify for copyright protection, a work must be original to the author * * * mean[ing] only that the work was independently created by the author (as opposed to copied from other works), and that it possesses at least some minimal degree of creativity." *Feist Publ'ns, Inc. v. Rural Tel. Serv. Co.,* 499 U.S. 340, 345 (1991). A sculptural work's creativity derives from the combination of texture, color, size, and shape, as well as the particular verse inscribed and the way the verse is presented. It means nothing that these elements may not be individually entitled to protection; "all creative works draw on the common wellspring that is the public domain. In this pool are not only elemental 'raw materials,' like colors, letters, descriptive facts, and the catalogue of standard geometric forms, but also earlier works of art that, due to the passage of time or for other reasons, are no longer copyright protected." *Tufenkian Import/Export Ventures, Inc. v. Einstein Moomjy, Inc.,* 338 F.3d 127, 132 (2d Cir. 2003). When an author combines these elements and adds his or her own imaginative spark, creation occurs, and the author is entitled to protection for the result. *Feist,* 499 U.S. at 345. This is true even when the author contributes only a minimal amount of creativity. *Id.* at 348 (factual compilations may be copyrightable when the author "chooses which facts to include, in what order to place them, and how to arrange the collected data so that they may be used effectively by readers."); *see also Reader's Digest Ass'n, Inc. v. Conservative Digest, Inc.,* 821 F.2d 800, 806 (D.C. Cir. 1987) (although no element of magazine cover—ordinary lines, typefaces, and colors—is entitled to copyright protection, the distinctive arrangement is entitled to protection as a graphic work).

Here, Kay Berry claims that it selected an inspirational poem from the public domain, adapted that poem to make it visually and rhythmically appealing, and then cast it on its own sculptural work. For these reasons, as well as those set forth above, we conclude that this quantum of creativity is sufficient to qualify for copyright protection. *See Feist,* 499 U.S. at 348.

C. *Kay Berry's Copyright Registration Does Not Extend to an Idea*

We next turn our attention to the second part of the infringement inquiry— whether Appellees improperly copied Sculpture No. 646, and specifically, whether the expression Kay Berry seeks to protect has merged with an unprotectible idea. Copying refers to the act of infringing any of the exclusive rights that accrue to the owner of a valid copyright, as set forth at 17 U.S.C. § 106, "including the rights to distribute and reproduce copyrighted material." *Ford Motor Co. v. Summit Motor Prods.,* 930 F.2d 277, 291 (3d Cir. 1991). It may be demonstrated by showing that the defendant had access to the copyrighted work and that the original and allegedly infringing works share substantial similarities. *Id.; see also Dam Things from Denmark v. Russ Berrie & Co.,* 290 F.3d 548, 561 (3d Cir. 2002); *Whelan Assocs., Inc. v. Jaslow Dental Lab.,* 797 F.2d 1222, 1231–32 (3d Cir. 1986).

"Substantial similarity," in turn, is further broken down into two considerations: "(1) whether the defendant copied from the plaintiff's work and (2) whether the copying, if proven, went so far as to constitute an improper appropriation." *Atari, Inc. v. North American Philips Consumer Elecs. Corp.,* 672 F.2d 607, 614 (7th Cir. 1982). "First, the fact-finder must decide whether there is sufficient similarity between the two works in question to conclude that the alleged infringer used the copyrighted work in making his own." *Whelan Assocs.,* 797 F.2d at 1232. A showing of substantial similarity in this sense, coupled with evidence that the infringing author had access to the original work, permits a fact-finder to infer that the infringing work is not itself original, but rather is based on the original. At this stage of the inquiry, expert testimony is permissible to help reveal the similarities that a lay person might not ordinarily perceive. *Id.* Direct evidence of copying or an admission by the infringing author would satisfy this test as well. *Dam Things from Denmark,* 290 F.3d at 562.

"Not all copying, however, is copyright infringement." *Feist,* 499 U.S. at 361. Even if actual copying is proven, "the fact-finder must decide without the aid of expert testimony, but with the perspective of the 'lay observer,' whether the copying was 'illicit,' or 'an unlawful appropriation' of the copyrighted work." *Whelan Assocs.,* 797 F.2d at 1232. The focus in this second step is "whether the substantial similarities relate to protectible material." *Dam Things from Denmark,* 290 F.3d at 562. "Phrased in an alternative fashion, it must be shown that copying went so far as to constitute improper appropriation, the test being the response of the ordinary lay person." *Universal Athletic Sales Co. v. Salkeld,* 511 F.2d 904, 907 (3d Cir. 1975) (citing *Arnstein v. Porter,* 154 F.2d 464 (2d Cir. 1946)). Part of this inquiry involves distinguishing between the author's expression and the idea or theme that he or she seeks to convey or explore.

It is a fundamental premise of copyright law that an author can protect only the expression of an idea, but not the idea itself. *See* 17 U.S.C. § 102(b); *see also Baker v. Selden,* 101 U.S. 99, 103–04 (1880); *Franklin Mint Corp. v. Nat'l Wildlife Art Exch., Inc.,* 575 F.2d 62, 64–65 (3d Cir.1978). Thus, an author may base his work on the same inspiration as that of an earlier work, but he may not "'copy the copy.'" *Franklin Mint,* 575 F.2d at 65 (quoting *Bleistein v. Donaldson Lithographing Co.,* 188 U.S. 239, 249 (1903)). When determining whether two works are sub-stantially similar, a fact-finder must determine whether the later work is similar because it appropriates the unique expressions of the original author, or merely because it contains elements that would be expected when two works express the same idea or explore the same theme.

Nearly every work involves a blend of idea and expression. Because an author can only demonstrate substantial similarity by referencing those aspects of his work that embody his creative contribution, he will have a more difficult time proving infringement if his work contains only a minimal amount of original ex-pression. *Universal Athletic Sales,* 511 F.2d at 908 ("[B]etween the extremes of conceded creativity and independent efforts amounting to no more than the triv-ial, the test of appropriation necessarily varies."). As one court has explained:

> [A] copyright on a work which bears practically a photographic likeness to the natural article ... is likely to prove a relatively weak copyright. This is not to say that, as a matter of law, infringement of such a copyright cannot be inferred from mere similarity of appearance, but only that the plaintiff's burden will be that much more difficult to sustain because of the intrinsic similarities of the copyrighted and accused works.

First Am. Artificial Flowers, Inc. v. Joseph Markovits, Inc., 342 F. Supp. 178, 186 (S.D.N.Y. 1972). The First Circuit has endorsed this view, explaining that when there is only a limited number of ways to express an idea "the burden of proof is heavy on the plaintiff who may have to show 'near identity' between the works at issue." *Concrete Mach. Co. v. Classic Lawn Ornaments, Inc.,* 843 F.2d 600, 606–07 (1st Cir. 1988) (citing *Sid & Marty Krofft Television v. McDonald's Corp.,* 562 F.2d 1157, 1167 (9th Cir. 1977)).

In some instances, there may come a point when an author's expression be-comes indistinguishable from the idea he seeks to convey, such that the two merge. *Educ. Testing Servs. v. Katzman,* 793 F.2d 533, 539 (3d Cir. 1986). In these circum-stances, no protection is available for the expression; otherwise, the copyright owner could effectively acquire a monopoly on the underlying art or the idea itself. *Id.* Merger is rare, however, and is generally found in works with a utilitarian func-tion, *id.* This Court has never found an instance in which a completely aesthetic expression merged into an idea. For instance, in *Masquerade Novelty, Inc. v. Unique Indus.,* 912 F.2d 663, 666 (3d Cir. 1990), this Court dealt with a case in which the plaintiff sued the defendant for infringing its copyright on humorous nose masks designed to resemble the noses, snouts, and beaks of different animals. Although the issue of substantial similarity was not before the Court, we recognized that "copyrights protect only expressions of ideas and not ideas themselves," and explained:

> By holding that Masquerade's nose masks are copyrightable, we do not intimate that it has the exclusive right to make nose masks representing pig, elephant and parrot noses. On remand, it will be Masquerade's burden to show that Unique's nose masks incorporate copies, in the copyright law sense, of Masquerade's sculptures, rather than sculptures that derive their similarity to Masquerade's sculptures merely from the commonality of the animal subjects both represent.

Masquerade Novelty, 912 F.2d at 671–72 (internal citation omitted).

Here, as in *Masquerade Novelty,* the protectible originality of the allegedly infringed work is to be found, if at all, solely in its appearance. Kay Berry claims that Appellees infringed its copyright not by using the same public domain poem, or inscribing a rock with text, but by copying the specific combination of elements it employed to give Sculpture No. 646 its unique look. Although the evidentiary burden upon it is high, for the reasons we have discussed, we conclude that Kay Berry is entitled to the opportunity to demonstrate that the Memory Stone is neither a unique creation, nor the unavoidable expression of a common idea, but rather an impermissible copy of Sculpture No. 646. We will therefore reverse and remand to the District Court.

Notes & Questions

1. The Seventh Circuit provides a good articulation of the starting point for understanding the elements of a prima facie case of infringement:

> Fundamentally, proving the basic tort of infringement simply requires the plaintiff to show that the defendant had an actual opportunity to copy the original (this is because independent creation is a defense to copyright infringement), and that the two works share enough unique features to give rise to a breach of the duty not to copy another's work.

Peters v. West, 692 F.3d 629, 633-34 (7th Cir. 2012).

The first element, sometimes referred to as "actual copying" or "copying-in-fact," is important because independent creation is a defense in copyright: if the defendant independently created her work without copying from the plaintiff's work, that does not constitute infringement. In that way, copyright protection is like trade secret protection and unlike patent law. With an extremely narrow exception that is not important here, patent law grants the patentee an exclusive right to the claimed invention even against others who independently create the same invention without copying from the patentee.[*] The copying-in-fact element of an infringement case puts some burden on the copyright owner to prove that it is more likely than not that the defendant copied from the plaintiff's copyrighted

---

[*] The exception is a tightly drawn prior-commercial-use defense in 35 U.S.C. § 273.

work. Showing access and a sufficient level of similarity between the two works can meet this initial burden. At the same time, once the plaintiff has demonstrated adequate levels of access and substantial similarity, the burden shifts to the defendant to provide convincing evidence of independent creation. That is why the Seventh Circuit refers to the "defense" of independent creation. Typically a defendant attempts to explain that the similarity is attributable to something other than copying. A defendant might, for example, demonstrate that the points of similarity are common in a particular genre. The ultimate burden of persuasion, however, remains on the plaintiff.

2. A characteristic that copyright shares with patent is its strict liability nature. To prevail on a claim of infringement the plaintiff need only demonstrate that the defendant engaged in an act encompassed within one of the rights granted by the Copyright Act. The copying-in-fact requirement itself does not require a mental state either; to commit infringement the defendant does not need to be actively aware that she is copying from a copyrighted work. In some instances a court may refer to this as "subconscious infringement." *See Three Boys Music Corp. v. Bolton*, 212 F.3d 477, 483 (9th Cir. 2000). A prominent example of subconscious infringement is *ABKCO Music, Inc v. Harrisongs Music, Ltd.*, 722 F.2d 988 (2d Cir. 1983), where the Second Circuit affirmed a jury's verdict that former Beatle George Harrison, in writing the song "My Sweet Lord," subconsciously copied The Chiffons' "He's So Fine." Harrison admitted hearing "He's So Fine" in 1963, when it was number one on the Billboard charts in the United States for five weeks and one of the top 30 hits in England for seven weeks. *See id.* at 998. Given that all of an author's life experiences inform the creative process, how should copyright draw the line between permissible inspiration and impermissible copying?

3. Now focus on the second of the fundamental elements that the Seventh Circuit identifies: that the defendant's work shares "enough unique features to give rise to a breach of the duty not to copy another's work." Often courts refer to this as demonstrating "improper appropriation," and it boils down to demonstrating that the defendant's work contains enough of the copyrightable expression from the plaintiff's work to be wrongful. In general, the copying must rise above a "de minimis" threshold. The legal maxim "de minimis non curat lex " refers to the concept that "the law does not concern itself with trifles." If the copying is de minimis it is considered so "trivial" as to fall below the quantitative threshold of substantial similarity and thus the copying is not actionable. *Compare Gottlieb Development LLC v. Paramount Pictures Corp.*, 590 F. Supp. 2d 625 (S.D.N.Y. 2008) (finding the appearance of a copyrighted graphic on a pinball machine in the background of a movie scene to be de minimis), *with Ringgold v. Black Entm't T.V. Inc.*, 126 F.3d 70, 75 (2d Cir. 1997) (finding the appearance of a copyrighted poster in scenes from an episode of a television series to exceed the de minimis threshold). How much is too little for the law to be concerned? What benchmarks or touchstones might make sense? One test often employed is whether the "average audience would recognize the appropriation." *VMG SalSoul, LLC v. Ciccone*, 824 F.3d 871, 874 (9th Cir. 2016) (rejecting an infringement claim concerning allegations of copying a .23 second sample of a sound recording). What are the benefits of an

"average audience" vantage point? What are the shortcoming? And what role should expertise play in assessing whether the amount of appropriation is wrongful?

4. Articulating the standards for how much material is too much only provides the framework for making the determination. The determination itself is based on the works at issue, and, if the copyrighted work takes elements from the public domain, those public domain works can also be important. The improper appropriation inquiry examines only the protected material from the copyrighted work that is included in the allegedly infringing work. The unprotected elements are "filtered out" and not included. The unprotected elements vary depending on the nature of the copyrighted work. For example, in a literary work or a play, stock characters or standard literary devices, referred to as "scenes a faire," are not protected. In computer programs the aspects of the program that are dictated by external constraints, such as the operating system on which the program is designed to operate, are not protectable and therefore are not considered when determining whether the defendant has copied a sufficient amount of the plaintiff's work to be adjudged an infringer. *See Computer Associates Int'l v. Altai Inc.*, 982 F.2d 693 (2d Cir. 1992). How is a court supposed to engage in the filtering inquiry? Given the wide range of works subject to copyright protection, when might it be appropriate to use expert witnesses?

5. When may the issue of improper appropriation be decided as a matter of law? In *Peters v. West*, the plaintiff's complaint is dismissed under Fed. R. Civ. Pro. 12(b)(6) because he has not even stated a plausible claim of infringement; the works are too dissimilar. In *Kay Berry, Inc. v. Taylor*, the defendant moved for summary judgment, but the Third Circuit holds that plaintiff has sufficient evidence to defeat that motion. The court notes that plaintiff's burden is high—why is that the case? What, exactly, does the court state the plaintiff will need to prove? On remand, could the plaintiff use an expert witness to meet its high burden? Or can expert testimony only be used to prove copying-in-fact?

B. THE § 106 RIGHTS

The Copyright Act specifies that the rights granted to a copyright owner in § 106 are divisible; they may be conveyed and owned separately. 17 U.S.C. § 201(d). This subsection provides a brief introduction to each of the different § 106 rights.

1. The Reproduction Right

The right to reproduce the work in copies granted to copyright owners in § 106(1) is the most fundamental right of a copyright owner and the right that most people understand intuitively. The reach of this exclusive right is governed by the standards for proving "copying" that are introduced in *Peters v. West* and *Kay Berry, Inc. v. Taylor*, set forth in subsection A. Additionally, the statutory definition of "copies," described in Section I of this Chapter in the context of what constitutes a fixation of a work, applies to determining infringement of the reproduction right as well. The Copyright Act defines copies as: "material objects, other than phonorecords, in which a work is fixed by any method now known or later developed, and from which the work can be perceived, reproduced, or otherwise communicated, either directly or with the aid of a machine or device." 17 U.S.C. § 101. "Copies" therefore can include reproductions on paper, film, wood, plastic, clay, etc. The exclusion of "phonorecords" from the definition of copies is merely meant to set apart the tangible objects in which *sounds* are embodied. Section 106(1) grants the copyright owner the right to control the reproduction of the work in both "copies" and "phonorecords." Thus, a recording of a recitation of David Sedaris' collection of essays, *When You Are Engulfed in Flames,* would implicate the § 106(1) right of the copyright owner in that work.

Copies also include digital files, no matter what format. When someone burns a copy to a disc, loads a copy onto a portable storage device, or sends a file attached to an email, he is making copies as defined by the Copyright Act. If what the file contains is someone else's copyrighted work, the mere act of reproduction constitutes an exercise of the right granted to the copyright owner in § 106(1). In the case of digital reproduction, it is unlikely that the answer to the question whether any particular act of digital copying constitutes infringement will turn on the subtleties of the two prongs of a prima facie case of infringement explored in subsection A. Verbatim copying of the entire original is plain to see. Also, as described at the outset of this Section, copyright infringement is a strict liability offense, so an individual's lack of awareness that the attached file is copyrighted, or that copying it might constitute a violation of a legal right, will not matter in determining whether infringement occurred. Instead, the determination of infringement in the digital context will likely rest on application of the limitations and exceptions that might apply, as well as potential arguments about implied licenses.

There are a few specific limitations that permit copying of entire works. For example, a broadcasting entity can make "ephemeral copies" of a work that is being broadcast, so long as the ephemeral copies are destroyed within a set time period. *See* 17 U.S.C. § 112. Webcasting companies can also make certain copies. *Id.* § 114. Individuals, however, are not granted a specific right to make digital copies.[*] As we explore in Section V, the fair use limitation encompasses some of the digital copying that individuals engage in, but fair use's full reach is uncertain.

Another example of a specific limitation on the reproduction right involves a compulsory license. Section 115 of the Copyright Act permits someone other than the copyright owner to record and distribute "mechanical copies" of musical works that have been previously recorded and distributed, commonly known as "covers." This compulsory license makes it possible for new artists to release their own recordings of old "hits," even over the objection of the owner of the copyright in the musical work. The license requires payment to the copyright owner of the underlying musical work set at a rate that is periodically adjusted. Currently that rate is 9.1 cents, or 1.75 cents per minute of playing time or fraction

---

[*] Section 1008, enacted as part of the Audio Home Recording Act, provides an exemption from infringement liability for copies made using "digital audio recording devices" or "analog" devices. However, the definition of those specific terms excludes general-purpose computers. *See* 17 U.S.C. § 1001; *see also Recording Industry Association of American v. Diamond Multimedia Sys.*, 180 F.3d 1072 (9th Cir. 1999).

thereof, whichever amount is larger, for each copy that is distributed. 37 C.F.R. § 255.3(m). Digital distributions are included in the § 115 compulsory license scheme. *Id.* at § 255.5.

A final example is an express limit that Congress enacted when it added the definition of "computer programs" to § 101 the Copyright Act. The limit, set forth in § 117, provides that "it is not an infringement for the owner of a copy of a computer program to make or authorize the making of another copy or adaptation of that computer program" when necessary to "the utilization of the computer program" or "for archival purposes only." 17 U.S.C. § 117.

2. The Right to Prepare Derivative Works

In addition to the right to control the reproduction of a copyrighted work in copies or phonorecords, the Copyright Act grants to a copyright owner the exclusive right to "prepare derivative works based on the copyrighted work." 17 U.S.C. § 106(2). The statute defines a derivative work as

> a work based upon one or more preexisting works, such as a translation, musical arrangement, dramatization, fictionalization, motion picture version, sound recording, art reproduction, abridgment, condensation, or any other form in which a work may be recast, transformed, or adapted. A work consisting of editorial revisions, annotations, elaborations, or other modifications which, as a whole, represent an original work of authorship, is a "derivative work."

17 U.S.C. § 101. Many of the concrete examples given in the definition are familiar: a translation of a work from one language to another, or a movie based on a book. It is the final phrase of the first sentence that widens the possibilities of what may constitute a derivative work: "any other form in which a work may be recast, transformed, or adapted." *Id.* The breadth of works potentially encompassed by that phrase has led courts to find many items to be "derivative works." For example, the following works have been held to be derivative works:

- a quiz book based upon a television series, *Castle Rock Entertainment v. Carol Publishing Group, Inc.*, 150 F.3d 132 (2d Cir. 1998);
- a decorative tile containing a purchased authorized copy of the work epoxied to the tile and encased in a resin, *Mirage Editions, Inc. v. Albuquerque A.R.T. Company*, 856 F.2d 1341 (9th Cir. 1988), *but see, Lee v. A.R.T. Company*, 125 F.3d 580 (7th Cir. 1997) (rejecting the analysis in *Mirage*);
- a computer program that creates new levels of the plaintiff's copyrighted video games by containing instructions that cause the computer program to retrieve images contained in authorized copies of the plaintiff's copyrighted software, *Micro Star v. FormGen Inc.*, 154 F.3d 1107 (9th Cir. 1998);
- a decorative plate containing a hand-painted image from the motion picture "The Wizard of Oz," *Gracen v. Bradford Exchange,* 698 F.2d 300, 305 (7th Cir. 1983);
- a catalog photograph of a child's toy train, *Schrock v. Learning Curve International, Inc.*, 586 F.3d 513 (7th Cir. 2009);

- three-dimensional inflatable costumes based upon cartoon characters, such as the Pillsbury Doughboy and Geoffrey the Giraffe, *Entertainment Research Group, Inc. v. Genesis Creative Group, Inc.,* 122 F.3d 1211 (9th Cir. 1997); and
- a guitar modeled on Prince's copyrighted name symbol, *Pickett v. Prince,* 207 F.3d 402 (7th Cir. 2000).

Each case alleging infringement of the derivative work right turns on the particular facts of that case. The right to prepare derivative works is not dependent on a subsequent sale of that derivative work. The mere preparation is what gives rise to potential liability. In many cases the question whether, in creating the alleged derivative work, the defendant has copied "enough" of the copyrightable elements of plaintiffs work must be decided by the court. Often the courts will turn to the familiar language of "substantial similarity" to make the determination. *Cf.* Pamela Samuelson, *The Quest for a Sound Conception of Copyright's Derivative Work Right,* 101 Geo. L.J. 1505 (2013) (suggesting the statutory language and legislative history support using the examples given in the definition to appropriately confine the derivative work right).

When might it matter whether the defendant has engaged in creating a reproduction versus a derivative work? Sometimes a license agreement may permit one activity but not the other. In such a case, determining the line between reproduction and derivative work may matter. Additionally, the Copyright Act has special rules for derivative works created pursuant to a transfer that is subsequently terminated under the termination of transfer provisions. For a discussion of termination of transfer provisions, see Section III.D. There are a few other places in the Copyright Act where derivative works are treated differently than reproductions. *See, e.g.,* 17 U.S.C. §§ 104A(d)(3), 304(a)(4)(A).

Notes & Questions

1. Is it wise to grant a copyright owner the right to prepare derivative works in addition to the right to reproduce the work? What might justify granting such a right? From a utilitarian perspective does the derivative work right make sense? How about from a natural rights perspective? Does a personhood justification require a copyright owner be able to control the creation of derivative works?

2. The Copyright Act treats sound recordings differently in a variety of ways. One important distinction is that for a copyright in a sound recording neither the reproduction right nor the derivative work right encompasses a sound-*alike* recording. Specifically, the statute provides that "[t]he exclusive rights of the owner of copyright in a sound recording under clauses (2) and (3) of section 106 do not extend to the making or duplication of another sound recording that consists entirely of an independent fixation of other sounds, even though such sounds imitate or simulate those in the copyrighted sound recording." 17 U.S.C. § 114(b). Why treat sound recordings differently? Beyond sound-alike recordings, what other types of "substantially similar" copies of sound recordings might be within the rights of a copyright owner?

Be careful when thinking through copyright issues related to sound recordings or musical works. There often are two different copyrights involved: one in the musical work (the notes and lyrics) and one in the sound recording. While a sound-alike recording does not infringe upon the sound recording copyright, it may nonetheless infringe upon the musical work copyright.

3. While a copyright owner is given the exclusive right to prepare derivative works based on the copyrighted work, derivative works themselves are eligible for copyright protection as new works. Specifically, the Copyright Act provides:

> The subject matter of copyright as specified by section 102 includes compilations and derivative works, but protection for a work employing preexisting material in which copyright subsists does not extend to any part of the work in which such material has been used unlawfully.

17 U.S.C. § 103. Courts have interpreted the second half of this sentence to mean that if a new work reproduces too much of a copyrighted work, such that it qualifies as a derivative work, and such use is unauthorized, the creator of that new work is not entitled to copyright protection. *See, e.g., Pickett v. Prince*, 207 F.3d 402 (7th Cir. 2000). Courts have concluded that allowing the second creator to obtain a copyright would give that creator "considerable power to interfere with the creation of subsequent derivative works from the same underlying work." *Gracen v. Bradford Exchange*, 698 F.2d 300, 305 (7th Cir. 1983). Note, however, that if the use is authorized, either by the copyright owner of the underlying work or by a limitation contained in the Copyright Act such as fair use, the creator of the derivative work is entitled to a copyright. *See Schrock v. Learning Curve International, Inc.*, 586 F.3d 513 (7th Cir. 2009) (finding photographer of copyrighted toys entitled to a copyright in photos taken with authorization of the copyright owner of the toys).

4. Recall that in patent law an inventor who builds on a pre-existing invention and creates an improvement that is novel and non-obvious is entitled to obtain a patent on that improvement, even without the authorization of the underlying patent owner. This is referred to as a "blocking patent." One who builds on a pre-existing copyrighted work and creates a new work that contains original material is not only *not* entitled to a copyright but is, instead, an infringer. Why is the rule different for copyrighted works? What might explain this different rule choice made by Congress? *See* Mark Lemley, *The Economics of Improvements in Intellectual Property Law*, 75 Tex. L. Rev. 989 (1997).

5. In many of the cases that you have read, the courts cite one important treatise on copyright law—*Nimmer on Copyright*—as authority for one or another legal proposition. Often courts will rely on this treatise as authoritative, even though it takes some positions that are open to fair debate, and are actively debated, among copyright scholars. Is it appropriate for a treatise, written by a private individual, to have such sway? *See* Anne Bartow, *The Hegemony of the Copyright Treatise*, 73 U. Cin. L. Rev. 1 (2004). In 2014 the American Law Institute announced a project to draft the first *Restatement of Copyright Law*.

3. The Right to Distribute Copies

Section 106(3) grants to the copyright owner the exclusive right to distribute copies of the copyrighted work to the public. Is it necessary for the copyright owner to possess both a right to reproduce the work in copies and a separate right to control the distribution of copies of the work? Imagine what would occur if the copyright owner did not possess the distribution right. Sometimes it is hard to obtain evidence of the act of copying, but the act of selling the copies is more likely to have witnesses. However, violation of the distribution right does not require a sale of copy; free give-aways are encompassed within the right as well.

A legal doctrine known as "the first sale doctrine," codified in 17 U.S.C. § 109, significantly limits the § 106(3) exclusive right to distribute copies of the copyrighted work. The equivalent doctrine in patent law is known as "exhaustion," and the underlying premise is identical. Once the copyright owner has parted with title to a "copy" (a material object in which the copyrighted work is embodied), the copyright owner cannot restrict the subsequent transfer of that copy. Specifically, § 109 provides:

> Notwithstanding the provisions of section 106(3), the owner of a particular copy or phonorecord lawfully made under this title, or any person authorized by such owner, is entitled, without the authority of the copyright owner, to sell or otherwise dispose of the possession of that copy or phonorecord.

17 U.S.C. § 109. The first sale doctrine permits everything from sales of used DVDs at garage sales and church rummage sales, to the longstanding tradition of used book stores. Section 109 makes these activities clearly lawful. *See Kirtsaeng v. John Wiley & Sons, Inc.*, 113 S. Ct. 1351 (2013) (discussing the longstanding, fundamental nature of the first sale doctrine). Used book sales are something that publishers take into account when pricing their books. For example, law school textbooks are often resold by students, and the publishers do not get a percentage of the resale price. Therefore the price of the initial sale is set at a level that takes into account the fact that new copies compete against used copies of the same title, which publishers think of as lost revenue. This is one of the many factors leading to law school casebooks costing $200 or more.

In the digital age, however, the first sale doctrine does not provide purchasers with the same level of freedom to "resell" the "lawfully made" copies that they purchased. Unless a user who purchases digital copies of works is selling her computer, she cannot pass her digital copies on to a third person without making reproductions of the works. The first sale doctrine limits *only* the distribution right, not the reproduction right. For an excellent discussion of the policies that animate the first sale doctrine and challenges faced by digital networks, *see* Anthony Reese, *The First Sale Doctrine in the Era of Digital Networks*, 44 B.C. L. Rev. 577 (2003).

In 1984 Congress narrowed the scope of the first sale doctrine for sound recordings and in 1990 it added computer programs to that limitation. For these types of works, for-profit "rental lease or lending" of copies or phonorecords expressly are not within the first sale doctrine. 17 U.S.C. § 109(b). This explains why any store can rent movies without fear of copyright infringement liability, but those stores cannot rent computer programs or music CDs.[*]

Notes & Questions

1. Prior to the 1976 Act the first sale doctrine was not contained in the statute, but rather was a judicial doctrine. In *Bobbs-Merrill Co. v. Straus*, 210 U.S. 339 (1908), the Supreme Court confronted a book publisher that had inserted the following notice on all copies of the novel *The Castaway*:

> The price of this book at retail is one dollar net. No dealer is licensed to sell it at a less price, and a sale at a less price will be treated as an infringement of the copyright.

When the defendant sold copies of the book for under $1, the publisher sued for copyright infringement. The Court held that the copyright statute did not give the publishers the right "to impose, by notice, . . . a limitation at which the book shall be sold at retail by future purchasers." *Id.* at 350. Fast forward to today: Many casebook publishers routinely send free "review copies" to professors for consideration in adoptions. These free copies are sometimes sold by professors to used book buyers, who then sell the copies to students at a lower price than the suggested retail price, cutting into the publishers' market for sales. Sometimes publishers will print notices on those books stating "Professor Review Copy Not For Sale." *See e.g.* Golstein & Reese, *Copyright, Patent, Trademark and Related State Doctrines, Cases and Materials on the Law of Intellectual Property* (6th ed. Foundation Press 2008). *Bobbs-Merrill* holds that such a notice should have no legal effect. Why do you think some publishers include such notices anyway?

Bobbs-Merrill left open the possibility that such restrictions could be imposed through contract. Should a copyright owner be able to control the terms of a subsequent resale of a copy of the copyrighted work? Should a copyright owner, by contract, be able to restrict resale entirely?

2. Why do you think Congress excluded for-profit rentals of computer software and sound recordings but not movies?

3. Importantly, the Copyright Act is clear that transfer of ownership of a copy that embodies a copyrighted work does not convey any rights in the underlying copyright. 17 U.S.C. § 202. Additionally, § 109 specifies that the rights granted to owners of copies are granted "notwithstanding the provisions of section 106(3)." Therefore, when someone has purchased a copy of a music CD, for example, §109 does not give them the right to make copies of that CD. Such copying would be an exercise of the §106(1) reproduction right, not the 106(3) distribution right. While perhaps fair use or an assertion of implied license might make certain copying not an infringement, the first sale doctrine as codified in §109 does not provide a defense in that context.

4. The first sale doctrine also does not limit the right to prepare derivative

---

[*] The statute provides a special carve-out for video games, explaining why rentals of video games is a popular business. 17 U.S.C. § 109(b)(1)(B)(ii).

works. Thus, if someone takes a copy of a work and adds new authorship to that copy, perhaps by annotating the work through highlighting passages and writing marginalia, a copyright owner could assert infringement of his derivative work right and the defendant could not point to § 109 as a defense. What if someone purchases a small artistic print embodied in a greeting card or contained in a book, mounts that card to a ceramic tile, and covers the entire tile with a transparent epoxy resin. Should that constitute an infringement? Does it matter whether the creator of the tile sells that tile? Compare *Mirage Editions, Inc. v. Albuquerque A.R.T. Company*, 856 F.2d 1341 (9th Cir. 1988) (defendant created a derivative work of image contained on book page and the first sale doctrine did not excuse the infringement) with *Lee v. A.R.T. Company*, 125 F.3d 580 (7th Cir. 1997) (same defendant did not create a derivative work of image on greeting card and the first sale doctrine permitted the sale of the tile).

4. Public Performance and Public Display

Section 106(4), (5) and (6) grant copyright owners of certain types of works the exclusive rights to publicly perform and publicly display their works. Review the statutory language quoted on the first page of this Section. The statute defines many of the terms relevant to understanding the scope of the public performance and public display rights. What it means to "perform" a work or to "display" a work are both defined in § 101. Section 101 also contains the following definition:

> To perform or display a work "publicly" means–
> (1) to perform or display it at a place open to the public or at any place where a substantial number of persons outside of a normal circle of a family and its social acquaintances is gathered; or
> (2) to transmit or otherwise communicate a performance or display of the work to a place specified by clause (1) or to the public, by means of any device or process, whether the members of the public capable of receiving the performance or display receive it in the same place or in separate places and at the same time or at different times.

Id. The first category of what constitutes a public performance is "a place open to the public." The second category, commonly referred to as a semi-public place, is determined by the size and composition of the audience. The legislative history indicates the second category was added to expand the concept of public performance by including those places that, although not open to the public at large, are accessible to a significant number of people. *See* H.R. Rep. No. 1476, 94th Cong., 2d Sess. 64, reprinted in 1976 U.S. Code Cong. & Ad. News 5659, 5677-78. Thus, if a place is "open to the public," the size and composition of the audience are irrelevant. However, if the place is not "open to the public," the size and composition of the audience will be determinative. Performing a copyrighted work in a private home, for example singing in the shower, does not implicate the public performance right of a copyright owner (unless a large number of people have gathered in the home to listen!).

The second definition of what constitutes a public performance involves a transmission or other communication. To "transmit" a performance or display is "to communicate it by any device or process whereby images or sounds are received beyond the place from which they are sent." 17 U.S.C. § 101. Understanding the boundaries of what constitutes a public performance under this second clause has taken on increased significance with the continued advancement of digital technologies.

The public display right relies on the same definition of "publicness." To "display" a work means "to show a copy of it, either directly or by means of a film, slide, television image, or any other device or process or, in the case of a motion picture or other audiovisual work, to show individual images nonsequentially." 17 U.S.C. § 101. Unlike the public performance right, the public display right has an important limitation related to the first sale doctrine:

> Notwithstanding the provisions of section 106(5), the owner of a particular copy lawfully made under this title, or any person authorized by such owner, is entitled, without the authority of the copyright owner, to display that copy publicly, either directly or by the projection of no more than one image at a time, to viewers present at the place where the copy is located.

17 U.S.C. § 109(c). This provision allows art patrons to hang their collections in lobbies of public buildings, for example, without triggering copyright infringement liability.

Notes & Questions

1. Consider videos posted on YouTube. Those videos are seen by individuals located at disparate locations and are watched individually at different times, often in the privacy of the home. Is YouTube engaging in a public performance of those videos? While uploading a video to YouTube creates a copy, the streaming technology used to "play" the video involves buffering. Buffering can be set so that it does not result in creating a complete copy of the work on the viewer's computer. Depending on the settings and the technology used should internet performances of videos constitute reproductions or public performances? Are the categories of rights created in the 1976 Act consistent with the technological reality of the twenty-first century?

2. ASCAP and BMI are called "collective rights organizations" (CROs). As CROs, ASCAP and BMI obtain authorization from a variety of copyright owners to license the public performance rights in works owned by those copyright owners. The CROs issue collective licenses for the public performance rights in musical works so that an entity that engages in public performances of large numbers of copyrighted musical works can obtain a license in a single transaction. ASCAP and BMI then remit royalties to the copyright owners, retaining a percentage for themselves. However, these CROs have been granted the rights to license only the public performance of the works. The division of the rights of a copyright owner is particularly pronounced in the music industry and has further increased the complexity of the law in this industry. For further exploration of the music industry, see Lydia Pallas Loren, *The Dual Narratives in the Landscape of Music Copyright*, 52 HOUSTON L. REV. 537 (2014).

3. When a song is played on an over-the-air radio broadcast or through a webcast, that constitutes a public performance of both the musical work and the sound recording, two separate copyrights. The radio station, however, only has to have a license to publicly perform the musical work (typically obtained from CROs ASCAP and BMI) because the Copyright Act does not grant sound recording copyright owners a general public performance right. 17 U.S.C. § 106(4). Instead, sound recording copyright owners are granted a more limited public performance right, added in 1995: the right to publicly perform the work by means of a digital audio transmission. 17 U.S.C. § 106(6). A webcaster, however, is engaging in public performances by such digital means and therefore needs to be concerned about the public performance rights of *both* the musical work copyright owners *and* the sound recording copyright owners. Should traditional over-the-air radio stations and webcasting radio stations be treated differently? If so, why? While the public performance right for the musical work can be licensed from ASCAP and BMI, the Copyright Act contains a statutory license regime for certain public performances of sound recordings. The licenses and revenue for this statutory license are handled by SoundExchange.

4. Section 110 of the Act sets out a series of specific limitations on both the public performance and public display rights. Those limitations range from one permitting teachers and students to perform copyrighted work in the course of face to face teaching activities, § 110(1), to one allowing the performance of a non-dramatic musical work "in the course of an annual agricultural or horticultural fair or exhibition," § 110(6). Special care must be taken to remain within the boundaries of the various § 110 exceptions because they are quite detailed and specific.

C. Contemporary Cases

Learning the breadth of the § 106 rights granted to copyright owners can be daunting. Below are two cases that examine the contours of some of the section § 106 rights.

Perfect 10, Inc. v. Amazon.com Inc.
508 F.3d 1146 (9th Cir. 2007)

IKUTA, J.:

In this appeal, we consider a copyright owner's efforts to stop an Internet search engine from facilitating access to infringing images. . . .

I

Background

Google's computers, along with millions of others, are connected to networks known collectively as the "Internet." "The Internet is a world-wide network of networks . . . all sharing a common communications technology." *Religious Tech. Ctr. v. Netcom On-Line Commc'n Servs.*, 923 F. Supp. 1231, 1238 n.1 (N.D. Cal. 1995). Computer owners can provide information stored on their computers to other users connected to the Internet through a medium called a webpage. A webpage consists of text interspersed with instructions written in Hypertext

Markup Language ("HTML") that is stored in a computer. No images are stored on a webpage; rather, the HTML instructions on the webpage provide an address for where the images are stored, whether in the webpage publisher's computer or some other computer. In general, webpages are publicly available and can be accessed by computers connected to the Internet through the use of a web browser.

Google operates a search engine, a software program that automatically accesses thousands of websites (collections of webpages) and indexes them within a database stored on Google's computers. When a Google user accesses the Google website and types in a search query, Google's software searches its database for websites responsive to that search query. Google then sends relevant information from its index of websites to the user's computer. Google's search engines can provide results in the form of text, images, or videos.

The Google search engine that provides responses in the form of images is called "Google Image Search." In response to a search query, Google Image Search identifies text in its database responsive to the query and then communicates to users the images associated with the relevant text. Google's software cannot recognize and index the images themselves. Google Image Search provides search results as a webpage of small images called "thumbnails," which are stored in Google's servers. The thumbnail images are reduced, lower-resolution versions of full-sized images stored on third-party computers.

When a user clicks on a thumbnail image, the user's browser program interprets HTML instructions on Google's webpage. These HTML instructions direct the user's browser to cause a rectangular area (a "window") to appear on the user's computer screen. The window has two separate areas of information. The browser fills the top section of the screen with information from the Google webpage, including the thumbnail image and text. The HTML instructions also give the user's browser the address of the website publisher's computer that stores the full-size version of the thumbnail.[2] By following the HTML instructions to access the third-party webpage, the user's browser connects to the website publisher's computer, downloads the full-size image, and makes the image appear at the bottom of the window on the user's screen. Google does not store the images that fill this lower part of the window and does not communicate the images to the user; Google simply provides HTML instructions directing a user's browser to access a third-party website. However, the top part of the window (containing the information from the Google webpage) appears to frame and comment on the bottom part of the window. Thus, the user's window appears to be filled with a single integrated presentation of the full-size image, but it is actually an image from a third-party website framed by information from Google's website. The process by which the webpage directs a user's browser to incorporate content from different computers into a single window is referred to as "in-line linking." *Kelly v. Arriba*

---

[2] The website publisher may not actually store the photographic images used on its webpages in its own computer, but may provide HTML instructions directing the user's browser to some further computer that stores the image. Because this distinction does not affect our analysis, for convenience, we will assume that the website publisher stores all images used on its webpages in the website publisher's own computer.

Soft Corp., 336 F.3d 811, 816 (9th Cir. 2003). The term "framing" refers to the process by which information from one computer appears to frame and annotate the in-line linked content from another computer.

. . . .

In addition to its search engine operations, Google generates revenue through a business program called "AdSense." Under this program, the owner of a website can register with Google to become an AdSense "partner." The website owner then places HTML instructions on its webpages that signal Google's server to place advertising on the webpages that is relevant to the webpages' content. Google's computer program selects the advertising automatically by means of an algorithm. AdSense participants agree to share the revenues that flow from such advertising with Google.

. . . .

Perfect 10 markets and sells copyrighted images of nude models. Among other enterprises, it operates a subscription website on the Internet. Subscribers pay a monthly fee to view Perfect 10 images in a "members' area" of the site. Subscribers must use a password to log into the members' area. Google does not include these password-protected images from the members' area in Google's index or database. Perfect 10 has also licensed Fonestarz Media Limited to sell and distribute Perfect 10's reduced-size copyrighted images for download and use on cell phones.

Some website publishers republish Perfect 10's images on the Internet without authorization. Once this occurs, Google's search engine may automatically index the webpages containing these images and provide thumbnail versions of images in response to user inquiries. When a user clicks on the thumbnail image returned by Google's search engine, the user's browser accesses the third-party webpage and in-line links to the full-sized infringing image stored on the website publisher's computer. This image appears, in its original context, on the lower portion of the window on the user's computer screen framed by information from Google's webpage.

Procedural History

. . . Perfect 10 sought a preliminary injunction to prevent . . . [Google] from "copying, reproducing, distributing, publicly displaying, adapting or otherwise infringing, or contributing to the infringement" of Perfect 10's photographs; linking to websites that provide full-size infringing versions of Perfect 10's photographs; and infringing Perfect 10's username/password combinations. [The district court granted in part and denied in part the preliminary injunction against Google] . . . .

III

Direct Infringement

Perfect 10 claims that Google's search engine program directly infringes two exclusive rights granted to copyright holders: its display rights and its distribution rights. "Plaintiffs must satisfy two requirements to present a prima facie case of direct infringement: (1) they must show ownership of the allegedly infringed material and (2) they must demonstrate that the alleged infringers violate at least one exclusive right granted to copyright holders under 17 U.S.C. § 106." [*A & M Records, Inc. v. Napster, Inc.*, 239 F.3d 1004, 1013 (9th Cir. 2001)]; *see* 17 U.S.C. § 501(a). Even if a plaintiff satisfies these two requirements and makes a prima facie case of direct infringement, the defendant may avoid liability if it can establish that its use of the images is a "fair use" as set forth in 17 U.S.C. § 107.[*] . . .

    . . . .

A. Display Right

In considering whether Perfect 10 made a prima facie case of violation of its display right, the district court reasoned that a computer owner that stores an image as electronic information and serves that electronic information directly to the user ("i.e., physically sending ones and zeroes over the [I]nternet to the user's browser," [*Perfect 10 v. Google, Inc.*, 416 F. Supp. 2d 828, 839 (C.D. Cal. 2006)]) is displaying the electronic information in violation of a copyright holder's exclusive display right. *Id.* at 843-45; *see* 17 U.S.C. § 106(5). Conversely, the owner of a computer that does not store and serve the electronic information to a user is not displaying that information, even if such owner in-line links to or frames the electronic information. *Perfect 10*, 416 F. Supp. 2d at 843-45. The district court referred to this test as the "server test." *Id.* at 838-39.

Applying the server test, the district court concluded that Perfect 10 was likely to succeed in its claim that Google's thumbnails constituted direct infringement but was unlikely to succeed in its claim that Google's in-line linking to full-size infringing images constituted a direct infringement. *Id.* at 843-45. As explained below, because this analysis comports with the language of the Copyright Act, we agree with the district court's resolution of both these issues.

We have not previously addressed the question when a computer displays a copyrighted work for purposes of section 106(5). Section 106(5) states that a copyright owner has the exclusive right "to display the copyrighted work publicly." The Copyright Act explains that "display" means "to show a copy of it, either directly or by means of a film, slide, television image, or any other device or process. . . ." 17 U.S.C. § 101. Section 101 defines "copies" as "material objects, other than phonorecords, in which a work is fixed by any method now known or later developed, and from which the work can be perceived, reproduced, or otherwise communicated, either directly or with the aid of a machine or device." *Id.* Finally, the Copyright Act provides that "[a] work is 'fixed' in a tangible medium of expression when its embodiment in a copy . . . is sufficiently permanent or stable to permit it to be perceived, reproduced, or otherwise communicated for a period of more than transitory duration." *Id.*

We must now apply these definitions to the facts of this case. A photographic image is a work that is "'fixed' in a tangible medium of expression," for purposes

---

[*] [*Eds. Note*: Section V of this Chapter explores the fair use doctrine and includes the Ninth Circuit's fair use analysis in this case.]

of the Copyright Act, when embodied (i.e., stored) in a computer's server (or hard disk, or other storage device). The image stored in the computer is the "copy" of the work for purposes of copyright law. The computer owner shows a copy "by means of a . . . device or process" when the owner uses the computer to fill the computer screen with the photographic image stored on that computer, or by communicating the stored image electronically to another person's computer. 17 U.S.C. § 101. In sum, based on the plain language of the statute, a person displays a photographic image by using a computer to fill a computer screen with a copy of the photographic image fixed in the computer's memory. There is no dispute that Google's computers store thumbnail versions of Perfect 10's copyrighted images and communicate copies of those thumbnails to Google's users.[6] Therefore, Perfect 10 has made a prima facie case that Google's communication of its stored thumbnail images directly infringes Perfect 10's display right.

Google does not, however, display a copy of full-size infringing photographic images for purposes of the Copyright Act when Google frames in-line linked images that appear on a user's computer screen. Because Google's computers do not store the photographic images, Google does not have a copy of the images for purposes of the Copyright Act. In other words, Google does not have any "material objects . . . in which a work is fixed . . . and from which the work can be perceived, reproduced, or otherwise communicated" and thus cannot communicate a copy. 17 U.S.C. § 101.

Instead of communicating a copy of the image, Google provides HTML instructions that direct a user's browser to a website publisher's computer that stores the full-size photographic image. Providing these HTML instructions is not equivalent to showing a copy. First, the HTML instructions are lines of text, not a photographic image. Second, HTML instructions do not themselves cause infringing images to appear on the user's computer screen. The HTML merely gives the address of the image to the user's browser. The browser then interacts with the computer that stores the infringing image. It is this interaction that causes an infringing image to appear on the user's computer screen. Google may facilitate the user's access to infringing images. However, such assistance raises only contributory liability issues, and does not constitute direct infringement of the copyright owner's display rights.

Perfect 10 argues that Google displays a copy of the full-size images by framing the full-size images, which gives the impression that Google is showing the image within a single Google webpage. While in-line linking and framing may cause some computer users to believe they are viewing a single Google webpage, the Copyright Act, unlike the Trademark Act, does not protect a copyright holder against

---

[6] Because Google initiates and controls the storage and communication of these thumbnail images, we do not address whether an entity that merely passively owns and manages an Internet bulletin board or similar system violates a copyright owner's display and distribution rights when the users of the bulletin board or similar system post infringing works. *Cf. CoStar Group, Inc. v. LoopNet, Inc.*, 373 F.3d 544 (4th Cir. 2004).

acts that cause consumer confusion. *Cf.* 15 U.S.C. § 1114(1) (providing that a person who uses a trademark in a manner likely to cause confusion shall be liable in a civil action to the trademark registrant).[7]

Nor does our ruling that a computer owner does not display a copy of an image when it communicates only the HTML address of the copy erroneously collapse the display right in section 106(5) into the reproduction right set forth in section 106(1). Nothing in the Copyright Act prevents the various rights protected in section 106 from overlapping. Indeed, under some circumstances, more than one right must be infringed in order for an infringement claim to arise. For example, a "Game Genie" device that allowed a player to alter features of a Nintendo computer game did not infringe Nintendo's right to prepare derivative works because the Game Genie did not incorporate any portion of the game itself. *See Lewis Galoob Toys, Inc. v. Nintendo of Am., Inc.*, 964 F.2d 965, 967 (9th Cir. 1992). We held that a copyright holder's right to create derivative works is not infringed unless the alleged derivative work "incorporate[s] a protected work in some concrete or permanent 'form.'" *Id.* In other words, in some contexts, the claimant must be able to claim infringement of its reproduction right in order to claim infringement of its right to prepare derivative works.

. . . .

B. Distribution Right

The district court also concluded that Perfect 10 would not likely prevail on its claim that Google directly infringed Perfect 10's right to distribute its full-size images. *Perfect 10*, 416 F. Supp. 2d at 844-45. The district court reasoned that distribution requires an "actual dissemination" of a copy. *Id.* at 844. Because Google did not communicate the full-size images to the user's computer, Google did not distribute these images. *Id.*

Again, the district court's conclusion on this point is consistent with the language of the Copyright Act. Section 106(3) provides that the copyright owner has the exclusive right "to distribute copies or phonorecords of the copyrighted work to the public by sale or other transfer of ownership, or by rental, lease, or lending." 17 U.S.C. § 106(3). As noted, "copies" means "material objects . . . in which a work is fixed." 17 U.S.C. § 101. The Supreme Court has indicated that in the electronic context, copies may be distributed electronically. *See N.Y. Times Co. v. Tasini*, 533 U.S. 483, 498 (2001) (a computer database program distributed copies of newspaper articles stored in its computerized database by selling copies of those articles through its database service). Google's search engine communicates

---

[7] Perfect 10 also argues that Google violates Perfect 10's right to display full-size images because Google's in-line linking meets the Copyright Act's definition of "to perform or display a work 'publicly.'" 17 U.S.C. § 101. This phrase means "to transmit or otherwise communicate a performance or display of the work to * * * the public, by means of any device or process, whether the members of the public capable of receiving the performance or display receive it in the same place or in separate places and at the same time or at different times." *Id.* Perfect 10 is mistaken. Google's activities do not meet this definition because Google transmits or communicates only an address which directs a user's browser to the location where a copy of the full-size image is displayed. Google does not communicate a display of the work itself.

HTML instructions that tell a user's browser where to find full-size images on a website publisher's computer, but Google does not itself distribute copies of the infringing photographs. It is the website publisher's computer that distributes copies of the images by transmitting the photographic image electronically to the user's computer. As in *Tasini*, the user can then obtain copies by downloading the photo or printing it.

Perfect 10 incorrectly relies on *Hotaling v. Church of Jesus Christ of Latter-Day Saints* and *Napster* for the proposition that merely making images "available" violates the copyright owner's distribution right. *Hotaling v. Church of Jesus Christ of Latter-Day Saints*, 118 F.3d 199 (4th Cir. 1997); *Napster*, 239 F.3d 1004. *Hotaling* held that the owner of a collection of works who makes them available to the public may be deemed to have distributed copies of the works. Similarly, the distribution rights of the plaintiff copyright owners were infringed by Napster users (private individuals with collections of music files stored on their home computers) when they used the Napster software to make their collections available to all other Napster users.

This "deemed distribution" rule does not apply to Google. Unlike the participants in the Napster system or the library in *Hotaling*, Google does not own a collection of Perfect 10's full-size images and does not communicate these images to the computers of people using Google's search engine. Though Google indexes these images, it does not have a collection of stored full-size images it makes available to the public. Google therefore cannot be deemed to distribute copies of these images under the reasoning of *Napster* or *Hotaling*. Accordingly, the district court correctly concluded that Perfect 10 does not have a likelihood of success in proving that Google violates Perfect 10's distribution rights with respect to full-size images.

. . . .

Notes & Questions

1. Perfect 10 asserted that Google infringed both the public display right and the public distribution right. Engaging in any activities that are within the exclusive rights granted to a copyright owner in § 106 constitutes infringement, unless some valid defense provides otherwise. There is no added advantage to demonstrating more than one type of infringement. The court notes that the § 106 rights can overlap and that sometimes "more than one right must be infringed in order for an infringement claim to arise." Is that overlap of rights a good policy, given copyright law's goals? Is the overlap of rights merely a consequence of changing technologies that were not envisioned when Congress passed the 1976 Copyright Act?

2. Perfect 10 asserted that Google infringed the distribution right by making the full-size images available. Whether infringement of the distribution right requires actual distribution or whether it is sufficient to prove that the defendant made copies *available* for distribution is a contested issue. *See Capitol Records, Inc. v. Thomas*, 579 F. Supp. 2d 1210 (D. Minn. 2008) (discussing the arguments and

concluding that the statute requires proof of actual distribution). What facts in the case permit the *Perfect 10* court to sidestep this particular question?

3. In *Perfect 10* the Ninth Circuit notes that liability based on Google *assisting* users' access to infringing copies or facilitating such access is not a basis for the direct liability that plaintiff alleged. Instead, such assistance might give rise to indirect liability through the doctrines of contributory or vicarious infringement. Section VI explores the distinction between those different theories of liability. For now, focus on one aspect of direct liability: volition. A defendant may be held directly liable only if he has engaged in volitional conduct that violates the Act.

American Broadcasting Cos. v. Aereo, Inc.
134 S. Ct. 2498 (2014)

BREYER, JUSTICE:

....

[handwritten note: Aereo is performing publicly and infringing]

I

A

For a monthly fee, Aereo offers subscribers broadcast television programming over the Internet, virtually as the programming is being broadcast. Much of this programming is made up of copyrighted works. Aereo neither owns the copyright in those works nor holds a license from the copyright owners to perform those works publicly.

Aereo's system is made up of servers, transcoders, and thousands of dime-sized antennas housed in a central warehouse. It works roughly as follows: First, when a subscriber wants to watch a show that is currently being broadcast, he visits Aereo's website and selects, from a list of the local programming, the show he wishes to see.

Second, one of Aereo's servers selects an antenna, which it dedicates to the use of that subscriber (and that subscriber alone) for the duration of the selected show. A server then tunes the antenna to the over-the-air broadcast carrying the show. The antenna begins to receive the broadcast, and an Aereo transcoder translates the signals received into data that can be transmitted over the Internet.

Third, rather than directly send the data to the subscriber, a server saves the data in a subscriber-specific folder on Aereo's hard drive. In other words, Aereo's system creates a subscriber-specific copy—that is, a "personal" copy—of the subscriber's program of choice.

Fourth, once several seconds of programming have been saved, Aereo's server begins to stream the saved copy of the show to the subscriber over the Internet. (The subscriber may instead direct Aereo to stream the program at a later time, but that aspect of Aereo's service is not before us.) The subscriber can watch the streamed program on the screen of his personal computer, tablet, smart phone, Internet-connected television, or other Internet-connected device. The streaming continues, a mere few seconds behind the over-the-air broadcast, until the subscriber has received the entire show.

Aereo emphasizes that the data that its system streams to each subscriber are the data from his own personal copy, made from the broadcast signals received by the particular antenna allotted to him. Its system does not transmit data saved in one subscriber's folder to any other subscriber. When two subscribers wish to watch the same program, Aereo's system activates two separate antennas and saves two separate copies of the program in two separate folders. It then streams the show to the subscribers through two separate transmissions—each from the subscriber's personal copy.

B

Petitioners are television producers, marketers, distributors, and broadcasters who own the copyrights in many of the programs that Aereo's system streams to its subscribers. They brought suit against Aereo for copyright infringement in Federal District Court. They sought a preliminary injunction, arguing that Aereo was infringing their right to "perform" their works "publicly," as the Transmit Clause defines those terms. The District Court denied the preliminary injunction. Relying on prior Circuit precedent, a divided panel of the Second Circuit affirmed. In the Second Circuit's view, Aereo does not perform publicly within the meaning of the [Copyright Act's] Transmit Clause because it does not transmit "to the public." Rather, each time Aereo streams a program to a subscriber, it sends a private transmission that is available only to that subscriber. . . .

II

This case requires us to answer two questions: First, in operating in the manner described above, does Aereo "perform" at all? And second, if so, does Aereo do so "publicly"? We address these distinct questions in turn.

Does Aereo "perform"? Phrased another way, does Aereo "transmit * * * a performance" when a subscriber watches a show using Aereo's system, or is it only the subscriber who transmits? In Aereo's view, it does not perform. It does no more than supply equipment that "emulate[s] the operation of a home antenna and [digital video recorder (DVR)]." Brief for Respondent 41. Like a home antenna and DVR, Aereo's equipment simply responds to its subscribers' directives. So it is only the subscribers who "perform" when they use Aereo's equipment to stream television programs to themselves.

Considered alone, the language of the Act does not clearly indicate when an entity "perform[s]" (or "transmit[s]") and when it merely supplies equipment that allows others to do so. But when read in light of its purpose, the Act is unmistakable: An entity that engages in activities like Aereo's performs.

A

History makes plain that one of Congress' primary purposes in amending the Copyright Act in 1976 was to overturn this Court's determination that community antenna television (CATV) systems (the precursors of modern cable systems) fell outside the Act's scope. In *Fortnightly Corp. v. United Artists Television, Inc.*, 392 U.S. 390 (1968), the Court considered a CATV system that carried local television broadcasting, much of which was copyrighted, to its subscribers in two cities. The CATV provider placed antennas on hills above the cities and used coaxial cables to carry the signals received by the antennas to the home television sets of its subscribers[below, whose antennas could not receive the broadcast signals unaided]. The system amplified and modulated the signals in order to improve their strength and efficiently transmit them to subscribers. A subscriber "could choose any of the . . . programs he wished to view by simply turning the knob on his own television set." *Id.* at 392. The CATV provider "neither edited the programs received nor originated any programs of its own." *Id.*

Asked to decide whether the CATV provider infringed copyright holders' exclusive right to perform their works publicly, the Court held that the provider did not "perform" at all. The Court drew a line: "Broadcasters perform. Viewers do not perform." 392 U.S. at 398. And a CATV provider "falls on the viewer's side of the line." *Id.* at 399. The Court reasoned that CATV providers were unlike broadcasters:

> Broadcasters select the programs to be viewed; CATV systems simply carry, without editing, whatever programs they receive. Broadcasters procure programs and propagate them to the public; CATV systems receive programs that have been released to the public and carry them by private channels to additional viewers.

Id. at 400. Instead, CATV providers were more like viewers, for "the basic function [their] equipment serves is little different from that served by the equipment generally furnished by" viewers. *Id.* at 399. "Essentially," the Court said, "a CATV system no more than enhances the viewer's capacity to receive the broadcaster's signals [by] provid[ing] a well-located antenna with an efficient connection to the viewer's television set." *Id.* Viewers do not become performers by using "amplifying equipment," and a CATV provider should not be treated differently for providing viewers the same equipment. *Id.* at 398–400.

In *Teleprompter Corp. v. Columbia Broadcasting Sys.*, 415 U.S. 394 (1974), the Court considered the copyright liability of a CATV provider that carried broadcast television programming into subscribers' homes from hundreds of miles away. Although the Court recognized that a viewer might not be able to afford amplifying equipment that would provide access to those distant signals, it nonetheless found that the CATV provider was more like a viewer than a broadcaster. *Id.* at 408–409. It explained: "The reception and rechanneling of [broadcast television signals] for simultaneous viewing is essentially a viewer function, irrespective of the distance between the broadcasting station and the ultimate viewer." *Id.* at 408. . . .

     . . . .

B

In 1976 Congress amended the Copyright Act in large part to reject the Court's holdings in *Fortnightly* and *Teleprompter*. See H.R. Rep. No. 94-1476, pp. 86–87 (1976) (hereinafter H.R. Rep.) (The 1976 amendments "completely overturned" this Court's narrow construction of the Act in *Fortnightly* and *Teleprompter*). Congress enacted new language that erased the Court's line between broadcaster and

^perform - definition

viewer, in respect to "perform[ing]" a work. The amended statute clarifies that to "perform" an audiovisual work means "to show its images in any sequence or to make the sounds accompanying it audible." §101; *see id.* (defining "[a]udiovisual works" as "works that consist of a series of related images which are intrinsically intended to be shown by the use of machines * * * , together with accompanying sounds"). Under this new language, both the broadcaster and the viewer of a television program "perform," because they both show the program's images and make audible the program's sounds. See H.R. Rep. at 63 ("[A] broadcasting network is performing when it transmits [a singer's performance of a song] * * * and any individual is performing whenever he or she * * * communicates the performance by turning on a receiving set").

Congress also enacted the Transmit Clause, which specifies that an entity performs publicly when it "transmit[s] * * * a performance * * * to the public." §101. Cable system activities, like those of the CATV systems in *Fortnightly* and *Teleprompter*, lie at the heart of the activities that Congress intended this language to cover. The Clause thus makes clear that an entity that acts like a CATV system itself performs, even if when doing so, it simply enhances viewers' ability to receive broadcast television signals.

Congress further created a new section of the Act to regulate cable companies' public performances of copyrighted works. Section 111 creates a complex, highly detailed compulsory licensing scheme that sets out the conditions, including the payment of compulsory fees, under which cable systems may retransmit broadcasts.

Congress made these three changes to achieve a similar end: to bring the activities of cable systems within the scope of the Copyright Act.

C

This history makes clear that Aereo is not simply an equipment provider. Rather, Aereo, and not just its subscribers, "perform[s]" (or "transmit[s]"). Aereo's activities are substantially similar to those of the CATV companies that Congress amended the Act to reach. Aereo sells a service that allows subscribers to watch television programs, many of which are copyrighted, almost as they are being broadcast. In providing this service, Aereo uses its own equipment, housed in a centralized warehouse, outside of its users' homes. . . .

Aereo's equipment may serve a "viewer function"; it may enhance the viewer's ability to receive a broadcaster's programs. It may even emulate equipment a viewer could use at home. But the same was true of the equipment that was before the Court, and ultimately before Congress, in *Fortnightly* and *Teleprompter*.

We recognize, and Aereo and the dissent emphasize, one particular difference between Aereo's system and the cable systems at issue in *Fortnightly* and *Teleprompter*. The systems in those cases transmitted constantly; they sent continuous programming to each subscriber's television set. In contrast, Aereo's system remains inert until a subscriber indicates that she wants to watch a program. Only at that moment, in automatic response to the subscriber's request, does Aereo's system activate an antenna and begin to transmit the requested program.

This is a critical difference, says the dissent. It means that Aereo's subscribers, not Aereo, "selec[t] the copyrighted content" that is "perform[ed]," and for that reason they, not Aereo, "transmit" the performance. Aereo is thus like "a copy shop that provides its patrons with a library card." A copy shop is not directly liable whenever a patron uses the shop's machines to "reproduce" copyrighted materials found in that library. See § 106(1) ("exclusive righ[t] * * * to reproduce the copyrighted work"). And by the same token, Aereo should not be directly liable whenever its patrons use its equipment to "transmit" copyrighted television programs to their screens.

In our view, however, the dissent's copy shop argument, in whatever form, makes too much out of too little. Given Aereo's overwhelming likeness to the cable companies targeted by the 1976 amendments, this sole technological difference between Aereo and traditional cable companies does not make a critical difference here. The subscribers of the *Fortnightly* and *Teleprompter* cable systems also selected what programs to display on their receiving sets. Indeed, as we explained in *Fortnightly*, such a subscriber "could choose any of the * * * programs he wished to view by simply turning the knob on his own television set." 392 U.S. at 392. The same is true of an Aereo subscriber. Of course, in *Fortnightly* the television signals, in a sense, lurked behind the screen, ready to emerge when the subscriber turned the knob. Here the signals pursue their ordinary course of travel through the universe until today's "turn of the knob"—a click on a website—activates machinery that intercepts and reroutes them to Aereo's subscribers over the Internet. But this difference means nothing to the subscriber. It means nothing to the broadcaster. We do not see how this single difference, invisible to subscriber and broadcaster alike, could transform a system that is for all practical purposes a traditional cable system into "a copy shop that provides its patrons with a library card."

In other cases involving different kinds of service or technology providers, a user's involvement in the operation of the provider's equipment and selection of the content transmitted may well bear on whether the provider performs within the meaning of the Act. But the many similarities between Aereo and cable companies, considered in light of Congress' basic purposes in amending the Copyright Act, convince us that this difference is not critical here. We conclude that Aereo is not just an equipment supplier and that Aereo "perform[s]." *— Aereo performs*

H

III

Next, we must consider whether Aereo performs petitioners' works "publicly," within the meaning of the Transmit Clause. Under the Clause, an entity performs a work publicly when it "transmit[s] * * * a performance * * * of the work * * * to the public." § 101. Aereo denies that it satisfies this definition. It reasons as follows: First, the "performance" it "transmit[s]" is the performance created by its act of transmitting. And second, because each of these performances is capable of being received by one and only one subscriber, Aereo transmits privately, not publicly. Even assuming Aereo's first argument is correct, its second does not follow.

We begin with Aereo's first argument. What performance does Aereo transmit? Under the Act, "[t]o 'transmit' a performance * * * is to communicate it by any device or process whereby images or sounds are received beyond the place from which they are sent." *Id.* And "[t]o 'perform'" an audiovisual work means "to show its images in any sequence or to make the sounds accompanying it audible." *Id.*

Petitioners say Aereo transmits a prior performance of their works. Thus when Aereo retransmits a network's prior broadcast, the underlying broadcast (itself a performance) is the performance that Aereo transmits. Aereo, as discussed above, says the performance it transmits is the new performance created by its act of transmitting. That performance comes into existence when Aereo streams the sounds and images of a broadcast program to a subscriber's screen.

We assume *arguendo* that Aereo's first argument is correct. Thus, for present purposes, to transmit a performance of (at least) an audiovisual work means to communicate contemporaneously visible images and contemporaneously audible sounds of the work. *Cf. United States v. American Soc. of Composers, Authors and Publishers*, 627 F.3d 64, 73 (2d Cir. 2010) (holding that a download of a work is not a performance because the data transmitted are not "contemporaneously perceptible"). When an Aereo subscriber selects a program to watch, Aereo streams the program over the Internet to that subscriber. Aereo thereby "communicate[s]" to the subscriber, by means of a "device or process," the work's images and sounds. §101. And those images and sounds are contemporaneously visible and audible on the subscriber's computer (or other Internet-connected device). So under our assumed definition, Aereo transmits a performance whenever its subscribers watch a program.

But what about the Clause's further requirement that Aereo transmit a performance "to the public"? As we have said, an Aereo subscriber receives broadcast television signals with an antenna dedicated to him alone. Aereo's system makes from those signals a personal copy of the selected program. It streams the content of the copy to the same subscriber and to no one else. One and only one subscriber has the ability to see and hear each Aereo transmission. The fact that each transmission is to only one subscriber, in Aereo's view, means that it does not transmit a performance "to the public."

In terms of the Act's purposes, these differences do not distinguish Aereo's system from cable systems, which do perform "publicly." Viewed in terms of Congress' regulatory objectives, why should any of these technological differences matter? They concern the behind-the-scenes way in which Aereo delivers television programming to its viewers' screens. They do not render Aereo's commercial objective any different from that of cable companies. Nor do they significantly alter the viewing experience of Aereo's subscribers. Why would a subscriber who wishes to watch a television show care much whether images and sounds are delivered to his screen via a large multi subscriber antenna or one small dedicated antenna, whether they arrive instantaneously or after a few seconds' delay, or whether they are transmitted directly or after a personal copy is made? And why, if Aereo is right, could not modern CATV systems simply continue the same commercial and consumer-oriented activities, free of copyright restrictions, provided they substitute such new technologies for old? Congress would as much have intended to protect a copyright holder from the unlicensed activities of Aereo as from those of cable companies.

The text of the Clause effectuates Congress' intent. Aereo's argument to the contrary relies on the premise that "to transmit * * * a performance" means to make a single transmission. But the Clause suggests that an entity may transmit a performance through multiple, discrete transmissions. That is because one can "transmit" or "communicate" something through a set of actions. Thus one can transmit a message to one's friends, irrespective of whether one sends separate identical e-mails to each friend or a single e-mail to all at once. So can an elected official communicate an idea, slogan, or speech to her constituents, regardless of whether she communicates that idea, slogan, or speech during individual phone calls to each constituent or in a public square.

The fact that a singular noun ("a performance") follows the words "to transmit" does not suggest the contrary. One can sing a song to his family, whether he sings the same song one-on-one or in front of all together. Similarly, one's colleagues may watch a performance of a particular play—say, this season's modern-dress version of "Measure for Measure"—whether they do so at separate or at the same showings. By the same principle, an entity may transmit a performance through one or several transmissions, where the performance is of the same work.

The Transmit Clause must permit this interpretation, for it provides that one may transmit a performance to the public "whether the members of the public capable of receiving the performance * * * receive it * * * at the same time or at different times." § 101. Were the words "to transmit * * * a performance" limited to a single act of communication, members of the public could not receive the performance communicated "at different times."

Therefore, in light of the purpose and text of the Clause, we conclude that when an entity communicates the same contemporaneously perceptible images and sounds to multiple people, it transmits a performance to them regardless of the number of discrete communications it makes.

We do not see how the fact that Aereo transmits via personal copies of programs could make a difference. The Act applies to transmissions "by means of any device or process." *Id.* And retransmitting a television program using user-specific copies is a "process" of transmitting a performance. A "cop[y]" of a work is simply a "material objec[t] * * * in which a work is fixed * * * and from which the work can be perceived, reproduced, or otherwise communicated." *Id.* So whether Aereo transmits from the same or separate copies, it performs the same work; it shows the same images and makes audible the same sounds. Therefore, when

Aereo streams the same television program to multiple subscribers, it "transmit[s] * * * a performance" to all of them.

Moreover, the subscribers to whom Aereo transmits television programs constitute "the public." Aereo communicates the same contemporaneously perceptible images and sounds to a large number of people who are unrelated and unknown to each other. This matters because, although the Act does not define "the public," it specifies that an entity performs publicly when it performs at "any place where a substantial number of persons outside of a normal circle of a family and its social acquaintances is gathered." *Id.* The Act thereby suggests that "the public" consists of a large group of people outside of a family and friends.

Neither the record nor Aereo suggests that Aereo's subscribers receive performances in their capacities as owners or possessors of the underlying works. This is relevant because when an entity performs to a set of people, whether they constitute "the public" often depends upon their relationship to the underlying work. When, for example, a valet parking attendant returns cars to their drivers, we would not say that the parking service provides cars "to the public." We would say that it provides the cars to their owners. We would say that a car dealership, on the other hand, does provide cars to the public, for it sells cars to individuals who lack a pre-existing relationship to the cars. Similarly, an entity that transmits a performance to individuals in their capacities as owners or possessors does not perform to "the public," whereas an entity like Aereo that transmits to large numbers of paying subscribers who lack any prior relationship to the works does so perform.

. . . .

IV

Aereo and many of its supporting *amici* argue that to apply the Transmit Clause to Aereo's conduct will impose copyright liability on other technologies, including new technologies, that Congress could not possibly have wanted to reach. We agree that Congress, while intending the Transmit Clause to apply broadly to cable companies and their equivalents, did not intend to discourage or to control the emergence or use of different kinds of technologies. But we do not believe that our limited holding today will have that effect.

For one thing, the history of cable broadcast transmissions that led to the enactment of the Transmit Clause informs our conclusion that Aereo "perform[s]," but it does not determine whether different kinds of providers in different contexts also "perform." For another, an entity only transmits a performance when it communicates contemporaneously perceptible images and sounds of a work.

Further, we have interpreted the term "the public" to apply to a group of individuals acting as ordinary members of the public who pay primarily to watch broadcast television programs, many of which are copyrighted. We have said that it does not extend to those who act as owners or possessors of the relevant product. And we have not considered whether the public performance right is infringed when the user of a service pays primarily for something other than the transmission of copyrighted works, such as the remote storage of content. See

Brief for United States as *Amicus Curiae* 31 (distinguishing cloud based storage services because they "offer consumers more numerous and convenient means of playing back copies that the consumers have *already* lawfully acquired" (emphasis in original)). In addition, an entity does not transmit to the public if it does not transmit to a substantial number of people outside of a family and its social circle.

We also note that courts often apply a statute's highly general language in light of the statute's basic purposes. Finally, the doctrine of "fair use" can help to prevent inappropriate or inequitable applications of the Clause. *See Sony Corp. v. Universal City Studios, Inc.*, 464 U.S. 417 (1984).

We cannot now answer more precisely how the Transmit Clause or other provisions of the Copyright Act will apply to technologies not before us. We agree with the Solicitor General that "[q]uestions involving cloud computing, [remote storage] DVRs, and other novel issues not before the Court, as to which 'Congress has not plainly marked [the] course,' should await a case in which they are squarely presented." Brief for United States as *Amicus Curiae* 34 (quoting *Sony*, 464 U.S. at 431 (alteration in original)). And we note that, to the extent commercial actors or other interested entities may be concerned with the relationship between the development and use of such technologies and the Copyright Act, they are of course free to seek action from Congress. *Cf.* Digital Millennium Copyright Act, 17 U. S. C. §512.

. . . .

SCALIA, JUSTICE, dissenting:

. . . .

III. Guilt By Resemblance

The Court's conclusion that Aereo performs boils down to the following syllogism: (1) Congress amended the Act to overrule our decisions holding that cable systems do not perform when they retransmit over-the-air broadcasts; (2) Aereo looks a lot like a cable system; therefore (3) Aereo performs. . . .

. . . .

The rationale for the Court's ad hoc rule for cable system lookalikes is so broad that it renders nearly a third of the Court's opinion superfluous. Part II of the opinion concludes that Aereo performs because it resembles a cable company, and Congress amended the Act in 1976 "to bring the activities of cable systems within [its] scope." Part III of the opinion purports to address separately the question whether Aereo performs "publicly." Trouble is, that question cannot remain open if Congress's supposed intent to regulate whatever looks like a cable company must be given legal effect (as the Court says in Part II). The Act reaches only public performances, see § 106(4), so Congress could not have regulated "the activities of cable systems" without deeming their retransmissions public performances. The upshot is this: If Aereo's similarity to a cable company means that it performs, then by necessity that same characteristic means that it does so publicly, and Part III of the Court's opinion discusses an issue that is no longer relevant— though discussing it certainly gives the opinion the "feel" of real textual analysis.

Making matters worse, the Court provides no criteria for determining when its cable-TV-lookalike rule applies. Must a defendant offer access to live television to qualify? If similarity to cable-television service is the measure, then the answer must be yes. But consider the implications of that answer: Aereo would be free to do exactly what it is doing right now so long as it built mandatory time shifting into its "watch" function.[6] Aereo would not be providing live television if it made subscribers wait to tune in until after a show's live broadcast ended. A subscriber could watch the 7 p.m. airing of a 1-hour program any time after 8 p.m. Assuming the Court does not intend to adopt such a do-nothing rule (though it very well may), there must be some other means of identifying who is and is not subject to its guilt-by-resemblance regime.

Two other criteria come to mind. One would cover any automated service that captures and stores live television broadcasts at a user's direction. That can't be right, since it is exactly what remote storage digital video recorders (RS–DVRs) do, and the Court insists that its "limited holding" does not decide the fate of those devices. The other potential benchmark is the one offered by the Government: The cable-TV-lookalike rule embraces any entity that "operates an integrated system, substantially dependent on physical equipment that is used in common by [its] subscribers." Brief for United States as *Amicus Curiae* 20. The Court sensibly avoids that approach because it would sweep in Internet service providers and a host of other entities that quite obviously do not perform.

That leaves as the criterion of cable-TV-resemblance nothing but th'ol' totality-of-the-circumstances test (which is not a test at all but merely assertion of an intent to perform test-free, ad hoc, case-by-case evaluation). It will take years, perhaps decades, to determine which automated systems now in existence are governed by the traditional volitional-conduct test and which get the Aereo treatment. (And automated systems now in contemplation will have to take their chances.) The Court vows that its ruling will not affect cloud-storage providers and cable television systems, but it cannot deliver on that promise given the imprecision of its result-driven rule. . . .

* * *

I share the Court's evident feeling that what Aereo is doing (or enabling to be done) to the Networks' copyrighted programming ought not to be allowed. But perhaps we need not distort the Copyright Act to forbid it. As discussed at the outset,[*] Aereo's secondary liability for performance infringement is yet to be determined, as is its primary and secondary liability for reproduction infringement. If that does not suffice, then (assuming one shares the majority's estimation of right and wrong) what we have before us must be considered a "loophole" in the

---

[6] Broadcasts accessible through the "watch" function are technically not live because Aereo's servers take anywhere from a few seconds to a few minutes to begin transmitting data to a subscriber's device. But the resulting delay is so brief that it cannot reasonably be classified as time shifting.

[*] [*Eds. Note*: A main thrust of Justice Scalia's dissent is the difference between direct liability for copyright infringement and indirect or "secondary" liablity. We take up that topic, and that portion of Justice Scalia's dissent, in Section VI.]

law. It is not the role of this Court to identify and plug loopholes. It is the role of good lawyers to identify and exploit them, and the role of Congress to eliminate them if it wishes. Congress can do that, I may add, in a much more targeted, better informed, and less disruptive fashion than the crude "looks-like-cable-TV" solution the Court invents today.

We came within one vote of declaring the VCR contraband 30 years ago in *Sony*. *See* 464 U.S. at 441, n.21. The dissent in that case was driven in part by the plaintiffs' prediction that VCR technology would wreak all manner of havoc in the television and movie industries. *See id.* at 483 (opinion of Blackmun, J.); *see also* Brief for CBS, Inc., as *Amicus Curiae*, O.T. 1982, No. 81–1687, p. 2 (arguing that VCRs "directly threatened" the bottom line of "[e]very broadcaster").

The Networks make similarly dire predictions about Aereo. We are told that nothing less than "the very existence of broadcast television as we know it" is at stake. Brief for Petitioners 39. Aereo and its *amici* dispute those forecasts and make a few of their own, suggesting that a decision in the Networks' favor will stifle technological innovation and imperil billions of dollars of investments in cloud-storage services. We are in no position to judge the validity of those self-interested claims or to foresee the path of future technological development. *See Sony*, 464 U.S. at 430–31; *see also Grokster*, 545 U.S. at 958 (Breyer, J., concurring). Hence, the proper course is not to bend and twist the Act's terms in an effort to produce a just outcome, but to apply the law as it stands and leave to Congress the task of deciding whether the Copyright Act needs an upgrade. I conclude, as the Court concluded in *Sony*: "It may well be that Congress will take a fresh look at this new technology, just as it so often has examined other innovations in the past. But it is not our job to apply laws that have not yet been written. Applying the copyright statute, as it now reads, to the facts as they have been developed in this case, the judgment of the Court of Appeals must be [affirmed]." 464 U.S. at 456.

I respectfully dissent.

Notes & Questions

1. Compare the approach to analyzing the scope of § 106 rights used by the Ninth Circuit in *Perfect 10* to that employed by the majority in *Aereo*. In what ways are they similar? In what ways are they different?

2. In *Aereo*, the majority assumes that to transmit a performance means "to communicate contemporaneously visible images and contemporaneously audible sounds of the work." It cites as support for that proposition a case involving ringtones downloaded to cellphones. *United States v. American Soc. of Composers, Authors and Publishers*, 627 F. 3d 64 (2d Cir. 2010). In that case, ASCAP had argued that downloading ringtones constituted both a distribution of a copy *and* a public performance. The cellphone company was already paying for authorization to engage in the distribution of copies. ASCAP, as the entity that licenses public performances of musical works, could insist upon payment only if the downloads were also public performances. The Second Circuit rejected ASCAP's argument. Can a download also be a public performance? What if the device that is receiving

the download is configured to begin playing as soon as it has received sufficient content?

3. The majority places significance on the Aereo system users' lack of relationship to the underlying work. The Court states that an entity would not be publicly performing a work if it were transmitting "to individuals in their capacities as owners or possessors." What is the relationship that the public has to free over-the-air television? Are viewers, who have been invited to watch publicly broadcast television, in a position to have a relationship with the work? Is it the lack of a *prior* relationship with the work that matters?

4. In part, the Court's focus on the user's relationship, or lack of it, with a work manifests the concern, raised by many *amici*, that a determination that Aereo was directly liable for infringement would suggest a similar liability for entities that provide remote digital storage (storage in "the cloud"). Imagine, for example, that one hundred different people all have cloud storage accounts with the same provider, and that each person stores a copy of the same song—say, *Uptown Funk* by Bruno Mars—in their respective storage spaces. If each person then plays that song, the cloud storage provider is engaged in an activity that is quite similar to what Aereo had done. How else does the majority seek to ensure its decision does not determine liability for cloud storage providers? Should such providers be seen as publicly performing works?

5. The majority notes that the 1976 Act implemented a complicated statutory license regime for cable retransmissions. Following the decision in *Aereo*, a natural question is whether an Internet-based service that achieves the same results as a cable company can use the statutory license. The Second Circuit had already concluded, before *Aereo*, that it cannot:

> The legislative history indicates that Congress enacted § 111 with the intent to address the issue of poor television reception, or, more specifically, to mitigate the difficulties that certain communities and households faced in receiving over-the-air broadcast signals by enabling the expansion of cable systems.
>
> . . . .
>
> Extending § 111's compulsory license to Internet retransmissions, moreover, would not fulfill or further Congress's statutory purpose. Internet retransmission services are not seeking to address issues of reception and remote access to over-the-air television signals. They provide not a local but a nationwide (arguably international) service. . . .

WPIX, Inc. v. ivi, Inc., 691 F.3d 275, 282 (2d Cir. 2012). On remand in the *Aereo* case itself, the district court cited *WPIX* in granting an injunction against Aereo. 2014 WL 5393867 (S.D.N.Y 2014). If Aereo is infringing because it is "just like" a cable company, should Aereo be entitled to rely on the other aspects of the Copyright Act that apply to cable companies? How can it be enough like a cable company to be held liable for infringement but not enough like a cable company to invoke the cable company license?

6. Justice Scalia raises a recurring question for this course: Who should drive change in innovation policy, Congress or the courts? Once the Supreme Court grants review in a case about the scope of an i.p. statute, it must make a decision, one way or the other. But in making its decision—which, from Congress's perspective, may be right, or may be an error—it is placing the burden on the losing party to seek congressional change in the law. In *Fortnightly* and *Teleprompter* the Court sided with the then-new technology companies (at the time, cable television providers), against the copyright owners. Later, Congress changed the statute. In *Aereo* the Court ruled in favor of copyright owners, allied with the incumbent television networks. If Congress thinks *Aereo* is mistaken, it must reduce the scope of copyright rights that, according to the Supreme Court, have existed since the late 1970s. Would that reduction in copyright rights require that Congress compensate the copyright owners, on the theory that the reduction is a Fifth Amendment taking of property? Had *Aereo* been decided the way Justice Scalia urged, and Congress had concluded that *that* holding (against the copyright owners) was mistaken, a change to the statute would not seem to involve a reduction in anyone's property rights. Should the Court take into account the ease with which Congress can fix its mistakes about the scope of the rights that i.p. statutes create? For an argument that it should, see Joseph Scott Miller, *Error Costs & IP Law*, 2014 U. Ill. L. Rev. 175.

D. § 106A Rights – The Visual Artist Rights Act ("VARA")

In addition to granting copyright owners rights in § 106, the Copyright Act provides a further set of rights to "the author of a work of visual art." The § 106A rights were added with the passage of the Visual Artist Rights Act of 1990, and they are often referred to as the "moral rights" of attribution and integrity. The definition of the term "work of visual art" is quite detailed and includes "a painting, drawing, print or sculpture existing in a single copy, [or] in a limited edition of 200 copies or fewer that are signed and consecutively numbered by the author . . . ." 17 U.S.C. § 101. The definition also specifies the types of works that are *not* included, such as a poster, applied art, a motion picture, merchandising items or advertising, and many other types of works. Importantly, works made for hire are expressly exempted from coverage.

For works that qualify, the attribution right that § 106A grants is the right: "to claim authorship of that work, and . . . to prevent the use of his or her name as the author of any work of visual art which he or she did not create" as well as "the right to prevent the use of his or her name as the author of the work of visual art in the event of a distortion, mutilation, or other modification of the work which would be prejudicial to his or her honor or reputation." 17 U.S.C. § 106A(a).

The right of integrity that the Copyright Act grants is more specific:

[S]ubject to the limitations set forth in section 113(d), [the author of a work of visual art] shall have the right—
(A) to prevent any intentional distortion, mutilation, or other modification of that work which would be prejudicial to his or her honor or reputation, and any intentional distortion, mutilation, or modification of that work is a

violation of that right, and

(B) to prevent any destruction of a work of recognized stature, and any intentional or grossly negligent destruction of that work is a violation of that right.

17 U.S.C. § 106A(a)(3). Courts have recognized claims for violations of 106A in a variety of contexts. *See, e.g., Martin v. City of Indianapolis,* 192 F.3d 608 (7th Cir. 1999) (finding violation when large outdoor stainless steel sculpture was destroyed by city as part of an urban renewal project).

Notes & Questions

1. Being credited as the creator of a work is important to many authors. Currently the Copyright Act grants a right of attribution only to authors of works of visual art (as defined by the statute). Should the Copyright Act grant a right of attribution to all authors?

2. The rights granted to authors in § 106A are separate from the economic rights granted in § 106 and separate from the personal property rights of the owner of a particular copy of the work. 17 U.S.C. § 106A(e)(1). These rights are personal to the author; they cannot be transferred, although they can be waived. 17 U.S.C. § 106A(e)(1). Imagine you are a business lawyer performing due diligence on a real estate transaction involving a large office complex that includes two works commissioned in 1995 by the previous owners: a large sculpture and a separate mosaic wall panel in the lobby. What issues would you need to raise in the context of the transaction? What options would new owners face if they desired to remove the sculpture or mosaic? Note that the statute contains special provisions relating to works of visual art that are incorporated into buildings. *See* 17 U.S.C. § 113(d). Note also that the rights granted by § 106A apply to works either created after June 1, 1991 (the effective date of the Visual Artist Rights Act) or created before that date but title to which the author transfers after that date. Thus, authors of older works, even older works of significant stature, are not granted any protection for their works under § 106A. Finally, for works created after the effective date of the act, the rights granted in § 106A endure for the life of the author. 17 U.S.C. § 106A(d).

3. Protection for moral rights is far more robust in other countries. In fact, the meager protection for moral rights in the United States is considered one of the major differences in global copyright law. Should the U.S grant moral rights protection to a wider variety of creative works? *See generally* Roberta Rosenthal Kwall, *The Soul of Creativity: Forging a Moral Rights Law for the United States* (2010).

V. Fair Use—The "Breathing Space Within the Confines of Copyright"

Section 106 makes clear that all of the rights granted to a copyright owner are "subject to" the express statutory limitations contained in sections 107 through 122. In the previous Section of this Chapter we discussed some of those limitations, such as the first sale doctrine codified in § 109, and statutory licenses contained in §§ 111 (for cable transmission of broadcast tv) and 115 (for cover tunes). Here we explore the important limitation known as *fair use*, embodied in § 107. Recognize that because the scope of the rights granted to a copyright owner are shaped not only by the grant of the rights in § 106, but also by the limitations on those rights, as you read the fair use cases you are continuing to develop your understanding of the rights of a copyright owner.

Fair use is an important limitation on the rights of a copyright owner for several reasons. First, fair use is a limitation that applies to *all* of the rights of a copyright owner—from the reproduction right to the digital performance right for sound recordings. Second, fair use takes on many different guises depending on the right at issue and the activities of the alleged infringer. Scholars have attempted to categorize these different faces of fair use, *see, e.g.*, Pamela Samuelson, *Unbundling Fair Uses*, 77 Fordham L. Rev. 2537 (2009); Michael Madison, *A Pattern-Oriented Approach to Fair Use,* 45 Wm. & Mary L. Rev. 1525 (2004), but fair use continues to evolve, and litigants seek to apply the fair use doctrine to new uses for copyrighted works. Third, the doctrine itself invites its own evolution, permitting courts to judge on a case-by-case basis whether a particular use should be categorized as a fair use or as an infringing one. The cases that follow will paint only a partial portrait of the many faces of fair use.

The Supreme Court has addressed the fair use doctrine three different times. We explore the Supreme Court's most recent decision on the doctrine first. This decision is the only fair use decision issued by a unanimous Court. The statute specifies four factors for the courts to consider in making a fair use determination. It may be helpful to begin to make a list of the types of facts that are relevant under each of the statutory factors. Be sure to include on your list the ways in which the Court identifies how the factors are related to one another.

A. Parody as a Potential Fair Use

Campbell v. Acuff-Rose Music, Inc.
510 U.S. 569 (1994)

[handwritten: 2 live crew wins, mostly on prong 1.]

Souter, Justice:

. . . .

In 1964, Roy Orbison and William Dees wrote a rock ballad called "Oh, Pretty Woman" and assigned their rights in it to respondent Acuff-Rose Music, Inc. . . .

Petitioners Luther R. Campbell, Christopher Wongwon, Mark Ross, and David Hobbs are collectively known as 2 Live Crew, a popular rap music group. In 1989, Campbell wrote a song entitled "Pretty Woman," which he later described in an affidavit as intended, "through comical lyrics, to satirize the original work * * * ." On July 5, 1989, 2 Live Crew's manager informed Acuff-Rose that 2 Live Crew had written a parody of "Oh, Pretty Woman," that they would afford all credit for ownership and authorship of the original song to Acuff-Rose, Dees, and Orbison, and that they were willing to pay a fee for the use they wished to make of it. Enclosed with the letter were a copy of the lyrics and a recording of 2 Live

Crew's song. Acuff-Rose's agent refused permission, stating that "I am aware of the success enjoyed by 'The 2 Live Crews', but I must inform you that we cannot permit the use of a parody of 'Oh, Pretty Woman.'" Nonetheless, in June or July 1989, 2 Live Crew released records, cassette tapes, and compact discs of "Pretty Woman" in a collection of songs entitled "As Clean As They Wanna Be." The albums and compact discs identify the authors of "Pretty Woman" as Orbison and Dees and its publisher as Acuff-Rose.

Almost a year later, after nearly a quarter of a million copies of the recording had been sold, Acuff-Rose sued 2 Live Crew and its record company, Luke Skyywalker Records, for copyright infringement. The District Court granted summary judgment for 2 Live Crew . . . . The District Court . . . held that 2 Live Crew's song made fair use of Orbison's original.

The Court of Appeals for the Sixth Circuit reversed and remanded. 972 F.2d 1429, 1439 (1992). . . . [T]he court concluded that [defendants'] "blatantly commercial purpose . . . prevents this parody from being a fair use."

We granted certiorari to determine whether 2 Live Crew's commercial parody could be a fair use.

II

It is uncontested here that 2 Live Crew's song would be an infringement of Acuff-Rose's rights in "Oh, Pretty Woman," under the Copyright Act of 1976, 17 U.S.C. § 106, but for a finding of fair use through parody. From the infancy of copyright protection, some opportunity for fair use of copyrighted materials has been thought necessary to fulfill copyright's very purpose, "[t]o promote the Progress of Science and useful Arts . . . ." U.S. Const., Art. I, § 8, cl. 8. For as Justice Story explained, "[i]n truth, in literature, in science and in art, there are, and can be, few, if any, things, which in an abstract sense, are strictly new and original throughout. Every book in literature, science and art, borrows, and must necessarily borrow, and use much which was well known and used before." *Emerson v. Davies,* 8 F. Cas. 615, 619 (No. 4,436) (CCD Mass. 1845). Similarly, Lord Ellenborough expressed the inherent tension in the need simultaneously to protect copyrighted material and to allow others to build upon it when he wrote, "while I shall think myself bound to secure every man in the enjoyment of his copy-right, one must not put manacles upon science." *Carey v. Kearsley,* 170 Eng. Rep. 679, 681 (K.B. 1803). In copyright cases brought under the Statute of Anne of 1710, English courts held that in some instances "fair abridgements" would not infringe an author's rights, see W. Patry, *The Fair Use Privilege in Copyright Law* 6-17 (1985) (hereinafter Patry); Leval, *Toward a Fair Use Standard,* 103 Harv. L. Rev. 1105 (1990) (hereinafter Leval), and although the First Congress enacted our initial copyright statute, Act of May 31, 1790, 1 Stat. 124, without any explicit reference to "fair use," as it later came to be known, the doctrine was recognized by the American courts nonetheless.

In *Folsom v. Marsh,* 9 F. Cas. 342 (No. 4,901) (CCD Mass. 1841), Justice Story distilled the essence of law and methodology from the earlier cases: "look to the nature and objects of the selections made, the quantity and value of the materials used, and the degree in which the use may prejudice the sale, or diminish the profits, or supersede the objects, of the original work." *Id.* at 348. Thus expressed, fair use remained exclusively judge-made doctrine until the passage of the 1976 Copyright Act, in which Justice Story's summary is discernible:

§ 107. Limitations on exclusive rights: Fair use

Notwithstanding the provisions of sections 106 and 106A, the fair use of a copyrighted work, including such use by reproduction in copies or phonorecords or by any other means specified by that section, for purposes such as criticism, comment, news reporting, teaching (including multiple copies for classroom use), scholarship, or research, is not an infringement of copyright. In determining whether the use made of a work in any particular case is a fair use the factors to be considered shall include—

(1) the purpose and character of the use, including whether such use is of a commercial nature or is for non-profit educational purposes;

(2) the nature of the copyrighted work;

(3) the amount and substantiality of the portion used in relation to the copyrighted work as a whole; and

(4) the effect of the use upon the potential market for or value of the copyrighted work.

The fact that a work is unpublished shall not itself bar a finding of fair use if such finding is made upon consideration of all the above factors."

17 U.S.C. § 107.

Congress meant § 107 "to restate the present judicial doctrine of fair use, not to change, narrow, or enlarge it in any way" and intended that courts continue the common-law tradition of fair use adjudication. H.R. Rep. No. 94-1476, p. 66 (1976) (hereinafter House Report); S. Rep. No. 94-473, p. 62 (1975) (hereinafter Senate Report). The fair use doctrine thus "permits [and requires] courts to avoid rigid application of the copyright statute when, on occasion, it would stifle the very creativity which that law is designed to foster." *Stewart v. Abend,* 495 U.S. 207, 236 (1990).

The task is not to be simplified with bright-line rules, for the statute, like the doctrine it recognizes, calls for case-by-case analysis. *Harper & Row [Publishers, Inc. v. Nation Enterprises],* 471 U.S. [417], 560 [(1985)]. The text employs the terms "including" and "such as" in the preamble paragraph to indicate the "illustrative and not limitative" function of the examples given, § 101, which thus provide only general guidance about the sorts of copying that courts and Congress most commonly had found to be fair uses.[9] Nor may the four statutory factors be treated in isolation, one from another. All are to be explored, and the results weighed to-

---

[9] See Senate Report, p. 62 ("[W]hether a use referred to in the first sentence of section 107 is a fair use in a particular case will depend upon the application of the determinative factors").

gether, in light of the purposes of copyright. See Leval 1110-1111; Patry & Perl-mutter, *Fair Use Misconstrued: Profit, Presumptions, and Parody*, 11 Cardozo Arts & Ent. L. J. 667, 685-687 (1993) (hereinafter Patry & Perlmutter).[10]

A

The first factor in a fair use enquiry is "the purpose and character of the use, including whether such use is of a commercial nature or is for nonprofit educational purposes." § 107(1). This factor draws on Justice Story's formulation, "the nature and objects of the selections made." *Folsom v. Marsh, supra,* at 348. The enquiry here may be guided by the examples given in the preamble to § 107, looking to whether the use is for criticism, or comment, or news reporting, and the like, see § 107. The central purpose of this investigation is to see, in Justice Story's words, whether the new work merely "supersede[s] the objects" of the original creation, *Folsom v. Marsh, supra,* at 348; accord, *Harper & Row, supra,* at 562 ("supplanting" the original), or instead adds something new, with a further purpose or different character, altering the first with new expression, meaning, or message; it asks, in other words, whether and to what extent the new work is "transformative." Leval 1111. Although such transformative use is not absolutely necessary for a finding of fair use,[11] the goal of copyright, to promote science and the arts, is generally furthered by the creation of transformative works. Such works thus lie at the heart of the fair use doctrine's guarantee of breathing space within the confines of copyright, and the more transformative the new work, the less will be the significance of other factors, like commercialism, that may weigh against a finding of fair use.

This Court has only once before even considered whether parody may be fair use, and that time issued no opinion because of the Court's equal division. *Benny v. Loew's Inc.,* 239 F.2d 532 (9th Cir. 1956), aff'd *sub nom. Columbia Broadcasting System, Inc. v. Loew's Inc.,* 356 U.S. 43 (1958). Suffice it to say now that parody has an obvious claim to transformative value, as Acuff-Rose itself does not deny. Like less ostensibly humorous forms of criticism, it can provide social benefit, by shedding light on an earlier work, and, in the process, creating a new one. We thus line up with the courts that have held that parody, like other comment or criticism, may claim fair use under § 107. See, *e. g., Fisher v. Dees,* 794 F.2d 432 (9th Cir. 1986) ("When Sonny Sniffs Glue," a parody of "When Sunny Gets Blue," is fair use); *Elsmere Music, Inc. v. National Broadcasting Co.,* 482 F. Supp. 741 (S.D.N.Y.), aff'd, 623 F.2d 252 (2d Cir. 1980) ("I Love Sodom," a "Saturday Night Live" television parody of "I Love New York," is fair use).

---

[10] Because the fair use enquiry often requires close questions of judgment as to the extent of permissible borrowing in cases involving parodies (or other critical works), courts may also wish to bear in mind that the goals of the copyright law, "to stimulate the creation and publication of edifying matter," Leval 1134, are not always best served by automatically granting injunctive relief when parodists are found to have gone beyond the bounds of fair use. See 17 U.S.C. § 502(a) (court "*may* * * * grant * * * injunctions on such terms as it may deem reasonable to prevent or restrain infringement") (emphasis added).

[11] The obvious statutory exception to this focus on transformative uses is the straight reproduction of multiple copies for classroom distribution.

The germ of parody lies in the definition of the Greek *parodeia*, . . . as "a song sung alongside another." 972 F.2d at 1440, quoting 7 *Encyclopedia Britannica* 768 (15th ed. 1975). Modern dictionaries accordingly describe a parody as a "literary or artistic work that imitates the characteristic style of an author or a work for comic effect or ridicule,"[12] or as a "composition in prose or verse in which the characteristic turns of thought and phrase in an author or class of authors are imitated in such a way as to make them appear ridiculous."[13] For the purposes of copyright law, the nub of the definitions, and the heart of any parodist's claim to quote from existing material, is the use of some elements of a prior author's composition to create a new one that, at least in part, comments on that author's works. If, on the contrary, the commentary has no critical bearing on the substance or style of the original composition, which the alleged infringer merely uses to get attention or to avoid the drudgery in working up something fresh, the claim to fairness in borrowing from another's work diminishes accordingly (if it does not vanish), and other factors, like the extent of its commerciality, loom larger.[14] Parody needs to mimic an original to make its point, and so has some claim to use the creation of its victim's (or collective victims') imagination, whereas satire can stand on its own two feet and so requires justification for the very act of borrowing.[15]

The fact that parody can claim legitimacy for some appropriation does not, of course, tell either parodist or judge much about where to draw the line. Like a book review quoting the copyrighted material criticized, parody may or may not be fair use, and petitioners' suggestion that any parodic use is presumptively fair has no more justification in law or fact than the equally hopeful claim that any use for news reporting should be presumed fair, see *Harper & Row*, 471 U.S. at 561. The Act has no hint of an evidentiary preference for parodists over their victims, and no workable presumption for parody could take account of the fact that parody often shades into satire when society is lampooned through its creative artifacts, or that a work may contain both parodic and nonparodic elements. Accordingly, parody, like any other use, has to work its way through the relevant factors, and be judged case by case, in light of the ends of the copyright law.

---

[12] American Heritage Dictionary 1317 (3d ed. 1992).

[13] 11 Oxford English Dictionary 247 (2d ed. 1989).

[14] A parody that more loosely targets an original than the parody presented here may still be sufficiently aimed at an original work to come within our analysis of parody. If a parody whose wide dissemination in the market runs the risk of serving as a substitute for the original or licensed derivatives (see *infra* discussing factor four), it is more incumbent on one claiming fair use to establish the extent of transformation and the parody's critical relationship to the original. By contrast, when there is little or no risk of market substitution, whether because of the large extent of transformation of the earlier work, the new work's minimal distribution in the market, the small extent to which it borrows from an original, or other factors, taking parodic aim at an original is a less critical factor in the analysis, and looser forms of parody may be found to be fair use, as may satire with lesser justification for the borrowing than would otherwise be required.

[15] Satire has been defined as a work "in which prevalent follies or vices are assailed with ridicule," 14 Oxford English Dictionary at 500, or are "attacked through irony, derision, or wit," American Heritage Dictionary at 1604.

Here, the District Court held, and the Court of Appeals assumed, that 2 Live Crew's "Pretty Woman" contains parody, commenting on and criticizing the original work, whatever it may have to say about society at large. As the District Court remarked, the words of 2 Live Crew's song copy the original's first line, but then "quickly degenerat[e] into a play on words, substituting predictable lyrics with shocking ones . . . [that] derisively demonstrat[e] how bland and banal the Orbison song seems to them." 754 F. Supp., at 1155 (footnote omitted). Judge Nelson, dissenting below, came to the same conclusion, that the 2 Live Crew song "was clearly intended to ridicule the white-bread original" and "reminds us that sexual congress with nameless streetwalkers is not necessarily the stuff of romance and is not necessarily without its consequences. The singers (there are several) have the same thing on their minds as did the lonely man with the nasal voice, but here there is no hint of wine and roses." 972 F.2d at 1442. Although the majority below had difficulty discerning any criticism of the original in 2 Live Crew's song, it assumed for purposes of its opinion that there was some.

We have less difficulty in finding that critical element in 2 Live Crew's song than the Court of Appeals did, although having found it we will not take the further step of evaluating its quality. The threshold question when fair use is raised in defense of parody is whether a parodic character may reasonably be perceived.[16] Whether, going beyond that, parody is in good taste or bad does not and should not matter to fair use. As Justice Holmes explained, "[i]t would be a dangerous undertaking for persons trained only to the law to constitute themselves final judges of the worth of [a work], outside of the narrowest and most obvious limits. At the one extreme some works of genius would be sure to miss appreciation. Their very novelty would make them repulsive until the public had learned the new language in which their author spoke." *Bleistein v. Donaldson Lithographing Co.,* 188 U.S. 239, 251 (1903) (circus posters have copyright protection); cf. *Yankee Publishing Inc. v. News America Publishing, Inc.,* 809 F. Supp. 267, 280 (SDNY 1992) (Leval, J.) ("First Amendment protections do not apply only to those who speak clearly, whose jokes are funny, and whose parodies succeed") (trademark case).

While we might not assign a high rank to the parodic element here, we think it fair to say that 2 Live Crew's song reasonably could be perceived as commenting on the original or criticizing it, to some degree. 2 Live Crew juxtaposes the romantic musings of a man whose fantasy comes true, with degrading taunts, a bawdy demand for sex, and a sigh of relief from paternal responsibility. The later words can be taken as a comment on the naivete of the original of an earlier day, as a rejection of its sentiment that ignores the ugliness of street life and the debasement that it signifies. It is this joinder of reference and ridicule that marks off the

---

[16] The only further judgment, indeed, that a court may pass on a work goes to an assessment of whether the parodic element is slight or great, and the copying small or extensive in relation to the parodic element, for a work with slight parodic element and extensive copying will be more likely to merely "supersede the objects" of the original. See *infra* . . . discussing factors three and four.

author's choice of parody from the other types of comment and criticism that traditionally have had a claim to fair use protection as transformative works.[17]

The Court of Appeals, however, immediately cut short the enquiry into 2 Live Crew's fair use claim by confining its treatment of the first factor essentially to one relevant fact, the commercial nature of the use. The court then inflated the significance of this fact by applying a presumption ostensibly culled from *Sony,* that "every commercial use of copyrighted material is presumptively . . . unfair . . . ." *Sony* [*Corp. v. Universal City Studios, Inc.*], 464 U.S. [417], 451 [(1984)]. In giving virtually dispositive weight to the commercial nature of the parody, the Court of Appeals erred.

The language of the statute makes clear that the commercial or nonprofit educational purpose of a work is only one element of the first factor enquiry into its purpose and character. Section 107(1) uses the term "including" to begin the dependent clause referring to commercial use, and the main clause speaks of a broader investigation into "purpose and character." As we explained in *Harper & Row,* Congress resisted attempts to narrow the ambit of this traditional enquiry by adopting categories of presumptively fair use, and it urged courts to preserve the breadth of their traditionally ample view of the universe of relevant evidence. 471 U.S. at 561; House Report, p. 66. Accordingly, the mere fact that a use is educational and not for profit does not insulate it from a finding of infringement, any more than the commercial character of a use bars a finding of fairness. If, indeed, commerciality carried presumptive force against a finding of fairness, the presumption would swallow nearly all of the illustrative uses listed in the preamble paragraph of § 107, including news reporting, comment, criticism, teaching, scholarship, and research, since these activities "are generally conducted for profit in this country." *Harper & Row,* 471 U.S. at 592 (Brennan, J., dissenting). Congress could not have intended such a rule, which certainly is not inferable from the common-law cases, arising as they did from the world of letters in which Samuel Johnson could pronounce that "[n]o man but a blockhead ever wrote, except for money." 3 *Boswell's Life of Johnson* 19 (G. Hill ed. 1934).

Sony itself called for no hard evidentiary presumption. There, we emphasized the need for a "sensitive balancing of interests," 464 U.S. at 455, n. 40, noted that Congress had "eschewed a rigid, bright-line approach to fair use," *id.* at 449, n. 31, and stated that the commercial or nonprofit educational character of a work is "not conclusive," *id.* at 448-449, but rather a fact to be "weighed along with other[s] in fair use decisions," *id.* at 449, n. 32 (quoting House Report, p. 66). The Court of Appeals's elevation of one sentence from *Sony* to a *per se* rule thus runs as much counter to *Sony* itself as to the long common-law tradition of fair use adjudication. Rather, as we explained in *Harper & Row, Sony* stands for the proposition that the "fact that a publication was commercial as opposed to nonprofit is a

---

[17] We note in passing that 2 Live Crew need not label their whole album, or even this song, a parody in order to claim fair use protection, nor should 2 Live Crew be penalized for this being its first parodic essay. Parody serves its goals whether labeled or not, and there is no reason to require parody to state the obvious (or even the reasonably perceived). See Patry & Perlmutter 716-717.

separate factor that tends to weigh against a finding of fair use." 471 U.S. at 562. But that is all, and the fact that even the force of that tendency will vary with the context is a further reason against elevating commerciality to hard presumptive significance. The use, for example, of a copyrighted work to advertise a product, even in a parody, will be entitled to less indulgence under the first factor of the fair use enquiry than the sale of a parody for its own sake, let alone one performed a single time by students in school.

B

The second statutory factor, "the nature of the copyrighted work," § 107(2) . . . calls for recognition that some works are closer to the core of intended copyright protection than others, with the consequence that fair use is more difficult to establish when the former works are copied. We agree with both the District Court and the Court of Appeals that the Orbison original's creative expression for public dissemination falls within the core of the copyright's protective purposes. This fact, however, is not much help in this case, or ever likely to help much in separating the fair use sheep from the infringing goats in a parody case, since parodies almost invariably copy publicly known, expressive works.

C

The third factor asks whether "the amount and substantiality of the portion used in relation to the copyrighted work as a whole," § 107(3), . . . are reasonable in relation to the purpose of the copying. Here, attention turns to the persuasiveness of a parodist's justification for the particular copying done, and the enquiry will harken back to the first of the statutory factors, for, as in prior cases, we recognize that the extent of permissible copying varies with the purpose and character of the use. The facts bearing on this factor will also tend to address the fourth, by revealing the degree to which the parody may serve as a market substitute for the original or potentially licensed derivatives. *See* Leval 1123.

The District Court considered the song's parodic purpose in finding that 2 Live Crew had not helped themselves overmuch. The Court of Appeals disagreed, stating that "[w]hile it may not be inappropriate to find that no more was taken than necessary, the copying was qualitatively substantial. . . . We conclude that taking the heart of the original and making it the heart of a new work was to purloin a substantial portion of the essence of the original." 972 F.2d at 1438.

The Court of Appeals is of course correct that this factor calls for thought not only about the quantity of the materials used, but about their quality and importance, too. . . . We also agree with the Court of Appeals that whether "a substantial portion of the infringing work was copied verbatim" from the copyrighted work is a relevant question, see *id.* at 565, for it may reveal a dearth of transformative character or purpose under the first factor, or a greater likelihood of market harm under the fourth; a work composed primarily of an original, particularly its heart, with little added or changed, is more likely to be a merely superseding use, fulfilling demand for the original.

Where we part company with the court below is in applying these guides to parody, and in particular to parody in the song before us. Parody presents a difficult case. Parody's humor, or in any event its comment, necessarily springs from recognizable allusion to its object through distorted imitation. Its art lies in the tension between a known original and its parodic twin. When parody takes aim at a particular original work, the parody must be able to "conjure up" at least enough of that original to make the object of its critical wit recognizable. What makes for this recognition is quotation of the original's most distinctive or memorable features, which the parodist can be sure the audience will know. Once enough has been taken to assure identification, how much more is reasonable will depend, say, on the extent to which the song's overriding purpose and character is to parody the original or, in contrast, the likelihood that the parody may serve as a market substitute for the original. But using some characteristic features cannot be avoided.

We think the Court of Appeals was insufficiently appreciative of parody's need for the recognizable sight or sound when it ruled 2 Live Crew's use unreasonable as a matter of law. It is true, of course, that 2 Live Crew copied the characteristic opening bass riff (or musical phrase) of the original, and true that the words of the first line copy the Orbison lyrics. But if quotation of the opening riff and the first line may be said to go to the "heart" of the original, the heart is also what most readily conjures up the song for parody, and it is the heart at which parody takes aim. Copying does not become excessive in relation to parodic purpose merely because the portion taken was the original's heart. If 2 Live Crew had copied a significantly less memorable part of the original, it is difficult to see how its parodic character would have come through. See *Fisher v. Dees*, 794 F.2d at 439.

This is not, of course, to say that anyone who calls himself a parodist can skim the cream and get away scot free. In parody, as in news reporting, context is everything, and the question of fairness asks what else the parodist did besides go to the heart of the original. It is significant that 2 Live Crew not only copied the first line of the original, but thereafter departed markedly from the Orbison lyrics for its own ends. 2 Live Crew not only copied the bass riff and repeated it, but also produced otherwise distinctive sounds, interposing "scraper" noise, over-laying the music with solos in different keys, and altering the drum beat. This is not a case, then, where "a substantial portion" of the parody itself is composed of a "verbatim" copying of the original. It is not, that is, a case where the parody is so insubstantial, as compared to the copying, that the third factor must be resolved as a matter of law against the parodists.

Suffice it to say here that, as to the lyrics, we think the Court of Appeals correctly suggested that "no more was taken than necessary," but just for that reason, we fail to see how the copying can be excessive in relation to its parodic purpose, even if the portion taken is the original's "heart." As to the music, we express no opinion whether repetition of the bass riff is excessive copying, and we remand to permit evaluation of the amount taken, in light of the song's parodic purpose and character, its transformative elements, and considerations of the potential for market substitution sketched more fully below.

D

The fourth fair use factor is "the effect of the use upon the potential market for or value of the copyrighted work." § 107(4). It requires courts to consider not only the extent of market harm caused by the particular actions of the alleged infringer, but also "whether unrestricted and widespread conduct of the sort engaged in by the defendant . . . would result in a substantially adverse impact on the potential market" for the original. Nimmer § 13.05[A][4], p. 13-102.61. The enquiry "must take account not only of harm to the original but also of harm to the market for derivative works." *Harper & Row, supra,* at 568.

Since fair use is an affirmative defense, its proponent would have difficulty carrying the burden of demonstrating fair use without favorable evidence about relevant markets.[21] In moving for summary judgment, 2 Live Crew left themselves at just such a disadvantage when they failed to address the effect on the market for rap derivatives, and confined themselves to uncontroverted submissions that there was no likely effect on the market for the original. They did not, however, thereby subject themselves to the evidentiary presumption applied by the Court of Appeals. In assessing the likelihood of significant market harm, the Court of Appeals quoted from language in *Sony* that " '[i]f the intended use is for commercial gain, that likelihood may be presumed. But if it is for a noncommercial purpose, the likelihood must be demonstrated.' " 972 F.2d at 1438, quoting *Sony,* 464 U.S. at 451. The court reasoned that because "the use of the copyrighted work is wholly commercial, . . . we presume that a likelihood of future harm to Acuff-Rose exists." 972 F.2d at 1438. In so doing, the court resolved the fourth factor against 2 Live Crew, just as it had the first, by applying a presumption about the effect of commercial use, a presumption which as applied here we hold to be error.

No "presumption" or inference of market harm that might find support in *Sony* is applicable to a case involving something beyond mere duplication for commercial purposes. *Sony*'s discussion of a presumption contrasts a context of verbatim copying of the original in its entirety for commercial purposes, with the noncommercial context of *Sony* itself (home copying of television programming). In the former circumstances, what *Sony* said simply makes common sense: when a commercial use amounts to mere duplication of the entirety of an original, it clearly "supersede[s] the objects," *Folsom v. Marsh, supra,* at 348, of the original and serves as a market replacement for it, making it likely that cognizable market harm to the original will occur. *Sony, supra,* at 451. But when, on the contrary, the second use is transformative, market substitution is at least less certain, and market harm may not be so readily inferred. Indeed, as to parody pure and simple, it

---

[21] Even favorable evidence, without more, is no guarantee of fairness. Judge Leval gives the example of the film producer's appropriation of a composer's previously unknown song that turns the song into a commercial success; the boon to the song does not make the film's simple copying fair. Leval 1124, n. 84. This factor, no less than the other three, may be addressed only through a "sensitive balancing of interests." *Sony,* 464 U.S. at 455 n.40. Market harm is a matter of degree, and the importance of this factor will vary, not only with the amount of harm, but also with the relative strength of the showing on the other factors.

is more likely that the new work will not affect the market for the original in a way cognizable under this factor, that is, by acting as a substitute for it ("supersed[ing] [its] objects"). This is so because the parody and the original usually serve different market functions.

We do not, of course, suggest that a parody may not harm the market at all, but when a lethal parody, like a scathing theater review, kills demand for the original, it does not produce a harm cognizable under the Copyright Act. Because "parody may quite legitimately aim at garroting the original, destroying it commercially as well as artistically," B. Kaplan, *An Unhurried View of Copyright* 69 (1967), the role of the courts is to distinguish between "[b]iting criticism [that merely] suppresses demand [and] copyright infringement[, which] usurps it." *Fisher v. Dees,* 794 F.2d at 438.

This distinction between potentially remediable displacement and unremediable disparagement is reflected in the rule that there is no protectible derivative market for criticism. The market for potential derivative uses includes only those that creators of original works would in general develop or license others to develop. Yet the unlikelihood that creators of imaginative works will license critical reviews or lampoons of their own productions removes such uses from the very notion of a potential licensing market. "People ask . . . for criticism, but they only want praise." S. Maugham, *Of Human Bondage* 241 (Penguin ed. 1992). Thus, to the extent that the opinion below may be read to have considered harm to the market for parodies of "Oh, Pretty Woman," the court erred.

In explaining why the law recognizes no derivative market for critical works, including parody, we have, of course, been speaking of the later work as if it had nothing but a critical aspect (*i.e.,* "parody pure and simple," *supra,* at 591). But the later work may have a more complex character, with effects not only in the arena of criticism but also in protectible markets for derivative works, too. In that sort of case, the law looks beyond the criticism to the other elements of the work, as it does here. 2 Live Crew's song comprises not only parody but also rap music, and the derivative market for rap music is a proper focus of enquiry. Evidence of substantial harm to it would weigh against a finding of fair use, because the licensing of derivatives is an important economic incentive to the creation of originals. Of course, the only harm to derivatives that need concern us, as discussed above, is the harm of market substitution. The fact that a parody may impair the market for derivative uses by the very effectiveness of its critical commentary is no more relevant under copyright than the like threat to the original market.

Although 2 Live Crew submitted uncontroverted affidavits on the question of market harm to the original, neither they, nor Acuff-Rose, introduced evidence or affidavits addressing the likely effect of 2 Live Crew's parodic rap song on the market for a nonparody, rap version of "Oh, Pretty Woman." And while Acuff-Rose would have us find evidence of a rap market in the very facts that 2 Live Crew recorded a rap parody of "Oh, Pretty Woman" and another rap group sought a license to record a rap derivative, there was no evidence that a potential rap market was harmed in any way by 2 Live Crew's parody, rap version. The fact that 2 Live Crew's parody sold as part of a collection of rap songs says very little about the parody's effect on a market for a rap version of the original, either of the music alone or of the music with its lyrics. The District Court essentially passed on this issue, observing that Acuff-Rose is free to record "whatever version of the original it desires," 754 F. Supp., at 1158; the Court of Appeals went the other way by erroneous presumption. Contrary to each treatment, it is impossible to deal with the fourth factor except by recognizing that a silent record on an important factor bearing on fair use disentitled the proponent of the defense, 2 Live Crew, to summary judgment. The evidentiary hole will doubtless be plugged on remand.

III

It was error for the Court of Appeals to conclude that the commercial nature of 2 Live Crew's parody of "Oh, Pretty Woman" rendered it presumptively unfair. No such evidentiary presumption is available to address either the first factor, the character and purpose of the use, or the fourth, market harm, in determining whether a transformative use, such as parody, is a fair one. The court also erred in holding that 2 Live Crew had necessarily copied excessively from the Orbison original, considering the parodic purpose of the use. We therefore reverse the judgment of the Court of Appeals and remand the case for further proceedings consistent with this opinion.

KENNEDY, JUSTICE, concurring:

I agree that remand is appropriate and join the opinion of the Court, with these further observations about the fair use analysis of parody.

. . . .

. . . More than arguable parodic content should be required to deem a would-be parody a fair use. Fair use is an affirmative defense, so doubts about whether a given use is fair should not be resolved in favor of the self-proclaimed parodist. We should not make it easy for musicians to exploit existing works and then later claim that their rendition was a valuable commentary on the original. Almost any revamped modern version of a familiar composition can be construed as a "comment on the naivete of the original," because of the difference in style and because it will be amusing to hear how the old tune sounds in the new genre. Just the thought of a rap version of Beethoven's Fifth Symphony or "Achy Breaky Heart" is bound to make people smile. If we allow any weak transformation to qualify as parody, however, we weaken the protection of copyright. And under-protection of copyright disserves the goals of copyright just as much as overprotection, by reducing the financial incentive to create.

The Court decides it is "fair to say that 2 Live Crew's song reasonably could be perceived as commenting on the original or criticizing it, to some degree." While I am not so assured that 2 Live Crew's song is a legitimate parody, the Court's treatment of the remaining factors leaves room for the District Court to determine on remand that the song is not a fair use. As future courts apply our fair use analysis, they must take care to ensure that not just any commercial takeoff is rationalized *post hoc* as a parody.

With these observations, I join the opinion of the Court.

APPENDIX A

"Oh, Pretty Woman" by Roy Orbison and William Dees

Pretty Woman, walking down the street,
Pretty Woman, the kind I like to meet,
Pretty Woman, I don't believe you,
   you're not the truth,
No one could look as good as you
Mercy

Pretty Woman, won't you pardon me,
Pretty Woman, I couldn't help but see,
Pretty Woman, that you look lovely as can be
Are you lonely just like me?

Pretty Woman, stop a while,
Pretty Woman, talk a while,
Pretty Woman give your smile to me
Pretty Woman, yeah, yeah, yeah
Pretty Woman, look my way,
Pretty Woman, say you'll stay with me
'Cause I need you, I'll treat you right
Come to me baby, Be mine tonight

Pretty Woman, don't walk on by,
Pretty Woman, don't make me cry,
Pretty Woman, don't walk away,
Hey, O. K.
If that's the way it must be, O. K.
I guess I'll go on home, it's late
There'll be tomorrow night, but wait!

What do I see
Is she walking back to me?
Yeah, she's walking back to me!
Oh, Pretty Woman.

APPENDIX B

"Pretty Woman" as recorded by 2 Live Crew

Pretty woman walkin' down the street
Pretty woman girl you look so sweet
Pretty woman you bring me down to that knee
Pretty woman you make me wanna beg please
Oh, pretty woman

Big hairy woman you need to shave that stuff
Big hairy woman you know I bet it's tough
Big hairy woman all that hair it ain't legit 'Cause you look like 'Cousin It'
Big hairy woman

Bald headed woman girl your hair won't grow
Bald headed woman you got a teeny weeny afro
Bald headed woman you know your hair could look nice
Bald headed woman first you got to roll it with rice
Bald headed woman here, let me get this hunk of biz for ya
Ya know what I'm saying you look better than rice a roni
Oh bald headed woman

Big hairy woman come on in
And don't forget your bald headed friend
Hey pretty woman let the boys Jump in

Two timin' woman girl you know you ain't right
Two timin' woman you's out with my boy last night
Two timin' woman that takes a load off my mind
Two timin' woman now I know the baby ain't mine
Oh, two timin' woman
Oh pretty woman.

Notes & Questions

1. The Court begins its opinion by noting that it is "uncontested" that 2 Live Crew's song would be an infringement of Acuff-Rose's right. Which of the 106 rights is infringed? Why do you think that 2 Live Crew conceded there was a prima facie case of infringement? With the question of prima facie infringement out of the way, the court turns its attention to the fair use argument. The court indicates that fair use is an affirmative defense, placing the burden on the defendant to prove facts supporting the defense. Given the language of the statute—that the rights granted in § 106 are "subject to" fair use rights, and that § 107 fair uses are "not an infringement of copyright" and are recognized "notwithstanding the provisions of section 106"—is it clear that Congress intended defendants to have to prove fair use? Is there at least an argument to be made that the absence of fair use

should be part of the plaintiff's case in chief? *See* Lydia Pallas Loren, *Fair Use: An Affirmative Defense?* 90 Wash. L. Rev. 685 (2015).

2. In his concurrence Justice Kennedy implores courts not to resolve doubts in favor of the putative parodist. He relies on the view that fair use is an affirmative defense. Indeed, if a defendant does not carry her burden on a defense, the plaintiff should prevail. Does this change how you feel about characterizing fair use as an affirmative defense? Recall the discussion of cost of errors at the beginning of this book. Does the cost of errors council in favor of deciding close cases for the copyright owner, as Justice Kennedy desires, or for the defendant?

3. With the burden placed on 2 Live Crew to prove fair use, what will it need to show on remand to prevail? How might it attempt to make that showing? On remand the parties settled, with 2 Live Crew taking a license on undisclosed terms. Stan Soocher, *They Fought the Law: Rock Music Goes to Court* 189-90 (1999). What incentives did both parties have to enter into the settlement agreement? What does the fact of settlement say about the viability of a fair use defense for parody? How would you advise a client who wants to engage in a parody?

4. While it is not accurate to say "parody is fair use," it is accurate to say that parody weighs in favor of a finding of fair use. How much in favor depends on a variety of other facts, as you see in *Campbell*. What about satire? Does satire have no claim to fair use? Consider groups such as The Capital Steps, artists like Weird Al Yankovic, or satirists like John Stewart and Stephen Colbert. Should they obtain permission before creating and performing their works? Pay close attention to footnote 14 in the Court's opinion.

5. The court emphasizes that the fair use analysis is to be conducted case by case. In this way the fair use doctrine constitutes *ex post* ordering by courts. When a dispute arises, the court will weigh the four factors set out in the statute and determine whether the use is fair. In conducting its analysis, the Court tells us, no presumptions are appropriate. Case-by-case analysis provides extreme flexibility for courts, but how about when advising clients? Given the case-by-case nature of the inquiry, and the lack of bright-line rules, it can be difficult to provide clients with an assessment of the lawfulness of their activities with a high level of confidence. Professor Larry Lessig has asserted that all that fair use guarantees is a right to hire a lawyer. Lawrence Lessig, *Free Culture* (2004). However, the uncertainty inherent in the fair use doctrine applies to both sides of an infringement claim — the copyright owner and the alleged infringer. Do we want to encourage such a dynamic in potential fair use cases?

The Copyright Act contains a fee shifting provision, permitting the prevailing party, whether it be plaintiff or defendant, to seek to recover attorney's fees. 17 U.S.C. § 505. Might the availability of fee shifting affect a party's willingness to litigate, as opposed to settle, a dispute? *See, e.g., Brownmark Films LLC v. Comedy Partners*, No. 10-CV-1013-JPS, 2011 WL 6002961, at *6 (E.D. Wis. Nov. 30, 2011) (awarding attorney's fees because "defendants' fair-use argument was very strong, and Brownmark's legal position was objectively unreasonable.").

6. The Court uses the word "transformative" or variations of it many times throughout the opinion. How far does the conception of "transformation" stretch? Does the copyrighted work itself need to be transformed or is a transformative "use" of the work all that is needed? In *Bill Graham Archives v. Dorling Kindersley Ltd.*, 448 F.3d 605 (2d Cir. 2006), the Second Circuit described the use of reproductions of entire concert posters in the coffee table book *Grateful Dead: The Illustrated Trip* as having a "transformative purpose." The court held that the "purpose in using the copyrighted images [in the book] is plainly different from the original purpose for which they were created." Noting that the posters were themselves "historical artifacts" that "represent the actual occurrence of Grateful Dead concert events," the court concluded that the use of seven separate posters was fair use. *Id.* at 609-10. When analyzing the first factor, is it appropriate to focus on a transformative *purpose* in addition to a transformative use?

7. The fourth factor examines the economic effect of the defendant's use on the market for the copyrighted work, but the Court is clear that we can include only certain types of harm. What harms are cognizable under this factor? Another way to consider the inquiry is to ask, what are the "relevant markets" that one should examine to determine harm? The market for the original is a relevant market, but how about markets for "derivative works"? If a use is fair use, should one consider a market for that type of use? What are *Campbell*'s "relevant markets"?

The Court supports its insistence that one must consider the markets for derivative works when examining the fourth factor, market harm, by quoting from its prior decision in *Harper & Row Publishers v. Nation Enters.*, 471 U.S. 539 (1985). In *Harper & Row* the defendant, *The Nation* magazine, had secretly obtained a pre-release copy of President Gerald Ford's memoir and published an article that included attention-grabbing quotations from the memoir. *Time* magazine had previously licensed from Ford's publisher, Harper & Row, the right to serialize the memoir in return for set payments. As a result of *The Nation*'s scoop, *Time* exercised a right it had under the license to cancel the project without making its second of two payments. Serialization of books was, at the time, an established market—the licensing market for pre-publication excerpts. The Court mentioned considering adverse effects on "the *potential* market for the copyrighted work" and harm to "any part of the normal market." *Id.* at 568 (quoting *Sony Corp. of America v. Universal City Studios, Inc.*, 464 U. S. 417, 484 (1984), and S. Rep. No. 94-473, p. 65 (1975)). The Court also highlighted that the inquiry into the first factor should consider "whether the user stands to profit from exploitation of the copyrighted materials without paying the customary price." *Id.* at 562. Determining which markets count in a fair use analysis (which are "normal" or "customary") can be an extremely important inquiry, especially as technology creates new opportunities for exploitation of previously created copyrighted works.

8. In footnote 10 the Court makes reference to the proper remedy if infringement is found in a close fair use case, indicating that an injunction may not be appropriate. This possibility took on greater force following the Court's examination of injunctions in patent cases, *eBay Inc. v. MercExchange, L.L.C.*, 547 U.S. 388 (2006), and is explored more in Chapter 7.

B. Transformative Uses

The open-ended nature of the fair use doctrine makes it an important tool for calibrating copyright protection when technologies change, and one of the most critical doctrines to understand in times of rapid technological change.

As new modes of creation and dissemination develop, and new models for earning money from "works of authorship" appear, these new models are often challenged by the established industries. Because fair use is meant to act as a safety valve, ensuring that copyright protection is not used to stifle innovation, the doctrine is often called upon to strike balances in the face of innovative technologies, inevitably breaking new ground for the doctrine. Using the fair use doctrine as the first line of defense in these scenarios can be extremely challenging for courts.

Perfect 10, Inc. v. Amazon, Inc.
508 F.3d 1146 (9th Cir. 2007)

Ikuta, Judge:

[Review the facts of this case above, in Section IV. After concluding that the plaintiff had demonstrated a likelihood of success on its claim that Google's thumbnail images infringed its display right, the court turned to Google's fair use argument.]

In applying the fair use analysis in this case, we are guided by *Kelly v. Arriba Soft Corp.*, which considered substantially the same use of copyrighted photographic images as is at issue here. *See* 336 F.3d 811. In *Kelly*, a photographer brought a direct infringement claim against Arriba, the operator of an Internet search engine. The search engine provided thumbnail versions of the photographer's images in response to search queries. We held that Arriba's use of thumbnail images was a fair use primarily based on the transformative nature of a search engine and its benefit to the public. We also concluded that Arriba's use of the thumbnail images did not harm the photographer's market for his image.

In this case, the district court determined that Google's use of thumbnails was not a fair use and distinguished *Kelly*. We consider these distinctions in the context of the four-factor fair use analysis, remaining mindful that Perfect 10 has the burden of proving that it will successfully challenge any evidence Google presents to support its affirmative defense.

Purpose and character of the use. The first factor, 17 U.S.C. § 107(1), requires a court to consider "the purpose and character of the use, including whether such use is of a commercial nature or is for nonprofit educational purposes." The central purpose of this inquiry is to determine whether and to what extent the new work is "transformative." *Campbell*, 510 U.S. at 579. A work is "transformative" when the new work does not "merely supersede the objects of the original creation" but rather "adds something new, with a further purpose or different character, altering the first with new expression, meaning, or message." *Id.* Conversely, if the new work "supersede[s] the use of the original," the use is likely not a fair use. *Harper & Row Publishers v. Nation Enters.*, 471 U.S. 539, 550-51 (1985); see also *Wall Data Inc. v. L.A. County Sheriff's Dep't*, 447 F.3d 769, 778-82 (9th Cir.

2006) (using a copy to save the cost of buying additional copies of a computer program was not a fair use).

As noted in *Campbell*, a "transformative work" is one that alters the original work "with new expression, meaning, or message." *Campbell*, 510 U.S. at 579. "A use is considered transformative only where a defendant changes a plaintiff's copyrighted work or uses the plaintiff's copyrighted work in a different context such that the plaintiff's work is transformed into a new creation." *Wall Data*, 447 F.3d at 778.

Google's use of thumbnails is highly transformative. In *Kelly*, we concluded that Arriba's use of thumbnails was transformative because "Arriba's use of the images serve[d] a different function than Kelly's use—improving access to information on the [I]nternet versus artistic expression." *Kelly*, 336 F.3d at 819. Although an image may have been created originally to serve an entertainment, aesthetic, or informative function, a search engine transforms the image into a pointer directing a user to a source of information. Just as a "parody has an obvious claim to transformative value" because "it can provide social benefit, by shedding light on an earlier work, and, in the process, creating a new one," *Campbell*, 510 U.S. at 579, a search engine provides social benefit by incorporating an original work into a new work, namely, an electronic reference tool. Indeed, a search engine may be more transformative than a parody because a search engine provides an entirely new use for the original work, while a parody typically has the same entertainment purpose as the original work. . . .

The fact that Google incorporates the entire Perfect 10 image into the search engine results does not diminish the transformative nature of Google's use. As the district court correctly noted, we determined in *Kelly* that even making an exact copy of a work may be transformative so long as the copy serves a different function than the original work. . . . Here, Google uses Perfect 10's images in a new context to serve a different purpose.

The district court nevertheless determined that Google's use of thumbnail images was less transformative than Arriba's use of thumbnails in *Kelly* because Google's use of thumbnails superseded Perfect 10's right to sell its reduced-size images for use on cell phones. *See Perfect 10*, 416 F. Supp. 2d at 849. The district court stated that "mobile users can download and save the thumbnails displayed by Google Image Search onto their phones," and concluded "to the extent that users may choose to download free images to their phone rather than purchase [Perfect 10's] reduced-size images, Google's use supersedes [Perfect 10's]." *Id.*

Additionally, the district court determined that the commercial nature of Google's use weighed against its transformative nature. Although *Kelly* held that the commercial use of the photographer's images by Arriba's search engine was less exploitative than typical commercial use, and thus weighed only slightly against a finding of fair use, the district court here distinguished *Kelly* on the ground that some website owners in the AdSense program had infringing Perfect 10 images on their websites. The district court held that because Google's thumbnails "lead users to sites that directly benefit Google's bottom line," the AdSense program increased the commercial nature of Google's use of Perfect 10's images. *Id.* at 847.

In conducting our case-specific analysis . . . in light of the purposes of copyright, *Campbell,* 510 U.S. at 581, we must weigh Google's superseding and commercial uses of thumbnail images against Google's significant transformative use, as well as the extent to which Google's search engine promotes the purposes of copyright and serves the interests of the public. Although the district court acknowledged the "truism that search engines such as Google Image Search provide great value to the public," *Perfect 10*, 416 F. Supp. 2d at 848- 49, the district court did not expressly consider whether this value outweighed the significance of Google's superseding use or the commercial nature of Google's use. *Id.* at 849. The Supreme Court, however, has directed us to be mindful of the extent to which a use promotes the purposes of copyright and serves the interests of the public.

We note that the superseding use in this case is not significant at present: the district court did not find that any downloads for mobile phone use had taken place. Moreover, while Google's use of thumbnails to direct users to AdSense partners containing infringing content adds a commercial dimension that did not exist in *Kelly*, the district court did not determine that this commercial element was significant. The district court stated that Google's AdSense programs as a whole contributed "$630 million, or 46% of total revenues" to Google's bottom line, but noted that this figure did not "break down the much smaller amount attributable to websites that contain infringing content." *Id.* at 847 & n.12 (internal quotation omitted).

We conclude that the significantly transformative nature of Google's search engine, particularly in light of its public benefit, outweighs Google's superseding and commercial uses of the thumbnails in this case. In reaching this conclusion, we note the importance of analyzing fair use flexibly in light of new circumstances. We are also mindful of the Supreme Court's direction that "the more transformative the new work, the less will be the significance of other factors, like commercialism, that may weigh against a finding of fair use." *Campbell*, 510 U.S. at 579.

Accordingly, we disagree with the district court's conclusion that because Google's use of the thumbnails could supersede Perfect 10's cell phone download use and because the use was more commercial than Arriba's, this fair use factor weighed "slightly" in favor of Perfect 10. Instead, we conclude that the transformative nature of Google's use is more significant than any incidental superseding use or the minor commercial aspects of Google's search engine and website. Therefore, the district court erred in determining this factor weighed in favor of Perfect 10.

The nature of the copyrighted work. . . .

Here, the district court found that Perfect 10's images were creative but also previously published. The right of first publication is "the author's right to control the first public appearance of his expression." *Harper & Row*, 471 U.S. at 564. Because this right encompasses "the choices of when, where, and in what form first to publish a work," *id.*, an author exercises and exhausts this one-time right by publishing the work in any medium. Once Perfect 10 has exploited this commercially valuable right of first publication by putting its images on the Internet for paid subscribers, Perfect 10 is no longer entitled to the enhanced protection available for an unpublished work. Accordingly the district court did not err in holding that this factor weighed only slightly in favor of Perfect 10.

The amount and substantiality of the portion used. . . . In *Kelly*, we held Arriba's use of the entire photographic image was reasonable in light of the purpose of a search engine. *Kelly*, 336 F.3d at 821. Specifically, we noted, "[i]t was necessary for Arriba to copy the entire image to allow users to recognize the image and decide whether to pursue more information about the image or the originating [website]. If Arriba only copied part of the image, it would be more difficult to identify it, thereby reducing the usefulness of the visual search engine." *Id.* Accordingly, we concluded that this factor did not weigh in favor of either party. *Id.* Because the same analysis applies to Google's use of Perfect 10's image, the district court did not err in finding that this factor favored neither party.

Effect of use on the market. The fourth factor is "the effect of the use upon the potential market for or value of the copyrighted work." 17 U.S.C. § 107(4). In *Kelly*, we concluded that Arriba's use of the thumbnail images did not harm the market for the photographer's full-size images. We reasoned that because thumbnails were not a substitute for the full-sized images, they did not harm the photographer's ability to sell or license his full-sized images. The district court here followed *Kelly*'s reasoning, holding that Google's use of thumbnails did not hurt Perfect 10's market for full-size images.

Perfect 10 argues that the district court erred because the likelihood of market harm may be presumed if the intended use of an image is for commercial gain. However, this presumption does not arise when a work is transformative because "market substitution is at least less certain, and market harm may not be so readily inferred." *Campbell*, 510 U.S. at 591. As previously discussed, Google's use of thumbnails for search engine purposes is highly transformative. Because market harm cannot be presumed, and because Perfect 10 has not introduced evidence that Google's thumbnails would harm Perfect 10's existing or potential market for full-size images, we reject this argument.

Perfect 10 also has a market for reduced-size images, an issue not considered in *Kelly*. The district court held that "Google's use of thumbnails likely does harm the potential market for the downloading of [Perfect 10's] reduced-size images onto cell phones." *Perfect 10*, 416 F. Supp. 2d at 851. The district court reasoned that persons who can obtain Perfect 10 images free of charge from Google are less likely to pay for a download, and the availability of Google's thumbnail images would harm Perfect 10's market for cell phone downloads. As we discussed above, the district court did not make a finding that Google users have downloaded thumbnail images for cell phone use. This potential harm to Perfect 10's market remains hypothetical. We conclude that this factor favors neither party.

Having undertaken a case-specific analysis of all four factors, we now weigh these factors together "in light of the purposes of copyright." *Campbell*, 510 U.S.

at 578. We note that Perfect 10 has the burden of proving that it would defeat Google's affirmative fair use defense. In this case, Google has put Perfect 10's thumbnail images (along with millions of other thumbnail images) to a use fundamentally different than the use intended by Perfect 10. In doing so, Google has provided a significant benefit to the public. Weighing this significant transformative use against the unproven use of Google's thumbnails for cell phone downloads, and considering the other fair use factors, all in light of the purpose of copyright, we conclude that Google's use of Perfect 10's thumbnails is a fair use. Because the district court here "found facts sufficient to evaluate each of the statutory factors . . . [we] need not remand for further fact-finding." *Harper & Row*, 471 U.S. at 560. We conclude that Perfect 10 is unlikely to be able to overcome Google's fair use defense and, accordingly, we vacate the preliminary injunction regarding Google's use of thumbnail images.

Notes & Questions

1. Has the Ninth Circuit stretched the notion of "transformative" use too far? Was such stretching an inevitable consequence of the Supreme Court's emphasis on transformative use in the *Campbell* case?

2. Should the operators of search engines ever be concerned about copyright infringement claims? After *Perfect 10* is there evidence that a plaintiff could present that would demonstrate a negative effect on the market for or value of the plaintiff's work?

Authors Guild, Inc. v. HathiTrust
755 F.3d 87 (2d Cir. 2014)

PARKER, JUDGE:

Beginning in 2004, several research universities including the University of Michigan, the University of California at Berkeley, Cornell University, and the University of Indiana agreed to allow Google to electronically scan the books in their collections. In October 2008, thirteen universities announced plans to create a repository for the digital copies and founded an organization called HathiTrust to set up and operate the HathiTrust Digital Library (or "HDL"). Colleges, universities, and other nonprofit institutions became members of HathiTrust and made the books in their collections available for inclusion in the HDL. HathiTrust currently has 80 member institutions and the HDL contains digital copies of more than ten million works, published over many centuries, written in a multitude of languages, covering almost every subject imaginable. . . .

. . . HathiTrust allows the general public to search for particular terms across all digital copies in the repository. Unless the copyright holder authorizes broader use, the search results show only the page numbers on which the search term is found within the work and the number of times the term appears on each page. The HDL does not display to the user any text from the underlying copyrighted work (either in "snippet" form or otherwise). Consequently, the user is not able to view either the page on which the term appears or any other portion of the book.

Below is an example of the results a user might see after running an HDL full-text search:

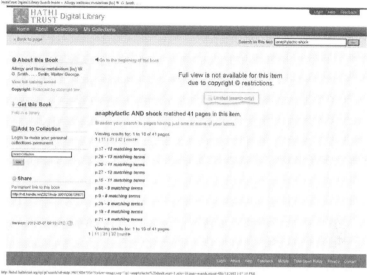

The HDL stores digital copies of the works in four different locations. One copy is stored on its primary server in Michigan, one on its secondary server in Indiana, and two on separate backup tapes at the University of Michigan.[3] Each copy contains the full text of the work, in a machine readable format, as well as the *images* of each page in the work as they appear in the print version. . . .

This case began when twenty authors and authors' associations (collectively, the "Authors") sued HathiTrust, one of its member universities, and the presidents of four other member universities (collectively, the "Libraries") for copyright infringement seeking declaratory and injunctive relief. . . .

The District Court's Opinion

The district court granted the Libraries' . . . motions for summary judgment on the infringement claims on the basis that the . . . uses permitted by the HDL were fair uses. In this assessment, the district court gave considerable weight to what it found to be the "transformative" nature of the . . . uses and to what it described as the HDL's "invaluable" contribution to the advancement of knowledge. The district court explained:

> Although I recognize that the facts here may on some levels be without precedent, I am convinced that they fall safely within the protection of fair use such that there is no genuine issue of material fact. I cannot imagine a definition of fair use that would not encompass the transformative uses

---

[3] Separate from the HDL, one copy is also kept by Google. Google's use of its copy is the subject of a separate lawsuit currently pending in this Court. *See Authors Guild, Inc. v. Google, Inc.,* 721 F.3d 132 (2d Cir. 2013), *on remand,* 954 F. Supp. 2d 282 (S.D.N.Y. 2013), *appeal docketed,* No. 13–4829 (2d Cir. Dec. 23, 2013).

made by [the HDL] and would require that I terminate this invaluable contribution to the progress of science and cultivation of the arts . . . .

[*Authors Guild, Inc. v. HathiTrust,* 902 F. Supp. 2d 445, 464 (S.D.N.Y. 2012).]

. . . .

Discussion

. . . .

I. Fair Use
A.

. . . .

. . . Section 107 requires a court to consider four nonexclusive factors which are to be weighed together to assess whether a particular use is fair . . . .

An important focus of the first factor is whether the use is "transformative." A use is transformative if it does something more than repackage or republish the original copyrighted work. The inquiry is whether the work "adds something new, with a further purpose or different character, altering the first with new expression, meaning or message * * * ." *Campbell,* 510 U.S. at 579. "[T]he more transformative the new work, the less will be the significance of other factors * * * that may weigh against a finding of fair use." *Id.* Contrary to what the district court implied, a use does not become transformative by making an "invaluable contribution to the progress of science and cultivation of the arts." *HathiTrust,* 902 F. Supp. 2d at 464. Added value or utility is not the test: a transformative work is one that serves a new and different function from the original work and is not a substitute for it. . . .

. . . .

B.

. . . .

It is not disputed that, in order to perform a full-text search of books, the Libraries must first create digital copies of the entire books. Importantly, . . . the HDL does not allow users to view any portion of the books they are searching. Consequently, in providing this service, the HDL does not add into circulation any new, human-readable copies of any books. Instead, the HDL simply permits users to "word search"—that is, to locate where specific words or phrases appear in the digitized books. Applying the relevant factors, we conclude that this use is a fair use.

i.

Turning to the first factor, we conclude that the creation of a full-text searchable database is a quintessentially transformative use. As the example . . ., *supra,* demonstrates, the result of a word search is different in purpose, character, expression, meaning, and message from the page (and the book) from which it is drawn. Indeed, we can discern little or no resemblance between the original text and the results of the HDL full-text search.

There is no evidence that the Authors write with the purpose of enabling text searches of their books. Consequently, the full-text search function does not "supersede[] the objects [or purposes] of the original creation," *Campbell*, 510 U.S. at 579. The HDL does not . . . merely recast "an original work into a new mode of presentation," *Castle Rock Entm't, Inc. v. Carol Publ'g Grp.*, 150 F.3d 132, 143 (2d Cir. 1998). Instead, by enabling full-text search, the HDL adds to the original something new with a different purpose and a different character.

Full-text search adds a great deal more to the copyrighted works at issue than did the transformative uses we approved in several other cases. For example, in *Cariou v. Prince,* we found that certain photograph collages were transformative, even though the collages were cast in the same medium as the copyrighted photographs. 714 F.3d at 706. Similarly, in *Bill Graham Archives v. Dorling Kindersley Ltd.,* we held that it was a transformative use to include in a biography copyrighted concert photos, even though the photos were unaltered (except for being reduced in size). 448 F.3d 605, 609–11 (2d Cir. 2006).

Cases from other Circuits reinforce this conclusion. In *Perfect 10, Inc.,* the Ninth Circuit held that the use of copyrighted thumbnail images in internet search results was transformative because the thumbnail copies served a different function from the original copyrighted images. 508 F.3d at 1165. And in *A.V. ex rel. Vanderhye v. iParadigms, LLC,* a company created electronic copies of unaltered student papers for use in connection with a computer program that detects plagiarism. Even though the electronic copies made no "substantive alteration to" the copyrighted student essays, the Fourth Circuit held that plagiarism detection constituted a transformative use of the copyrighted works. 562 F.3d 630, 639–40.

ii.

The second fair-use factor—the nature of the copyrighted work—is not dispositive. The HDL permits the full-text search of every type of work imaginable. Consequently, there is no dispute that the works at issue are of the type that the copyright laws value and seek to protect. However, "this factor 'may be of limited usefulness where,' as here, 'the creative work * * * is being used for a transformative purpose." *Cariou,* 714 F.3d at 710 (quoting *Bill Graham Archives,* 448 F.3d at 612). Accordingly, our fair-use analysis hinges on the other three factors.

iii.

The third factor asks whether the copying used more of the copyrighted work than necessary and whether the copying was excessive. . . . "[T]here are no absolute rules as to how much of a copyrighted work may be copied and still be considered a fair use." *Maxtone–Graham v. Burtchaell,* 803 F.2d 1253, 1263 (2d Cir. 1986). "[T]he extent of permissible copying varies with the purpose and character of the use." *Campbell,* 510 U.S. at 586–87. The crux of the inquiry is whether "no more was taken than necessary." *Id.* at 589. For some purposes, it may be necessary to copy the entire copyrighted work, in which case Factor Three does not weigh against a finding of fair use.

In order to enable the full-text search function, the Libraries . . . created digital copies of all the books in their collections. Because it was reasonably necessary for the HDL to make use of the entirety of the works in order to enable the full-text search function, we do not believe the copying was excessive.

The Authors also contend that the copying is excessive because the HDL creates and maintains copies of the works at four different locations. But the record demonstrates that these copies are also reasonably necessary in order to facilitate the HDL's legitimate uses. In particular, the HDL's services are offered to patrons through two servers, one at the University of Michigan (the primary server) and an identical one at the University of Indiana (the "mirror" server). Both servers contain copies of the digital works at issue. According to the HDL executive director, the "existence of a[n] [identical] mirror site allows for balancing the load of user web traffic to avoid overburdening a single site, and each site acts as a back-up of the HDL collection in the event that one site were to cease operation (for example, due to failure caused by a disaster, or even as a result of routine maintenance)." To further guard against the risk of data loss, the HDL stores copies of the works on two encrypted backup tapes, which are disconnected from the internet and are placed in separate secure locations on the University of Michigan campus. The HDL creates these backup tapes so that the data could be restored in "the event of a disaster causing large-scale data loss" to the primary and mirror servers.

We have no reason to think that these copies are excessive or unreasonable in relation to the purposes identified by the Libraries and permitted by the law of copyright. In sum, even viewing the evidence in the light most favorable to the Authors, the record demonstrates that these copies are reasonably necessary to facilitate the services HDL provides to the public and to mitigate the risk of disaster or data loss. Accordingly, we conclude that this factor favors the Libraries.

iv.

The fourth factor requires us to consider "the effect of the use upon the potential market for or value of the copyrighted work," 17 U.S.C. § 107(4), and, in particular, whether the secondary use "usurps the market of the original work," *NXIVM Corp. [v. Ross Inst.*, 364 F.3d 471, 482 (2d Cir. 2004)].

The Libraries contend that the full-text-search use poses no harm to any existing or potential traditional market and point to the fact that, in discovery, the Authors admitted that they were unable to identify "any specific, quantifiable past harm, or any documents relating to any such past harm," resulting from any of the Libraries' uses of their works (including full-text search). Defs.-Appellees' Br. 38 (citing Pls.' Resps. to Interrogs.). The district court agreed with this contention, as do we.

At the outset, it is important to recall that the Factor Four analysis is concerned with only one type of economic injury to a copyright holder: the harm that results because the secondary use serves as a substitute for the original work. *See Campbell,* 510 U.S. at 591 ("cognizable market harm" is limited to "market substitution"). In other words, under Factor Four, any economic "harm" caused by transformative uses does not count because such uses, by definition, do not serve as substitutes for the original work.

To illustrate why this is so, consider how copyright law treats book reviews. Book reviews often contain quotations of copyrighted material to illustrate the reviewer's points and substantiate his criticisms; this is a paradigmatic fair use. And a negative book review can cause a degree of economic injury to the author by dissuading readers from purchasing copies of her book, even when the review does not serve as a substitute for the original. But, obviously, in that case, the author has no cause for complaint under Factor Four: The only market harms that count are the ones that are caused because the secondary use serves as a substitute for the original, not when the secondary use is transformative (as in quotations in a book review). *See Campbell,* 510 U.S. at 591–92 ("[W]hen a lethal parody, like a scathing theater review, kills demand for the original, it does not produce a harm cognizable under the Copyright Act.").

The Authors assert two reasons why the full-text-search function harms their traditional markets. The first is a "lost sale" theory which posits that a market for licensing books for digital search could possibly develop in the future, and the HDL impairs the emergence of such a market because it allows patrons to search books without any need for a license. Thus, according to the Authors, every copy employed by the HDL in generating full-text searches represents a lost opportunity to license the book for search.

This theory of market harm does not work under Factor Four, because the full-text search function does not serve as a substitute for the books that are being searched. Thus, it is irrelevant that the Libraries might be willing to purchase licenses in order to engage in this transformative use (if the use were deemed unfair). Lost licensing revenue counts under Factor Four only when the use serves as a substitute for the original and the full-text-search use does not.

Next, the Authors assert that the HDL creates the risk of a security breach which might impose irreparable damage on the Authors and their works. In particular, the Authors speculate that, if hackers were able to obtain unauthorized access to the books stored at the HDL, the full text of these tens of millions of books might be distributed worldwide without restriction, "decimat[ing]" the traditional market for those works.

The record before us documents the extensive security measures the Libraries have undertaken to safeguard against the risk of a data breach. . . .

This showing of the security measures taken by the Libraries is essentially unrebutted. Consequently, we see no basis in the record on which to conclude that a security breach is likely to occur, much less one that would result in the public release of the specific copyrighted works belonging to any of the plaintiffs in this case. *Cf. Clapper v. Amnesty Int'l USA,* 133 S. Ct. 1138, 1143, 1149 (2013) (risk of future harm must be "certainly impending," rather than merely "conjectural" or "hypothetical," to constitute a cognizable injury-in-fact); *Sony Corp.,* 464 U.S. at 453–54 (concluding that time-shifting using a Betamax is fair use because the copyright owners' "prediction that live television or movie audiences will decrease" was merely "speculative"). Factor Four thus favors a finding of fair use.

Without foreclosing a future claim based on circumstances not now predictable, and based on a different record, we hold that the balance of relevant factors in this case favors the Libraries. In sum, we conclude that the doctrine of fair use allows the Libraries to digitize copyrighted works for the purpose of permitting full-text searches.

. . . .

Notes & Questions

1. The Copyright Act expressly treats libraries and archives differently. Section 108 authorizes libraries to engage in a variety of activities that would constitute infringement if conducted by others. The plaintiffs argued that the HDL exceeded the bounds of what § 108 permits and thus should be found to be an infringement without consideration of the libraries' fair use defense. The court rejected that argument, noting the "savings clause" in § 108: "Nothing in this section . . . in any way affects the right of fair use as provided by section 107 . . . ." 17 U.S.C. § 108(f)(4).

Many of the digital copies contained in the HDL were created by Google, in partnership with the libraries. In return for access to their collections, Google provided the libraries with digital copies. Google used the digital copies it obtained to create what is known as Google Book Search (GBS). GBS permits the public to search books via the internet and returns search results that include "snippets" of approximately 1/8th of a page of a book containing the target text string. The Authors Guild also sued Google. In affirming a judgment in Google's favor, the Second Circuit emphasized limitations Google had placed on the snippets:

> These include the small size of the snippets (normally one eighth of a page), the blacklisting of one snippet per page and of one page in every ten, the fact that no more than three snippets are shown—and no more than one per page—for each term searched, and the fact that the same snippets are shown for a searched term no matter how many times, or from how many different computers, the term is searched. In addition, Google does not provide snippet view for types of books, such as dictionaries and cookbooks, for which viewing a small segment is likely to satisfy the searcher's need.
>
> . . .

Authors Guild v. Google, Inc., 804 F.3d 202, 222 (2d Cir. 2015), *cert. denied*, 136 S. Ct. 1658 (2016). The court also saw "no reason why Google's overall profit motivation should prevail as a reason for denying fair use over its highly convincing transformative purpose, together with the absence of significant substitutive competition, as reasons for granting fair use." *Id.* at 219. Because of the limitations Google had implemented on the snippet views, copyright holders were not threatened with "any significant harm to the value of their copyrights or diminish[ment of] their harvest of copyright revenue." *Id.* at 224.

2. How did HathiTrust demonstrate a lack of market harm under the fourth factor? The plaintiffs had argued that lost licensing fees should be taken into account. Why did the court reject that argument?

3. The use of lost licensing revenue in fair use cases can be tricky and, some-times, the argument can seem circular. In an earlier case, the Second Circuit had considered lost licensing fees as a component of market harm in a fair use analysis. In *American Geophysical Union v. Texaco, Inc.*, 60 F.3d 913 (2d Cir. 1994), the court concluded that the photocopying practices of research scientists at Texaco to fur-ther their research needs was not fair use, based in part on lost licensing fees. The publishers in that case had created the Copyright Clearance Center (the CCC) and authorized the CCC to grant blanket licenses to business and nonprofit enti-ties. Texaco had argued that "whether the publishers can demand a fee for per-mission to make photocopies is the very question that the fair use trial is supposed to answer." The Second Circuit held:

> Despite Texaco's claims to the contrary, it is not unsound to conclude that the right to seek payment for a particular use tends to become legally cognizable under the fourth fair use factor when the means for paying for such a use is made easier. This notion is not inherently troubling: it is sen-sible that a particular unauthorized use should be considered "more fair" when there is no ready market or means to pay for the use, while such an unauthorized use should be considered "less fair" when there is a ready market or means to pay for the use. The vice of circular reasoning arises only if the availability of payment is conclusive against fair use. Whatever the situation may have been previously, before the development of a market for institutional users to obtain licenses to photocopy articles, it is now ap-propriate to consider the loss of licensing revenues in evaluating "the effect of the use upon the potential market for or value of" journal articles.

Id. at 930-31. The court also noted a that there are limits on the concept of "po-tential licensing revenues" that can be considered: only licensing revenues from "traditional, reasonable, or likely to be developed markets" should be considered. *Id.* at 930. In dissent, Judge Jacobs focused on this part of the court's decision:

> In this case the only harm to a market is to the supposed market in pho-tocopy licenses. The CCC scheme is neither traditional nor reasonable; and its development into a real market is subject to substantial impediments. There is a circularity to the problem: the market will not crystallize unless courts reject the fair use argument that Texaco presents; but, under the statutory test, we cannot declare a use to be an infringement unless (assum-ing other factors also weigh in favor of the secondary user) there is a market to be harmed.

American Geophysical Union, 60 F.3d at 937. What do you think of this "lost licens-ing" revenue debate? Is the way that the Second Circuit handles plaintiff's lost licensing revenue claim in the *HathiTrust* case consistent with its ruling in *Texaco*?

C. FAIR USE OF COMPUTER SOFTWARE: REVERSE ENGINEERING

Another facet of fair use involves computer software. In order to make pro-grams that are compatible with existing operating systems or other programs, cer-tain technical interface information must be known. The best way to obtain that information is to examine the source code of the software. Many computer programs, however, are distributed in object code only with the source code kept as a closely guarded secret (indeed, sometimes, a trade secret). With great effort, the source code can be obtained through reverse engineering (or at least a sufficient amount of the source code containing the technical information necessary to write interoperable programs). Inevitably, that reverse engineering involves making multiple copies of the entire software program.

Courts have held that the intermediate copies made in the process of reverse engineering, also known as disassembly, are not infringements of the copyrights in the computer software. Specifically, in *Sega Enterprises Ltd. v. Accolade, Inc.*, 977 F.2d 1510 (9th Cir. 1992), the Ninth Circuit held "that where disassembly is the only way to gain access to the ideas and functional elements embodied in a copyrighted computer program and where there is a legitimate reason for seeking such access, disassembly is a fair use of the copyrighted work, as a matter of law." In *Sega* the legitimate reason was for the defendant to create its own video games that operated on the Sega Genesis video game platform. Other decisions have followed *Sega. See, e.g. Sony Computer Entertainment, Inc. v. Connectix Corp.*, 203 F.3d 596 (9th Cir. 2000).

D. Personal Use

Sony Corp. v. Universal City Studios, Inc.
464 U.S. 417 (1984)

Stevens, Justice:

Petitioners manufacture and sell home video tape recorders. Respondents own the copyrights on some of the television programs that are broadcast on the public airwaves. Some members of the general public use video tape recorders sold by petitioners to record some of these broadcasts, as well as a large number of other broadcasts. The question presented is whether the sale of petitioners' copying equipment to the general public violates any of the rights conferred upon respondents by the Copyright Act.

Respondents commenced this copyright infringement action against petitioners in the United States District Court for the Central District of California in 1976. Respondents alleged that some individuals had used Betamax video tape recorders (VTR's) to record some of respondents' copyrighted works which had been exhibited on commercially sponsored television and contended that these individuals had thereby infringed respondents' copyrights. Respondents further maintained that petitioners were liable for the copyright infringement allegedly committed by Betamax consumers because of petitioners' marketing of the Betamax VTR's. Respondents sought no relief against any Betamax consumer. Instead, they sought money damages and an equitable accounting of profits from petitioners, as well as an injunction against the manufacture and marketing of Betamax VTR's.

After a lengthy trial, the District Court denied respondents all the relief they sought and entered judgment for petitioners. 480 F. Supp. 429 (1979). The United

States Court of Appeals for the Ninth Circuit reversed the District Court's judgment on respondent's copyright claim, holding petitioners liable for contributory infringement and ordering the District Court to fashion appropriate relief. 659 F.2d 963 (1981). We granted certiorari; since we had not completed our study of the case last Term, we ordered reargument. We now reverse.

. . . [The District Court] findings reveal that the average member of the public uses a VTR principally to record a program he cannot view as it is being televised and then to watch it once at a later time. This practice, known as "time-shifting," enlarges the television viewing audience. For that reason, a significant amount of television programming may be used in this manner without objection from the owners of the copyrights on the programs. For the same reason, even the two respondents in this case, who do assert objections to time-shifting in this litigation, were unable to prove that the practice has impaired the commercial value of their copyrights or has created any likelihood of future harm. . . .

[*Eds. Note*: After a detailed review of the District Court's factual findings, the Court established a test for determining when a manufacturer of an article of commerce can be held liable for infringement engaged in by third parties using that article of commerce. We will explore that aspect of the Court's ruling in Section VII of this Chapter concerning third party liability for infringement. The Court held that the "sale of copying equipment, like the sale of other articles of commerce, does not constitute contributory infringement if the product is widely used for legitimate, unobjectionable purposes. Indeed, it need merely be capable of substantial noninfringing uses." The Court then turned to the task of determining whether VTRs were "capable of substantial noninfringing use." In addition to copyright owners that testified they did not object to individuals using VTRs to "time-shift" their programs (among others, Fred Rogers, of *Mr. Rogers' Neighborhood* fame, testified at trial), the Court also examined whether, even without copyright owners' authorization, individuals who time-shifted programs were infringers or fair users.]

[Section 107] identifies various factors that enable a Court to apply an "equitable rule of reason" analysis to particular claims of infringement. Although not conclusive, the first factor requires that "the commercial or nonprofit character of an activity" be weighed in any fair use decision. If the Betamax were used to make copies for a commercial or profit-making purpose, such use would presumptively be unfair. The contrary presumption is appropriate here, however, because the District Court's findings plainly establish that time-shifting for private home use must be characterized as a noncommercial, nonprofit activity. Moreover, when one considers the nature of a televised copyrighted audiovisual work, see 17 U.S.C. § 107(2), and that timeshifting merely enables a viewer to see such a work which he had been invited to witness in its entirety free of charge, the fact that the entire work is reproduced, see *id.* at § 107(3), does not have its ordinary effect of militating against a finding of fair use.[33]

---

[33] It has been suggested that "consumptive uses of copyrights by home VTR users are commercial even if the consumer does not sell the homemade tape because the consumer will not buy tapes

This is not, however, the end of the inquiry because Congress has also directed us to consider "the effect of the use upon the potential market for or value of the copyrighted work." *Id.* at § 107(4). The purpose of copyright is to create incentives for creative effort. Even copying for noncommercial purposes may impair the copyright holder's ability to obtain the rewards that Congress intended him to have. But a use that has no demonstrable effect upon the potential market for, or the value of, the copyrighted work need not be prohibited in order to protect the author's incentive to create. The prohibition of such noncommercial uses would merely inhibit access to ideas without any countervailing benefit.

Thus, although every commercial use of copyrighted material is presumptively an unfair exploitation of the monopoly privilege that belongs to the owner of the copyright, noncommercial uses are a different matter. A challenge to a noncommercial use of a copyrighted work requires proof either that the particular use is harmful, or that if it should become widespread, it would adversely affect the potential market for the copyrighted work. Actual present harm need not be shown; such a requirement would leave the copyright holder with no defense against predictable damage. Nor is it necessary to show with certainty that future harm will result. What is necessary is a showing by a preponderance of the evidence that *some* meaningful likelihood of future harm exists. If the intended use is for commercial gain, that likelihood may be presumed. But if it is for a noncommercial purpose, the likelihood must be demonstrated.

In this case, respondents failed to carry their burden with regard to home time-shifting. The District Court described respondents' evidence as follows:

> Plaintiffs' experts admitted at several points in the trial that the time-shifting without librarying would result in 'not a great deal of harm.' Plaintiffs' greatest concern about time-shifting is with 'a point of important philosophy that transcends even commercial judgment.' They fear that with any Betamax usage, 'invisible boundaries' are passed: 'the copyright owner has lost control over his program.'

480 F. Supp. at 467. . . .

---

separately sold by the copyright holder." Home Recording of Copyrighted Works: Hearing before Subcommittee on Courts, Civil Liberties and the Administration of Justice of the House Committee on the Judiciary, 97th Congress, 2d Session, pt. 2, p. 1250 (1982) (memorandum of Prof. Laurence H. Tribe). Furthermore, "[t]he error in excusing such theft as noncommercial," we are told, "can be seen by simple analogy: jewel theft is not converted into a noncommercial veniality if stolen jewels are simply worn rather than sold." *Id.* The premise and the analogy are indeed simple, but they add nothing to the argument. The use to which stolen jewelry is put is quite irrelevant in determining whether depriving its true owner of his present possessory interest in it is venial; because of the nature of the item and the true owner's interests in physical possession of it, the law finds the taking objectionable even if the thief does not use the item at all. Theft of a particular item of personal property of course may have commercial significance, for the thief deprives the owner of his right to sell that particular item to any individual. Timeshifting does not even remotely entail comparable consequences to the copyright owner. Moreover, the timeshifter no more steals the program by watching it once than does the live viewer, and the live viewer is no more likely to buy pre-recorded videotapes than is the timeshifter. Indeed, no live viewer would buy a pre-recorded videotape if he did not have access to a VTR.

There was no need for the District Court to say much about past harm. "Plaintiffs have admitted that no actual harm to their copyrights has occurred to date." *Id.* at 451.

On the question of potential future harm from time-shifting, the District Court offered a more detailed analysis of the evidence. It rejected respondents' "fear that persons 'watching' the original telecast of a program will not be measured in the live audience and the ratings and revenues will decrease," by observing that current measurement technology allows the Betamax audience to be reflected. *Id.* at 466.[36] It rejected respondents' prediction "that live television or movie audiences will decrease as more people watch Betamax tapes as an alternative," with the observation that "[t]here is no factual basis for [the underlying] assumption." *Id.* It rejected respondents' "fear that time-shifting will reduce audiences for telecast reruns," and concluded instead that "given current market practices, this should aid plaintiffs rather than harm them." *Id.*[38] And it declared that respondents' suggestion "that theater or film rental exhibition of a program will suffer because of time-shift recording of that program" "lacks merit." 480 F. Supp. at 467.

After completing that review, the District Court restated its overall conclusion several times, in several different ways. "Harm from time-shifting is speculative and, at best, minimal." *Id.* "The audience benefits from the time-shifting capability have already been discussed. It is not implausible that benefits could also accrue to plaintiffs, broadcasters, and advertisers, as the Betamax makes it possible for more persons to view their broadcasts." *Id.* "No likelihood of harm was shown at trial, and plaintiffs admitted that there had been no actual harm to date." *Id.* at 468-469. "Testimony at trial suggested that Betamax may require adjustments in marketing strategy, but it did not establish even a likelihood of harm." *Id.* at 469. "Television production by plaintiffs today is more profitable than it has ever been,

---

[36] . . . In a separate section, the District Court rejected plaintiffs' suggestion that the commercial attractiveness of television broadcasts would be diminished because Betamax owners would use the pause button or fast-forward control to avoid viewing advertisements:

"It must be remembered, however, that to omit commercials, Betamax owners must view the program, including the commercials, while recording. To avoid commercials during playback, the viewer must fast-forward and, for the most part, guess as to when the commercial has passed. For most recordings, either practice may be too tedious. As defendants' survey showed, 92% of the programs were recorded with commercials and only 25% of the owners fast-forward through them. Advertisers will have to make the same kinds of judgments they do now about whether persons viewing televised programs actually watch the advertisements which interrupt them."

Id. at 468.

[38] "The underlying assumptions here are particularly difficult to accept. Plaintiffs explain that the Betamax increases access to the original televised material and that the more people there are in this original audience, the fewer people the rerun will attract. Yet current marketing practices, including the success of syndication, show just the opposite. Today, the larger the audience for the original telecast, the higher the price plaintiffs can demand from broadcasters from rerun rights. There is no survey within the knowledge of this court to show that the rerun audience is comprised of persons who have not seen the program. In any event, if ratings can reflect Betamax recording, original audiences may increase and, given market practices, this should aid plaintiffs rather than harm them." *Id.*

and, in five weeks of trial, there was no concrete evidence to suggest that the Beta-max will change the studios' financial picture." *Id.*

The District Court's conclusions are buttressed by the fact that to the extent time-shifting expands public access to freely broadcast television programs, it yields societal benefits. Earlier this year, in *Community Television of Southern California v. Gottfried,* 103 S. Ct. 885, 891-892 (1983), we acknowledged the public interest in making television broadcasting more available. Concededly, that interest is not unlimited. But it supports an interpretation of the concept of "fair use" that requires the copyright holder to demonstrate some likelihood of harm before he may condemn a private act of time-shifting as a violation of federal law.

When these factors are all weighed in the "equitable rule of reason" balance, we must conclude that this record amply supports the District Court's conclusion that home time-shifting is fair use. In light of the findings of the District Court regarding the state of the empirical data, it is clear that the Court of Appeals erred in holding that the statute as presently written bars such conduct.[40]

In summary, the record and findings of the District Court lead us to two conclusions. First, Sony demonstrated a significant likelihood that substantial numbers of copyright holders who license their works for broadcast on free television would not object to having their broadcasts time-shifted by private viewers. And

---

[40] The Court of Appeals chose not to engage in any "equitable rule of reason" analysis in this case. Instead, it assumed that the category of "fair use" is rigidly circumscribed by a requirement that every such use must be "productive." It therefore concluded that copying a television program merely to enable the viewer to receive information or entertainment that he would otherwise miss because of a personal scheduling conflict could never be fair use. That understanding of "fair use" was erroneous.

Congress has plainly instructed us that fair use analysis calls for a sensitive balancing of interests. The distinction between "productive" and "unproductive" uses may be helpful in calibrating the balance, but it cannot be wholly determinative. Although copying to promote a scholarly endeavor certainly has a stronger claim to fair use than copying to avoid interrupting a poker game, the question is not simply two-dimensional. For one thing, it is not true that all copyrights are fungible. Some copyrights govern material with broad potential secondary markets. Such material may well have a broader claim to protection because of the greater potential for commercial harm. Copying a news broadcast may have a stronger claim to fair use than copying a motion picture. And, of course, not all uses are fungible. Copying for commercial gain has a much weaker claim to fair use than copying for personal enrichment. But the notion of social "productivity" cannot be a complete answer to this analysis. A teacher who copies to prepare lecture notes is clearly productive. But so is a teacher who copies for the sake of broadening his personal understanding of his specialty. Or a legislator who copies for the sake of broadening her understanding of what her constituents are watching; or a constituent who copies a news program to help make a decision on how to vote.

Making a copy of a copyrighted work for the convenience of a blind person is expressly identified by the House Committee Report as an example of fair use, with no suggestion that anything more than a purpose to entertain or to inform need motivate the copying. In a hospital setting, using a VTR to enable a patient to see programs he would otherwise miss has no productive purpose other than contributing to the psychological well-being of the patient. Virtually any time-shifting that increases viewer access to television programming may result in a comparable benefit. The statutory language does not identify any dichotomy between productive and nonproductive time-shifting, but does require consideration of the economic consequences of copying.

second, respondents failed to demonstrate that time-shifting would cause any likelihood of nonminimal harm to the potential market for, or the value of, their copyrighted works. The Betamax is, therefore, capable of substantial noninfringing uses. Sony's sale of such equipment to the general public does not constitute contributory infringement of respondent's copyrights.

V

"The direction of Art. I is that *Congress* shall have the power to promote the progress of science and the useful arts. When, as here, the Constitution is permissive, the sign of how far Congress has chosen to go can come only from Congress." *Deepsouth Packing Co. v. Laitram Corp.,* 406 U.S. 518, 530 (1972).

One may search the Copyright Act in vain for any sign that the elected representatives of the millions of people who watch television every day have made it unlawful to copy a program for later viewing at home, or have enacted a flat prohibition against the sale of machines that make such copying possible.

It may well be that Congress will take a fresh look at this new technology, just as it so often has examined other innovations in the past. But it is not our job to apply laws that have not yet been written. Applying the copyright statute, as it now reads, to the facts as they have been developed in this case, the judgment of the Court of Appeals must be reversed.

BLACKMUN, JUSTICE, dissenting (with Justice Marshall, Powell and Rehnquist):

. . . .

The purpose of copyright protection, in the words of the Constitution, is to "promote the Progress of Science and useful Arts." Copyright is based on the belief that by granting authors the exclusive rights to reproduce their works, they are given an incentive to create, and that "encouragement of individual effort by personal gain is the best way to advance public welfare through the talents of authors and inventors in 'Science and the useful Arts.'" *Mazer v. Stein,* 347 U.S. 201, 219 (1954). The monopoly created by copyright thus rewards the individual author in order to benefit the public.

There are situations, nevertheless, in which strict enforcement of this monopoly would inhibit the very "Progress of Science and useful Arts" that copyright is intended to promote. An obvious example is the researcher or scholar whose own work depends on the ability to refer to and to quote the work of prior scholars. Obviously, no author could create a new work if he were first required to repeat the research of every author who had gone before him.[28] The scholar, like the ordinary user, of course could be left to bargain with each copyright owner for permission to quote from or refer to prior works. But there is a crucial difference between the scholar and the ordinary user. When the ordinary user decides that the owner's price is too high, and forgoes use of the work, only the individual is the

---

[28] "The world goes ahead because each of us builds on the work of our predecessors. 'A dwarf standing on the shoulders of a giant can see farther than the giant himself.'" Chafee, *Reflections on the Law of Copyright: I,* 45 Colum. L. Rev. 503, 511 (1945).

loser. When the scholar forgoes the use of a prior work, not only does his own work suffer, but the public is deprived of his contribution to knowledge. The scholar's work, in other words, produces external benefits from which everyone profits. In such a case, the fair use doctrine acts as a form of subsidy—albeit at the first author's expense—to permit the second author to make limited use of the first author's work for the public good. See Latman Fair Use Study 31; Gordon, *Fair Use as Market Failure: A Structural Analysis of the* Betamax *Case and its Predecessors,* 82 Colum. L. Rev. 1600, 1630 (1982).

A similar subsidy may be appropriate in a range of areas other than pure scholarship. The situations in which fair use is most commonly recognized are listed in § 107 itself; fair use may be found when a work is used "for purposes such as criticism, comment, news reporting, teaching, . . . scholarship, or research." The House and Senate Reports expand on this list somewhat,[29] and other examples may be found in the case law. Each of these uses, however, reflects a common theme: each is a productive use, resulting in some added benefit to the public beyond that produced by the first author's work. The fair use doctrine, in other words, permits works to be used for "socially laudable purposes." See Copyright Office, *Briefing Papers on Current Issues,* reprinted in 1975 House Hearings 2051, 2055. I am aware of no case in which the reproduction of a copyrighted work for the sole benefit of the user has been held to be fair use.

I do not suggest, of course, that every productive use is a fair use. A finding of fair use still must depend on the facts of the individual case, and on whether, under the circumstances, it is reasonable to expect the user to bargain with the copyright owner for use of the work. The fair use doctrine must strike a balance between the dual risks created by the copyright system: on the one hand, that depriving authors of their monopoly will reduce their incentive to create, and, on the other, that granting authors a complete monopoly will reduce the creative ability of others. The inquiry is necessarily a flexible one, and the endless variety of situations that may arise precludes the formulation of exact rules. But when a user reproduces an entire work and uses it for its original purpose, with no added benefit to the public, the doctrine of fair use usually does not apply. There is then no need whatsoever to provide the ordinary user with a fair use subsidy at the author's expense.

The making of a videotape recording for home viewing is an ordinary rather than a productive use of the Studios' copyrighted works. . . . Copyright gives the

---

[29] . . . [T]he Senate and House Reports give examples of possible fair uses:

quotation of excerpts in a review or criticism for purposes of illustration or comment; quotation of short passages in a scholarly or technical work, for illustration or clarification of the author's observations; use in a parody of some of the content of the work parodied; summary of an address or article, with brief quotations, in a news report; reproduction by a library of a portion of a work to replace part of a damaged copy; reproduction by a teacher or student of a small part of a work to illustrate a lesson; reproduction of a work in legislative or judicial proceedings or reports; incidental and fortuitous reproduction, in a newsreel or broadcast, of a work located in the scene of an event being recorded.

1975 Senate Report 61-62; 1976 House Report 65, U.S. Code Cong. & Admin. News 1976, p. 5678.

author a right to limit or even to cut off access to his work. A VTR recording creates no public benefit sufficient to justify limiting this right. Nor is this right extinguished by the copyright owner's choice to make the work available over the airwaves. Section 106 of the 1976 Act grants the copyright owner the exclusive right to control the performance and the reproduction of his work, and the fact that he has licensed a single television performance is really irrelevant to the existence of his right to control its reproduction. Although a television broadcast may be free to the viewer, this fact is equally irrelevant; a book borrowed from the public library may not be copied any more freely than a book that is purchased.

It may be tempting, as, in my view, the Court today is tempted, to stretch the doctrine of fair use so as to permit unfettered use of this new technology in order to increase access to television programming. But such an extension risks eroding the very basis of copyright law, by depriving authors of control over their works and consequently of their incentive to create. . . .

. . . .

I recognize, nevertheless, that there are situations where permitting even an unproductive use would have no effect on the author's incentive to create, that is, where the use would not affect the value of, or the market for, the author's work. Photocopying an old newspaper clipping to send to a friend may be an example; pinning a quotation on one's bulletin board may be another. In each of these cases, the effect on the author is truly de minimis. Thus, even though these uses provide no benefit to the public at large, no purpose is served by preserving the author's monopoly, and the use may be regarded as fair.

Courts should move with caution, however, in depriving authors of protection from unproductive "ordinary" uses. As has been noted above, even in the case of a productive use, § 107(4) requires consideration of "the effect of the use upon the *potential* market for or value of the copyrighted work" (emphasis added). "[A] particular use which may seem to have little or no economic impact on the author's rights today can assume tremendous importance in times to come." Register's Supplementary Report 14. Although such a use may seem harmless when viewed in isolation, "[i]solated instances of minor infringements, when multiplied many times, become in the aggregate a major inroad on copyright that must be prevented." 1975 Senate Report 65.

I therefore conclude that, at least when the proposed use is an unproductive one, a copyright owner need prove only a potential for harm to the market for or the value of the copyrighted work. Proof of actual harm, or even probable harm, may be impossible in an area where the effect of a new technology is speculative, and requiring such proof would present the "real danger . . . of confining the scope of an author's rights on the basis of the present technology so that, as the years go by, his copyright loses much of its value because of unforeseen technical advances." Register's Supplementary Report 14. Infringement thus would be found if the copyright owner demonstrates a reasonable possibility that harm will result from the proposed use. When the use is one that creates no benefit to the public at large, copyright protection should not be denied on the basis that a new technology that may result in harm has not yet done so.

The Studios have identified a number of ways in which VTR recording could damage their copyrights. VTR recording could reduce their ability to market their works in movie theaters and through the rental or sale of pre-recorded videotapes or videodiscs; it also could reduce their rerun audience, and consequently the license fees available to them for repeated showings. Moreover, advertisers may be willing to pay for only "live" viewing audiences, if they believe VTR viewers will delete commercials or if rating services are unable to measure VTR use; if this is the case, VTR recording could reduce the license fees the Studios are able to charge even for first-run showings. Library-building may raise the potential for each of the types of harm identified by the Studios, and time-shifting may raise the potential for substantial harm as well.

. . . .

In this case, the Studios and their amici demonstrate that the advent of the VTR technology created a potential market for their copyrighted programs. That market consists of those persons who find it impossible or inconvenient to watch the programs at the time they are broadcast, and who wish to watch them at other times. These persons are willing to pay for the privilege of watching copyrighted work at their convenience, as is evidenced by the fact that they are willing to pay for VTRs and tapes; undoubtedly, most also would be willing to pay some kind of royalty to copyright holders. The Studios correctly argue that they have been deprived of the ability to exploit this sizable market.

It is thus apparent from the record and from the findings of the District Court that time-shifting does have a substantial adverse effect upon the "potential market for" the Studios' copyrighted works. Accordingly, even under the formulation of the fair use doctrine advanced by Sony, time-shifting cannot be deemed a fair use.

. . . .

Notes & Questions

1. As the Court indicates in the opening paragraphs of its opinion, the Court heard argument in the case twice. Justice Thurgood Marshall's papers, released after his death, show that after the first hearing the straw vote among the Justices was 5-4 in favor of a determination of infringement. As opinions were circulated, the shifting majority required re-argument of the case, which ultimately resulted in a 5-4 decision in favor of fair use.

As with the other two Supreme Court fair use decisions, this one involved reversals at each level of appeal; the District Court found for defendants, the Court of Appeals reversed, and the Supreme Court, in turn, reversed the Court of Appeals. What explains the flip-flop pattern in these decisions?

2. After *Sony*, lower courts applied a set of presumptions depending on whether the use at issue was characterized as commercial or non-commercial. As we explored in Section V, however, in *Campbell* the Court has clearly stated that

presuming no fair use from the defendant's commercial gain is inappropriate. What if the use is characterized as non-commercial—would it be permissible to apply a presumption? Is a presumption just another way to identify who has the burden on a particular issue? If a use is non-commercial, is it appropriate to require the plaintiff to demonstrate some actual harm or, perhaps, likely potential harm? Do the majority and the dissent disagree about what facts a copyright owner must demonstrate to prevail, or do they disagree about the facts proven by the plaintiffs in this case?

3. Time-shifting occurs in the privacy of individual homes. How much did the private nature of the copying affect the Court's decision? If you were to exempt private personal-use copying from infringement liability, how would you limit the exemption so as not to undermine the incentive copyright law seeks to provide?

The Court highlights that plaintiffs did not seek relief from any individual user of a VTR. Instead, they sought relief from the manufacturers of the VTRs. The status of personal use copying, whether it is infringement or fair use, is often litigated in the context of third-party liability asserted against an equipment maker or service provider. This factual context can affect the analysis. For example, peer to peer file sharing of music was held to be infringement and not fair use in the context of a lawsuit against the original Napster (before Napster changed to a subscription model). *A & M Records, Inc. v. Napster, Inc.*, 239 F.3d 1004 (9th Cir. 2001). Is peer to peer file sharing personal use or public use?

4. In part the *Sony* case is about what happens when new technologies disrupt old business models. Equipping individuals to record television programs threatened the advertiser-driven model of broadcast television. Jack Valenti, president of the industry group known as the Motion Picture Association of America (MPAA), testified before Congress that "the VCR is to the American film producer and the American public as the Boston Strangler is to the woman home alone." Home Recording of Copyrighted Works: Hearing on H.R. 4783, H.R. 4794, H.R. 4808, H.R. 5250, H.R. 5488, and H.R. 5705 Before the Subcomm. on Courts, Civil Liberties, and the Admin. of Justice, of the H. Comm. on the Judiciary, 97th Cong. (1982). Did the existence of VCRs result in the death of the American film industry? Did it create business pressure toward subscription-based premium cable TV channels? What can this experience teach us about new technologies and copyright protection?

When the courts apply the Copyright Act to new technologies, they are determining on whom the burden will fall to convince Congress to change the statute. In *Sony* the majority concluded its opinion with an indication that if copyright owners were unhappy with the result they should seek relief from Congress. No changes were made to the statute related to the *Sony* decision. Does this mean the case was correctly decided?

Fair Use and Market Failure

One way to view fair use is through the lens of law and economics. In that context, fair use is a mechanism to cope with market failures. Generally, those who favor a minimally-regulated market over a greatly-regulated market believe that, so long as the boundaries of the legal entitlements are clear, market-based transactions will move resources to their highest-value uses. Regulation should step in only when there is likely to be market failure.

Markets can fail for different reasons, but two significant reasons are transaction costs and externalities. *Transaction costs*: Imagine a user who desires to make a copy of a particular work. The user may value the ability to make that copy at $5. Let's assume that the copyright owner would permit the use at any amount above $3. If the costs of seeking out the copyright owner, negotiating for permission, and monitoring performance of the license—what economists refer to as "transaction costs"—are higher than $2, the bargain will not be struck and the market will have failed. We call this a market failure because, but for the transaction costs, the parties would have been able to reach a deal. The copyright owner does not realize the $3 gain, and the user does not realize a benefit valued at $5. Fair use could step in to permit the use to occur. When the problem of transaction costs has been reduced through a functioning licensing system, there is less of a case for a market-failure-based justification for fair use. For example, in *Harper & Row* there was a functioning market for serializations of forthcoming novels— Time and Harper & Row had already reached an agreement for such serialization. The Nation did not "pay the customary price" within that functioning market and thus its claim to fair use was diminished.

Externalities: Economic analysis of law often focuses on negative externalities—costs cast on others—such as the pollution caused by a factory. Absent a legal rule to the contrary, the pollution imposes costs, such as adverse health effects, on people who are external to the basic decision to pollute. In the fair use context we instead focus on positive externalities, also known as spillovers. *See* Brett M. Frischmann & Mark A. Lemley, *Spillovers*, 107 Colum. L. Rev. 257 (2007) For example, consider an article a teacher desires to reproduce for her students. Let's say the copyright owner would be willing to accept $3 to permit that use. When a teacher reproduces an article for her students, the students obtain some benefit, but additionally, society as a whole benefits, by having a more well-educated citizenry. Those benefits are external to the use of the article in the class. The students and the teachers might be willing to pay, say $1 for that article, but if all of the external value of the use could be factored in, the offered payment might exceed $3. In this scenario, a bargained-for exchange will not be reached between the teacher/students and the copyright owner. Again, fair use could step in and permit that use to occur. For example, in *Campbell* the external benefits to society of having the ability to engage in critical commentary on creative works makes the claim of fair use in the parody context quite strong.

Notes & Questions

1. The *Sony* dissent focuses on what it perceives to be a lack of external benefits for "time-shifting." Using the tools of law and economics, is there a way to explain the majority's result? The dissent characterizes fair use as a kind of subsidy for uses that have high external benefits. Would it be accurate, at the same time, to

characterize copyright rights themselves as a tax on the public meant to provide an incentive to create and disseminate works of authorship?

2. Does a market failure rationale based on high transaction costs explain the outcomes in the *Perfect 10* case or in the *HathiTrust* case?

E. An Alternative Means of Creating "Breathing Space": Open Source and Creative Commons Licenses

At the same time that the internet and digital technology significantly increase access to copyrighted works, copyright law provides copyright owners control over how those works can be used. That control is present whether the copyright owner has registered the work and whether she has indicated an intent to claim copyright by affixing a notice to copies of the work. In its most recent case addressing fair use, the Supreme Court indicated that the doctrine provides a "guarantee of breathing space within the confines of copyright." *Campbell v. Acuff-Rose, Music, Inc.,* 510 U.S. 569, 579 (1994). Another way to create breathing space is for individual copyright owners to indicate a willingness to permit uses that might otherwise constitute infringement. While it is possible to abandon a copyright through an express indication of abandonment, what if a copyright owner does not want to completely abandon copyright protections, but would like to permit some uses to occur? Licenses present one possibility to achieve this result.

Using blanket licenses to grant wider use-rights to the public began with what is known as the open source or free software movement. The best known open source license is the GNU General Public License (GNU GPL), under which the GNU/Linux computer operating system is distributed. The GNU GPL authorizes others to use, modify, and redistribute the software programs and to create and distribute new programs based on the initial ones. A condition of the license requires that, if those new programs (which the Copyright Act classifies as derivative works) are distributed, they must be distributed subject to the same GNU GPL. The GNU GPL also requires that the program's source code must be distributed along with the object code, thus avoiding having to use fair use as a means of "excusing" the copies that are made in the process of reverse engineering a computer program. Open source licenses like the GNU GPL are also called "copyleft" licenses. The GNU GPL and other open source licenses are available at <http://www.fsf.org/licensing/licenses/>.

Inspired in part by the success of the open source movement in software, an organization called Creative Commons developed licensing tools for all types of creative works, and it did so with a similar aim: to use the private rights of copyright and contract "to create public goods." http://creativecomons.org/about/history. Copyright owners can use Creative Commons licenses to declare what rights they are reserving and what rights they are granting to users. As a condition for the different uses granted, all of the Creative Commons licenses require attribution—crediting the creator of the work. Each license characteristic has a corresponding symbol that can be placed on the owner's work. The attribution condition symbol looks like this:

Three additional conditions are available for a copyright owner to select when determining which Creative Commons license to use for his work:

 Noncommercial: others may copy, distribute, publicly display, and publicly perform the work and derivative works based upon it, but for noncommercial purposes only.

 No Derivative: others may copy, distribute, publicly display, and publicly perform only verbatim copies of the work, not derivative works based upon it.

 Share Alike: others are allowed to distribute derivative works only under a license identical to the license that governs the original work.

In addition to providing licensing tools for authors and copyright owners, Creative Commons provides a search function that facilitates discovery of creative works for which the copyright owner has already granted users certain use rights. If you are looking for a picture to enhance a presentation, or for a sound recording to use for a website, you can search for works with the licensing terms you need.

Use of Creative-Commons-licensed works requires compliance with the license terms. For example, if someone uses a Creative-Commons-licensed photo but fails to provide the required attribution of authorship, he cannot rely on a defense of "licensed used" if the owner of the photo copyright sues him for infringement. *See Jacobsen v. Katzer*, 535 F.3d 1373 (Fed. Cir. 2008).

VI. Secondary Liability and Para-Copyright

As you learned in Section IV, copyright infringement is a strict liability offense; there is no mental-state requirement. A direct infringer is one who causes the infringing copies to be made or distributed, or who engages in a public performance or public display. A defendant may be held directly liable only if he has engaged in volitional conduct that violates a copyright owner's rights. But what about others who may have assisted the infringer in engaging in that conduct? Should they be held liable for infringement as well? Sometimes the assistance comes in the form of technological tools that help the direct infringer engage in the infringing conduct. Should the mere distribution of those tools give rise to liability? What if the tools have many uses that have nothing to do with copyright infringement?

A. Secondary Liability

1. Contributory and Vicarious Infringement

Fonovisa, Inc. v. Cherry Auction, Inc.

76 F.3d 259 (9th Cir. 1996)

Schroeder, Judge:

. . . .

The plaintiff and appellant is Fonovisa, Inc., a California corporation that owns copyrights and trademarks to Latin/Hispanic music recordings. Fonovisa filed this action in district court against defendant-appellee, Cherry Auction, Inc., and its individual operators (collectively "Cherry Auction"). For purposes of this appeal, it is undisputed that Cherry Auction operates a swap meet in Fresno, California, similar to many other swap meets in this country where customers come to purchase various merchandise from individual vendors. The vendors pay a daily rental fee to the swap meet operators in exchange for booth space. Cherry Auction supplies parking, conducts advertising and retains the right to exclude any vendor for any reason, at any time, and thus can exclude vendors for patent and trademark infringement. In addition, Cherry Auction receives an entrance fee from each customer who attends the swap meet.

There is also no dispute for purposes of this appeal that Cherry Auction and its operators were aware that vendors in their swap meet were selling counterfeit recordings in violation of Fonovisa's trademarks and copyrights. Indeed, it is alleged that in 1991, the Fresno County Sheriff's Department raided the Cherry Auction swap meet and seized more than 38,000 counterfeit recordings. The following year, after finding that vendors at the Cherry Auction swap meet were still selling counterfeit recordings, the Sheriff sent a letter notifying Cherry Auction of the on-going sales of infringing materials, and reminding Cherry Auction that they had agreed to provide the Sheriff with identifying information from each vendor. In addition, in 1993, Fonovisa itself sent an investigator to the Cherry Auction site and observed sales of counterfeit recordings.

Fonovisa filed its original complaint in the district court on February 25, 1993, and on March 22, 1994, the district court granted defendants' motion to dismiss pursuant to Federal Rule of Civil Procedure 12(b)(6). In this appeal, Fonovisa does not challenge the district court's dismissal of its claim for direct copyright infringement, but does appeal the dismissal of its claims for contributory copyright infringement [and] vicarious copyright infringement . . . .

The copyright claims are brought pursuant to 17 U.S.C. §§ 101 et seq. Although the Copyright Act does not expressly impose liability on anyone other than direct infringers, courts have long recognized that in certain circumstances, vicarious or contributory liability will be imposed. *See Sony Corp. of America v. Universal City Studios, Inc.*, 464 U.S. 417, 435 (1984) (explaining that "vicarious liability is imposed in virtually all areas of the law, and the concept of contributory infringement is merely a species of the broader problem of identifying circumstances in which it is just to hold one individually accountable for the actions of another").

. . . .

Vicarious Copyright Infringement

The concept of vicarious copyright liability was developed in the Second Circuit as an outgrowth of the agency principles of respondeat superior. The landmark case on vicarious liability for sales of counterfeit recordings is *Shapiro, Bernstein and Co. v. H.L. Green Co.*, 316 F.2d 304 (2d Cir. 1963). In *Shapiro*, the court was faced with a copyright infringement suit against the owner of a chain of department stores where a concessionaire was selling counterfeit recordings. Noting that the normal agency rule of respondeat superior imposes liability on an employer for copyright infringements by an employee, the court endeavored to fashion a principle for enforcing copyrights against a defendant whose economic interests were intertwined with the direct infringer's, but who did not actually employ the direct infringer.

The *Shapiro* court looked at the two lines of cases it perceived as most clearly relevant. In one line of cases, the landlord-tenant cases, the courts had held that a landlord who lacked knowledge of the infringing acts of its tenant and who exercised no control over the leased premises was not liable for infringing sales by its tenant. *See e.g. Deutsch v. Arnold*, 98 F.2d 686 (2d Cir. 1938). In the other line of cases, the so-called "dance hall cases," the operator of an entertainment venue was held liable for infringing performances when the operator (1) could control the premises and (2) obtained a direct financial benefit from the audience, who paid to enjoy the infringing performance. *See e.g. Buck v. Jewell-LaSalle Realty Co.*, 283 U.S. 191, 198-199 (1931).

From those two lines of cases, the *Shapiro* court determined that the relationship between the store owner and the concessionaire in the case before it was closer to the dance-hall model than to the landlord-tenant model. It imposed liability even though the defendant was unaware of the infringement. *Shapiro* deemed the imposition of vicarious liability neither unduly harsh nor unfair because the store proprietor had the power to cease the conduct of the concessionaire, and because the proprietor derived an obvious and direct financial benefit from the infringement. The test was more clearly articulated in a later Second Circuit case as follows: "even in the absence of an employer-employee relationship one may be vicariously liable if he has the right and ability to supervise the infringing activity and also has a direct financial interest in such activities." *Gershwin Publishing Corp. v. Columbia Artists Management, Inc.*, 443 F.2d 1159, 1162 (2d Cir. 1971). The most recent and comprehensive discussion of the evolution of the doctrine of vicarious liability for copyright infringement is contained in Judge Keeton's opinion in *Polygram Intern. Pub., Inc. v. Nevada/TIG, Inc.*, 855 F. Supp. 1314 (D. Mass. 1984).

The district court in this case agreed with defendant Cherry Auction that Fonovisa did not, as a matter of law, meet either the control or the financial benefit prong of the vicarious copyright infringement test articulated in *Gershwin, supra*. Rather, the district court concluded that based on the pleadings, "Cherry Auction neither supervised nor profited from the vendors' sales." 847 F. Supp. at 1496. In the district court's view, with respect to both control and financial benefit, Cherry Auction was in the same position as an absentee landlord who has surrendered its exclusive right of occupancy in its leased property to its tenants.

This analogy to absentee landlord is not in accord with the facts as alleged in the district court and which we, for purposes of appeal, must accept. The allegations below were that vendors occupied small booths within premises that Cherry Auction controlled and patrolled. According to the complaint, Cherry Auction had the right to terminate vendors for any reason whatsoever and through that right had the ability to control the activities of vendors on the premises. In addition, Cherry Auction promoted the swap meet and controlled the access of customers to the swap meet area. In terms of control, the allegations before us are strikingly similar to those in *Shapiro* and *Gershwin*.

In *Shapiro*, for example, the court focused on the formal licensing agreement between defendant department store and the direct infringer-concessionaire. There, the concessionaire selling the bootleg recordings had a licensing agreement with the department store (H.L. Green Company) that required the concessionaire and its employees to "abide by, observe and obey all regulations promulgated from time to time by the H.L. Green Company," and H.L. Green Company had the "unreviewable discretion" to discharge the concessionaires' employees. In practice, H.L. Green Company was not actively involved in the sale of records and the concessionaire controlled and supervised the individual employees. *Id.* Nevertheless, H.L. Green's ability to police its concessionaire—which parallels Cherry Auction's ability to police its vendors under Cherry Auction's similarly broad contract with its vendors—was sufficient to satisfy the control requirement. *Id.* at 308.

In *Gershwin*, the defendant lacked the formal, contractual ability to control the direct infringer. Nevertheless, because of defendant's "pervasive participation in the formation and direction" of the direct infringers, including promoting them (i.e. creating an audience for them), the court found that defendants were in a position to police the direct infringers and held that the control element was satisfied. As the promoter and organizer of the swap meet, Cherry Auction wields the same level of control over the direct infringers as did the *Gershwin* defendant.

The district court's dismissal of the vicarious liability claim in this case was therefore not justified on the ground that the complaint failed to allege sufficient control.

We next consider the issue of financial benefit. The plaintiff's allegations encompass many substantive benefits to Cherry Auction from the infringing sales. These include the payment of a daily rental fee by each of the infringing vendors; a direct payment to Cherry Auction by each customer in the form of an admission fee, and incidental payments for parking, food and other services by customers seeking to purchase infringing recordings.

Cherry Auction nevertheless contends that these benefits cannot satisfy the financial benefit prong of vicarious liability because a commission, directly tied to the sale of particular infringing items, is required. They ask that we restrict the financial benefit prong to the precise facts presented in *Shapiro*, where defendant H.L. Green Company received a 10 or 12 per cent commission from the direct infringers' gross receipts. Cherry Auction points to the low daily rental fee paid by each vendor, discounting all other financial benefits flowing to the swap meet, and asks that we hold that the swap meet is materially similar to a mere landlord. The facts alleged by Fonovisa, however, reflect that the defendants reap substantial financial benefits from admission fees, concession stand sales and parking fees, all of which flow directly from customers who want to buy the counterfeit recordings at bargain basement prices. The plaintiff has sufficiently alleged direct financial benefit.

Our conclusion is fortified by the continuing line of cases, starting with the dance hall cases, imposing vicarious liability on the operator of a business where infringing performances enhance the attractiveness of the venue to potential customers. In *Polygram*, for example, direct infringers were participants in a trade show who used infringing music to communicate with attendees and to cultivate interest in their wares. 855 F. Supp. at 1332. The court held that the trade show participants "derived a significant financial benefit from the attention" that attendees paid to the infringing music. *Id.; See also Famous Music Corp. v. Bay State Harness Horse Racing and Breeding Ass'n*, 554 F.2d 1213, 1214 (1st Cir. 1977) (race track owner vicariously liable for band that entertained patrons who were not "absorbed in watching the races"); *Shapiro*, 316 F.2d at 307 (dance hall cases hold proprietor liable where infringing "activities provide the proprietor with a source of customers and enhanced income"). In this case, the sale of pirated recordings at the Cherry Auction swap meet is a "draw" for customers, as was the performance of pirated music in the dance hall cases and their progeny.

Plaintiffs have stated a claim for vicarious copyright infringement.

Contributory Copyright Infringement

Contributory infringement originates in tort law and stems from the notion that one who directly contributes to another's infringement should be held accountable. Contributory infringement has been described as an outgrowth of enterprise liability, and imposes liability where one person knowingly contributes to the infringing conduct of another. The classic statement of the doctrine is in *Gershwin*, 443 F.2d 1159, 1162: "[O]ne who, with knowledge of the infringing activity, induces, causes or materially contributes to the infringing conduct of another, may be held liable as a 'contributory' infringer."

There is no question that plaintiff adequately alleged the element of knowledge in this case. The disputed issue is whether plaintiff adequately alleged that Cherry Auction materially contributed to the infringing activity. We have little difficulty in holding that the allegations in this case are sufficient to show material contribution to the infringing activity. Indeed, it would be difficult for the infringing activity to take place in the massive quantities alleged without the support services provided by the swap meet. These services include, *inter alia*, the provision of space, utilities, parking, advertising, plumbing, and customers.

Here again Cherry Auction asks us to ignore all aspects of the enterprise described by the plaintiffs, to concentrate solely on the rental of space, and to hold that the swap meet provides nothing more. Yet Cherry Auction actively strives to provide the environment and the market for counterfeit recording sales to thrive. Its participation in the sales cannot be termed "passive," as Cherry Auction would prefer.

The district court apparently took the view that contribution to infringement should be limited to circumstances in which the defendant "expressly promoted or encouraged the sale of counterfeit products, or in some manner protected the identity of the infringers." 847 F. Supp. 1492, 1496. Given the allegations that the local sheriff lawfully requested that Cherry Auction gather and share basic, identifying information about its vendors, and that Cherry Auction failed to comply, the defendant appears to qualify within the last portion of the district court's own standard that posits liability for protecting infringers' identities. Moreover, we agree with the Third Circuit's analysis in *Columbia Pictures Industries, Inc. v. Aveco, Inc.*, 800 F.2d 59 (3rd Cir. 1986) that providing the site and facilities for known infringing activity is sufficient to establish contributory liability. See 2 William F. Patry, COPYRIGHT LAW & PRACTICE 1147 ("Merely providing the means for infringement may be sufficient" to incur contributory copyright liability).

   . . . .

Notes & Questions

1. As the Ninth Circuit makes clear in the *Fonovisa* case, there are at least two distinct doctrines of secondary liability for copyright infringement: contributory infringement and vicarious liability. Be sure to identify the requirements needed to show liability under each theory separately. Additionally, there must be direct infringement. What is the evidence of direct infringement in *Fonovisa*? Which §106 right(s) was/were infringed?

2. Vicarious liability requires a right and ability to control the infringing activity as well as a direct financial benefit. How direct was the financial benefit in *Fonovisa*? Why did the court reject the defendant's proposed standard for what constitutes a direct financial benefit? What about credit card companies that process payments for infringing products, particularly through web sites that host infringing content; without the credit card processing services the web sites may not be financially viable and thus would cease operations. Should the credit card companies be held vicariously liable? *See Perfect 10, Inc. v. Visa International Service, Assoc.*, 494 F.3d 788 (9th Cir. 2007) (rejecting claims of copyright infringement liability asserted against Visa). Some courts have expressed caution about extending the rationale of *Fonovisa* too far: "Without the requirement that the counterfeit goods provide the main customer 'draw' to the venue, *Fonovisa* would provide essentially for the limitless expansion of vicarious liability into spheres wholly unintended by the court." *Adobe Systems, Inc. v. Canus Productions, Inc.*, 173 F. Supp. 2d 1044, 1051 (C.D. Cal. 2001).

3. Liability for contributory infringement requires knowledge of the infringing activity and material contribution to that activity. According to the Ninth Circuit, what constitutes material contribution? At one point the court quotes from a treatise on copyright law written by William Patry indicating that "merely providing means [for infringement] may be sufficient" to prove contributory infringement (together with the requisite knowledge). Such a standard would be extremely broad. For example, much infringement happens through the use of computers and the internet. Should Intel (the manufacturer and distributor of computer processor technology) and Microsoft (the creator and distributor of computer operating systems) be liable for that infringing activity? How about internet service

providers? Running computers requires electricity—should the electric company be liable too? Is there a point at which the level of causation is too attenuated? How would you define that point? Does the knowledge requirement address all of the concerns that might arise with such a broad standard? What happens if a copyright owner sends a letter notifying these companies of infringement occurring on the internet—is the requisite knowledge now present? *See Sony Discos, Inc. v. E.J.C. Family Partnership*, 2010 WL 1270342 (S.D. Tex. 2010) (expressing caution about stretching the requirements of contributory infringement "to the point where they have become empty gestures").

4. The possibility of secondary liability for internet activity led Congress to adopt "safe harbor" provisions for internet service providers, web hosting services, and on-line search engines as part of the Digital Millennium Copyright Act in 1998. These provisions are currently codified in § 512 of the Act. If an internet company desires to maintain the protection from certain remedies that these safe harbors provide, they must not have knowledge of infringing activity and they must, in certain circumstances, remove infringing material upon receipt of a notification from a copyright owner. These "notice and takedown" provisions are explored later in this Section.

2. Secondary Liability for Device Manufacturers

Sony Corp. v. Universal City Studios, Inc.
464 U.S. 417 (1984)

STEVENS, JUSTICE:

[Review the facts of this case above, in Section VI.]

. . . .

II

. . . .

From its beginning, the law of copyright has developed in response to significant changes in technology. Indeed, it was the invention of a new form of copying equipment—the printing press—that gave rise to the original need for copyright protection. Repeatedly, as new developments have occurred in this country, it has been the Congress that has fashioned the new rules that new technology made necessary. Thus, long before the enactment of the Copyright Act of 1909, it was settled that the protection given to copyrights is wholly statutory. *Wheaton v. Peters*, 33 U.S. (8 Peters) 591, 661-662 (1834). The remedies for infringement "are only those prescribed by Congress." *Thompson v. Hubbard*, 131 U.S. 123, 151 (1889).

The judiciary's reluctance to expand the protections afforded by the copyright without explicit legislative guidance is a recurring theme. *See, e.g. Teleprompter Corp. v. CBS*, 415 U.S. 394 (1974); *Fortnightly Corp. v. United Artists*, 392 U.S. 390 (1968); *White-Smith Music Publishing Co. v. Apollo Co.*, 209 U.S. 1 (1908); *Williams and Wilkins v. United States*, 487 F.2d 1345 (1973), *affirmed by an equally divided court*, 420 U.S. 376 (1975). Sound policy, as well as history, supports our consistent deference to Congress when major technological innovations alter the market for copyrighted materials. Congress has the constitutional authority and the institutional ability to accommodate fully the varied permutations of competing interests that are inevitably implicated by such new technology.

In a case like this, in which Congress has not plainly marked our course, we must be circumspect in construing the scope of rights created by a legislative enactment which never contemplated such a calculus of interests. In doing so, we are guided by Justice Stewart's exposition of the correct approach to ambiguities in the law of copyright:

> "The limited scope of the copyright holder's statutory monopoly, like the limited copyright duration required by the Constitution, reflects a balance of competing claims upon the public interest: Creative work is to be encouraged and rewarded, but private motivation must ultimately serve the cause of promoting broad public availability of literature, music, and the other arts. The immediate effect of our copyright law is to secure a fair return for an 'author's' creative labor. But the ultimate aim is, by this incentive, to stimulate artistic creativity for the general public good. 'The sole interest of the United States and the primary object in conferring the monopoly,' this Court has said, 'lie in the general benefits derived by the public from the labors of authors.' *Fox Film Corp. v. Doyal,* 286 U.S. 123, 127. See *Kendall v. Winsor,* 21 How. 322, 327-328; *Grant v. Raymond,* 6 Pet. 218, 241-242. When technological change has rendered its literal terms ambiguous, the Copyright Act must be construed in light of this basic purpose." *Twentieth Century Music Corp. v. Aiken,* 422 U.S. 151, 156 (footnotes omitted).

Copyright protection "subsists * * * in original works of authorship fixed in any tangible medium of expression." 17 U.S.C. § 102(a). This protection has never accorded the copyright owner complete control over all possible uses of his work. Rather, the Copyright Act grants the copyright holder "exclusive" rights to use and to authorize the use of his work in five qualified ways, including reproduction of the copyrighted work in copies. *Id.* § 106. All reproductions of the work, however, are not within the exclusive domain of the copyright owner; some are in the public domain. Any individual may reproduce a copyrighted work for a "fair use;" the copyright owner does not possess the exclusive right to such a use. Compare *id.* § 106 with *id.* § 107.

. . . .

The two respondents in this case do not seek relief against the Betamax users who have allegedly infringed their copyrights. Moreover, this is not a class action on behalf of all copyright owners who license their works for television broadcast, and respondents have no right to invoke whatever rights other copyright holders may have to bring infringement actions based on Betamax copying of their works. As was made clear by their own evidence, the copying of the respondents' programs represents a small portion of the total use of VTR's. It is, however, the taping of respondents own copyrighted programs that provides them with standing to charge Sony with contributory infringement. To prevail, they have the burden of proving that users of the Betamax have infringed their copyrights and that Sony should be held responsible for that infringement.

III

The Copyright Act does not expressly render anyone liable for infringement committed by another. In contrast, the Patent Act expressly brands anyone who "actively induces infringement of a patent" as an infringer, 35 U.S.C. § 271(b), and further imposes liability on certain individuals labeled "contributory" infringers, *id.* § 271(c). The absence of such express language in the copyright statute does not preclude the imposition of liability for copyright infringements on certain parties who have not themselves engaged in the infringing activity. For vicarious liability is imposed in virtually all areas of the law, and the concept of contributory infringement is merely a species of the broader problem of identifying the circumstances in which it is just to hold one individual accountable for the actions of another.

. . . Petitioners in the instant case do not supply Betamax consumers with respondents' works; respondents do. Petitioners supply a piece of equipment that is generally capable of copying the entire range of programs that may be televised: those that are uncopyrighted, those that are copyrighted but may be copied without objection from the copyright holder, and those that the copyright holder would prefer not to have copied. The Betamax can be used to make authorized or unauthorized uses of copyrighted works . . . .

Justice Holmes stated that the producer [of an infringing film] had "contributed" to the infringement of the copyright, and the label "contributory infringement" has been applied in a number of lower court copyright cases involving an ongoing relationship between the direct infringer and the contributory infringer at the time the infringing conduct occurred. In such cases . . . the "contributory" infringer was in a position to control the use of copyrighted works by others and had authorized the use without permission from the copyright owner. This case, however, plainly does not fall in that category. The only contact between Sony and the users of the Betamax that is disclosed by this record occurred at the moment of sale. The District Court expressly found that "no employee of Sony . . . had either direct involvement with the allegedly infringing activity or direct contact with purchasers of Betamax who recorded copyrighted works off-the-air." 480 F. Supp., at 460. And it further found that "there was no evidence that any of the copies made by Griffiths[*] or the other individual witnesses in this suit were influenced or encouraged by [Sony's] advertisements." *Id.* If vicarious liability is to be imposed on petitioners in this case, it must rest on the fact that they have sold equipment with constructive knowledge of the fact that their customers may use that equipment to make unauthorized copies of copyrighted material. There is no precedent in the law of copyright for the imposition of vicarious liability on such a theory. The closest analogy is provided by the patent law cases to which it is

---

*[*Eds. Note*: Griffiths was an individual who had been named as a defendant, although the plaintiffs did not seek any relief against him.]

appropriate to refer because of the historic kinship between patent law and copyright law.

In the Patent Code both the concept of infringement and the concept of contributory infringement are expressly defined by statute. The prohibition against contributory infringement is confined to the knowing sale of a component especially made for use in connection with a particular patent. There is no suggestion in the statute that one patentee may object to the sale of a product that might be used in connection with other patents. Moreover, the Act expressly provides that the sale of a "staple article or commodity of commerce suitable for substantial noninfringing use" is not contributory infringement.

When a charge of contributory infringement is predicated entirely on the sale of an article of commerce that is used by the purchaser to infringe a patent, the public interest in access to that article of commerce is necessarily implicated. A finding of contributory infringement does not, of course, remove the article from the market altogether; it does, however, give the patentee effective control over the sale of that item. Indeed, a finding of contributory infringement is normally the functional equivalent of holding that the disputed article is within the monopoly granted to the patentee.

For that reason, in contributory infringement cases arising under the patent laws the Court has always recognized the critical importance of not allowing the patentee to extend his monopoly beyond the limits of his specific grant. These cases deny the patentee any right to control the distribution of unpatented articles unless they are "unsuited for any commercial noninfringing use." *Dawson Chemical Co. v. Rohm & Hass Co.,* 448 U.S. 176, 198 (1980). Unless a commodity "has no use except through practice of the patented method," *id.,* the patentee has no right to claim that its distribution constitutes contributory infringement. "To form the basis for contributory infringement the item must almost be uniquely suited as a component of the patented invention." P. Rosenberg, *Patent Law Fundamentals* § 17.02[2] (1982). "[A] sale of an article which though adapted to an infringing use is also adapted to other and lawful uses, is not enough to make the seller a contributory infringer. Such a rule would block the wheels of commerce." *Henry v. A.B. Dick Co.,* 224 U.S. 1, 48 (1912), overruled on other grounds, *Motion Picture Patents Co. v. Universal Film Mfg. Co.,* 243 U.S. 502, 517 (1917).

We recognize there are substantial differences between the patent and copyright laws. But in both areas the contributory infringement doctrine is grounded on the recognition that adequate protection of a monopoly may require the courts to look beyond actual duplication of a device or publication to the products or activities that make such duplication possible. The staple article of commerce doctrine must strike a balance between a copyright holder's legitimate demand for effective—not merely symbolic—protection of the statutory monopoly, and the rights of others freely to engage in substantially unrelated areas of commerce. Accordingly, the sale of copying equipment, like the sale of other articles of commerce, does not constitute contributory infringement if the product is widely used for legitimate, unobjectionable purposes. Indeed, it need merely be capable of substantial noninfringing uses.

IV

The question is thus whether the Betamax is capable of commercially significant noninfringing uses. In order to resolve that question, we need not explore *all* the different potential uses of the machine and determine whether or not they would constitute infringement. Rather, we need only consider whether on the basis of the facts as found by the district court a significant number of them would be non-infringing. Moreover, in order to resolve this case we need not give precise content to the question of how much use is commercially significant. For one potential use of the Betamax plainly satisfies this standard, however it is understood: private, noncommercial time-shifting in the home. It does so both (A) because respondents have no right to prevent other copyright holders from authorizing it for their programs, and (B) because the District Court's factual findings reveal that even the unauthorized home time-shifting of respondents' programs is legitimate fair use.

. . . .

Notes & Questions

1. In Section V we explored the portion of the *Sony* opinion in which five Justices held that time-shifting of over-the-air broadcast television programs was fair use. In this portion of the opinion those five Justices conclude that the manufacture and sale of VCRs does not constitute contributory infringement because of that substantial non-infringing use. Assume that there were individuals who used the VCRs to infringe. Why does the Court not hold the defendants liable for that infringing conduct? Was there no evidence of direct infringement? Was there no evidence of knowledge on the part of the defendant of that infringing conduct? The Court is not as clear as the Ninth Circuit was in the *Fonovisa* decision concerning the difference between contributory and vicarious infringement. Does this lack of precision about these two different doctrines of secondary liability help explain the outcome in *Sony?*

2. If secondary liability had been found in this case, what would have been the appropriate remedy? If Sony had entered into a license agreement with the plaintiffs in order to be able to continue selling its VTRs, would the right people have been receiving compensation? What about other copyright owners of televised programs? Would it make sense to collect a royalty on copying devices used in recording devices (for example Mp3 players) or on blank media (for example CDs)? If so, to whom should the royalty be distributed? Should such a royalty obligation be imposed by courts, or is Congress the proper body to implement such a system?

3. Congress codified the rules for imposing secondary liability in the Patent Act, from which the Court borrows the standard for Copyright. Are the Patent Act and the Copyright Act sufficiently similar to justify the Court's borrowing? Or, should the Court wait for Congress to impose any kind of secondary liability?

3. Inducing Infringement

In the late 1990s and early 2000s, a technology for "sharing" digital files became ubiquitous on the world wide web. Labeled as peer-to-peer ("p2p") technology, the first iterations used a centralized server that facilitated the indexing of the files that were available on users' individual computers. Copyright owners promptly brought suit against the entities that controlled the centralized servers and that distributed the software that made p2p sharing possible. The Ninth Circuit determined that "sharing" copyrighted files constituted infringement—both the "upload" (infringing reproduction and distribution rights) and the "download" (infringing reproduction)—and that the purveyors of the technology were liable for infringement on theories of secondary liability. *A & M Records, Inc. v. Napster, Inc.*, 239 F.3d 1004 (9th Cir. 2001). The court held that "absent specific information which identifies infringing activity, a computer system operator cannot be liable for contributory infringement merely because the structure of the system allows for the exchange of copyrighted material." *Id.* at 1020-21. However, the court held that Napster had both specific knowledge of infringing files and the ability to "purge" the infringing files. *Id.* The court also ruled that Napster could be held vicariously liable for infringement, because it received a direct financial benefit from advertising sales and had both the right and the ability to control the infringement by filtering or otherwise blocking the exchange of infringing files. *Id.* at 1023-24. The Seventh Circuit also ruled against a peer-to-peer technology company. *In re Aimster Copyright Litigation*, 334 F.3d 643 (7th Cir. 2003).

The next phase of the technological development involved p2p software that did not rely on a centralized server to facilitate the "sharing" process.

MGM Studios Inc. v. Grokster Ltd.,

545 U.S. 913 (2005)

SOUTER, JUSTICE:

. . . .

Respondents, Grokster, Ltd., and StreamCast Networks, Inc., defendants in the trial court, distribute free software products that allow computer users to share electronic files through peer-to-peer networks, so called because users' computers communicate directly with each other, not through central servers. The advantage of peer-to-peer networks over information networks of other types shows up in their substantial and growing popularity. Because they need no central computer server to mediate the exchange of information or files among users, the high-bandwidth communications capacity for a server may be dispensed with, and the need for costly server storage space is eliminated. Since copies of a file (particularly a popular one) are available on many users' computers, file requests and retrievals may be faster than on other types of networks, and since file exchanges do not travel through a server, communications can take place between any computers that remain connected to the network without risk that a glitch in the server will disable the network in its entirety. Given these benefits in security, cost, and efficiency, peer-to-peer networks are employed to store and distribute electronic files by universities, government agencies, corporations, and libraries, among others.

Other users of peer-to-peer networks include individual recipients of Grokster's and StreamCast's software, and although the networks that they enjoy through using the software can be used to share any type of digital file, they have prominently employed those networks in sharing copyrighted music and video files without authorization. A group of copyright holders (MGM for short, but including motion picture studios, recording companies, songwriters, and music publishers) sued Grokster and StreamCast for their users' copyright infringements, alleging that they knowingly and intentionally distributed their software to enable users to reproduce and distribute the copyrighted works in violation of the Copyright Act. MGM sought damages and an injunction.

. . . .

Although Grokster and StreamCast do not . . . know when particular files are copied, a few searches using their software would show what is available on the networks the software reaches. MGM commissioned a statistician to conduct a systematic search, and his study showed that nearly 90% of the files available for download . . . were copyrighted works. Grokster and StreamCast dispute this figure, raising methodological problems and arguing that free copying even of copyrighted works may be authorized by the rightholders. They also argue that potential noninfringing uses of their software are significant in kind, even if infrequent in practice. Some musical performers, for example, have gained new audiences by distributing their copyrighted works for free across peer-to-peer networks, and some distributors of unprotected content have used peer-to-peer networks to disseminate files, Shakespeare being an example. Indeed, StreamCast has given Morpheus users the opportunity to download the briefs in this very case, though their popularity has not been quantified.

. . . MGM's evidence gives reason to think that the vast majority of users' downloads are acts of infringement, and because well over 100 million copies of the software in question are known to have been downloaded, and billions of files are shared across the [defendants' p2p] networks each month, the probable scope of copyright infringement is staggering.

Grokster and StreamCast concede the infringement in most downloads, Brief for Respondents 10, n. 6, and it is uncontested that they are aware that users employ their software primarily to download copyrighted files, even if the decentralized . . . networks fail to reveal which files are being copied, and when. From time to time, moreover, the companies have learned about their users' infringement directly, as from users who have sent e-mail to each company with questions about playing copyrighted movies they had downloaded, to whom the companies have responded with guidance. And MGM notified the companies of 8 million copyrighted files that could be obtained using their software.

Grokster and StreamCast are not, however, merely passive recipients of information about infringing use. The record is replete with evidence that from the moment Grokster and StreamCast began to distribute their free software, each one clearly voiced the objective that recipients use it to download copyrighted works, and each took active steps to encourage infringement.

After the notorious file-sharing service, Napster, was sued by copyright holders for facilitation of copyright infringement, StreamCast gave away a software program of a kind known as OpenNap, designed as compatible with the Napster program and open to Napster users for downloading files from other Napster and OpenNap users' computers . . . .

. . . Internal company documents indicate that StreamCast hoped to attract large numbers of former Napster users if that company was shut down by court order or otherwise, and that StreamCast planned to be the next Napster. A kit developed by StreamCast to be delivered to advertisers, for example, contained press articles about StreamCast's potential to capture former Napster users, and it introduced itself to some potential advertisers as a company "which is similar to what Napster was." It broadcast banner advertisements to users of other Napster-compatible software, urging them to adopt its OpenNap. An internal e-mail from a company executive stated: " 'We have put this network in place so that when Napster pulls the plug on their free service . . . or if the Court orders them shut down prior to that . . . we will be positioned to capture the flood of their 32 million users that will be actively looking for an alternative.' "

Thus, StreamCast developed promotional materials to market its service as the best Napster alternative. One proposed advertisement read: "Napster Inc. has announced that it will soon begin charging you a fee. That's if the courts don't order it shut down first. What will you do to get around it?" Another proposed ad touted StreamCast's software as the "#1 alternative to Napster" and asked "[w]hen the lights went off at Napster . . . where did the users go?"[7] StreamCast even planned to flaunt the illegal uses of its software; when it launched the OpenNap network, the chief technology officer of the company averred that "[t]he goal is to get in trouble with the law and get sued. It's the best way to get in the new[s]."

. . . .

StreamCast's executives monitored the number of songs by certain commercial artists available on their networks, and an internal communication indicates they aimed to have a larger number of copyrighted songs available on their networks than other file-sharing networks. The point . . . would be to attract users of a mind to infringe, just as it would be with their promotional materials developed showing copyrighted songs as examples of the kinds of files available through Morpheus. Morpheus in fact allowed users to search specifically for "Top 40" songs, which were inevitably copyrighted. Similarly, Grokster sent users a newsletter promoting its ability to provide particular, popular copyrighted materials.

---

[7] The record makes clear that StreamCast developed these promotional materials but not whether it released them to the public. Even if these advertisements were not released to the public and do not show encouragement to infringe, they illuminate StreamCast's purposes.

In addition to this evidence of express promotion, marketing, and intent to promote further, the business models employed by Grokster and StreamCast confirm that their principal object was use of their software to download copyrighted works. Grokster and StreamCast receive no revenue from users, who obtain the software itself for nothing. Instead, both companies generate income by selling advertising space, and they stream the advertising to Grokster and Morpheus users while they are employing the programs. As the number of users of each program increases, advertising opportunities become worth more. While there is doubtless some demand for free Shakespeare, the evidence shows that substantive volume is a function of free access to copyrighted work. Users seeking Top 40 songs, for example, or the latest release by Modest Mouse, are certain to be far more numerous than those seeking a free Decameron, and Grokster and StreamCast translated that demand into dollars.

Finally, there is no evidence that either company made an effort to filter copyrighted material from users' downloads or otherwise impede the sharing of copyrighted files. Although Grokster appears to have sent e-mails warning users about infringing content when it received threatening notice from the copyright holders, it never blocked anyone from continuing to use its software to share copyrighted files. StreamCast not only rejected another company's offer of help to monitor infringement, but blocked the Internet Protocol addresses of entities it believed were trying to engage in such monitoring on its networks.

B

After discovery, the parties on each side of the case cross-moved for summary judgment.... The District Court held that those who used the Grokster and Morpheus software to download copyrighted media files directly infringed MGM's copyrights, a conclusion not contested on appeal, but the court nonetheless granted summary judgment in favor of Grokster and StreamCast as to any liability arising from distribution of the then current versions of their software. Distributing that software gave rise to no liability in the court's view, because its use did not provide the distributors with actual knowledge of specific acts of infringement.

The Court of Appeals affirmed. . . .

II

A

MGM and many of the *amici* fault the Court of Appeals's holding for upsetting a sound balance between the respective values of supporting creative pursuits through copyright protection and promoting innovation in new communication technologies by limiting the incidence of liability for copyright infringement. The more artistic protection is favored, the more technological innovation may be discouraged; the administration of copyright law is an exercise in managing the trade off.

The tension between the two values is the subject of this case, with its claim that digital distribution of copyrighted material threatens copyright holders as never before, because every copy is identical to the original, copying is easy, and many people (especially the young) use file-sharing software to download copyrighted works. This very breadth of the software's use may well draw the public directly into the debate over copyright policy, and the indications are that the ease of copying songs or movies using software like Grokster's and Napster's is fostering disdain for copyright protection. As the case has been presented to us, these fears are said to be offset by the different concern that imposing liability, not only on infringers but on distributors of software based on its potential for unlawful use, could limit further development of beneficial technologies.[8]

The argument for imposing indirect liability in this case is, however, a powerful one, given the number of infringing downloads that occur every day using StreamCast's and Grokster's software. When a widely shared service or product is used to commit infringement, it may be impossible to enforce rights in the protected work effectively against all direct infringers, the only practical alternative being to go against the distributor of the copying device for secondary liability on a theory of contributory or vicarious infringement.

One infringes contributorily by intentionally inducing or encouraging direct infringement, and infringes vicariously by profiting from direct infringement while declining to exercise a right to stop or limit it.[9] Although "[t]he Copyright Act does not expressly render anyone liable for infringement committed by another," *Sony Corp. v. Universal City Studios*, 464 U.S. at 434, these doctrines of secondary liability emerged from common law principles and are well established in the law.

B

. . . In *Sony Corp.* v. *Universal City Studios*, this Court addressed a claim that secondary liability for infringement can arise from the very distribution of a commercial product . . . . There was no evidence that Sony had expressed an object of bringing about taping in violation of copyright or had taken active steps to increase its profits from unlawful taping. *Id.* at 438. Although Sony's advertisements urged consumers to buy the VCR to " 'record favorite shows' " or " 'build a library' " of recorded programs, *id.* at 459 (Blackmun, J., dissenting), neither of these uses was necessarily infringing, *id.* at 424, 454—55.

On those facts, with no evidence of stated or indicated intent to promote infringing uses, the only conceivable basis for imposing liability was on a theory of contributory infringement arising from its sale of VCRs to consumers with

---

[8] The mutual exclusivity of these values should not be overstated, however. On the one hand technological innovators, including those writing filesharing computer programs, may wish for effective copyright protections for their work. (StreamCast itself was urged by an associate to "get [its] technology written down and [its intellectual property] protected.") On the other hand the widespread distribution of creative works through improved technologies may enable the synthesis of new works or generate audiences for emerging artists.

[9] We stated in *Sony* that "the lines between direct infringement, contributory infringement and vicarious liability are not clearly drawn . . ." *id.* at 435, n.17. In the present case MGM has argued a vicarious liability theory, which allows imposition of liability when the defendant profits directly from the infringement and has a right and ability to supervise the direct infringer, even if the defendant initially lacks knowledge of the infringement. Because we resolve the case based on an inducement theory, there is no need to analyze separately MGM's vicarious liability theory.

knowledge that some would use them to infringe. *Id*. at 439. But because the VCR was "capable of commercially significant noninfringing uses," we held the manufacturer could not be faulted solely on the basis of its distribution. *Id*. at 442.

. . . .

. . . We do not revisit *Sony* further, as MGM requests, to add a more quantified description of the point of balance between protection and commerce when liability rests solely on distribution with knowledge that unlawful use will occur. It is enough to note that the Ninth Circuit's judgment rested on an erroneous understanding of *Sony* and to leave further consideration of the *Sony* rule for a day when that may be required.

C

Sony's rule limits imputing culpable intent as a matter of law from the characteristics or uses of a distributed product. But nothing in *Sony* requires courts to ignore evidence of intent if there is such evidence, and the case was never meant to foreclose rules of fault-based liability derived from the common law.[10] Thus, where evidence goes beyond a product's characteristics or the knowledge that it may be put to infringing uses, and shows statements or actions directed to promoting infringement, *Sony*'s staple-article rule will not preclude liability.

The classic case of direct evidence of unlawful purpose occurs when one induces commission of infringement by another, or "entic[es] or persuad[es] another" to infringe, *Black's Law Dictionary* 790 (8th ed. 2004), as by advertising. Thus at common law a copyright or patent defendant who "not only expected but invoked [infringing use] by advertisement" was liable for infringement "on principles recognized in every part of the law." *Kalem Co.* v. *Harper Brothers*, 222 U.S. at 62–63 (copyright infringement).

The rule on inducement of infringement as developed in the early cases is no different today. Evidence of "active steps * * * taken to encourage direct infringement," *Oak Indus.* v. *Zenith Electronics Corp.*, 697 F. Supp. 988, 992 (N.D. Ill. 1988), such as advertising an infringing use or instructing how to engage in an infringing use, show an affirmative intent that the product be used to infringe, and a showing that infringement was encouraged overcomes the law's reluctance to find liability when a defendant merely sells a commercial product suitable for some lawful use.

For the same reasons that *Sony* took the staple-article doctrine of patent law as a model for its copyright safe-harbor rule, the inducement rule, too, is a sensible one for copyright. We adopt it here, holding that one who distributes a device with the object of promoting its use to infringe copyright, as shown by clear expression or other affirmative steps taken to foster infringement, is liable for the resulting acts of infringement by third parties. We are, of course, mindful of the need to keep from trenching on regular commerce or discouraging the development of technologies with lawful and unlawful potential. Accordingly, just as *Sony* did not

---

[10] Nor does the Patent Act's exemption from liability for those who distribute a staple article of commerce, 35 U.S.C. § 271(c), extend to those who induce patent infringement, §271(b).

find intentional inducement despite the knowledge of the VCR manufacturer that its device could be used to infringe, mere knowledge of infringing potential or of actual infringing uses would not be enough here to subject a distributor to liability. Nor would ordinary acts incident to product distribution, such as offering customers technical support or product updates, support liability in themselves. The inducement rule, instead, premises liability on purposeful, culpable expression and conduct, and thus does nothing to compromise legitimate commerce or discourage innovation having a lawful promise.

III

A

The only apparent question about treating MGM's evidence as sufficient to withstand summary judgment under the theory of inducement goes to the need on MGM's part to adduce evidence that StreamCast and Grokster communicated an inducing message to their software users. The classic instance of inducement is by advertisement or solicitation that broadcasts a message designed to stimulate others to commit violations. MGM claims that such a message is shown here. It is undisputed that StreamCast beamed onto the computer screens of users of Napster-compatible programs ads urging the adoption of its OpenNap program, which was designed, as its name implied, to invite the custom of patrons of Napster, then under attack in the courts for facilitating massive infringement. . . . Grokster distributed an electronic newsletter containing links to articles promoting its software's ability to access popular copyrighted music. . . . And both companies communicated a clear message by responding affirmatively to requests for help in locating and playing copyrighted materials.

. . . Here, the summary judgment record is replete with other evidence that Grokster and StreamCast, unlike the manufacturer and distributor in *Sony*, acted with a purpose to cause copyright violations by use of software suitable for illegal use.

Three features of this evidence of intent are particularly notable. First, each company showed itself to be aiming to satisfy a known source of demand for copyright infringement, the market comprising former Napster users. StreamCast's internal documents made constant reference to Napster, it initially distributed its Morpheus software through an OpenNap program compatible with Napster, it advertised its OpenNap program to Napster users, and its Morpheus software functions as Napster did except that it could be used to distribute more kinds of files, including copyrighted movies and software programs. Grokster's name is apparently derived from Napster, it too initially offered an OpenNap program, its software's function is likewise comparable to Napster's, and it attempted to divert queries for Napster onto its own Web site. Grokster and StreamCast's efforts to supply services to former Napster users, deprived of a mechanism to copy and distribute what were overwhelmingly infringing files, indicate a principal, if not exclusive, intent on the part of each to bring about infringement.

Second, this evidence of unlawful objective is given added significance by MGM's showing that neither company attempted to develop filtering tools or other mechanisms to diminish the infringing activity using their software. While the Ninth Circuit treated the defendants' failure to develop such tools as irrelevant because they lacked an independent duty to monitor their users' activity, we think this evidence underscores Grokster's and StreamCast's intentional facilitation of their users' infringement.[12]

Third, there is a further complement to the direct evidence of unlawful objective. It is useful to recall that StreamCast and Grokster make money by selling advertising space, by directing ads to the screens of computers employing their software. As the record shows, the more the software is used, the more ads are sent out and the greater the advertising revenue becomes. Since the extent of the software's use determines the gain to the distributors, the commercial sense of their enterprise turns on high-volume use, which the record shows is infringing.[13] This evidence alone would not justify an inference of unlawful intent, but viewed in the context of the entire record its import is clear.

The unlawful objective is unmistakable.

B

In addition to intent to bring about infringement and distribution of a device suitable for infringing use, the inducement theory of course requires evidence of actual infringement by recipients of the device, the software in this case. As the account of the facts indicates, there is evidence of infringement on a gigantic scale, and there is no serious issue of the adequacy of MGM's showing on this point in order to survive the companies' summary judgment requests. Although an exact calculation of infringing use, as a basis for a claim of damages, is subject to dispute, there is no question that the summary judgment evidence is at least adequate to entitle MGM to go forward with claims for damages and equitable relief.

. . . .

GINSBURG, JUSTICE, concurring (with Chief Justice Rehnquist and Justice Kennedy):

I concur in the Court's decision . . . and write separately to clarify why I conclude that the Court of Appeals misperceived, and hence misapplied, our holding in *Sony Corp. v. Universal City Studios, Inc.,* 464 U.S. 417 (1984). . . .

At bottom, however labeled, the question in this case is whether Grokster and StreamCast are liable for the direct infringing acts of others. Liability under our jurisprudence may be predicated on actively encouraging (or inducing) infringement through specific acts (as the Court's opinion develops) or on distributing a product distributees use to infringe copyrights, if the product is not capable of

---

[12] Of course, in the absence of other evidence of intent, a court would be unable to find contributory infringement liability merely based on a failure to take affirmative steps to prevent infringement, if the device otherwise was capable of substantial noninfringing uses. Such a holding would tread too close to the *Sony* safe harbor.

[13] . . . [T]he distribution of a product can itself give rise to liability where evidence shows that the distributor intended and encouraged the product to be used to infringe. In such a case, the culpable act is not merely the encouragement of infringement but also the distribution of the tool intended for infringing use.

"substantial" or "commercially significant" noninfringing uses. *Sony*, 464 U.S. at 442....

. . . .

. . . [T]here was no need in *Sony* to "give precise content to the question of how much [actual or potential] use is commercially significant." *Id*. Further development was left for later days and cases.

. . . .

This case differs markedly from *Sony*. Here, there has been no finding of any fair use and little beyond anecdotal evidence of noninfringing uses. In finding the Grokster and StreamCast software products capable of substantial noninfringing uses, the District Court and the Court of Appeals appear to have relied largely on declarations submitted by the defendants....

. . . Review of these declarations reveals mostly anecdotal evidence, sometimes obtained second-hand, of authorized copyrighted works or public domain works available online and shared through peer-to-peer networks, and general statements about the benefits of peer-to-peer technology.

. . . These declarations do not support summary judgment in the face of evidence, proffered by MGM, of overwhelming use of Grokster's and StreamCast's software for infringement.

Even if the absolute number of noninfringing files copied using the Grokster and StreamCast software is large, it does not follow that the products are therefore put to substantial noninfringing uses and are thus immune from liability. The number of noninfringing copies may be reflective of, and dwarfed by, the huge total volume of files shared. Further, the District Court and the Court of Appeals did not sharply distinguish between uses of Grokster's and StreamCast's software products (which this case is about) and uses of peer-to-peer technology generally (which this case is not about).

. . . .

If, on remand, the case is not resolved on summary judgment in favor of MGM based on Grokster and StreamCast actively inducing infringement, the Court of Appeals, I would emphasize, should reconsider, on a fuller record, its interpretation of *Sony*'s product distribution holding.

BREYER, JUSTICE, concurring (with Justices Stevens and O'Connor):

I agree with the Court that the distributor of a dual-use technology may be liable for the infringing activities of third parties where he or she actively seeks to advance the infringement. I further agree that, in light of our holding today, we need not now "revisit" *Sony*. Other Members of the Court, however, take up the *Sony* question: whether Grokster's product is "capable of 'substantial' or 'commercially significant' noninfringing uses." (Ginsburg, J., concurring). And they answer that question by stating that the Court of Appeals was wrong when it granted summary judgment on the issue in Grokster's favor. I write to explain why I disagree with them on this matter.

I

. . . .

A

. . . Sony knew many customers would use its VCRs to engage in unauthorized copying and "'library-building.'" *Id.* at 458-459 (Blackmun, J., dissenting). But that fact, said the Court, was insufficient to make Sony itself an infringer. And the Court ultimately held that Sony was not liable for its customers' acts of infringement.

In reaching this conclusion, the Court recognized the need for the law, in fixing *secondary* copyright liability, to "strike a balance between a copyright holder's legitimate demand for effective—not merely symbolic—protection of the statutory monopoly, and the rights of others freely to engage in substantially unrelated areas of commerce." *Id.* at 442. . . .

. . . The Court had before it a survey (commissioned by the District Court and then prepared by the respondents) showing that roughly 9% of all VCR recordings were of the type—namely, religious, educational, and sports programming— owned by producers and distributors testifying on Sony's behalf who did not object to time-shifting. . . .

The Court found that the magnitude of authorized programming was "significant," and it also noted the "significant potential for future authorized copying." 464 U.S. at 444. The Court supported this conclusion by referencing the trial testimony of professional sports league officials and a religious broadcasting representative. *Id.* at 444, and n.24. It also discussed (1) a Los Angeles educational station affiliated with the Public Broadcasting Service that made many of its programs available for home taping, and (2) Mr. Rogers' Neighborhood, a widely watched children's program. *Id.* at 445. On the basis of this testimony and other similar evidence, the Court determined that producers of this kind had authorized duplication of their copyrighted programs "in significant enough numbers to create a *substantial* market for a noninfringing use of the" VCR. *Id.* at 447, n.28 (emphasis added).

The Court, in using the key word "substantial," indicated that these circumstances alone constituted a sufficient basis for rejecting the imposition of secondary liability. Nonetheless, the Court buttressed its conclusion by finding separately that, in any event, *un*authorized time-shifting often constituted not infringement, but "fair use."

B

When measured against *Sony's* underlying evidence and analysis, the evidence now before us shows that Grokster passes *Sony's* test—that is, whether the company's product is capable of substantial or commercially significant noninfringing uses. For one thing, petitioners' (hereinafter MGM) own expert declared that 75% of current files available on Grokster are infringing and 15% are "likely infringing." That leaves some number of files near 10% that apparently are noninfringing, a figure very similar to the 9% or so of authorized time-shifting uses of the VCR that the Court faced in *Sony*.

As in *Sony,* witnesses here explained the nature of the noninfringing files on Grokster's network without detailed quantification. Those files include:

– Authorized copies of music by artists such as Wilco, Janis Ian, Pearl Jam, Dave Matthews, John Mayer, and others.

– Free electronic books and other works from various online publishers, including Project Gutenberg.

– Public domain and authorized software, such as WinZip 8.1.

– Licensed music videos and television and movie segments distributed via digital video packaging with the permission of the copyright holder.

The nature of these and other lawfully swapped files is such that it is reasonable to infer quantities of current lawful use roughly approximate to those at issue in *Sony.* . . .

Importantly, *Sony* also used the word "capable," asking whether the product is *"capable of"* substantial noninfringing uses. Its language and analysis suggest that a figure like 10%, if fixed for all time, might well prove insufficient, but that such a figure serves as an adequate foundation where there is a reasonable prospect of expanded legitimate uses over time. See *id.* (noting a "significant potential for future authorized copying"). And its language also indicates the appropriateness of looking to potential future uses of the product to determine its "capability."

Here the record reveals a significant future market for noninfringing uses of Grokster-type peer-to-peer software. Such software permits the exchange of *any* sort of digital file—whether that file does, or does not, contain copyrighted material. As more and more uncopyrighted information is stored in swappable form, it seems a likely inference that lawful peer-to-peer sharing will become increasingly prevalent.

And that is just what is happening. Such legitimate noninfringing uses are coming to include the swapping of: *research information* (the initial purpose of many peer-to-peer networks); *public domain films* (*e.g.,* those owned by the Prelinger Archive); *historical recordings and digital educational materials* (*e.g.,* those stored on the Internet Archive); *digital photos* (OurPictures, for example, is starting a P2P photo-swapping service); *"shareware" and "freeware"* (*e.g.,* Linux and certain Windows software); *secure licensed music and movie files* (Intent MediaWorks, for example, protects licensed content sent across P2P networks); *news broadcasts past and present* (the BBC Creative Archive lets users "rip, mix and share the BBC"); *user-created audio and video files* (including "podcasts" that may be distributed through P2P software); *and all manner of free "open content" works collected by Creative Commons* (one can search for Creative Commons material on StreamCast). I can find nothing in the record that suggests that this course of events will *not* continue to flow naturally as a consequence of the character of the software taken together with the foreseeable development of the Internet and of information technology.

There may be other now-unforeseen noninfringing uses that develop for peer-to-peer software, just as the home-video rental industry (unmentioned in *Sony*) developed for the VCR. But the foreseeable development of such uses, when taken together with an estimated 10% noninfringing material, is sufficient to meet *Sony*'s standard. . . . [T]here are no facts asserted by MGM in its summary judgment filings that lead me to believe the outcome after a trial here could be any different. The lower courts reached the same conclusion.

Of course, Grokster itself may not want to develop these other noninfringing uses. But *Sony*'s standard seeks to protect not the Groksters of this world (which in any event may well be liable . . .), but the development of technology more generally. And Grokster's desires in this respect are beside the point.

II

The real question here, I believe, is not whether the record evidence satisfies *Sony*. . . .

Instead, the real question is whether we should modify the *Sony* standard, as MGM requests, or interpret *Sony* more strictly, as I believe Justice Ginsburg's approach would do in practice.

As I have said, *Sony* itself sought to "strike a balance between a copyright holder's legitimate demand for effective—not merely symbolic—protection of the statutory monopoly, and the rights of others freely to engage in substantially unrelated areas of commerce." *Id.* at 442. Thus, to determine whether modification, or a strict interpretation, of *Sony* is needed, I would ask whether MGM has shown that *Sony* incorrectly balanced copyright and new-technology interests. In particular: (1) Has *Sony* (as I interpret it) worked to protect new technology? (2) If so, would modification or strict interpretation significantly weaken that protection? (3) If so, would new or necessary copyright-related benefits outweigh any such weakening?

. . . .

[Justice Breyer concludes that *Sony* has protected new technology by, *inter alia*, being clear and forward looking, and that while a smaller *Sony* zone would "provide greater revenue security for copyright holders," it is doubtful that the gains for copyright would exceed the losses to technological innovation.]

Notes & Questions

1. As the Court notes, some musicians authorized the sharing of their copyrighted works. Was the Court's focus on defendants' awareness that copyrighted works were being shared the right focus? Isn't it the unauthorized sharing of copyrighted works rather than the status of the works as copyrighted that should matter? Many of the musicians that authorize sharing are seeking to gain an audience without the support and promotion of a major record label. P2P software is critical to their business model. Is the Court barring the creation and distribution of P2P software? If not, what must distributors of such software do to remain within the bounds of lawful activity, at least as far as copyright is concerned?

2. The Court is clearly concerned with innovation having a lawful purpose. The Court requires evidence of affirmative acts inducing infringement that are independent of design and distribution of the product. What were the independent affirmative acts in this case?

3. In many ways, one of the battles being fought in *Grokster* concerns the obligations of copyright owners versus technology providers to stop infringement from occurring. Do copyright owners have to notify technology companies of each instance of infringement before there is an obligation to assist in stopping the infringement? Or, should technology companies have some obligations to design their technologies to reduce infringement? What does the Court say about the defendants' design choice to decline to include elements that would assist in reducing infringing use of the software? How much pressure does that put on companies when designing products or services?

4. In *Sony* the Court borrowed the contributory infringement doctrine from the Patent Act. In *Grokster* the Court again looks to patent law for the boundaries of an inducement theory of liability. Is such borrowing from patent law appropriate?

5. Balancing innovation in technology and incentives for the creation of original works of authorship is central to the *Sony* framework. What are the differences in opinion between Justice Ginsburg and Justice Breyer? Do they disagree about the facts of this case or about the fundamental policy balance, or both?

B. The Line Between Direct and Secondary Liability

Now that you have an understanding of the doctrines of secondary liability, consider again the case of *American Broadcasting Cos. v. Aereo, Inc.*, 134 S. Ct. 2498 (2014). The majority opinion, which you read in Section IV.C, concluded that Aereo was directly liable for publicly performing television broadcasts with its internet service. In dissent, Justice Scalia argued that the Court had gone astray.

American Broadcasting Cos. v. Aereo, Inc.

134 S. Ct. 2498 (2014)

Scalia, Justice, dissenting:

. . . .

I. Legal Standard

There are two types of liability for copyright infringement: direct and secondary. As its name suggests, the former applies when an actor personally engages in infringing conduct. *See Sony Corp. v. Universal City Studios, Inc.*, 464 U.S. 417, 433 (1984). Secondary liability, by contrast, is a means of holding defendants responsible for infringement by third parties, even when the defendants "have not themselves engaged in the infringing activity." *Id.* at 435. It applies when a defendant "intentionally induc[es] or encourag[es]" infringing acts by others or profits from such acts "while declining to exercise a right to stop or limit [them]." *Metro-Goldwyn-Mayer Studios Inc. v. Grokster, Ltd.*, 545 U.S. 913, 930 (2005).

Most suits against equipment manufacturers and service providers involve secondary-liability claims. For example, when movie studios sued to block the sale of

Sony's Betamax videocassette recorder (VCR), they argued that Sony was liable because its customers were making unauthorized copies. Record labels and movie studios relied on a similar theory when they sued Grokster and StreamCast, two providers of peer-to-peer file-sharing software.

This suit, or rather the portion of it before us here, is fundamentally different. The Networks claim that Aereo directly infringes their public-performance right. Accordingly, the Networks must prove that Aereo "perform[s]" copyrighted works, § 106(4), when its subscribers log in, select a channel, and push the "watch" button. That process undoubtedly results in a performance; the question is who does the performing. If Aereo's subscribers perform but Aereo does not, the claim necessarily fails.

The Networks' claim is governed by a simple but profoundly important rule: A defendant may be held directly liable only if it has engaged in volitional conduct that violates the Act. . . . And since the Act makes it unlawful to copy or perform copyrighted works, not to copy or perform in general, see § 501(a), the volitional-act requirement demands conduct directed to the plaintiff's copyrighted material. Every Court of Appeals to have considered an automated-service provider's direct liability for copyright infringement has adopted that rule. Although we have not opined on the issue, our cases are fully consistent with a volitional-conduct requirement. . . .

The volitional-conduct requirement is not at issue in most direct-infringement cases; the usual point of dispute is whether the defendant's conduct is infringing (e.g., Does the defendant's design copy the plaintiff's?), rather than whether the defendant has acted at all (e.g., Did this defendant create the infringing design?). But it comes right to the fore when a direct-infringement claim is lodged against a defendant who does nothing more than operate an automated, user-controlled system. Internet-service providers are a prime example. When one user sends data to another, the provider's equipment facilitates the transfer automatically. Does that mean that the provider is directly liable when the transmission happens to result in the "reproduc[tion]," § 106(1), of a copyrighted work? It does not. The provider's system is "totally indifferent to the material's content," whereas courts require "some aspect of volition" directed at the copyrighted material before direct liability may be imposed. [*CoStar Group, Inc. v. LoopNet, Inc.*, 373 F.3d 544, 550–51 (4th Cir. 2004)].[2] The defendant may be held directly liable only if the defendant itself "trespassed on the exclusive domain of the copyright owner." *Id.* at 550. Most of the time that issue will come down to who selects the copyrighted content: the defendant or its customers.

A comparison between copy shops and video-on-demand services illustrates the point. A copy shop rents out photocopiers on a per-use basis. One customer might copy his 10-year-old's drawings—a perfectly lawful thing to do—while another might duplicate a famous artist's copyrighted photographs—a use clearly

---

[2] Congress has enacted several safe-harbor provisions applicable to automated network processes, *see, e.g.*, 17 U.S.C. § 512(a)–(b), but those provisions do not foreclose "any other defense," § 512(l), including a volitional-conduct defense.

prohibited by § 106(1). Either way, the customer chooses the content and activates the copying function; the photocopier does nothing except in response to the customer's commands. Because the shop plays no role in selecting the content, it cannot be held directly liable when a customer makes an infringing copy.

Video-on-demand services, like photocopiers, respond automatically to user input, but they differ in one crucial respect: They choose the content. When a user signs in to Netflix, for example, "thousands of * * * movies [and] TV episodes" carefully curated by Netflix are "available to watch instantly." That selection and arrangement by the service provider constitutes a volitional act directed to specific copyrighted works and thus serves as a basis for direct liability.

The distinction between direct and secondary liability would collapse if there were not a clear rule for determining whether the defendant committed the infringing act. The volitional-conduct requirement supplies that rule; its purpose is not to excuse defendants from accountability, but to channel the claims against them into the correct analytical track. See Brief for 36 Intellectual Property and Copyright Law Professors as *Amici Curiae* 7. Thus, in the example given above, the fact that the copy shop does not choose the content simply means that its culpability will be assessed using secondary-liability rules rather than direct-liability rules.

II. Application to Aereo

So which is Aereo: the copy shop or the video-on-demand service? In truth, it is neither. Rather, it is akin to a copyshop that provides its patrons with a library card. Aereo offers access to an automated system consisting of routers, servers, transcoders, and dime-sized antennae. Like a photocopier or VCR, that system lies dormant until a subscriber activates it. When a subscriber selects a program, Aereo's system picks up the relevant broadcast signal, translates its audio and video components into digital data, stores the data in a user-specific file, and transmits that file's contents to the subscriber via the Internet—at which point the subscriber's laptop, tablet, or other device displays the broadcast just as an ordinary television would. The result of that process fits the statutory definition of a performance to a tee: The subscriber's device "show[s]" the broadcast's "images" and "make[s] the sounds accompanying" the broadcast "audible." § 101. The only question is whether those performances are the product of Aereo's volitional conduct.

They are not. Unlike video-on-demand services, Aereo does not provide a prearranged assortment of movies and television shows. Rather, it assigns each subscriber an antenna that—like a library card—can be used to obtain whatever broadcasts are freely available. Some of those broadcasts are copyrighted; others are in the public domain. The key point is that subscribers call all the shots: Aereo's automated system does not relay any program, copyrighted or not, until a subscriber selects the program and tells Aereo to relay it. Aereo's operation of that system is a volitional act and a but-for cause of the resulting performances, but, as in the case of the copy shop, that degree of involvement is not enough for direct liability.

In sum, Aereo does not "perform" for the sole and simple reason that it does not make the choice of content. And because Aereo does not perform, it cannot be held directly liable for infringing the Networks' public-performance right. That conclusion does not necessarily mean that Aereo's service complies with the Copyright Act. Quite the contrary. The Networks' complaint alleges that Aereo is directly *and* secondarily liable for infringing their public performance rights (§ 106(4)) *and also* their reproduction rights (§ 106(1)). Their request for a preliminary injunction—the only issue before this Court—is based exclusively on the direct-liability portion of the public-performance claim . . . . Affirming the judgment below would merely return this case to the lower courts for consideration of the Networks' remaining claims.

. . . .

Notes & Questions

1. As Justice Scalia recognizes, the Supreme Court has never expressly formalized a separate volitional element to a claim of direct liability. Does that volition element make sense? Recall from your first-year torts course the distinction between a "but-for" cause and a proximate cause. Typically a plaintiff must demonstrate proximate cause in order to show the defendant is liable; but-for cause is insufficient. Computers make a great deal of infringement possible—but for computers, that infringement would not occur. But is that a sufficient reason to hold the manufacturers or distributors of computers directly liable for that infringement? Similarly, a software creator could design an application to filter out content that is likely to infringe. Does choosing to design the software without that filtering component mean the software creator should be liable for infringement that occurs?

2. Congress did not codify any doctrines of secondary liability in the Copyright Act. Instead, courts have created and shaped those doctrines, borrowing both from patent law and from general principles of agency and enterprise liability. Might the majority in *Aereo* be unwilling to force the plaintiff to rely on theories of secondary liability out of concern for the non-statutory nature of the doctrine?

3. In *Aereo* both the majority and the dissent refer to the Digital Millennium Copyright Act and, specifically, to § 512. That section provides an exemption from both direct and contributory liability for certain activities engaged in by "on-line service providers" (OSPs). The majority points to this provision as an example of how Congress can accommodate technologic advances with amendments to the statute. Justice Scalia notes the § 512 exemption but then highlights that the exemption does "not foreclose 'any other defense,' § 512(l), including a volitional-conduct defense." *American Broadcasting Cos. v. Aereo, Inc.*, 134 S. Ct. 2498, 2513 n.2 (2014) (Scalia, dissenting). Is there a reliable guidepost, in the Copyright Act itself, to point one toward the sounder inference about what Congress intended here? Or is the inquiry grounded on some other policy?

C. On-line Service Provider Liability: Notice and Take-Down

On-line service providers, whether they be Comcast, Facebook, Twitter, or your academic institution, are continually storing and distributing content. As you now know, most of that content is copyrighted. Some of it is infringing. When should an OSP be liable for the content it stores and distributes? As you saw in *Perfect 10 v. Amazon* in Section IV.C, the Ninth Circuit employed a "server test:" if the infringing content was located on the defendant's equipment, and internet users could access that copy, the plaintiff had made out a case of prima facie infringement. The requirement of volition for direct liability might permit the defendant to avoid direct liability: for example, if someone posts an infringing video on YouTube, no affirmative, voluntary act on YouTube's part that is specific to that video led to that infringement. But indirect liability does not have the same volition element. Even if the infringing material is located on the OSP's equipment only as a result of the independent actions of an internet user, the OSP certainly is "contributing" to the infringement that is occurring by providing the very equipment that makes the infringement possible. *See Religious Tech. Ctr. v. Netcom On-Line Commc'n Servs.*, 923 F. Supp. 1231 (N.D. Cal. 1995). Knowledge of the infringement is required for contributory liability, but what if the OSP receives a notice from a copyright owner asserting infringement? If the OSP allows the material to stay on the OSPs site, should the OSP now be contributorily liable?

Uncertainty about their liability exposure in the early days of the world wide web led OSPs to seek assistance from Congress in clarifying their liability. At the same time, the major content owning industries wanted OSPs to have some obligation to help reduce infringement on the internet. The result was a compromise, enacted in 1998 as part of the Digital Millennium Copyright Act that created four separate safe harbors from infringement liability for online service providers.

The four separate safe harbors are for (1) transitory digital network communications, (2) system caching, (3) information residing on systems at the direction of users, and (4) providing information location tools, such as hypertext links. 17 U.S.C. § 512 (a)-(d). Each safe harbor has specific requirements that must be met in order to obtain and maintain the protection against monetary damages that the safe harbors provide. Failing to comply with the requirements does not render a service provider liable for copyright infringement, it only renders the protection of the safe harbor unavailable. *See Perfect 10 v. CCBill*, 488 F.3d 1102, 1109 (9th Cir. 2007).

The safe harbors identified as (3) and (4) above create a mechanism for potentially obtaining removal of material that a copyright owner believes is infringing from a website or a search engine, without having to file a lawsuit or prove a case of infringement. The obligations of these safe harbors are triggered when a copyright owner sends a "takedown notice" to an on-line service provider. The DMCA requires that copyright owners provide the certain information in a takedown notice. 17 U.S.C. § 512(c)(3)(A).

In order to maintain the protection of the safe harbor, when an on-line service provider receives a proper notice from a copyright owner it must "expeditiously . . . remove, or disable access to, " the allegedly infringing material. 17 U.S.C. § 512(c)(A)(iii). Thus, by sending a notice of infringement to the on-line service provider, a copyright owner can obtain a significant remedy: removal from the internet of material it asserts is infringing. In the context of widespread user-generated content posted on websites such as YouTube, Facebook, and Flickr, this is an important remedial tool for copyright owners. The takedown obligation is not an independent obligation; failing to take the alleged infringing material down does not automatically make the service provider liable. If the service provider does remove the material, however, it avoids the possibility of either direct or contributory liability.

In order to qualify for safe-harbor protection, in addition to responding expeditiously when a copyright owner sends a takedown notice, an on-line service provider must designate an agent to receive such notifications with the copyright office. 17 U.S.C. § 512(c)(2), 37 CFR § 201.38. A list of designated agents is available on the copyright office website, http://www.copyright.gov/onlinesp. Designating an agent is not a mere formality–it is an eligibility requirement to be able to invoke the safe harbor protection. *See BWP Media USA, Inc. v. Hollywood Fan Sites LLC*, 2015 WL 3971750 (S.D.N.Y. June 30, 2015) (the § 512(c) safe harbor does not apply to any infringement before the date on which the defendant filed the designated-agent notice with the Copyright Office).

In order to qualify for safe-harbor protection, an online service provider must also adopt, and reasonably implement, a policy to terminate repeat infringers. 17 U.S.C. § 512 (i)(1)(A). Courts are divided on what constitutes a "repeat infringer." *Compare Corbis v. Amazon*, 351 F. Supp. 2d 1090, 1105 n.9 (W.D. Wash. 2004) (repeat takedown notices not conclusive proof), *with Perfect 10 v. CCBill*, 340 F. Supp. 2d 1077, 1088 (C.D. Cal. 2004) (multiple notices requires termination under § 512(i)). Additionally, an on-line service provider must not have actual knowledge of specific identifiable infringing activity on its system and it must not be aware of facts or circumstances from which such infringing activity is apparent. *Viacom Int'l, Inc. v. YouTube*, 676 F.3d 19 (2d Cir. 2012) (interpreting the requirements of § 512(c) as applied to YouTube).

No court reviews the takedown request to determine the validity of the copyright owner's copyright or whether the material is, in fact, infringing. The on-line service provider could engage in its own analysis of whether the material is infringing, or the copyright is valid, but it earns the safe-harbor protection whether it undertakes those analyses or not. Thus the incentives are such that the on-line service provider is more likely to remove the material and let the copyright owner and the individual who posted the material take their dispute to court, if either party so desires.

The mechanism for the copyright owner to take the individual to court is, of course, a suit alleging copyright infringement. Even with the material removed from the web, there still was past activity that may have amounted to infringement. If the copyright owner desires more relief than simply removal from the internet, such a lawsuit will be necessary. Section 512(h) provides a special subpoena power to assist the copyright owner in identifying the individual responsible for the allegedly infringing material. *See* 17 U.S.C. § 512(h).

Section 512 also provides the on-line service provider with protection from a lawsuit brought by an individual that has had her material removed, so long as the on-line service provider notifies her that a copyright owner requested removal and also complies with any timely counter-notification received from the individual. 17 U.S.C. § 512(g). A user whose material has been removed can send a counter-notice to the provider and the on-line service provider is required re-post the material to maintain protection against a user-filed lawsuit. However, if the copyright owner files a lawsuit against the individual within 10-14 business days, and notifies the provider of that lawsuit, the provider is not required to repost the allegedly infringing material. *Id.* The requirements for a counter-notice are contained in § 512(g)(3):

> . . . a counter-notification must be a written communication provided to the service provider's designated agent that includes substantially the following:
>
> (A) A physical or electronic signature of the subscriber.
>
> (B) Identification of the material that has been removed or to which access has been disabled and the location at which the material appeared before it was removed or access to it was disabled.
>
> (C) A statement under penalty of perjury that the subscriber has a good faith belief that the material was removed or disabled as a result of mistake or misidentification of the material to be removed or disabled.
>
> (D) The subscriber's name, address, and telephone number, and a statement that the subscriber consents to the jurisdiction of Federal District Court for the judicial district in which the address is located, or if the subscriber's address is outside of the United States, for any judicial district in which the service provider may be found, and that the subscriber will accept service of process from the person who provided notification under subsection (c)(1)(C) or an agent of such person.

17 U.S.C. § 512(g)(3). Clearly, the counter-notice greatly aids the copyright owner in filing a lawsuit.

The potential for significant abuse of takedown notices is somewhat tempered in the statute by providing a cause of action for "material misrepresentations." 17 U.S.C. § 512(f). The Ninth Circuit has held that "fair use is 'authorized by the law' and a copyright holder must consider the existence of fair use before sending a takedown notification . . . ." *Lenz v. Universal Music Corp.*, 815 F.3d 1145, 1153 (9th Cir. 2016). This requirement stems from the fact that the notice sent by a copyright owner must contain "[a] statement that the complaining party has a good faith belief that use of the material in the manner complained of is not authorized by the copyright owner, its agent, or the law." 17 U.S.C. § 512(c)(3)(A)(v).

Notes & Questions

1. What happens if an OSP decides not to seek the protection of the safe harbor and has material that a copyright owner thinks is infringing—if the copyright owner sues the OSP, what is the theory of liability? What would the copyright owner need to prove? What if, prior to suing, the copyright owner sends the OSP an email identifying the infringing content that is available on that OSPs site and the OSP does not remove the material—if the copyright owner sues the OSP, what is the theory of liability? Working through these scenarios helps understand the value of the safe harbor.

2. As one of the largest OSPs, YouTube receives millions of takedown notices each year, *i.e.*, thousands *per day*. Many of those notices are meant to be § 512 compliant. Separate from the § 512 notice and takedown regime, YouTube instituted a "content ID" program through which copyright owners can submit reference data for identification of their copyrighted content. As content is uploaded to YouTube, it is checked against this reference data. If a match occurs the video gets a "Content ID claim." Once that happens, "[i]t's up to copyright owners to decide whether or not others can reuse their original material. In many cases, copyright owners allow the use of their content in YouTube videos in exchange for putting ads on those videos." https://support.google.com/youtube/answer/6013276. Copyright owners can request removal of content so identified or, in some cases, can seek to monetize the content through advertising associated with that video and/or track the video's viewership statistics. https://support.google.com/youtube/answer/2797370. Any of these actions can be country-specific; for example, "a video may be monetized in one country and blocked or tracked in another." *Id*. Should sites that host user content have to provide this type of filtering?

D. DIGITAL RIGHTS MANAGEMENT (DRM) AND PARA-COPYRIGHT

In response to the ease of reproduction and distribution of expressive works in the digital environment, some copyright owners have employed technological measures to prevent unauthorized copying and distribution of their works. This technology, often referred to as Digital Rights Management or DRM, provides some protection against copying that does not rely on legal protections or federally granted copyrights.

In 1998 congress passed the Digital Millennium Copyright Act, which granted federal prohibitions against circumvention of the technological protection measures employed by copyright owners. These prohibitions are sometimes referred to as "para-copyright"—alongside of copyright—because they are not, strictly speaking, copyright protection. Instead, these protections make it unlawful both to circumvent certain types of DRM and to distribute or "traffic in" tools that are designed or marketed for circumventing certain kinds of DRM. They are found in 17 U.S.C. § 1201. While a full study of the protections for copyright owners provided in § 1201 cannot be undertaken in a survey course, it is important to be aware that these legal protections exist.

The first significant case that interpreted § 1201 involved an encryption program known as Content Scramble System, or CSS, which the motion picture industry used as a standard form of copy-prevention on DVDs. A decryption algorithm, called DeCSS, was authored and released on the internet. The Motion Picture Association of America promptly sent cease and desist letters and then sued the operators of many web sites seeking to have DeCSS removed from the internet. Following a bench trial, the district court held in favor of the plaintiffs, concluding that making DeCSS available on a website constituted a violation of § 1201 and that none of the exceptions within § 1201 applied. The district court also rejected the defendant's fair use argument:

> The use of technological means of controlling access to a copyrighted work may affect the ability to make fair uses of the work. Focusing specifically on the facts of this case, the application of CSS to encrypt a copyrighted motion picture requires the use of a compliant DVD player to view or listen to the movie. Perhaps more significantly, it prevents exact copying of either the video or the audio portion of all or any part of the film. This latter point means that certain uses that might qualify as "fair" for purposes of copyright infringement—for example, the preparation by a film studies professor of a single CD-ROM or tape containing two scenes from different movies in order to illustrate a point in a lecture on cinematography, as opposed to showing relevant parts of two different DVDs—would be difficult or impossible absent circumvention of the CSS encryption. Defendants therefore argue that the DMCA cannot properly be construed to make it difficult or impossible to make any fair use of plaintiffs' copyrighted works and that the statute therefore does not reach their activities, which are simply a means to enable users of DeCSS to make such fair uses.
>
> Defendants have focused on a significant point. [Technical protection measures] such as CSS do involve some risk of preventing lawful as well as unlawful uses of copyrighted material. Congress, however, clearly faced up to and dealt with this question in enacting the DMCA.
>
> . . . .
>
> The question here is whether the possibility of noninfringing fair use by someone who gains access to a protected copyrighted work through a circumvention technology distributed by the defendants saves the defendants from liability under Section 1201. But nothing in Section 1201 so suggests. By prohibiting the provision of circumvention technology, the DMCA fundamentally altered the landscape. A given device or piece of technology might have "a substantial noninfringing use, and hence be immune from attack under *Sony's* construction of the Copyright Act—but nonetheless still be subject to suppression under Section 1201." [RealNetworks, Inc. v. Streambox, Inc., No. 2:99CV02070, 2000 WL 127311, *8 (W.D.Wash. Jan.18, 2000).] Indeed, Congress explicitly noted that Section 1201 does not incorporate *Sony*.
>
> . . . Defendants' statutory fair use argument therefore is entirely without merit.

Universal City Studios, Inc. v. Reimerdes, 111 F. Supp. 2d 294 (S.D.N.Y. 2000), *aff'd sub nom. Universal City Studios, Inc. v. Corley*, 273 F.3d 429 (2d Cir. 2001). Given our extensive discussion of the fair use doctrine in this Chapter, it is important to appreciate the threat to the "breathing space" of fair use in copyright created by DRM technologies and the new digital para-copyright created by § 1201.

CHAPTER 5: TRADEMARK LAW

I. TRADEMARK PROTECTION

We all encounter multiple trademarks every day. The coffee we drink in the morning, the bike we ride to work, the websites from which we retrieve our email, and the knives we use to prepare our dinner—all are marketed using trademarks that help us know something about those products and services. That something is the item's source, which the mark communicates to us. Given the depth of our consumer culture, we all likely think we're reasonably sophisticated in navigating brands in the marketplace. But few consumers have a full appreciation for the legal doctrines that make a trademark a reliable and valuable source identifier.

"The protection of trade-marks is a law's recognition of the psychological function of symbols." *Mishawaka Rubber Woolen Co. v. S.S. Kresge Co.*, 316 U.S. 203, 205 (1942). Consumers make mental associations between trademarks and the sources of the products and services they purchase. Traditional trademark protection grants the owner of a trademark a legal right to stop others from using a mark that confuses consumers, essentially by drawing upon that psychological association in the minds of consumers. A feedback mechanism is also at play in the strength of the signaling power of trademarks: once the law provides protection, the association that individuals make becomes stronger. In the United States, for example, where trademark protection is quite robust, consumers routinely expect that almost any use of a trademark is controlled by, or at least licensed by, the trademark owner. This expectation, in turn, has led to stronger legal protection for trademarks, as we will explore in this Chapter.

Trademarks are protected by both state and federal law. This Chapter focuses almost entirely on federal protection because most trademark owners prefer to bring claims under the federal trademark law, known as the Lanham Act, 15 U.S.C. §§ 1051 et seq. Congress has provided that the federal district courts may hear trademark infringement claims brought under the Lanham Act, and that plaintiffs may join state law claims with those federal claims. 28 U.S.C. §§ 1338, 1367.

A copy of the Lanham Act is available from CALI's eLangdell Press:
http://www.cali.org/books/united-states-trademark-law
A pdf copy may also be obtained from the U.S. Patent & Trademark Office:
http://www.uspto.gov/sites/default/files/documents/tmlaw.pdf

In many ways, trademark law is quite different from patent and copyright protection. As we have described in Chapters 3 and 4, patent and copyright promote unconventionality through the standards of nonobviousness and originality that they set. Trademark law, on the other hand, promotes conventionality; it encourages people to create conventions—mental associations that words and symbols identify the source of goods or services. Trademark protection is based on the use of a mark in commerce; it is not based on creation, as with copyright, or invention, as with patents. As with copyright and patent protection, a federal statute grants certain rights; but unlike copyright and patent, trademarks are also protected by state law. Trademark infringement actions can be brought in federal court, although, unlike patent and copyright litigation, subject matter jurisdiction over trademark cases is not exclusive in the federal courts. 28 U.S.C. § 1338. Thus, trademark actions can be brought in state court as well. The Constitution grants Congress the authority to adopt the copyright and patent laws specifically in the Progress Clause, Art. I, sec. 8 cl. 8. The Supreme Court has held that the Progress Clause does not give Congress the authority to protect trademarks. *The Trademark Cases*, 100 U.S. 82 (1879). Instead, the authority to grant national trademark protection is based on the Commerce Clause, Art. I, sec. 8 cl. 3. This difference in constitutional authority indicates that different policy choices animate legal recognition for trademark owners.

A. THEORETICAL UNDERPINNINGS

Traditional trademark protection prohibits a competitor from using a word or symbol in association with its products or services in a way that is likely to cause consumers to be confused about the source of those products or services. Such confusing use would cause harm to the consumers who thought they were getting a product from a different source, and would also cause harm to the owner of the trademark by diverting those sales to the competitor. Consider the company Apple, Inc., maker of the *iPad* table computer. If Acme Co. began selling a tablet computer called an *iPad*, some consumers might purchase an *iPad* made by Acme, thinking they were actually buying an *iPad* made by Apple. These consumers may experience disappointment, Apple will have experienced lost sales, and, if the Acme product is inferior, Apple's reputation might also be harmed. At the same time, Acme gains sales it might not otherwise have made. Examining each of these consequences helps us understand the theoretical justifications for protecting trademarks.

First, consider the consumer. Imagine a world in which trademarks don't exist. Each purchase a consumer makes would likely be preceded by research on the part of the consumer to determine the contents and, to the extent possible, the quality of the product and the source of the product. These search costs would be significant, certainly in the aggregate of purchases made by an individual. With trademarks, a consumer can minimize those search costs. For example, once you have decided which brand of toothpaste you prefer, you can reliably find that brand on store shelves without having to read the details of the label or inquire of the store owner where each different tube of toothpaste comes from. If a friend tells you about a toothpaste that has superior qualities (be it taste, whitening power, or tube squeezability), all you need to remember is the name, and you can try it for yourself. Similarly, an advertisement might entice you to try a new product and all you need to remember is the trademark name. Once you've decided that you like that toothpaste, you can quickly and efficiently purchase that same toothpaste in the future. If a competitor could sell toothpaste with the same or even a confusingly similar name, you would need to be extra vigilant in your purchases in order to get what you really wanted. Because trademark law prevents competitors from using confusingly similar marks, that vigilance is not required. Protecting consumers from confusion and reducing consumer search costs is one of the underlying goals of trademark law.

Next, consider the potential harm to the trademark owner. When a competitor uses a mark that confuses a consumer, the trademark owner is likely to face two kinds of harm. First, the trademark owner may suffer a loss in sales. The consumer may purchase the competitor's product thinking that he is purchasing a product made by the trademark owner. Courts will sometimes refer to competitors' inappropriate sales-diverting use of the trademarks of others as "passing off" or "palming off." Second, the trademark owner may suffer a reputational harm. If the competitor's products are not similar in quality, the consumer may come to believe that this new quality is more representative of the trademark owner's products, and this distorted belief may affect the consumer's willingness to purchase the trademark owner's goods in the future. Both of these reputation-related harms affect the trademark owner's willingness to invest in creating and distributing products that are of a consistent, and hopefully high, quality—in other words, investing in building up the "goodwill" associated with the mark. If such investments can be easily undermined and turned to a competitor's advantage by the simple expedient of a confusingly similar mark, there is less incentive to make the initial investment. Trademark law seeks to guard against both of these harms by permitting the trademark owner to stop others from using a mark in a way that is likely to confuse consumers about the source (or origin) of a good. In so doing, trademark law provides an incentive for providers of goods and services to develop and consistently use trademarks to help consumers identify their goods and services. Trademark law also provides an incentive for producers to invest in creating goodwill that is associated with those marks stemming from consistent and high-quality goods or services.

Finally, consider the profit that the competitor earned through palming off. This profit was gained by free riding on the goodwill developed by the trademark owner. While we favor competition in the marketplace, we also insist upon honest and fair dealing, including identifying products and services in a way that does not confuse consumers. Through protection of trademarks, the law ultimately helps to regulate business ethics and to foster fair and honest competition.

B. What Can Be a Trademark?

The Lanham Act provides:

The term trademark includes any word, name, symbol, or device, or any combination thereof—
  (1) used by a person, or
  (2) which a person has a bona fide intention to use in commerce and applies to register on the principal register established by this chapter,
to identify and distinguish his or her goods, including a unique product, from those manufactured or sold by others and to indicate the source of the goods even if that source is unknown.

15 U.S.C. § 1127. The statute defines the term "service mark" similarly, except that service marks are used to identify and distinguish services, not goods. Most people use the term "trademark" to indicate either a trademark or a service mark, and we will do the same throughout this Chapter. A trade name, on the other hand, is a name used to identify a person's business. Trade names are not protected by traditional trademark law, except to the extent that they are used as a trademark.

As the statute makes clear, it is not the *creation* of a trademark that gives rise to protection, but rather it is *use* of that mark as a source identifier in the marketplace that entitles the user to protection. The statute also requires use of the mark, or at least a bona fide intention to use the mark, to make the mark eligible for federal registration. A second requirement contained in the statutory definition of a "trademark" is that the "word, name, symbol, or device" be used to "to identify and distinguish" the mark owner's goods. The next case addresses what kinds of marks meet this requirement.

E.T. Browne Drug Co. v. Cococare Products, Inc.
538 F.3d 185 (3d Cir. 2008)

invalid trademark

Ambro, Judge:

This case involves a dispute between two manufacturers of personal care and beauty products that contain cocoa butter. E.T. Browne Drug Co. claims that it has a protected trademark interest under the Lanham Act in the term "Cocoa Butter Formula," which features prominently on its products. Cococare Products disputes the validity of this asserted trademark.[*] The District Court entered summary judgment in Cococare's favor after concluding that the term is generic and thus may not receive protection from the trademark laws. We agree that Browne has not demonstrated that "Cocoa Butter Formula" is a protectable trademark, but reach that conclusion by a different path. . . .   — H

Cococare disputes E.T. trademark

---

[*] [*Eds. Note*: We provide the images in the case, which we found using Google Image Search. The court's opinion does not contain any images.]

I. Background and Procedural History

Browne, a New Jersey corporation, markets personal care and beauty products containing cocoa butter under the brand name "Palmer's." The "Palmer's" line of cocoa butter products is the sales leader among personal care and beauty products containing cocoa butter. The packaging containing those products displays "Palmer's" and "Cocoa Butter Formula." "Palmer's Cocoa Butter Formula" is on the principal register of the United States Patent & Trademark Office, and thus this term is presumptively valid as a trademark. See 15 U.S.C. § 1057(b). In contrast, "Cocoa Butter Formula" is on the PTO's supplemental register but not on the principal register. The statutory presumption of validity accordingly does not attach to that term. See § 1094.

Cococare, a New Jersey corporation, also sells personal care and beauty products containing cocoa butter, although its sales are far smaller than those of Browne. In 1994, it introduced new products formulated with cocoa butter and Vitamin E, labeling them "Cococare Cocoa Butter Formula." This use of "Cocoa Butter Formula" gave rise to its dispute with Browne.

. . . Browne first objected to Cococare's use of the term "Cocoa Butter Formula" in 2002 after it became aware of a product flyer from a seller of Cococare's products.

Browne then brought suit in the United States District Court for the District of New Jersey after a cease-and-desist letter sent to Cococare failed to cause it to stop using the contested term. Browne alleged, inter alia, that Cococare had violated the Lanham Act and equivalent New Jersey law by its use of the term "Cocoa Butter Formula." Cococare . . . moved for summary judgment on the grounds that "Cocoa Butter Formula" is not a protectable trademark because it is a generic term, [and] that Browne's claims . . . should be dismissed because Cococare's use of "Cocoa Butter Formula" was a fair use of a product descriptor. Browne cross-moved for summary judgment on the genericness and fair use issues . . . .

The District Court concluded that "Cocoa Butter Formula" is a generic term and entered summary judgment in favor of Cococare. . . .

. . . .

. . . Our review is plenary. *Berner Int'l Corp. v. Mars Sales Co.*, 987 F.2d 975, 978 (3d Cir. 1993). On appeal from a grant of summary judgment, our Court exercises the same standard of review as the District Court and considers whether genuine issues of material fact exist that preclude entry of summary judgment. *Id.*

III. Discussion

A. The Protectability of the Mark "Cocoa Butter Formula"

To establish trademark infringement in violation of the Lanham Act, 15 U.S.C. § 1114, a plaintiff must prove that (1) the mark is valid and legally protectable, (2) it owns the mark, and (3) the defendant's use of the mark is likely to create confusion concerning the origin of goods or services. *Freedom Card, Inc. v. JPMorgan*

Chase & Co., 432 F.3d 463, 469-70 (3d Cir. 2005). We thus begin our analysis by asking whether "Cocoa Butter Formula" is a valid, legally protectable trademark.

Terms asserted as trademarks may fall in four categories:

> [1] arbitrary (or fanciful) terms, which bear no logical or suggestive relation to the actual characteristics of the goods; [2] suggestive terms, which suggest rather than describe the characteristics of the goods; [3] descriptive terms, which describe a characteristic or ingredient of the article to which it refers[;] and [4] generic terms, which function as the common descriptive name of a product class.

A.J. Canfield Co. v. Honickman, 808 F.2d 291, 296 (3d Cir. 1986). The Lanham Act protects only some of these categories of terms. Working backward, it provides no protection for generic terms because a first-user of a term "cannot deprive competing manufacturers of the product of the right to call an article by its name." *Id.* at 297; *see also Park 'N Fly, Inc. v. Dollar Park and Fly, Inc.*, 469 U.S. 189, 194 (1985) ("Generic terms are not registrable and a registered mark may be cancelled at any time on the ground[] it has become generic."). In contrast, the Lanham Act protects descriptive terms if they have acquired secondary meaning associating the term with the claimant. *Canfield*, 808 F.2d at 292-93, 296. Finally, trademark law protects suggestive and arbitrary or fanciful terms without any showing of secondary meaning.

Browne has the burden in this case of proving the existence of a protectable mark because "Cocoa Butter Formula" does not appear on the PTO's principal register. *Canfield*, 808 F.2d at 297. It contends that "Cocoa Butter Formula" should receive protection from the trademark laws as a descriptive term that has acquired secondary meaning. Cococare responds that the term should receive no protection because it is generic. The parties thus pose a difficult question of trademark law. *See id.* at 296 ("Courts and commentators have recognized the difficulties of distinguishing between suggestive, descriptive, and generic marks.").

Whether "Cococare Butter Formula" is generic or descriptive, and whether that term has acquired secondary meaning, are questions of fact. *See id.* at 307 n.24 (noting that "Courts of Appeals have generally held that a designation's level of inherent distinctiveness is a question of fact"); *Dranoff-Perlstein Assocs. v. Sklar*, 967 F.2d 852, 862 (3d Cir. 1992) (identifying secondary meaning as a question of fact). We therefore ask whether Browne has presented sufficient evidence to create a genuine issue of material fact as to those questions. See Fed. R. Civ. P. 56(c).

. . . [W]e conclude that Browne has produced evidence sufficient to create a genuine issue of material fact on its claim that the term "Cocoa Butter Formula" is not generic, but descriptive. But as we also conclude that Browne has failed to produce sufficient evidence to create a genuine issue of material fact as to whether the term has acquired secondary meaning, we reach the same result as the District Court—"Cocoa Butter Formula" may not receive the protections of the Lanham Act.

1. *Is "Cocoa Butter Formula" Generic?*

a. *The Primary Significance Test and the Limited Circumstances in Which* Canfield'*s Alternative Test Applies*

This appeal raises the initial question of the proper test under which to evaluate whether the term "Cocoa Butter Formula" is generic and thus not protectable as a trademark. "The jurisprudence of genericness revolves around the primary significance test, which inquires whether the primary significance of a term in the minds of the consuming public is the product or the producer." *Canfield*, 808 F.2d at 292-93. We ask "whether consumers think the term represents the generic name of the product [or service] or a mark indicating merely one source of that product [or service]." *Dranoff-Perlstein Assocs.*, 967 F.2d at 859 (alterations in original). If the term refers to the product (*i.e.*, the genus), the term is generic. If, on the other hand, it refers to one source or producer of that product, the term is not generic (*i.e.*, it is descriptive, suggestive, or arbitrary or fanciful). To give an example, "Cola" is generic because it refers to a product, whereas "Pepsi Cola" is not generic because it refers to the producer. To repeat, Cococare contends that "Cocoa Butter Formula" is generic whereas Browne argues it is descriptive.

The District Court did not apply the primary significance test. It instead applied an alternative test stated in *Canfield*. Neither party disputed that approach. We conclude, however, that the District Court should not have ventured beyond the primary significance test to any alternative gloss.

Canfield addressed situations in which a manufacturer created a new product and it was not clear if it also had created a new product genus. It involved a dispute over the term "Diet Chocolate Fudge Soda." "[A] fundamental question * * * [was] whether chocolate soda or chocolate fudge soda is the relevant product genus for evaluating genericness." *Id.* at 298-99. The primary significance test could not answer that question, we reasoned, since it applied "only after we have determined the relevant genus." *Id.* at 299.

Our Court concluded that the following rule would help us fill in this gap of identifying the appropriate genus for analysis: "If a producer introduces a product that [1] differs from an established product class in a particular characteristic, and [2] uses a common descriptive term of that characteristic as the name of the product, then the product should be considered its own genus." *Id.* at 305-06. In those circumstances, "[w]hether the term that identifies the product is generic then depends on the competitors' need to use it. At the least, if no commonly used alternative effectively communicates the same functional information, the term that denotes the product is generic." *Id.* at 306. *See generally Genesee Brewing Co. v. Stroh Brewing Co.*, 124 F.3d 137, 145 (2d Cir. 1997) (discussing *Canfield* and describing its test as a complement to, rather than a rejection of, the primary significance test when a court cannot readily determine the genus of a new product).

Canfield does not control here for a simple reason: this case does not pose the question addressed in *Canfield*. The "question * * * at the core" of *Canfield* was whether "the relevant product category or genus for purposes of evaluating genericness is chocolate soda or chocolate fudge soda." *Id.* at 293. We do not face a comparable question, as the parties before us do not dispute whether we should use an existing genus or a new genus in our analysis. They instead agree, with only insignificant quibbles over wording, that "Cocoa Butter Skin Care Products" or an equivalent term defines the category.

To understand why this distinction matters, we return to the principles underlying *Canfield*. It addressed a weakness in the primary significance test—the presumption that a court knows the product's genus. In most cases, that genus will be obvious, even for new products. A slight change in a detergent's formula, for example, likely will not create a new product genus. Problems may arise, however, if a product differs from existing products in what *Canfield* calls a "particular characteristic." Examples may include the addition of a new flavor or a new featured ingredient (such as honey in the "Honey Brown Ale" at issue in *Genesee Brewing Co.*). The manufacturer then likely has created a new type of product. That manufacturer may well need to use descriptive terms in the product name to identify the product to consumers. This raises the question of the proper genus for the Court's genericness analysis: the established product class or a new product class that modifies the established product class with the new characteristic.

Canfield addressed this problem by articulating a test that supplies the proper genus for a genericness analysis. Its test applies when a manufacturer uses the following equation: name of new product = name of the established product class ("Diet Chocolate Soda" in *Canfield*) + name of the new characteristic ("Fudge" in *Canfield*). *See id.* at 305-06.

The established product class in our case is "Skin Care Products" or "Lotion." The new characteristic is "Cocoa Butter." Browne could have called its new products "Cocoa Butter Skin Care Products" or "Cocoa Butter Skin Care Lotion." Use of these terms would have satisfied *Canfield*'s equation (name of the new characteristic + name of the established product class) and triggered its test. *Canfield* stated that the primary significance test would not have been useful because the genericness determination would have depended on the unresolved threshold definition of the genus ("Cocoa Butter Skin Care Products" v. "Skin Care Products").

Of course, when it introduced skin-care products containing cocoa butter (*i.e.*, adding a new characteristic), Browne did not label those products with the term "Cocoa Butter Skin Care Products" or "Cocoa Butter Lotion." Instead, it used "Cocoa Butter Formula." This term does not frustrate the application of the primary significance test because it does not raise the question whether to use "Cocoa Butter Formula" or "Formula" as the proper genus for our analysis. Cococare also has not suggested that "Formula" identifies the established product class. Browne does not make baby formula after all, or sell algorithms or recipes. Nor does any record evidence suggest that consumers use "Formula" to describe the skin care product category. Browne's use of a different equation to name its product ("Cocoa Butter Formula" = name of the new characteristic ("Cocoa Butter") + a term not describing the established product class ("Formula")) does not bring into play the weakness in the primary significance test that *Canfield* addressed because it does not raise the question of the proper genus for our analysis. Applying

Canfield here amounts to attempting to remedy a non-existent problem. We therefore will evaluate the genericness of the term "Cocoa Butter Formula" under the primary significance test only.

b. Is the Term "Cocoa Butter Formula" Generic Under the Primary Significance Test?

"[T]he primary significance test * * * inquires whether the primary significance of a term in the minds of the consuming public is the product or the producer." *Id.* at 292-93.[6] We appl[y] that test . . . , asking whether the evidence demonstrated that the term at issue primarily signified the product genus to consumers. [T]he meaning of "Cocoa Butter Formula" should be "evaluated by examining its meaning to the relevant consuming public." *Id.* at 981. That evaluation requires looking at the mark as a whole, not dissecting it into various parts. *Id.*[7] We therefore inquire whether the consuming public understands "Cocoa Butter Formula" to refer to a product genus or to a producer. We ask specifically if the evidence submitted by Browne creates a genuine issue of material fact as to its contention that the consuming public does not understand the term "Cocoa Butter Formula" to refer to a product genus (*i.e.*, that it is not generic, but descriptive in this case).

Plaintiffs seeking to establish the descriptiveness of a mark often use one of two types of survey evidence. J. Thomas McCarthy, *McCarthy on Trademarks & Unfair Competition* (4th ed. 2008), describes a "Teflon survey" as "essentially a mini-course in the generic versus trademark distinction, followed by a test." 2 *McCarthy on Trademarks* § 12:16. That survey runs a participant through a number of terms (such as "washing machine" and "Chevrolet"), asking whether they are common names or brand names. After the participant grasps the distinction, the survey asks the participant to categorize a number of terms, including the term at issue. *Id.* (discussing survey created for *E.I. DuPont de Nemours & Co. v. Yoshida Int'l, Inc.*, 393 F. Supp. 502 (E.D.N.Y. 1975)).

A "Thermos survey," on the other hand, asks the respondent how he or she would ask for the product at issue. If, to use the term under dispute in the case from which the survey gets its name, the respondents largely say the brand name ("Thermos") rather than the initial product category name ("Vacuum Bottle"), the survey provides evidence that the brand name ("Thermos") has become a generic term for the product category. 2 *McCarthy on Trademarks* § 12:15 (discussing survey used in *American Thermos Prods. v. Aladdin Indus., Inc.*, 207 F. Supp. 9 (D. Conn. 1962)). To put this in the terms of the primary significance test, the term would be generic because the consumers would be using it to refer to the product category rather than a producer who makes products within that product category.

---

[6] The public need not know the identity of the producer for the primary significance of the term to be the producer. *Canfield*, 808 F.2d at 300 (citing S. Rep. No. 98-627, at 5 (1984), *U.S. Code Cong. & Admin. News* 1984, pp. 5718, 5722).

[7] Under this analysis, the District Court should not have broken up the term for purposes of its analysis and considered "Cocoa Butter" separately from "Formula."

Browne conducted a survey in this case that generally adheres to the "Thermos survey" model. The survey posed a number of open-ended questions asking respondents to "identify or describe the product category" in which its products fall. It asked each of the 154 valid respondents "[w]hat word or words would you use to identify or describe a skin care product which contains cocoa butter?" and "[i]f you needed to identify or describe a skin care product containing cocoa butter, what word or words would you use instead of or in addition to just saying cocoa butter, if any?" Neither "Cocoa Butter Formula" nor any form of the word "Formula" appeared among the respondents' answers.

The District Court appears to have admitted the survey. Cococare does not suggest on this appeal that we should exclude it. Nor do we see a reason to do so. We therefore ask whether the survey evidence creates a genuine issue of material fact as to whether "Cocoa Butter Formula" is not generic, but descriptive.

The District Court concluded that the survey had "little or no probative value" and "should be afforded little or no weight." We understand the Court to have made that determination within the summary judgment framework rather than to have engaged in any inappropriate weighing of the evidence at the summary judgment stage. *See, e.g.*, *Universal City Studios, Inc. v. Nintendo Co.*, 746 F.2d 112, 118 (2d Cir. 1984) (holding a survey to be "so badly flawed that it cannot be used to demonstrate the existence of a question of fact"). . . .

. . . .

The District Court faulted the survey for what it perceived as two errors. It criticized the survey for not using the term "Palmer's," believing that this omission "undermine[d] Browne's theory of the case" and made the questions "flawed and misleading." The Court also believed that the survey contained leading questions. For example, it considered the question "What word or words would you use to identify or describe a skin care product which contains cocoa butter?" to be highly suggestive in order to evoke a specific response.

We do not agree these questions were so misplaced. The survey was intended to reveal whether customers use the word "Cocoa Butter Formula" to describe cocoa butter skin care products or lotions. The parties' genericness dispute turns on that question. We thus steer away from the [district court's] criticism that "some of the survey questions use two of the three words (cocoa butter) of Browne's source identifier (cocoa butter formula)." We struggle to conjure how Browne could have pursued the core genericness inquiry without doing so.[10] . . . .

We also do not perceive any reason for Browne to have included the word "Palmer's" in its survey. This litigation focuses on the term "Cocoa Butter Formula," not on the registered trademark "Palmer's Cocoa Butter Formula." The

---

[10] Cococare presents expert evidence indicating that "Cocoa Butter" may have discouraged the use of those words in the answers. But this does not deprive the survey of probative value. It merely creates a question as to how much weight a jury should give the survey.

inclusion of the word "Palmer's" in the survey would have confused matters and would have taken the survey outside the "Thermos survey" model.[12]

Browne's survey does have non-trivial flaws, however. Only 30% of valid respondents used a noun identifying the product genus (*e.g.*, lotion, cream). The majority of respondents either answered with an adjective describing the product class (*e.g.*, healing, moisturizing) or did not answer. This suggests that the questions confused many respondents. The survey may have caused this confusion by deviating from the standard "Thermos survey" model by asking respondents for terms describing the products in addition to asking (as a "Thermos survey" should) for terms identifying the products. The survey likely would have been strongest if it had asked respondents, as the "Thermos survey" also did, how they would ask at a store for the type of product at issue.

These flaws nonetheless do not deprive the survey of probative value. The survey raises a reasonable inference that "Cocoa Butter Formula" does not describe the product genus in the opinion of the 46 respondents who described the product class. A reasonable jury could rely on that inference to conclude that consumers do not use the words "Cocoa Butter Formula" to describe the category of skin care products containing cocoa butter. Cococare could attack the inference at trial, but that does not stop the survey from creating a genuine issue of material fact. It is premature for us now to conclude that the survey does not provide probative evidence that "Cocoa Butter Formula" is not generic. Browne could have performed a better survey. Indeed, it might have rued the survey's design flaws after a trial. But the survey is strong enough to allow Browne to survive summary judgment on the genericness issue.

Browne also points to evidence that competitors use terms other than "Cocoa Butter Formula." This evidence tends to prove that the term is not generic. *See Canfield*, 808 F.2d at 306 n.20 ("Courts have long focused on the availability of commonly used alternatives in deciding whether a term is generic.") (citing *Holzapfel's Compositions Co. v. Rahtjen's American Composition Co.*, 183 U.S. 1 (1901)). It indicates that other competitors (and thus Cococare) did not need to use the word "Formula" to communicate to consumers the type of products that it sold. This bolsters our conclusion that a genuine issue of material fact exists on the question of genericness,[13] and thus the District Court should not have entered

---

[12] Cococare does not distinguish in its opening brief between "Teflon surveys" and "Thermos surveys." It implicitly argues, however, that Browne should have conducted the former rather than the latter when it contends that "[t]he proper question would have asked survey respondents in a straightforward fashion whether the designation 'cocoa butter formula' identified a brand, or a common name." Cococare does not explain why Browne had to conduct a "Teflon survey" rather than a "Thermos survey."

[13] In contrast, we struggle to perceive the relevance ascribed by Browne to the fact that dictionaries do not include the term "Cocoa Butter Formula." Not all generic terms appear in dictionaries, a fact so obvious Browne makes no effort to disprove it.

summary judgment that "Cocoa Butter Formula" is generic.[14] Instead, it should have proceeded to a secondary meaning analysis.

2. Assuming "Cocoa Butter Formula" is Descriptive, Has it Acquired Secondary Meaning?

Because a genuine issue of material fact exists on the question whether "Cocoa Butter Formula" is generic, the parties normally would need to proceed to trial to resolve that issue. Even assuming that Browne prevailed and proved "Cocoa Butter Formula" to be descriptive, it also would need to show that the term had acquired secondary meaning which associated it with Browne.

Cococare moved for summary judgment on the basis that Browne had not produced evidence creating a genuine issue of material fact on the secondary meaning question. The District Court set out in its opinion how a secondary meaning analysis would proceed and laid out relevant factors from our case law. But it did not go further. . . . The parties presented the secondary meaning question to the District Court, however, and we have their arguments before us. We thus resolve that question because of our interest in judicial economy, and affirm on this basis the District Court's entry of summary judgment. *See Nicini v. Morra*, 212 F.3d 798, 805 (3d Cir. 2000) (stating that we may affirm on any ground supported by the record).

Secondary meaning is a new and additional meaning that attaches to a word or symbol that is not inherently distinctive. *See generally* 2 *McCarthy on Trademarks* § 15:1. We have explained:

> Secondary meaning exists when the trademark is interpreted by the consuming public to be not only an identification of the product, but also a representation of the product's origin. Secondary meaning is generally established through extensive advertising which creates in the mind of consumers an association between different products bearing the same mark. This association suggests that the products originate from a single source. Once a trademark which could not otherwise have exclusive appropriation achieves secondary meaning, competitors can be prevented from using a similar mark. The purpose of this rule is to minimize confusion of the public as to the origin of the product and to avoid diversion of customers misled by a similar mark.

---

[14] The District Court appeared to reason that the registration of "Cocoa Butter Formula" on the PTO's supplemental register rather than its principal register weakens Browne's claim that the term is descriptive. We know of no support for that view. As the District Court correctly explained elsewhere, the validity of a term that does not appear on the principal register and is not distinctive (*i.e.*, one that is descriptive rather than suggestive, or arbitrary or fanciful) depends on its acquisition of secondary meaning. *See Berner*, 987 F.2d at 979 (explaining that arbitrary or suggestive terms are distinctive and automatically qualify for trademark protection, while descriptive terms only receive trademark protection after a showing of secondary meaning).

We also note that Browne is not entitled to summary judgment in its favor on the question of genericness. The weaknesses in the genericness survey alone create a genuine issue of material fact on that point.

Scott Paper Co. v. Scott's Liquid Gold, Inc., 589 F.2d 1225, 1228 (3d Cir. 1978).

We have identified an eleven-item, non-exhaustive list of factors relevant to the factual determination whether a term has acquired secondary meaning:

> (1) the extent of sales and advertising leading to buyer association; (2) length of use; (3) exclusivity of use; (4) the fact of copying; (5) customer surveys; (6) customer testimony; (7) the use of the mark in trade journals; (8) the size of the company; (9) the number of sales; (10) the number of customers; and, (11) actual confusion.

Commerce Nat'l Ins. Services, Inc. v. Commerce Ins. Agency, Inc., 214 F.3d 432, 438 (3d Cir. 2000). "[T]he evidentiary bar must be placed somewhat higher" when the challenged term is particularly descriptive. *Id.* at 441.

Browne's proffered showing of secondary meaning includes the following evidence:

> • its use and promotion of the term "Cocoa Butter Formula" continuously for 20 years;
>
> • the substantial amounts of money it has spent promoting the term "Cocoa Butter Formula;"
>
> • the nature and quality of the advertising in support of the term "Cocoa Butter Formula;"
>
> • Cococare's alleged intent to copy the term "Cocoa Butter Formula;" and
>
> • the increase in the sales of products bearing the term "Cocoa Butter Formula."

This evidence may seem, at first blush, to support Browne's claim that the term "Cocoa Butter Formula" has gained secondary meaning. But serious flaws cause it to fail to create a genuine issue of material fact on the question of secondary meaning.

The evidence's core deficiency is that while it shows Browne used the term "Cocoa Butter Formula" on many occasions over a long period of time, it does not show Browne succeeded in creating secondary meaning in the minds of consumers. Although the evidence leaves no doubt that Browne hoped the term would acquire secondary meaning, nothing shows that it achieved this goal. Jurors would have to make a leap of faith to conclude that the term gained secondary meaning because the record fails to provide meaningful support. A jury could evaluate the quality of the advertising or consider the rise in product sales, but it would have to guess what lasting impression the advertising left in the mind of consumers or what portion of Browne's revenue growth it caused.

We indicated in *Commerce National Insurance Services* that a plaintiff might establish secondary meaning through evidence of advertising and sales growth. *See id.* at 438. A plaintiff could create a reasonable inference, for example, that a term had gained secondary meaning by showing that it had appeared for a long period of time in a prevalent advertising campaign. Evidence of revenue growth simultaneous with such marketing would strengthen that inference, particularly if supported by evidence of other factors among those we listed in *Commerce National*.

This case, however, differs in an important way from such an example. Browne has introduced no evidence indicating that it ever used "Cocoa Butter Formula" as a standalone term in marketing or packaging. Instead, it always used the term connected with the "Palmer's," forming the phrase "Palmer's Cocoa Butter Formula." For example, Browne's lotion bottles bore logos with "Palmer's" immediately above the words "Cocoa Butter Formula," creating one visual presentation for the consumer. The marketing and sales evidence thus likely would raise a reasonable inference that "Palmer's Cocoa Butter Formula" has gained secondary meaning in the minds of the public.[16]

But Browne wants to do something more complicated: it wants to establish that a portion ("Cocoa Butter Formula") of the larger term ("Palmer's Cocoa Butter Formula") has acquired an independent secondary meaning. Nothing in the record would allow a jury to evaluate the strength of the term "Cocoa Butter Formula" independently from the larger term including "Palmer's." We thus conclude, under the specific circumstances presented by this case, that the marketing and sales evidence provided by Browne does not create a reasonable inference that "Cocoa Butter Formula" has acquired secondary meaning.

Nor does Browne's asserted evidence of Cococare's intent to copy the term "Cocoa Butter Formula" create a genuine issue of material fact as to secondary meaning. This evidence pertains almost exclusively to trade dress (*i.e.*, the overall appearance of labels, wrappers, and containers used in packaging a product), an issue not presented by this case. Browne only identifies one piece of evidence that conceivably suggests an intent to copy. Cococare's founder testified that he may have known about Browne's use of "Cocoa Butter Formula" when Cococare began using that term on its own products. But he also testified that he decided to use the word "Formula" because it is a standard descriptor in the cosmetics industry. He never testified that he copied Browne and nothing in the record suggests that he did. Even viewing this evidence in the light most favorable to Browne, we cannot discern how a reasonable jury could conclude that Cococare intended to copy Browne's use of the term "Cocoa Butter Formula."

Browne could have overcome these deficiencies in its evidence by conducting a secondary meaning survey in the same way it conducted its genericness survey. It could have used survey evidence to show that "Cocoa Butter Formula" had acquired secondary meaning in the minds of consumers, thus creating a genuine issue of material fact on that issue. We never have held, and do not hold today, that a party seeking to establish secondary meaning must submit a survey on that point. However, Browne's failure to conduct a secondary meaning survey leaves it without evidence of any sort in this case of the secondary meaning of the term "Cocoa Butter Formula."

---

[16] This case does not put that question at issue. The parties do not dispute that Browne has a valid, registered trademark in the term "Palmer's Cocoa Butter Formula." . . . .

We thus conclude that Browne has failed to identify evidence creating a genuine issue of material fact on the question whether "Cocoa Butter Formula" has acquired secondary meaning. Cococare is entitled to entry of summary judgment on the basis that Browne lacks a protectable trademark interest in the term "Cocoa Butter Formula."

. . . .

Notes & Question

1. The *Cococare* court holds that there is a genuine fact dispute about whether "Cocoa Butter Formula" is a *generic* term, and thus not protectable as a mark at all, or is instead a potentially protectable *descriptive* mark as applied to the skin care products at issue. Having raised the prospect that the mark is descriptive, the court requires Browne Drug to show that the mark has *secondary meaning* as a source indicator as a threshold requirement before excluding others from making confusingly similar use of the phrase. The phrase "secondary meaning" is a bit of a misnomer, if one thinks of it strictly as a reference to importance. It makes more sense to think of the source-identifying meaning as second *in time*. To prove secondary meaning, one "must show that the *primary* significance of the term in the minds of the consuming public is not the product but the producer."[*] *Kellogg Co. v. National Biscuit Co.*, 305 U.S. 111, 118 (1938) (emphasis added). To quote Prof. Glynn Lunney, this producer-fostered source-indicating meaning is "second in time, but first in mind." And as *Cococare* states, one can use a wide range of both direct and circumstantial evidence to make the required showing that the mark has *acquired distinctiveness*. You will see that many courts use the less confusing term "acquired distinctiveness," instead of "secondary meaning."

2. While marks classified as descriptive must be shown to have acquired distinctiveness, marks that are classified as suggestive, arbitrary or fanciful are said to be "inherently distinctive." Being inherently distinctive, these marks are eligible for protection upon a demonstration of use by the entity asserting rights in the mark. Why are marks in these categories eligible for protection without any need to show further evidence that the mark does signal to consumers something about the producer and not the product? Remember, it is a word or symbol's capacity to serve that "psychological function" that courts are looking for when making the initial determination whether a mark is protectable. As the statute defines a trademark, it must be able "to identify and distinguish his or her goods, including a unique product, from those manufactured or sold by others and to indicate the source of the goods, even if that source is unknown."

3. The *Cococare* court lays out the traditional classification of terms asserted as trademarks. In ascending order of strength, they are: (i) generic, (ii) descriptive, (iii) suggestive, (iv) arbitrary, and (v) fanciful. This classification is often called the "hierarchy of marks." The concept of tiers of strength is particularly apt when dealing with word marks, though courts use it with other marks as well. In *Cococare*

---

[*] Consumers don't actually have to know precisely *who* the producer is by name; the statutory definition of "trademark" indicates that the source of the goods can be unknown. 15 U.S.C. § 1127.

the court combined the fourth and fifth categories, but there is a distinction: *arbitrary* marks use words or symbols that already have a familiar meaning but are used to mark unrelated goods or services (*e.g.*, "Ivory" for soap), whereas *fanciful* marks are newly invented words or symbols (*e.g.*, "Xerox" for copiers).

No one argues, in *Cococare*, that "Cocoa Butter Formula" is a *suggestive* mark, rather than a descriptive one. The common test courts use to tell one from the other helps show why Browne Drug did not attempt the argument (even though a finding of suggestiveness would have made its mark protectable without any additional evidence of acquired distinctiveness). Specifically, to determine if a mark is suggestive, courts employ an "imagination test," asking if a consumer needs to use her imagination to link the mark to the good: "suggestive terms require the buyer to use thought, imagination, or perception to connect the mark with the goods, whereas descriptive terms directly convey to the buyer the ingredients, qualities, or characteristics of the product." *Hornady Mfg. v. DoubleTap, Inc.*, 746 F.3d 995, 1007 (10th Cir. 2014); *see also POM Wonderful LLC v. Hubbard*, 775 F.3d 1118, 1126 (9th Cir. 2014) ("Unlike descriptive marks, which define qualities or characteristics of a product in a straightforward way, suggestive marks convey impressions of goods that require the consumer to use imagination or any type of multistage reasoning to understand the mark's significance."); *Stoncor Group, Inc. v. Specialty Coatings, Inc.*, 759 F.3d 1327, 1332-33 (Fed. Cir. 2014) ("A suggestive mark requires imagination, thought and perception to reach a conclusion as to the nature of the goods, while a merely descriptive mark forthwith conveys an immediate idea of the ingredients, qualities or characteristics of the goods."); *Welding Servs., Inc. v. Forman*, 509 F.3d 1351, 1357-58 (11th Cir. 2007) ("A suggestive mark refers to some characteristic of the goods, but requires a leap of the imagination to get from the mark to the product.").

4. Try your hand at classifying the following marks, keeping in mind that the classifications are, as one court once put it, more like guidelines than pigeonholes.

> a. "Apple" for fruit commonly known as apples
> b. "Apple" for computers
> c. "Kleenex" for facial tissues
> d. "Jolt" for energy drinks
> e. "Semaphore Press" for digital publications
> f. "Off" for insect repellent
> g. "Slinky" for women's clothing

5. A generic term cannot be a mark at all—it is not protected by trademark law. The word "Peach" used in connection with the sale of peaches is a clear-cut example—it's the name of the good, not the name of a particular source of the good. Competitors need to be able to use the word "peach" if they are selling peaches. Indeed, to allow one source to control sole use of that word for that good would interfere with legitimate competition. Trademark law is thus sensitive to language's different functions. "Peach" used to sell peaches can function to tell us what the item is, but cannot function to distinguish the peaches of one orchard from those of another.

6. Escalator, aspirin, thermos, cellophane. All are words that were, when first created, strong trademarks. And all came, over time, to signify the type of good, rather than the good's particular source. When that happens to a once-famous mark, some refer to it as the mark committing "genericide." Quite a bit is at stake for marks that have this problem. Xerox Corporation faces this challenge with the word "Xerox" being used synonymously with "photocopying" or with "photocopies." One of the steps it has taken is to run advertisements like the one below.

The text of this advertisement, which appeared in the February 2008 issue of the *ABA Journal*, reads as follows:

> If you use 'Xerox' the way you use 'zipper,' our trademark could be left wide open. There's a new way to look at it.
>
> No one likes to leave their name open to misuse. Which is what happens when you use our name in a generic manner. Basically you're putting it in a compromising position which could cause it to lose its trademark status. That's what happened to the name 'zipper' years ago. So when you use our name, please use it as an adjective to identify our products and services, such as 'Xerox copiers.' Never as a verb: 'to Xerox' in place of 'to copy,' or as a noun: 'Xeroxes' in place of 'copies.' Now that you're aware of all this, that should just about zip things up. Thanks."

The potential for genericide is acute for unique products. Trademark owners who are introducing innovative products can combat this by providing a generic name for the product as well as the trademark—that way the public can use the generic name when identifying competing products, and when a competitor introduces a competing product there is the generic name available for its use. Examples of this strategy include *Palm* personal organizers, *Scotch* transparent tape, *iPod* portable digital music players, *TiVo* digital video recorders, and *rollerblade* in-line skates. If the Otis Elevator Company, inventor of the escalator, had promoted the product as a "moving stairway," *escalator* might still be a trademark. *See Haughton Elevator Co. v. Seeberger*, 85 U.S.P.Q. 80 (Com'r Pat. & Trademarks 1950). The problem of genericide can be heightened for innovative products subject to patent protection. For the period of patent protection, the consuming public encounters the unique good with only one mark, and thus may come to think of the mark as the name for that type of good.

C. Introduction to "Likelihood of Confusion"

Demonstrating the protectable nature of a trademark does not mean that the owner can automatically stop others from using that mark. In Sections III and V we explore the requirements for proving infringement and the potential defenses that can be raised. A basic requirement of proving infringement is demonstrating that what the defendant is doing—the mark that they are using and the way that they are using it—is likely to confuse the consumer about the source of a good or service.

Top Tobacco, L.P v. North Atlantic Operating Company, Inc

no infringement
no confusion
b/c
509 F.3d 380 (7th Cir. 2007)

EASTERBROOK, JUDGE:

This case illustrates the power of pictures. One glance is enough to decide the appeal.

Top Tobacco, L.P., sells tobacco to people who want to roll cigarettes by hand or make them using a cranked machine. This is known as the roll-your-own, make-your own or RYO/MYO business. Top Tobacco and its predecessors have been in this segment of the cigarette market for more than 100 years, and the mark TOP®, printed above a drawing of a spinning top, is well known among merchants and customers of cigarette tobacco. North Atlantic Operating Company and its prede-

cessors also have been in the roll-your-own, make-your-own business for more than 100 years, though initially only as manufacturers of cigarette paper. Not until 1999 did North Atlantic bring its own tobacco to market. The redesigned can that it introduced in 2001 bears the phrase Fresh-Top™ Canister. Top Tobacco maintains in this suit under the Lanham Act that none of its rivals may use the word "top" as a trademark.

Trademarks are designed to inform potential buyers who makes the goods on sale. Knowledge of origin may convey information about a product's attributes and quality, and consistent attribution of origin is vital when vendors' reputations matter. Without a way to know who makes what, reputations cannot be created and evaluated, and the process of competition will be less effective. See generally William M. Landes & Richard A. Posner, *The Economic Structure of Intellectual Property Law* 166–209 (2003).

Top Tobacco insists that it has exclusive rights to the word "top" for use on tobacco in this market. But many words have multiple meanings: "Top" may mean the best, or a spinning toy, or a can's lid. Top Tobacco uses the word "top" in the second sense and may hope that consumers will hear the first as well; North Atlantic uses the word in its third sense, to refer to a pull-tab design that keeps tobacco fresh. If English used different words to encode these different meanings, there could not be a trademark problem. Because our language gives the word "top" so many different meanings, however, there is a potential for confusion. But no one who saw these cans side by side could be confused about who makes which[.]

The phrase "Fresh-Top Canister" on North American's can does not stand out; no consumer could miss the difference between Top Tobacco's TOP brand, with a spinning top, and North Atlantic's ZIG-ZAG® brand, with a picture of a Zouave soldier. The trade dress (including colors and typography) of each producer's can is distinctive. . . .

The left panel shows the can as it was between 2001 and 2004, when Fresh-Top Canister was on the front (right under "Classic American Blend"), and the two right panels show the can as it was from 2004 through 2006, when the phrase Fresh-Top Canister was on the side. The phrase was removed in 2006 when North Atlantic replaced the aluminum pull-tab design with a plastic lid. (This change does not make the case moot, because the possibility of damages remains.)

The district court granted summary judgment for the defendants, 2007 U.S. Dist. LEXIS 2838 (N.D. Ill. Jan. 4, 2007), and the pictures show why. It is next to impossible to believe that any consumer, however careless, would confuse these products. "Next to impossible" doesn't mean "absolutely impossible"; judges are not perceptual psychologists or marketing experts and may misunderstand how trade dress affects purchasing decisions. But the pictures are all we have. Top Tobacco did not conduct a survey of consumers' reactions to the cans and did not produce an affidavit from even a single consumer or merchant demonstrating confusion.

What Top Tobacco wants us to do is to ignore the pictures and the lack of any reason to believe that any one ever has been befuddled. Like other courts, this circuit has articulated a multi-factor approach to assessing the probability of confusion. These factors include whether the trademarks use the same word, whether they sound alike, and so on. Top Tobacco insists that "Fresh Top" is spelled and sounds the same as fresh "TOP", and thus it traipses through the list. It conveniently omits the fact that the phrase on the ZIGZAG can is "Fresh-Top Canister", with "Fresh-Top" serving as a phrasal adjective modifying the word "canister" rather than as the product's brand. But it's unnecessary to belabor the point. A list of factors designed as *proxies* for the likelihood of confusion can't supersede the statutory inquiry. If we know for sure that consumers are not confused about a product's origin, there is no need to consult even a single proxy.

Top Tobacco says that merchants may have been confused, because a few of the price lists that North Atlantic sent to its wholesalers and retailers omitted the ZIGZAG brand and gave prices for a "6 oz. Fresh-Top™ Can" and a ".75 oz. Pocket Pouch™". Yet all of these lists prominently include the seller's name (North Atlantic or National Tobacco), and if any commercial buyer thought that North Atlantic was selling the TOP brand the record does not contain a shred of evidence to that effect.

      . . . .

Notes & Questions

1. Likelihood of confusion is a key concept in trademark law. Section III will explore the factors that courts use to evaluate whether there is a likelihood of confusion. For now focus on the core concept. Consider the first justification for trademark protection: protecting consumers from deception and reducing consumer search costs. Prohibiting any confusion about a product or service that is significant enough to be characterized as deception would be consistent with this underlying goal. Prohibiting confusion that relates to the consumer's assumptions concerning source of a product or a service would also be consistent with the goal of reducing search costs. In addition to protecting consumers, trademark law is also designed to protect a business' good will from usurpation by another, thereby encouraging investment in creating that good will. With this justification in mind, any type of consumer confusion that seeks to profit from the good will of another should be relevant when determining infringement.

2. Trademark litigants use surveys for a variety of purposes, as the *Cococare* case indicates. Often survey evidence is used in the context of demonstrating whether a "likelihood of confusion" does or does not exist. In a recent opinion by Judge Posner, the Seventh Circuit explained the skepticism courts have towards this type of evidence:

> Consumer surveys conducted by party-hired expert witnesses are prone to bias. There is such a wide choice of survey designs, none foolproof, involving such issues as sample selection and size, presentation of the allegedly confusing products to the consumers involved in the survey, and phrasing of questions in a way that is intended to elicit the surveyor's desired response—confusion or lack thereof—from the survey respondents. See Robert H. Thornburg, *Trademark Surveys: Development of Computer-Based Survey Methods*, 4 John Marshall Rev. Intellectual Property L. 91, 97 (2005); Michael Rappeport, *Litigation Surveys—Social 'Science' as Evidence*, 92 Trademark Rep. 957, 960–61 (2002); Jacob Jacoby, *Experimental Design and the Selection of Controls in Trademark and Deceptive Advertising Surveys*, 92 Trademark Rep. 890, 890 (2002). Among the problems identified by the academic literature are the following: when a consumer is a survey respondent, this changes the normal environment in which he or she encounters, compares, and reacts to trademarks; a survey that produces results contrary to the interest of the party that sponsored the survey may be suppressed and thus never become a part of the trial record; and the expert

witnesses who conduct surveys in aid of litigation are likely to be biased in favor of the party that hired and is paying them, usually generously.

Kraft Foods Group Brands LLC v. Cracker Barrel Old Country Store, Inc., 735 F.3d 735, 741-42 (7th Cir. 2013). How significant was the absence of a survey in *Top Tobacco*? How did the parties use survey evidence in *Cococare*?

D. FEDERAL REGISTRATION

The Lanham Act permits trademark owners to register their marks. Registration is not required for protection; one may sue for the infringement of an unregistered mark under § 43 of the Act. 15 U.S.C. § 1125.* State law, both statutory and common law, can also provide protection for unregistered marks. Additionally, many states do permit the registration of trade marks, trade names, and dba's (doing business as). State registration has limited benefits, particularly for interstate actors. Federal registration requires use of the mark in interstate commerce. An application can be filed based upon an intent to use the mark, but the registration will not issue until actual use of the mark is made by the applicant and demonstrated to the Trademark Office.

Federal registration on the Principal Register brings with it many benefits, making registration an important step for mark owners interested in legal protection for their marks. First, registration provides prima-facie evidence of the validity of the mark. 15 U.S.C. § 1115. When a trademark owner sues based upon a registered mark the burden is shifted to the defendant to prove a problem with the mark that would prevent protection. Second, registration provides prima-facie evidence of the registrant's ownership of the mark, again shifting the burden to the challenger to prove otherwise.

A third important function of registration is to provide nation-wide notice to others of the registrant's rights in the mark and provide constructive "use" nationwide. This notice and constructive use can be extremely important in determining priority of use. Recall that trademark rights arise from "use" of the mark in commerce. The first entity to use a mark in commerce is called the "senior user." Someone who comes later in time is called the "junior user." A senior user who is also the first to register in the federal system can stop a junior user from using a confusingly similar mark in a way that is likely to cause confusion. This is true even if the junior user was unaware of the senior user's use of the mark. The junior user could have (and indeed, should have) checked the federal registrations to determine if the mark was registered by someone else. If, on the other hand, the junior user is the first to register the mark, a senior user can be geographically confined in the subsequent use of the mark—even though its use is "senior." Because rights flow from use of a mark, the senior user will be able to continue using the mark in the geographic area in which it was operating prior to the junior user's

---

* Note that the section numbers in the Lanham Act correspond to different section numbers in Title 15 of the U.S. Code. This results from the Lanham Act beginning at § 1051 in Title 15. Among experienced practitioners, some sections, such as § 43, are more often referred to by their Lanham Act section number than by their Title 15 number.

registration of the mark. The registered junior user obtains superior rights in all other areas as a result of federal registration.

A fourth benefit of protection is that U.S. Customs & Border Protection (CBP), a bureau of the Department of Homeland Security, maintains a trademark recordation system for marks registered at the United States Patent and Trademark Office. CBP officers can access the recordation database at each of the 317 ports of entry to the United States, 19 CFR §§ 133.1 et seq., and can stop the importation of infringing products.

A fifth benefit of registration is the ability to defend against a state law claim of trademark dilution, a cause of action we explore in Section IV. Specifically, the federal dilution statute provides that ownership of a registered mark "shall be a complete bar to an action" based on either common law or state statutory law seeking to prevent dilution. 15 U.S.C. § 1125(c)(6).

The final benefit of having a mark registered on the Principal Register is the ability to make the mark "incontestable." We explore the benefits of incontestability below.

The federal registration system consists of two registers: the Principal Register and the Supplemental Register. The Supplemental Register is a lesser register; it does not confer the benefits identified above. *See* 15 U.S.C. §§ 1094, 1096. The Supplemental Register is used by applicants who are unable to obtain registration on the Principal Register at the time of filing, in the hopes of ultimately gaining registration on that register. Recall that, in *Cococare*, Browne Drug had put "Cocoa Butter Formula" on the Supplemental Register. This process and the types of marks that end up having to use this Supplemental Register are explained below.

1. Registrable and Unregistrable Marks

Section 2 of the Lanham Act specifies the marks that are "registrable on the principal register" by listing the types of marks that are not eligible for registration. 15 U.S.C. § 1052. Specifically, the section provides that "[n]o trademark by which the goods of the applicant may be distinguished from the goods of others shall be refused registration . . . on account of its nature unless . . . ," *id.*, and then lists exclusions. *Some* of the exclusions include:

> (a) a mark that "consists of or comprises immoral, deceptive, or scandalous matter . . . ;"
> (b) a mark that "consists of or comprises a flag or coat of arms or other insignia of the United States, or of any State or municipality, or of any foreign nation;"
> (c) a mark that "consists of or comprises a name, portrait, or signature identifying a particular individual except by his written consent;"
> (d) a mark that "consists of or comprises a mark which so resembles a mark registered in the Patent and Trademark Office or a mark or trade name previously used in the United States by another and not abandoned, as to be likely, when used on or in connection with the goods of the applicant, to cause confusion, or to cause mistake, or to deceive;"
> (e₁) marks that are "primarily geographically deceptively misdescriptive,"

or that as a whole are "functional;" or

(e₂) marks that are "descriptive," "deceptively misdescriptive," "primarily geographically descriptive," or "primarily merely a surname."

See 15 U.S.C. § 1052. Each one of these exclusions from registration, and others contained in § 2, have their own tests and legal issues. This Section merely provides an overview of some of the issues.

Determining whether a mark is disqualified from registration under (d), because the use of the mark on the goods of the applicant would be likely to cause confusion, employs the same test that is used to determine whether a particular mark infringes: likelihood of confusion. The one difference is that in the context of registration, the trademark office considers the use of the mark only on the goods as described in the application. The applicant has control over the drafting of the description of the goods or services identified in the application. Thus, care in drafting that description is warranted. Only the goods or services on which the applicant uses or intends to use the mark should be listed. Over-claiming can lead not only to problems under § 1052(d) but also can be the basis of a fraud claim which can result in the cancellation of the entire registration. *Standard Knitting, Ltd. v. Toyota Jidosha Kabushiki Kaisha*, 77 U.S.P.Q.2d 1917 (T.T.A.B. 2006).

While the marks identified in (e₁) above can never be registered, the exclusions listed in (e₂) can be registered if the applicant can show that the mark "has become distinctive of the applicant's goods in commerce." 15 U.S.C. § 1052(f). Thus, the lessons you learned in *Cococare* about how to demonstrate acquired distinctiveness can come into play when seeking to obtain a registration for a mark that the Trademark Office classifies as "descriptive." *See, e.g., In re STEELBUILDING.COM*, 415 F.3d 1293, 1300-01 (Fed. Cir. 2005) (analyzing acquired distinctiveness on appeal from a refusal to register a mark). The statute also provides that proof of substantially exclusive and continuous use of the mark "for the five years before the date on which the claim of distinctiveness is made" constitutes prima-facie evidence of acquired distinctiveness. During those five years of use, a mark owner may register the mark on the Supplemental Register. Indeed if the registration is rejected as descriptive with insufficient evidence of secondary meaning, the mark owner can register the mark on the Supplemental Register and, five years later, seek registration on the Principal Register.

Notes & Question

1. In Chapter 3, Section IV, we noted U.S. patent law's lack of a moral utility doctrine. Note that, as enacted, the Lanham Act bars registration for marks that consist of or comprise "immoral . . . or scandalous matter." 15 U.S.C. § 1052(a). Is there a valid governmental interest in declining to lend official approval to marks in commerce that the consuming public of the day finds shocking or offensive?

Section 1052(a), as written, also contained a bar against the registration of a mark that "may disparage . . . persons, living or dead, institutions, beliefs, or national symbols, or bring them into contempt, or disrepute." The Supreme Court, affirming the Federal Circuit's determination, struck down this bar as a violation of the First Amendment. *See Matal v. Tam*, 137 S. Ct. 1744 (2017). The case involved a rock band with Asian-American members that had sought to register its name, The Slants, as a mark. Simon Tam, the band's lead singer, "chose this moniker in order to 'reclaim' and 'take ownership' of stereotypes about people of Asian ethnicity." *Id.* at 1754. Though all eight justices agreed that the disparagement bar could not survive First Amendment scrutiny, they differed, 4-4, on the precise rationale for this conclusion. A common theme in the two plurality opinions, however, is that the offensiveness of private* speech is not a good ground for government to formally disadvantage that speech. Justice Alito, in the Court's principal opinion, rejected the contention that "[t]he Government has an interest in preventing speech expressing ideas that offend," concluding that "that idea strikes at the heart of the First Amendment." *Id.* at 1764 (Part IV, plurality portion). Similarly, Justice Kennedy relied heavily on the predicate that, aside from settled exceptions such as fraud, defamation, and incitement to violence, "it is a fundamental principle of the First Amendment that the government may not punish or suppress speech based on disapproval of the ideas or perspectives the speech conveys." *Id.* at 1765 (Kennedy, J., plurality opinion).

The ultimate constitutional fate of the § 1052(a) scandalous-matter bar, which *Matal v. Tam* did *not* formally adjudicate, depends on how broadly we are to apply the principle the Court stated at the start of the decision: "this provision . . . offends a bedrock First Amendment principle: Speech may not be banned on the ground that it expresses ideas that offend." *Id.* at 1751. The Federal Circuit, for its part, recently struck down the scandalous-matter bar on First Amendment grounds, applying *Tam*. *In re Brunetti*, 877 F.3d 1330, 1348-57 (Fed. Cir. 2017).

2. Marks can consist of just words, or of words and images, including colors. The different parts of a mark contribute to its signaling function in the minds of consumers. For example, the distinct script lettering used by the Coca-Cola Company has a level of association in the minds of consumers such that when they see words other than "Coca-Cola" in that distinct script they still make the association with the Coca-Cola Company. *Coca-Cola Co. v. Gemini Rising, Inc.*, 346 F. Supp. 1183 (E.D.N.Y. 1972) (enjoining defendant's use of the words "enjoy cocaine" in the distinctive script on posters). What one registers for protection determines the scope of the *registered* mark, but remember that it is use that gives rise to protection. Consider the trademark of the company that publishes this book, Semaphore Press, Inc. The company has filed for protection for the word mark "Semaphore Press," but it has not yet filed for protection for the stylized way those words are used on its website. If another entity were to sell law related books using the phrase "Semi-4 Press," the publisher would likely sue for registered mark infringement and would likely prevail. If another publishing company were to use different words, but stylized them in the same colors and font, and

---

* In a part of the decision that was unanimous (*Tam*, 137 S. Ct. at 1757-60), the Court—rejecting the PTO's arguments to the contrary—concluded that "[t]rademarks are private, not government, speech." *Id.* at 1760.

with a flag-like image extending from one letter, Semaphore Press would consider a suit for unregistered mark infringement.

3. A word of caution concerning trademarks that include geographic names. The use of geographic names as a trademark or as part of a trademark is fraught with complicated trademark issues, particularly in the area of wine and spirits. Because of the complexity of these issues, we leave them for a full-semester course in trademark law.

2. The Registration Process

The federal registration of a trademark begins with the filing of an application with the Patent and Trademark Office. The application must include a description and a drawing of the mark, as well as a description of the goods or services with which the applicant uses or intends to use the mark. Goods or services are organized into classifications under an international classification system. The filing fee as of June 2018 is $400 per mark, per classification of goods or services (reduced to $275 for electronically submitted applications).

Once the Trademark Office receives the application it is examined for compliance with the statutory requirements. The examination process is not as detailed as the patent examination process we discussed in Chapter 3, but it is a thorough examination that can involve correspondence back and forth between the applicant (or the applicant's attorney) and the examiner. If the application is ultimately rejected the applicant can appeal that rejection to the Trademark Trial and Appeal Board (TTAB) and then to the Federal Circuit, or can sue in federal district court. 15 U.S.C. § 1071.

If the application is accepted for registration on the Principal Register, the mark will be "published for opposition," meaning that the mark will be listed in the Official Gazette (published on-line <http://www.uspto.gov/web/trademarks/tmog/>). Anyone who might be harmed by the issuance of the registration can file an opposition to the registration with the Trademark Office within 30 days of publication. If such opposition is filed, the dispute between the registrant and the party opposing registration is heard through the Trademark Office, with appeal rights to the federal courts. The most common type of opposition filed is by a registered mark owner who believes that the use of the mark proposed for registration, when used on the goods or services described in the publication notice, is likely to cause confusion among consumers, 15 U.S.C. § 1052(d)—the same test used to determine infringement that we explore in Section III.

Because of the examination process and the opportunity for other mark owners to oppose a registration, trademark lawyers routinely conduct searches of pre-existing trademarks prior to filing a trademark application. It is better to evaluate the prospects of obtaining a registration prior to incurring the costs associated with filing and examination, not to mention the marketing costs a company will incur once it adopts a mark for a product of line of goods.

If no opposition is filed within the opposition period, or if the opposition is resolved favorably to the registrant, the Trademark Office issues a certificate of registration. Because trademark rights are based upon use of the mark in commerce, if the application was based on an "intent to use" the mark, an applicant must show actual use in order for the Trademark Office to issue the registration. Use is demonstrated by submitting affidavits with specimens, such as hang tags or labels, to the Trademark Office.

3. Notice, Renewal, and Duration of a Registered Mark

Prior to registration, a trademark owner is free to use the notice "TM" or, less common "SM", to indicate the trademark or servicemark owner is asserting rights in the mark but has not yet obtained a federal registration. Once a mark is registered the trademark owner may use the symbol "®" in connection with the mark. The Lanham Act provides that, if a registrant fails to include such notice of the registration of the mark, "no profits and no damages shall be recovered under the provisions of this chapter unless the defendant had actual notice of the registration." 15 U.S.C. § 1111.

Section 8 of the Lanham Act provides that a registration stays in effect for ten years. The registration can be renewed for subsequent 10-year periods indefinitely. However, to obtain the full first ten years of protection, a trademark owner must file an affidavit of continued use with specimens showing current use of the mark between the fifth and sixth year following registration. 15 U.S.C. § 1058. Many novice trademark owners are unaware of this requirement and neglect to file the required affidavit resulting in a loss of their mark's registration status. For trademark attorneys, docketing software provides reminders of such filing requirements for their different clients.

4. Incontestability

At the time that the affidavit of continued use is filed, trademark owners typically file for "incontestability" under § 15 of the Act. The two affidavits are typically combined into one "section 8 & 15 affidavit." Section 15 requires that the trademark owner swear to several different facts concerning continued use. If incontestability is achieved, the grounds for challenging the protectability of a mark or for seeking cancellation of the mark are far more limited. See 15 U.S.C. §§ 1064, 1065. One of the most important consequences of a mark achieving incontestable status is that a defendant can no longer challenge the protectable status of the mark on the grounds that it is descriptive and lacks secondary meaning. *Park'N Fly, Inc. v. Dollar Park & Fly, Inc.*, 469 U.S. 189 (1985). Challenges based on genericness are still permitted. *Id.*

5. Cancellation of a Registered Mark

Prior to a mark achieving incontestable status, it is possible to challenge a registration and seek to have the registration cancelled. Within five years of registration, any person who believes that he is or will be damaged by registration may seek to cancel a mark. § 14(a), 15 U.S.C. § 1064(a). A mark may be canceled at any time for certain specified grounds, including that it was obtained fraudulently or has become generic. § 14(c), 15 U.S.C. § 1064(c).

6. Certification Marks and Collective Marks

In addition to standard trademarks and servicemarks, federal law provides for registration of two unique types of marks: collective marks and certification marks. Collective marks can either be marks used by "members of a cooperative, an association, or other collective group or organization" such as agricultural co-operatives, unions or trade organizations, or marks used by members of such an organization to identify and distinguish their goods or services. 15 U.S.C. § 1127. For example, "realtor" is a collective service mark owned by the National Association of Realtors. Only members of that association are permitted to use that term. Collective marks are subject to the same general trademark rules regarding protectability and, if registered, their registrations may be cancelled if the mark becomes generic. *See, e.g., Zimmerman v. National Association of Realtors*, 70 U.S.P.Q.2d 1425 (T.T.A.B. 2004) (rejecting challenge that the marks "realtor" and "realtors" are generic when used in connection with real estate sales services).

The Lanham Act defines certification marks as marks used "to certify regional or other origin, material, mode of manufacture, quality, accuracy, or other characteristics of such person's goods or services or that the work or labor on the goods or services was performed by members of a union or other organization." 15 U.S.C. § 1127. As you can see from the definition, a certification mark does not indicate the producer of the good or service but rather that the goods or services meet certain criteria. A familiar example is the certification mark "UL" owned by Underwriters Laboratory, which certifies electrical appliances as meeting certain safety standards. The owner of the mark is required to permit all entities that meet the certification standards to use the mark. Indeed, certification marks are subject to cancellation if the owner "discriminately refuses to certify or to continue to certify the goods or services of any person who maintains the standards or conditions which such mark certifies." 15 U.S.C. § 1064(5). Owners of certification marks are prohibited from using the mark on their own goods or services, such use being grounds for cancellation. *Id.*

One example of a certification mark is the dolphin safe mark (depicted here), owned by Earth Island Institute which monitors tuna fishing companies around the world for their fishing methods and the effect on dolphins. Earth Island Institute states that in order for tuna to be considered "Dolphin Safe", it must be produced according to the following standards:

1. No intentional chasing, netting, or encirclement of dolphins during an entire tuna fishing trip;
2. No use of drift gill nets to catch tuna;
3. No accidental killing or serious injury to any dolphins during net sets;
4. No mixing of dolphin-safe and dolphin-deadly tuna in individual boat wells (for accidental kill of dolphins), or in processing or storage facilities; and
5. Each trip in the Eastern Tropical Pacific Ocean (ETP) by vessels 400

gross tons and above must have an independent observer on board attesting
to the compliance with points (1) through (4) above.

See <http://www.earthisland.org/immp/Dol_Safe_Standard.html>. Consumers
desiring to purchase tuna but concerned about dolphin well-being can look for the
symbol as a way to reduce their search costs for tuna that comports with their
ethics. Other "dolphin safe" labels exist, including one controlled by the U.S. Department of Commerce pursuant to the Dolphin Protection Consumer Information Act, see 16 U.S.C.A. § 1385. Should other entities be permitted to use the
phrase "dolphin safe" on their products? As part of their own trademarks? As part
of an alternative certification mark? The answers to these questions depend
largely on two inquiries. The first inquiry concerns material we explored in this
Section—the classification of the phrase "dolphin safe"—is it generic, descriptive, or suggestive? The second inquiry involves understanding the standard for
trademark infringement. We turn to that inquiry in Section III.

II. "Device" Marks and Trade Dress Protection

The Lanham Act defines a trademark as "any word, name, symbol, or device,
or any combination thereof." 15 U.S.C. § 1127. Just what *is* a "symbol" or a "device" that can function as a trademark?

A. Color as a Trademark

Qualitex Co. v. Jacobson Prods. Co.

514 U.S. 159 (1995)

Breyer, Justice:

. . . .

The case before us grows out of petitioner Qualitex Company's use (since the
1950's) of a special shade of green-gold color on the pads that it makes and sells
to dry cleaning firms for use on dry cleaning presses. In 1989 respondent Jacobson
Products (a Qualitex rival) began to sell its own press pads to dry cleaning firms;
and it colored those pads a similar green-gold. In 1991 Qualitex registered the special green-gold color on press pads with the Patent and Trademark Office as a
trademark. Registration No. 1,633,711 (Feb. 5, 1991). Qualitex subsequently added
a trademark infringement count, 15 U.S.C. § 1114(1), to an unfair competition
claim, § 1125(a), in a lawsuit it had already filed challenging Jacobson's use of the
green-gold color.

Qualitex won the lawsuit in the District Court. But, the Court of Appeals for
the Ninth Circuit set aside the judgment in Qualitex's favor on the trademark infringement claim because, in that Circuit's view, the Lanham Act does not permit
Qualitex, or anyone else, to register "color
alone" as a trademark. 13 F.3d 1297, 1300,
1302 (1994).

The courts of appeals have differed as
to whether or not the law recognizes the

use of color alone as a trademark. Compare *NutraSweet Co. v. Stadt Corp.*, 917 F.2d 1024, 1028 (7th Cir. 1990) (absolute prohibition against protection of color alone), with *In re Owens-Corning Fiberglas Corp.*, 774 F.2d 1116, 1128 (Fed. Cir. 1985) (allowing registration of color pink for fiberglass insulation), and *Master Distributors, Inc. v. Pako Corp.*, 986 F.2d 219, 224 (8th Cir. 1993) (declining to establish *per se* prohibition against protecting color alone as a trademark). Therefore, this Court granted certiorari. We now hold that there is no rule absolutely barring the use of color alone, and we reverse the judgment of the Ninth Circuit.

II

The Lanham Act gives a seller or producer the exclusive right to "register" a trademark, 15 U.S.C. § 1052, and to prevent his or her competitors from using that trademark, § 1114(1). Both the language of the Act and the basic underlying principles of trademark law would seem to include color within the universe of things that can qualify as a trademark. The language of the Lanham Act describes that universe in the broadest of terms. It says that trademarks "includ[e] any word, name, symbol, or device, or any combination thereof." § 1127. Since human beings might use as a "symbol" or "device" almost anything at all that is capable of carrying meaning, this language, read literally, is not restrictive. The courts and the [PTO] have authorized for use as a mark a particular shape (of a Coca-Cola bottle), a particular sound (of NBC's three chimes), and even a particular scent (of plumeria blossoms on sewing thread). See, *e.g.*, Registration No. 696,147 (Apr. 12, 1960); Registration Nos. 523,616 (Apr. 4, 1950) and 916,522 (July 13, 1971); *In re Clarke*, 17 U.S.P.Q.2d 1238, 1240 (TTAB 1990). If a shape, a sound, and a fragrance can act as symbols why, one might ask, can a color not do the same?

A color is also capable of satisfying the more important part of the statutory definition of a trademark, which requires that a person "us[e]" or "inten[d] to use" the mark

> to identify and distinguish his or her goods, including a unique product, from those manufactured or sold by others and to indicate the source of the goods, even if that source is unknown.

15 U.S.C. § 1127.

True, a product's color is unlike "fanciful," "arbitrary," or "suggestive" words or designs, which almost *automatically* tell a customer that they refer to a brand. The imaginary word "Suntost," or the words "Suntost Marmalade," on a jar of orange jam immediately would signal a brand or a product "source"; the jam's orange color does not do so. But, over time, customers may come to treat a particular color on a product or its packaging (say, a color that in context seems unusual, such as pink on a firm's insulating material or red on the head of a large industrial bolt) as signifying a brand. And, if so, that color would have come to identify and distinguish the goods—*i.e.* "to indicate" their "source"—much in the way that descriptive words on a product (say, "Trim" on nail clippers or "Car-Freshner" on deodorizer) can come to indicate a product's origin. In this circumstance, trademark law says that the word (*e.g.*, "Trim"), although not inherently distinctive, has developed "secondary meaning." See *Inwood Laboratories, Inc. v.*

Ives Laboratories, Inc., 456 U.S. 844, 851, n.11 (1982) ("secondary meaning" is acquired when "in the minds of the public, the primary significance of a product feature . . . is to identify the source of the product rather than the product itself"). Again, one might ask, if trademark law permits a descriptive word with secondary meaning to act as a mark, why would it not permit a color, under similar circumstances, to do the same?

We cannot find in the basic objectives of trademark law any obvious theoretical objection to the use of color alone as a trademark, where that color has attained "secondary meaning" and therefore identifies and distinguishes a particular brand (and thus indicates its "source"). In principle, trademark law, by preventing others from copying a source-identifying mark, "reduce[s] the customer's costs of shopping and making purchasing decisions," 1 J. McCarthy, McCarthy on Trademarks and Unfair Competition § 2.01[2], p. 2-3 (3d ed. 1994) (hereinafter McCarthy), for it quickly and easily assures a potential customer that *this* item—the item with this mark—is made by the same producer as other similarly marked items that he or she liked (or disliked) in the past. At the same time, the law helps assure a producer that it (and not an imitating competitor) will reap the financial, reputation-related rewards associated with a desirable product. The law thereby "encourage[s] the production of quality products," *ibid.,* and simultaneously discourages those who hope to sell inferior products by capitalizing on a consumer's inability quickly to evaluate the quality of an item offered for sale. It is the source-distinguishing ability of a mark—not its ontological status as color, shape, fragrance, word, or sign—that permits it to serve these basic purposes. And, for that reason, it is difficult to find, in basic trademark objectives, a reason to disqualify absolutely the use of a color as a mark.

Neither can we find a principled objection to the use of color as a mark in the important "functionality" doctrine of trademark law. The functionality doctrine prevents trademark law, which seeks to promote competition by protecting a firm's reputation, from instead inhibiting legitimate competition by allowing a producer to control a useful product feature. It is the province of patent law, not trademark law, to encourage invention by granting inventors a monopoly over new product designs or functions for a limited time, after which competitors are free to use the innovation. If a product's functional features could be used as trademarks, however, a monopoly over such features could be obtained without regard to whether they qualify as patents and could be extended forever (because trademarks may be renewed in perpetuity). See *Kellogg Co. v. National Biscuit Co.,* 305 U.S. 111, 119-120 (1938) (Brandeis, J.); *Inwood Laboratories, Inc.,* 456 U.S. at 863 (White, J., concurring in result) ("A functional characteristic is 'an important ingredient in the commercial success of the product,' and, after expiration of a patent, it is no more the property of the originator than the product itself"). Functionality doctrine therefore would require, to take an imaginary example, that even if customers have come to identify the special illumination-enhancing shape of a new patented light bulb with a particular manufacturer, the manufacturer may not use that shape as a trademark, for doing so, after the patent had expired, would impede competition—not by protecting the reputation of the original bulb maker, but by frustrating competitors' legitimate efforts to produce an equivalent illumi-nation-enhancing bulb. See, *e.g., Kellogg Co.*, 305 U.S. at 119-120 (trademark law cannot be used to extend monopoly over "pillow" shape of shredded wheat bis-cuit after the patent for that shape had expired). This Court consequently has ex-plained that, "[i]n general terms, a product feature is functional," and cannot serve as a trademark, "if it is essential to the use or purpose of the article or if it affects the cost or quality of the article," that is, if exclusive use of the feature would put competitors at a significant non-reputation-related disadvantage. *In-wood Laboratories, Inc.*, 456 U.S. at 850, n.10. Although sometimes color plays an important role (unrelated to source identification) in making a product more de-sirable, sometimes it does not. And, this latter fact—the fact that sometimes color is not essential to a product's use or purpose and does not affect cost or quality—indicates that the doctrine of "functionality" does not create an absolute bar to the use of color alone as a mark. See *Owens-Corning*, 774 F.2d, at 1123 (pink color of insulation in wall "performs no nontrademark function").

It would seem, then, that color alone, at least sometimes, can meet the basic legal requirements for use as a trademark. It can act as a symbol that distinguishes a firm's goods and identifies their source, without serving any other significant function. See U.S. Dept. of Commerce, Patent and Trademark Office, Trademark Manual of Examining Procedure § 1202.04(e), p. 1202-13 (2d ed., May 1993) (hereinafter PTO Manual) (approving trademark registration of color alone where it "has become distinctive of the applicant's goods in commerce," provided that "there is [no] competitive need for colors to remain available in the industry" and the color is not "functional"). Indeed, the District Court, in this case, entered findings (accepted by the Ninth Circuit) that show Qualitex's green-gold press pad color has met these requirements. The green-gold color acts as a symbol. Having developed secondary meaning (for customers identified the green-gold color as Qualitex's), it identifies the press pads' source. And, the green-gold color serves no other function. (Although it is important to use *some* color on press pads to avoid noticeable stains, the [district] court found "no competitive need in the press pad industry for the green-gold color, since other colors are equally usable.") Accordingly, unless there is some special reason that convincingly militates against the use of color alone as a trademark, trademark law would protect Quali-tex's use of the green-gold color on its press pads.

III

Respondent Jacobson Products says that there are four special reasons why the law should forbid the use of color alone as a trademark. We . . . find them unper-suasive.

First, Jacobson says that, if the law permits the use of color as a trademark, it will produce uncertainty and unresolvable court disputes about what shades of a color a competitor may lawfully use. Because lighting (morning sun, twilight mist) will affect perceptions of protected color, competitors and courts will suffer from "shade confusion" as they try to decide whether use of a similar color on a similar product does, or does not, confuse customers and thereby infringe a trademark. Jacobson adds that the "shade confusion" problem is "more difficult" and "far different from" the "determination of the similarity of words or symbols." Brief for Respondent 22.

We do not believe, however, that color, in this respect, is special. Courts traditionally decide quite difficult questions about whether two words or phrases or symbols are sufficiently similar, in context, to confuse buyers. They have had to compare, for example, such words as "Bonamine" and "Dramamine" (motion-sickness remedies); "Huggies" and "Dougies" (diapers); "Cheracol" and "Syrocol" (cough syrup); "Cyclone" and "Tornado" (wire fences); and "Mattres" and "1-800-Mattres" (mattress franchisor telephone numbers). Legal standards exist to guide courts in making such comparisons. See, *e.g.,* 2 McCarthy § 15.08; 1 McCarthy §§ 11.24-11.25 ("[S]trong" marks, with greater secondary meaning, receive broader protection than "weak" marks). We do not see why courts could not apply those standards to a color, replicating, if necessary, lighting conditions under which a colored product is normally sold. Indeed, courts already have done so in cases where a trademark consists of a color plus a design, *i.e.,* a colored symbol such as a gold stripe (around a sewer pipe), a yellow strand of wire rope, or a "brilliant yellow" band (on ampules). See, *e.g., Youngstown Sheet & Tube Co. v. Tallman Conduit Co.,* 149 U.S.P.Q. 656, 657 (TTAB 1966); *Amsted Industries, Inc. v. West Coast Wire Rope & Rigging Inc.,* 2 U.S.P.Q.2d 1755, 1760 (TTAB 1987); *In re Hodes-Lange Corp.,* 167 U.S.P.Q. 255, 256 (TTAB 1970).

Second, Jacobson argues, as have others, that colors are in limited supply. Jacobson claims that, if one of many competitors can appropriate a particular color for use as a trademark, and each competitor then tries to do the same, the supply of colors will soon be depleted. Put in its strongest form, this argument would concede that "[h]undreds of color pigments are manufactured and thousands of colors can be obtained by mixing." L. Cheskin, Colors: What They Can Do For You 47 (1947). But, it would add that, in the context of a particular product, only some colors are usable. By the time one discards colors that, say, for reasons of customer appeal, are not usable, and adds the shades that competitors cannot use lest they risk infringing a similar, registered shade, then one is left with only a handful of possible colors. . . . [A] competitor's inability to find a suitable color will put that competitor at a significant disadvantage.

This argument is unpersuasive, however, largely because it relies on an occasional problem to justify a blanket prohibition. When a color serves as a mark, normally alternative colors will likely be available for similar use by others. Moreover, if that is not so—if a "color depletion" or "color scarcity" problem does arise—the trademark doctrine of "functionality" normally would seem available to prevent the anticompetitive consequences that Jacobson's argument posits, thereby minimizing that argument's practical force.

The functionality doctrine, as we have said, forbids the use of a product's feature as a trademark where doing so will put a competitor at a significant disadvantage because the feature is "essential to the use or purpose of the article" or "affects [its] cost or quality." *Inwood Laboratories, Inc.,* 456 U.S. at 850, n.10. The functionality doctrine thus protects competitors against a disadvantage (unrelated to recognition or reputation) that trademark protection might otherwise impose, namely their inability reasonably to replicate important non-reputation-related product features. For example, this Court has written that competitors might be free to copy the color of a medical pill where that color serves to identify the kind of medication (*e.g.,* a type of blood medicine) in addition to its source. See *id.* at 853, 858, n. 20 ("[S]ome patients commingle medications in a container and rely on color to differentiate one from another"); see also J. Ginsburg *et al.,* Trademark and Unfair Competition Law 194-195 (1991) (noting that drug color cases "have more to do with public health policy" regarding generic drug substitution "than with trademark law"). And, the federal courts have demonstrated that they can apply this doctrine in a careful and reasoned manner, with sensitivity to the effect on competition. Although we need not comment on the merits of specific cases, we note that lower courts have permitted competitors to copy the green color of farm machinery (because customers wanted their farm equipment to match) and have barred the use of black as a trademark on outboard boat motors (because black has the special functional attributes of decreasing the apparent size of the motor and ensuring compatibility with many different boat colors). See *Deere & Co. v. Farmhand, Inc.,* 560 F.Supp. 85, 98 (S.D. Iowa 1982), *aff'd,* 721 F.2d 253 (8th Cir. 1983); *Brunswick Corp. v. British Seagull Ltd.,* 35 F.3d 1527, 1532 (Fed. Cir. 1994); see also *Nor-Am Chemical v. O.M. Scott & Sons Co.,* 4 U.S.P.Q.2d 1316, 1320 (E.D. Pa. 1987) (blue color of fertilizer held functional because it indicated the presence of nitrogen). The Restatement (Third) of Unfair Competition adds that, if a design's "aesthetic value" lies in its ability to "confe[r] a significant benefit that cannot practically be duplicated by the use of alternative designs," then the design is "functional." Restatement (Third) of Unfair Competition § 17, Comment *c,* pp. 175-176 (1995). The "ultimate test of aesthetic functionality," it explains, "is whether the recognition of trademark rights would significantly hinder competition." *Id.* at 176.

The upshot is that, where a color serves a significant nontrademark function—whether to distinguish a heart pill from a digestive medicine or to satisfy the "noble instinct for giving the right touch of beauty to common and necessary things," G.K. Chesterton, Simplicity and Tolstoy 61 (1912)—courts will examine whether its use as a mark would permit one competitor (or a group) to interfere with legitimate (nontrademark-related) competition through actual or potential exclusive use of an important product ingredient. That examination should not discourage firms from creating aesthetically pleasing mark designs, for it is open to their competitors to do the same. But, ordinarily, it should prevent the anticompetitive consequences of Jacobson's hypothetical "color depletion" argument, when, and if, the circumstances of a particular case threaten "color depletion."

Third, Jacobson points to many older cases—including Supreme Court cases—in support of its position. . . .

These Supreme Court cases, however, interpreted trademark law as it existed *before* 1946, when Congress enacted the Lanham Act. The Lanham Act significantly changed and liberalized the common law to "dispense with mere technical prohibitions," S. Rep. No. 1333, 79th Cong., 2d Sess., 3 (1946), most notably, by permitting trademark registration of descriptive words (say, "U-Build-It" model airplanes) where they had acquired "secondary meaning." See *Abercrombie & Fitch Co.,* 537 F.2d at 9 (Friendly, J.). The Lanham Act extended protection to descriptive marks by making clear that (with certain explicit exceptions not relevant here), "nothing . . . shall prevent the registration of a mark used by the applicant which has become distinctive of the applicant's goods in commerce." 15 U.S.C. § 1052(f). This language permits an ordinary word, normally used for a nontrademark purpose (*e.g.,* description), to act as a trademark where it has gained "secondary meaning." Its logic would appear to apply to color as well. . . .

. . . .

Fourth, Jacobson argues that there is no need to permit color alone to function as a trademark because a firm already may use color as part of a trademark, say, as a colored circle or colored letter or colored word, and may rely upon "trade dress" protection, under § 43(a) of the Lanham Act, if a competitor copies its color and thereby causes consumer confusion : . . . , see 15 U.S.C. § 1125(a). The first part of this argument begs the question. One can understand why a firm might find it difficult to place a usable symbol or word on a product (say, a large industrial bolt that customers normally see from a distance); and, in such instances, a firm might want to use color, pure and simple, instead of color as part of a design. Neither is the second portion of the argument convincing. Trademark law helps the holder of a mark in many ways that "trade dress" protection does not. See 15 U.S.C. § 1124 (ability to prevent importation of confusingly similar goods); § 1072 (constructive notice of ownership); § 1065 (incontestible status); § 1057(b) (prima facie evidence of validity and ownership). Thus, one can easily find reasons why the law might provide trademark protection in addition to trade dress protection.

IV

Having determined that a color may sometimes meet the basic legal requirements for use as a trademark and that respondent Jacobson's arguments do not justify a special legal rule preventing color alone from serving as a trademark (and, in light of the District Court's here undisputed findings that Qualitex's use of the green-gold color on its press pads meets the basic trademark requirements), we conclude that the Ninth Circuit erred in barring Qualitex's use of color as a trademark. For these reasons, the judgment of the Ninth Circuit is reversed.

Notes & Questions

1. *Qualitex* makes it clear that color alone can be protected as a trademark so long as it has acquired distinctiveness (a.k.a. secondary meaning). Why can't color be "inherently distinctive"? Is it because consumers don't think color alone can signal source? Or are we uncomfortable permitting one seller of goods to prohibit others from using a particular color without placing some burden on that seller to demonstrate entitlement to protection? Does it matter that the color was used on the product itself versus on the packaging? Consider the blue packaging in which the trademarked sweetener, Nutra-Sweet, is sold. The product itself is white, but the package is blue. Should the use of color on packaging be treated differently? The distinction between product design and product packaging is an important

one following the Supreme Court's decision in *Wal-Mart v. Samara Bros.*, reproduced in the subsection B, below.

2. The functionality doctrine is an important limitation on color alone being eligible for trademark protection. What, exactly, does the court say about determining functionality? Would the color blue used for an automatic toilet bowl cleaner, such as Ty-D-Bol, be functional? What about the silver color of steel chain-link fences?

3. The PTO guidelines require an applicant for a color mark to demonstrate that the color is not functional *and* that there is no "competitive need for the colors to remain available in the industry." Are these really two separate showings? The Court mentions "aesthetic" functionality. Is that the same thing as the competitive need for the color to remain available in the industry?

B. TRADE DRESS PROTECTION

The defendant in *Qualitex* argued that color alone did not need separate protection because color could be protected as part of a trademark or as part of "trade dress." While the Court rejected this argument, understanding the protection available for "trade dress" is an important part of learning about trademark law. The next case, decided by the Supreme Court before *Qualitex*, addresses trade dress.

Two Pesos, Inc. v. Taco Cabana, Inc.
505 U.S. 763 (1992)

WHITE, JUSTICE:

The issue in this case is whether the trade dress[1] of a restaurant may be protected under § 43(a) of the [Lanham Act], 15 U.S.C. § 1125(a), based on a finding of inherent distinctiveness, without proof that the trade dress has secondary meaning.

Respondent Taco Cabana, Inc., operates a chain of fast-food restaurants in Texas. The restaurants serve Mexican food. The first Taco Cabana restaurant was opened in San Antonio in September 1978, and five more restaurants had been opened in San Antonio by 1985. Taco Cabana describes its Mexican trade dress as

> a festive eating atmosphere having interior dining and patio areas decorated with artifacts, bright colors, paintings and murals. The patio includes interior and exterior areas with the interior patio capable of being sealed off from the outside patio by overhead garage doors. The stepped exterior of

---

[1] The District Court instructed the jury: "'[T]rade dress' is the total image of the business. Taco Cabana's trade dress may include the shape and general appearance of the exterior of the restaurant, the identifying sign, the interior kitchen floor plan, the decor, the menu, the equipment used to serve food, the servers' uniforms and other features reflecting on the total image of the restaurant." The Court of Appeals accepted this definition and quoted from *Blue Bell Bio-Medical v. Cin-Bad, Inc.*, 864 F.2d 1253, 1256 (5th Cir. 1989): "The 'trade dress' of a product is essentially its total image and overall appearance." It "involves the total image of a product and may include features such as size, shape, color or color combinations, texture, graphics, or even particular sales techniques." *John H. Harland Co. v. Clarke Checks, Inc.*, 711 F.2d 966, 980 (11th Cir. 1983).

the building is a festive and vivid color scheme using top border paint and neon stripes. Bright awnings and umbrellas continue the theme.

932 F.2d 1113, 1117 (5th Cir. 1991).

In December 1985, a Two Pesos, Inc., restaurant was opened in Houston. Two Pesos adopted a motif very similar to the foregoing description of Taco Cabana's trade dress. Two Pesos restaurants expanded rapidly in Houston and other markets, but did not enter San Antonio. In 1986, Taco Cabana entered the Houston and Austin markets and expanded into other Texas cities, including Dallas and El Paso where Two Pesos was also doing business.

In 1987, Taco Cabana sued Two Pesos in the United States District Court for the Southern District of Texas for trade dress infringement under § 43(a) of the Lanham Act, 15 U.S.C. § 1125(a), and for theft of trade secrets under Texas common law. The case was tried to a jury, which was instructed to return its verdict in the form of answers to five questions propounded by the trial judge. The jury's answers were: Taco Cabana has a trade dress; taken as a whole, the trade dress is nonfunctional; the trade dress is inherently distinctive;[3] the trade dress has not acquired a secondary meaning in the Texas market; and the alleged infringement creates a likelihood of confusion on the part of ordinary customers as to the source or association of the restaurant's goods or services. Because, as the jury was told, Taco Cabana's trade dress was protected if it either was inherently distinctive or had acquired a secondary meaning, judgment was entered awarding damages to Taco Cabana. In the course of calculating damages, the trial court held that Two Pesos had intentionally and deliberately infringed Taco Cabana's trade dress.

The Court of Appeals ruled that the instructions adequately stated the applicable law and that the evidence supported the jury's findings. In particular, the Court of Appeals rejected petitioner's argument that a finding of no secondary meaning contradicted a finding of inherent distinctiveness.

In so holding, the court below followed precedent in the Fifth Circuit. In *Chevron Chemical Co. v. Voluntary Purchasing Groups, Inc.*, 659 F.2d 695, 702 (5th Cir. 1981), the court noted that trademark law requires a demonstration of secondary meaning only when the claimed trademark is not sufficiently distinctive of itself to identify the producer; the court held that the same principles should apply to protection of trade dresses. The Court of Appeals noted that this approach conflicts with decisions of other courts . . . that § 43(a) protects unregistered trademarks or designs only where secondary meaning is shown. We granted certiorari to resolve the conflict among the Courts of Appeals on the question whether trade

---

[3] The instructions were that to be found inherently distinctive, the trade dress must not be descriptive.

dress which is inherently distinctive is protectable under § 43(a) without a showing that it has acquired secondary meaning.[6] We find that it is, and we therefore
affirm.

II

The Lanham Act was intended to make "actionable the deceptive and misleading use of marks" and "to protect persons engaged in . . . commerce against unfair
competition." § 45, 15 U.S.C. § 1127. Section 43(a) "prohibits a broader range of
practices than does § 32," which applies to registered marks, *Inwood Laboratories,
Inc. v. Ives Laboratories, Inc.*, 456 U.S. 844, 858 (1982), but it is common ground
that § 43(a) protects qualifying unregistered trademarks and that the general principles qualifying a mark for registration under § 2 of the Lanham Act are for the
most part applicable in determining whether an unregistered mark is entitled to
protection under § 43(a).

. . . .

The general rule regarding distinctiveness is clear: an identifying mark is distinctive and capable of being protected if it *either* (1) is inherently distinctive *or* (2)
has acquired distinctiveness through secondary meaning. It is also clear that eligibility for protection under § 43(a) depends on nonfunctionality. It is, of course,
also undisputed that liability under § 43(a) requires proof of the likelihood of confusion.

. . . There is no persuasive reason to apply to trade dress a general requirement
of secondary meaning which is at odds with the principles generally applicable to
infringement suits under § 43(a). . . .

Petitioner argues that the jury's finding that the trade dress has not acquired a
secondary meaning shows conclusively that the trade dress is not inherently distinctive. The Court of Appeals' disposition of this issue was sound:

> . . . While the necessarily imperfect (and often prohibitively difficult) meth-
> ods for assessing secondary meaning address the empirical question of cur-
> rent consumer association, the legal recognition of an inherently distinctive
> trademark or trade dress acknowledges the owner's legitimate proprietary
> interest in its unique and valuable informational device, regardless of
> whether substantial consumer association yet bestows the additional em-
> pirical protection of secondary meaning.

932 F.2d at 1120, n. 7.

Although petitioner makes the above argument, it appears to concede elsewhere in its briefing that it is possible for a trade dress, even a restaurant trade
dress, to be inherently distinctive and thus eligible for protection under § 43(a).
Recognizing that a general requirement of secondary meaning imposes "an unfair
prospect of theft [or] financial loss" on the developer of fanciful or arbitrary trade

---

[6] We limited our grant of certiorari to the above question on which there is a conflict. We did not
grant certiorari on the second question presented by the petition, which challenged the Court of
Appeals' acceptance of the jury's finding that Taco Cabana's trade dress was not functional.

dress at the outset of its use, petitioner suggests that such trade dress should receive limited protection without proof of secondary meaning. Petitioner argues that such protection should be only temporary and subject to defeasance when over time the dress has failed to acquire a secondary meaning. This approach is also vulnerable for the reasons given by the Court of Appeals. If temporary protection is available from the earliest use of the trade dress, it must be because it is neither functional nor descriptive but an inherently distinctive dress that is capable of identifying a particular source of the product. Such a trade dress, or mark, is not subject to copying by concerns that have an equal opportunity to choose their own inherently distinctive trade dress. To terminate protection for failure to gain secondary meaning over some unspecified time could not be based on the failure of the dress to retain its fanciful, arbitrary, or suggestive nature, but on the failure of the user of the dress to be successful enough in the marketplace. This is not a valid basis to find a dress or mark ineligible for protection. The user of such a trade dress should be able to maintain what competitive position it has and continue to seek wider identification among potential customers.

. . . .

It would be a different matter if there were textual basis in § 43(a) for treating inherently distinctive verbal or symbolic trademarks differently from inherently distinctive trade dress. But there is none. The section does not mention trademarks or trade dress, whether they be called generic, descriptive, suggestive, arbitrary, fanciful, or functional. Nor does the concept of secondary meaning appear in the text of § 43(a). Where secondary meaning does appear in the statute, 15 U.S.C. § 1052, it is a requirement that applies only to merely descriptive marks and not to inherently distinctive ones. We see no basis for requiring secondary meaning for inherently distinctive trade dress protection under § 43(a) but not for other distinctive words, symbols, or devices capable of identifying a producer's product.

Engrafting onto § 43(a) a requirement of secondary meaning for inherently distinctive trade dress also would undermine the purposes of the Lanham Act. Protection of trade dress, no less than of trademarks, serves the Act's purpose to "secure to the owner of the mark the goodwill of his business and to protect the ability of consumers to distinguish among competing producers. National protection of trademarks is desirable, Congress concluded, because trademarks foster competition and the maintenance of quality by securing to the producer the benefits of good reputation." *Park'N Fly,* [*Inc. v. Dollar Park & Fly, Inc.,*] 469 U.S. [189,] 198 [(1985)]. By making more difficult the identification of a producer with its product, a secondary meaning requirement for a nondescriptive trade dress would hinder improving or maintaining the producer's competitive position.

Suggestions that under the Fifth Circuit's law, the initial user of any shape or design would cut off competition from products of like design and shape are not persuasive. Only nonfunctional, distinctive trade dress is protected under § 43(a). The Fifth Circuit holds that a design is legally functional, and thus unprotectable, if it is one of a limited number of equally efficient options available to competitors and free competition would be unduly hindered by according the design trademark protection. *See Sicilia Di R. Biebow & Co. v. Cox*, 732 F.2d 417, 426 (5th Cir. 1984). This serves to assure that competition will not be stifled by the exhaustion of a limited number of trade dresses.

On the other hand, adding a secondary meaning requirement could have anti-competitive effects, creating particular burdens on the start-up of small companies. It would present special difficulties for a business, such as respondent, that seeks to start a new product in a limited area and then expand into new markets. Denying protection for inherently distinctive nonfunctional trade dress until after secondary meaning has been established would allow a competitor, which has not adopted a distinctive trade dress of its own, to appropriate the originator's dress in other markets and to deter the originator from expanding into and competing in these areas.

As noted above, petitioner concedes that protecting an inherently distinctive trade dress from its inception may be critical to new entrants to the market and that withholding protection until secondary meaning has been established would be contrary to the goals of the Lanham Act. Petitioner specifically suggests, however, that the solution is to dispense with the requirement of secondary meaning for a reasonable, but brief period at the outset of the use of a trade dress. If § 43(a) does not require secondary meaning at the outset of a business' adoption of trade dress, there is no basis in the statute to support the suggestion that such a requirement comes into being after some unspecified time.

III

We agree with the Court of Appeals that proof of secondary meaning is not required to prevail on a claim under § 43(a) of the Lanham Act where the trade dress at issue is inherently distinctive, and accordingly the judgment of that court is affirmed.

. . . .

STEVENS, JUSTICE, concurring in the judgment:

As the Court notes in its opinion, the text of § 43(a) of the Lanham Act, 15 U.S.C. § 1125(a), "does not mention trademarks or trade dress." Nevertheless, the Court interprets this section as having created a federal cause of action for infringement of an unregistered trademark or trade dress and concludes that such a mark or dress should receive essentially the same protection as those that are registered. Although I agree with the Court's conclusion, I think it is important to recognize that the meaning of the text has been transformed by the federal courts over the past few decades. I agree with this transformation, even though it marks a departure from the original text, because it is consistent with the purposes of the statute and has recently been endorsed by Congress.

It is appropriate to begin with the relevant text of § 43(a). Section 43(a) provides a federal remedy for using either "a false designation of origin" or a "false description or representation" in connection with any goods or services. The full text of the section makes it clear that the word "origin" refers to the geographic location in which the goods originated, and in fact, the phrase "false designation of origin" was understood to be limited to false advertising of geographic origin.

For example, the "false designation of origin" language contained in the statute makes it unlawful to represent that California oranges came from Florida, or vice versa.

For a number of years after the 1946 enactment of the Lanham Act, a "false description or representation," like "a false designation of origin," was construed narrowly. The phrase encompassed two kinds of wrongs: false advertising[4] and the common-law tort of "passing off." False advertising meant representing that goods or services possessed characteristics that they did not actually have and passing off meant representing one's goods as those of another. Neither "secondary meaning" nor "inherent distinctiveness" had anything to do with false advertising, but proof of secondary meaning was an element of the common-law passing-off cause of action.

II

Over time, the Circuits have expanded the categories of "false designation of origin" and "false description or representation." One treatise[6] identified the Court of Appeals for the Sixth Circuit as the first to broaden the meaning of "origin" to include "origin of source or manufacture" in addition to geographic origin.[7] Another early case, described as unique among the Circuit cases because it was so "forward-looking,"[8] interpreted the "false description or representation" language to mean more than mere "palming off." *L'Aiglon Apparel, Inc. v. Lana Lobell, Inc.*, 214 F.2d 649 (3rd Cir. 1954). The court explained: "We find nothing in the legislative history of the Lanham Act to justify the view that [§ 43(a)] is merely declarative of existing law. . . . It seems to us that Congress has defined a statutory civil wrong of false representation of goods in commerce and has given a broad class of suitors injured or likely to be injured by such wrong the right to relief in the federal courts." *Id.* at 651. Judge Clark, writing a concurrence in 1956, presciently observed: "Indeed, there is indication here and elsewhere that the bar has not yet realized the potential impact of this statutory provision [§ 43(a)]." *Maternally Yours, Inc. v. Your Maternity Shop, Inc.*, 234 F.2d 538, 546 (2d Cir. 1956). Although some have criticized the expansion as unwise, it is now "a firmly embedded reality."[10] The United States Trade[mark] Association Trademark Review Commission noted this transformation with approval: "Section 43(a) is an enig-ma, but a very popular one. Narrowly drawn and intended to reach

---

[4] The deleterious effects of false advertising were described by one commentator as follows: "[A] campaign of false advertising may completely discredit the product of an industry, destroy the confidence of consumers and impair a communal or trade good will. Less tangible but nevertheless real is the injury suffered by the honest dealer who finds it necessary to meet the price competition of inferior goods, glamorously misdescribed by the unscrupulous merchant. The competition of a liar is always dangerous even though the exact injury may not be susceptible of precise proof." Handler, *Unfair Competition*, 21 Iowa L. Rev. 175, 193 (1936).

[6] [2 J. McCarthy, Trademarks and Unfair Competition(2d ed. 1984) (McCarthy) at], § 27:3, p. 345.

[7] Federal-Mogul-Bower Bearings, Inc. v. Azoff, 313 F.2d 405, 408 (6th Cir. 1963).

[8] Derenberg, [*Federal Unfair Competition Law at the End of the First Decade of the Lanham Act: Prologue or Epilogue?*,] 32 N.Y.U. L. Rev. [1029,] 1047, 1049 [(1957)].

[10] 2 McCarthy § 27:3, p. 345.

false designations or representations as to the geographical origin of products, the section has been widely interpreted to create, in essence, a federal law of unfair competition. . . . It has definitely eliminated a gap in unfair competition law, and its vitality is showing no signs of age."[11]

Today, it is less significant whether the infringement falls under "false designation of origin" or "false description or representation" because in either case § 43(a) may be invoked. The federal courts are in agreement that § 43(a) creates a federal cause of action for trademark and trade dress infringement claims. . . .

III

. . . .

Congress has revisited this statute from time to time, and has accepted the "judicial legislation" that has created this federal cause of action. Recently, for example, in the Trademark Law Revision Act of 1988, Pub.L. 100-667, 102 Stat. 3935, Congress codified the judicial interpretation of § 43(a), giving its imprimatur to a growing body of case law from the Circuits that had expanded the section beyond its original language.

. . . Congress broadened the language of § 43(a) to make explicit that the provision prohibits "any word, term, name, symbol, or device, or any combination thereof" that is "likely to cause confusion, or to cause mistake, or to deceive as to the affiliation, connection, or association of such person with another person, or as to the origin, sponsorship, or approval of his or her goods, services, or commercial activities by another person." 15 U.S.C. § 1125(a). That language makes clear that a confusingly similar trade dress is actionable under § 43(a), without necessary reference to "falsity." . . . Congress explicitly extended to any violation of § 43(a) the basic Lanham Act remedial provisions whose text previously covered only registered trademarks. The aim of the amendments was to apply the same protections to unregistered marks as were already afforded to registered marks. See S. Rep. No. 100-515, p. 40 (1988). These steps buttress the conclusion that § 43(a) is properly understood to provide protection in accordance with the standards for registration in § 2. These aspects of the 1988 legislation bolster the claim that an inherently distinctive trade dress may be protected under § 43(a) without proof of secondary meaning.

IV

In light of the general consensus among the Courts of Appeals that have actually addressed the question, and the steps on the part of Congress to codify that consensus, *stare decisis* concerns persuade me to join the Court's conclusion that secondary meaning is not required to establish a trade dress violation under § 43(a) once inherent distinctiveness has been established. Accordingly, I concur in the judgment, but not in the opinion of the Court.

---

[11] The United States Trademark Association Trademark Review Commission Report and Recommendations to USTA President and Board of Directors, 77 Trademark Rep. 375, 426 (1987).

Notes & Questions

1. In addition to being a case concerning trade dress, *Two Pesos* involves a service, a restaurant, as opposed to a product. As we explored in Section I, service marks are a type of mark that can be registered under the Lanham Act. It is difficult, although not impossible, to register a trade dress. For example, Pommery Mustard is sold in a crockery jar sealed with a red ribbon. A registration for the use of a red ribbon can be obtained. Another example, of course, is the green-gold color at issue in *Qualitex*, registered on the principal register by Qualitex. The protection at issue in *Two Pesos* involves unregistered trade dress. What level of protection is obtained for that trade dress? What else would be gained by registration? As Justice Stevens makes clear in his concurrence, the protection afforded by § 43(a) has grown significantly over time. Currently § 43(a) provides:

> (1) Any person who, on or in connection with any goods or services, or any container for goods, uses in commerce any word, term, name, symbol, or device, or any combination thereof, or any false designation of origin, false or misleading description of fact, or false or misleading representation of fact, which—
>
>> (A) is likely to cause confusion, or to cause mistake, or to deceive as to the affiliation, connection, or association of such person with another person, or as to the origin, sponsorship, or approval of his or her goods, services, or commercial activities by another person, or
>>
>> (B) in commercial advertising or promotion, misrepresents the nature, characteristics, qualities, or geographic origin of his or her or another person's goods, services, or commercial activities,
>
> shall be liable in a civil action by any person who believes that he or she is or is likely to be damaged by such act.
>
> . . . .
>
> (3) In a civil action for trade dress infringement under this chapter for trade dress not registered on the principal register, the person who asserts trade dress protection has the burden of proving that the matter sought to be protected is not functional.

The expansion, over time, of the protection afforded by § 43 has continued with additions to § 43 that protect both registered and unregistered marks against dilution of a trade mark, § 43(c), and against "cybersquatting," § 43(d). These protections will be explored later in this Chapter.

2. Taco Cabana did not sue Two Pesos when Two Pesos first started using a similar "look" for its restaurants. Taco Cabana waited until it expanded into an overlapping geographic area. Why did the company wait? Taco Cabana was not suing pursuant to a registered trademark and thus, as you learned in Section I, Taco Cabana did not enjoy the benefit of nationwide rights that comes with federal registration. Additionally, as you will see in Section III, when conducting a likelihood of confusion analysis, overlapping in the same market is an important consideration.

3. In *Two Pesos* the defendant suggested an approach to trade dress that would provide a limited protection for trade dress subject to defeasance if, after a set period of time, that trade dress does not acquire distinctiveness in the minds of consumers. The Court rejected this approach as inconsistent with the statute. Congress could adopt such an approach if it so chose. Would such an approach be wise?

4. *Qualitex* was decided after *Two Pesos*. Why allow trade dress protection immediately (if the trade dress is inherently distinctive) but not permit any color to be immediately protectable? The Court indicates that "[e]ngrafting onto § 43(a) a requirement of secondary meaning for inherently distinctive trade dress also would undermine the purposes of the Lanham Act. Protection of trade dress, no less than of trademarks, serves the Act's purpose to 'secure to the owner of the mark the goodwill of his business and to protect the ability of consumers to distinguish among competing producers.'" Why treat trade dress more favorably than color? If one equates the protection of "device" marks with words, following the holdings of *Qualitex* and *Two Pesos*, colors are always classified as descriptive, and hence require a showing of secondary meaning, while some trade dress can be suggestive, arbitrary or fanciful and thus receive protection without demonstrating secondary meaning. Is the trade dress at issue in *Two Pesos* different in type than the color mark at issue in *Qualitex*? Keep this possibility in mind as you read the next case.

Wal-Mart Stores, Inc. v. Samara Brothers, Inc.
529 U.S. 205 (2000)

SCALIA, JUSTICE: — copying baby clothes from another brand

In this case, we decide under what circumstances a product's design is distinctive, and therefore protectible, in an action for infringement of unregistered trade dress under § 43(a) of the [Lanham Act].

Respondent Samara Brothers, Inc., designs and manufactures children's clothing. Its primary product is a line of spring/summer one-piece seersucker outfits decorated with appliques of hearts, flowers, fruits, and the like. A number of chain stores, including JCPenney, sell this line of clothing under contract with Samara.

Petitioner Wal-Mart Stores, Inc., is one of the nation's best known retailers, selling among other things children's clothing. In 1995, Wal-Mart contracted with one of its suppliers, Judy-Philippine, Inc., to manufacture a line of children's outfits for sale in the 1996 spring/summer season. Wal-Mart sent Judy-Philippine photographs of a number of garments from Samara's line, on which Judy-Philippine's garments were to be based; Judy-Philippine duly copied, with only minor modifications, 16 of Samara's garments, many of which contained copyrighted elements. In 1996, Wal-Mart briskly sold the so-called knockoffs, generating more than $1.15 million in gross profits.

In June 1996, a buyer for JCPenney called a representative at Samara to complain that she had seen Samara garments on sale at Wal-Mart for a lower price than JCPenney was allowed to charge under its contract with Samara. The Samara

— not copyright b/c not a work of art

representative told the buyer that Samara did not supply its clothing to Wal-Mart. Their suspicions aroused, however, Samara officials launched an investigation, which disclosed that Wal-Mart and several other major retailers—Kmart, Caldor, Hills, and Goody's—were selling the knockoffs of Samara's outfits produced by Judy-Philippine.

After sending cease-and-desist letters, Samara brought this action in the United States District Court for the Southern District of New York against Wal-Mart, Judy-Philippine, Kmart, Caldor, Hills, and Goody's for copyright infringement under federal law, consumer fraud and unfair competition under New York law, and—most relevant for our purposes—infringement of unregistered trade dress under § 43(a) of the Lanham Act, 15 U.S.C. § 1125(a). All of the defendants except Wal-Mart settled before trial.

After a weeklong trial, the jury found in favor of Samara on all of its claims. Wal-Mart then renewed a motion for judgment as a matter of law, claiming, *inter alia*, that there was insufficient evidence to support a conclusion that Samara's clothing designs could be legally protected as distinctive trade dress for purposes of § 43(a). The District Court denied the motion and awarded Samara damages, interest, costs, and fees totaling almost $1.6 million, together with injunctive relief. The Second Circuit affirmed the denial of the motion for judgment as a matter of law, and we granted certiorari.

II

The Lanham Act provides for the registration of trademarks, which it defines in § 45 to include "any word, name, symbol, or device, or any combination thereof [used or intended to be used] to identify and distinguish [a producer's] goods * * * from those manufactured or sold by others and to indicate the source of the goods * * *." 15 U.S.C. § 1127. Registration of a mark under § 2 of the Act, 15 U.S.C. § 1052, enables the owner to sue an infringer under § 32, 15 U.S.C. § 1114; it also entitles the owner to a presumption that its mark is valid, see § 7(b), 15 U.S.C. § 1057(b), and ordinarily renders the registered mark incontestable after five years of continuous use, see § 15, 15 U.S.C. § 1065. In addition to protecting registered marks, the Lanham Act, in § 43(a), gives a producer a cause of action for the use by any person of "any word, term, name, symbol, or device, or any combination thereof * * * which * * * is likely to cause confusion * * * as to the origin, sponsorship, or approval of his or her goods * * *." 15 U.S.C. § 1125(a). It is the latter provision that is at issue in this case.

The breadth of the definition of marks registrable under § 2, and of the confusion-producing elements recited as actionable by § 43(a), has been held to embrace not just word marks, such as "Nike," and symbol marks, such as Nike's "swoosh" symbol, but also "trade dress"—a category that originally included only the packaging, or "dressing," of a product, but in recent years has been expanded by many courts of appeals to encompass the design of a product. See, e.g. , *Ashley Furniture Industries, Inc. v. Sangiacomo N. A., Ltd.*, 187 F.3d 363 (4th Cir. 1999) (bedroom furniture); *Knitwaves, Inc. v. Lollytogs, Ltd.*, 71 F.3d 996 (2d Cir. 1995) (sweaters); *Stuart Hall Co., Inc. v. Ampad Corp.*, 51 F.3d 780 (8th Cir. 1995) (notebooks).

These courts have assumed, often without discussion, that trade dress constitutes a "symbol" or "device" for purposes of the relevant sections, and we conclude likewise. "Since human beings might use as a 'symbol' or 'device' almost anything at all that is capable of carrying meaning, this language, read literally, is not restrictive." *Qualitex Co. v. Jacobson Products Co.*, 514 U.S. 159, 162 (1995). This reading of § 2 and § 43(a) is buttressed by a recently added subsection of § 43(a), § 43(a)(3), which refers specifically to "civil action[s] for trade dress infringement under this chapter for trade dress not registered on the principal register." 15 U.S.C. § 1125(a)(3).

The text of § 43(a) provides little guidance as to the circumstances under which unregistered trade dress may be protected. It does require that a producer show that the allegedly infringing feature is not "functional," see § 43(a)(3), and is likely to cause confusion with the product for which protection is sought, see § 43(a)(1)(A), 15 U.S.C. § 1125(a)(1)(A). Nothing in § 43(a) explicitly requires a producer to show that its trade dress is distinctive, but courts have universally imposed that requirement, since without distinctiveness the [accused] trade dress would not "cause confusion . . . as to the origin, sponsorship, or approval of [the] goods," as the section requires. Distinctiveness is, moreover, an explicit prerequisite for registration of trade dress under § 2, and "the general principles qualifying a mark for registration under § 2 of the Lanham Act are for the most part applicable in determining whether an unregistered mark is entitled to protection under 43(a)." *Two Pesos, Inc. v. Taco Cabana, Inc.*, 505 U.S. 763, 768 (1992).

In evaluating the distinctiveness of a mark under § 2 (and therefore, by analogy, under § 43(a)), courts have held that a mark can be distinctive in one of two ways. First, a mark is inherently distinctive if "[its] intrinsic nature serves to identify a particular source." *Id.* . . . Second, a mark has acquired distinctiveness, even if it is not inherently distinctive, if it has developed secondary meaning, which occurs when, "in the minds of the public, the primary significance of a [mark] is to identify the source of the product rather than the product itself." *Inwood Laboratories, Inc. v. Ives Laboratories, Inc.*, 456 U.S. 844, 851, n. 11 (1982).

The judicial differentiation between marks that are inherently distinctive and those that have developed secondary meaning has solid foundation in the statute itself. Section 2 requires that registration be granted to any trademark "by which the goods of the applicant may be distinguished from the goods of others"—subject to various limited exceptions. 15 U.S.C. § 1052. It also provides, again with limited exceptions, that "nothing in this chapter shall prevent the registration of a mark used by the applicant which has become distinctive of the applicant's goods in commerce"—that is, which is not inherently distinctive but has become so only through secondary meaning. § 2(f), 15 U.S.C. § 1052(f). Nothing in § 2, however, demands the conclusion that every category of mark necessarily includes some marks "by which the goods of the applicant may be distinguished from the goods of others" without secondary meaning—that in every category some marks are inherently distinctive.

Indeed, with respect to at least one category of mark—colors—we have held that no mark can ever be inherently distinctive. *See Qualitex*, 514 U.S. at 162-163.

... We held that a color could be protected as a trademark, but only upon a showing of secondary meaning. Reasoning by analogy to the *Abercrombie & Fitch* test developed for word marks, we noted that a product's color is unlike a "fanciful," "arbitrary," or "suggestive" mark, since it does not "almost automatically tell a customer that [it] refer[s] to a brand," *ibid.*, and does not "immediately * * * signal a brand or a product 'source,'" *id.* at 163. However, we noted that, "over time, customers may come to treat a particular color on a product or its packaging * * * as signifying a brand." *Id.* at 162-163. Because a color, like a "descriptive" word mark, could eventually "come to indicate a product's origin," we concluded that it could be protected upon a showing of secondary meaning. *Id.*

It seems to us that design, like color, is not inherently distinctive. The attribution of inherent distinctiveness to certain categories of word marks and product packaging derives from the fact that the very purpose of attaching a particular word to a product, or encasing it in a distinctive packaging, is most often to identify the source of the product. Although the words and packaging can serve subsidiary functions—a suggestive word mark (such as "Tide" for laundry detergent), for instance, may invoke positive connotations in the consumer's mind, and a garish form of packaging (such as Tide's squat, brightly decorated plastic bottles for its liquid laundry detergent) may attract an otherwise indifferent consumer's attention on a crowded store shelf—their predominant function remains source identification. Consumers are therefore predisposed to regard those symbols as indication of the producer, which is why such symbols "almost automatically tell a customer that they refer to a brand," *id.* at 162-163, and "immediately * * * signal a brand or a product 'source,'" *id.* at 163. And where it is not reasonable to assume consumer predisposition to take an affixed word or packaging as indication of source—where, for example, the affixed word is descriptive of the product ("Tasty" bread) or of a geographic origin ("Georgia" peaches)—inherent distinctiveness will not be found. That is why the statute generally excludes, from those word marks that can be registered as inherently distinctive, words that are "merely descriptive" of the goods, § 2(e)(1), 15 U.S.C. § 1052(e)(1), or "primarily geographically descriptive of them," see § 2(e)(2), 15 U.S.C. § 1052(e)(2). In the case of product design, as in the case of color, we think consumer predisposition to equate the feature with the source does not exist. Consumers are aware of the reality that, almost invariably, even the most unusual of product designs—such as a cocktail shaker shaped like a penguin—is intended not to identify the source, but to render the product itself more useful or more appealing.

The fact that product design almost invariably serves purposes other than source identification not only renders inherent distinctiveness problematic; it also renders application of an inherent-distinctiveness principle more harmful to other consumer interests. Consumers should not be deprived of the benefits of competition with regard to the utilitarian and esthetic purposes that product design ordinarily serves by a rule of law that facilitates plausible threats of suit against new entrants based upon alleged inherent distinctiveness. How easy it is to mount a plausible suit depends, of course, upon the clarity of the test for inherent distinctiveness, and where product design is concerned we have little confidence that a reasonably clear test can be devised. Respondent and the United States as amicus curiae urge us to adopt for product design relevant portions of the test formulated by the Court of Customs and Patent Appeals for product packaging in *Seabrook Foods, Inc. v. Bar-Well Foods, Ltd.*, 568 F.2d 1342 (1977). That opinion, in determining the inherent distinctiveness of a product's packaging, considered, among other things, "whether it was a 'common' basic shape or design, whether it was unique or unusual in a particular field, [and] whether it was a mere refinement of a commonly-adopted and well-known form of ornamentation for a particular class of goods viewed by the public as a dress or ornamentation for the goods." *Id.* at 1344. Such a test would rarely provide the basis for summary disposition of an anticompetitive strike suit. Indeed, at oral argument, counsel for the United States quite understandably would not give a definitive answer as to whether the test was met in this very case, saying only that "[t]his is a very difficult case for that purpose." Tr. of Oral Arg. 19.

It is true, of course, that the person seeking to exclude new entrants would have to establish the nonfunctionality of the design feature, see § 43(a)(3), 15 U.S.C. § 1125(a)(3)—a showing that may involve consideration of its esthetic appeal, see *Qualitex*, 514 U.S. at 170. Competition is deterred, however, not merely by successful suit but by the plausible threat of successful suit, and given the unlikelihood of inherently source-identifying design, the game of allowing suit based upon alleged inherent distinctiveness seems to us not worth the candle. That is especially so since the producer can ordinarily obtain protection for a design that is inherently source identifying (if any such exists), but that does not yet have secondary meaning, by securing a design patent or a copyright for the design—as, indeed, respondent did for certain elements of the designs in this case. The availability of these other protections greatly reduces any harm to the producer that might ensue from our conclusion that a product design cannot be protected under § 43(a) without a showing of secondary meaning.

Respondent contends that our decision in *Two Pesos* forecloses a conclusion that product-design trade dress can never be inherently distinctive. . . . *Two Pesos* unquestionably establishes the legal principle that trade dress can be inherently distinctive, but it does not establish that product-design trade dress can be. *Two Pesos* is inapposite to our holding here because the trade dress at issue, the decor of a restaurant, seems to us not to constitute product design. It was either product packaging—which, as we have discussed, normally is taken by the consumer to indicate origin—or else some *tertium quid* that is akin to product packaging and has no bearing on the present case.

Respondent replies that this manner of distinguishing *Two Pesos* will force courts to draw difficult lines between product-design and product-packaging trade dress. There will indeed be some hard cases at the margin: a classic glass Coca-Cola bottle, for instance, may constitute packaging for those consumers who drink the Coke and then discard the bottle, but may constitute the product itself for those consumers who are bottle collectors, or part of the product itself for those consumers who buy Coke in the classic glass bottle, rather than a can, because they think it more stylish to drink from the former. We believe, however, that the frequency and the difficulty of having to distinguish between product design and product packaging will be much less than the frequency and the difficulty of having to decide when a product design is inherently distinctive. To the extent there are close cases, we believe that courts should err on the side of caution and classify ambiguous trade dress as product design, thereby requiring secondary meaning. The very closeness will suggest the existence of relatively small utility in adopting an inherent-distinctiveness principle, and relatively great consumer benefit in requiring a demonstration of secondary meaning.

* * *

We hold that, in an action for infringement of unregistered trade dress under 43(a) of the Lanham Act, a product's design is distinctive, and therefore protectible, only upon a showing of secondary meaning. The judgment of the Second Circuit is reversed, and the case is remanded for further proceedings consistent with this opinion.

Notes & Questions

1. You have now read three Supreme Court decisions, decided within a relatively short time span (at least for the Court), concerning protection for symbols other than words. The Court requires a showing of secondary meaning for color alone and for product design, but not for the trade dress of a restaurant or for product packaging (so long as it is inherently distinctive). Do the lines that the Court has drawn make sense? In *Wal-Mart*, the Court bases its distinctions on its understanding that consumers are predisposed to regard symbols and product packaging as indications of the producer of a product, but are not predisposed to regard the shape of a product as a source indicator. Let's assume, for the sake of argument, that the Court is correct. Is the consumer predisposition on which the Court relies shaped by the strength of the legal protection we give to symbols and product packaging? If so, over time, will consumers begin to regard product shape or design as signaling source, as more producers prove secondary meaning? How likely is it that producers will be able to make that showing?

2. In *Wal-Mart* the court is concerned with the anti-competitive effect that strike suits could have if product design could be protected without requiring a showing of secondary meaning. Is the anti-competitive threat of such suits significantly lower for the overall look of a restaurant? For the color of a product? Distinguishing *Two Pesos*, the Court says restaurant décor is more similar to product packaging and thus it can be inherently distinctive. Are consumers predisposed to expect a restaurant's décor to signal the services' source? Or is the décor meant to be a pleasant environment in which to have a meal, or to evoke the region on the planet from which the food is supposed to emanate or be inspired? If it's the latter, doesn't the décor become part of the dining experience, part of the product or service itself? If that is true, décor is more like product design than product packaging. In close cases, the Court tells us that courts should err on the side of requiring a showing of secondary meaning. Is the décor of a restaurant such a "close case," or does *Two Pesos* foreclose that argument?

3. The court acknowledges that another purpose of packaging can be to catch the eye of the consumer on a crowded shelf. Does that goal conflict with or assist in obtaining trademark protection?

4. Samara Brothers also asserted a claim for copyright infringement. Although Samara Brothers failed in its trademark claim, the Supreme Court left standing the lower courts' copyright rulings. The jury awarded Samara Brothers $912,856.77 for the copyright infringement. Wal-Mart had argued that the appliqués of strawberries, daisies, hearts and tulips lacked sufficient originality to be protected by copyright. In affirming the finding of infringement by the District Court, the Second Circuit relied on the presumption of validity dictated by Samara Brothers' copyright registrations. *Samara Bros., Inc. v. Wal-Mart Stores, Inc.*, 165 F.2d 120, 132 (2d Cir. 1998). Noting that the protection for works depicting familiar objects is "very narrow," the identical copying by the defendant was important to the finding of infringement.

C. Functionality

In trademark law, the line between protectable and unprotectable product design depends not only on a demonstration of secondary meaning but also on a showing that the design is not functional. In each of the cases in this Section the Supreme Court formulated its holdings by relying, in part, on the doctrine that functional trade dress is not protectable. In *Qualitex* the Court provided the following articulation of what it means for a product feature to be functional:

> "[i]n general terms, a product feature is functional," and cannot serve as a trademark, "if it is essential to the use or purpose of the article or if it affects the cost or quality of the article," that is, if exclusive use of the feature would put competitors at a significant non-reputation-related disadvantage.

Qualitex, 514 U.S. at 165 (quoting *Inwood Laboratories, Inc.*, 456 U.S. at 850, n. 10.). In discussing what is sometimes referred to as the "aesthetic functionality" doctrine, the Court was guided by the common law:

> The Restatement (Third) of Unfair Competition adds that, if a design's "aesthetic value" lies in its ability to "confe[r] a significant benefit that cannot practically be duplicated by the use of alternative designs," then the design is "functional." Restatement (Third) of Unfair Competition § 17, Comment *c*, pp. 175-176 (1995). The "ultimate test of aesthetic functionality," it explains, "is whether the recognition of trademark rights would significantly hinder competition." *Id.*, at 176.

Qualitex, 514 U.S. at 170.

In *Two Pesos*, the Court cited approvingly the Fifth Circuit's articulation of the functionality doctrine:

> The Fifth Circuit holds that a design is legally functional, and thus unprotectable, if it is one of a limited number of equally efficient options available to competitors and free competition would be unduly hindered by according the design trademark protection. See *Sicilia Di R. Biebow & Co. v. Cox*,

732 F.2d 417, 426 (CA5 1984). This serves to assure that competition will not be stifled by the exhaustion of a limited number of trade dresses.

Two Pesos, 505 U.S. at 775.

One concern the functionality doctrine addresses is a desire to avoid "back-door" patents. This is a concern that you have seen before. For example, in copyright law the protection for "useful articles" is constrained by the separability doctrine and in trade secret law a competitor can use the secret if the competitor obtains it through lawful means, such as through reverse engineering or independently creating the product. In the case below the Court expressly addresses the functionality doctrine's relationship to patent protection.

TrafFix Devices, Inc. v. Marketing Displays, Inc.
532 U.S. 23 (2001)

Kennedy, Justice:

Temporary road signs with warnings like "Road Work Ahead" or "Left Shoulder Closed" must withstand strong gusts of wind. An inventor named Robert Sarkisian obtained two utility patents for a mechanism built upon two springs (the dual-spring design) to keep these and other outdoor signs upright despite adverse wind conditions. The holder of the now-expired Sarkisian patents, respondent Marketing Displays, Inc. (MDI), established a successful business in the manufacture and sale of sign stands incorporating the patented feature. MDI's stands for road signs were recognizable to buyers and users (it says) because the dual-spring design was visible near the base of the sign.

This litigation followed after the patents expired and a competitor, TrafFix Devices, Inc., sold sign stands with a visible spring mechanism that looked like MDI's. MDI and TrafFix products looked alike because they were. When TrafFix started in business, it sent an MDI product abroad to have it reverse engineered, that is to say copied. Complicating matters, TrafFix marketed its sign stands under a name similar to MDI's. MDI used the name "WindMaster," while TrafFix, its new competitor, used "WindBuster."

MDI brought suit under the [Lanham Act] against TrafFix for trademark infringement (based on the similar names), trade dress infringement (based on the copied dual-spring design) and unfair competition. TrafFix counterclaimed on antitrust theories. After the United States District Court for the Eastern District of Michigan considered cross-motions for summary judgment, MDI prevailed on its trademark claim for the confusing similarity of names and was held not liable on the antitrust counterclaim; and those two rulings, affirmed by the Court of Appeals, are not before us.

We are concerned with the trade dress question. The District Court ruled against MDI on its trade dress claim. 971 F. Supp. 262 (E.D. Mich. 1997). After determining that the one element of MDI's trade dress at issue was the dual-spring design, it held that "no reasonable trier of fact could determine that MDI has established secondary meaning" in its alleged trade dress. In other words, consumers did not associate the look of the dual-spring design with MDI. As a second, independent reason to grant summary judgment in favor of TrafFix, the District Court determined the dual-spring design was functional. On this rationale secondary meaning is irrelevant because there can be no trade dress protection in any event. In ruling on the functional aspect of the design, the District Court noted that Sixth Circuit precedent indicated that the burden was on MDI to prove that its trade dress was nonfunctional, and not on TrafFix to show that it was functional (a rule since adopted by Congress, see 15 U.S.C. § 1125(a)(3)), and then went on to consider MDI's arguments that the dual-spring design was subject to trade dress protection. Finding none of MDI's contentions persuasive, the District Court concluded MDI had not "proffered sufficient evidence which would enable a reasonable trier of fact to find that MDI's vertical dual-spring design is non-functional." Summary judgment was entered against MDI on its trade dress claims.

The Court of Appeals for the Sixth Circuit reversed the trade dress ruling. 200 F.3d 929 (1999). The Court of Appeals held the District Court had erred in ruling MDI failed to show a genuine issue of material fact regarding whether it had secondary meaning in its alleged trade dress and had erred further in determining that MDI could not prevail in any event because the alleged trade dress was in fact a functional product configuration. The Court of Appeals suggested the District Court committed legal error by looking only to the dual-spring design when evaluating MDI's trade dress. Basic to its reasoning was the Court of Appeals' observation that it took "little imagination to conceive of a hidden dual-spring mechanism or a tri or quad-spring mechanism that might avoid infringing [MDI's] trade dress." The Court of Appeals explained that "[i]f TrafFix or another competitor chooses to use [MDI's] dual-spring design, then it will have to find some other way to set its sign apart to avoid infringing [MDI's] trade dress." It was not sufficient, according to the Court of Appeals, that allowing exclusive use of a particular feature such as the dual-spring design in the guise of trade dress would "hinde[r] competition somewhat." Rather, "[e]xclusive use of a feature must 'put competitors at a significant non-reputation-related disadvantage' before trade dress protection is denied on functionality grounds." ([Q]uoting *Qualitex Co. v. Jacobson Products Co.*, 514 U.S. 159, 165 (1995)). In its criticism of the District Court's ruling on the trade dress question, the Court of Appeals took note of a split among Courts of Appeals in various other Circuits on the issue whether the existence of an expired utility patent forecloses the possibility of the patentee's claiming trade dress protection in the product's design. Compare *Sunbeam Products, Inc. v. West Bend Co.*, 123 F.3d 246 (5th Cir. 1997) (holding that trade dress protection is not foreclosed), *Thomas & Betts Corp. v. Panduit Corp.*, 138 F.3d 277 (7th Cir. 1998) (same), and *Midwest Industries, Inc. v. Karavan Trailers, Inc.*, 175 F.3d 1356 (Fed. Cir. 1999) (same), with *Vornado Air Circulation Systems, Inc. v. Duracraft Corp.*, 58 F.3d 1498, 1500 (10th Cir. 1995) ("Where a product configuration is a significant inventive component of an invention covered by a utility patent * * * it cannot receive trade dress protection"). To resolve the conflict, we granted certiorari.

II

. . . .

Trade dress protection must subsist with the recognition that in many instances there is no prohibition against copying goods and products. In general, unless an intellectual property right such as a patent or copyright protects an item, it will be subject to copying. As the Court has explained, copying is not always discouraged or disfavored by the laws which preserve our competitive economy. *Bonito Boats, Inc. v. Thunder Craft Boats, Inc.*, 489 U.S. 141, 160 (1989). Allowing competitors to copy will have salutary effects in many instances. "Reverse engineering of chemical and mechanical articles in the public domain often leads to significant advances in technology." *Id.*

The principal question in this case is the effect of an expired patent on a claim of trade dress infringement. A prior patent, we conclude, has vital significance in resolving the trade dress claim. A utility patent is strong evidence that the features therein claimed are functional. If trade dress protection is sought for those features the strong evidence of functionality based on the previous patent adds great weight to the statutory presumption that features are deemed functional until proved otherwise by the party seeking trade dress protection. Where the expired patent claimed the features in question, one who seeks to establish trade dress protection must carry the heavy burden of showing that the feature is not functional, for instance by showing that it is merely an ornamental, incidental, or arbitrary aspect of the device.

In the case before us, the central advance claimed in the expired utility patents (the Sarkisian patents) is the dual-spring design; and the dual-spring design is the essential feature of the trade dress MDI now seeks to establish and to protect. The rule we have explained bars the trade dress claim, for MDI did not, and cannot, carry the burden of overcoming the strong evidentiary inference of functionality based on the disclosure of the dual-spring design in the claims of the expired patents.

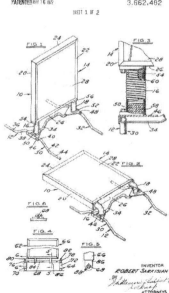

The dual springs shown in the Sarkisian patents were well apart (at either end of a frame for holding a rectangular sign when one full side is the base) while the dual springs at issue here are close together (in a frame designed to hold a sign by one of its corners). As the District Court recognized, this makes little difference. The point is that the springs are necessary to the operation of the device. The fact that the springs in this very different-looking device fall within the claims of the patents is illustrated by MDI's own position in earlier litigation. In the late 1970's, MDI engaged in a long-running intellectual property battle with a company known as Winn-Proof. Although the precise claims of the Sarkisian patents cover sign stands with springs "spaced apart," U.S. Patent

No. 3,646,696, col. 4; U.S. Patent No. 3,662,482, col. 4, the Winn-Proof sign stands (with springs much like the sign stands at issue here) were found to infringe the patents by the United States District Court for the District of Oregon, and the Court of Appeals for the Ninth Circuit affirmed the judgment. *Sarkisian v. Winn-Proof Corp.*, 697 F.2d 1313 (1983). Although the Winn-Proof traffic sign stand (with dual springs close together) did not appear, then, to infringe the literal terms of the patent claims (which called for "spaced apart" springs), the Winn-Proof sign stand was found to infringe the patents under the doctrine of equivalents . . . . In light of this past ruling—a ruling procured at MDI's own insistence—it must be concluded the products here at issue would have been covered by the claims of the expired patents.

The rationale for the rule that the disclosure of a feature in the claims of a utility patent constitutes strong evidence of functionality is well illustrated in this case. The dual-spring design serves the important purpose of keeping the sign upright even in heavy wind conditions; and, as confirmed by the statements in the expired patents, it does so in a unique and useful manner. As the specification of one of the patents recites, prior art "devices, in practice, will topple under the force of a strong wind." U.S. Patent No. 3,662,482, col. 1. The dual-spring design allows sign stands to resist toppling in strong winds. Using a dual-spring design rather than a single spring achieves important operational advantages. For example, the specifications of the patents note that the "use of a pair of springs * * * as opposed to the use of a single spring to support the frame structure prevents canting or twisting of the sign around a vertical axis," and that, if not prevented, twisting "may cause damage to the spring structure and may result in tipping of the device." U.S. Patent No. 3,646,696, col. 3. In the course of patent prosecution, it was said that "[t]he use of a pair of spring connections as opposed to a single spring connection * * * forms an important part of this combination" because it "forc[es] the sign frame to tip along the longitudinal axis of the elongated ground-engaging members." App. 218. The dual-spring design affects the cost of the device as well; it was acknowledged that the device "could use three springs but this would unnecessarily increase the cost of the device." App. 217. These statements made in the patent applications and in the course of procuring the patents demonstrate the functionality of the design. MDI does not assert that any of these representations are mistaken or inaccurate, and this is further strong evidence of the functionality of the dual-spring design.

III

In finding for MDI on the trade dress issue the Court of Appeals gave insufficient recognition to the importance of the expired utility patents, and their evidentiary significance, in establishing the functionality of the device. The error likely was caused by its misinterpretation of trade dress principles in other respects. As we have noted, even if there has been no previous utility patent the party asserting trade dress has the burden to establish the nonfunctionality of alleged trade dress features. MDI could not meet this burden. Discussing trademarks, we have said " '[i]n general terms, a product feature is functional,' and cannot serve as a trademark, 'if it is essential to the use or purpose of the article or if it affects the cost or quality of the article.'" *Qualitex*, 514 U.S. at 165 (quoting *Inwood Laboratories, Inc. v. Ives Laboratories, Inc.*, 456 U.S. 844, 850, n. 10 (1982)). Expanding upon the meaning of this phrase, we have observed that a functional feature is one the "exclusive use of [which] would put competitors at a significant non-reputation-related disadvantage." 514 U.S. at 165. The Court of Appeals in the instant case seemed to interpret this language to mean that a necessary test for functionality is "whether the particular product configuration is a competitive necessity." 200 F.3d at 940.

This was incorrect as a comprehensive definition. As explained in *Qualitex* and *Inwood* a feature is also functional when it is essential to the use or purpose of the device or when it affects the cost or quality of the device. The *Qualitex* decision did not purport to displace this traditional rule. Instead, it quoted the rule as *Inwood* had set it forth. It is proper to inquire into a "significant non-reputation-related disadvantage" in cases of aesthetic functionality, the question involved in *Qualitex*. Where the design is functional under the *Inwood* formulation there is no need to proceed further to consider if there is a competitive necessity for the feature. In *Qualitex*, by contrast, aesthetic functionality was the central question, there having been no indication that the green-gold color of the laundry press pad had any bearing on the use or purpose of the product or its cost or quality.

The Court has allowed trade dress protection to certain product features that are inherently distinctive. *Two Pesos*, 505 U.S. at 774. In *Two Pesos*, however, the Court at the outset made the explicit analytic assumption that the trade dress features in question (decorations and other features to evoke a Mexican theme in a restaurant) were not functional. *Id.* at 767, n. 6. The trade dress in those cases did not bar competitors from copying functional product design features. In the instant case, beyond serving the purpose of informing consumers that the sign stands are made by MDI (assuming it does so), the dual-spring design provides a unique and useful mechanism to resist the force of the wind. Functionality having been established, whether MDI's dual-spring design has acquired secondary meaning need not be considered.

There is no need, furthermore, to engage, as did the Court of Appeals, in speculation about other design possibilities, such as using three or four springs which might serve the same purpose. Here, the functionality of the spring design means that competitors need not explore whether other spring juxtapositions might be used. The dual-spring design is not an arbitrary flourish in the configuration of MDI's product; it is the reason the device works. Other designs need not be attempted.

Because the dual-spring design is functional, it is unnecessary for competitors to explore designs to hide the springs, say by using a box or framework to cover them, as suggested by the Court of Appeals. The dual-spring design assures the user the device will work. If buyers are assured the product serves its purpose by seeing the operative mechanism that in itself serves an important market need. It would be at cross-purposes to those objectives, and something of a paradox, were we to require the manufacturer to conceal the very item the user seeks.

In a case where a manufacturer seeks to protect arbitrary, incidental, or ornamental aspects of features of a product found in the patent claims, such as arbitrary curves in the legs or an ornamental pattern painted on the springs, a different result might obtain. There the manufacturer could perhaps prove that those aspects do not serve a purpose within the terms of the utility patent. The inquiry into whether such features, asserted to be trade dress, are functional by reason of their inclusion in the claims of an expired utility patent could be aided by going beyond the claims and examining the patent and its prosecution history to see if the feature in question is shown as a useful part of the invention. No such claim is made here, however. MDI in essence seeks protection for the dual-spring design alone. The asserted trade dress consists simply of the dual-spring design, four legs, a base, an upright, and a sign. MDI has pointed to nothing arbitrary about the components of its device or the way they are assembled. The Lanham Act does not exist to reward manufacturers for their innovation in creating a particular device; that is the purpose of the patent law and its period of exclusivity. The Lanham Act, furthermore, does not protect trade dress in a functional design simply because an investment has been made to encourage the public to associate a particular functional feature with a single manufacturer or seller. The Court of Appeals erred in viewing MDI as possessing the right to exclude competitors from using a design identical to MDI's and to require those competitors to adopt a different design simply to avoid copying it. MDI cannot gain the exclusive right to produce sign stands using the dual-spring design by asserting that consumers associate it with the look of the invention itself. Whether a utility patent has expired or there has been no utility patent at all, a product design which has a particular appearance may be functional because it is "essential to the use or purpose of the article" or "affects the cost or quality of the article." *Inwood*, 456 U.S. at 850, n. 10.

TrafFix and some of its amici argue that the Patent Clause of the Constitution, Art. I, 8, cl. 8, of its own force, prohibits the holder of an expired utility patent from claiming trade dress protection. We need not resolve this question. If, despite the rule that functional features may not be the subject of trade dress protection, a case arises in which trade dress becomes the practical equivalent of an expired utility patent, that will be time enough to consider the matter. The judgment of the Court of Appeals is reversed, and the case is remanded for further proceedings consistent with this opinion.

Notes & Questions

1. The Supreme Court makes clear in *TrafFix Devices* that even if a producer can demonstrate secondary meaning for the design of a product, if that design is functional it will not be protected as trade dress. When one company obtains a patent on a particular configuration of elements, it is the only company that can make and sell that configuration for almost 20 years. Under such circumstances, it would be understandable for the public to come to associate that configuration with a particular source, i.e., for that configuration to acquire distinctivness. Why

isn't the Court more concerned about protecting the public from consumer confusion? What is it that overrides our general desire to prevent confusing consumers?

2. In *TrafFix* the Court is concerned with the background norm of not just competition, but competition through copying. The Court cites its opinion in *Bonito Boats, Inc. v. Thunder Craft Boats, Inc.*, 489 U.S. 141, 160 (1989), to support the idea that copying can lead to new advances in the state of the art. Is permitting competitors to copy a successful design, such as the design at issue in *TrafFix,* likely to lead to advances in the art? What other patent doctrines contribute to the Court's decision?

3. The Court establishes a "heavy presumption" of functionality for a design covered by an expired utility patent. What must a producer demonstrate to be able to overcome that presumption? If an aspect of a product design is part of a patent claim, how likely is it that the producer will be able to meet that burden?

4. Recall the useful article doctrine in copyright law. One can imagine a party that fails to secure copyright protection for features of its useful article turning to a product-design trade dress theory as an alternative means of protection. In an influential case asserting multiple intellectual property rights in the widely used Ribbon Rack bike rack (which you can see at www.ribbonrack.com, and which you have almost certainly seen hundreds of times in your life), as part of a decision issued before any of the Supreme Court's cases included in this Section, the U.S. Court of Appeals for the Second Circuit addressed the district court's grant of defendant's motion for summary judgment on the trade dress claim:

> There are numerous alternative bicycle rack constructions. The nature, price, and utility of these constructions are material issues of fact not suitable for determination by summary judgment. For example, while it is true that the materials used by Brandir are standard-size pipes, we have no way of knowing whether the particular size and weight of the pipes used is the best, the most economical, or the only available size and weight pipe in the marketplace. We would rather think the opposite might be the case. So, too, with the dimension of the bends being dictated by a standard formula corresponding to the pipe size; it could be that there are many standard radii and that the particular radius of Brandir's RIBBON Rack actually required new tooling. This issue of functionality on remand should be viewed in terms of bicycle racks generally and not one-piece undulating bicycle racks specifically.

Brandir Intern., Inc. v. Cascade Pacific Lumber Co., 834 F.2d 1142, 1148 (2nd Cir. 1987).

How would the analysis of trade dress protection for the Ribbon Rack differ, if at all, after the Supreme Court decisions that you have read in this Section?

5. Following the approach in *TrafFix,* which treats the various tests for functionality in trademark as disjunctive (functional if *a,* or *b,* or *c*), how likely is it that

a trademark-eligible product design will remain after the hunt to root out any functional feature(s) in that design? The use of three separate options to show functionality appears to make it very likely that most product designs will be excluded from trademark. Why make the trigger for exclusion from protection relatively easy to trip?

6. The *TrafFix* case involved an expired utility patent. What about an expired design patent? After the design patent expires, should others be permitted to use the ornamental features claimed in the now-expired design patent? What if those features had come to signal the source of the product? In *Kellogg Co. v. National Biscuit Co.*, 305 U.S. 111 (1938), the plaintiff had possessed both utility and design patents related to the shape of shredded wheat breakfast cereal. After the expiration of the patents, plaintiff sued Kellogg, a competitor, for violation of trademark law. The Court held for the defendant:

> Where an article may be manufactured by all, a particular manufacturer can no more assert exclusive rights in a form in which the public has become accustomed to see the article and which, in the minds of the public, is primarily associated with the article rather than a particular producer, than it can in the case of a name with similar connections in the public mind. Kellogg Company was free to use the pillow-shaped form, subject only to the obligation to identify its product lest it be mistaken for that of the plaintiff.

Id. at 120. Does this suggest a downside of electing to obtain a design patent on certain aspects of product design?

III. TRADEMARK INFRINGEMENT

Having introduced the types of marks that qualify for protection, we now turn to the rights that the federal statute grants to trademark owners. In this Section we begin with the right to prevent infringement, and in Section IV, we explore the right owners of famous marks have to prevent dilution.

Section 32 of the Lanham Act, 15 U.S.C. § 1114, defines the infringement cause of action for registered trademarks:

> (1) Any person who shall, without the consent of the registrant—
> (a) use in commerce any reproduction, counterfeit, copy, or colorable imitation of a registered mark in connection with the sale, offering for sale, distribution, or advertising of any goods or services on or in connection with which such use is likely to cause confusion, or to cause mistake, or to deceive; or
> (b) reproduce, counterfeit, copy, or colorably imitate a registered mark and apply such reproduction, counterfeit, copy, or colorable imitation to labels, signs, prints, packages, wrappers, receptacles or advertisements intended to be used in commerce upon or in connection with the sale, offering for sale, distribution, or advertising of goods or services on or in connection with which such use is likely to cause confusion, or to cause mistake, or to deceive,

shall be liable in a civil action by the registrant. . . .

Additionally, as you learned in Section II, § 43(a) of the Lanham Act, 15 U.S.C. § 1125, permits unregistered mark owners to bring claims for infringement as well. Registered mark owners can also assert claims under § 43(a). Specifically, § 43(a) provides:

> (a) (1) Any person who, on or in connection with any goods or services, or any container for goods, uses in commerce any word, term, name, symbol, or device, or any combination thereof, or any false designation of origin, false or misleading description of fact, or false or misleading representation of fact, which—
>
> > (A) is likely to cause confusion, or to cause mistake, or to deceive as to the affiliation, connection, or association of such person with another person, or as to the origin, sponsorship, or approval of his or her goods, services, or commercial activities by another person, or
> >
> > (B) in commercial advertising or promotion, misrepresents the nature, characteristics, qualities, or geographic origin of his or her or another person's goods, services, or commercial activities,
>
> shall be liable in a civil action by any person who believes that he or she is or is likely to be damaged by such act.

The fundamental test for infringement is whether there is a likelihood of confusion. But confusion as to what? Section 32 does not specify the nature of the confusion that must be present to constitute infringement, while § 43(a) identifies two types of confusion: confusion as to "affiliation, connection, or association of such person with another person"; and confusion as to "the origin, sponsorship, or approval of his or her goods, services, or commercial activities by another person." In Section I we identified the underlying policy goals and justifications for trademark protection. How should these goals help determine the type of "confusion" we are interested in preventing?

A. Registered Mark Infringement

We begin our exploration of infringement with an inquiry into registered mark infringement under 15 U.S.C. § 1114, § 32 of the Lanham Act.

Kellogg Co. v. Toucan Golf, Inc.
337 F.3d 616 (6th Cir. 2003)

Suhrheinrich, Judge:

. . . .

I. Facts

Kellogg, a Delaware corporation based in Battle Creek, Michigan, is the largest producer of breakfast cereal in the world. On July 24, 1963, Kellogg first introduced Toucan Sam on boxes of "Froot Loops" cereal. Kellogg has used Toucan Sam on Froot Loops boxes, and in every print and television advertisement for the cereal, since. Toucan Sam is an anthropomorphic cartoon toucan. He is short and stout and walks upright. He is nearly always smiling with a pleasant and cheery demeanor, but looking nothing similar to a real toucan. He has a royal and powder blue body and an elongated and oversized striped beak, colored shades of orange, red, pink, and black. He has human features, such as fingers and toes, and only exhibits his wings while flying. Moreover, in television advertisements over the past forty years, Toucan Sam has been given a voice. He speaks with a British accent, allowing him to fervently sing the praises of the cereal he represents, and to entice several generations of children to "follow his nose" because "it always knows" where to find the Froot Loops.

Kellogg is the holder of five federally-registered Toucan Sam marks at issue in this case. The first was registered on August 18, 1964, under United States Patent and Trademark Office (USPTO) Reg. No. 775,496, and consists of a simplistic toucan design, drawn with an exaggerated, striped beak, standing in profile with hands on hips and smiling, as reproduced [here].

The second mark was registered March 20, 1984, under USPTO Reg. No. 1,270,940, and consists of an updated version of the same toucan, standing and smiling with his mouth open widely; and pointing his left index finger upward[.]

The third mark is for the word mark, "Toucan Sam." This mark was registered on June 18, 1985, under USPTO Reg. No. 1,343,023. The fourth mark, registered on June 21, 1994, under USPTO Reg. No. 1,840,746, is a shaded drawing of Toucan Sam flying, with wings spread, and smiling.

The fifth mark, registered January 31, 1995, under USPTO Reg. No. 1,876,803, is essentially the same drawing as in the fourth mark, except unshaded . . . .

Together the five registrations indicate that Kellogg's marks are for use in the breakfast cereal industry, and on clothing.

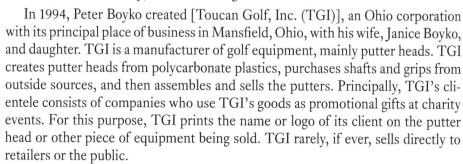

In 1994, Peter Boyko created [Toucan Golf, Inc. (TGI)], an Ohio corporation with its principal place of business in Mansfield, Ohio, with his wife, Janice Boyko, and daughter. TGI is a manufacturer of golf equipment, mainly putter heads. TGI creates putter heads from polycarbonate plastics, purchases shafts and grips from outside sources, and then assembles and sells the putters. Principally, TGI's clientele consists of companies who use TGI's goods as promotional gifts at charity events. For this purpose, TGI prints the name or logo of its client on the putter head or other piece of equipment being sold. TGI rarely, if ever, sells directly to retailers or the public.

TGI likewise uses a toucan drawing, known as "GolfBird" or "Lady GolfBird," to represent its products. TGI has placed this logo on letterhead, business cards, its web site, and even on the outside of its building in Mansfield.

 GolfBird has a multi-colored body, and TGI displays GolfBird in a myriad of color schemes for different purposes. Invariably, however, she has a long, narrow, yellow beak with a black tip, not disproportionate to or unlike that of a real toucan. GolfBird is always seen perched upon a golf iron as if it were a tree branch. She has no human features whatsoever, and resembles a real toucan in all aspects except, perhaps, her variable body coloring[.]

TGI has not registered its GolfBird logo with the USPTO. . . . [Kellogg filed suit in the United States District Court for the Western District of Michigan.] In its complaint, Kellogg . . . claimed that TGI's use of the word mark "Toucan Gold" created a likelihood of confusion among consumers with respect to Kellogg's Toucan Sam word mark. Kellogg added a likelihood of confusion claim with respect to the GolfBird logo as well. . . . On September 6, 2001, after a four day bench trial, the district court dismissed Kellogg's complaint. . . . [Kellogg appealed.]

. . . .

III. Analysis

. . . Kellogg asserts that there is a Lanham Act violation because there exists a likelihood that consumers will be confused as to the source of TGI's products. . . .

A. Likelihood of Confusion

. . . .

This Court has established an eight-part test for determining when a likelihood of confusion exists between the origins of two products. *Therma-Scan, Inc. v. Thermoscan, Inc.*, 295 F.3d 623, 629-30 (6th Cir. 2002); *Daddy's Junky Music Store, Inc. v. Big Daddy's Family Music Center*, 109 F.3d 275, 280 (6th Cir. 1997). The factors are: (1) the strength of the plaintiff's mark; (2) the relatedness of the goods or services offered by the parties; (3) similarity of the marks; (4) any evidence of actual confusion; (5) the marketing channels used by the parties; (6) the probable degree of purchaser care and sophistication; (7) the defendant's intent; and (8) the likelihood of either party expanding its product line using the marks. *Therma-Scan*, 295 F.3d at 630; *Daddy's Junky Music Stores*, 109 F.3d at 280. Not all of these factors will be relevant in every case, and "[t]he ultimate question remains whether relevant consumers are likely to believe that the products or services offered by the parties are affiliated in some way." *Homeowners Group, Inc. v. Home Mktg. Specialists, Inc.*, 931 F.2d 1100, 1107 (6th Cir. 1991). Thus, the question here, as in all trademark cases, is whether we believe consumers of TGI's golf equipment are likely to think it was manufactured by Kellogg. None of the factors is dispositive, but the factors guide us in our ultimate determination.

1. Strength of Kellogg's Marks

The first factor of the test focuses on the distinctiveness of a mark and the public's ability to recognize it. In *Daddy's Junky Music Stores*, we recognized a spectrum of distinctiveness for trademarks, ranging from "generic" to "fanciful."

Daddy's Junky Music Stores, 109 F.3d at 280-81. For example, the word "cereal" is generic, whereas the names "Xerox" and "Kodak" are fanciful, having been completely fabricated by the trademark holders.

We find the "Toucan Sam" word mark and logo each to be fanciful. Kellogg completely created the name "Toucan Sam." Kellogg also completely fabricated Toucan Sam's logo design. He does not resemble a real toucan. His unique shape, coloring, size, and demeanor are entirely the creation of Kellogg, and not reminiscent of anything seen in the wild. Therefore, as a logo, he is also a fanciful mark and distinctive.

In further support of the strength of its Toucan Sam marks, Kellogg has submitted survey information indicating that 94% of Americans recognize Toucan Sam, and 81% of children who recognize him correspond him with Froot Loops. Moreover, Kellogg has submitted extensive records detailing the massive amount of time, money, and effort expended in regard to the marketing of Toucan Sam and Froot Loops. We need not delve into Kellogg's records; we find the fact that Kellogg is the largest cereal maker in the world, that Froot Loops is one of its best selling cereals, and that Toucan Sam has appeared in every print and television advertisement for Froot Loops since 1963 enough to establish that Toucan Sam is visually recognizable by an overwhelming cross-section of American consumers. Coupling that with his distinctiveness, Toucan Sam is a very strong mark.

2. Relatedness of the Products

In consideration of the second factor, we must examine the relatedness of the goods and services offered by each party. We have established three benchmarks regarding the relatedness of parties' goods and services. First, if the parties compete directly, confusion is likely if the marks are sufficiently similar; second, if the goods and services are somewhat related, but not competitive, then the likelihood of confusion will turn on other factors; finally, if the products are unrelated, confusion is highly unlikely.

TGI makes golf equipment, mainly putter heads. TGI also sells bag tags, divot tools, and full sets of clubs, but has never sold any merchandise unrelated to golf. *Therma-Scan*, 295 F.3d at 632; *Daddy's Junky Music Stores*, 109 F.3d at 282.

Kellogg is primarily a producer of breakfast cereal, but has branched off from cereal and sold products in other industries on a limited basis. It has also at times licensed its name and characters to outside companies. Kellogg asserts before this Court that it has sufficiently entered the golf equipment industry. In support of this claim, Kellogg presents a catalog, wherein it offers for sale golf balls and golf shirts on which is imprinted the picture of Toucan Sam. Moreover, Kellogg has presented a mass-marketed 1982 animated television advertisement wherein Toucan Sam is portrayed soliciting his Froot Loops on a golf course, and interacting with a golf playing bear. Kellogg claims these materials indicate that the Toucan Sam marks are related not only to the manufacture of breakfast cereal, but to the golf equipment industry as well.

However, Kellogg, although it is the largest producer of breakfast cereal nationally, has not presented evidence that its golf "equipment" has been marketed nationally. The golf balls and shirts are available on a limited basis, either through the aforementioned catalog— which is not widely distributed— or through select local theme stores, such as Kellogg's own "Cereal City" in Battle Creek, Michigan. Moreover, the commercial in which Toucan Sam plays golf is nonetheless an advertisement for Froot Loops, not golf equipment. The district court found that Kellogg's presence in the golf industry was insignificant, and nothing more than a marketing tool to further boost sales of its cereal. We agree. We find that one thirty second advertisement does not render Toucan Sam a golfer, nor does a novelty catalog make Kellogg a player in the golfing industry. In any event, trademark law is grounded on a likelihood of confusion standard. We find that no consumer would associate Kellogg with top-line golf equipment based on Kellogg's extremely limited licensing of its characters on novelty items. We also believe that if any consumers ever did associate Kellogg and Toucan Sam with golf based on the 1982 commercial, it is highly unlikely that they would still do so twenty years after the advertisement last aired. We find the parties' products completely unrelated. And under the benchmarks established in this Circuit, the second factor therefore supports a conclusion that confusion is not likely to occur. *See Therma-Scan*, 295 F.3d at 632 (stating that confusion is highly unlikely where goods are completely unrelated).

3. Similarity of the Marks

Kellogg argues that it can prove a likelihood of confusion notwithstanding the unrelatedness of the goods. It has presented several cases to demonstrate that courts have held for trademark owners relying heavily on the similarity of the marks, even where the parties' goods were in different product markets. *See, e.g.*, *Recot*, 214 F.3d at 1328 (finding likelihood of confusion between "Frito Lay" and "Fido Lay" even though one is used for snack chips and one is used for dog food); *Hunt Foods & Indus., Inc. v. Gerson Stewart Corp.*, 367 F.2d 431, 435 (C.C.P.A. 1966) (holding "Hunt's" for canned goods and "Hunt" for cleaning products confusingly similar); *American Sugar Refining Co. v. Andreassen*, 296 F.2d 783, 784 (C.C.P.A. 1961) (finding "Domino" for sugar and "Domino" for pet food confusingly similar); *Yale Elec. Corp. v. Robertson*, 26 F.2d 972, 974 (2d Cir. 1928) (finding "Yale" for flashlights and locks confusingly similar); *Quality Inns Int'l, Inc. v. McDonald's Corp.*, 695 F. Supp. 198, 221-22 (D. Md. 1988) (finding similarity between "McSleep Inn" and McDonald's' trademarks); *John Walker & Sons, Ltd. v. Bethea*, 305 F. Supp. 1302, 1307-08 (D.S.C. 1969) (finding "Johnnie Walker" whiskey and "Johnny Walker" hotels confusingly similar). But each of these cases is distinguishable. In some of the cases cited by Kellogg, the courts *did* find that the goods were related. *See, e.g.*, *Recot*, 214 F.3d at 1328 (finding that some snack chip makers might also make dog food); *Hunt Foods*, 367 F.2d at 434 (finding a relationship between the respective products); *American Sugar Refining Co.*, 296 F.2d at 784 (finding goods related because both are sold at grocery stores); *Yale Elec. Corp.*, 26 F.2d at 974 (finding locks and flashlights related because "the trade has so classed them"). In the other cases cited by Kellogg, the names, as well as other marks, were either not only similar, but substantially identical, *see John Walker & Sons*, 305 F. Supp. at 1307-08 (comparing "Johnnie Walker" whiskey to "Johnny

Walker" hotels and finding infringement where defendant also used same color scheme and same script); or the similar portion of the senior mark was both famous and fanciful, and thus so distinctive that its use would transcend its market.[2] *Cf. Recot*, 214 F.3d at 1328 (stating that "Frito Lay" word mark "casts a 'long shadow which competitors must avoid'") (citations omitted); *Quality Inns*, 695 F. Supp. at 216-21 (intimating that the prefix mark "Mc" used by McDonald's is highly distinctive in regard to anything but surnames).

But here, the parties' goods are completely unrelated, and the "Toucan Sam" and "Toucan Gold" word marks are similar only in that they each contain the common word "toucan." Although the name "Toucan Sam" is itself fanciful and distinctive, use of the word "toucan" for cereal is merely arbitrary. Kellogg has taken an everyday word and applied it to a setting where it is not naturally placed. *See, e.g., Daddy's Junky Music Stores*, 109 F.3d at 280-81 (recognizing distinctiveness spectrum and stating that a mark is arbitrary when it is an everyday name or thing mismatched to the product it represents, such as "Camel" for cigarettes or "Apple" for computers). As opposed to a fanciful mark, an arbitrary mark is distinctive only within its product market and entitled to little or no protection outside of that area. *See, e.g., Amstar Corp. v. Domino's Pizza, Inc.*, 615 F.2d 252, 260 (5th Cir. 1980) (implying that plaintiff's arbitrary term "Domino" is entitled to no protection outside of the sugar and condiments market). Thus, unlike the *Recot*, *John Walker & Sons*, and *Quality Inns* cases, here TGI has not used any distinctive portion of Kellogg's word mark at all. Admittedly, we would have a far different case had TGI attempted to use a mark such as "Toucan Sam Gold" for its line of products, because the "Toucan Sam" word mark, in its entirety, is fanciful and likely transcends its market in the same way "Frito Lay" and the "Mc" prefix do. *Cf. Recot*, 214 F.3d at 1328; *Quality Inns*, 695 F.Supp. at 216-21. Kellogg has not cornered the market on all potential uses of the common bird name "toucan" in commerce, only on uses of "Toucan Sam." In regard to the word marks, TGI's apparently similar use is therefore not enough to overcome the unrelatedness of the goods.

As for the logos, the actual Toucan Sam design is fanciful. Hence, in step with cases like *Recot*, if TGI's GolfBird is similar to Toucan Sam's design, there may be a Lanham Act violation in spite of the unrelated goods. But we find GolfBird dissimilar to Toucan Sam. GolfBird resembles a real toucan. She has the look and proportions of a toucan that one would encounter in the wild. Toucan Sam is anthropomorphic, with a discolored, misshaped beak. His body type is not the same as that of a real toucan; and he smiles and has several other human features. We therefore find no similarity between Toucan Sam and GolfBird.

---

[2] It is also of note that in each of the cases cited by Kellogg, the infringed upon trademark was the actual name of the senior user's product. Here, Kellogg claims that TGI has infringed only upon the name of a character that represents Kellogg's product. This would again be a different case if TGI had named itself "Froot Loops Golf" or some derivative thereof.

4. The Other Confusion Factors

The other five factors can be disposed of quickly. Kellogg has presented no evidence of actual customer confusion. Thus, we need not consider that factor.

The parties do not use similar avenues of commerce. Kellogg distributes Froot Loops through regular wholesale and retail channels. Kellogg advertises its product nationally on television and in print. Conversely, TGI distributes its product primarily at trade shows and over the internet. TGI does not sell its golf equipment via retail outlets or advertise on television or radio. *Cf. Hunt Foods*, 367 F.2d at 435 (finding same channels of commerce because both goods are sold at grocery stores).

TGI's clientele is primarily, and almost exclusively, comprised of corporations and wealthy golfers.[3] We find each of these groups to be sufficiently sophisticated, so as not to believe that Kellogg, a cereal company, has manufactured a golf club named "Toucan Gold." Moreover, we find the two industries sufficiently separate, so that there will rarely, if ever, exist a consumer who is looking for Kellogg's product in the golf equipment market.

Next, there is no evidence to suggest that Boyko chose his toucan marks in order to dishonestly trade on Kellogg's marks. Again, the goods are so unrelated as to dispose of this factor with little discussion. Boyko testified that he chose the name "toucan" because of any bird's obvious connection to the game of golf, as evidenced through golfing terms such as "eagle," "birdie," and "albatross." The district court found his testimony on this issue credible, and Kellogg has presented no evidence to cause us to doubt that Boyko's intent was not dishonorable.

Lastly, there is no evidence to suggest that TGI has any desire to enter the cereal game, or that Kellogg has any plan to begin manufacturing golf equipment on a full-scale basis. As stated above, we do not believe Kellogg's limited licensing of golf balls and golf shirts with a Toucan Sam logo, nor the single 1982 advertisement wherein Toucan Sam parades around a golf course, announces Kellogg's entry into the golf market, or its intention to do so.

Accordingly, we find no likelihood of confusion between TGI's use of its marks—the word mark Toucan Gold" and its GolfBird logo; and Kellogg's marks—the word mark "Toucan Sam" and the Toucan Sam design. In fact, the only of the eight factors we find in favor of Kellogg is the strength of its marks. The products sold by each party are wholly unrelated; the similarity between the word marks or the bird designs is not enough to overcome this unrelatedness; and TGI's clientele is not the sort to believe that Kellogg now manufactures golf clubs. We affirm the decision of the district court and find no likelihood of confusion.

    . . . .

---

[3] A set of Toucan Gold clubs costs $1500.

Notes & Questions

1. Kellogg triggered this litigation in an attempt to prevent TGI from registering the word mark "Toucan Gold" on the principal register. TGI had filed an intent to use application for the mark on "golf clubs and golf putters." When the mark was published for opposition, Kellogg opposed TGI's registration asserting, *inter alia*, that TGI's use of the mark would infringe upon its marks and thus could not be registered by TGI. One of the types of marks that is not registerable is one that "so resembles a mark registered in the Patent and Trademark Office, or a mark or trade name previously used in the United States by another and not abandoned, as to be likely, when used on or in connection with the goods of the applicant, to cause confusion, or to cause mistake, or to deceive . . . ." 15 U.S.C. § 1052(d). The test to determine the registrability of a mark is identical to the test to determine infringement. Thus, one cannot register a mark the use of which would infringe another's mark.

The Trademark Trial and Appeal Board (TTAB) rejected the Kellogg's opposition, finding no likelihood of confusion. At that point, Kellogg faced two options: appeal the TTAB's decision to the Federal Circuit or file a lawsuit in federal district court. *See* 15 U.S.C. § 1071. When appealing to the Federal Circuit, the TTAB's conclusions of law are reviewed *de novo* and its findings of facts are reviewed to determine whether there is "substantial evidence" to support the findings. *In re Thrifty, Inc.,* 274 F.3d 1349, 1350 (Fed. Cir. 2001). When challenging a TTAB decision by filing a lawsuit in federal district court, the court reviews the TTAB decision *de novo* and parties may present new evidence to the district court. *See Dickinson v. Zurko,* 526 U.S. 150, 164 (1999). In addition to the difference in standards of review and evidence, might there be other reasons Kellogg chose the strategy it did? In what district court did Kellogg file suit?

2. The Sixth Circuit sets out an eight-part test for determining the ultimate question of likelihood of confusion. Each of the circuits has articulated slightly differently the set of factors to be examined, although all are in agreement that the factors are designed to help the court with the underlying question central to trademark infringement: likelihood of confusion as to the origin of the products. While the Sixth Circuit identified eight factors, it only provided separate analysis of the first three, indicating that the other "five factors can be disposed of quickly." Why did the court believe those other factors did not warrant separate attention? The fluidity of the multi-factor test makes predicting outcomes in trademark litigation difficult and, for the student new to trademark law, it presents difficulties in understanding how the different factors are weighed. Reading more opinions is the best way to build your understanding of the test for trademark infringement.

Experience Hendrix, LLC. v. Electric Hendrix, LLC.
2008 WL 3243896 (W.D. Wa)

ZILLY, JUDGE:

. . . .

This case involves the ongoing and expensive litigation saga between two factions of the Jimi Hendrix family. In this lawsuit, the Authentic Hendrix and Electric Hendrix companies run by Janie Hendrix (Jimi Hendrix's stepsister) (collectively "Authentic Hendrix" or "Plaintiffs") are suing Craig Dieffenbach individually and the companies formed by Craig Dieffenbach (collectively "Hendrix Electric" or "Defendants"), for infringement of Plaintiffs' trademarks in connection with selling, bottling, and marketing vodka as "Hendrix Electric," "Jimi Hendrix Electric," or "Jimi Hendrix Electric Vodka." . . .

Jimi Hendrix died intestate in 1970 in London, England, and his estimated $80 million estate went to his father, Al Hendrix, according to the intestate laws of New York. In 1995, Al Hendrix established and incorporated Experience Hendrix LLC and Authentic Hendrix LLC, and by written agreements, Al Hendrix assigned his rights, personally and as sole heir to the estate of Jimi Hendrix, to Plaintiffs. Plaintiffs now license and sell Jimi Hendrix related products, including Jimi Hendrix music, as well as branded T-shirts, guitars, guitar accessories, and posters. Plaintiffs also operate a web site where Jimi Hendrix goods are sold.

Defendant Dieffenbach has a relationship with the Jimi Hendrix family through Jimi's brother, Leon Hendrix. Mr. Dieffenbach helped underwrite the James Marshall Hendrix Foundation ("the Foundation") which Leon Hendrix founded. Mr. Dieffenbach also assisted Leon Hendrix in preserving the original Jimi Hendrix family home when it was in danger of being torn down. Dieffenbach is the CEO of Electric Hendrix and Electric Hendrix Spirits, Limited Liability Companies.

Mr. Dieffenbach has had previous litigation with Plaintiffs when Plaintiffs sued the Foundation and certain individuals including Mr. Dieffenbach over the use of Jimi Hendrix's name . . . . In that case, the defendants moved for partial summary judgment on the publicity right claims plaintiffs had asserted. This Court granted the motion for partial summary judgment and held that "no right of publicity descended to Al Hendrix in 1970." ("the *Foundation* Order"). . . . The Ninth Circuit affirmed . . . . *Experience Hendrix LLC v. James Marshall Hendrix Found.*, 240 Fed. Appx. 739 (9th Cir. 2007).

In late 2005, Mr. Dieffenbach filed trademark registration applications for JIMI HENDRIX ELECTRIC, JIMI HENDRIX ELECTRIC VODKA, HENDRIX ELECTRIC VODKA, HENDRIX ELECTRIC, and a design featuring a Jimi Hendrix likeness with and without "Jimi" written over a portion of the hair ("the Electric marks"). . . .

Plaintiffs own at least 48 registered trademarks including the following five marks that have become incontestable (collectively the "incontestable marks"):

(1) AUTHENTIC HENDRIX, registration no. 2,245,408, for online ordering services of music, apparel, and memorabilia related to the music industry;

(2) EXPERIENCE HENDRIX registration no. 2,245,409, for print materials such as magazines and posters regarding the music industry;

(3) The design ("AUTHENTIC HENDRIX BUST") registration no. 2,250,912, for use on musical sound recordings;

(4) JIMI HENDRIX, registration no. 2,245,408, for use on clothing such as T-shirts, jackets, and caps; and

(5) JIMI HENDRIX registration no. 2,322,761, for use in entertainment services.

Plaintiffs' contestable marks which have been registered for less than 5 years include, in summary:

(1) Variations of the AUTHENTIC HENDRIX BUST, for use on a number of products including ceramic tiles, key chains, afghans, printed matter, etc.;

(2) ELECTRIC LADYLAND for use on air fresheners, books, key chains, etc.;

(3) HENDRIX and JIMI HENDRIX on compact discs, stickers, key chains, T-shirts, buttons, watches, wall clocks, entertainment services, etc.;

(4) The design ("AUTHENTIC HENDRIX SIGNATURE") for use on printed matter, key chains, clothing, sports bottles, and entertainment services;

(5) JIMI HENDRIX ELECRIC GUITAR COMPETITION and JIMI HENDRIX ELECTRIC GUITAR FESTIVAL for use on posters and pins; and

(6) The design ("AUTHENTIC I AM EXPERIENCED") for use on online retail store services.

Plaintiffs' contend that the following unregistered marks used by Defendants ["Electric Marks"] infringe Plaintiffs' incontestable marks as follows:

(1) JIMI HENDRIX ELECTRIC, JIMI HENDRIX ELECTRIC VODKA, HENDRIX ELECTRIC, and HENDRIX ELECTRIC VODKA for distilled spirits excluding gin; and

(2) The design for distilled spirits excluding gin (with and without Jimi signature)[.]

The Electric marks were used on vodka and promotional material such as shot glasses and T-shirts.

Discussion

[Both parties filed cross motions for summary judgment, each seeking a declaratory judgment on the issue of infringement.]

. . . .

I. Publicity Rights

A threshold issue in this case is what effect the *Foundation* Order has on Plaintiffs' trademark rights. The *Foundation* Order held that on Jimi Hendrix's death his publicity (privacy) rights expired and did not pass to Al Hendrix. Publicity Rights under New York law are governed by New York Civil Rights Law § 51. The primary thrust of Defendants' arguments is that because the Jimi Hendrix publicity rights expired, Plaintiffs could not seek trademark protection for the use of the name or likeness of Jimi Hendrix.

Defendants contend that the effect of the *Foundation* Order was to foreclose any possibility of Plaintiffs asserting rights in the name or likeness of Jimi Hendrix, including those derived from trademarks based on Jimi Hendrix's fame. The *Foundation* Order does not compel this result. The *Foundation* Order can only be read to say that no rights in the commercial value of Jimi Hendrix's name, likeness or other indicia of identity descended to Plaintiffs by operation of New York law.

Trademark rights and publicity rights are distinct rights that are analyzed under different standards. The inquiry under a right of publicity action is whether there was a commercial appropriation of one's identity without consent. *See* J. Thomas McCarthy, *The Rights of Publicity and Privacy* § 1:3 (2d ed. 2008). The proper inquiry when analyzing a trademark right in a famous name is whether the mark in question will falsely suggest a connection with the famous person. 15 U.S.C. § 1052(a). The right of publicity or privacy under state law is not necessary to acquire valid trademark rights under federal law. The Court must therefore analyze Plaintiffs' trademark claims under federal law.

II. Plaintiffs' Trademarks

The Lanham Act allows the holder of a protectable trademark to prevent unauthorized persons from, "us[ing] in commerce any . . . registered mark in connection with the sale, offering for sale, distribution, or advertising of any goods or services" which is likely to cause confusion. 15 U.S.C. § 1114(1)(a). Plaintiffs must prove two elements for their trademark claims: (1) ownership of a valid trademark, and (2) Defendants' use of a similar mark which is likely to cause customer confusion. 15 U.S.C. §§ 1114(1)(a), 1114(1)(b). . . . Plaintiffs' ownership of its registered marks is presumed by registration of the marks and is not disputed by Defendants. 15 U.S.C. § 1057(b).

Trademarks are traditionally a word, phrase, or graphic design used to identify the origin of goods or the sponsorship of goods. *Mattel, Inc. v. MCA Records, Inc.*, 296 F.3d 894, 900 (9th Cir. 2002). The law permits the use of the name of a person as a trademark. Defendants cannot seriously challenge this legal principle. In fact, Defendants have attempted to register marks using the Hendrix name and image.

A. Validity of Plaintiffs' Incontestable Trademarks

Registration of a trademark establishes prima facie evidence of the validity of the mark, and of the registrant's exclusive ownership of the mark and right to exclusive use of the mark in commerce, subject to the goods or services listed in the registration and any conditions or limitations noted on the registration. 15 U.S.C.

§ 1057(b). A registered trademark provides a presumption of validity that shifts the burden to the alleged infringer to prove otherwise.

. . . .

Five of Plaintiffs' marks have attained incontestable status under 15 U.S.C. § 1065: AUTHENTIC HENDRIX, EXPERIENCE HENDRIX, the AUTHENTIC HENDRIX BUST, and both uses of JIMI HENDRIX. . . .

. . . .

[The court rejected defendants' challenges to the validity of the plaintiffs' marks on theories of descriptiveness, fraud on the PTO, abandonment, and unclean hands.]

B. Infringement

. . . .

Trademark infringement of a plaintiff's mark occurs when a defendant's mark is used in commerce and that mark is likely to cause consumer confusion between plaintiff's and defendant's marks. 15 U.S.C. § 1114. The test for likelihood of confusion is the likely reaction of typical buyers including those who are ignorant, unthinking, and credulous. *Fleischmann Distilling Corp. v. Maier Brewing Co.*, 314 F.2d 149, 156 (9th Cir. 1963).

Likelihood of confusion is analyzed using the eight so-called *Sleekcraft* factors: (1) strength of the mark, (2) proximity or relatedness of goods, (3) similarity of the marks, (4) evidence of actual confusion, (5) convergence of marketing channels, (6) degree of care customers are likely to use when purchasing goods of the type in question, (7) intent of defendant in selecting the mark, and (8) likelihood the parties will expand their business lines. *White v. Samsung Electronics America, Inc.*, 971 F.2d 1395, 1400 (9th Cir. 1992). These factors are a non-exclusive set helpful in making an ultimate factual determination of the likelihood of confusion, but they are not "requirements or hoops that a district court need jump through to make a determination." *Eclipse Associates Ltd., v. Data General Corp.*, 894 F.2d 1114, 1118 (9th Cir. 1990).

i. First Sleekcraft Factor: Strength of Plaintiffs' Marks

Trademark law seeks to promote identification of a good's source so that consumers can make an informed purchasing choice. Marks are given varying degrees of protection based on the strength of the mark, and the strength of the mark is determined by the ability to signify to the consumer the source or sponsorship of the goods. Stronger marks are marks more likely to be remembered and associated by consumers with the mark owner, and stronger marks are given greater protection. An incontestable mark must still be analyzed to determine the strength of the mark.

Plaintiffs' marks all contain the name Hendrix. A person's name is a descriptive trademark and requires secondary meaning in order to be afforded trademark protection. When analyzing the strength of a name used as a trademark, the particular facts of the name and the goods must be analyzed. A famous person's name

may have secondary meaning if it is likely to be recognized by prospective purchasers.

> If a mark consists of the name of an historical figure or other noted person
> and is likely to be recognized as such by prospective purchasers, secondary
> meaning ordinarily will not be required. Unless consumers are likely to be-
> lieve that the named person is connected with the goods, services, or busi-
> ness, the use will be understood as arbitrary or suggestive. DA VINCI on
> jewelry, for example, is an inherently distinctive designation.

Restatement (Third) of Unfair Competition, § 14 cmt. e (citations omitted).

Whether Jimi Hendrix's name is likely to be recognized by prospective purchasers is not in dispute. Defendants admit that they chose to market Hendrix-
related goods because of the recognition of the Hendrix name in the market place.
See Electric Hendrix marketing material ("Jimi Hendrix is a world-wide iconic
legend. . . . Jimi's name, image, and music is [sic] recognized by hundreds of millions of people throughout the world."). Defendants argue that other trademarks
use the Hendrix name. Similar registrations that are in actual use in the relevant
market can show a weakness in a trademark. However, Defendants have not shown
that any other registrations containing the word "Hendrix" are in actual use or
that they have any recognition in the relevant marketplace. These similar registrations do not indicate a weakness of Plaintiffs marks. The Court holds that the
name Jimi Hendrix is a famous name and that it has acquired a secondary meaning. Accordingly, the Court holds that the Plaintiffs' incontestable Hendrix-related marks are strong marks and this factor can only support infringement.

Additionally, the AUTHENTIC HENDRIX BUST is a high-contrast black
and white silhouette image with the word EXPERIENCE arched above the bust
and the word HENDRIX below the bust. The arbitrary nature of this mark's features, such as text placement, artistic choices in image style, and overall feel, also
make this a strong mark even apart from the strength of the secondary meaning
associated with HENDRIX.

ii. Second Sleekcraft Factor: Proximity or Relatedness of Goods

Goods are related if consumers are likely to associate the two products lines.
Survivor Media, Inc., v. Survivor Productions, 406 F.3d 625, 633 (9th Cir. 2005).
Compl[e]mentary goods, such as wine and cheese or cheese and salami, are related goods under this analysis. *See E. & J. Gallo Winery,* 967 F.2d at 1291; *White,*
971 F.2d at 1400 (television performances related to VCRs); *but see Levi Strauss &
Co. v. Blue Bell, Inc.,* 778 F.2d 1352, 1359-60 (9th Cir. 1985) (pants and shirts are
not related goods). Defendants assert that they are only selling vodka, which is not
a good that is related to the types of goods Plaintiffs' marks cover, such as key
chains, T-shirts, and memorabilia. However, it is undisputed that Defendants
have created and distributed promotional items such as beverage glasses, T-shirts,
and posters. The undisputed evidence of collisions of identical goods, such as the
T-shirts distributed by both parties, clearly demonstrates that some of the products are in identical categories of goods. Defendants argue that these items are
merely promotional, such as when LEXUS automobiles and LEXIS legal research

services each produce T-shirts bearing their marks for promotion of their products, and like LEXUS and LEXIS, Plaintiffs' and Defendants' marks should be allowed to co-exist on promotional items. The analogy, however, does not apply to the Hendrix-related items because while items like caps and shirts may be promotional for Defendants, they are the actual products for Plaintiffs. Lexus could not produce a promotional legal research service without fear of confusing customers, just as Lexis could not produce a promotional automobile without fear of confusing customers.

Additionally in this particular case, people purchasing both Plaintiffs' and Defendants' goods are primarily motivated to purchase the goods due to the goods' relationship with Jimi Hendrix. The goods share a defined market of Jimi Hendrix fans and are therefore in close proximity when fans search for Hendrix-related items. Defendants also argue that Plaintiffs have vowed to never produce alcohol or related products such as shot glasses or martini glasses. This is certainly Plaintiffs' stated position on alcohol-related products; however the test for relatedness of goods measures the consumer's likely association of product lines. Defendants have not produced any evidence to show that consumers are aware of Plaintiffs' position or that such awareness would prevent consumers from associating the goods.

For these reasons, the Court holds that Plaintiffs' and Defendants' goods are in close proximity and related, and this factor can only support infringement.

iii. Third Sleekcraft Factor: Similarity of the Marks

When considering the similarity of marks, the marks should be evaluated as they appear in the market place as a whole and the appearance, sound, and meaning of the marks should be analyzed individually.

For comparison purposes, Plaintiffs' incontestable marks and Defendants' marks fall into two categories: text marks containing the word "Hendrix," and design marks featuring images of Jimi Hendrix.

Plaintiffs' Incontestable Text Marks	Defendants' Text Marks
AUTHENTIC HENDRIX	HENDRIX ELECTRIC
EXPERIENCE HENDRIX	HENDRIX ELECTRIC VODKA
JIMI HENDRIX	JIMI HENDRIX ELECTRIC
	JIMI HENDRIX ELECTRIC VODKA

Defendants' longest mark is JIMI HENDRIX ELECTRIC VODKA, and the other marks are shortened versions of that mark. All of these marks share the word HENDRIX in common with Plaintiffs' marks, and two of them share the word JIMI as well. This similarity is hardly surprising given that both Plaintiffs and Defendants seek to call to mind Jimi Hendrix, and therefore share at least that much similarity. The primary distinguishing characteristics of Defendants' marks are the use of the word ELECTRIC and the placement of ELECTRIC after HENDRIX. It is notable, however, that Jimi Hendrix's band was called "The Jimi Hendrix Experience," and it was actually Plaintiffs who differentiated their mark from the band name by placing EXPERIENCE before Hendrix. Defendants' choice of placement of ELECTRIC is less distinguishing because it follows the same pattern as the original band name. The sound of these marks is similar in that they share the same root of HENDRIX or JIMI HENDRIX and the words ELECTRIC and EXPERIENCE have similar sounds. The appearance of these marks is similar, especially in that EXPERIENCE HENDRIX, and HENDRIX ELECTRIC have the word HENDRIX paired with an eight or nine letter word beginning with E and ending with "CE" or "IC." The appearance of these marks is therefore similar. The sounds, while having some specific differences, share an overall similarity.

The overall meaning of the marks is more difficult to determine; however, there is similarity between the words EXPERIENCE and ELECTRIC in that they both call to mind a distinct primary meaning and they also refer to the music of Jimi Hendrix. EXPERIENCE is a reference to the band name, The Jimi Hendrix Experience, and ELECTRIC is a reference to the particular musical talent that made Jimi famous, his electric guitar playing in a manner completely distinct from acoustic guitar playing. In addition, Electric is a reference to Jimi's song "Electric Ladyland." As a whole, both Plaintiffs' and Defendants' marks primarily convey the same impression of a musical association with Jimi Hendrix. For these reasons, the Court holds the text marks of Plaintiffs and Defendants are substantially similar and this factor favors the Plaintiffs.

The design marks also share a number of similarities.

Plaintiffs' incontestable design mark   *Defendants' design mark*

Two notable similarities are immediately apparent in the appearance of the design marks: the use of the word HENDRIX directly underneath the picture, and the rendering of the image of Jimi in a high-contrast silhouette-like picture that is mostly black with white details. Both images feature Jimi's distinctive afro, although Defendants' design is a larger, older version of Jimi's hair. There are other differences as well. Defendants chose to place their text completely below the image where Plaintiffs chose to surround the image. Defendants' slanted and stylized text is different than Plaintiffs' block font, although both use all capitals. Overall, however, the marks convey the same dominant impression due to the similarities of the images, the predominant placement of the word HENDRIX, and identical references to Jimi Hendrix. For these reasons the design marks have an overall general similarity and this factor weighs in favor of Plaintiffs.

iv. Fourth Sleekcraft Factor: Evidence of Actual Confusion

Evidence of actual confusion is persuasive evidence that future confusion is likely. Actual confusion evidence is often difficult to obtain and its absence is often given little weight as a result. Plaintiffs have provided emails indicating that at least some consumers were confused by Defendants' products. Defendants dispute whether the emailers were potential customers or people closely associated with Plaintiffs who were not confused, but were instead alerting Plaintiffs to Defendants' products. The Lee email, however, does show confusion between Plaintiffs and Defendants.[10] Defendants' argument that this email is from an associated contractor doesn't negate this evidence; in fact, the relatedness of the party to Plaintiffs only shows that even someone who is relatively familiar with Plaintiffs would be confused. This showing of actual confusion indicates a substantial likelihood of confusion and therefore infringement.

v. Fifth Sleekcraft Factor: Convergence of Marketing Channels

Similarity of marketing channels can increase the likelihood that consumers will be confused. *AMF Inc. v. Sleekcraft Boats*, 599 F.2d 341, 353 (9th Cir. 1979). Both parties in this case sell their goods directly to consumers via web sites, authentichendrix.com for Plaintiffs and houseofhendrix.com for Defendants. Advertising and selling goods on the internet does not necessarily indicate similar marketing channels when almost everything is available online. Here however, the similarity of the web sites, and the chances that consumers performing Jimi Hendrix related searches are likely to see both web sites, strongly indicates that the web-based marketing of both companies are related channels. Both parties also advertise in locations where Jimi Hendrix fans are likely to be exposed to the advertising because these fans are the target market for both parties. Because the marketing channels are related, this factor favors Plaintiffs.

vi. Sixth Sleekcraft Factor: Degree of Customer Care

The degree of care customers use in a given market is a factor in determining whether the consumer will be confused by similar marks. Plaintiffs and Defendants agree that consumers of T-shirts and ball caps are likely to exercise relatively little care, but Defendants assert that purchasers of Defendants' premium vodka are likely to exercise a higher degree of care. Purchasers of vodka products purchase them in a number of locales, including shots from the bar and bottles from the store. While full-bottle purchasers are likely to exercise more care, purchases at the bar are more likely to exercise relatively little care. However, even the more cautious shopper in this case is unlikely to use the type of care found to be relevant in other trademark cases in reducing the likelihood of confusion. *Compare Sleekcraft*, 599 F.2d at 353-54 (purchasers of expensive boats likely to exercise more care), *with Fleischmann Distilling Corp.*, 314 F.2d at 161 (purchasers of alcoholic beverages would not notice small difference in labels at point of purchase). Under the facts of this case, the degree of care customers would use does not alleviate any likelihood of confusion that would arise from the similarity of the marks, and as a result, this factor cannot help Defendants.

---

[10] "Bob, . . . You got great PR in today's Wall Street Journal with the article on the front page about the explosion of vodka sales in the U.S. and the introduction of Hendrix Electric vodka." Wilson Decl., ex. M (email addressed to Plaintiffs from business associate).

vii. Seventh Sleekcraft Factor: Intent of Defendant

A defendant who knowingly crafts his mark to emulate a plaintiff's mark is presumed to have accomplished his purpose and this intent weighs heavily in favor of a plaintiff. *Sleekcraft*, 599 F.2d at 354. Defendants concede they intended to create a mark that associated their goods with the persona of Jimi Hendrix. Defendants may have believed that the *Foundation* Order announced an open season on Jimi's identity, but this mistaken belief does not provide them safe harbor. Defendants admit that they were aware of Plaintiffs' marks and registrations. Defendants also compared their design mark to Plaintiffs' while Defendants were still in the design stage and considered crafting portions of their design to look like Plaintiffs'.[11] Defendants argue that they compared the marks in an attempt to avoid confusion, however they cite no evidence that such was the case, and at summary judgment Defendants are obligated to produce more than argument to create a genuine issue of material fact. The intent of Defendants to emulate Plaintiffs' marks and trade off of the fame of Jimi Hendrix, a name Plaintiffs have invested with further good will after Jimi's death, strongly favors the Plaintiffs.

viii. Eighth Sleekcraft Factor: Likelihood of Expansion

Where there is a strong possibility of expansion into conflicting markets, this factor favors infringement. Here, Defendants admit they contemplated expansion into Plaintiffs' market, licensed merchandise such as posters and T-shirts, but abandoned the expansion. Plaintiffs admitted they have refused to enter the alcoholic beverage market or any related market. This factor, however, is not particularly relevant to the analysis at hand because the goods of both parties are already in directly conflicting areas in some cases and somewhat related and compl[e]mentary areas in others, so no further expansion is required in order for the overlapping trademarks to cause customer confusion.

C. Infringement Conclusion—Incontestable Marks

. . . Although summary judgment in trademark cases is disfavored, summary judgment is appropriate when the undisputed facts lead to only one conclusion.

Consumers are accustomed to celebrity endorsements, including endorsements by the family of deceased celebrities. This would be a different case were Defendants and Plaintiffs starting out at the same time and each trying to build a brand around Jimi Hendrix. In contrast, here Plaintiffs are the senior user, have incontestable marks and are a longstanding source of Jimi Hendrix related items.

Defendants' opposition . . . rests on the assumption that Plaintiffs' lack of ownership of any publicity rights in Jimi Hendrix limited the scope of protection they secured through their trademarks. As a matter of law, no such limitation exists and Plaintiffs have created valid trademarks that do not violate the prohibitions against misrepresentation or use of the trademarks. Plaintiffs have legally acquired the right to use Jimi Hendrix's fame to sell their products, and they are entitled to all

---

[11] See First Wilson Decl., ex. E, (email to Craig Dieffenbach containing Defendants signature mark and Plaintiffs' signature mark; email from Craig Dieffenbach suggesting they make their signature look more like Plaintiffs').

the rights of a trademark holder. . . . Defendants have produced no evidence that any of the *Sleekcraft* factors would indicate anything other than a likelihood of confusion between Plaintiffs' and Defendants' marks. The most probative *Sleekcraft* factor applicable to this particular case is the intent of Defendants. Defendants intended to trade off the good will associated with Jimi Hendrix and intended to have a mark similar to Plaintiffs.

The Court holds that Defendants' marks create a substantial likelihood of confusion with Plaintiffs' incontestable marks. Having already held that the marks are valid and owned by Plaintiffs, the Court holds that Defendants have infringed Plaintiffs' trademark rights with respect to Plaintiffs' incontestable trademarks.[13]
. . .

. . . .

Notes & Questions

1. Compare the eight factors the Sixth Circuit identified in the *Kellogg* case with the eight factors the *Hendrix* court identifies as the appropriate Ninth Circuit test. How do they differ? Which factors were most probative in the *Hendrix* case? Which were most probative in *Kellogg*? These factor-based tests for determining whether a likelihood of confusion exists are used in all of the different circuits, with each circuit court's formulation of the factors stated slightly differently. These same factor-based tests are also typically used in cases where the plaintiff asserts trade dress infringement.

2. Is it appropriate to use the defendant's intent or "bad faith" as a factor in assessing likelihood of confusion? The Fourth Circuit applies "a presumption of likelihood of consumer confusion" when there is "intentional copying" of a plaintiff's trade dress or trademark by a defendant and the copier intends "to exploit the good will created by an already registered trademark." *Shakespeare Co. v. Silstar Corp. of Am., Inc.*, 110 F.3d 234, 239 (4th Cir. 1997). The court justifies this presumption because "one who tries to deceive the public should hardly be allowed to prove that the public has not in fact been deceived." *Id.* Importantly, this presumption does not apply if there is no evidence that the defendant attempted to pass off its own goods as those of the plaintiffs. *See Rosetta Stone v. Google, Inc.*, 676 F.3d 144, 155-56 and n.5 (4th Cir. 2012).

3. In an examination of 331 district court opinions decided between 2000 and 2004, one scholar concludes that there is a significant inter-circuit and inter-district variation in plaintiff win rates. *See* Barton Beebe, *An Empirical Study of the Multifactor Tests for Trademark Infringement*, 94 Cal. L. Rev. 1581 (2006). His survey identifies different tests that range from five to 13 factors. *Id.* at 1591 (presenting all factors in tabular form). Professor Beebe argues that his data "shows that judges employ *fast and frugal* heuristics to short-circuit the multifactor test. Perhaps as an expression of their cognitive limitations, but more likely as an expression of their cognitive ingenuity, judges rely upon a few factors or combinations of

---

[13] Because Plaintiffs are entitled to summary judgment on the incontestable marks, the Court does not address Plaintiffs' motion with respect to the remaining marks at this time.

factors to make their decisions. The rest of the factors are at best redundant and at worst irrelevant." *Id.* at 1586. He asserts that "[i]n theory, the multifactor test is a full-fledged balancing test. In practice, it is a complex of per se rules." *Id.* at 1614. Are there any per se rules, or at least "benchmarks," that the *Kellogg* court identifies? Are there any per se rules in the *Hendrix* decision?

Recall the *Top Tobacco* opinion from Section I. What reasons did Judge Easterbrook give for not analyzing factors to determine whether there was a "likelihood of confusion"? Now that you have read two cases that fully utilize factors (some circuits call them "digits of confusion"), what are the pros and cons of such an approach?

4. The *Hendrix* case explores the relationship between trademark rights and the right of publicity. As the court makes clear, those two different rights have different standards for obtaining and maintaining rights, as well as different durations. A supplemental chapter exploring the right of publicity is available on the Semaphore Press download page for this book. If Jimi Hendrix were still alive, the legal arguments presented in this case would have been very different. The parties subsequently settled the dispute. Pursuant to a stipulated Supplemental Judgment and Permanent Injunction, defendants were required to pay plaintiff $3.2 million and withdraw all of their vodka products from the market. *See* Sara Jean Green, *Judge orders Hendrix Vodka pulled from shelves, $3.2 million settlement,* The Seattle Times, Feb. 18, 2009.

5. In *Hendrix* the plaintiffs owned several marks that were incontestable. How did the inconstestable status of the marks influence the outcome?

While defendants' arguments concerning descriptiveness of plaintiffs' incontestable marks were rejected automatically, in a portion of the opinion not reproduced above, the court addressed defendant's other challenges to the validity of plaintiffs' marks: fraud, abandonment, and unclean hands. The abandonment argument was based on plaintiffs' licensing practices. A mark is deemed abandoned when it no longer serves the purpose of indicating the source of the goods. 15 U.S.C. § 1127. When a mark owner permits others to use the mark without exercising any quality control over the goods or services with which the mark is used, the mark owner is said to be engaged in "naked licensing." Because there are no rights in a trademark alone (sometimes phrased as "in gross"), given that the rights are to the goodwill that the trademark is associated with or represents, naked licensing or naked transfers are considered a form of abandonment. The type of quality control required to prevent abandonment varies with the circumstances, but having product samples sent by licensees to the trademark owner at regular intervals is an important example of quality control. *See Barcamerica International USA Trust v. Tyfield Importers, Inc.*, 289 F.3d 589, 595-598 (9th Cir. 2002).

6. What type of confusion was the court looking for in *Hendrix*? How would you characterize the confusion shown in the Lee email?

In addition to point of sale confusion, another type of confusion exists: reverse confusion. This type of confusion occurs when the registered mark owner, or senior user, is less well known than the junior user. In that case the risk is that the consuming public will be confused into believing that the source of the mark owner's goods or services actually is the defendant. Courts generally accept reverse confusion as a cognizable harm in trademark law. *See, e.g., Playmakers LLC v. ESPN, Inc.*, 376 F.3d 894, 897 (9th Cir. 2004) (concerning use by ESPN's television series of "Play-Makers" a registered mark owned by plaintiff for sports agent services).

B. Claims Under § 43(a)

Owners of registered trademarks can sue for infringement under § 32 of the Lanham Act, 15 U.S.C. § 1114. Registered-mark owners may also sue for infringement under § 43(a), as may unregistered-mark owners. As you learned in the first part of this Section, the fundamental test for infringement is whether there is a likelihood of confusion. Review the relevant provisions of §§ 32 and 43(a) set out at the beginning of this Section. Specifically what type of confusion must be shown to prove infringement? Section 32 does not specify the nature of the confusion that must be present to constitute infringement, while § 43(a) identifies two types of confusion: confusion as to "affiliation, connection, or association of such person with another person"; and confusion as to "the origin, sponsorship, or approval of his or her goods, services, or commercial activities by another person." The next case focuses on a different type of confusion that § 43(a) makes actionable.

King of the Mountain Sports, Inc. v. Chrysler Corp.
185 F.3d 1084 (10th Cir. 1999)

Tacha, Judge:

   . . . .

I. Background

Plaintiff [King of the Mountain Sports, Inc. ("KOM")] sells camouflage-patterned, natural fiber, outdoor clothing and related mountaineering accessories. It obtained a federal registration for its first stylized trademark on June 4, 1991, and for a second stylized mark on September 28, 1993. "[Plaintiff's] first mark consisted of the words 'King of the Mountain Sports Inc.' in Gothic lettering super-

imposed on the outline of a mountain and enclosed by a thin, rectangular border." *King of the Mountain Sports, Inc. v. Chrysler Corp.*, 968 F. Supp. 568, 570 (D. Colo. 1997).

The second mark consists of the words "King of the Mountain" in Gothic lettering with an outline of a mountain in the background. As used, plaintiff's marks employ dark lettering against drab background colors such as the blacks, browns,

and greens found in camouflage. The Gothic lettering is horizontally oriented in one line and only the "K" of King and "M" of Mountain are capitalized.

Plaintiff primarily markets its products to hunters, fishers, campers, and hikers but notes that its customers use its products for all cold weather outdoor activities. It also contends that the downhill skiing and snowboarding apparel markets are a logical expansion area for its product line.

In 1995, defendant Eclipse Television and Sports Marketing LLC ("Eclipse") purchased from defendant Eclipse Television and Sports Marketing, Inc. ("Eclipse California") the right to contract with defendant Chrysler to use the "Jeep KING OF THE MOUNTAIN DOWNHILL SERIES" logo. "Defendants' primary logo consists of the word 'Jeep' in largest, purple type above the words 'KING OF THE MOUNTAIN' in smaller, blue type, and the words 'DOWN-HILL SERIES' in even smaller, red type at the bottom of the logo." *King of the Mountain Sports*, 968 F. Supp. at 570. Other than the word "Jeep," all words consist entirely of capital letters and "are superimposed over a blue outline of a mountain with a picture of a red ski racer in a tucked position and a series of red and orange straight lines stretched out behind him to suggest the speed with which he is racing." *Id.*

Defendants' logo, however, does not always conform to this description. For example, on banners and scoreboards, defendants have employed a bright blue background with the word "Jeep" on either side of "KING OF THE MOUNTAIN" in stark white, bold, capital letters. Also, television listings have simply referred to the event as "Skiing: King of the Mountain Downhill Series."

Defendants use the logo to identify and promote a series of televised downhill ski races. To this end, they have placed the logo, for example, on billboards, banners, clothing apparel, and in magazines. Defendants have given away several items of clothing featuring the logo to participants, television commentators, and spectators at the skiing events. Defendant Bogner manufactures the ski jackets upon which defendants' logo appears.

Plaintiff objects to defendants' use of the phrase "king of the mountain" in the name of its ski event. It argues that defendants' use of the phrase creates a likelihood of sponsorship confusion. Plaintiff fears that its consumers will believe that KOM sponsors, or is otherwise associated with, the downhill skiing event.

Defendants filed a summary judgment motion in district court arguing, as to the trademark infringement claims, that no likelihood of confusions exists. After thoroughly analyzing the appropriate factors, the district court agreed with defendants and entered summary judgment against KOM.

. . . .

II. Discussion

Likelihood of confusion forms the gravamen for a trademark infringement action. Section 43(a) of the Lanham Act declares that any person who uses in commerce any word, term, name, symbol, or device, . . . which . . . is likely to cause confusion, or to cause mistake, or to deceive as to the affiliation, connection, or association of such person with another person, or as to the origin, sponsorship, or approval of his or her goods, services, or commercial activities by another person . . . shall be liable in a civil action by any person who believes that he or she is or is likely to be damaged by such act.

Id. 1125(a). . . .

. . . [I]n order to prevail on appeal, plaintiff must show that a genuine issue of material fact exists regarding whether defendants' use of its logo would likely cause confusion about the sponsorship of the downhill ski race. The Tenth Circuit has identified six factors, derived from the *Restatement of Torts* 729 (1938), that aid in determining whether a likelihood of confusion exists between two marks:

(a) the degree of similarity between the marks;
(b) the intent of the alleged infringer in adopting its mark;
(c) evidence of actual confusion;
(d) the relation in use and the manner of marketing between the goods or services marketed by the competing parties;
(e) the degree of care likely to be exercised by purchasers; and
(f) the strength or weakness of the marks.

First Savings Bank, 101 F.3d at 652. "This list is not exhaustive. All of the factors are interrelated, and no one factor is dispositive." *Universal Money Ctrs.*, 22 F.3d at 1530. While we consider these factors to determine whether a likelihood of confusion exists regardless of whether the trademark infringement suit involves source or sponsorship confusion, the weight afforded to some of the factors differs when applied in these separate contexts. In both confusion of source and confusion of sponsorship cases, the similarity of the marks factor constitutes the heart of our analysis. *See Heartsprings v. Heartspring,* 143 F.3d 550, 554 (10th Cir. 1998) (recognizing that the key inquiry in a source confusion case is "whether the consumer is likely to be deceived or confused by the similarity of the marks." (quoting *Two Pesos, Inc. v. Taco Cabana, Inc.*, 505 U.S. 763, 780 (1992))); *Wendt v. Host Int'l, Inc.*, 125 F.3d 806, 812 (9th Cir. 1998) (emphasizing the primacy of the similarity of mark factor in a confusion of endorsement analysis). However, in the rare, pure sponsorship action, other factors—such as the relation in use and the manner of marketing between the goods or services and the degree of care likely to be exercised by purchasers—have little importance. Bearing this in mind, if, as in any case, the examination of the various factors establishes a genuine issue of material fact regarding the likelihood of sponsorship confusion, summary judgment is not appropriate.

A. Similarity Between the Marks

We test the degree of similarity between marks on three levels: sight, sound, and meaning. We do not consider these factors in isolation. Instead, we must examine them "in the context of the marks as a whole as they are encountered by consumers in the marketplace." *Beer Nuts, [Inc. v. Clover Club Foods Co.*, 805 F.2d

920, 925 (10th Cir. 1986)]. Furthermore, we do not engage in a "'side-by-side' comparison." *Universal Money Ctrs.*, 22 F.3d at 1531. Rather, "the court must determine whether the alleged infringing mark will be confusing to the public when singly presented." *Id.* We give the similarities of the marks more weight than the differences. See *id.*

In this case, plaintiff argues that because defendants employ the phrase "King of the Mountain," the similarity factor weighs in favor of it and creates a genuine issue of material fact regarding likelihood of confusion. We disagree. Comparing the stylized trademark of plaintiff with the ways in which defendants have used the phrase "king of the mountain," we hold that no reasonable jury could find similarity. Even assuming that the phrase "king of the mountain" constitutes the dominant portion of defendants' logo, as plaintiff argues, the marks as a whole are not confusingly similar. Although defendants' logo employs, in part, the same phrase as plaintiff's mark and therefore might sound somewhat similar, the sight and sense of meaning invoked by defendants' logo and plaintiff's stylized mark differ drastically. First, plaintiff's stylized mark is in Gothic lettering, with only the first letter of "King" and "Mountain" capitalized, against a solitary mountain ridge outline. It appears in camouflage colors, consistent with plaintiff's marketing toward hunters. In short, the mark is quite understated. The opposite is true regarding the ways in which defendants have used the phrase "king of the mountain." Defendant's primary logo is quite colorful, done in brilliant blue, purple, red, and orange. The letters consist of bold capitals, and the words are vertically oriented. The mountain outline in the background has a different shape than that in plaintiff's mark, and the logo depicts a skier with red and orange lines shooting out from behind him to suggest the great speed at which he flies down the mountainside. Thus, the visual impact of the plaintiff's mark and defendants' logo differs dramatically. Moreover, as the district court noted,

> the marks of plaintiff and defendants do not convey the same meaning or stimulate the same mental reaction. Plaintiff's marks are simple, reserved, and dignified. Indeed, one of plaintiff's complaints is that it does not want to be associated with the 'glitzy' ski races promoted by defendants. Defendants' mark is bright and attention-grabbing, connoting the fun and speed associated with ski racing.

King of the Mountain Sports, Inc. v. Chrysler Corp., 968 F. Supp. 568, 573 (D. Colo. 1997). Thus, the first and most important factor, similarity of the marks, weighs heavily in defendants' favor.

B. Intent of the Defendants

The proper focus under this factor "is whether defendant had the intent to derive benefit from the reputation or goodwill of plaintiff." *Jordache Enters., Inc. v. Hogg Wyld, Ltd.*, 828 F.2d 1482, 1485 (10th Cir. 1987). Plaintiff argues that although it has no direct evidence of bad intent, a jury could infer defendants' intent to derive the benefit and goodwill of KOM's mark because they failed to conduct a full trademark search before using the phrase "king of the mountain." We disagree. As the district court noted, "Plaintiff has presented no evidence to suggest that the defendants were even aware of plaintiff's existence, let alone that they intentionally attempted to trade on plaintiff's reputation or goodwill." *King of the Mountain Sports*, 968 F. Supp. at 574. Instead, uncontested evidence indicates:

> (1) defendants did not know of plaintiff or its trademarks when they designed their logo; (2) defendants knew that no other competitor in the ski-race industry used the term "King of the Mountain;" and (3) defendants incorporated the phrase "King of the Mountain" to describe the goal of the ski racers competing in the event to be the "king of the mountain" in downhill ski racing.

Id. Because the undisputed "evidence indicates [that the] defendant[s] did not intend to derive benefits from . . . plaintiff's existing mark, this factor weighs against the likelihood of confusion." *Heartsprings*, 143 F.3d at 556.

C. Similarity in Products/Services and Manner of Marketing

Typically, "[t]he greater the similarity between the products and services, the greater the likelihood of confusion." *Universal Money Ctrs.*, 22 F.3d at 1532. This is undoubtedly true when the action pertains to source or affiliation confusion. For example, *Beer Nuts, Inc. v. Clover Club Foods Co.*, 805 F.2d 920, 926 (10th Cir. 1986), involved two competing companies marketing very similar goods—sweetened salted peanuts—in the same manner. In that case, we found that this factor added strength to the position that a reasonable consumer would likely be confused as to the source of the peanuts. *See id.* However, in a case involving pure sponsorship confusion, the parties may have little similarity in their products or manner of marketing. Here, KOM's clothing is only marginally related to defendant's ski event. This disconnection greatly reduces the relevance of the similarity of products factor. We therefore find it provides no support of plaintiff's claim.

D. Degree of Care Exercised by Purchasers

"A consumer exercising a high degree of care in selecting a product reduces the likelihood of confusing similar trade names." *Heartsprings*, 143 F.3d at 557. Plaintiff argues, however, that this factor rarely reduces the risk of sponsorship confusion. We agree. The harm plaintiff seeks to remedy in this case is loss of reputation and goodwill stemming from its perceived sponsorship of the ski event. The care with which consumers select a product does not impact the association they may make regarding the sponsorship of an event. Therefore, even if plaintiff's current and potential customers exercise a high degree of care, it would have little impact on our determination regarding likelihood of confusion in this case.

E. Actual Confusion

Although plaintiff need not set forth evidence of actual confusion to prevail in a trademark infringement action, "[a]ctual confusion in the marketplace is often considered the best evidence of likelihood of confusion." *Universal Money Ctrs.*, 22 F.3d at 1534. However, "isolated instances of actual confusion [may] be de minimis." [*Id.*] at 1535; see also McCarthy, supra, 23:14 ("Evidence of actual confusion of a very limited scope may be dismissed as de minimis: Probable confusion cannot be shown by pointing out that at someplace, at some time, someone made a false identification." (internal quotation marks omitted)). "De minimis evidence of actual confusion does not establish the existence of a genuine issue of material fact regarding the likelihood of confusion." *Universal Money Ctrs.*, 22 F.3d at 1535. After carefully reviewing the record, we have found that plaintiff has put into evidence, at most, only seven examples of actual confusion. This handful of anecdotal evidence is de minimis and does not support a finding of a genuine issue of material fact as to the likelihood of confusion, especially in light of the complete lack of similarity between the defendants' uses and plaintiff's mark. *See id.* at 1535-36 ("The de minimis evidence of actual confusion is especially undermined in this case by the sheer lack of similarity between the marks.").

F. Strength of Plaintiff's Mark

The stronger a trademark, the more likely that encroachment upon it will lead to sponsorship confusion. *See First Sav. Bank*, 101 F.3d at 653. "A strong trademark is one that is rarely used by parties other than the owner of the trademark, while a weak trademark is one that is often used by other parties." *Id.* To assess the relative strength of a mark, one must consider the two aspects of strength: (1) "Conceptual Strength: the placement of the mark on the [distinctiveness or fanciful-suggestive-descriptive] spectrum;" and (2) "Commercial Strength: the marketplace recognition value of the mark." McCarthy, *supra*, 11:83.

. . . Viewed in the light most favorable to plaintiff, its stylized mark is at least suggestive on the conceptual strength spectrum. Moreover, again drawing all inferences in favor of plaintiff, we assume that the mark has great commercial strength in the hunting apparel market. Therefore, plaintiff's mark is quite strong.

The strength of plaintiff's mark cannot outweigh the other factors, however. Given the great dissimilarity between its mark as a whole and the ways in which defendants have used the phrase, "king of the mountain," as well as our above analysis of the other factors, we find that no reasonable juror could find likelihood of confusion between plaintiff's and defendants' marks. Accordingly, we hold that there exists no genuine issue of material fact as to the likelihood of confusion.

. . . .

Notes & Questions

1. In the *King of the Mountain* case the court indicates that the likelihood of confusion factors are weighed differently in a case asserting confusion as to endorsement or sponsorship. Why might that be appropriate? Should there be a distinct set of factors for a court to consider depending on the nature of the confusion that the plaintiff alleges? Do consumers think of these types of confusion categories in distinct ways, or is it more of a continuum?

2. Each of the cases that you have read in this Section involved a combination of word marks and design marks. Determining whether there is a likelihood of confusion is tough enough with basic word marks, let alone with marks that combine images and words. Even words that sound very different can have confusingly similar meanings. *See e.g. Hancock v. American Steel & Wire Co.*, 203 F.2d 737 (CCPA 1953) (involving marks "Cyclone" and "Tornado" for fences), *Jellibeans,*

Inc. v. Skating Clubs of Georgia, Inc., 716 F.2d 833 (11th Cir. 1983) (involving the marks "jellibeans" and "lollipops" for skating rinks). Overall consumer impression is important, but so is the communicative impact of a mark in the spoken language. When a mark combines images and words, the spoken language conveys only part of the mark's meaning.

C. Confusion as to "Origin"

Dastar Corp. v. Twentieth Century Fox Film Corp.
539 U.S. 23 (2003)

Scalia, Justice:

. . . .

In 1948, three and a half years after the German surrender at Reims, General Dwight D. Eisenhower completed *Crusade in Europe*, his written account of the allied campaign in Europe during World War II. Doubleday published the book, registered it with the Copyright Office in 1948, and granted exclusive television rights to an affiliate of respondent Twentieth Century Fox Film Corporation (Fox). Fox, in turn, arranged for Time, Inc., to produce a television series, also called *Crusade in Europe*, based on the book, and Time assigned its copyright in the series to Fox. The television series, consisting of 26 episodes, was first broadcast in 1949. It combined a soundtrack based on a narration of the book with film footage from the United States Army, Navy, and Coast Guard, the British Ministry of Information and War Office, the National Film Board of Canada, and unidentified "Newsreel Pool Cameramen." In 1975, Doubleday renewed the copyright on the book as the "'proprietor of copyright in a work made for hire.'" Fox, however, did not renew the copyright on the *Crusade* television series, which expired in 1977, leaving the television series in the public domain.

. . . .

Enter petitioner Dastar. In 1995, Dastar decided to expand its product line from music compact discs to videos. Anticipating renewed interest in World War II on the 50th anniversary of the war's end, Dastar released a video set entitled *World War II Campaigns in Europe*. To make *Campaigns*, Dastar purchased eight beta cam tapes of the *original* version of the Crusade television series, which is in the public domain, copied them, and then edited the series. Dastar's *Campaigns* series is slightly more than half as long as the original *Crusade* television series. Dastar substituted a new opening sequence, credit page, and final closing for those of the *Crusade* television series; inserted new chapter-title sequences and narrated chapter introductions; moved the "recap" in the *Crusade* television series to the beginning and retitled it as a "preview"; and removed references to and images of the book. Dastar created new packaging for its *Campaigns* series and (as already noted) a new title.

Dastar manufactured and sold the *Campaigns* video set as its own product. The advertising states: "Produced and Distributed by: *Entertainment Distributing*" (which is owned by Dastar), and makes no reference to the *Crusade* television se-

ries. Similarly, the screen credits state "DASTAR CORP presents" and "an EN-
TERTAINMENT DISTRIBUTING Production," and list as executive producer,
producer, and associate producer, employees of Dastar. The *Campaigns* videos
themselves also make no reference to the *Crusade* television series, New Line's
Crusade videotapes, or the book. . . .

In 1998, respondents Fox, SFM, and New Line brought this action alleging that
Dastar's sale of its *Campaigns* video set infringes Doubleday's copyright in General Eisenhower's book and, thus, their exclusive television rights in the book. Respondents later amended their complaint to add claims that Dastar's sale of *Campaigns* "without proper credit" to the *Crusade* television series constitutes "reverse passing off"[1] in violation of § 43(a) of the Lanham Act, 15 U.S.C. § 1125(a),
and in violation of state unfair-competition law. On cross-motions for summary
judgment, the District Court found for respondents on all three counts . . . .

The Court of Appeals for the Ninth Circuit affirmed the judgment for respondents on the Lanham Act claim, but reversed as to the copyright claim and remanded. 34 Fed. Appx. 312, 316 (2002). . . . With respect to the Lanham Act claim,
the Court of Appeals . . . concluded that "Dastar's 'bodily appropriation' of Fox's
original [television] series is sufficient to establish the reverse passing off." *Ibid.* . . .
.

II

The Lanham Act was intended to make "actionable the deceptive and misleading use of marks," and "to protect persons engaged in . . . commerce against unfair
competition." 15 U.S.C. § 1127. While much of the Lanham Act addresses the registration, use, and infringement of trademarks and related marks, § 43(a) . . . is one
of the few provisions that goes beyond trademark protection. . . . As the Second
Circuit accurately observed with regard to the original enactment, however—and
as remains true after the 1988 revision—§ 43(a) "does not have boundless application as a remedy for unfair trade practices," *Alfred Dunhill, Ltd.* v. *Interstate Cigar Co.*, 499 F.2d 232, 237 (1974). "[B]ecause of its inherently limited wording,
§ 43(a) can never be a federal 'codification' of the overall law of 'unfair competition,'" 4 J. McCarthy, *Trademarks and Unfair Competition* §27:7, p. 27-14 (4th ed.
2002) (McCarthy), but can apply only to certain unfair trade practices prohibited
by its text.

. . . .

. . . [T]he gravamen of respondents' claim is that, in marketing and selling
Campaigns as its own product without acknowledging its nearly wholesale reliance
on the *Crusade* television series, Dastar has made a "false designation of origin,
false or misleading description of fact, or false or misleading representation of fact,
which . . . is likely to cause confusion . . . as to the origin . . . of his or her goods."
See, *e.g.*, Brief for Respondents 8, 11. That claim would undoubtedly be sustained

---

[1] Passing off (or palming off, as it is sometimes called) occurs when a producer misrepresents his
own goods or services as someone else's. "Reverse passing off," as its name implies, is the opposite:
The producer misrepresents someone else's goods or services as his own.

if Dastar had bought some of New Line's *Crusade* videotapes and merely repackaged them as its own. Dastar's alleged wrongdoing, however, is vastly different: it took a creative work in the public domain—the *Crusade* television series—copied it, made modifications (arguably minor), and produced its very own series of videotapes. If "origin" refers only to the manufacturer or producer of the physical "goods" that are made available to the public (in this case the videotapes), Dastar was the origin. If, however, "origin" includes the creator of the underlying work that Dastar copied, then someone else (perhaps Fox) was the origin of Dastar's product. At bottom, we must decide what § 43(a)(1)(A) of the Lanham Act means by the "origin" of "goods."

III

The dictionary definition of "origin" is "[t]he fact or process of coming into being from a source," and "[t]hat from which anything primarily proceeds; source." *Webster's New International Dictionary* 1720-1721 (2d ed. 1949). And the dictionary definition of "goods" (as relevant here) is "[w]ares; merchandise." *Id.* at 1079. We think the most natural understanding of the "origin" of "goods"— the source of wares—is the producer of the tangible product sold in the marketplace, in this case the physical *Campaigns* videotape sold by Dastar. The concept might be stretched (as it was under the original version of § 43(a)) to include not only the actual producer, but also the trademark owner who commissioned or assumed responsibility for ("stood behind") production of the physical product. But as used in the Lanham Act, the phrase "origin of goods" is in our view incapable of connoting the person or entity that originated the ideas or communications that "goods" embody or contain. Such an extension would not only stretch the text, but it would be out of accord with the history and purpose of the Lanham Act and inconsistent with precedent.

Section 43(a) of the Lanham Act prohibits actions like trademark infringement that deceive consumers and impair a producer's goodwill. . . . The consumer who buys a branded product does not automatically assume that the brand-name company is the same entity that came up with the idea for the product, or designed the product—and typically does not care whether it is. The words of the Lanham Act should not be stretched to cover matters that are typically of no consequence to purchasers.

It could be argued, perhaps, that the reality of purchaser concern is different for what might be called a communicative product—one that is valued not primarily for its physical qualities, such as a hammer, but for the intellectual content that it conveys, such as a book or, as here, a video. The purchaser of a novel is interested not merely, if at all, in the identity of the producer of the physical tome (the publisher), but also, and indeed primarily, in the identity of the creator of the story it conveys (the author). And the author, of course, has at least as much interest in avoiding passing-off (or reverse passing-off) of his creation as does the publisher. For such a communicative product (the argument goes) "origin of goods" in § 43(a) must be deemed to include not merely the producer of the physical item (the publishing house Farrar, Straus and Giroux, or the video producer

Dastar) but also the creator of the content that the physical item conveys (the author Tom Wolfe, or—assertedly—respondents).

The problem with this argument according special treatment to communicative products is that it causes the Lanham Act to conflict with the law of copyright, which addresses that subject specifically. The right to copy, and to copy without attribution, once a copyright has expired, like "the right to make [an article whose patent has expired]—including the right to make it in precisely the shape it carried when patented—passes to the public." *Sears, Roebuck & Co.* v. *Stiffel Co.*, 376 U.S. 225, 230 (1964). "In general, unless an intellectual property right such as a patent or copyright protects an item, it will be subject to copying." *TrafFix Devices, Inc.* v. *Marketing Displays, Inc.*, 532 U.S. 23, 29 (2001). The rights of a patentee or copyright holder are part of a "carefully crafted bargain," *Bonito Boats, Inc.* v. *Thunder Craft Boats, Inc.*, 489 U.S. 141, 150-151 (1989), under which, once the patent or copyright monopoly has expired, the public may use the invention or work at will and without attribution. Thus, in construing the Lanham Act, we have been "careful to caution against misuse or over-extension" of trademark and related protections into areas traditionally occupied by patent or copyright. *TrafFix*, 532 U.S. at 29. "The Lanham Act," we have said, "does not exist to reward manufacturers for their innovation in creating a particular device; that is the purpose of the patent law and its period of exclusivity." *Id.* at 34. Federal trademark law "has no necessary relation to invention or discovery," *Trade-Mark Cases,* 100 U.S. 82, 94 (1879), but rather, by preventing competitors from copying "a source-identifying mark," "reduce[s] the customer's costs of shopping and making purchasing decisions," and "helps assure a producer that it (and not an imitating competitor) will reap the financial, reputation-related rewards associated with a desirable product," *Qualitex Co.* v. *Jacobson Products Co.*, 514 U.S. 159, 163-164 (1995) (internal quotation marks and citation omitted). Assuming for the sake of argument that Dastar's representation of itself as the "Producer" of its videos amounted to a representation that it originated the creative work conveyed by the videos, allowing a cause of action under §43(a) for that representation would create a species of mutant copyright law that limits the public's "federal right to 'copy and to use,' " expired copyrights, *Bonito Boats*, *supra*, at 165.

When Congress has wished to create such an addition to the law of copyright, it has done so with much more specificity than the Lanham Act's ambiguous use of "origin." The Visual Artists Rights Act of 1990, § 603(a), 104 Stat. 5128, provides that the author of an artistic work "shall have the right … to claim authorship of that work." 17 U.S.C. § 106A(a)(1)(A). That express right of attribution is carefully limited and focused: It attaches only to specified "work[s] of visual art," § 101, is personal to the artist, §§ 106A(b) and (e), and endures only for "the life of the author," at § 106A(d)(1). Recognizing in § 43(a) a cause of action for misrepresentation of authorship of noncopyrighted works (visual or otherwise) would render these limitations superfluous. A statutory interpretation that renders another statute superfluous is of course to be avoided.

Reading "origin" in § 43(a) to require attribution of uncopyrighted materials would pose serious practical problems. Without a copyrighted work as the base-point, the word "origin" has no discernable limits. . . . In many cases, figuring out who is in the line of "origin" would be no simple task. Indeed, in the present case it is far from clear that respondents have that status. Neither SFM nor New Line had anything to do with the production of the Crusade television series—they merely were licensed to distribute the video version. While Fox might have a claim to being in the line of origin, its involvement with the creation of the television series was limited at best. Time, Inc., was the principal if not the exclusive creator, albeit under arrangement with Fox. And of course it was neither Fox nor Time, Inc., that shot the film used in the Crusade television series. Rather, that footage came from the United States Army, Navy, and Coast Guard, the British Ministry of Information and War Office, the National Film Board of Canada, and unidentified "Newsreel Pool Cameramen." If anyone has a claim to being the *original* creator of the material used in both the Crusade television series and the Campaigns videotapes, it would be those groups, rather than Fox. We do not think the Lanham Act requires this search for the source of the Nile and all its tributaries.

Another practical difficulty of adopting a special definition of "origin" for communicative products is that it places the manufacturers of those products in a difficult position. On the one hand, they would face Lanham Act liability for *failing* to credit the creator of a work on which their lawful copies are based; and on the other hand they could face Lanham Act liability for *crediting* the creator if that should be regarded as implying the creator's "sponsorship or approval" of the copy, 15 U.S.C. § 1125(a)(1)(A). . . .

Finally, reading § 43(a) of the Lanham Act as creating a cause of action for, in effect, plagiarism—the use of otherwise unprotected works and inventions without attribution—would be hard to reconcile with our previous decisions. For example, in *Wal-Mart Stores, Inc.* v. *Samara Brothers, Inc.*, 529 U.S. 205 (2000), we considered whether product-design trade dress can ever be inherently distinctive. . . . We concluded that the designs could not be protected under § 43(a) without a showing that they had acquired "secondary meaning," *id*. at 214, so that they "'identify the source of the product rather than the product itself,'" *id*. at 211. This carefully considered limitation would be entirely pointless if the "original" producer could turn around and pursue a reverse-passing-off claim under exactly the same provision of the Lanham Act. Samara would merely have had to argue that it was the "origin" of the designs that Wal-Mart was selling as its own line. It was not, because "origin of goods" in the Lanham Act referred to the producer of the clothes, and not the producer of the (potentially) copyrightable or patentable designs that the clothes embodied.

. . . .

In sum, reading the phrase "origin of goods" in the Lanham Act in accordance with the Act's common-law foundations (which were *not* designed to protect originality or creativity), and in light of the copyright and patent laws (which *were*), we conclude that the phrase refers to the producer of the tangible goods that are offered for sale, and not to the author of any idea, concept, or communication embodied in those goods. To hold otherwise would be akin to finding that § 43(a) created a species of perpetual patent and copyright, which Congress may not do. See *Eldred* v. *Ashcroft*, 537 U.S. 186, 208 (2003).

The creative talent of the sort that lay behind the *Campaigns* videos is not left without protection. The original film footage used in the *Crusade* television series could have been copyrighted, see 17 U.S.C. § 102(a)(6), as was copyrighted (as a compilation) the *Crusade* television series, even though it included material from the public domain, see § 103(a). Had Fox renewed the copyright in the Crusade television series, it would have had an easy claim of copyright infringement. And respondents' contention that *Campaigns* infringes Doubleday's copyright in General Eisenhower's book is still a live question on remand. . . .

. . . .

Notes & Questions

1. *Dastar* is not your typical trademark infringement case. The defendant was not using a mark that was in any way similar to marks owned by the plaintiffs. Instead, the claim was almost one of false advertising, of confusing or misleading consumers through representations of the origins of the product. The Supreme Court interprets the phrase "origin of the goods" in § 43(a) to refer to the producer of the tangible goods. You have encountered the distinction between the tangible goods that embody intangible assets throughout this course. What intangible asset was the plaintiff seeking to protect?

2. The copyright in the television series had expired. Does the holding apply to situations in which the copyright has not expired? Courts that have addressed the issue have concluded that it does. *See, e.g., General Universal Sys. v. Lee*, 379 F.3d 131 (5th Cir. 2004); *Zyla v. Wadsworth*, 360 F.3d 243 (1st Cir. 2004).

3. As the Court notes, plaintiffs also asserted a copyright claim. The sticking point was whether General Eisenhower had prepared his manuscript as a work made for hire under the 1909 Copyright Act. The Supreme Court stated in a footnote, not reproduced above, that it was expressing "no opinion as to whether petitioner's product would infringe a valid copyright in General Eisenhower's book." On remand, the district court found that Eisenhower's book *was* a work made for hire, the copyright in that work had been properly renewed, and Dastar was liable for infringement of the copyright in the Eisenhower book. *See Twentieth Century Fox Film Corp. v. Entertainment Distrib.*, 429 F.3d 869 (9th Cir. 2005).

D. INITIAL INTEREST CONFUSION

Generally, trademark law is meant to protect consumers from confusion concerning the origin or source of a product, or at least as to affiliation or sponsorship. Does it matter, however, when that confusion occurs? Should the law guard against labels or packaging that might initially confuse a consumer as he grabs a product off a shelf, but as to which the confusion is promptly dispelled before purchase through prominent labeling about the "real" source? For that matter, should the law guard against placing products of the same type near one another on store shelves, which creates this risk of momentary confusion in the first place? A quick visit to the nearest grocery store suggests it's a perfectly permissible practice. *See* Eric Goldman, *Brand Spillovers*, 22 Harv. J.L. & Tech. (2009). This confusion theory, known as "initial interest confusion," is one that courts sometimes consider. It has taken on increased prominence in the internet era. This is so because a domain name might use another's mark, causing momentary confusion, even if seeing the website to which it corresponds would immediately dispel any confusion.

Courts had long recognized the potential harm to the mark owner that results when a competitor attracts the attention of the consumer even if by the time of sale any confusion that the consumer might have initially experienced has dissipated, sometimes referring to the activity as a type of "bait and switch" (although the "switch" happened, and the consumer was aware of the switch, before any purchase was completed).

> The following example demonstrates why protection against initial-interest confusion might make sense under the right circumstances: Suppose that you are taking a long roadtrip, you have become very hungry, and you are keeping an eye out for a McDonald's, which is your fast-food restaurant of choice. Soon you spot a 'McDonald's' sign by an exit. You take the exit and follow the signs, looking forward to your favorite McDonald's hamburger. But—behold—it's a Burger King. The signs were misleading. You are not so fond of Burger King but, having already made the detour and loath to waste even more time, you reluctantly buy a Whopper and get on with your trip. One does not have to be an economist to see that such a deceitful creation of an initial interest is harmful to consumer interests, brand-development incentives, and efficient allocation of capital, even if the confusion is ultimately dissipated by the time of purchase.

Groeneveld Transport Efficiency, Inc. v. Lubecore Int'l, Inc., 730 F.3d 494, 518 (6th Cir. 2013). *See also Dorr-Oliver, Inc. v. Fluid-Quip, Inc.*, 94 F.3d 376, 382 (7th Cir. 1996); *Mobil Oil Corp. v. Pegasus Petroleum Corp.*, 818 F.2d 254, 260 (2d Cir. 1987).

As the example from *Groeneveld Transport* demonstrates, the social harm from such bait & switch tactics is greatest when the consumer's self-help cost—retracing one's steps until one is back off the hook—is high enough to prevent that self-help, to the commercial benefit of the wrongdoer. But when the self-help cost is as low as moving over a few feet on the store shelf or hitting the "back" button on one's browser, the claim of harm rings hollow. *See generally* Eric Goldman, *Deregulating Relevancy in Internet Trademark Law*, 54 Emory L.J. 507 (2005). The Courts of Appeals are divided on the initial-interest theory: some embrace it, *e.g.*, *McNeil Nutrionals, LLC v. Heartland Sweeteners, LLC*, 511 F.3d 350, 358 (3d Cir. 2007); *Nissan Motor Co. v. Nissan Computer Corp.*, 378 F.3d 1002, 1018 (9th Cir. 2004); others reject it, *e.g.*, *Lamparello v. Falwell*, 420 F.3d 309, 316 (4th Cir. 2005); *Gibson Guitar Corp. v. Paul Reed Smith Guitars*, 423 F.3d 539, 552 (6th Cir. 2005). The Supreme Court has done neither.

E. CYBERSQUATTING

Squatting, when applied to real property, refers to occupying property belonging to someone else—a form of trespass. Labeling the practice of registering a domain name consisting of a trademark that belongs to someone else "cybersquatting" aided mark owners in successfully lobbying for specific legislation targeting this behavior: The Anti-cybersquatting Consumer Protection Act, Pub. L. No. 106-113 (1999), often referred to by its acronym "ACPA."

ACPA, codified at § 43(d) of the Lanham Act, gives the owner of trademark a cause of action against a person who "has a bad faith intent to profit from that mark . . . and registers, traffics in, or uses a domain name that—(I) in the case of a mark that is distinctive at the time of registration of the domain name, is identical or confusingly similar to that mark . . . ." 15 U.S.C. § 1125(d)(1)(A). The statute is clear that liability exists "without regard to the goods or services of the parties." *Id.* The statute also provides an illustrative list of factors a court may consider "[i]n determining whether a person has a bad faith intent" to profit in the manner prohibited. That list is as follows:

(I) the trademark or other intellectual property rights of the person, if any, in the domain name;

(II) the extent to which the domain name consists of the legal name of the person or a name that is otherwise commonly used to identify that person;

(III) the person's prior use, if any, of the domain name in connection with the bona fide offering of any goods or services;

(IV) the person's bona fide noncommercial or fair use of the mark in a site accessible under the domain name;

(V) the person's intent to divert consumers from the mark owner's online location to a site accessible under the domain name that could harm the goodwill represented by the mark, either for commercial gain or with the intent to tarnish or disparage the mark, by creating a likelihood of confusion as to the source, sponsorship, affiliation, or endorsement of the site;

(VI) the person's offer to transfer, sell, or otherwise assign the domain name to the mark owner or any third party for financial gain without having used, or having an intent to use, the domain name in the bona fide offering of any goods or services, or the person's prior conduct indicating a pattern of such conduct;

(VII) the person's provision of material and misleading false contact information when applying for the registration of the domain name, the person's intentional failure to maintain accurate contact information, or the person's prior conduct indicating a pattern of such conduct;

(VIII) the person's registration or acquisition of multiple domain names which the person knows are identical or confusingly similar to marks of others that are distinctive at the time of registration of such domain names, or dilutive of famous marks of others that are famous at the time of registration of such domain names, without regard to the goods or services of the parties; and

(IX) the extent to which the mark incorporated in the person's domain name registration is or is not distinctive and famous within the meaning of subsection (c)(1) of section 43.

15 U.S.C. § 1125(d)(1)(B)(i). Below is a series of trial and appellate decisions involving a single cybersquatting dispute. As you read the excerpts, pay attention to how the courts conduct the inquiry into the presence or absence of the required "bad faith intent to profit."

Carnivale v. Staub Design LLC
754 F. Supp. 2d 652 (D. Del. 2010)

ROBINSON, JUDGE:

. . . .

The parties do not contest the following facts. Plaintiff is an architect from Staten Island, New York. Defendant Staub Design, LLC is a Delaware limited liability corporation that was formed on February 9, 2005; its principal place of business is in Arlington, Virginia. Staub Design is a residential design company focused on the application of autoclaved aerated concrete ("AAC"), a lightweight building material. Defendants John and David Staub are principals of Staub Design.

Plaintiff wrote a book entitled "The Affordable House," which was copyrighted on January 26, 1996. On March 15, 1996, plaintiff published his book on the internet at www.affordablehouse.com. Plaintiff registered the domain name www.affordablehouse.com on July 20, 1998 and has renewed it continuously since that date. Plaintiff's website contains home plan designs and excerpts from his book, and he uses his website to advertise blueprints and copies of his book for sale.

In 2000, defendants John and David Staub built a house in South Dakota using AAC as a building material. Defendants referred to this house as "The Affordable House" and described the house as an embodiment of the "concept of The Affordable House," representing a house that is "affordable on many angles."

In the spring of 2004, defendants brainstormed a list of potential domain names for a website they planned to create. During their research of various domain names, defendants learned that plaintiff had previously registered the domain name www.affordablehouse.com. On May 7, 2004, defendant John Staub registered the domain name www.theaffordablehouse.com under his name and contact information and has renewed the registration using his updated contact information since that date. Defendants registered several other domain names as well, but none of these other domain names resembled plaintiff's book title and domain name.

Beginning in December of 2004, defendants posted information on AAC at www.theaffordablehouse.com. Defendants used the slogan "The Affordable House—a project of Staub Design LLC" at the top of each page on their website and on booths at renewable energy conferences held between 2005 and 2007. The website provided information and links to resources about AAC as well as background on Staub Design's focus on the application of AAC. Although defendants did not directly offer anything for sale at www.theaffordablehouse.com, they directed viewers to contact John and David Staub with inquiries regarding AAC.

Plaintiff filed his application for registration of the mark THE AFFORDABLE HOUSE on January 4, 2005, certifying use of the mark in commerce since March 15, 1996. On February 14, 2006, plaintiff secured U.S. Trademark and Service Mark Registration No. 3,058,545 for "architectural plans and specifications" and "on-line retail store services featuring books and sets of blue prints" . . . . Defendants have not licensed the mark THE AFFORDABLE HOUSE.

Plaintiff discovered defendants' website . . . in March 2007 and sent defendants a cease and desist letter, asserting his rights to the mark THE AFFORDABLE HOUSE and stating his belief that defendants' use of the domain name . . . constituted infringement. On May 3, 2007, defendants moved the content of www.theaffordablehouse.com to www.staubdesian.com [sic, www.staubdesign.com] in response to plaintiff's letter. Defendants did not offer to sell the allegedly infringing domain name to plaintiff.

On May 16, 2007, defendants filed a petition for cancellation of plaintiff's registration with the [PTO's] Trademark Trial and Appeal Board ("TTAB"), Cancellation No. 92047553. The TTAB has suspended the cancellation proceeding
. . . .

    . . . .

Carnivale v. Staub Design LLC
456 Fed. Appx. 104 (3d Cir. 2012)

RENDELL, JUDGE:

Staub Design LLC, John Staub, and David Staub (together, the Staubs) appeal a judgment the District Court entered against them on appellee David John Carnivale's Anticybersquatting Consumer Protection Act, or ACPA, claim. See 15 U.S.C. § 1125(d). In a summary judgment ruling and a decision after a bench trial, the District Court found that: (1) Carnivale's trademark, "The Affordable House," was distinctive; (2) the Staubs' domain name, www.theaffordablehouse.com, was identical or confusingly similar to Carnivale's mark; and (3) in registering their domain name, the Staubs acted with a bad faith intent to profit from Carnivale's mark. We conclude that the District Court clearly erred in evaluating two of the factors pertinent to its conclusion that the Staubs acted with a bad faith intent to profit, and, therefore, will reverse and remand for further proceedings consistent with this opinion.

    . . . .

The ACPA lists nine, non-exclusive factors that a court should consider in determining whether a defendant possessed the requisite bad faith intent to profit from the use of the plaintiff's mark. See 15 U.S.C. § 1125(d)(1)(B)(i). In this case, the District Court concluded that factors one through five and nine weighed in favor of a finding of bad faith, and that factors six through eight weighed against such a conclusion. Weighing all of the factors "qualitatively in light of the circumstances of the case as a whole," the District Court found that the Staubs acted in bad faith. 754 F. Supp. 2d 652, 658-61 (D. Del. 2010). We agree with the Staubs that the District Court's analysis with respect to factors five, the defendant's intent to divert consumers, and nine, the extent to which the plaintiff's mark is distinctive or famous, was flawed.

A.

The fifth statutory bad-faith factor is:

> the [defendant's] intent to divert consumers from the mark owner's online location to a site accessible under the domain name that could harm the goodwill represented by the mark, either for commercial gain or with the intent to tarnish or disparage the mark, by creating a likelihood of confusion as to the source, sponsorship, affiliation, or endorsement of the site.

15 U.S.C. § 1125(d)(1)(B)(i)(V). The District Court concluded that this factor weighed against the Staubs, as follows:

> Defendants' knowing and wholesale inclusion of the mark THE AFFORD-ABLE HOUSE in the domain name implies that defendants may have sought to divert customers away from plaintiff's website. Moreover, both plaintiff and defendants are in the business of designing house plans and likely compete for at least part of their client bases. The similarities between the parties' domain names and the nature of their businesses suggest that defendants might be motivated to divert web traffic away from plaintiff's website as a potential competitor.

That analysis is not grounded in the record. Rather than relying on actual evidence presented at trial, the District Court hypothesized that the Staubs "*may* have sought to divert customers away from plaintiff's website," that Carnivale and the Staubs "*likely* compete for *at least part of* their client bases," and that "similarities" between the parties' businesses "*suggest* that defendants *might* be motivated to divert web traffic away from plaintiff's website." (Emphases added). While we understand and agree with the District Court's assertion that a plaintiff rarely will uncover *direct* evidence of an intent to divert consumers, it does not follow that this factor may be satisfied without *any* specific facts from which intent to divert could be inferred, such as evidence tending to show actual customer overlap or competition between the plaintiff's and defendant's businesses.

Further, that the Staubs "may have" or "might have been motivated" to divert customers away from Carnivale's website is not enough to tilt this factor in favor of Carnivale, who, as the plaintiff, bears the burden of proof. On remand, the District Court should consider whether the evidence in the record supports a finding that the Staubs did intend to divert customers from Carnivale.

B.

The ninth statutory factor is "the extent to which the mark incorporated in the [defendant's] domain name registration is or is not distinctive and famous within the meaning of subsection (c) of this section."[2] 15 U.S.C. § 1125(d)(1)(B)(i)(IX). The District Court relied solely on its earlier determination, at summary judgment, that Carnivale's mark is "inherently distinctive" to conclude that this factor supported a finding of bad faith.

The District Court also clearly erred in this regard. We do not read the ninth factor, concerning "the extent to which" the relevant mark "is or is not distinctive and famous," as merely duplicative of the element requiring that a mark be distinctive or famous to qualify for protection under the ACPA. *See* 15 U.S.C. § 1125(d)(1)(A)(ii)(I)-(III); *see also Shields v. Zuccarini*, 254 F.3d 476, 482 (3d Cir. 2001). Instead, the statutory language suggests that the District Court must separately assess the extent of the distinctive or famous nature of the mark.

Obviously, the District Court's determination at summary judgment that Carnivale's mark is "distinctive" must be considered as part of the factor nine analysis.[3] But the District Court also should have evaluated how strong or distinctive Carnivale's mark was. In this regard, the record suggests that Carnivale's mark, even if legally distinctive, is relatively weak. Although Carnivale used the phrase "The Affordable House" before the Staubs registered their website, his use does not appear to have been exclusive—the Staubs presented evidence of numerous other websites using the same phrase in similar ways. The phrase "The Affordable House" is descriptive, at best, and the evidence Carnivale submitted to establish secondary meaning does not include customer surveys, customer testimony, evidence of actual confusion, or several of the other criteria we typically consider in determining secondary meaning. *See Browne Drug Co. v. Cococare Prods., Inc.*, 538 F.3d 185, 199 (3d Cir. 2008). Moreover, at the time the Staubs registered their domain name, Carnivale had not yet applied to register his trademark, and there was no indication on his website that he claimed exclusive rights to the phrase "The Affordable House." On remand, the District Court should evaluate the extent to which Carnivale's mark is distinctive or famous in light of this evidence and the record as a whole.

C.

As the District Court properly recognized, applying the ACPA's bad-faith factors "is a holistic, not mechanical, exercise." *Green v. Fornario*, 486 F.3d 100, 106 (3d Cir. 2007). Thus, although we conclude that the District Court erred in evaluating factors five and nine, we leave to the District Court the task of reanalyzing those factors and determining how any changes affect the overall balance of the factors in this case.

---

[2] Subsection (c) of the statute refers to marks that are distinctive, "inherently or through acquired distinctiveness," 15 U.S.C. § 1125(c)(1), and defines a "famous" mark as one that is "widely recognized by the general consuming public of the United States as a designation of source of the goods or services of the mark's owner," *id.* § 1125(c)(2)(A).

[3] We do not here disturb the District Court's ruling that Carnivale's mark is distinctive, but we read that ruling narrowly. In our view, the District Court based its decision more on the Staubs' failures to make the proper arguments and to present legally relevant evidence to counter Carnivale's summary judgment motion than on the legal merits of the question.

. . . .

Carnivale v. Staub Design LLC
547 Fed. Appx. 114 (3d Cir. 2013)

McKEE, JUDGE:

Staub Design . . . appeal[s] a judgment [on remand] in favor of David John Carnivale for violation of the Anticybersquatting Consumer Protection Act ("ACPA")[, including an award of statutory damages of $25,000, 15 U.S.C. § 1117(d), and an order prohibiting Staub Design from resuming use of the disputed domain name]. For the reasons set forth below, we will affirm the District Court's judgment.

. . . .

. . . We are only concerned with the third element [of an ACPA claim]: Staub Design's bad faith intent to profit from Carnivale's mark. See 15 U.S.C. § 1125(d)(1)(A)(i).

. . . .

As to factor five, although we agreed with the District Court that because "intent is rarely discernible directly, it must typically be inferred from pertinent facts and circumstances," *Audi AG v. D'Amato*, 469 F.3d 534, 549 (6th Cir. 2006), we nevertheless directed the District Court to consider whether the record supported an inference that the Appellants intended to divert consumers.

In doing so, the District Court highlighted testimony in which John Staub admitted to viewing Carnivale's website, www.affordablehouse.com. That website was replete with home building-related content, and the District Court pointed to evidence that Appellants registered their domain name, www.theaffordable-house.com, a short time after John Staub viewed Carnival's website. Moreover, the District Court refused to credit John Staub's claim that his website was intended to serve a purely educational function. Rather, the court ruled to the contrary based on evidence that supported a finding of a profit motive. From that evidence, the District Court inferred that Staub Design sought to ride the coattails of Carnivale's long-established website—an indicator of bad faith.

On this record, we are not left with a definite and firm conviction that Staub Design's intent was unrelated to consumer diversion. This is especially true when we consider that Staub Design's domain name is the entirety of Carnivale's mark, "The Affordable House."

As to factor nine, we directed the District Court to assess the extent to which the mark is distinctiveness. Secondary meaning is also relevant to this inquiry. *See E.T. Browne Drug Co. v. Cococare Products, Inc.*, 538 F.3d 185, 199 (3d Cir. 2008). On remand, the District Court considered these factors and again concluded that "the ninth factor slightly supports a finding of bad faith."

Although the District Court's analysis of factor nine on remand is not without some shortcomings, we cannot conclude that its findings were clearly erroneous.

Further, while the "unique circumstances" cited by the District Court[*] are not exceedingly indicative of bad faith, we find no clear error with their inclusion in the holistic approach. Moreover, when we accord the proper amount of deference to the District Court's balancing of both statutory and non-statutory factors, we are not left with a definite and firm conviction that a mistake has been committed.

. . . .

Notes & Questions

1. As you can tell from the *Carnivale* dispute, bad faith intent to profit is the heart of a cybersquatting claim. Are the factors the statute identifies helpful in determining the presence or absence of bad faith? What steps would you advise a client to take to minimize its legal risk if it wanted to set up, *e.g.*, a gripe site?

2. The ACPA expressly states that "[b]ad faith intent . . . shall not be found in any case in which the court determines that the person believed and had reasonable grounds to believe that the use of the domain name was a fair use or otherwise lawful." § 1125(d)(1)(B)(ii). How might a defendant prove both the subjective and objective elements required by this subsection?

3. The cybersquatting statute expressly rules out consideration of one of the core components of a traditional likelihood-of-confusion analysis. Specifically, it provides for liability "without regard to the goods or services of the parties." 15 U.S.C. § 1125(d)(1)(A). Why define liability in a way that disregards the presence or absence of competitive injury, or the likelihood of consumer confusion about the website's source? Is this a form of initial-interest confusion, in fact if not in name? Does the "bad faith" requirement sufficiently guard against using cybersquatting to harass a vigorous competitor?

4. The ACPA, in a separately codified section, also provides statutory protection for people's names that are registered as domain names:

Any person who registers a domain name that consists of the name of another living person, or a name substantially and confusingly similar thereto, without that person's consent, with the specific intent to profit from such name by selling the domain name for financial gain to that person or any third party, shall be liable in a civil action by such person.

---

*[*Eds. Note*: The district court's remand decision includes the following: "To round out its holistic analysis of the unique circumstances in this case, this court also considers the following. Defendants went out of their way to use plaintiffs domain name in its entirety in their own domain name after viewing it online. While defendants asserted that they did not wish to spend money on the domain name disagreement, defendants filed a petition for cancellation of plaintiff's registration with the United States Patent and Trademark Office Trademark Trial and Appeal Board, Cancellation No. 92047553, on May 16, 2007. Defendants assert that they are pursuing this case on principal, wanting only to educate the public about a new building material. Defendants' alleged principled stand does not ring true in light of their actions or testimony. Other domain names would easily accomplish defendants' alleged goal of educating the public. Defendants' testimony failed to substantiate their claims that the website was for educational not business purposes, and that the domain name theaffordablehouse.com was a key factor in accomplishing this educational goal." *Carnivale v. Staub Design LLC*, Civ. No. 08-764, 2012 WL 6814251, at *5 (D. Del. Jan. 7, 2013).]

15 U.S.C. § 8131(1)(a). Notice that an individual does not need to prove trademark rights in his or her name, and that no proof of bad faith intent to profit is required, although proof of "specific intent to profit" by selling the domain name *is* required. In what way is that different? What might justify the different protection for individual names?

The remedies available for violations of § 8131 are injunctive relief, including the forfeiture or cancellation of the domain name or the transfer of the domain name to the plaintiff, as well as the potential for an award costs and attorneys fees to the prevailing party at the discretion of the court. The statute does not provide for an award of damages for claims. 15 U.S.C. § 8131(2).

5. The internet is a global phenomenon. A registrant or an owner of a domain name might be located outside of the United States. The ACPA provides for a mechanism to proceed against a domain name "in rem," although damages cannot be recovered when proceeding in rem. See 15 U.S.C. § 1125(d)(2).

In addition to the legal rights granted to trademark owners under U.S. trademark law, the governing entity that controls the domain name system, the Internet Corporation for Assigned Names and Numbers (ICANN), instituted an arbitration system that mark owners can use for certain generic top level domains. ICANN adopted the Uniform Dispute Resolution Policy (UDRP) in 1999 and it is now applicable to most of the generic top level domains, such as .com and .org. Under the UDRP a mark owner can institute a dispute resolution process that can result in the transfer of the domain name to the mark owner. No award of damages is possible. All proceedings take place via written documents submitted electronically. Under the UDRP, a domain name owner can opt out of the proceedings by identifying a jurisdiction in which he is willing to litigate ownership issues. For more information on the UDRP see <http://www.icann.org/en/udrp/>.

IV. TRADEMARK DILUTION

Section III introduced you to the basics involved in proving trademark infringement, focusing on the central requirement of "likelihood of confusion." You saw that the type of confusion necessary to prevail could vary depending on the circumstances. Sometimes the plaintiff asserted that consumers would be confused as to the source of the defendant's goods. Other times the allegation was confusion about the relationship between the mark owner and the alleged infringer. When the products or services sold by the mark owner and the alleged infringer vary significantly, it is much harder to demonstrate the likelihood of confusion necessary for the plaintiff to prevail, although confusion as to endorsement or sponsorship sometimes may be a viable cause of action.

The use of a similar mark on goods that are significantly different from those sold by the mark owner might cause two different types of harm. The first type of harm arises not from consumers being confused, but rather from consumers having a weaker connection in their minds between a trademark and the mark owner after they encounter the similar mark. This type of harm is referred to as "blurring." A strong mark's signaling power is weakened by the appearance of the same, or a similar, mark on completely unrelated goods or services. The second type of harm can arise if the product sold by the junior user is of shoddy or inferior quality, or in some other way is distasteful or unsavory. Consumers, although not confused, nonetheless in their minds might downgrade what the trademark means. This type of harm is referred to as "tarnishment." In trademark law, both blurring and tarnishment are categories of trademark dilution.

For many years, protection against dilution was available only under state law. Massachusetts was the first state to grant protection against dilution, passing its law in 1947. By 1996, 26 states had enacted anti-dilution statutes. Some states granted the protection to all marks, while others granted dilution protection only to famous marks. In 1995 Congress determined that federal law should protect owners of "famous" trademarks against the dilution of their marks and passed the Federal Trademark Dilution Act. In 2006 Congress amended the statute with the Trademark Dilution Revision Act. Federal dilution protection is codified in § 43(c) of the Lanham Act, 15 U.S.C. § 1125(c).

A. Likelihood of Dilution

Starbucks Corp. v. Wolfe's Borough Coffee, Inc.
588 F.3d 97 (2d Cir. 2009)

Miner, Judge:

[Wolfe's Borough Coffee, Inc. does business under the name Black Bear Micro Roastery ("Black Bear"). Starbucks sued Black Bear, alleging, among other things, dilution in violation of the Lanham Act. Black Bear won after a bench trial. This appeal followed.]

. . . .

I. BACKGROUND

A. Preliminary Facts

Starbucks, a company primarily engaged in the sale of coffee products, was founded in Seattle, Washington in 1971. Since its founding, Starbucks has grown to over 8,700 retail locations in the United States, Canada, and 34 foreign countries and territories. In addition to operating its retail stores, Starbucks supplies its coffees to hundreds of restaurants, supermarkets, airlines, sport and entertainment venues, motion picture theaters, hotels, and cruise ship lines. Starbucks also maintains an internet site that generates over 350,000 "hits" per week from visitors.

In conducting all of its commercial activities, Starbucks prominently displays its registered "Starbucks" marks (the "Starbucks Marks") on its products and areas of business. The Starbucks Marks include, inter alia, the tradename "Starbucks" and its logo, which is circular and generally contains a graphic of a mermaid-like siren encompassed by the phrase "Starbucks Coffee." Starbucks "has been the subject of U.S. trademark registrations continuously since 1985" and has approximately 60 U.S. trademark registrations. Starbucks also has foreign trademark registrations in 130 countries.

From fiscal years 2000 to 2003, Starbucks spent over $136 million on advertising, promotion, and marketing activities. These promotional activities included television and radio commercials, print advertising, and in-store displays, and "prominently feature[d] (or, in the case of radio, mention[ed]) the Starbucks Marks, which Starbucks considers to be critical to the maintenance of its positive public image and identity." Starbucks also enhanced its commercial presence by permitting the use of its products and retail stores in Hollywood films and popular television programs. These films and programs contained scenes in which the Starbucks Marks were also "prominently displayed."

As may be expected from its spending "substantial time, effort and money advertising and promoting the Starbucks Marks throughout the United States and elsewhere," Starbucks devotes "substantial effort to policing its registered Starbucks Marks." Starbucks "has a regular practice of using watch services and other methods to identify potential infringers of the Starbucks Marks," and it "routinely sends cease and desist letters and, if necessary, commences litigation in support of these efforts."

Black Bear, also a company engaged in the sale of coffee products, has its principal place of business in Tuftonboro, New Hampshire. In contrast to Starbucks, Black Bear is a relatively small company owned by Jim Clark and his wife. It is a family-run business that "manufactures and sells * * * roasted coffee beans and related goods via mail order, internet order, and at a limited number of New England supermarkets." Black Bear also sold coffee products from a retail outlet called "The Den," in Portsmouth, New Hampshire. To help operate its business, Black Bear hires some part-time employees, such as "one girl who comes in two days a week and helps with packaging," but Black Bear is otherwise operated by Mr. and Mrs. Jim Clark, with the occasional help of their two daughters.

In April 1997, Black Bear began selling a "dark roasted blend" of coffee called "Charbucks Blend" and later "Mister Charbucks" (together, the "Charbucks Marks"). Charbucks Blend was sold in a packaging that showed a picture of a black bear above the large font "BLACK BEAR MICRO ROASTERY." The package informed consumers that the coffee was roasted and "Air Quenched" in New Hampshire and, in fairly large font, that "You wanted it dark ... You've got it dark!" Mister Charbucks was sold in a packaging that showed a picture of a man walking above the large font "Mister Charbucks." The package also informed consumers that the coffee was roasted in New Hampshire by "The Black Bear Micro Roastery" and that the coffee was "ROASTED TO THE EXTREME ... FOR THOSE WHO LIKE THE EXTREME."

Not long after making its first sale of Charbucks Blend, in August 1997, Starbucks demanded that Black Bear cease use of the Charbucks Marks. Having felt wrongly threatened by Starbucks, and believing that "[w]e hadn't done anything wrong," Black Bear ultimately decided to continue selling its "Charbucks Blend" and "Mister Charbucks." Mr. Clark later testified, "[m]y main objection was that basically this was a large corporation coming at me and saying, telling us what to do, and, oh, by the way you're going to pay for it, too. * * * [S]ome of the requests that they were making were really off the wall."

B. Complaint and Trial

After failed negotiations with Black Bear, on July 2, 2001, Starbucks filed a complaint in the District Court, alleging trademark dilution in violation of 15 U.S.C. §§ 1125(c), 1127; trademark infringement in violation of 15 U.S.C. § 1114(1); unfair competition in violation of 15 U.S.C. § 1125(a); (as well as parallel state law causes of action). . . .

A two-day bench trial was held . . . . Among the evidence proffered during trial, Starbucks introduced the testimony of Warren J. Mitofsky ("Dr. Mitofsky"), a scientist in the field of consumer research and polling. His testimony explained the results of his survey, which concluded in part that "[t]he number one association of the name 'Charbucks' in the minds of consumers is with the brand 'Starbucks' * * * [and that] [t]he name 'Charbucks' creates many negative associations in the mind of the consumer when it comes to describing coffee." Dr. Mitofsky testified that the surveyed sample of persons were "designed to be representative of the United States" and that he believed a telephone survey of 600 adults in the United States would "do a good job of random sampling." Dr. Mitofsky summarized the scope of his survey: "Well, if you want to know the reaction to the name Charbucks, then the telephone is perfectly adequate. If you want to measure the reaction or the familiarity with other visual cues, then it's not the right method."

On December 22, 2005, the District Court issued an opinion and order ruling in favor of Black Bear and dismissing Starbucks' complaint. Among its findings, the court determined that there was neither actual dilution to establish a violation of the federal trademark laws nor any likelihood of dilution to establish a violation of New York's trademark laws. The court also found that Starbucks failed to prove its trademark infringement and unfair competition claims because there was no likelihood that consumers would confuse the Charbucks Marks for the Starbucks Marks.

C. Subsequent Proceedings

Starbucks appealed the District Court's judgment, and, while the appeal was pending, Congress amended the trademark laws by passing the Trademark Dilution Revision Act of 2005 (the "TDRA"). The TDRA was in response to the Supreme Court's decision in *Moseley v. V Secret Catalogue, Inc.*, 537 U.S. 418, 433 (2003), in which the Supreme Court held that the Federal Trademark Dilution Act required a showing of "actual dilution" in order to establish a dilution claim. The TDRA amended the Federal Trademark Dilution Act to provide, *inter alia*, that the owner of a famous, distinctive mark is entitled to an injunction against the use of a mark that is "likely" to cause dilution of the famous mark. 15 U.S.C. § 1125(c)(1). In light of the change in law, we vacated the judgment of the District Court and remanded for further proceedings. . . .

On remand, . . . the District Court again entered judgment in favor of Black Bear, finding—for substantially the same reasons detailed in the court's December 2005 decision but with additional analysis . . . .

. . . .

II. DISCUSSION

. . . .

Under federal law, an owner of a "famous, distinctive mark" is entitled to an "injunction against the user of a mark that is 'likely to cause dilution' of the famous mark." *Starbucks Corp. v. Wolfe's Borough Coffee, Inc.*, 477 F.3d 765, 766 (2d Cir. 2007) (quoting 15 U.S.C. § 1125(c)(1)). Although the requirement that the mark be "famous" and "distinctive" significantly limits the pool of marks that may receive dilution protection, *see Savin Corp. v. Savin Group*, 391 F.3d 439, 449 (2d Cir. 2004), that the Starbucks Marks are "famous" within the meaning of 15 U.S.C. § 1125(c) is not disputed by the parties in this case. Rather, the focus of this appeal is on dilution itself. . . .

1. Dilution by Blurring

Dilution by blurring is an "association arising from the similarity between a mark or trade name and a famous mark that impairs the distinctiveness of the famous mark," 15 U.S.C. § 1125(c)(2)(B), and may be found "regardless of the presence or absence of actual or likely confusion, of competition, or of actual economic injury," 15 U.S.C. § 1125(c)(1); *see also Deere & Co. v. MTD Products, Inc.*, 41 F.3d 39, 43 (2d Cir. 1994); *Nabisco, Inc. v. PF Brands, Inc.*, 191 F.3d 208, 219 (2d Cir. 1999). Some classic examples of blurring include "hypothetical anomalies as Dupont shoes, Buick aspirin tablets, Schlitz varnish, Kodak pianos, Bulova gowns, and so forth." See *Mead Data Cent., Inc. v. Toyota Motor Sales, U.S.A., Inc.*, 875 F.2d 1026, 1031 (2d Cir. 1989) (internal quotation marks omitted); *see also id.* (stating that the primary concern in blurring actions is preventing "the whittling away of an established trademark's selling power through its unauthorized use by others.").

Federal law specifies six non-exhaustive factors for the courts to consider in determining whether there is dilution by blurring:

(i) The degree of similarity between the mark or trade name and the famous mark.

(ii) The degree of inherent or acquired distinctiveness of the famous mark.

(iii) The extent to which the owner of the famous mark is engaging in substantially exclusive use of the mark.

(iv) The degree of recognition of the famous mark.

(v) Whether the user of the mark or trade name intended to create an association with the famous mark.

(vi) Any actual association between the mark or trade name and the famous mark.

15 U.S.C. § 1125(c)(2)(B)(i)-(vi). The District Court found that the second, third, and fourth factors favored Starbucks, and those findings are not challenged in this appeal.

With respect to the first factor—the degree of similarity between the marks—the District Court did not clearly err in finding that the Charbucks Marks were minimally similar to the Starbucks Marks. Although "Ch"arbucks is similar to

"St"arbucks in sound and spelling, it is evident from the record that the Charbucks Marks—as they are presented to consumers—are minimally similar to the Starbucks Marks. The Charbucks line of products are presented as either "Mister Charbucks" or "Charbucks Blend" in packaging that displays the "Black Bear" name in no subtle manner, and the packaging also makes clear that Black Bear is a "Micro Roastery" located in New Hampshire. Moreover, Black Bear's package design for Charbucks coffee is "different in imagery, color, and format from Starbucks' logo and signage." For example, either a graphic of a bear or a male person is associated with Charbucks, and those marks are not comparable to the Starbucks graphic of a siren in pose, shape, art-style, gender, or overall impression. Indeed, the Starbucks siren appears nowhere on the Charbucks package. To the extent the Charbucks Marks are presented to the public through Black Bear's website, the dissimilarity between the marks is still evident as the Charbucks brand of coffee is accompanied by Black Bear's domain name, www.blackbearcoffee.com, and other products, such as shirts and cups, displaying Black Bear's name.

Furthermore, we note that it is unlikely that "Charbucks" will appear to consumers outside the context of its normal use, since "Charbucks" is not directly identifiable with the actual product, *i.e.*, coffee beans. *Cf. Nabisco v. PF Brands, Inc.*, 191 F.3d 208, 213, 218 (2d Cir. 1999) (observing that Pepperidge Farm's famous "goldfish" mark may be identified outside of the packaging because the goldfish cracker itself is the famous mark). The term "Charbucks" appears only on the packaging and on Black Bear's website—both mediums in which the Charbucks Marks' similarity with Starbucks is demonstratively minimal—and Starbucks has not identified any other method by which the Charbucks Marks likely would be presented to the public outside the context of its normal use. *Cf. id.* at 218 ("[M]any consumers of [the defendant's] crackers will not see the box; they will find goldfish-shaped cheddar cheese crackers served in a dish at a bar or restaurant or friend's house, looking very much like the familiar Pepperidge Farm Goldfish product."). To be sure, consumers may simply refer to "Mister Charbucks" or "Charbucks Blend" in conversation; however, it was not clearly erroneous for the District Court to find that the "Mister" prefix or "Blend" suffix lessened the similarity between the Charbucks Marks and the Starbucks Marks in the court's overall assessment of similarity.

. . . Starbucks asserts that the District Court should have ignored the term "Mister" or "Blend" before or after "Charbucks" in assessing the "degree of similarity" factor because those terms are generic and "too weak to serve a brand-identifying function." This argument to ignore relevant evidence is unfounded in the law. . . . And in any event, even if the core term "Charbucks" were used to identify a product as a stand-alone term, such finding would not be dispositive of the District Court's overall assessment of the degree of similarity. In this case, the District Court's reasons for a finding of minimal similarity between the Charbucks Marks and the Starbucks Marks were well supported by the record . . . .

Upon its finding that the marks were not substantially similar, however, the District Court concluded that "[t]his dissimilarity alone is sufficient to defeat [Starbucks'] blurring claim, and in any event, this factor at a minimum weighs strongly against [Starbucks] in the dilution analysis." We conclude that the District Court erred to the extent it required "substantial" similarity between the marks, and, in this connection, we note that the court may also have placed undue significance on the similarity factor in determining the likelihood of dilution in its alternative analysis.

Prior to the TDRA, this Court has held that "[a] plaintiff cannot prevail on a state or federal dilution claim unless the marks at issue are 'very' or 'substantially similar.'" *Playtex Prods., Inc.* [*v. Georgia-Pacific Corp.*], 390 F.3d [158,] 167 [(2d Cir. 2004)]. . . .

The post-TDRA federal dilution statute, however, provides us with a compelling reason to discard the "substantially similar" requirement for federal trademark dilution actions. The current federal statute defines dilution by blurring as an "association arising from the similarity between a mark * * * and a famous mark that impairs the distinctiveness of the famous mark," and the statute lists six non-exhaustive factors for determining the existence of an actionable claim for blurring. 15 U.S.C. § 1125(c)(2)(B). Although "similarity" is an integral element in the definition of "blurring," we find it significant that the federal dilution statute does not use the words "very" or "substantial" in connection with the similarity factor to be considered in examining a federal dilution claim. *See* 15 U.S.C. § 1125(c).

Indeed, one of the six statutory factors informing the inquiry as to whether the allegedly diluting mark "impairs the distinctiveness of the famous mark" is "[t]he *degree* of similarity between the mark or trade name and the famous mark." 15 U.S.C. § 1125(c)(2)(B)(i) (emphasis added). Consideration of a "degree" of similarity as a factor in determining the likelihood of dilution does not lend itself to a requirement that the similarity between the subject marks must be "substantial" for a dilution claim to succeed. Moreover, were we to adhere to a substantial similarity requirement for all dilution by blurring claims, the significance of the remaining five factors would be materially diminished because they would have no relevance unless the degree of similarity between the marks are initially determined to be "substantial." . . . Accordingly, the District Court erred to the extent it focused on the absence of "substantial similarity" between the Charbucks Marks and the Starbucks Marks to dispose of Starbucks' dilution claim. We note that the court's error likely affected its view of the importance of the other factors in analyzing the blurring claim, which must ultimately focus on whether an association, arising from the similarity between the subject marks, "impairs the distinctiveness of the famous mark." 15 U.S.C. § 1125(c)(2)(B).

Turning to the remaining two disputed factors—(1) whether the user of the mark intended to create an association with the famous mark, and (2) whether there is evidence of any actual association between the mark and the famous mark—we conclude that the District Court also erred in considering these factors.

The District Court determined that Black Bear possessed the requisite intent to associate Charbucks with Starbucks but that this factor did not weigh in favor of Starbucks because Black Bear did not act in "bad faith." The determination of an "intent to associate," however, does not require the additional consideration of whether bad faith corresponded with that intent. The plain language of section 1125(c) requires only the consideration of "[w]hether the user of the mark or trade name intended to create an association with the famous mark." See 15 U.S.C. § 1125(c)(2)(B)(v). Thus, where, as here, the allegedly diluting mark was created with an intent to associate with the famous mark, this factor favors a finding of a likelihood of dilution.

The District Court also determined that there was not an "actual association" favoring Starbucks in the dilution analysis. Starbucks, however, submitted the results of a telephone survey where 3.1% of 600 consumers responded that Starbucks was the possible source of Charbucks. The survey also showed that 30.5% of consumers responded "Starbucks" to the question: "[w]hat is the first thing that comes to mind when you hear the name 'Charbucks.'" In rejecting Starbucks' claim of actual association, the District Court referred to evidence supporting the absence of "actual confusion" to conclude that "the evidence is insufficient to make the * * * factor weigh in [Starbucks'] favor to any significant degree." (internal quotation marks and original alteration omitted). This was error, as the absence of actual or even of a likelihood of confusion does not undermine evidence of trademark dilution. See 15 U.S.C. § 1125(c)(2)(B); *accord Nabisco*, 191 F.3d at 221 (stating that while a showing of consumer confusion is relevant in determining dilution by blurring, the absence of confusion "has no probative value" in the dilution analysis).

Accordingly, in light of the foregoing, we remand to the District Court for consideration of Starbucks' claim of trademark dilution by blurring . . . .

2. Dilution by Tarnishment

Dilution by tarnishment is an "association arising from the similarity between a mark or trade name and a famous mark that harms the reputation of the famous mark." 15 U.S.C. § 1125(c)(2)(C). "A trademark may be tarnished when it is linked to products of shoddy quality, or is portrayed in an unwholesome or unsavory context, with the result that the public will associate the lack of quality or lack of prestige in the defendant's goods with the plaintiff's unrelated goods." *Hormel Foods Corp. v. Jim Henson Productions, Inc.*, 73 F.3d 497, 507 (2d Cir. 1996) (internal quotation marks omitted). A trademark may also be diluted by tarnishment if the mark loses its ability to serve as a "wholesome identifier" of plaintiff's product. *Id.*; *accord Chemical Corp. v. Anheuser-Busch, Inc.*, 306 F.2d 433 (5th Cir. 1962) (finding that use of exterminator's slogan "where there's life, ... there's Bugs" tarnished the use of beer company's slogan "where there's life, ... there's Bud."); *Steinway & Sons v. Robert Demars & Friends*, 210 U.S.P.Q. 954 (C.D. Cal. 1981) (finding that use of "STEIN-WAY CO." to sell clip-on beverage handles tarnished high-end musical instrument company's use of its name of "STEINWAY & SONS"); *Eastman Kodak Co. v. Rakow*, 739 F. Supp. 116 (W.D.N.Y. 1989) (finding that comedian's stage name "Kodak" tarnished the mark of the Eastman Kodak Company because the comedian's act "includes humor that relates to bodily functions and sex * * * and * * * uses crude, off-color language repeatedly" (internal quotation marks omitted)); *Dallas Cowboys Cheerleaders, Inc. v. Pussycat Cinema, Ltd.*, 467 F. Supp. 366 (S.D.N.Y.) (finding that pornographic depiction of a Dallas Cowboys

Cheerleader-style cheerleader in an adult film tarnished the professional mark of the Dallas Cowboys Cheerleaders), *aff'd*, 604 F.2d 200 (2d Cir. 1979).

Starbucks argues that the District Court "erred by failing to find that 'Charbucks' damages the positive reputation of Starbucks by evoking both 'Starbucks' and negative impressions in consumers, including the image of bitter, over-roasted coffee." Starbucks reasons that it has shown dilution by tarnishment because, pursuant to its survey, (1) 30.5% of persons surveyed "immediately associated 'Charbucks' with 'Starbucks'"; and (2) 62% of those surveyed who associated "Charbucks" with "Starbucks" "indicated that they would have a negative impression" of a "coffee named 'Charbucks.'" We are unpersuaded by Starbucks' reasoning.

To the extent Starbucks relies on the survey, a mere association between "Charbucks" and "Starbucks," coupled with a negative impression of the name "Charbucks," is insufficient to establish a likelihood of dilution by tarnishment. That a consumer may associate a negative-sounding junior mark with a famous mark says little of whether the consumer views the junior mark as harming the reputation of the famous mark. The more relevant question, for purposes of tarnishment, would have been how a hypothetical coffee named either "Mister Charbucks" or "Charbucks Blend" would affect the positive impressions about the coffee sold by Starbucks. We will not assume that a purportedly negative-sounding junior mark will likely harm the reputation of the famous mark by mere association when the survey conducted by the party claiming dilution could have easily enlightened us on the matter. Indeed, it may even have been that "Charbucks" would strengthen the positive impressions of Starbucks because it brings to the attention of consumers that the "Char" is absent in "Star"bucks, and, therefore, of the two "bucks," Starbucks is the "un-charred" and more appealing product. Juxtaposition may bring to light more appealing aspects of a name that otherwise would not have been brought to the attention of ordinary observers.

Starbucks also argues that "Charbucks" is a pejorative term for Starbucks' coffee, and, therefore, the Charbucks "name has negative associations that consumers are likely to associate with Starbucks' coffee." Although the term "Charbucks" was once used pejoratively during the so-called "coffee-wars"[5] in Boston, Massachusetts, Black Bear is not propagating that negative meaning but, rather, is redefining "Charbucks" to promote a positive image for its brand of coffee. Black Bear sells "Charbucks" as its own product, and, consistent with its intent on profiting from selling Charbucks, the Charbucks line of coffee is of "[v]ery high quality. It's our life. We put everything into it." In short, Black Bear is promoting "Charbucks" and not referring to it in a way as to harm the reputation of Starbucks' coffees. *Cf. Deere & Co.*, 41 F.3d at 45 (stating that the likelihood of dilution by tarnishment means "the possibility that consumers will come to attribute unfavorable characteristics to a mark and ultimately associate the mark with inferior goods and services").

---

[5] The name "Charbucks" was used publicly by the owner of a former chain of coffee bars, "Coffee Connection," to describe what the owner believed was the "over-roasted" type of coffee Starbucks was serving.

Moreover, that the Charbucks line of coffee is marketed as a product of "[v]ery high quality"—as Starbucks also purports its coffee to be—is inconsistent with the concept of "tarnishment." *See Hormel Foods Corp.*, 73 F.3d at 507 (citing cases finding tarnishment where challenged marks were either "seamy" or substantially of lesser quality than the famous mark). Certainly, the similarity between Charbucks and Starbucks in that they are both "[v]ery high quality" coffees may be relevant in determining dilution, *see* 15 U.S.C. 1125(c)(2)(B), (c)(2)(C), but such similarity in this case undercuts the claim that Charbucks harms the reputation of Starbucks. *See Deere & Co.*, 41 F.3d at 43 ("'Tarnishment' generally arises when the plaintiff's trademark is linked to products of shoddy quality, or is portrayed in an unwholesome or unsavory context likely to evoke unflattering thoughts about the owner's product."). Accordingly, we conclude that the District Court did not err in rejecting Starbucks' claim of dilution by tarnishment.

. . . .

Starbucks Corp. v. Wolfe's Borough Coffee, Inc.
2011 WL 6747431 (S.D.N.Y. 2011)

SWAIN, JUDGE:

. . . .

The one remaining question on remand is whether Defendant's use of its "Mister Charbucks," "Mr. Charbucks" and "Charbucks Blend" marks (the "Charbucks Marks") for one of its blended coffee products is likely to dilute Plaintiffs "Starbucks" marks by blurring. . . .

. . . .

The Statutory Factors

At this stage of the litigation, there is no dispute that four of the six factors weigh in Plaintiff's favor. They are: the distinctiveness of Plaintiff's marks, Plaintiff's exclusivity of use, the high degree of recognition of Plaintiff's marks, and Defendant's intent to associate its marks with the Plaintiff's marks. On remand, this Court focuses on the degree of similarity of the marks and the evidence of actual association between the marks.

Similarity of the Marks

When determining similarity for purposes of a dilution claim, courts must consider "the differences in the way the [marks] are presented" in commerce. *Starbucks IV*, 588 F.3d at 106 (citing *Playtex Products. Inc. v. Georgia-Pacific Corp.*, 390 F.3d 158, 167-68 (2d Cir. 2004)). Here, there is no evidence that Charbucks is ever used as a standalone term, and it is unlikely that Charbucks "will appear to consumers outside the context of its normal use." *Starbucks IV*, 588 F.3d at 106. In commerce, the term Charbucks is always preceded or followed by the terms "Mister," "Mr." or "Blend." Defendant uses the Charbucks marks in conjunction with its Black Bear mark, a large black bear, or the figure of a walking man above the words "Black Bear Micro Roastery." *Starbucks IV*, 588 F.3d at 106. These marks are not similar to Plaintiff's highly recognizable siren mark, which does not appear on the Charbucks product packaging. *Id.* Further, Defendant's packaging, which uses an entirely different color scheme from that employed by Starbucks, identifies Black Bear as a "Micro Roastery" located in New Hampshire. *Id.* Where the Charbucks marks are used on Black Bear's website, they are accompanied by Black Bear's domain name, www.blackbearcoffee.com. *Id.* Thus, although the term "Ch"arbucks is similar to "St"arbucks "in sound and spelling" when compared out of context, the marks are only minimally similar as they are presented in commerce. *Id.*

Plaintiff cites to decisions finding other marks sufficiently similar to the Starbucks marks to create a likelihood of dilution. Plaintiff argues that a finding of dilution is likewise warranted in the instant case. (See Pl.s Opening Brief on Second Remand, pg. 5 n.6, citing *Bell v. Starbucks U.S. Brands Corp.*, 389 F. Supp. 2d 766 (S.D. Tex. 2005) (prohibiting use of "Starbock" and "Star Bock" marks for beer), *aff'd* 205 Fed. Appx. 289 (5th Cir. 2006), and *Starbucks Corp. v. Lundberg*, No. Civ. 02-948-HA, 2005 WL 3183858 (D. Or. Nov. 29, 2005) (finding "extensive and obvious" similarities between "Sambuck's" and "Starbucks'").) In those cases, however, the Court found dilution where the junior marks were used on their own, without contextual features distinguishing the junior mark from the senior mark. *See, e.g., Bell*, 389 F. Supp. 2d at 780 (finding that "Plaintiff's use of the words 'Star Bock' and 'Starbock' alone violate numerous state and federal laws" but that "the 'Star Bock Beer' logo with the 'Born in Galveston' wording as shown in Plaintiff's Exhibit 1, does not violate any trademark or unfair competition laws"). Here, the Charbucks marks are used exclusively with terms "Mister," "Mr." or "Blend" and in contexts dissimilar from the contexts in which the Starbucks marks are used. The Court will not "ignore relevant evidence" of such distinguishing contextual features. *Starbucks IV*, 588 F.3d at 107. The minimal degree of similarity between the marks as they are used in commerce weighs in Defendant's favor.

Actual Association with the Famous Mark

Plaintiff has proffered, as evidence of actual association, the results of a telephonic survey in which respondents were asked to react to the terms "Charbucks" and "Starbucks." Of the 600 respondents surveyed, 30.5% said that they associated the term "Charbucks" with "Starbucks," and 9% said they associated the term "Charbucks" with coffee. When asked to name a company or store that they thought might "offer a product called 'Charbucks,'" 3.1% of respondents said Starbucks. These results constitute evidence of actual association.

The results of Plaintiff's survey show some association between the terms Charbucks and Starbucks. However, the survey did not measure how consumers would react to the Charbucks marks as they are actually packaged and presented in commerce. Further, the percentage of respondents who indicated a mental association between the marks is relatively small. In the cases relied on by Plaintiff, survey respondents typically made an association between the marks between 70% and 90% of the time. *See e.g., Visa Intern.*, 590 F. Supp. 2d at 1319 (73% of respondents said EVISA reminded them of Visa); *Nike, Inc. v. Nikepal Intern., Inc.*, No., 2:05-cv-1468-GEB-JFM, 2007 WL 2782030, *4 (E.D. Cal. Sep. 18, 2007) (87% of respondents said Nikepal reminded them of Nike); *Lundberg*, 2005 WL 3183858, at *8 (85% of respondents thought of Starbucks when shown "Sambuck's Coffee-house" and 70% said they thought of Starbucks because the marks were so similar). Here, even stand-alone use of the core term "Charbucks" drew only a 30.5% association response.

Plaintiff invokes the Ninth Circuit's decision in *Jada Toys v. Mattel. Inc.* to demonstrate that lower survey numbers (28%) have been found significant. *Jada Toys v. Mattel, Inc.*, 518 F.3d 628, 636 (9th Cir. 2008). In *Jada Toys*, the survey asked respondents who they thought "puts out or makes" a product called HOT RIGZ. *Id.* Twenty-eight percent of those responding said that they thought it was either made by Mattel, by the company that makes HOT WHEELS, or that whoever made it required Mattel's permission to do so. *Id.* The Ninth Circuit found that these survey results showed "significant evidence of actual association." *Id.* By contrast, when asked a similar question, only 3.1% of those responding to the survey in this case said that they thought Plaintiff offered a product called Charbucks. While *Jada Toys* does confirm that association numbers in the lowest third can be significant, it does little to bolster Starbucks' argument that a single-digit source confusion indicator produced by a survey that did not present the relevant terms in context is probative of a likelihood of dilution by blurring.

The Court finds, after careful consideration of the survey results and methodology, that the actual association factor weighs no more than minimally in Plaintiff's favor.

Likelihood of Dilution by Blurring

The ultimate analytical question before the Court is not simply whether there has been an association between the marks. As the Second Circuit explained in *Starbucks IV*, the ultimate analytical question presented by a dilution-by-blurring claim is whether there is an association, arising from the similarity of the relevant marks, that impairs the distinctiveness of the famous mark. 588 F.3d at 109. The Court evaluates the non-exclusive statutory factors in light of that ultimate question.

The Court is also mindful of the purposes and core principles of trademark law when analyzing a blurring claim. It is settled law that trademarks do not create a "right-in-gross" or an unlimited right at large. *American Footwear Corp. v. General Footwear Co. Ltd.*, 609 F.2d 655, 663 (2d Cir. 1979); see also 4 McCarthy on Trademarks § 24:11 (4th Ed. 2010) (collecting cases). Federal anti-dilution law should not be read to "prohibit all uses of a distinctive mark that the owner prefers not be made." *Nabisco Inc. v. P.F. Brands. Inc.*, 191 F.3d 208, 224 n.6; see also 4 McCarthy on Trademarks § 24:67 ("[N]o antidilution law should be so interpreted and applied as to result in granting the owner of a famous mark the automatic right to exclude any and all uses of similar marks in all product or service lines.") Antidilution law has been called "a scalpel, not a battle axe," and should be applied with care after rigorous evidentiary examination by the courts. 4 McCarthy § 24:67.

As previously explained, the distinctiveness, recognition, and exclusivity of use factors weigh in Plaintiff's favor. Indeed, Plaintiff's evidence on all three of these factors is strong. None of the three, however, is dependent on any consideration of the nature of the challenged marks or any defendant's use of any challenged mark. Thus, although these factors are significant insofar as they establish clearly Plaintiff's right to protection of its marks against dilution, they are not informative as to whether any association arising from similarity of the marks used by Defendant to Plaintiff's marks is likely to impair the distinctiveness of Plaintiff's marks.

A fourth factor—intent to associate—also weighs in Plaintiff's favor, as Defendant's principal testified during trial that, by using the term Charbucks, he meant to evoke an image of dark-roasted coffee of the type offered by Starbucks.

Similarity of the marks and association between the marks are obviously important factors. The statutory language leaves no doubt in this regard—dilution "is association arising from the similarity between a mark or trade name and a famous mark that impairs the distinctiveness of the famous mark." 15 U.S.C.A. § 1125 (c)(2)(B) (West 2009). It is thus appropriate to examine carefully, in considering the significance of both the evidence of similarity and the evidence of actual association, the degree to which any likelihood of dilution by blurring has been shown to arise from similarity between Defendant's marks and those of Plaintiff. As explained above, the marks being compared in this case are only minimally similar as they are presented in commerce, and the evidence of association weighs no more than minimally in Plaintiff's favor.

After considering all of the evidence and noting the dissimilarity of the marks as used in commerce, the weakness of the survey evidence, and the fact that consumers encounter Defendant's Charbucks term only in conjunction with other marks unique to Defendant, the Court holds that the Charbucks marks are only weakly associated with the minimally similar Starbucks marks and, thus, are not likely to impair the distinctiveness of the famous Starbucks marks. In other words, Plaintiff has failed to carry its burden of proving that Defendant's use of its marks, as evidenced on the record before the Court, is likely to cause dilution by blurring.

. . . .

Notes & Questions

1. What is the policy justification for protecting a mark from blurring? Certainly mark owners would like to have exclusive control over their marks and prohibit anyone from using the mark for any reason without the mark owner's permission. But would such control make for good policy? Judge Posner offers one search-cost-based account:

> [T]here is concern that consumer search costs will rise if a trademark becomes associated with a variety of unrelated products. Suppose an upscale restaurant calls itself "Tiffany." There is little danger that the consuming public will think it's dealing with a branch of the Tiffany jewelry store if it patronizes this restaurant. But when consumers next see the name "Tiffany" they may think about both the restaurant and the jewelry store, and if so the efficacy of the name as an identifier of the store will be diminished. Consumers will have to think harder—incur as it were a higher imagination

cost—to recognize the name as the name of the store.

Ty Inc. v. Perryman, 306 F.3d 509, 511 (7th Cir. 2002). Does the increased search cost rationale justify the scope of protection afforded by a dilution cause of action?

2. To determine whether the defendant's mark is likely to cause dilution through blurring, the statute directs courts to consider "all relevant factors" and then lists six specific ones, although not indicating any particular weight to assign to any particular factor. Review that list.

How similar should the marks have to be to weigh in favor of the plaintiff? The Second Circuit held that it was error for the district court to require substantial similarity between the marks. At one point the Second Circuit notes that marks are "minimally similar." Should "minimal similarity" between the marks be a threshold requirement for a claim of trademark dilution? Recall the Sixth Circuit's "benchmarks" in the infringement context, discussed in *Kellogg Co. v. Toucan Golf, Inc.*, 337 F.3d 616 (6th Cir. 2003), excerpted in Section III. Might a benchmark approach work in the dilution context? If so, how might you formulate such benchmarks?

How does the intent of the defendant to create an association with the famous mark help show a likelihood of blurring? Is this similar to examining the intent of the defendant when analyzing a likelihood of confusion? Should the presence or absence of the defendant's "bad faith" be more relevant for an infringement claim than for dilution?

A court is also directed to consider evidence of "actual association" to assist in the determination of blurring, but the Supreme Court has held that blurring "is not a necessary consequence of a mental association" in the mind of the consumer between an accused mark and a famous mark. *Moseley v. V Secret Catalogue, Inc.*, 537 U.S. 418, 434 (2003). Thus, mere proof of actual association is not enough. What else should a plaintiff prove to make the evidence of actual association meaningful in this context?

3. You have probably heard stories of people that have had to change their business name or some activity that they might have been conducting because of an objection by a trademark owner. By now you can begin to see why trademark owners aggressively assert their trademark rights. While trademark law does not impose a "duty to police" third-party use of a mark, in what way does the federal dilution statute encourage zealous assertions of unauthorized use of trademarks?

While the trademark laws might encourage vigilance by trademark owners, sometimes that vigilance can seem a bit over the top. For example, consider the assertion by Chick-fil-A that a t-shirt urging "Eat More Kale" infringed and diluted its "Eat Mor Chicken" mark. Jess Bidgood, *Chicken Chain Says Stop, but T-Shirt Maker Balks*, N.Y. Times Dec. 4, 2011, at A12. Similarly, a church in Washington created t-shirts to encourage congregants to attend services with the image above, using the name of the church, "Mount Olivet," and the slogan "do the pew."

PepsiCo, the owner of the mark "Mountain Dew" for beverages, sent a cease and desist letter, explaining the problem with the church's design:

> The MOUNT OLIVET logo closely resembles PepsiCo, Inc.'s registered MOUNTAIN DEW trademarks in both stylization and color. Such use wrongly trades on the goodwill of our brand and could lead to consumer confusion by creating the impression of a sponsorship or association with MOUNTAIN DEW soft drinks. In addition, by mimicking our famous MOUNTAIN DEW logo and federally-registered DO THE DEW advertising slogan, the distinctiveness of our trademarks are diluted.
>
> I trust you will not proceed with producing and selling the t-shirts.

Michael Atkins, PepsiCo Shines in Gentle Demand Letter to Church, *available at* http://seattletrademarklawyer.com/blog/2008/7/2/pepsico-shines-in-gentle-demand-letter-to-church.html (last visited July 5, 2011). Which type of dilution, blurring or tarnishment, do you think PepsiCo was referring to? What harm would PepsiCo risk by forbearance—not sending a letter at all—in a situation such as this?

4. In *Starbucks* we again see the use of surveys. What were the surveys attempting to show in that case? A major hurdle in constructing a solid survey is using a proper "stimulus" to frame the survey question. Mitofsky conducted his survey by phone and thus didn't *show* respondents the Charbucks mark in the commercial context in which Black Bear used it. Is this how most consumers would have encountered the accused mark? Was asking aloud, with a pronunciation chosen by Mitofsky, a form of leading question? *See*, Irina Manta, *In Search of Validity: A New Model for the Context and Procedural Treatment of Trademark Infringement Surveys*, 24 Cardozo Arts & Ent. L.J. 1027 (2007) (discussing standards for survey stimuli in trademark infringement cases).

5. Should trade dress also be protected from dilution? The Levi Strauss company asserted a claim for dilution (and infringement) of the

 "Arcuate" stitching pattern used on the back pocket of its jeans since 1873 (shown on the left). Abercrombie began using the stitching pattern shown below (on the right). The Ninth Circuit reversed the district court decision against Levi because it
had incorrectly required the marks to be "identical or nearly identical." *Levi Strauss & Co. v. Abercrombie & Fitch Trading Co.*, 633 F.3d 1158 (9th Cir. 2011). The case was remanded for further proceedings. What evidence should Levi consider introducing to support its claim? What evidence should Abercrombie consider introducing?

6. Many cases addressing dilution, including the *Starbucks* case excerpted above, do not have the fact pattern that the advocates for antidilution statutes had used to justify creating this new cause of action. Traditionally conceived, dilution protection was to provide at least some mark owners with a way to prevent the use

of an established mark on very different products—like the use of Tiffany for a restaurant or Kodak for pianos. When those marks are used on such different products or services, consumers are unlikely to think that Kodak is now in the piano business or Tiffany is running a restaurant. In that context an infringement claim, requiring proof of a likelihood of confusion, would not be successful. However, both trademark law and consumer expectations have evolved, perhaps in something of a feedback loop. As trademark owners have pushed for protection against confusion as to sponsorship, leading Congress to revise § 43(a) in 1988 to expressly protect against such confusion, the consuming public may have become more likely to expect that when a mark is used on a wildly different product there is likely some sponsorship or approval by the mark owner. *See* James Gibson, *Risk Aversion and Rights Accretion in Intellectual Property Law*, 116 YALE L.J. 882 (2007). Does this mean that protection against dilution is unnecessary?

While the protection against dilution, traditionally conceived, permitted mark owners to sue for use of a similar mark on very different products, increasingly mark owners add dilution claims to their complaints alleging trademark infringement—even though the products the defendant is selling are strongly similar to those of the mark owner. When the defendant's products are similar to the mark owner's products, and the mark owner fails to demonstrate a likelihood of confusion, should a court interpret dilution law to prevent defendant's use of the mark anyway? Even if the ultimate outcome in the dilution cases is not favorable to the plaintiffs, have dilution claims become a way for owners of strong marks to harass competitors? Or, should those who seek to compete with famous brands be required to select marks that are not even "minimally similar," thereby providing famous mark owner a "wide berth"?

7. In addition to dilution by blurring, federal law also guards against dilution through tarnishment. In one opinion, Judge Posner uses as an example of tarnshiment a "striptease joint" that adopts the name "Tiffany." Consumers will not be confused and think that the jewelry store is affiliated with the striptease joint, "[b]ut because of the inveterate tendency of the human mind to proceed by association, every time they think of the word 'Tiffany' their image of the fancy jewelry store will be tarnished by the association of the word with the strip joint." *Ty Inc. v. Perryman*, 306 F.3d 509, 511 (7th Cir. 2002).

A tarnishment case decided under New York law that is often cited—including by the Second Circuit in *Starbucks*—involved the logo consisting of a two-dimensional silhouette of a male deer, used in connection with John Deere tractors. *Deere & Co. v. MTD Prods.*, 41 F.3d 39, 41 (2d Cir. 1994). Defendant sold lawn tractors using the trademark "Yard-Man" and had created a television commercial using the deer logo to comedic effect, as described by the court:

> [T]he deer in the Commercial Logo is animated and assumes various poses. Specifically, the . . . deer looks over its shoulder, jumps through the logo frame (which breaks into pieces and tumbles to the ground), hops to a pinging noise, and, as a two-dimensional cartoon, runs, in apparent fear, as it is pursued by the Yard-Man lawn tractor and a barking dog. [The trial judge]

described the dog as "recognizable as a breed that is short in stature," and in the commercial the fleeing deer appears to be even smaller than the dog.

Id. at 41. Interpreting New York's anti-dilution statute, the Second Circuit affirmed the district court's grant of a preliminary injunction. The court was particularly concerned with the competing nature of the defendant's product being advertised:

> Whether the use of the mark is to identify a competing product in an informative comparative ad, to make a comment, or to spoof the mark to enliven the advertisement for a noncompeting or a competing product, the scope of protection under a dilution statute must take into account the degree to which the mark is altered and the nature of the alteration. Not every alteration will constitute dilution, and more leeway for alterations is appropriate in the context of satiric expression and humorous ads for noncompeting products. But some alterations have the potential to so lessen the selling power of a distinctive mark that they are appropriately proscribed by a dilution statute. Dilution of this sort is more likely to be found when the alterations are made by a competitor with both an incentive to diminish the favorable attributes of the mark and an ample opportunity to promote its products in ways that make no significant alteration.
>
> We need not attempt to predict how New York will delineate the scope of its dilution statute in all of the various contexts in which an accurate depiction of a distinctive mark might be used, nor need we decide how variations of such a mark should be treated in different contexts. Some variations might well be de minimis, and the context in which even substantial variations occur may well have such meritorious purposes that any diminution in the identifying and selling power of the mark need not be condemned as dilution.
>
> Wherever New York will ultimately draw the line, we can be reasonably confident that the MTD commercial challenged in this case crosses it. The commercial takes a static image of a graceful, full-size deer-symbolizing Deere's substance and strength—and portrays, in an animated version, a deer that appears smaller than a small dog and scampers away from the dog and a lawn tractor, looking over its shoulder in apparent fear. Alterations of that sort, accomplished for the sole purpose of promoting a competing product, are properly found to be within New York's concept of dilution because they risk the possibility that consumers will come to attribute unfavorable characteristics to a mark and ultimately associate the mark with inferior goods and services.

Id. at 45.

While all of trademark law creates tensions with the First Amendment, the tensions are most intense as to tarnishment. We revisit the First Amendment tensions with trademark law in Section V.

B. The Fame Requirement

The federal statute is clear: dilution is actionable "regardless of the presence or absence of actual or likely confusion, of competition, or of actual economic injury." The potential for dilution law to unmoor trademark protection from its consumer protection rationale and cause troubling conflicts with the First Amendment places significant pressure on the gatekeeping requirement that, in order to obtain protection against dilution, the mark must be "famous." Indeed, as the Second Circuit notes in *Starbucks*, "the requirement that the mark be 'famous' . . . significantly limits the pool of marks that may receive dilution protection." The defendant did not contest that the Starbucks mark was famous. Just how famous does a mark need to be to qualify for federal dilution protection?

Board of Regents, University of Texas v. KST Elec.
550 F. Supp. 2d 657 (W.D. Tex. 2008)

YEAKEL, J.:

[The opinion in this case was authored by Magistrate Judge Austin as a report and recommendation under 28 U.S.C. §636(b); FRCP 72; and Loc. R.W.D. Tex. Appx. C, 1(d), and was accepted and approved by District Court Judge Yeakel.]

. . . .

In this case the University of Texas ("UT") is suing KST Electric ("KST") for a number of state and federal trademark claims, alleging that several logos developed and used by KST infringe on UT's registered trademark that depicts its mascot, a longhorn steer, in silhouette (referred to by UT as its "longhorn silhouette logo" or LSL).

KST was started by Kenneth and Suanna Tumlinson in 1994. They are avid fans of the University of Texas athletics and have had season tickets to the football games for many years. In 1998, KST designed what the Court will refer to as the Longhorn Lightning Bolt Logo (or "LLB Logo"). The logo's design consists of a longhorn silhouette with a "K" on the left cheek area of the longhorn, an "S" on the right cheek area, a "lightening bolt T" in the face of the silhouette, and the words "ELECTRIC, LTD." in the space between the horns.

In March 2002, when UT asserts it learned of the LLB Logo . . . UT asked KST to cease and desist using that logo. KST refused. Eventually, in December 2006 UT filed suit. KST now moves for summary judgment on a number of affirmative defenses and on the merits of some of UT's claims.

[The court first addressed and rejected KST's affirmative defenses of laches, statute of limitations, and acquiescence. The court also denied KST's motion for summary judgment on UT's trademark infringement and

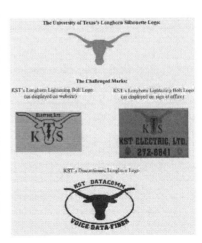

unfair competition claims. It then turned to the dilution claim. —*Eds.*]

. . . .

KST argues that it should be granted summary judgment on UT's federal dilution claim because UT has not provided any evidence that the longhorn silhouette logo is famous for purposes of the Trademark Dilution Revision Act ("TDRA"). On October 6, 2006, President Bush signed into law the Trademark Dilution Revision Act of 2006 (TDRA), which amended the Federal Trademark Dilution Act (FTDA).[13] Under the TDRA, "the owner of a famous mark . . . shall be entitled to an injunction against another person who, at any time after the owner's mark has become famous, commences use of a mark or trade name in commerce that is likely to cause dilution by blurring or dilution by tarnishment of the famous mark, regardless of the presence or absence of actual or likely confusion, of competition, or of actual economic injury." 15 U.S.C. § 1125(c)(1).

Specifically, to state a dilution claim under the TDRA, a plaintiff must show:

(1) that the plaintiff owns a famous mark that is distinctive;
(2) that the defendant has commenced using a mark in commerce that allegedly is diluting the famous mark;
(3) that a similarity between the defendant's mark and the famous mark gives rise to an association between the marks; and
(4) that the association is likely to impair the distinctiveness of the famous mark or likely to harm the reputation of the famous mark.

See Louis Vuitton Malletier S.A. v. Haute Dggity Dog, LLC, 507 F.3d 252, 264-65 (4th Cir. 2007). Under the TDRA, the owner of a famous mark has to establish a likelihood of dilution to obtain monetary relief on a federal dilution claim under the Lanham Act. *See Adidas America, Inc. v. Payless Shoesource,* 529 F. Supp. 2d 1215, 1244-45 (D. Or. 2007); *Malletier v. Dooney & Bourke, Inc.,* 500 F. Supp. 2d 276, 280 (S.D.N.Y. 2007); 15 U.S.C. § 1125(c)(5). The TDRA specifically requires that the mark be "*widely recognized by the general consuming public of the United States* as a designation of source of the goods or services of the mark's owner." 15 U.S.C. § 1125(c)(2)(A) (emphasis added); *Adidas America,* 529 F. Supp. 2d at 1244-45.

Dilution is a cause of action invented and reserved for a select class of marks — those marks with such powerful consumer associations that even non-competing uses can impinge on their value. *Avery Dennison Corp. v. Sumpton,* 189 F.3d 868, 875 (9th Cir. 1999) (citing the seminal trademark article by Frank L. Schechter, *The Rational Basis for Trademark Protection,* 40 Harv. L. Rev. 813 (1927), which proposed a cause of action for dilution). Dilution causes of action, much more so

---

[13] Specific changes to federal dilution law under the TDRA include: (1) the establishment of a "likelihood of dilution" standard for dilution claims, rather than an "actual dilution" standard; (2) a provision that noninherently distinctive marks may qualify for protection; (3) a reconfiguration of the factors used to determine whether a mark is famous for dilution purposes, including a rejection of dilution claims based on "niche" fame; (4) the specification of separate and explicit causes of action for dilution by blurring and dilution by tarnishment; and (5) an expanded set of exclusions. See 15 U.S.C. § 1125(c).

than infringement and unfair competition laws, tread very close to granting "rights in gross" in a trademark. *Id.*

The legislative history of the TDRA shows a similar concern. Congress passed the anti-dilution legislation because it sought to protect . . . famous marks from those "attempt[ing] to trade upon the goodwill and established renown of such marks," regardless of whether such use causes a likelihood of confusion about the product's origin. H.R. Rep. No. 104–374, at 3 (1995), reprinted in 1995 U.S.C.-C.A.N. 1029, 1030. The legislative history speaks of protecting those marks that have an "aura" and explains that the harm from dilution occurs "when the unauthorized use of a famous mark reduces the public's perception that the mark signifies something unique, singular, or particular." *Id.*; S. Rep. No. 100–515, at 7 (1988). For example, such harm occurs in the hypothetical cases of "DUPONT shoes, BUICK aspirin, and KODAK pianos," according to the legislative history. *Moseley v. V. Secret Catalogue, Inc.*, 537 U.S. 418, 431 (2003) (quoting H.R. Rep. No. 104–374, at 3 (1995), as reprinted in 1995 U.S.C.C.A.N. 1029, 1030). This is true even though a consumer would be unlikely to confuse the manufacturer of KODAK film with the hypothetical producer of KODAK pianos. In short, for purposes of § 1125(c), a mark usually will achieve broad-based fame only if a large portion of the general consuming public recognizes that mark. *Thane Intern., Inc. v. Trek Bicycle Corp.*, 305 F.3d 894, 911 (9th Cir. 2002). In other words, the mark must be a household name. *Id.*

Under the TDRA, four non-exclusive factors are relevant when determining whether a mark is sufficiently famous for anti-dilution protection:

(i) The duration, extent, and geographic reach of advertising and publicity of the mark, whether advertised or publicized by the owner or third parties;

(ii) The amount, volume, and geographic extent of sales of goods or services offered under the mark;

(iii) The extent of actual recognition of the mark;

(iv) Whether the mark was registered under the Act of March 3, 1881, or the Act of February 20, 1905, or on the principal register.

15 U.S.C. § 1125(c)(2).

Last things first. The LSL is registered, and has been for over 20 years according to UT. KST does not dispute this. However, as the leading commentator notes, "[o]ne cannot logically infer fame from the fact that a mark is one of the millions on the federal Register. On the other hand, one could logically infer lack of fame from a lack of registration . . . ." 4 J. Thomas McCarthy, *McCarthy On Trademarks And Unfair Competition*, § 24:106, at 24-293 (4th ed.).

KST's primary argument goes to the third issue, the actual recognition of the mark, and contends that UT's mark is not sufficiently recognized on a national level to be famous. They base this argument on their expert, Robert Klein, who conducted a national survey he contends demonstrates that only 5.8% of respondents in the United States "associated the UT registered longhorn logo with UT alone" and that "only 21.1% of respondents in Texas associated the UT registered longhorn logo with UT alone." As an initial matter, it simply does not matter, for purposes of a federal dilution cause of action, what the results are of a survey conducted of people in the Austin metropolitan area or the State of Texas. As noted above, the TDRA specifically requires that the mark be "*widely recognized by the general consuming public of the United States.*" 15 U.S.C. § 1125(c)(2)(A). The Court must presume that this is exactly why UT did not invoke, in its briefing on the dilution issue, the study it commissioned by a University of Houston professor that was conducted solely in the metropolitan Austin area.

Instead, UT, via its expert, attacks the methodology and assumptions of Klein's nationwide survey, while championing its evidence, detailed in full below, that UT has permeated the national consciousness primarily through the success of its football program (circumstantial evidence that touches on the first two famousness factors). Klein conducted his survey using a test group and a control group. For the test group he used the UT LSL that is registered with the Federal Register (however he showed it to participants in white with a white background, rather than the burnt orange color it almost invariably appears in, or against). For the control group he used the longhorn or "the cow's head" from the Longhorn World Championship Rodeo logo.[14] Klein contends that he used this alternative longhorn depiction because "[r]espondents who associate the 'control' symbol with UT are either guessing or indicating that any depiction of a longhorn steer will be associated with UT." Among the survey respondents in Texas—slightly less than half of the total survey respondents (the rest were from the different geographical regions across the U.S.)—who were shown the control logo, nearly 22% associated it with UT alone, and another 26.5% associated it with UT and another "company, organization, or place." Across the rest of the United States, only 5.3% in the control associated it with UT alone (and an additional 6.1% with UT and some other entity).

UT raises several arguments contesting the validity of Klein's results. First, it complains that Klein presented the LSL to the survey respondents in a biased fashion. The evidence shows that Klein did not test the LSL in the burnt orange color in which it almost invariably appears (apparently survey respondents were shown a "whited out" version of the LSL). When confronted with this fact—that the mark he was ostensibly attempting to test was not being shown to the general consuming public in the manner in which they would be exposed to in the marketplace—Klein stated UT's mark, the LSL, is used in a wide variety of contexts, and picking any one of them would be to bias the results. This response seems disingenuous. It is difficult to understand how it would bias the results to show the general consuming public the mark in the typical circumstance in which the public would confront that mark in the marketplace, rather than showing them it in a manner in which they would likely not confront it.

UT also correctly points out that Klein's coding of a number of survey responses seemed designed to favor KST's position in the litigation. For example, Klein did not count respondents who, upon seeing the LSL and being asked if they

---

[14] Note that in his survey Klein only used the longhorn portion of this graphic (i.e., not the entire logo).

associated that mark with any "company, organization, or place," responded by stating "Texas football" or "the Texas Longhorns" because the response did not make a reference to an academic institution. Klein also admitted to numerous errors in coding the responses. Most troublesome is that the coding errors almost invariably were to the benefit of KST and the detriment of UT. Klein was exacting when it came to what responses would be classified as validly associating the LSL with UT. To give just one example, Klein coded a response that stated that s/he associated the LSL with "Texas college" and the reason for that being that "That's the image on helmets" as not being associated with UT.

UT also contends that Klein's survey question was "leading" because it asked respondents whether they associated the LSL with any "company, organization, or place" (thus implicitly directing respondents to simply say "Texas," which would then be coded unfavorably to UT). UT also has pointed out flaws in the standard Klein used to determine whether a response should be coded as associating the LSL with UT alone (which is the only response which "counts" as demonstrating fame in Klein's survey).[16]

This is not the first time Klein has been criticized, *see Urban Outfitters, Inc. v. BCBG Max Azria Group, Inc.*, 511 F.Supp.2d 482, 499 (E.D. Pa. 2007) ("The Court finds that [Klein's] survey's use of the 'classic' control suffers from a significant shortcoming that limits its probative value in showing a likelihood of confusion between the brands."); *Avocados Plus, Inc. v. Johanns*, 2007 WL 172305, *3, No. 02–1798 (D.D.C. Jan. 23, 2007) ("In any event, the [Klein's] survey itself is fatally flawed"), and the Court hopes "that he will take these criticisms to heart in his next courtroom appearance." *Indianapolis Colts*, 34 F.3d at 415 (Judge Posner noting that one of the parties' expert's study "was a survey that consisted of three loaded questions asked in one Baltimore mall."). These criticisms significantly undermine Klein's conclusion, and the Court must, at this stage of the litigation, find that KST has failed to demonstrate through its summary judgment evidence that the LSL is not "famous" for dilution purposes.

. . . [B]ecause UT bears the burden of proof on the famousness issue at trial, this means that UT must submit sufficient evidence to demonstrate that there is at least a triable issue under the TDRA that the LSL is "famous."

In its response, UT offers evidence of famousness that at first blush appears impressive. Upon closer scrutiny, however, it is apparent that the evidence submitted is evidence of "niche" fame, which is a category of fame to which the TDRA explicitly does not apply. To summarize, UT's response contains evidence that UT football games are regularly nationally televised on ABC and ESPN, and the LSL is prominently featured as UT's logo during these broadcasts. Similarly,

---

[16] Although not raised by UT, the court also notes that Klein's sample size of 454 seems small for a national sample. For example, that sample size would not pass muster for a reputable national poll (like a Gallup or Zogby). *See* Frank Newport et. al, How Are Polls Conducted?, in WHERE AMERICA STANDS (Michael Golay, ed. 1997), available at http://media.gallup.com/PDF/FAQ/HowArePolls.pdf. Gallup and other major polling and survey organizations typically have a sample size of 1,000–1,500 individuals for national polls. *Id.* at 3.

the men's college basketball team's games have been televised nationally 97 times in the past five seasons. UT points to the Bowl Championship Series (BCS) Rose Bowl national championship game, in which UT beat the University of Southern California. That game was, at that point, the highest rated game in the eight-year history of the BCS and was also the highest rated college football game since 1987. Over 35 million people watched it nationwide. Along these same lines, the 2006 Alamo Bowl game between UT and Iowa was the most watched bowl game in ESPN's history, with nearly 9 million viewers. Spanning from 1963–2006, UT football players have been featured solely or as a part of the cover of Sports Illustrated ten times (although the logo is not featured prominently or totally visible on all these covers). A writer from SI.com (Sports Illustrated's website) named UT's football helmet as the number 1 non-letter, i.e., only logo, helmet. That same helmet was displayed on two separate Wheaties' boxes: one celebrating UT's national BCS win and the other "commemorating UT's rivalry game with Texas A&M." UT sells "official corporate sponsorships" for $250,000 each (at a minimum) to companies such as Coca-Cola, Dodge, Nike, Pizza Hut, State Farm, and Wells Fargo. The Collegiate Licensing Company (CLC), which licenses the merchandise for many if not most major universities, reported that UT holds the record for most royalties earned in a single year and has been the number one university for licensing royalties for the past two years. Coming in closely behind Notre Dame, Forbes recently valued UT's football program as the second most valuable in the country. Finally, retail sales of UT products in stores such as Wal-Mart and Target totaled nearly $400 million in 2005–06.[18]

UT contends that all of this is enough to at least create a fact issue at to whether the LSL is famous. In support of this they cite an Eastern District of Virginia case for the proposition that 41% recognition by the general consuming public as enough for Ringling Bros. Circus' mark "Greatest Show on Earth" to be deemed sufficiently famous. *Ringling Brothers-Barnum & Bailey Combined Shows, Inc. v. Utah Div. of Travel Development*, 955 F. Supp. 605, 613 & n. 4 (E.D. Va. 1997), *aff'd*, 170 F.3d 449 (4th Cir. 1999). The problem with this proposition is that this case was decided under the FTDA, which was amended by the TDRA in 2006. One of the central revisions of the TDRA was to make it more difficult to get the "rights in gross" that a famous mark is entitled to. *Avery Dennison*, 189 F.3d at 875. Getting rid of niche fame, the federal trademark law (as already noted) now requires that the mark be "widely recognized by the general consuming public of the United States." 15 U.S.C. § 1125(c)(2)(A).

And the central problem for UT is that its circumstantial evidence is largely evidence of niche market fame. Reading through the evidence, it is not at all clear that if one is not a college football fan (or, to a much lesser extent, college baseball or basketball fan) [one] would recognize the LSL as being associated with UT, as all of the evidence relates to the use of the logo in sporting events. The Court is

---

[18] UT notes a few other pieces of evidence—for example, a *Texas Monthly* cover story with the LSL on it—however they have little to no relevance to determining the knowledge of the general consuming public.

well aware that NCAA college football is a popular sport—the Court counts itself as a more than casual fan of Saturday afternoon football in the Fall—but this hardly equals a presence with the general consuming public (nearly the entire population of the United States). Simply because UT athletics have achieved a level of national prominence does not necessarily mean that the longhorn logo is so ubiquitous and well-known to stand toe-to-toe with Buick or KODAK.

Nor is it determinative that well-known corporations pay to sponsor UT. The University of Florida's media guide lists Nike as an official sponsor. *See* Florida Gator Football Media Guide, Table of Contents, *available at* http://www.gatorzone.com/football/ media/2007/pdf/2.pdf. Similarly, on the Florida Gator's home page for its football team there are ''Gator Links'' to SIRIUS satellite radio and radio behemoth Clear Channel. *See* http://www.gatorzone.com/football/ (last visited January 29, 2008). A visit to the University of Michigan's football team's website reveals sponsorships by AT&T and State Farm Insurance. *See* M Go Blue Football, http://www.mgoblue.com/Sport/Default.aspx?sid=516&id=12116.19. None of this is evidence that these programs' logos are nationally famous.

One of the major purposes of the TDRA was to restrict dilution causes of action to those few truly famous marks like Budweiser beer, *see Anheuser–Busch, Inc. v. Andy's Sportswear, Inc.*, 1996 WL 657219, No. C–96-2783, (N.D. Cal. Aug. 28, 1996) (Budweiser's mark ''unquestionably famous''); Camel cigarettes, *R.J. Reynolds Tobacco Co. v. Premium Tobacco Stores, Inc.*, 2004 WL 1613563, No. 99–C–1174 (N.D. Ill. July 19, 2004); Barbie Dolls, *Mattel, Inc. v. Jcom, Inc.*, 1998 WL 766711, No. 97 Civ. 7191(SS), (S.D.N.Y. Sep. 10, 1998), and the like. Despite the problems with Klein's study, the fact remains that UT has not created a genuine issue of material fact that the longhorn silhouette logo is ''a household name.'' *Thane Intern.*, 305 F.3d at 911. As one academic commentator put it, the ''TDRA is simply not intended to protect trademarks whose fame is at all in doubt.'' Barton Beebe, *A Defense of the New Federal Trademark Antidilution Law*, 16 Fordham Intell. Prop. Media & Ent. L.J. 1143, 1158 (2006); *see* also Marc L. Delflache et al, *Life After Moseley: The Trademark Dilution Act*, 16 Tex. Intell. Prop. L.J. 125, 142–43 (2007) (noting that the TDRA rejects niche fame). The leading trademarks treatise agrees. 4 J.Thomas McCarthy, *McCarthy On Trademarks And Unfair Competition*, § 24:104, at 24–280-81 (only ''truly eminent and widely recognized marks'' should be given the famous label). Because UT's evidence fails to demonstrate the extremely high level of recognition necessary to show ''fame'' under the TDRA, summary judgment is appropriate on this claim.

. . . .

Notes & Questions

1. What does it mean for a mark to be a "household name"? How does one prove that? It is tempting to use survey evidence in trademark cases, but surveys must be constructed carefully because they are prone to rejection by the courts on a variety of grounds. Additionally, lawyers must choose their experts carefully. Did the lawyers for KST do their due diligence in selecting their expert?

2. Review the decision in *Starbucks*. Why didn't the accused diluter challenge the premise that the Starbucks mark is famous? Is it possible for a non-profit entity to show famousness? What other kinds of evidence should the University have considered submitting? Would the nationwide distribution to high school students of catalogues prominently displaying the longhorn logo have assisted the University's claim? Or would that simply have been another example of niche fame?

Some examples of other marks that have been held to be famous include:

- LOUIS VUITTON (*Louis Vuitton Malletier S.A. v. Haute Diggity Dog, LLC*, 507 F.3d 252 (4th Cir. 2007));

- EBAY (*Perfumebay.com Inc. v. EBAY, Inc.*, 506 F.3d 1165 (9th Cir. 2007));

- AMERICA'S TEAM (*Dallas Cowboys Football Club, Ltd. v. America's Team Properties, Inc.*, 616 F. Supp. 2d 622 (N.D. Tex. 2009));

- VISA (*Visa Intern. Service Ass'n v. JSL Corp.*, 590 F. Supp. 2d 1306 (D. Nev. 2008));

- TIFFANY (*Tiffany (NJ) Inc. v. eBay, Inc.*, 576 F. Supp. 2d 463 (S.D.N.Y. 2008));

- NYC TRIATHLON (*New York City Triathlon, LLC v. NYC Triathlon Club, Inc.*, 704 F. Supp. 2d 305 (S.D.N.Y. 2010));

- VOLVO (*Volvo Trademark Holding AB v. Volvospares.com*, 2010 WL 1404175 (E.D.Va. 2010));

- CHEM-DRY (*Harris Research, Inc. v. Lydon*, 505 F. Supp. 2d 1161 (D. Utah 2007));

- VICTORIA'S SECRET (*V Secret Catalogue, Inc. v. Moseley*, 605 F.3d 382 (6th Cir. 2010));

- MARBLELIFE (*MarbleLife, Inc. v. Stone Resources, Inc.*, 759 F. Supp. 2d 552 (E.D. Pa. 2010));

- NEWPORT (*Lorillard Tobacco Co. v. Zoom Enterprises, Inc.*, 809 F. Supp. 2d 692 (E.D. Mich. 2011)); and,

- FORD logos (*Nat'l Bus. Forms & Printing, Inc. v. Ford Motor Co.*, 671 F.3d 526 (5th Cir. 2012)).

- BENTLEY (Bentley Motors Corp. v. McEntegart, 976 F. Supp. 2d 1297 (M.D. Fla. 2013))

- YELP (Yelp Inc. v. Catron, 70 F. Supp. 3d 1082 (N.D. Cal. 2014));

Marks that courts have concluded fail to meet the test for famousness include:

- LOKAR (*Gennie Shifter, LLC. v. Lokar, Inc.*, 2010 WL 126181 (D. Colo. 2010));

- Design and trade dress of the "Bellini Chair" (*Heller Inc. v. Design Within Reach, Inc.*, 2009 WL 2486054 (S.D.N.Y 2009));

- MOSCOW CATS THEATRE (*Kuklachev v. Gelfman*, 600 F. Supp. 2d 437 (E.D.N.Y. 2009));

- DIGISPLINT (*Silver Ring Splint Co. v. Digisplint, Inc.*, 567 F. Supp. 2d 847 (W.D. Va. 2008));

- RAAGA (*Vista India v. Raaga, LLC*, 501 F. Supp. 2d 605 (D.N.J. 2007));

- BAY STATE (*Bay State Sav. Bank v. Baystate Financial Services, LLC*, 484 F. Supp. 2d 205 (D. Mass. 2007));

- LINGO'S MARKET (*Lingo v. Lingo*, 785 F. Supp. 2d 443 (D. Del. 2011));

- TIMBER PRODUCTS (*Timber Products Inspection, Inc. v. Coastal Container Corp.*, 827 F. Supp. 2d 819 (W.D. Mich. 2011));

- COACH (Coach Services, Inc. v. Triumph Learning LLC, 668 F.3d 1356, 1366-67, 1376 (Fed. Cir. 2012));

- ERERLINA LAURICE (Harp v. Rahme, 984 F. Supp. 2d 398 (E.D. Pa. 2013));

- VAUGHN TRADE DRESS (Mike Vaughn Custom Sports, Inc. v. Piku, 15 F. Supp. 3d 735 (E.D. Mich. 2014));

- MOVIE MANIA (Movie Mania Metro, Inc. v. GZ DVD's Inc., 306 Mich. App. 594, 857 N.W.2d 677 (2014)); and,

- PROLACTO (Paleteria La Michoacana, Inc. v. Productos Lacteos Tocumbo S.A. De C.V., 69 F. Supp. 3d 175 (D.D.C. 2014)).

Some circuits had concluded, before the 2006 enactment of the TDRA, that the fame required could be market-specific, particularly if the alleged diluting use was directed at the same niche market. *See, e.g., Syndicate Sales, Inc. v. Hampshire Paper Corp.*, 192 F.3d 633, 640 (7th Cir. 1999); *Advantage Rent–A–Car, Inc. v. Enter. Rent–A–Car, Co.*, 238 F.3d 378, 380 (5th Cir. 2001) (mark "need only be famous within its specific industry or market"). Applying a niche-market fame test has led courts to conclude, even after the TDRA, that the following marks are famous and should be protected against dilution:

- PET SILK (Pet Silk, Inc. v. Jackson, 481 F. Supp. 2d 824 (S.D. Tex. 2007));

- GALVOTC and GA (*Galvotec Alloys, Inc. v. Gaus Anodes Int'l, LLC*, 2014 WL 6805458 (S.D. Tex. 2014).

3. In addition to proving famousness, a plaintiff must also prove that the mark is distinctive. Under what circumstances might a mark be famous but not be distinctive?

4. The decision in *KST Electric* involved a Report and Recommendation from a magistrate judge. The District Court adopted the Report and Recommendation without change, although the University of Texas did not object to the conclusion on famousness. This failure to object bars the University from attacking this conclusion on appeal. *Douglass v. United Servs. Auto. Ass'n,* 79 F.3d 1415 (5th Cir. 1996) (en banc).

5. If the University had been able to show famousness, what is the likelihood that it would have prevailed on its dilution claim? What other facts would you want to know to assess the University's likelihood of prevailing?

6. The owner of a *famous* mark, unlike others, can assert a cybersquatting claim, discussed above in Section III.E, against a defendant who "registers, traffics in, or uses a domain name that . . . is identical or confusingly similar to *or dilutive of* that mark." 15 U.S.C. § 1125(d)(1)(A) (emphasis added). To prevail, the mark owner will, of course, also need to show the defendant's "bad faith intent to profit" that is central to any cybersquatting claim.

V. COMPARATIVE ADVERTISING, FAIR USE, AND OTHER DEFENSES

The first trademark case in this Chapter involved the use of the mark "Cocoa Butter Formula" for beauty products that contained cocoa butter. In that case, the court determined that the phrase was descriptive and that the plaintiff had failed to show the required secondary meaning for the phrase to be protectable as a trademark. Even when a descriptive term or phrase has acquired distinctivenss and thus is protected as a mark, others remain free to use the term or phrase in its descriptive sense to characterize their own goods or services. This type of descriptive use is allowed so long as the term is not used in a source-identifying way. This defense is set forth in the statute. Specifically, the Lanham Act provides a defense to infringement if:

> the use of the name, term, or device charged to be an infringement is a use, *otherwise than as a mark* . . . of a term or device which is descriptive of and used fairly and in good faith only to describe the goods or services of such party, or their geographic origin;

15 U.S.C. § 1115(b)(4) (emphasis added).[*] This type of use by a defendant is sometimes called "descriptive fair use" and applies to marks that consist of words that describe a product or service, or an attribute of the product or service.

The Trademark Dilution Revision Act also provides several liability "exclusions":

> **(3) Exclusions.** The following shall not be actionable as dilution by blurring or dilution by tarnishment under this subsection:
> (A) Any fair use, including a nominative or descriptive fair use, or facilitation of such fair use, of a famous mark by another person *other than as a designation of source for the person's own goods or services*, including use in connection with—
> > (i) advertising or promotion that permits consumers to compare goods or services; or
> > (ii) identifying and parodying, criticizing, or commenting upon the famous mark owner or the goods or services of the famous mark owner.

---

[*] This defense is identified in the section of the statute concerning viable defenses for marks that have become incontestable. Courts have also expressly recognized this provision as a basis for this defense as to contestable marks. *M.B.H. Enterprises, Inc. v. WOKY, Inc.*, 633 F.2d 50, 52 n. 2 (7th Cir. 1980).

(B) All forms of news reporting and news commentary.

(C) Any noncommercial use of a mark.

15 U.S.C. § 1125(c) (emphasis added). As you read the cases in this Section, focus on the basis for the defense asserted by the defendant and the analysis employed by the court in determining whether the defense is applicable.

A. Advertising and Comparative Advertising

Smith v. Chanel, Inc.

402 F.2d 562 (9th Cir. 1968)

Browning, Judge:

Appellant R.G. Smith, doing business as Ta'Ron, Inc., advertised a fragrance called "Second Chance" as a duplicate of appellees' "Chanel No. 5," at a fraction of the latter's price. Appellees were granted a preliminary injunction prohibiting any reference to Chanel No. 5 in the promotion or sale of appellants' product. This appeal followed.

The action rests upon a single advertisement published in "Specialty Salesmen," a trade journal directed to wholesale purchasers. The advertisement offered "The Ta'Ron Line of Perfumes" for sale. It gave the seller's address as "Ta'Ron Inc., 26 Harbor Cove, Mill Valley, Calif." It stated that the Ta'Ron perfumes "duplicate 100% perfect the exact scent of the world's finest and most expensive perfumes and colognes at prices that will zoom sales to volumes you have never before experienced!" It repeated the claim of exact duplication in a variety of forms.

The advertisement suggested that a "Blindfold Test" be used "on skeptical prospects," challenging them to detect any difference between a well known fragrance and the Ta'Ron "duplicate." One suggested challenge was, "We dare you to try to detect any difference between Chanel #5 (25.00) and Ta'Ron's 2nd Chance. $7.00."

In an order blank printed as part of the advertisement each Ta'Ron fragrance was listed with the name of the well known fragrance which it purportedly duplicated immediately beneath. Below "Second Chance" appeared "*(Chanel #5)." The asterisk referred to a statement at the bottom of the form reading "Registered Trade Name of Original Fragrance House."

Appellees conceded below and concede here that appellants "have the right to copy, if they can, the unpatented formula of appellees' products." Moreover, for the purposes of these proceedings, appellees assume that "the products manufactured and advertised by [appellants] are *in fact* equivalents of those products manufactured by appellees." (Emphasis in original.) Finally, appellees disclaim any contention that the packaging or labeling of appellants' "Second Chance" is misleading or confusing.[4]

---

[4] Appellants' product was packaged differently from appellees', and the only words appearing on the outside of appellants' packages were "Second Chance Perfume by Ta'Ron." The same words ap-

I.

The principal question presented on this record is whether one who has copied an unpatented product sold under a trademark may use the trademark in his advertising to identify the product he has copied. We hold that he may, and that such advertising may not be enjoined under either the Lanham Act, or the common law of unfair competition, so long as it does not contain misrepresentations or create a reasonable likelihood that purchasers will be confused as to the source, identity, or sponsorship of the advertiser's product.

This conclusion is supported by . . . *Saxlehner v. Wagner*, 216 U.S. 375 (1910). . . .

In *Saxlehner* the copied product was a "bitter water" drawn from certain privately owned natural springs. The plaintiff sold the natural water under the name "Hunyadi Janos," a valid trademark. The defendant was enjoined from using plaintiff's trademark to designate defendant's "artificial" water, but was permitted to use it to identify plaintiff's natural water as the product which defendant was copying.

Justice Holmes wrote:

> We see no reason for disturbing the finding of the courts below that there was no unfair competition and no fraud. The real intent of the plaintiff's bill, it seems to us, is to extend the monopoly of such trademark or tradename as she may have to a monopoly of her type of bitter water, by preventing manufacturers from telling the public in a way that will be understood, what they are copying and trying to sell. But the plaintiff has no patent for the water, and the defendants have a right to reproduce it as nearly as they can. *They have a right to tell the public what they are doing, and to get whatever share they can in the popularity of the water by advertising that they are trying to make the same article, and think that they succeed. If they do not convey, but, on the contrary, exclude, the notion that they are selling the plaintiff's goods, it is a strong proposition that when the article has a well-known name they have not the right to explain by that name what they imitate.* By doing so, they are not trying to get the good will of the name, but the good will of the goods. 216 U.S. at 380-381 (citations omitted) (emphasis added.)

. . . .

. . . Since appellees' perfume was unpatented, appellants had a right to copy it, as appellees concede. There was a strong public interest in their doing so, "[f]or imitation is the life blood of competition. It is the unimpeded availability of substantially equivalent units that permits the normal operation of supply and demand to yield the fair price society must pay for a given commodity." *American Safety Table Co. v. Schreiber*, 269 F.2d 255, 272 (2d Cir. 1959). But this public benefit might be lost if appellants could not tell potential purchasers that appellants' product was the equivalent of appellees' product. "A competitor's chief weapon

---

peared on the front of appellants' bottles; the words "Ta'Ron trademark by International Fragrances, Inc., of Dallas and New York" appeared on the back.

is his ability to represent his product as being equivalent and cheaper * * *." Alexander, *Honesty and Competition*, 39 So. Cal. L. Rev. 1, 4 (1966). The most effective way (and, where complex chemical compositions sold under trade names are involved, often the only practical way) in which this can be done is to identify the copied article by its trademark or trade name. To prohibit use of a competitor's trademark for the sole purpose of identifying the competitor's product would bar effective communication of claims of equivalence. Assuming the equivalence of "Second Chance" and "Chanel No. 5," the public interest would not be served by a rule of law which would preclude sellers of "Second Chance" from advising consumers of the equivalence and thus effectively deprive consumers of knowledge that an identical product was being offered at one third the price.

As Justice Holmes wrote in *Saxlehner v. Wagner*, the practical effect of such a rule would be to extend the monopoly of the trademark to a monopoly of the product. The monopoly conferred by judicial protection of complete trademark exclusivity would not be preceded by examination and approval by a governmental body, as is the case with most other government-granted monopolies. Moreover, it would not be limited in time, but would be perpetual.

Against these considerations, two principal arguments are made for protection of trademark values other than source identification.

The first of these, as stated in the findings of the district court, is that the creation of the other values inherent in the trademark require "the expenditure of great effort, skill and ability," and that the competitor should not be permitted "to take a free ride" on the trademark owner's "widespread goodwill and reputation."

A large expenditure of money does not in itself create legally protectable rights. Appellees are not entitled to monopolize the public's desire for the unpatented product, even though they themselves created that desire at great effort and expense. As we have noted, the most effective way (and in some cases the only practical way) in which others may compete in satisfying the demand for the product is to produce it and tell the public they have done so, and if they could be barred from this effort appellees would have found a way to acquire a practical monopoly in the unpatented product to which they are not legally entitled.

Disapproval of the copyist's opportunism may be an understandable first reaction, "[b]ut this initial response to the problem has been curbed in deference to the greater public good." *American Safety Table Co. v. Schreiber*, 269 F.2d at 272. By taking his "free ride," the copyist, albeit unintentionally, serves an important public interest by offering comparable goods at lower prices. On the other hand, the trademark owner, perhaps equally without design, sacrifices public to personal interests by seeking immunity from the rigors of competition.

Moreover, appellees' reputation is not directly at stake. Appellants' advertisement makes it clear that the product they offer is their own. If it proves to be inferior, they, not appellees, will bear the burden of consumer disapproval.

The second major argument for extended trademark protection is that even in the absence of confusion as to source, use of the trademark of another "creates a serious threat to the uniqueness and distinctiveness" of the trademark, and "if continued would create a risk of making a generic or descriptive term of the words" of which the trademark is composed.

The contention has little weight in the context of this case. Appellants do not use appellees' trademark as a generic term. They employ it only to describe appellees' product, not to identify their own. They do not label their product "Ta'Ron's Chanel No. 5," as they might if appellees' trademark had come to be the common name for the product to which it is applied. Appellants' use does not challenge the distinctiveness of appellees' trademark, or appellees' exclusive right to employ that trademark to indicate source or sponsorship. For reasons already discussed, we think appellees are entitled to no more. The slight tendency to carry the mark into the common language which even this use may have is outweighed by the substantial value of such use in the maintenance of effective competition.

We are satisfied, therefore, that both authority and reason require a holding that in the absence of misrepresentation or confusion as to source or sponsorship a seller in promoting his own goods may use the trademark of another to identify the latter's goods. The district court's contrary conclusion cannot support the injunction.

. . . .

Notes & Questions

1. In a part of the opinion not reproduced above, the court reviewed some criticism of trademark in general:

> The object of much modern advertising is "to impregnate the atmosphere of the market with the drawing power of a congenial symbol," *Mishawaka Rubber & Woolen Mfg. Co. v. S. S. Kresge Co.*, 316 U.S. 203, 205 (1942), rather than to communicate information as to quality or price. The primary value of the modern trademark lies in the "conditioned reflex developed in the buyer by imaginative or often purely monotonous selling of the mark itself." Derring, *Trademarks on Noncompetitive Products*, 36 Or. L. Rev. 1, 2 (1956). To the extent that advertising of this type succeeds, it is suggested, the trademark is endowed with sales appeal independent of the quality or price of the product to which it is attached; economically irrational elements are introduced into consumer choices; and the trademark owner is insulated from the normal pressures of price and quality competition. In consequence the competitive system fails to perform its function of allocating available resources efficiently.
>
> Moreover, the economically irrelevant appeal of highly publicized trademarks is thought to constitute a barrier to the entry of new competition into the market. "[T]he presence of irrational consumer allegiances may constitute an effective barrier to entry. Consumer allegiances built over the years with intensive advertising, trademarks, trade names, copyrights and so forth extend substantial protection to firms already in the market. In some markets this barrier to entry may be insuperable." Papandreou, *The Economic Effects of Trademarks*, 44 Calif. L. Rev. 503, 508-09 (1956). High

barriers to entry tend, in turn, to produce "high excess profits and monop-olistic output restriction" and "probably * * * high and possibly excessive costs of sales promotion." J. Bain, *Barriers to New Competition* 203 (1955).

Smith v. Chanel, Inc., 402 F.2d at 566-67. The problems the court identifies are lessened by permitting comparative advertising, but do other aspects of the problems remain? Is this something that trademark law should be concerned with, or are these problems merely the costs of trademark protection that are outweighed by its benefits?

2. Store brands that imitate the trade dress of branded products could be thought of as comparative advertising, albeit of a quite aggressive type. As the court in *Smith v. Chanel* repeatedly cautions, comparative advertisements are permissible so long as the consumer is not confused. Comparative advertising cases thus can turn on the likelihood of confusion. Consider the example of sucralose products, sold by McNeil Nutritionals under the trademark "Splenda." This product and its competitors are sold in grocery stores, and each contains individual sweetener packets. Should McNeil's trademark and trade dress rights allow it to prevent the sale of the store-brand packages depicted below?

McNeil Industries instituted a lawsuit involving these packages and others. The district court refused to grant McNeil a preliminary injunction, noting the different colors used for different artificial sweeteners (red and pink for saccharin, blue for aspartame, and, now, yellow for sucralose) and consumers' high familiarity with store-brand products that seek to have shoppers compare their products with national brands by imitating components of the trade dress. *McNeil Nutritionals, LLC v. Heartland Sweeteners LLC*, 512 F. Supp. 2d 217 (E.D. Pa. 2007). The court held that consumers are not confused by this type of activity.

On appeal, the Third Circuit reversed in part. 511 F.3d 350 (3d Cir. 2007). While the court acknowledged that 90% of consumers were familiar with store brands, some of the packaging involved did not have prominent "store-specific signatures" and some of the packaging was too closely imitative of the overall packaging used by McNeil. The Third Circuit affirmed the district court's decision concerning the Safeway packaging (bottom left), but as to the Giant package (next page) the court concluded that McNeil had demonstrated a likelihood of

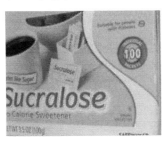

prevailing on the critical factor in an infringement case: likelihood of consumer confusion. "The danger in the District Court's result is that producers of store-brand products will be held to a lower standard of infringing behavior, that is, they effectively would acquire per se immunity as long as the store brand's name or logo appears somewhere on the allegedly infringing package, even when the name or logo is tiny." On remand, the district court concluded that McNeil had met the requirements for a preliminary injunction on the Giant package. *McNeil Nutritionals, LLC v. Heartland Sweeteners LLC*, 566 F. Supp. 2d 378 (E.D. Pa. 2008). Did the court strike the appropriate balance between protecting trademark owners and facilitating competition in non-patented products?

B. Descriptive Fair Use

KP Permanent Make-Up, Inc. v. Lasting Impression I, Inc.
543 U.S. 111 (2004)

SOUTER, JUSTICE:

. . . .

Each party to this case sells permanent makeup, a mixture of pigment and liquid for injection under the skin to camouflage injuries and modify nature's dispensations, and each has used some version of the term "micro color" (as one word or two, singular or plural) in marketing and selling its product. Petitioner KP Permanent Make-Up, Inc., claims to have used the single-word version since 1990 or 1991 on advertising flyers and since 1991 on pigment bottles. . . . In 1992, Lasting [Impression] applied to the United States Patent and Trademark Office (PTO) under 15 U.S.C. § 1051 for registration of a trademark consisting of the words "Micro Colors" in white letters separated by a green bar within a black square. The PTO registered the mark to Lasting in 1993, and in 1999 the registration became incontestable. § 1065.

It was also in 1999 that KP produced a 10-page advertising brochure using "microcolor" in a large, stylized typeface, provoking Lasting to demand that KP stop using the term. Instead, KP sued Lasting in the Central District of California, seeking, on more than one ground, a declaratory judgment that its language infringed no such exclusive right as Lasting claimed. Lasting counterclaimed, alleging, among other things, that KP had infringed Lasting's "Micro Colors" trademark.

KP sought summary judgment on the infringement counterclaim, based on the statutory affirmative defense of fair use, 15 U.S.C. § 1115(b)(4). After finding that Lasting had conceded that KP used the term only to describe its goods and not as a mark, the District Court held that KP was acting fairly and in good faith because undisputed facts showed that KP had employed the term "microcolor" continuously from a time before Lasting adopted the two-word, plural variant as a mark. Without enquiring whether the practice was likely to cause confusion, the court concluded that KP had made out its affirmative defense under § 1115(b)(4) and entered summary judgment for KP on Lasting's infringement claim.

On appeal, the Court of Appeals for the Ninth Circuit thought it was error for the District Court to have addressed the fair use defense without delving into the matter of possible confusion on the part of consumers about the origin of KP's goods. The reviewing court took the view that no use could be recognized as fair where any consumer confusion was probable, and although the court did not pointedly address the burden of proof, it appears to have placed it on KP to show absence of consumer confusion. Since it found there were disputed material facts relevant under the Circuit's eight-factor test for assessing the likelihood of confusion, it reversed the summary judgment and remanded the case.

We granted KP's petition for certiorari to address a disagreement among the Courts of Appeals on the significance of likely confusion for a fair use defense to a trademark infringement claim, and the obligation of a party defending on that ground to show that its use is unlikely to cause consumer confusion. . . .

II

. . . .

The holder of a registered mark (incontestable or not) has a civil action against anyone employing an imitation of it in commerce when "such use is likely to cause confusion, or to cause mistake, or to deceive." § 1114(1)(a). Although an incontestable registration is "conclusive evidence . . . of the registrant's exclusive right to use the . . . mark in commerce," § 1115(b), the plaintiff's success is still subject to "proof of infringement as defined in section 1114," *ibid*. And that, as just noted, requires a showing that the defendant's actual practice is likely to produce confusion in the minds of consumers about the origin of the goods or services in question. This plaintiff's burden has to be kept in mind when reading the relevant portion of the further provision for an affirmative defense of fair use, available to a party whose "use of the name, term, or device charged to be an infringement is a use, otherwise than as a mark, . . . of a term or device which is descriptive of and used fairly and in good faith only to describe the goods or services of such party, or their geographic origin . . . ." § 1115(b)(4).

Two points are evident. Section 1115(b) places a burden of proving likelihood of confusion (that is, infringement) on the party charging infringement even when relying on an incontestable registration. And Congress said nothing about likelihood of confusion in setting out the elements of the fair use defense in § 1115(b)(4).

Starting from these textual fixed points, it takes a long stretch to claim that a defense of fair use entails any burden to negate confusion. It is just not plausible that Congress would have used the descriptive phrase "likely to cause confusion, or to cause mistake, or to deceive" in § 1114 to describe the requirement that a markholder show likelihood of consumer confusion, but would have relied on the phrase "used fairly" in § 1115(b)(4) in a fit of terse drafting meant to place a defendant under a burden to negate confusion. " '[W]here Congress includes particular language in one section of a statute but omits it in another section of the same Act, it is generally presumed that Congress acts intentionally and purposely in the disparate inclusion or exclusion.' " *Russello* v. *United States,* 464 U.S. 16, 23 (1983) (quoting *United States* v. *Wong Kim Bo,* 472 F.2d 720, 722 (CA5 1972)) (alteration in original).[4]

Nor do we find much force in Lasting's suggestion that "used fairly" in § 1115(b)(4) is an oblique incorporation of a likelihood-of-confusion test developed in the common law of unfair competition. Lasting is certainly correct that some unfair competition cases would stress that use of a term by another in conducting its trade went too far in sowing confusion, and would either enjoin the use or order the defendant to include a disclaimer. But the common law of unfair competition also tolerated some degree of confusion from a descriptive use of words contained in another person's trademark. See, *e. g., William R. Warner & Co.* v. *Eli Lilly & Co.,* 265 U.S. 526, 528 (1924) (as to plaintiff's trademark claim, "[t]he use of a similar name by another to truthfully describe his own product does not constitute a legal or moral wrong, even if its effect be to cause the public to mistake the origin or ownership of the product"); *Canal Co.* v. *Clark,* 13 Wall. 311, 327 (1872) ("Purchasers may be mistaken, but they are not deceived by false representations, and equity will not enjoin against telling the truth"). While these cases are consistent with taking account of the likelihood of consumer confusion as one consideration in deciding whether a use is fair, see Part II-B, *infra,* they do not stand for the proposition that an assessment of confusion alone may be dispositive. Certainly one cannot get out of them any defense burden to negate it entirely.

Finally, a look at the typical course of litigation in an infringement action points up the incoherence of placing a burden to show nonconfusion on a defendant. If a plaintiff succeeds in making out a prima facie case of trademark infringement, including the element of likelihood of consumer confusion, the defendant may offer rebutting evidence to undercut the force of the plaintiff's evidence on this (or any) element, or raise an affirmative defense to bar relief even if the prima facie case is sound, or do both. But it would make no sense to give the defendant a defense of showing affirmatively that the plaintiff cannot succeed in proving some element (like confusion); all the defendant needs to do is to leave the factfinder unpersuaded that the plaintiff has carried its own burden on that point. A defendant has no need of a court's true belief when agnosticism will do. Put another way, it is only when a plaintiff has shown likely confusion by a preponderance of the evidence that a defendant could have any need of an affirmative defense, but under Lasting's theory the defense would be foreclosed in such a case. "[I]t defies logic to argue that a defense may not be asserted in the only situation where it even becomes relevant." *Shakespeare Co.* v. *Silstar Corp.,* 110 F.3d, at 243. Nor would it make sense to provide an affirmative defense of no confusion plus good faith, when merely rebutting the plaintiff's case on confusion would entitle the defendant to judgment, good faith or not.

   . . . .

---

[4] Not only that, but the failure to say anything about a defendant's burden on this point was almost certainly not an oversight, not after the House Subcommittee on Trademarks declined to forward a proposal to provide expressly as an element of the defense that a descriptive use be "'[un]likely to deceive the public.'" Hearings on H.R. 102 et al. before the Subcommittee on Trademarks of the House Committee on Patents, 77th Cong., 1st Sess., 167-168 (1941).

B

Since the burden of proving likelihood of confusion rests with the plaintiff, and the fair use defendant has no free-standing need to show confusion unlikely, it follows (contrary to the Court of Appeals's view) that some possibility of consumer confusion must be compatible with fair use, and so it is. The common law's tolerance of a certain degree of confusion on the part of consumers followed from the very fact that in cases like this one an originally descriptive term was selected to be used as a mark, not to mention the undesirability of allowing anyone to obtain a complete monopoly on use of a descriptive term simply by grabbing it first. *Canal Co.* v. *Clark,* 13 Wall., at 323-324, 327. The Lanham Act adopts a similar leniency, there being no indication that the statute was meant to deprive commercial speakers of the ordinary utility of descriptive words. "If any confusion results, that is a risk the plaintiff accepted when it decided to identify its product with a mark that uses a well known descriptive phrase." *Cosmetically Sealed Industries, Inc.* v. *Chesebrough-Pond's USA Co.,* 125 F.3d, at 30. See also *Park 'N Fly, Inc.* v. *Dollar Park & Fly, Inc.,* 469 U.S. 189, 201 (1985) (noting safeguards in Lanham Act to prevent commercial monopolization of language); *Car-Freshner Corp.* v. *S.C. Johnson & Son, Inc.,* 70 F.3d 267, 269 (CA2 1995) (noting importance of "protect[ing] the right of society at large to use words or images in their primary descriptive sense"). This right to describe is the reason that descriptive terms qualify for registration as trademarks only after taking on secondary meaning as "distinctive of the applicant's goods," 15 U.S.C. § 1052(f), with the registrant getting an exclusive right not in the original, descriptive sense, but only in the secondary one associated with the markholder's goods, 2 McCarthy § 11:45, p. 11-90 ("The only aspect of the mark which is given legal protection is that penumbra or fringe of secondary meaning which surrounds the old descriptive word").

. . . .

III

In sum, a plaintiff claiming infringement . . . must show likelihood of consumer confusion as part of the prima facie case, 15 U.S.C. § 1115(b), while the defendant has no independent burden to negate the likelihood of any confusion in raising the affirmative defense that a term is used descriptively, not as a mark, fairly, and in good faith, § 1115(b)(4).

Because we read the Court of Appeals as requiring KP to shoulder a burden on the issue of confusion, we vacate the judgment and remand the case for further proceedings consistent with this opinion.

Notes & Questions

1. Courts generally treat arguments of fair use as a defense to a claim of infringement. As a defense, it would be relevant only if a plaintiff can prove a prima facie case of infringement. However, when a defendant uses the words from the plaintiff's mark in a descriptive way, will consumers be confused? If descriptive uses are not confusing to consumers the plaintiff will fail in its prima facie case. Is

it more appropriate to consider descriptive fair use as part of the prima facie case or as an affirmative defense?

2. Comparative advertising is sometimes styled as a defense. However, as you saw in *Smith v. Chanel*, courts generally hold that comparative advertisements are permissible only if the consumer is not confused. Does the Supreme Court's decision in *KP Permanent* support an argument that some confusion might be permissible in the context of comparative advertising?

3. Another defense to trademark infringement involves the first sale or exhaustion doctrine: In order to infringe, a defendant must be using the mark to identify the source and distinguish *the defendant's* goods. Using a mark to accurately identify goods produced by the mark owner is *not* an infringement. In the case of the resale of goods, the origin of the goods has not changed. *See Enesco Corp. v. Price/Costco Inc.*, 146 F.3d 1083, 1085 (9th Cir. 1998). As long as there has been no material alteration in the product, the use of the mark in the context of reselling goods does not constitute infringement. *See, e.g., Davidoff & CIE, S.A. v. PLD Int'l Corp.*, 263 F.3d 1297 (11th Cir. 2001) (affirming finding that etching glass fragrance bottles to remove batch codes constituted a material alteration). We consider first sale doctrines in detail in Chapter 6.

C. Nominative / Noncommercial Fair Use

People other than the mark owner sometimes use the mark to refer to the products or services that the mark owner offers. For example, if a car mechanic specializes in repairing certain types of vehicles, she might advertise that she repairs "Toyotas" or "Volkswagens." Clearly, the mechanic is referring to another's product, but does so in order to accurately identify what she does. This type of use of a trademark has been held to constitute nominative fair use. *See Volkswagenwerk Aktiengesellschaft v. Church*, 411 F.2d 350 (9th Cir. 1969). The next case explores the contours of the nominative fair use defense.

Swarovski Aktiengesellschaft v. Building #19, Inc.
704 F.3d 44 (1st Cir. 2013)

Per Curiam:

. . . .

Building #19 is an off-price retail store that acquires products through secondary, non-traditional channels and then resells them at discounted prices in its 11 stores located throughout New Hampshire, Massachusetts and Rhode Island. The business is known by its motto, "Good Stuff Cheap," and aside from Swarovski crystal figurines, Building #19 has previously acquired and resold brand-name collectible merchandise made by companies such as Thomas Kinkade and M.I. Hummel. Building #19 spends millions of dollars on newspaper advertising in New England to market itself as a vendor of salvage, overstock and discontinued merchandise—these ads often feature descriptions of the advertised goods alongside humorous cartoons.

[Swarovski Aktiengesellschaft and Swarovski North America Limited (collectively, "Swarovski")] is a world-famous manufacturer and distributor of crystal, jewelry and other luxury products. It holds several registered federal trademarks for the mark "Swarovski," which it has used in the United States since at least 1969. Swarovski monitors and polices the use of its mark through a dedicated anti-infringement unit. It sells its crystal products online, in Swarovski retail stores (including stores in New Hampshire, Massachusetts and Rhode Island), in small independent retailers and in authorized national retailers such as Macy's, Bloomingdale's and Nordstrom. Swarovski merchandise may also be purchased through secondary channel re-sellers, eBay and other discount retailers.

In December 2011, Building #19 acquired a number of Swarovski crystal figurines, with a total retail value of approximately $500,000, from an insurer's salvage sale after a severe storm damaged the warehouse where they had been stored by their prior owner. The figurines were apparently unaltered and free from damage, and came boxed in their original packaging with the Swarovski Certificate of Authenticity. Although industry practice varies, in this case the salvor placed no restrictions on Building #19's ability to use the "Swarovski" name or product label. Swarovski itself never directly authorized Building #19 to use its trademark or name.

That same month, Building #19 conducted two, one-day-only sales of its Swarovski merchandise. It promoted the events through newspaper advertisements, one of which was headlined with the word "SWAROVSKI" in extra-large, capitalized, bold, and distinctive lettering. The bodies of the ads included cartoons of a tornado wrecking a warehouse, several pictures of crystal figurines, a list of the available items along with their prices and other text describing the details of the

sale. The name "Building #19" also appeared at the bottom of the advertisements in extra-large, capitalized, bold and distinctive lettering. At the events themselves, Building #19 set up a separate, roped-off display and sales area for the crystal along with various decorations featuring the word "Swarovski."

On December 7, 2011, Swarovski sent Building #19 a cease and desist letter objecting to the ads, and a week later, on December 14, Swarovski filed a complaint against Building #19 . . . alleg[ing] various claims for trademark infringement and unfair competition under the Lanham Act and Rhode Island statutory and common law. In response, Building #19 agreed to voluntarily refrain from further advertising or sale of the Swarovski crystal in its possession.

All was well until April 5, 2012, when Building #19 provided Swarovski with a copy of a proposed newspaper advertisement that it intended to run to

promote a Mother's Day sale of its remaining Swarovski crystal, and invited Swarovski to suggest any further steps it could take to avoid consumer confusion. The advertisement came crowned with a headline that read "ONE DAY EVENT 11AM to 8PM, Tornado Hits Warehouse containing GENUINE SWAROVSKI(R) CRYSTAL Collectibles." The word "Swarovski" appeared in extra-large, capitalized, bold and distinctive font. Like the previous ads, the body of the advertisement contained several images and text giving details on the sale. The name "Building #19" ran at the bottom of the add in extra-large, capitalized, bold and distinctive font. A disclaimer, in much smaller, unbolded font, also appeared near the bottom of the ad, reading: "Disclaimer: Building #19 is selling GENUINE SWAROVSKI(R) CRYSTAL products BUT Building #19 is NOT an authorized dealer, has no affiliation, connection or association with SWAROVSKI(R) and the standard SWAROVSKI(R) limited warranty is not available to our customers. (The lawyers told us we should add that big 'but'...)."

In response, Swarovski filed a motion for a preliminary injunction to forbid Building #19 from: (1) using the Swarovski trademark or name or any other marks or names or logos confusingly similar thereto in advertisements, in-store promotions, customer cards or signage of any kind, and (2) doing any act likely to induce the mistaken belief that Swarovski products or services are in any way affiliated, connected or associated with or sponsored by Building #19, including creating display areas tending to imply a "store within a store" event.

On May 1, 2012, the district court held an evidentiary hearing on Swarovski's motion, during which it heard arguments from counsel as well as witness testimony from several Building #19 employees and two Swarovski employees who attended the first crystal sale. After a brief recess, the district court acknowledged the need for an expedited determination, due to Building #19's wish to advertise before Mother's Day on May 13, and so the court departed from its usual practice of issuing a written decision and instead engaged "in a little more rough justice by providing an oral decision today." Ultimately, the court granted Swarovski's motion only to the extent that the capitalized word "Swarovski" at the top of the proposed advertisement could be no larger than the font used for the name "Swarovski" in the disclaimer at the bottom of the proposed advertisement. Neither party raised any objection to this decision at the time.

Building #19 now appeals the district court's order . . .

We begin with the district court's approach to the first prong of the preliminary injunction analysis, Swarovski's likelihood of success on its infringement claim against Building #19 . . .

To establish infringement successfully, a trademark plaintiff must demonstrate that the defendant used an imitation of its protected mark in commerce in a way that is "likely to cause confusion, or to cause mistake, or to deceive." 15 U.S.C. § 1114(1)(a) (2006). Historically, the subject of trademark "confusion" has been the source of the good or service to which the mark is attached. *See New Kids on the Block v. News Am. Publ'g, Inc.*, 971 F.2d 302, 305 (9th Cir. 1992). For instance, a shoddy crystal manufacturer might label its goods "Swarovski" or something similar in order to fool consumers into thinking they were buying the luxury product. "The typical situation in a trademark case involves the defendant's having passed off another's mark as its own or having used a similar name, confusing the public as to precisely whose goods are being sold." *Century 21 Real Estate Corp. v. LendingTree, Inc.*, 425 F.3d 211, 217 (3rd Cir. 2005). In this circuit, we evaluate the likelihood of such confusion through the eight-factor analysis laid out in *Pignons S.A. de Mecanique de Precision v. Polaroid Corp.*, 657 F.2d 482 (1st Cir. 1981).

But the "confusion" at issue in this case is of a different kind. Building #19 has not labeled its own products with the "Swarovski" mark, but instead wants to use the "Swarovski" mark to describe actual Swarovski crystal. Because some trade-marked products are so well-known and so unique, "many goods and services are effectively identifiable only by their trademarks." *New Kids on the Block*, 971 F.2d at 306. As the district court found, Swarovski crystal is among them. Unlike typical trademark infringement, where the defendant uses the plaintiff's mark to refer to the defendant's product, this so-called "nominative use" involves Building #19's use of Swarovski's mark to refer to Swarovski's own product.[3] *Id.* at 308.

The potential for confusion in a nominative use case is not one of source—here, the crystal really was manufactured by Swarovski—but rather one of endorsement or affiliation. The fear is that a consumer glancing at Building #19's proposed advertisement might mistakenly believe that Swarovski had some official association with the sale; perhaps that Swarovski sponsored the sale and so stood behind the goods as a direct seller, or that it had partnered with Building #19 in a way that might detract from its luxury status.

Because this kind of confusion does not implicate the traditional "source-identification function" of a trademark, other circuits, most notably the Ninth and the Third, have developed a distinct "nominative fair use" analysis to identify unlawful infringement in cases like this one. Although these courts differ on the precise articulation of the doctrine and on whether it should replace the standard likelihood-of-confusion analysis or should serve as an affirmative defense, they generally evaluate the lawfulness of a defendant's nominative use of a mark through the lens of three factors: (1) whether the plaintiff's product was identifiable without use of the mark; (2) whether the defendant used more of the mark than necessary; and (3) whether the defendant accurately portrayed the relationship between itself and the plaintiff. *See Toyota Motor Sales, U.S.A., Inc. v. Tabari*, 610 F.3d 1171, 1175-76 (9th Cir. 2010); *Century 21 Real Estate Corp.*, 425 F.3d at 222. In the First Circuit, we have recognized the "underlying principle" of nominative fair use, but like several other circuits, we have never endorsed any particular version of the

---

[3] "Nominative use" may also include situations where the defendant ultimately describes his own product, so long as his use of the mark is in reference to the plaintiff's product. For instance, a car mechanic who specializes in Volkswagens might put the word "Volkswagen" in an advertisement; he makes nominative use of the word in reference to another's product, but he does so in order to describe his own service. *See Century 21 Real Estate Corp.*, 425 F.3d at 214, 218 (citing *Volks-wagen-werk Aktiengesellschaft v. Church*, 411 F.2d 350 (9th Cir.1969)).

doctrine. *See Universal Commc'n Sys., Inc. v. Lycos, Inc.*, 478 F.3d 413, 424 (1st Cir. 2007).

Given the uncertainty in this area of the law, the district court made an admirable attempt to evaluate the likelihood that Swarovski would succeed on its infringement claim against Building #19. Ultimately, however, the district court's opinion, perhaps because it was delivered to provide an expedited resolution of the interlocutory motion contemporaneously with the hearing, did not include a finding on whether Building #19's use of the "Swarovski" mark in its proposed advertisement was likely to confuse consumers in order to support the issuance of the preliminary injunction. Nor are we able to infer such a finding from the court's reasoning.

There is a need for greater clarity on the matter of likely confusion in this case. A trademark holder's claim over his mark extends to uses of the mark "likely to cause confusion, or to cause mistake, or to deceive." 15 U.S.C. § 1114(1)(a). Swarovski may not charge infringement against all unauthorized uses of the "Swarovski" name, but only those uses likely to cause consumer confusion, mistake or deception. The Supreme Court has made clear that a trademark infringement action "requires a showing that the defendant's actual practice is likely to produce confusion in the minds of consumers," with the burden placed firmly on the plaintiff. *KP Permanent Make-Up, Inc. v. Lasting Impression I, Inc.*, 543 U.S. 111, 117-18 (2004). Without such a showing, no trademark infringement has occurred and so the trademark holder has no cause of action.

In this case, there is no indication from the record that the district court found a likelihood of confusion either under the traditional eight-factor *Pignons* test or under the nominative fair use test. The court first stated that Swarovski had tried to "shoehorn" this case into the analysis laid out in *Pignons*, but that *Pignons* "was a very different case from this." The court then applied the *Pignons* analysis, as it felt it was required to do by our precedent. . . . The court ultimately concluded that it was "not so sure that Plaintiff met its burden under the *Pignons* test." While this statement may not be a negative finding on the matter, it certainly does not indicate a positive finding that Swarovski had met its burden under the traditional *Pignons* confusion test.

The court next analyzed the advertisement through the three nominative fair use factors, as Building #19 urged it to do in its opposition motion to Swarovski's motion for a preliminary injunction, and which the court believed were more appropriate to the facts of this case. Yet the court never explained if it was using those factors to measure the likelihood of consumer confusion. The court described the three nominative fair use factors as: (1) whether "the product [was] readily identifiable without use of the mark," (2) whether the defendant "utilize[d] more of the mark than [was] necessary," and (3) whether the defendant "falsely suggested that [it] was sponsored or endorsed by the trademark holder." It found that the first and third factors weighed in favor of Building #19: Swarovski crystal was not readily identifiable without use of the "Swarovski" name, and the proposed ad made clear the product's origin, how it came into the hands of Building #19 and that the sale was not sponsored by Swarovski. Presumably, then, these factors did not suggest a likelihood of confusion to the district court.

In its application of the second factor, on which its decision against Building #19 must have turned, the district court stated that in the headline of the proposed advertisement, "the name Swarovski is larger than any other font in that ad. It's larger than the name Building #19. It's larger than the words 'one-day event.' It is clearly, I think, more use of the mark than is necessary to make the point." Nevertheless, the court apparently believed that the font and size of the word "Swarovski" as it appeared in the advertisement's disclaimer was sufficient to attract the attention of anyone reading the advertisement. The court concluded that "Defendant has some issues with its burden under the nominative fair use test because I think that they are wanting to use more of the mark than is necessary to describe the product."

This analysis does not mention consumer confusion, but merely decided that Building #19 used "more of the mark than necessary" to effectively communicate its message. But as we have explained, a trademark holder has no right to police "unnecessary" use of its mark. Whether necessary or not, a defendant's use of a mark must be confusing in the relevant statutory sense for a plaintiff to raise a viable infringement claim. *See Prestonettes, Inc. v. Coty*, 264 U.S. 359, 368 (1924) (Holmes, J.) ("When the mark is used in a way that does not deceive the public we see no such sanctity in the word as to prevent its being used to tell the truth. It is not taboo."); *Dow Jones & Co., Inc. v. Int'l Sec. Exch., Inc.*, 451 F.3d 295, 308 (2d Cir. 2006) ("While a trademark conveys an exclusive right to the use of a mark in commerce in the area reserved, that right generally does not prevent one who trades a branded product from accurately describing it by its brand name, so long as the trader does not create confusion by implying an affiliation with the owner of the product.").

The word "Swarovski" in the headline of Building #19's proposed advertisement might, in theory, have created confusion in a number of ways: it might have made the ad look like one from Swarovski rather than Building #19; it might have lead viewers to believe that Swarovski endorsed or sponsored the sale; or it might have suggested some official affiliation between Swarovski and Building #19. But the district court referenced none of these possibilities in its analysis; instead, it simply applied the second nominative fair use factor in order to restrict Building #19's use of the word "Swarovski" to the minimum font size necessary to convey its message.

Part of the difficulty may have stemmed from the fact that different circuits use the doctrine to measure different concepts. In the Ninth Circuit, the three factors are applied to determine confusion; they "replace[]" the traditional likelihood of confusion analysis "as the proper test for likely consumer confusion whenever defendant asserts to have referred to the trademarked good itself." *Toyota Motor Sales*, 610 F.3d at 1182 (internal quotation marks omitted). But in the Third Circuit, the three factors "demonstrate *fairness*"; once the plaintiff has proven likely

confusion through the traditional analysis, the defendant may invoke the nominative fair use factors to show that its use of the mark was fair. *Century 21 Real Estate
Corp.*, 425 F.3d at 222 (emphasis added).

We are especially wary here because the court's conclusion that Building #19
"has some issues with its burden under the nominative fair use test" suggests that
it had the Third Circuit's "affirmative defense" iteration of the doctrine in mind.
As we explained above, the Third Circuit applies the nominative fair use factors
to measure fairness, not confusion, and a defendant's failure to meet its burden
under that version of the test means nothing if the plaintiff has not first demonstrated a likelihood of confusion through the traditional analysis. *See id.* ("Once
plaintiff has met its burden of proving that confusion is likely, the burden then
shifts to defendant to show that its nominative use of plaintiff's mark is fair.").

Without at this time endorsing any particular approach to the nominative fair
use doctrine, it is enough to observe that whether the factors serve as the plaintiff's case-in-chief (as appears to be true in the Ninth Circuit) or as an affirmative
defense (as in the Third) a trademark defendant has no burden to prove anything
until the plaintiff has first met its responsibility to show infringement by demonstrating that the defendant's use of its mark is likely to confuse consumers. *Cf. KP
Permanent Make-Up, Inc.*, 543 U.S. at 120, 125 S.Ct. 542 ("[I]t is only when a plaintiff has shown likely confusion by a preponderance of the evidence that a defendant could have any need of an affirmative defense."). The district court erred by
granting Swarovski's request for a preliminary injunction without first finding that
Building #19's use of its mark in the proposed advertisement was likely to cause
confusion.

. . . .

Notes & Questions

1. It is a tricky exercise to balance the right of a trademark owner to prevent
uses of its mark that might confuse consumers, particularly about a sponsorship
or affiliation, and the rights of others to accurately identify products or services
that they sell. Do you think the First Circuit's framework will provide the right
balance? Consider the situation of compatible products, for example cases designed to fit an iPad, a trademark owned by the Apple Corporation. Should the
case manufacturer be free to advertise the use of the cases for iPads? What if the
case manufacturer wanted to use the distinctive apple silhouette? Should the manufacturer need to get a license from Apple for either of these uses? If the company
came to you seeking advice about whether a license was needed from Apple, how
would the Third Circuit's approach influence your analysis?

2. Must nominative fair use be treated similarly to descriptive fair use? While
the Supreme Court's analysis in *KP Permanent* concerned descriptive fair use, the
Court's approach clearly influenced the First Circuit. In what ways is nominative
fair use similar to descriptive fair use? In what ways is it different?

3. Both *KP Permanent* and *Swarovski* involved claims of trademark infringement. What if the plaintiffs had asserted claims for dilution? The Trademark Dilution Revision Act expressly provides that "[a]ny fair use, including a nominative or descriptive fair use . . . of a famous mark by another person other than as a designation of source for the person's own goods" shall not be actionable as dilution. 15 U.S.C. § 1125(c)(3). Would this defense be viable on the facts of *KP Permanent*? How about on the facts of *Swarovski*?

4. As you have seen throughout this Chapter, trademark owners object to a wide array of uses of "their" marks. It can be helpful to remember these words from the First Circuit: "a trademark holder has no right to police 'unnecessary' use of its mark. Whether necessary or not, a defendant's use of a mark must be confusing in the relevant statutory sense for a plaintiff to raise a viable infringement claim." Keep this in mind as you explore the next section on fair use.

D. Fair Use and Parody

Louis Vuitton Malletier S.A. v. Haute Diggity Dog, LLC
507 F.3d 252 (4th Cir. 2007)

Niemeyer, Judge:

Louis Vuitton Malletier S.A., a French corporation located in Paris, that manufactures luxury luggage, handbags, and accessories, commenced this action against Haute Diggity Dog, LLC, a Nevada corporation that manufactures and sells pet products nationally, alleging trademark infringement under 15 U.S.C. § 1114(1)(a) [and] trademark dilution under 15 U.S.C. § 1125(c)[.] . . .

. . . .

. . . In connection with the sale of its products, LVM has adopted trademarks and trade dress that are well recognized and have become famous and distinct. Indeed, in 2006, BusinessWeek ranked LOUIS VUITTON as the 17th "best brand" of all corporations in the world and the first "best brand" for any fashion business.

LVM has registered trademarks for "LOUIS VUITTON," in connection with luggage and ladies' handbags (the "LOUIS VUITTON mark"); for a stylized monogram of "LV," in connection with traveling bags and other goods (the "LV mark"); and for a monogram canvas design consisting of a canvas with repetitions of the LV mark along with four-pointed stars, four-pointed stars inset in curved diamonds, and four-pointed flowers inset in circles, in connection with traveling bags and other products (the "Monogram Canvas mark"). In 2002, LVM adopted a brightly-colored version of the Monogram Canvas mark in which the LV mark and the designs were of various colors and the background was white (the "Multicolor design"), created in collaboration with Japanese artist Takashi Murakami. For the Multicolor design, LVM obtained a copyright in 2004. In 2005, LVM adopted another design consisting of a canvas with repetitions of the LV mark and smiling cherries on a brown background (the "Cherry design").

As LVM points out, the Multicolor design and the Cherry design attracted immediate and extraordinary media attention and publicity in magazines such as

Vogue, W, Elle, Harper's Bazaar, Us Weekly, Life and Style, Travel & Leisure, People, In Style, and *Jane*. The press published photographs showing celebrities carrying these handbags, including Jennifer Lopez, Madonna, Eve, Elizabeth Hurley, Carmen Electra, and Anna Kournikova, among others. When the Multicolor design first appeared in 2003, the magazines typically reported, "The Murakami designs for Louis Vuitton, which were the hit of the summer, came with hefty price tags and a long waiting list." *People Magazine* said, "the wait list is in the thousands." The handbags retailed in the range of $995 for a medium handbag to $4500 for a large travel bag. The medium size handbag that appears to be the model for the "Chewy Vuiton" dog toy retailed for $1190. The Cherry design appeared in 2005, and the handbags including that design were priced similarly—in the range of $995 to $2740. LVM does not currently market products using the Cherry design.

The original LOUIS VUITTON, LV, and Monogram Canvas marks, however, have been used as identifiers of LVM products continuously since 1896.

During the period 2003-2005, LVM spent more than $48 million advertising products using its marks and designs, including more than $4 million for the Multicolor design. It sells its products exclusively in LVM stores and in its own in-store boutiques that are contained within department stores such as Saks Fifth Avenue, Bloomingdale's, Neiman Marcus, and Macy's. LVM also advertises its products on the Internet through the specific websites www.louisvuitton.com and www.eluxury.com.

Although better known for its handbags and luggage, LVM also markets a limited selection of luxury pet accessories—collars, leashes, and dog carriers—which bear the Monogram Canvas mark and the Multicolor design. These items range in price from approximately $200 to $1600. LVM does not make dog toys.

Haute Diggity Dog, LLC, which is a relatively small and relatively new business located in Nevada, manufactures and sells nationally—primarily through pet stores—a line of pet chew toys and beds whose names parody elegant high-end brands of products such as perfume, cars, shoes, sparkling wine, and handbags. These include—in addition to Chewy Vuiton (LOUIS VUITTON)—Chewnel No. 5 (Chanel No. 5), Furcedes (Mercedes), Jimmy Chew (Jimmy Choo), Dog Perignonn (Dom Perignon), Sniffany & Co. (Tiffany & Co.), and Dogior (Dior). The chew toys and pet beds are plush, made of polyester, and have a shape and design that loosely imitate the signature product of the targeted brand. They are mostly distributed and sold through pet stores, although one or two Macy's stores carries Haute Diggity Dog's products. The dog toys are generally sold for less than $20, although larger versions of some of Haute Diggity Dog's plush dog beds sell for more than $100.

Haute Diggity Dog's "Chewy Vuiton" dog toys, in particular, loosely resemble miniature handbags and undisputedly evoke LVM handbags of similar shape, design, and color. In lieu of the LOUIS VUITTON mark, the dog toy uses "Chewy Vuiton"; in lieu of the LV mark, it uses "CV"; and the other symbols and colors employed are imitations, but not exact ones, of those used in the LVM Multicolor and Cherry designs.

. . . .

II

LVM contends first that Haute Diggity Dog's marketing and sale of its "Chewy Vuiton" dog toys infringe its trademarks because the advertising and sale of the "Chewy Vuiton" dog toys is likely to cause confusion. See 15 U.S.C. § 1114(1)(a). LVM argues:

> The defendants in this case are using almost an exact imitation of the house mark VUITTON (merely omitting a second "T"), and they painstakingly copied Vuitton's Monogram design mark, right down to the exact arrangement and sequence of geometric symbols. They also used the same design marks, trade dress, and color combinations embodied in Vuitton's Monogram Multicolor and Monogram Cerises [Cherry] handbag collections. Moreover, HDD did not add any language to distinguish its products from Vuitton's, and its products are not "widely recognized."[23]

Haute Diggity Dog contends that there is no evidence of confusion, nor could a reasonable factfinder conclude that there is a likelihood of confusion, because it successfully markets its products as parodies of famous marks such as those of LVM. It asserts that "precisely because of the [famous] mark's fame and popularity . . . confusion is avoided, and it is this lack of confusion that a parodist depends upon to achieve the parody." Thus, responding to LVM's claims of trademark infringement, Haute Diggity Dog argues:

> The marks are undeniably similar in certain respects. There are visual and phonetic similarities. [Haute Diggity Dog] admits that the product name and design mimics LVM's and is based on the LVM marks. It is necessary for the pet products to conjure up the original designer mark for there to be a parody at all. However, a parody also relies on "equally obvious dissimilarit[ies] between the marks" to produce its desired effect.

Concluding that Haute Diggity Dog did not create any likelihood of confusion as a matter of law, the district court granted summary judgment to Haute Diggity Dog. We review its order de novo. *See CareFirst of Md., Inc. v. First Care, P.C.*, 434 F.3d 263, 267 (4th Cir. 2006).

To prove trademark infringement, LVM must show (1) that it owns a valid and protectable mark; (2) that Haute Diggity Dog uses a "reproduction, counterfeit, copy, or colorable imitation" of that mark in commerce and without LVM's consent; and (3) that Haute Diggity Dog's use is likely to cause confusion. 15 U.S.C. § 1114(1)(a); *Care-First*, 434 F.3d at 267. The validity and protectability of LVM's

---

[23] We take this argument to be that Haute Diggity Dog is copying too closely the marks and trade dress of LVM. But we reject the statement that LVM has a trademark consisting of the one word VUITTON. At oral argument, counsel for LVM conceded that the trademark is "LOUIS VUITTON," and it is always used in that manner rather than simply as "VUITTON." It appears that LVM has employed this technique to provide a more narrow, but irrelevant, comparison between its VUITTON and Haute Diggity Dog's "Vuiton." In resolving this case, however, we take LVM's arguments to compare "LOUIS VUITTON" with Haute Diggity Dog's "Chewy Vuiton."

marks are not at issue in this case, nor is the fact that Haute Diggity Dog uses a colorable imitation of LVM's mark. Therefore, we give the first two elements no further attention. To determine whether the "Chewy Vuiton" product line creates a likelihood of confusion, we have identified several nonexclusive factors to consider: (1) the strength or distinctiveness of the plaintiff's mark; (2) the similarity of the two marks; (3) the similarity of the goods or services the marks identify; (4) the similarity of the facilities the two parties use in their businesses; (5) the similarity of the advertising used by the two parties; (6) the defendant's intent; and (7) actual confusion. *See Pizzeria Uno Corp. v. Temple*, 747 F.2d 1522, 1527 (4th Cir. 1984). These *Pizzeria Uno* factors are not always weighted equally, and not all factors are relevant in every case.

Because Haute Diggity Dog's arguments with respect to the *Pizzeria Uno* factors depend to a great extent on whether its products and marks are successful parodies, we consider first whether Haute Diggity Dog's products, marks, and trade dress are indeed successful parodies of LVM's marks and trade dress.

For trademark purposes, "[a] 'parody' is defined as a simple form of entertainment conveyed by juxtaposing the irreverent representation of the trademark with the idealized image created by the mark's owner." *People for the Ethical Treatment of Animals v. Doughney* ("*PETA*"), 263 F.3d 359, 366 (4th Cir. 2001) (internal quotation marks omitted). "A parody must convey two simultaneous—and contradictory—messages: that it is the original, but also that it is not the original and is instead a parody." *Id.* (internal quotation marks and citation omitted). This second message must not only differentiate the alleged parody from the original but must also communicate some articulable element of satire, ridicule, joking, or amusement. Thus, "[a] parody relies upon a difference from the original mark, presumably a humorous difference, in order to produce its desired effect." *Jordache Enterprises, Inc. v. Hogg Wyld, Ltd.*, 828 F.2d 1482, 1486 (10th Cir. 1987) (finding the use of "Lardashe" jeans for larger women to be a successful and permissible parody of "Jordache" jeans).

When applying the *PETA* criteria to the facts of this case, we agree with the district court that the "Chewy Vuiton" dog toys are successful parodies of LVM handbags and the LVM marks and trade dress used in connection with the marketing and sale of those handbags. First, the pet chew toy is obviously an irreverent, and indeed intentional, representation of an LVM handbag, albeit much smaller and coarser. The dog toy is shaped roughly like a handbag; its name "Chewy Vuiton" sounds like and rhymes with LOUIS VUITTON; its monogram CV mimics LVM's LV mark; the repetitious design clearly imitates the design on the LVM handbag; and the coloring is similar. In short, the dog toy is a small, plush imitation of an LVM handbag carried by women, which invokes the marks and design of the handbag, albeit irreverently and incompletely. No one can doubt that LVM handbags are the target of the imitation by Haute Diggity Dog's "Chewy Vuiton" dog toys.

At the same time, no one can doubt also that the "Chewy Vuiton" dog toy is not the "idealized image" of the mark created by LVM. The differences are immediate, beginning with the fact that the "Chewy Vuiton" product is a dog toy, not an expensive, luxury LOUIS VUITTON handbag. The toy is smaller, it is plush, and virtually all of its designs differ. Thus, "Chewy Vuiton" is not LOUIS VUITTON ("Chewy" is not "LOUIS" and "Vuiton" is not "VUITTON," with its two Ts); CV is not LV; the designs on the dog toy are simplified and crude, not detailed and distinguished. The toys are inexpensive; the handbags are expensive and marketed to be expensive. And, of course, as a dog toy, one must buy it with pet supplies and cannot buy it at an exclusive LVM store or boutique within a department store. In short, the Haute Diggity Dog "Chewy Vuiton" dog toy undoubtedly and deliberately conjures up the famous LVM marks and trade dress, but at the same time, it communicates that it is not the LVM product.

Finally, the juxtaposition of the similar and dissimilar—the irreverent representation and the idealized image of an LVM handbag—immediately conveys a joking and amusing parody. The furry little "Chewy Vuiton" imitation, as something to be chewed by a dog, pokes fun at the elegance and expensiveness of a LOUIS VUITTON handbag, which must not be chewed by a dog. The LVM handbag is provided for the most elegant and well-to-do celebrity, to proudly display to the public and the press, whereas the imitation "Chewy Vuiton" "handbag" is designed to mock the celebrity and be used by a dog. The dog toy irreverently presents haute couture as an object for casual canine destruction. The satire is unmistakable. The dog toy is a comment on the rich and famous, on the LOUIS VUITTON name and related marks, and on conspicuous consumption in general. This parody is enhanced by the fact that "Chewy Vuiton" dog toys are sold with similar parodies of other famous and expensive brands—"Chewnel No. 5" targeting "Chanel No. 5"; "Dog Perignonn" targeting "Dom Perignon"; and "Sniffany & Co." targeting "Tiffany & Co."

We conclude that the *PETA* criteria are amply satisfied in this case and that the "Chewy Vuiton" dog toys convey "just enough of the original design to allow the consumer to appreciate the point of parody," but stop well short of appropriating the entire marks that LVM claims. *PETA*, 263 F.3d at 366 (quoting *Jordache*, 828 F.2d at 1486).

Finding that Haute Diggity Dog's parody is successful, however, does not end the inquiry into whether Haute Diggity Dog's "Chewy Vuiton" products create a likelihood of confusion. See 6 J. Thomas McCarthy, *Trademarks and Unfair Competition* § 31:153, at 262 (4th ed. 2007) ("There are confusing parodies and non-confusing parodies. All they have in common is an attempt at humor through the use of someone else's trademark"). The finding of a successful parody only influences the way in which the *Pizzeria Uno* factors are applied. Indeed, it becomes apparent that an effective parody will actually diminish the likelihood of confusion, while an ineffective parody does not. We now turn to the *Pizzeria Uno* factors.

A

As to the first *Pizzeria Uno* factor, the parties agree that LVM's marks are strong and widely recognized. They do not agree, however, as to the consequences of this fact. LVM maintains that a strong, famous mark is entitled, as a matter of law, to broad protection. While it is true that finding a mark to be strong and fa-mous usually favors the plaintiff in a trademark infringement case, the opposite may be true when a legitimate claim of parody is involved. As the district court observed, "In cases of parody, a strong mark's fame and popularity is precisely the mechanism by which likelihood of confusion is avoided." *Louis Vuitton Malletier*, 464 F. Supp. 2d at 499 (citing *Hormel Foods Corp. v. Jim Henson Prods., Inc.*, 73 F.3d 497, 503-04 (2d Cir. 1996); *Schieffelin & Co. v. Jack Co. of Boca, Inc.*, 850 F. Supp. 232, 248 (S.D.N.Y. 1994)). "An intent to parody is not an intent to confuse the public." *Jordache*, 828 F.2d at 1486.

We agree with the district court. It is a matter of common sense that the strength of a famous mark allows consumers immediately to perceive the target of the parody, while simultaneously allowing them to recognize the changes to the mark that make the parody funny or biting. *See Tommy Hilfiger Licensing, Inc. v. Nature Labs, LLC*, 221 F. Supp. 2d 410, 416 (S.D.N.Y. 2002) (noting that the strength of the "TOMMY HILFIGER" fashion mark did not favor the mark's owner in an infringement case against "TIMMY HOLEDIGGER" novelty pet perfume). In this case, precisely because LOUIS VUITTON is so strong a mark and so well recognized as a luxury handbag brand from LVM, consumers readily recognize that when they see a "Chewy Vuiton" pet toy, they see a parody. Thus, the strength of LVM's marks in this case does not help LVM establish a likelihood of confusion.

B

With respect to the second *Pizzeria Uno* factor, the similarities between the marks, the usage by Haute Diggity Dog again converts what might be a problem for Haute Diggity Dog into a disfavored conclusion for LVM.

Haute Diggity Dog concedes that its marks are and were designed to be some-what similar to LVM's marks. But that is the essence of a parody—the invocation of a famous mark in the consumer's mind, so long as the distinction between the marks is also readily recognized. While a trademark parody necessarily copies enough of the original design to bring it to mind as a target, a successful parody also distinguishes itself and, because of the implicit message communicated by the parody, allows the consumer to appreciate it.

In concluding that Haute Diggity Dog has a successful parody, we have im-pliedly concluded that Haute Diggity Dog appropriately mimicked a part of the LVM marks, but at the same time sufficiently distinguished its own product to communicate the satire. The differences are sufficiently obvious and the parody sufficiently blatant that a consumer encountering a "Chewy Vuiton" dog toy would not mistake its source or sponsorship on the basis of mark similarity.

This conclusion is reinforced when we consider how the parties actually use their marks in the marketplace. The record amply supports Haute Diggity Dog's contention that its "Chewy Vuiton" toys for dogs are generally sold alongside other pet products, as well as toys that parody other luxury brands, whereas LVM markets its handbags as a top-end luxury item to be purchased only in its own stores or in its own boutiques within department stores. These marketing channels further emphasize that "Chewy Vuiton" dog toys are not, in fact, LOUIS VUITTON products.

C

Nor does LVM find support from the third *Pizzeria Uno* factor, the similarity of the products themselves. It is obvious that a "Chewy Vuiton" plush imitation handbag, which does not open and is manufactured as a dog toy, is not a LOUIS VUITTON handbag sold by LVM. Even LVM's most proximate products—dog collars, leashes, and pet carriers—are fashion accessories, not dog toys. As Haute Diggity Dog points out, LVM does not make pet chew toys and likely does not intend to do so in the future. Even if LVM were to make dog toys in the future, the fact remains that the products at issue are not similar in any relevant respect, and this factor does not favor LVM.

D

The fourth and fifth *Pizzeria Uno* factors, relating to the similarity of facilities and advertising channels, have already been mentioned. LVM products are sold exclusively through its own stores or its own boutiques within department stores. It also sells its products on the Internet through an LVM-authorized website. In contrast, "Chewy Vuiton" products are sold primarily through traditional and Internet pet stores, although they might also be sold in some department stores. The record demonstrates that both LVM handbags and "Chewy Vuiton" dog toys are sold at a Macy's department store in New York. As a general matter, however, there is little overlap in the individual retail stores selling the brands.

Likewise with respect to advertising, there is little or no overlap. LVM markets LOUIS VUITTON handbags through high-end fashion magazines, while "Chewy Vuiton" products are advertised primarily through pet-supply channels.

The overlap in facilities and advertising demonstrated by the record is so minimal as to be practically nonexistent. "Chewy Vuiton" toys and LOUIS VUITTON products are neither sold nor advertised in the same way, and the de minimis overlap lends insignificant support to LVM on this factor.

E

The sixth factor, relating to Haute Diggity Dog's intent, again is neutralized by the fact that Haute Diggity Dog markets a parody of LVM products. As other courts have recognized, "An intent to parody is not an intent to confuse the public." *Jordache*, 828 F.2d at 1486. Despite Haute Diggity Dog's obvious intent to profit from its use of parodies, this action does not amount to a bad faith intent to create consumer confusion. To the contrary, the intent is to do just the opposite—to evoke a humorous, satirical association that distinguishes the products. This factor does not favor LVM.

F

On the actual confusion factor, it is well established that no actual confusion is required to prove a case of trademark infringement, although the presence of actual confusion can be persuasive evidence relating to a likelihood of confusion.

While LVM conceded in the district court that there was no evidence of actual confusion, on appeal it points to incidents where retailers misspelled "Chewy Vuiton" on invoices or order forms, using two Ts instead of one. Many of these invoices also reflect simultaneous orders for multiple types of Haute Diggity Dog parody products, which belies the notion that any actual confusion existed as to the source of "Chewy Vuiton" plush toys. The misspellings pointed out by LVM are far more likely in this context to indicate confusion over how to spell the product name than any confusion over the source or sponsorship of the "Chewy Vuiton" dog toys. We conclude that this factor favors Haute Diggity Dog.

In sum, the likelihood-of-confusion factors substantially favor Haute Diggity Dog. But consideration of these factors is only a proxy for the ultimate statutory test of whether Haute Diggity Dog's marketing, sale, and distribution of "Chewy Vuiton" dog toys is likely to cause confusion. Recognizing that "Chewy Vuiton" is an obvious parody and applying the *Pizzeria Uno* factors, we conclude that LVM has failed to demonstrate any likelihood of confusion. Accordingly, we affirm the district court's grant of summary judgment in favor of Haute Diggity Dog on the issue of trademark infringement.

III

LVM also contends that Haute Diggity Dog's advertising, sale, and distribution of the "Chewy Vuiton" dog toys dilutes its LOUIS VUITON, LV, and Monogram Canvas marks, which are famous and distinctive, in violation of the Trademark Dilution Revision Act of 2006 ("TDRA"), 15 U.S.C. § 1125(c). It argues, "Before the district court's decision, Vuitton's famous marks were unblurred by any third party trademark use." "Allowing defendants to become the first to use similar marks will obviously blur and dilute the Vuitton Marks." . . .

Haute Diggity Dog urges that, in applying the TDRA to the circumstances before us, we reject LVM's suggestion that a parody "automatically" gives rise to "actionable dilution." Haute Diggity Dog contends that only marks that are "identical or substantially similar" can give rise to actionable dilution, and its "Chewy Vuiton" marks are not identical or sufficiently similar to LVM's marks. It also argues that "[its] spoof, like other obvious parodies," "'tends to increase public identification' of [LVM's] mark with [LVM]," quoting *Jordache*, 828 F.2d at 1490, rather than impairing its distinctiveness, as the TDRA requires. . . .

. . . .

A

We address first LVM's claim for dilution by blurring.

The first three elements of a trademark dilution claim are not at issue in this case. LVM owns famous marks that are distinctive; Haute Diggity Dog has commenced using "Chewy Vuiton," "CV," and designs and colors that are allegedly diluting LVM's marks; and the similarity between Haute Diggity Dog's marks and LVM's marks gives rise to an association between the marks, albeit a parody. The issue for resolution is whether the association between Haute Diggity Dog's marks and LVM's marks is likely to impair the distinctiveness of LVM's famous marks.

In deciding this issue, the district court correctly outlined the six factors to be considered in determining whether dilution by blurring has been shown. See 15 U.S.C. § 1125(c)(2)(B). But in evaluating the facts of the case, the court did not directly apply those factors it enumerated. It held simply:

> [The famous mark's] strength is not likely to be blurred by a parody dog toy product. Instead of blurring Plaintiff's mark, the success of the parodic use depends upon the continued association with LOUIS VUITTON.

Louis Vuitton Malletier, 464 F. Supp. 2d at 505. The amicus supporting LVM's position in this case contends that the district court, by not applying the statutory factors, misapplied the TDRA to conclude that simply because Haute Diggity Dog's product was a parody meant that "there can be no association with the famous mark as a matter of law." Moreover, the amicus points out correctly that to rule in favor of Haute Diggity Dog, the district court was required to find that the "association" did not impair the distinctiveness of LVM's famous mark.

LVM goes further in its own brief, however, and contends:

> When a defendant uses an imitation of a famous mark in connection with related goods, a claim of parody cannot preclude liability for dilution.
>
> * * *
>
> The district court's opinion utterly ignores the substantial goodwill VUITTON has established in its famous marks through more than a century of exclusive use. Disregarding the clear Congressional mandate to protect such famous marks against dilution, the district court has granted [Haute Diggity Dog] permission to become the first company other than VUITTON to use imitations of the famous VUITTON Marks.

In short, LVM suggests that any use by a third person of an imitation of its famous marks dilutes the famous marks as a matter of law. This contention misconstrues the TDRA.

The TDRA prohibits a person from using a junior mark that is likely to dilute (by blurring) the famous mark, and blurring is defined to be an impairment to the famous mark's distinctiveness. "Distinctiveness" in turn refers to the public's recognition that the famous mark identifies a single source of the product using the famous mark.

To determine whether a junior mark is likely to dilute a famous mark through blurring, the TDRA directs the court to consider all factors relevant to the issue, including six factors that are enumerated in the statute:

> (i) The degree of similarity between the mark or trade name and the famous mark.
>
> (ii) The degree of inherent or acquired distinctiveness of the famous mark.
>
> (iii) The extent to which the owner of the famous mark is engaging in substantially exclusive use of the mark.
>
> (iv) The degree of recognition of the famous mark.

(v) Whether the user of the mark or trade name intended to create an association with the famous mark.

(vi) Any actual association between the mark or trade name and the famous mark.

15 U.S.C. § 1125(c)(2)(B). Not every factor will be relevant in every case, and not every blurring claim will require extensive discussion of the factors. But a trial court must offer a sufficient indication of which factors it has found persuasive and explain why they are persuasive so that the court's decision can be reviewed. The district court did not do this adequately in this case. Nonetheless, after we apply the factors as a matter of law, we reach the same conclusion reached by the district court.

We begin by noting that parody is not automatically a complete defense to a claim of dilution by blurring where the defendant uses the parody as its own designation of source, i.e., as a trademark. Although the TDRA does provide that fair use is a complete defense and allows that a parody can be considered fair use, it does not extend the fair use defense to parodies used as a trademark. As the statute provides:

> The following shall not be actionable as dilution by blurring or dilution by tarnishment under this subsection:
>
> (A) Any fair use . . . *other than as a designation of source for the person's own goods or services*, including use in connection with . . . parodying . . . .

15 U.S.C. § 1125(c)(3)(A)(ii) (emphasis added). Under the statute's plain language, parodying a famous mark is protected by the fair use defense only if the parody is not "a designation of source for the person's own goods or services."

The TDRA, however, does not require a court to ignore the existence of a parody that is used as a trademark, and it does not preclude a court from considering parody as part of the circumstances to be considered for determining whether the plaintiff has made out a claim for dilution by blurring. Indeed, the statute permits a court to consider "all relevant factors," including the six factors supplied in § 1125(c)(2)(B).

Thus, it would appear that a defendant's use of a mark as a parody is relevant to the overall question of whether the defendant's use is likely to impair the famous mark's distinctiveness. Moreover, the fact that the defendant uses its marks as a parody is specifically relevant to several of the listed factors. For example, factor (v) (whether the defendant intended to create an association with the famous mark) and factor (vi) (whether there exists an actual association between the defendant's mark and the famous mark) directly invite inquiries into the defendant's intent in using the parody, the defendant's actual use of the parody, and the effect that its use has on the famous mark. While a parody intentionally creates an association with the famous mark in order to be a parody, it also intentionally communicates, if it is successful, that it is not the famous mark, but rather a satire of the famous mark. *See PETA*, 263 F.3d at 366. That the defendant is using its mark as a parody is therefore relevant in the consideration of these statutory factors.

Similarly, factors (i), (ii), and (iv)—the degree of similarity between the two marks, the degree of distinctiveness of the famous mark, and its recognizability—are directly implicated by consideration of the fact that the defendant's mark is a successful parody. Indeed, by making the famous mark an object of the parody, a successful parody might actually enhance the famous mark's distinctiveness by making it an icon. The brunt of the joke becomes yet more famous. See *Hormel Foods*, 73 F.3d at 506 (observing that a successful parody "tends to increase public identification" of the famous mark with its source); *see also Yankee Publ'g Inc. v. News Am. Publ'g Inc.*, 809 F. Supp. 267, 272-82 (S.D.N.Y. 1992) (suggesting that a sufficiently obvious parody is unlikely to blur the targeted famous mark).

In sum, while a defendant's use of a parody as a mark does not support a "fair use" defense, it may be considered in determining whether the plaintiff-owner of a famous mark has proved its claim that the defendant's use of a parody mark is likely to impair the distinctiveness of the famous mark.

In the case before us, when considering factors (ii), (iii), and (iv), it is readily apparent, indeed conceded by Haute Diggity Dog, that LVM's marks are distinctive, famous, and strong. The LOUIS VUITTON mark is well known and is commonly identified as a brand of the great Parisian fashion house, Louis Vuitton Malletier. So too are its other marks and designs, which are invariably used with the LOUIS VUITTON mark. It may not be too strong to refer to these famous marks as icons of high fashion.

While the establishment of these facts satisfies essential elements of LVM's dilution claim, see 15 U.S.C. § 1125(c)(1), the facts impose on LVM an increased burden to demonstrate that the distinctiveness of its famous marks is likely to be impaired by a successful parody. Even as Haute Diggity Dog's parody mimics the famous mark, it communicates simultaneously that it is not the famous mark, but is only satirizing it. And because the famous mark is particularly strong and distinctive, it becomes more likely that a parody will not impair the distinctiveness of the mark. In short, as Haute Diggity Dog's "Chewy Vuiton" marks are a successful parody, we conclude that they will not blur the distinctiveness of the famous mark as a unique identifier of its source.

It is important to note, however, that this might not be true if the parody is so similar to the famous mark that it likely could be construed as actual use of the famous mark itself. Factor (i) directs an inquiry into the "degree of similarity between the junior mark and the famous mark. If Haute Diggity Dog used the actual marks of LVM (as a parody or otherwise), it could dilute LVM's marks by blurring, regardless of whether Haute Diggity Dog's use was confusingly similar, whether it was in competition with LVM, or whether LVM sustained actual injury. See 15 U.S.C. § 1125(c)(1). . . .

But in this case, Haute Diggity Dog mimicked the famous marks; it did not come so close to them as to destroy the success of its parody and, more importantly, to diminish the LVM marks' capacity to identify a single source. Haute Diggity Dog designed a pet chew toy to imitate and suggest, but not use, the marks of a high-fashion LOUIS VUITTON handbag. It used "Chewy Vuiton" to mimic

"LOUIS VUITTON"; it used "CV" to mimic "LV"; and it adopted imperfectly the items of LVM's designs. We conclude that these uses by Haute Diggity Dog were not so similar as to be likely to impair the distinctiveness of LVM's famous marks.

In a similar vein, when considering factors (v) and (vi), it becomes apparent that Haute Diggity Dog intentionally associated its marks, but only partially and certainly imperfectly, so as to convey the simultaneous message that it was not in fact a source of LVM products. Rather, as a parody, it separated itself from the LVM marks in order to make fun of them.

In sum, when considering the relevant factors to determine whether blurring is likely to occur in this case, we readily come to the conclusion, as did the district court, that LVM has failed to make out a case of trademark dilution by blurring by failing to establish that the distinctiveness of its marks was likely to be impaired by Haute Diggity Dog's marketing and sale of its "Chewy Vuiton" products.

B

LVM's claim for dilution by tarnishment does not require an extended discussion. To establish its claim for dilution by tarnishment, LVM must show, in lieu of blurring, that Haute Diggity Dog's use of the "Chewy Vuiton" mark on dog toys harms the reputation of the LOUIS VUITTON mark and LVM's other marks. LVM argues that the possibility that a dog could choke on a "Chewy Vuiton" toy causes this harm. LVM has, however, provided no record support for its assertion. It relies only on speculation about whether a dog could choke on the chew toys and a logical concession that a $10 dog toy made in China was of "inferior quality" to the $1190 LOUIS VUITTON handbag. The speculation begins with LVM's assertion in its brief that "defendant Woofie's admitted that 'Chewy Vuiton' products pose a choking hazard for some dogs. Having prejudged the defendant's mark to be a parody, the district court made light of this admission in its opinion, and utterly failed to give it the weight it deserved," citing to a page in the district court's opinion where the court states:

> At oral argument, plaintiff provided only a flimsy theory that a pet may some day choke on a Chewy Vuiton squeak toy and incite the wrath of a confused consumer against LOUIS VUITTON.

Louis Vuitton Malletier, 464 F. Supp. 2d at 505. The court was referring to counsel's statement during oral argument that the owner of Woofie's stated that "she would not sell this product to certain types of dogs because there is a danger they would tear it open and choke on it." There is no record support, however, that any dog has choked on a pet chew toy, such as a "Chewy Vuiton" toy, or that there is any basis from which to conclude that a dog would likely choke on such a toy.

We agree with the district court that LVM failed to demonstrate a claim for dilution by tarnishment.

. . . .

Notes & Questions

1. Sometimes, a picture (or two) is worth a thousand words:

2. LVM brought claims both for infringement and for dilution. The Lanham Act does not contain an express "parody" defense to trademark infringement. Thus, courts must either fashion a judge-made defense, or use the traditional likelihood of infringement factors. In addressing the infringement claim, should the fact that the defendant's use is parodic prompt the court to engage in a different analysis than one that walks through that circuit's "likelihood of confusion" factors? The Fourth Circuit uses the familiar factor-based approach. How does the fact that the court finds defendant's activity parodic affect each of the factors? Does that, essentially, amount to a different analysis for parodies? In *KP Permanent* the Court adopted a bifurcated approach that shifted the burden after the plaintiff had demonstrated a likelihood of confusion, permitting the defendant to then offer proof of the defense of descriptive fair use. Is there any reason to treat parody fair use differently? How does the nominative fair use in *Swarovski* compare?

3. According to the Fourth Circuit, creating a successful trademark parody involves conveying two contradictory messages at the same time. On the one hand, the use must convey association with the original. On the other, the use must convey one or more "elements of satire, ridicule, joking, or amusement." A parody that is sufficiently obvious, the court indicates, is "unlikely to blur the targeted famous mark." Should sufficiently obvious parodies then be expressly exempted from dilution claims without resort to the dilution factors identified in the statute?

4. Is the Fourth Circuit setting up a presumption of no dilution if a defendant can show successful parody? If so, is that presumption rebuttable? How would a plaintiff go about rebutting the presumption? What fact or facts would you need to change in this case to alter the outcome on the dilution claim?

5. Instead of a presumption approach, should courts employ a type of burden-shifting framework, similar to what you saw for infringement claims in *KP Permanent*? Should it matter that, unlike the way it frames infringement, the federal statute defining dilution expressly provides that the fair use of trademarks, including for parody, "shall not be actionable." 15 U.S.C. § 1125(c)(3)?

6. Review the facts of *Starbucks Corp. v. Wolfe's Borough Coffee, Inc.*, 588 F.3d 97 (2d Cir. 2009), excerpted in Section IV of this Chapter. In that case the Second Circuit rejected defendant's parody defense to the dilution claim:

> As [is] evident from the statutory language, Black Bear's use of the Charbucks Marks cannot qualify under the parody exception because the Charbucks Marks are used "as a designation of source for [Black Bear's] own goods[, *i.e.*, the Charbucks line of coffee]." See 15 U.S.C. § 1125(c)(3)(A). . . .
>
> Inasmuch as Black Bear's argument may be construed as advocating for consideration of parody in determining the likelihood of dilution by blurring—such as is recognized by the Fourth Circuit, *see* [*Louis Vuitton Malletier S.A. v. Haute Diggity Dog*] at 267—we need not adopt or reject *Louis Vuitton*'s parody holding. We conclude that Black Bear's use of the Charbucks Marks is not a parody of the kind which would favor Black Bear in the dilution analysis even if we were to adopt the Fourth Circuit's rule.
>
> . . . .
>
> Here, unlike in *Louis Vuitton*, Black Bear's use of the Charbucks Marks is, at most, a subtle satire of the Starbucks Marks. Although we recognize some humor in "Char"bucks as a reference to the dark roast of the Starbucks coffees, Black Bear's claim of humor fails to demonstrate such a clear parody as to qualify under the Fourth Circuit's rule. As the owner of Black Bear affirmed during his testimony, "[t]he inspiration for the term Charbucks comes directly from Starbucks' tendency to roast its products more darkly than that of other major roasters." The owner of Black Bear further testified that the Charbucks line of products "is the darkest roasted coffee that we do" and is of "[v]ery high quality." Thus, the Charbucks parody is promoted not as a satire or irreverent commentary of Starbucks but, rather, as a beacon to identify Charbucks as a coffee that competes at the same level and quality as Starbucks in producing dark-roasted coffees.
>
> Therefore, because the Charbucks Marks do not effect an "increase [in] public identification [of the Starbucks Marks with Starbucks]," the purported Charbucks parody plays no part in undermining a finding of dilution under the Fourth Circuit's rule. Accordingly, we conclude that Black Bear's incantation of parody does nothing to shield it from Starbucks' dilution claim in this case.

7. Professor Jessica Litman has argued that "[l]egal protection for trade symbols, in the absence of confusion, disserves competition and thus the consumer. It arrogates to the producer the entire value of cultural icons that we should more appropriately treat as collectively owned." Jessica Litman, *Breakfast with Batman: The Public Interest in the Advertising Age*, 108 Yale L.J. 1717, 1718 (1999). Is granting any protection against dilution a good idea? Currently federal protection is limited to "famous" marks, yet those are the very marks, the "household names," that have been imbued with meaning as a result of actions of the public, not solely because of the actions of the mark owners. Consider the "Eat More Kale" and "Do the Pew" t-shirt disputes described in the Notes & Questions at the end of Section

IV.A. Does the existence of a defense to dilution liability provide sufficient protection for the public interest at stake?

CHAPTER 6: IP LIMITS

No property gives unlimited rights. My fee simple in Greenacre does not give me a right to engage in nuisance behavior; tort law gives my neighbors a remedy against that. My ownership of a baseball bat does not give me a right to swing it into the headlight of your car; criminal laws prohibit that. My ownership of a gas station does not give me the right to agree with my competitor about the price we'll both charge for gas; antitrust laws forbid that. Property rights are always limited rights. And intellectual property rights are no different.

As you learned about the different types of intellectual property, you encountered limits inherent in the way intellectual property rights are defined. For example, the duration of all intellectual property rights are limited in some way: patents and copyrights expire after a term of years, trade secret rights end when the information is no longer secret, and trademarks last only so long as the mark is used in commerce as a source-identifying symbol. Defenses also limit intellectual property rights: fair use limits copyright and trademark, and a Patent Act safe harbor gives generic drug makers the ability to do drug testing that would otherwise be patent infringement so long as they do so to get FDA approval to enter the market. 35 U.S.C. § 271(e)(1).

Many limits constrain intellectual property rights. This chapter explores two additional limits that cut across all intellectual property types. One, preemption of state law by federal law, is a structural limit. Another, the first-sale principle, limits the scope of intellectual property rights.

I. STRUCTURAL LIMITS: PREEMPTION

Intellectual property law is a mix of state and national law. In some domains, such as trade secret law and the right of publicity, the applicable law is almost, or entirely, state law. In other domains, such as copyright law and patent law, the applicable law is almost, or entirely, federal law. And in trademark, state law and federal law offer much the same protection, but differ in the markets to which they apply (intrastate v. interstate). One can work through a good deal of these different domains without encountering any serious clash between state and national law. Clashes do, however, occur. When state and national laws conflict, how should courts respond? The answers courts have provided, in different subject-matter areas and at different times, make up the jurisprudence of *preemption*, a facet of our federalism.

The Constitution sets the parameters of preemption analysis, in part, by allocating powers among the national and state governments. It gives some powers,

such as coining money, to the national government exclusively.[1] It gives other powers, such as appointing presidential electors to the Electoral College, to the state
governments exclusively.[2] Much power, however, is shared. The Constitution
gives Congress the power "[t]o regulate commerce . . . among the several states"
(Art. I, § 8, cl. 3), and at the same time the states retain their traditional police
powers, *i.e.*, "th[e] general power of governing, possessed by the States but not by
the Federal Government." *National Federation of Independent Business v. Sebelius*,
132 S. Ct. 2566, 2578 (2012).[3] As a result, many items fall within overlapping state
and national regulatory programs. When they do, tensions between their respective dictates can arise. The outright impossibility of complying with both state and
national law, where one requires what the other forbids, is the most obvious example, but it is far from the most common. Low-grade tensions and frictions predominate. The Constitution, for its part, explicitly makes national law supreme in
a case of conflict:

> This Constitution, and the laws of the United States which shall be made
> in pursuance thereof; and all treaties made, or which shall be made, under
> the authority of the United States, *shall be the supreme law of the land*; and
> the judges in every state shall be bound thereby, anything in the Constitu-
> tion or laws of any State to the contrary notwithstanding.

Art. VI, cl. 2 (emphasis added). "Consistent with th[is] command," the Supreme
Court "ha[s] long recognized that state laws that conflict with federal law are
'without effect.'" *Altria Group, Inc. v. Wood*, 555 U.S. 70, 76 (2008).

"The formulaic aspect of preemption law consists of a judge-made categorization of different 'types' or varieties of preemption. Here the cases nearly always
begin with a root distinction between 'express' and 'implied' preemption."
Thomas W. Merrill, *Preemption and Institutional Choice*, 102 Nw. U. L. Rev. 727,
738 (2008). With *express* preemption, the federal statute in question expressly provides that it displaces state law in a particular fashion. For example, the federal
statutes regulating interstate motor carriers explicitly provide that "a State . . .
may not enact or enforce a law, regulation, or other provision having the force and

---

[1] On coining money, compare Article I, § 8, clause 5 (giving Congress the power "[t]o coin money
[and to] regulate the value thereof") with § 10 (providing that "[n]o state shall . . . coin money").

[2] *See* Art. II, § 1, cl. 2: "Each state shall appoint, in such manner as the Legislature thereof may
direct, a number of electors, equal to the whole number of Senators and Representatives to which
the State may be entitled in the Congress[.]"

[3] "The legal authority of the States is based on the concept of their sovereignty that existed just
before the Constitution was adopted. The concept, and the political reality, was that the States had
the complete sovereignty of independent nations. Accordingly, their statutory and common law—
and the civil law of states such as Louisiana—were conceived and administered as comprehending
all legal relationships arising in their respective territories." Geoffrey C. Hazard, Jr., *Quasi-Preemp-
tion: Nervous Breakdown In Our Constitutional System*, 84 Tul. L. Rev. 1143, 1146 (2010). The Tenth
Amendment confirms that this reservoir of state power exists: "The powers not delegated to the
United States by the Constitution, nor prohibited by it to the states, are reserved to the states respectively, or to the people." Amend. X; *see also Bond v. United States*, 131 S. Ct. 2355, 2366 (2011)
("The principles of limited national powers and state sovereignty are intertwined. While neither
originates in the Tenth Amendment, both are expressed by it.").

effect of law related to a price, route, or service of any motor carrier . . . with respect to the transportation of property." 49 U.S.C. § 14501(c)(1). Of course, the fact that Congress has been explicit does not prevent there being difficult questions about the reach of particular preemptive language. The motor carrier provision just mentioned has been the subject of multiple Supreme Court cases. *See, e.g., Dan's City Used Cars v. Pelkey*, 133 S. Ct. 1769 (2013); *Rowe v. New Hampshire Motor Transp. Assn.*, 552 U.S. 364 (2008). Today's Copyright Act, unlike the Patent Act or Lanham Act, includes an express preemption provision.

Implied preemption comes in two varieties—field preemption, and conflict preemption. In a case rejecting the contention that the federal Atomic Energy Act of 1954 preempted a California state law's storage and disposal requirements for spent fuel rods for nuclear reactors, the Supreme Court contrasted field and conflict preemption this way:

> Absent explicit preemptive language, Congress' intent to supersede state law altogether may be found from a scheme of federal regulation so pervasive as to make reasonable the inference that Congress left no room to supplement it, because the Act of Congress may touch a field in which the federal interest is so dominant that the federal system will be assumed to preclude enforcement of state laws on the same subject, or because the object sought to be obtained by the federal law and the character of obligations imposed by it may reveal the same purpose. Even where Congress has not entirely displaced state regulation in a specific area, state law is preempted to the extent that it actually conflicts with federal law. Such a conflict arises when "compliance with both federal and state regulations is a physical impossibility," *Florida Lime & Avocado Growers, Inc. v. Paul*, 373 U.S. 132, 142-143 (1963), or where state law "stands as an obstacle to the accomplishment and execution of the full purposes and objectives of Congress." *Hines v. Davidowitz*, 312 U.S. 52, 67 (1941).

Pacific Gas & Elec. Co. v. State Energy Resources Conservation & Dev. Comm'n, 461 U.S. 190, 203-04 (1983). *See also Hillman v. Maretta*, 133 S. Ct. 1943, 1949-50 (2013) (prescribing the same basic framework). Notably, a federal statute with an express preemption clause that does *not* preempt a challenged state law may nevertheless *impliedly* preempt that same state law.[4]

One can imagine a preemption jurisprudence that focused tightly on the narrowest grounds for displacing state with national law—express preemption provisions of unmistakable clarity, and cases of mutual impossibility (*i.e.*, one sovereign

---

[4] *See Altria Group, Inc. v. Good*, 555 U.S. 70, 76-77 (2008) ("If a federal law contains an express preemption clause, it does not immediately end the inquiry because the question of the substance and scope of Congress' displacement of state law still remains. Pre-emptive intent may also be inferred if the scope of the statute indicates that Congress intended federal law to occupy the legislative field, or if there is an actual conflict between state and federal law."); *Freightliner Corp. v. Myrick*, 514 U.S. 280, 287 (1995) ("According to respondents and the Court of Appeals . . . implied pre-emption cannot exist when Congress has chosen to include an express pre-emption clause in a statute. This argument is without merit.").

requires what the other forbids). But that has never been the Supreme Court's approach. Instead, whether construing an express preemption provision or considering a case of implied conflict or field preemption, the Supreme Court has engaged in a far-ranging inquiry in support of making a purpose-based "decision to displace state law in some area in order to advance perceived federal policy goals." Merrill, *Institutional Choice*, at 729. As you consider the materials below, keep in mind the varied policy goals that federal intellectual property laws embody.

A. An Early Benchmark

Federalism "adopts the principle that both the National and State Governments have elements of sovereignty the other is bound to respect." *Arizona v. United States*, 132 S. Ct. 2492, 2500 (2012). One result of this structure is that states can approach different questions of law and policy differently. States can differ in the way they regulate insurance markets or labor markets, including collective bargaining rights; in the way they provide contract and tort laws, both in statutes and in case law; in the way they raise money through taxes, and spend that money on infrastructure, education, and the like. We have seen differences in the way states approach intellectual property law, too. Some states require that, to qualify for protection, a trade secret not only be not generally known, but that it also be not readily ascertainable by proper means; but some states do not require that second showing. In short, the several states are largely free to experiment with laws and policies that make sense to them. Justice Brandeis, in what is now a famous metaphor, called the states laboratories of democracy:

> There must be power in the states and the nation to remould, through experimentation, our economic practices and institutions to meet changing social and economic needs. . . . Denial of the right to experiment may be fraught with serious consequences to the nation. It is one of the happy incidents of the federal system that a single courageous state may, if its citizens choose, serve as a laboratory; and try novel social and economic experiments without risk to the rest of the country.

New State Ice Co. v. Liebmann, 285 U.S. 262, 311 (1932) (Brandeis, J., dissenting from the invalidation of an Oklahoma state ice-sales licensing statute). But the Supreme Court has not always afforded states this flexibility. From the 1870s to the 1930s, the Supreme Court viewed many state legislative interventions in economic markets as illicit favors to interest groups, and thus voided them.

The principle that legislatures have wide latitude to experiment with market regulation has a rich, complex history in the U.S. We can only scratch the surface of that history here. For our purposes, the critical facts are these: In April 1938, the Supreme Court transformed the federal judiciary's relationship both to state law and to congressional statutes, dramatically restructuring federalism's architecture by reducing federal judicial control over national and state market regulation. In *Erie Railroad Co. v. Tompkins*, 304 U.S. 64 (1938), the Court held that federal courts hearing state-law claims are bound not only by state statutes, but also by state-court decisional law. And in *United States v. Carolene Products Co.*, 304 U.S. 144 (1938), decided the same day as *Erie*, the Court squarely rejected both Commerce Clause and Substantive Due Process challenges to a federal statute that banned the interstate shipment of filled milk (*i.e.*, a mixture of skim milk and a nonmilk fat, such as coconut oil). *Erie* and *Carolene Products* thus "announce[d] the inauguration of the modern era of federal judicial life": "Economic regulation was, subject to very weak constraints, to be the province of the state and federal legislatures." John H. Robinson, *The Compromise of '38 and the Federal Courts Today*, 73 Notre Dame L. Rev. 891, 893, 895 (1998). In doing so, the two cases cemented changes set in place the year before,[5] when the Court set aside what had been its own aggressive use of both the Due Process Clause to invalidate state-level market regulation and the Commerce Clause, narrowly construed, to invalidate federal-level market regulation.[6]

Amid these transformations, the Supreme Court decided the first modern intellectual property preemption case, *Kellogg v. Nabisco*, in November 1938. Seven months after *Erie* and *Carolene Products*, the Court had to resolve a putative clash of state law with national statutes, characterizing and shaping national policy in the process. The conventional analytical approach to preemption described above ("Express? Implied: field, or conflict?") had not yet emerged. Instead, the plaintiff simply brought a state-based unfair competition claim, and the Court considered the state-law claim foreclosed by the nature of federal utility and design patent protection. In doing so, the Court set a critical benchmark for the overriding importance of national intellectual property policy as a constraint on state law.

Justice Brandeis, who wrote the *Kellogg* opinion and had written *Erie*, retired in February 1939. The case is thus a bookend with *INS v. AP*, the 1918 case you encountered in the Introduction, in which Justice Brandeis so memorably dissented just a year and a half after his confirmation to the Court. Consider what themes are common to his *INS* dissent and his *Kellogg* opinion for the Court. Specifically, consider whether the fact that, as a matter of federal law, patents and copyrights decisively expire should limit the protection states can provide to formerly patented or copyrighted materials *after* those federal rights expire.

---

[5] *See NLRB v. Jones & Laughlin Steel Corp.*, 301 U.S. 1 (1937) (rejecting a Commerce Clause challenge to the National Labor Relations Act); *West Coast Hotel Co. v. Parrish*, 300 U.S. 379 (1937) (rejecting a Due Process challenge to a Washington state minimum wage law, and overruling *Adkins v. Children's Hospital*, 261 U.S. 525 (1923)); *see also Nebbia v. New York*, 291 U.S. 502 (1934) (rejecting a Due Process challenge to New York state milk price control law).

[6] That general federal common law and economic substantive due process should meet a common fate is no coincidence. For a masterful exposition of the mutually reinforcing conceptions of general federal common law and freedom-of-contract constitutionalism that were repudiated in *Erie* and *Carolene Products*, respectively, the interested reader should consult Robert Post, *Federalism in the Taft Court Era: Can It Be "Revived"?*, 51 Duke L.J. 1513 (2002).

Kellogg Co. v. National Biscuit Co.
305 U.S. 111 (1938)

BRANDEIS, JUSTICE:

This suit was brought in the federal court for Delaware[1] by National Biscuit Company against Kellogg Company to enjoin alleged unfair competition by the manufacture and sale of the breakfast food commonly known as shredded wheat. The competition was alleged to be unfair mainly because Kellogg Company uses, like the plaintiff, the name shredded wheat and, like the plaintiff, produces its biscuit in pillow-shaped form.

Shredded wheat is a product composed of whole wheat which has been boiled, partially dried, then drawn or pressed out into thin shreds and baked. The shredded wheat biscuit generally known is pillow-shaped in form. It was introduced in 1893 by Henry D. Perky, of Colorado; and he was connected until his death in 1908 with companies formed to make and market the article. Commercial success was not attained until the Natural Food Company built, in 1901, a large factory at Niagara Falls, New York. . . . [I]n 1930 its business and goodwill were acquired by National Biscuit Company.

Kellogg Company has been in the business of manufacturing breakfast food cereals since its organization in 1905. For a period commencing in 1912 and ending in 1919 it made a product whose form was somewhat like the product in question, but whose manufacture was different, the wheat being reduced to a dough before being pressed into shreds. For a short period in 1922 it manufactured the article in question. In 1927, it resumed manufacturing the product. In 1928, the plaintiff sued for alleged unfair competition two dealers in Kellogg shredded wheat biscuits. That suit was discontinued by stipulation in 1930. On June 11, 1932, the present suit was brought. . . .

In 1935, the District Court dismissed the bill. It found that the name "Shredded Wheat" is a term describing alike the product of the plaintiff and of the defendant; and that no passing off or deception had been shown. It held that upon the expiration of the Perky patent No. 548,086 issued October 15, 1895, the name of the patented article passed into the public domain. In 1936, the Circuit Court of Appeals affirmed that decree. Upon rehearing, it vacated, in 1937, its own decree and reversed that of the District Court, with direction "to enter a decree enjoining the defendant from the use of the name 'Shredded Wheat' as its tradename and from advertising or offering for sale its product in the form and shape of plaintiff's biscuit in violation of its trade-mark; and with further directions to order an accounting for damages and profits." In its opinion the court described the trade-mark as "consisting of a dish, containing two biscuits submerged in milk." . . .

---

[1] The federal jurisdiction rests on diversity of citizenship—National Biscuit Company being a New Jersey corporation and Kellogg Company a Delaware corporation. Most of the issues in the case involve questions of common law and hence are within the scope of *Erie R. Co. v. Tompkins*, 304 U.S. 64 (1938). But no claim has been made that the local law is any different from the general law on the subject, and both parties have relied almost entirely on federal precedents.

On January 5, 1938, the District Court entered its mandate in the exact language of the order of the Circuit Court of Appeals, and issued a permanent injunction. Shortly thereafter National Biscuit Company petitioned the Circuit Court of Appeals to recall its mandate "for purposes of clarification." It alleged that Kellogg Company was insisting, contrary to the court's intention, that . . . it was not enjoined from making its biscuit in the form and shape of the plaintiff's biscuit, nor from calling it "Shredded Wheat," unless at the same time it uses upon its cartons plaintiff's trade-mark consisting of a dish with two biscuits in it. On May 5, 1938, the Circuit Court of Appeals granted the petition for clarification . . . .

. . . .

The plaintiff concedes that it does not possess the exclusive right to make shredded wheat. But it claims the exclusive right to the trade name "Shredded Wheat" and the exclusive right to make shredded wheat biscuits pillow-shaped. It charges that the defendant, by using the name and shape, and otherwise, is passing off, or enabling others to pass off, Kellogg goods for those of the plaintiff. Kellogg Company denies that the plaintiff is entitled to the exclusive use of the name or of the pillow-shape; denies any passing off; asserts that it has used every reasonable effort to distinguish its product from that of the plaintiff; and contends that in honestly competing for a part of the market for shredded wheat it is exercising the common right freely to manufacture and sell an article of commerce unprotected by patent.

First. The plaintiff has no exclusive right to the use of the term "Shredded Wheat" as a trade name. For that is the generic term of the article, which describes it with a fair degree of accuracy; and is the term by which the biscuit in pillow-shaped form is generally known by the public. Since the term is generic, the original maker of the product acquired no exclusive right to use it. As Kellogg Company had the right to make the article, it had, also, the right to use the term by which the public knows it. Ever since 1894 the article has been known to the public as shredded wheat. For many years, there was no attempt to use the term "Shredded Wheat" as a trade-mark. When in 1905 plaintiff's predecessor, Natural Food Company, applied for registration of the words "Shredded Whole Wheat" as a trade-mark . . . [t]he Commissioner of Patents refused registration. The Court of Appeals of the District of Columbia affirmed his decision, holding that "these words accurately and aptly describe an article of food which * * * has been produced * * * for more than ten years * * * ."

Moreover, the name "Shredded Wheat," as well as the product, the process and the machinery employed in making it, has been dedicated to the public. The basic patent for the product and for the process of making it, and many other patents for special machinery to be used in making the article, issued to Perky. In those patents the term "shredded" is repeatedly used as descriptive of the product. The basic patent expired October 15, 1912; the others soon after. Since during the life of the patents "Shredded Wheat" was the general designation of the patented product, there passed to the public upon the expiration of the patent, not only the right to make the article as it was made during the patent period, but also the right to apply thereto the name by which it had become known. As was said in *Singer Mfg. Co. v. June Mfg. Co.*, 163 U.S. 169, 185 [(1896)]:

> It equally follows from the cessation of the monopoly and the falling of the patented device into the domain of things public, that along with the public ownership of the device there must also necessarily pass to the public the generic designation of the thing which has arisen during the monopoly * * * . To say otherwise would be to hold that although the public had acquired the device covered by the patent, yet the owner of the patent or the manufacturer of the patented thing had retained the designated name which was essentially necessary to vest the public with the full enjoyment of that which had become theirs by the disappearance of the monopoly.

It is contended that the plaintiff has the exclusive right to the name "Shredded Wheat," because those words acquired the "secondary meaning" of shredded wheat made at Niagara Falls by the plaintiff's predecessor. There is no basis here for applying the doctrine of secondary meaning. . . .

The plaintiff seems to contend that even if Kellogg Company acquired upon the expiration of the patents the right to use the name shredded wheat, the right was lost by delay. The argument is that Kellogg Company, although the largest producer of breakfast cereals in the country, did not seriously attempt to make shredded wheat, or to challenge plaintiff's right to that name until 1927, and that meanwhile plaintiff's predecessor had expended more than $17,000,000 in making the name a household word and identifying the product with its manufacture. Those facts are without legal significance. Kellogg Company's right was not one dependent upon diligent exercise. Like every other member of the public, it was, and remained, free to make shredded wheat when it chose to do so; and to call the product by its generic name. The only obligation resting upon Kellogg Company was to identify its own product lest it be mistaken for that of the plaintiff.

Second. The plaintiff has not the exclusive right to sell shredded wheat in the form of a pillow-shaped biscuit—the form in which the article became known to the public. That is the form in which shredded wheat was made under the basic patent. The patented machines used were designed to produce only the pillow-shaped biscuits. And a design patent was taken out to cover the pillow-shaped form. Hence, upon expiration of the patents the form, as well as the name, was dedicated to the public. As was said in *Singer Mfg. Co. v. June Mfg. Co.*, *supra*, p. 185:

> It is self evident that on the expiration of a patent the monopoly granted by it ceases to exist, and the right to make the thing formerly covered by the patent becomes public property. It is upon this condition that the patent is granted. It follows, as a matter of course, that on the termination of the patent there passes to the public the right to make the machine in the form in which it was constructed during the patent. We may, therefore, dismiss without further comment the complaint, as to the form in which the defendant made his machines.

Where an article may be manufactured by all, a particular manufacturer can no more assert exclusive rights in a form in which the public has become accustomed to see the article and which, in the minds of the public, is primarily associated with the article rather than a particular producer, than it can in the case of a name with similar connections in the public mind. Kellogg Company was free to use the pillow-shaped form, subject only to the obligation to identify its product lest it be mistaken for that of the plaintiff.

Third. The question remains whether Kellogg Company in exercising its right to use the name "Shredded Wheat" and the pillow-shaped biscuit, is doing so fairly. Fairness requires that it be done in a manner which reasonably distinguishes its product from that of plaintiff.

Each company sells its biscuits only in cartons. The standard Kellogg carton contains fifteen biscuits; the plaintiff's twelve. The Kellogg cartons are distinctive. They do not resemble those used by the plaintiff either in size, form, or color. And the difference in the labels is striking. The Kellogg cartons bear in bold script the names "Kellogg's Whole Wheat Biscuit" or "Kellogg's Shredded Whole Wheat Biscuit" so sized and spaced as to strike the eye as being a Kellogg product. It is true that on some of its cartons it had a picture of two shredded wheat biscuits in a bowl of milk which was quite similar to one of the plaintiff's registered trademarks. But the name Kellogg was so prominent on all of the defendant's cartons as to minimize the possibility of confusion.

Some hotels, restaurants, and lunchrooms serve biscuits not in cartons and guests so served may conceivably suppose that a Kellogg biscuit served is one of the plaintiff's make. But no person familiar with plaintiff's product would be misled. The Kellogg biscuit is about two-thirds the size of plaintiff's; and differs from it in appearance. Moreover, the field in which deception could be practiced is negligibly small. Only 2½ per cent of the Kellogg biscuits are sold to hotels, restaurants and lunchrooms. Of those so sold 98 per cent are sold in individual cartons containing two biscuits. These cartons are distinctive and bear prominently the Kellogg name. To put upon the individual biscuit some mark which would identify it as the Kellogg product is not commercially possible. Relatively few biscuits will be removed from the individual cartons before they reach the consumer. The obligation resting upon Kellogg Company is not to insure that every purchaser will know it to be the maker but to use every reasonable means to prevent confusion.

It is urged that all possibility of deception or confusion would be removed if Kellogg Company should refrain from using the name "Shredded Wheat" and adopt some form other than the pillow-shape. But the name and form are integral parts of the goodwill of the article. To share fully in the goodwill, it must use the name and the pillow-shape. And in the goodwill Kellogg Company is as free to share as the plaintiff. Moreover, the pillow-shape must be used for another reason. The evidence is persuasive that this form is functional—that the cost of the biscuit would be increased and its high quality lessened if some other form were substituted for the pillow-shape.

Kellogg Company is undoubtedly sharing in the goodwill of the article known as "Shredded Wheat"; and thus is sharing in a market which was created by the skill and judgment of plaintiff's predecessor and has been widely extended by vast expenditures in advertising persistently made. But that is not unfair. Sharing in the goodwill of an article unprotected by patent or trade-mark is the exercise of a right possessed by all—and in the free exercise of which the consuming public is deeply interested. There is no evidence of passing off or deception on the part of the Kellogg Company; and it has taken every reasonable precaution to prevent confusion or the practice of deception in the sale of its product.

. . . .

Notes & Questions

1. When Justice Brandeis dissented from the creation of a new general federal common law right against the misappropriation of hot news, in *INS v. AP*, 248 U.S. 215 (1918), his main argument was relative institutional competence:

[W]ith the increasing complexity of society, the public interest tends to become omnipresent; and the problems presented by new demands for justice cease to be simple. . . . Courts are ill-equipped to make the investigations which should precede a determination of the limitations which should be set upon any property right in news or of the circumstances under which news gathered by a private agency should be deemed affected with a public interest. Courts would be powerless to prescribe the detailed regulations essential to full enjoyment of the rights conferred or to introduce the machinery required for enforcement of such regulations. Considerations such as these should lead us to decline to establish a new rule of law in the effort to redress a newly disclosed wrong, although the propriety of some remedy appears to be clear.

Id. at 262-67. He was also quite clear, however, that intellectual property's default rules were decidedly against AP's misappropriation claim:

The general rule of law is, that the noblest of human productions—knowledge, truths ascertained, conceptions, and ideas—become, after voluntary communication to others, free as the air to common use. Upon these incorporeal productions the attribute of property is continued after such communication only in certain classes of cases where public policy has seemed to demand it. . . . The creations which are recognized as property by the common law are literary, dramatic, musical, and other artistic creations; and these have also protection under the copyright statutes. The inventions and discoveries upon which this attribute of property is conferred only by statute, are the few comprised within the patent law. . . . Thus it [has been] held that one may ordinarily make and sell anything in any form, may copy with exactness that which another has produced, or may otherwise use his ideas without his consent and without the payment of compensation, and yet not inflict a legal injury; and that ordinarily one is at perfect

liberty to find out, if he can by lawful means, trade secrets of another, however valuable, and then use the knowledge so acquired gainfully, although
it cost the original owner much in effort and in money to collect or produce.
. . . That competition is not unfair in a legal sense, merely because the profits gained are unearned, even if made at the expense of a rival, is shown by
many cases besides those referred to above. He who follows the pioneer into
a new market, or who engages in the manufacture of an article newly introduced by another, seeks profits due largely to the labor and expense of the
first adventurer; but the law sanctions, indeed encourages, the pursuit. He
who makes a city known through his product, must submit to sharing the
resultant trade with others who, perhaps for that reason, locate there later.

Id. at 250-59. How do these default rules compare to the competitive landscape
for shredded wheat cereal that Justice Brandeis describes, where Perky's utility
and design patents have expired?

2. The Court holds that the phrase "shredded wheat" is the generic name of
the disputed product, and is thus not eligible for protection as a Nabisco trademark. The Court also holds that the product's pillow shape is functional, and thus
also not eligible for protection as (what we would now call) a product design trade
dress. Did the Court have any need, then, to rely on the fact that the Perky patents
had expired, to hold for Kellogg? What does the Court's heavy reliance on, and
quotations from, its 1896 decision in *Singer* add to the policy picture the Court
establishes in *Kellogg*?

3. Under the Patent Act, patents expire. If state law provides a former patentee
with market exclusivity comparable to an expired right, has the existence of the
federal expiration date been undermined or frustrated?

4. Who, other than Kellogg, benefits from Kellogg's ability to "shar[e] in the
goodwill of the article known as 'Shredded Wheat'"? How, if at all, do *consumers*
benefit? The Court does assert that "[s]haring in the goodwill of an article unprotected by patent or trade-mark is the exercise of a right possessed by all—and in
the free exercise of which the consuming public is deeply interested." What does
the Court mean here?

5. Recall that in *Wal-Mart Stores v. Samara Bros.*, 529 U.S. 205, 216 (2000), the
Supreme Court held that a product design trade dress cannot be inherently distinctive; instead, the putative trade dress owner must show that the product design has acquired distinctivness as a source indicater. The *Wal-Mart* Court did not
discuss the *Kellogg* case, but perhaps it should have done so. How does *Wal-Mart*'s
holding about the conditions for product design trade dress protection support the
intellectual property policies *Kellogg* sets forth? How much should it matter that
the trade dress protection sought in that case was asserted under federal law, rather than state law?

B. Contemporary Cases

In the 75 years since *Kellogg*, intellectual property law has presented many preemption questions. Contemporary cases consider both implied and express preemption. We begin with implied preemption cases.

1. Implied Preemption

After *Kellogg*, the Supreme Court did not revisit implied preemption in an intellectual property case until *Sperry v. Florida*, 373 U.S. 379 (1963). In that case, the Supreme Court held that the Patent & Trademark Office bar expressly established under the Patent Act,[*] preempted a state bar's power to condemn a patent agent's work as the unauthorized practice of law. "[B]y virtue of the Supremacy Clause, Florida may not deny to those failing to meet its own [law bar] qualifications the right to perform the functions within the scope of the federal authority" to license patent attorneys. *Id.* at 385.

The following year, the Supreme Court decided a pair of cases similar, in some respects, to *Kellogg*—*Sears, Roebuck & Co. v. Stiffel Co.*, 376 U.S. 225 (1964); and *Compco Corp. v. Day-Brite Lighting, Inc.*, 376 U.S. 234 (1964). The various utility and design patents at issue in both cases had not expired, as in *Kellogg*; instead, they had been declared invalid in litigation. And in both cases, the Seventh Circuit had construed the state unfair competition law of Illinois to permit a design originator to prevail against a design copyist in the absence of any valid patent rights *and* without any showing that the product design in question had acquired distinctiveness as a source indicator (contrary to what the law of trade dress infringement now requires, recall *Wal-Mart Stores v. Samara Bros.*, 529 U.S. 205 (2000), Chapter 5, Part II.B). The Supreme Court reversed in both cases, concluding that a state unfair competition law of this sort squarely conflicted with federal law:

> In the present case the "pole lamp" sold by Stiffel has been held not to be entitled to the protection of either a mechanical or a design patent. An unpatentable article, like an article on which the patent has expired, is in the public domain and may be made and sold by whoever chooses to do so. What Sears did was to copy Stiffel's design and to sell lamps almost identical to those sold by Stiffel. This it had every right to do under the federal patent laws. That Stiffel originated the pole lamp and made it popular is immaterial. "Sharing in the goodwill of an article unprotected by patent or trade-mark is the exercise of a right possessed by all—and in the free exercise of which the consuming public is deeply interested." *Kellogg Co. v. National Biscuit Co.*, 305

---

[*] The authority to do so is currently codified at 35 U.S.C. § 2(b)(2)(D).

U.S. at 122. To allow a State by use of its law of unfair competition to prevent the copying of an article which represents too slight an advance to be patented would be to permit the State to block off from the public something which federal law has said belongs to the public. The result would be that while federal law grants only 14 or 17 years' protection to genuine inventions, see 35 U.S.C. §§ 154, 173, States could allow perpetual protection to articles too lacking in novelty to merit any patent at all under federal constitutional standards. This would be too great an encroachment on the federal patent system to be tolerated.

Sears, 376 U.S. at 231-32; *see also Compco*, 376 U.S. at 237-38 ("Here Day-Brite's fixture has been held not to be entitled to a design or mechanical patent. Under the federal patent laws it is, therefore, in the public domain and can be copied in every detail by whoever pleases.").

In the next noteworthy intellectual property implied-preemption case, *Kewanee Oil Co. v. Bicron Corp.*, 416 U.S. 470 (1974), the Supreme Court rejected the contention that the federal Patent Act preempts state trade secret laws. The Court viewed trade secret and patent protection as complementary rather than conflicting. First, trade secret law protects some matters as to which patents are not available. *Id.* at 483. Second, even as to subject matter that can be patented, the patent law "policy that matter once in the public domain must remain in the public domain is not incompatible with the existence of trade secret protection. By definition a trade secret has not been placed in the public domain." *Id.* at 484. Third, trade secret protection does not undermine the national policy in favor of invention disclosure embodied in patent law's core disclosure-for-protection *quid pro quo*. Trade secret law helps those who make inventions that, although patentable subject matter in general, certainly or likely fail one or more of the rigors limiting patent protection. *Id.* at 484-89. Such secrets are not serious candidates for disclosure anyway. And, the Court believed, those who invent plainly patentable items will recognize, and generally seek the benefit of, the stronger protection that patent law provides: "Trade secret law provides far weaker protection in many respects than the patent law. . . . The possibility that an inventor who believes his invention meets the standards of patentability will sit back, rely on trade secret law, and after one year of use forfeit any right to patent protection, is remote indeed." *Id.* at 489-90. *Kewanee* was the first Supreme Court case to reject a patent-law-based preemption challenge to a state intellectual property law. What do you think of the Court's assumption concerning the choice one might make between trade secret protection and patent protection?

The Supreme Court's most recent intellectual property preemption case is, once again, an implied preemption case. It involves the Patent Act and a Florida state statute regulating the use of boat hull designs.

Bonito Boats, Inc. v. Thunder Craft Boats, Inc.
489 U.S. 141 (1989)

O'CONNOR, JUSTICE:

We must decide today what limits the operation of the federal patent system places on the States' ability to offer substantial protection to utilitarian and design ideas which the patent laws leave otherwise unprotected. In *Interpart Corp. v. Italia*, 777 F.2d 678 (Fed. Cir. 1985), the Court of Appeals for the Federal Circuit concluded that a California law prohibiting the use of the "direct molding process" to duplicate unpatented articles posed no threat to the policies behind the federal patent laws. In this case, the Florida Supreme Court came to a contrary conclusion. It struck down a Florida statute which prohibits the use of the direct molding process to duplicate unpatented boat hulls, finding that the protection offered by the Florida law conflicted with the balance struck by Congress in the federal patent statute between the encouragement of invention and free competition in unpatented ideas. 515 So. 2d 220 (1987). We granted certiorari to resolve the conflict and we now affirm the judgment of the Florida Supreme Court.

I

In September 1976, petitioner Bonito Boats, Inc. (Bonito), a Florida corporation, developed a hull design for a fiberglass recreational boat which it marketed under the trade name Bonito Boat Model 5VBR. Designing the boat hull required substantial effort on the part of Bonito. A set of engineering drawings was prepared, from which a hardwood model was created. The hardwood model was then sprayed with fiberglass to create a mold, which then served to produce the finished fiberglass boats for sale. The 5VBR was placed on the market sometime in September 1976. There is no indication in the record that a patent application was ever filed for protection of the utilitarian or design aspects of the hull, or for the process by which the hull was manufactured. The 5VBR was favorably received by the boating public, and "a broad interstate market" developed for its sale.

In May 1983, after the Bonito 5VBR had been available to the public for over six years, the Florida Legislature enacted Fla. Stat. § 559.94. The statute makes "[i]t * * * unlawful for any person to use the direct molding process to duplicate for the purpose of sale any manufactured vessel hull or component part of a vessel made by another without the written permission of that other person." § 559.94(2). The statute also makes it unlawful for a person to "knowingly sell a vessel hull or component part of a vessel duplicated in violation of subsection (2)." § 559.94(3). Damages, injunctive relief, and attorney's fees are made available to "[a]ny person who suffers injury or damage as the result of a violation" of the statute. § 559.94(4). The statute was made applicable to vessel hulls or component parts duplicated through the use of direct molding after July 1, 1983. § 559.94(5).

On December 21, 1984, Bonito filed this action in the Circuit Court of Orange County, Florida. The complaint alleged that respondent here, Thunder Craft Boats, Inc. (Thunder Craft), a Tennessee corporation, had violated the Florida statute by using the direct molding process to duplicate the Bonito 5VBR fiberglass hull, and had knowingly sold such duplicates in violation of the Florida stat-ute. . . . Respondent filed a motion to dismiss the complaint, arguing that . . . the Florida statute conflicted with federal patent law and was therefore invalid under the Supremacy Clause of the Federal Constitution. The trial court granted re-spondent's motion, and a divided Court of Appeals affirmed the dismissal of pe-titioner's complaint.

On appeal, a sharply divided Florida Supreme Court agreed with the lower courts' conclusion that the Florida law impermissibly interfered with the scheme established by the federal patent laws. . . .

II

Article I, § 8, cl. 8, of the Constitution gives Congress the power "[t]o promote the Progress of Science and useful Arts, by securing for limited Times to Authors and Inventors the exclusive Right to their respective Writings and Discoveries." The Patent Clause itself reflects a balance between the need to encourage innova-tion and the avoidance of monopolies which stifle competition without any con-comitant advance in the "Progress of Science and useful Arts." As we have noted in the past, the Clause contains both a grant of power and certain limitations upon the exercise of that power. Congress may not create patent monopolies of unlim-ited duration, nor may it "authorize the issuance of patents whose effects are to remove existent knowledge from the public domain, or to restrict free access to materials already available." *Graham v. John Deere Co.*, 383 U.S. 1, 6 (1966).

From their inception, the federal patent laws have embodied a careful balance between the need to promote innovation and the recognition that imitation and refinement through imitation are both necessary to invention itself and the very lifeblood of a competitive economy. . . .

. . . .

. . . The novelty requirement[s] of patentability . . . operate . . . to exclude from consideration for patent protection knowledge that is already available to the pub-lic. They express a congressional determination that the creation of a monopoly in such information would not only serve no socially useful purpose, but would in fact injure the public by removing existing knowledge from public use. From the Patent Act of 1790 to the present day, the public sale of an unpatented article has acted as a complete bar to federal protection of the idea embodied in the article thus placed in public commerce.

. . . .

. . . [T]he federal patent scheme creates a limited opportunity to obtain a prop-erty right in an idea. Once an inventor has decided to lift the veil of secrecy from his work, he must choose the protection of a federal patent or the dedication of his idea to the public at large. As Judge Learned Hand once put it: "[I]t is a condition upon the inventor's right to a patent that he shall not exploit his discovery com-petitively after it is ready for patenting; he must content himself with either se-crecy or legal monopoly." *Metallizing Engineering Co. v. Kenyon Bearing & Auto Parts Co.*, 153 F. 2d 516, 520 (2d. Cir. 1946).

. . . Taken together, the novelty and nonobviousness requirements express a congressional determination that the purposes behind the Patent Clause are best served by free competition and exploitation of either that which is already available to the public or that which may be readily discerned from publicly available material. *See Aronson v. Quick Point Pencil Co.*, 440 U.S. 257, 262 (1979) ("[T]he stringent requirements for patent protection seek to ensure that ideas in the public domain remain there for the use of the public").

. . . The federal patent system thus embodies a carefully crafted bargain for encouraging the creation and disclosure of new, useful, and nonobvious advances in technology and design in return for the exclusive right to practice the invention for a period of years. . . .

The attractiveness of such a bargain, and its effectiveness in inducing creative effort and disclosure of the results of that effort, depend almost entirely on a backdrop of free competition in the exploitation of unpatented designs and innovations. The novelty and nonobviousness requirements of patentability embody a congressional understanding, implicit in the Patent Clause itself, that free exploitation of ideas will be the rule, to which the protection of a federal patent is the exception. Moreover, the ultimate goal of the patent system is to bring new designs and technologies into the public domain through disclosure. State law protection for techniques and designs whose disclosure has already been induced by market rewards may conflict with the very purpose of the patent laws by decreasing the range of ideas available as the building blocks of further innovation. The offer of federal protection from competitive exploitation of intellectual property would be rendered meaningless in a world where substantially similar state law protections were readily available. To a limited extent, the federal patent laws must determine not only what is protected, but also what is free for all to use. *Cf. Arkansas Electric Cooperative Corp. v. Arkansas Public Service Comm'n*, 461 U.S. 375, 384 (1983) ("[A] federal decision to forgo regulation in a given area may imply an authoritative federal determination that the area is best left *un*regulated, and in that event would have as much pre-emptive force as a decision *to* regulate") (emphasis in original).

Thus our past decisions have made clear that state regulation of intellectual property must yield to the extent that it clashes with the balance struck by Congress in our patent laws. The tension between the desire to freely exploit the full potential of our inventive resources and the need to create an incentive to deploy those resources is constant. Where it is clear how the patent laws strike that balance in a particular circumstance, that is not a judgment the States may second-guess. We have long held that after the expiration of a federal patent, the subject matter of the patent passes to the free use of the public as a matter of federal law. *See Kellogg Co. v. National Biscuit Co.*, 305 U.S. 111 (1938); *Singer Mfg. Co. v. June Mfg. Co.*, 163 U.S. 169 (1896). Where the public has paid the congressionally mandated price for disclosure, the States may not render the exchange fruitless by offering patent-like protection to the subject matter of the expired patent. "It is self-evident that on the expiration of a patent the monopoly created by it ceases to exist, and the right to make the thing formerly covered by the patent becomes public property." *Singer*, 163 U.S. at 185.

In our decisions in *Sears, Roebuck & Co. v. Stiffel Co.*, 376 U.S. 225 (1964), and *Compco Corp. v. Day-Brite Lighting, Inc.*, 376 U.S. 234 (1964), we found that publicly known design and utilitarian ideas which were unprotected by patent occupied much the same position as the subject matter of an expired patent. . . .

The pre-emptive sweep of our decisions in *Sears* and *Compco* has been the subject of heated scholarly and judicial debate. Read at their highest level of generality, the two decisions could be taken to stand for the proposition that the States are completely disabled from offering any form of protection to articles or processes which fall within the broad scope of patentable subject matter. Since the potentially patentable includes "anything under the sun that is made by man," *Diamond v. Chakrabarty*, 447 U.S. 303, 309 (1980), the broadest reading of *Sears* would prohibit the States from regulating the deceptive simulation of trade dress or the tortious appropriation of private information.

That the extrapolation of such a broad pre-emptive principle from *Sears* is inappropriate is clear from the balance struck in *Sears* itself. The *Sears* Court made it plain that the States "may protect businesses in the use of their trademarks, labels, or distinctive dress in the packaging of goods so as to prevent others, by imitating such markings, from misleading purchasers as to the source of the goods." *Sears*, 376 U.S. at 232. Trade dress is, of course, potentially the subject matter of design patents. Yet our decision in *Sears* clearly indicates that the States may place limited regulations on the circumstances in which such designs are used in order to prevent consumer confusion as to source. Thus, while *Sears* speaks in absolutist terms, its conclusion that the States may place some conditions on the use of trade dress indicates an implicit recognition that all state regulation of potentially patentable but unpatented subject matter is not ipso facto pre-empted by the federal patent laws.

What was implicit in our decision in *Sears*, we have made explicit in our subsequent decisions concerning the scope of federal pre-emption of state regulation of the subject matter of patent. Thus, in *Kewanee Oil Co. v. Bicron Corp.*, 416 U. S. 470 (1974), we held that state protection of trade secrets did not operate to frustrate the achievement of the congressional objectives served by the patent laws. Despite the fact that state law protection was available for ideas which clearly fell within the subject matter of patent, the Court concluded that the nature and degree of state protection did not conflict with the federal policies of encouragement of patentable invention and the prompt disclosure of such innovations.

. . . .

At the heart of *Sears* and *Compco* is the conclusion that the efficient operation of the federal patent system depends upon substantially free trade in publicly known, unpatented design and utilitarian conceptions. In *Sears*, the state law offered "the equivalent of a patent monopoly," 376 U.S. at 233, in the functional aspects of a product which had been placed in public commerce absent the protection of a valid patent. While, as noted above, our decisions since *Sears* have taken a decidedly less rigid view of the scope of federal pre-emption under the patent laws, we believe that the *Sears* Court correctly concluded that the States may not offer patent-like protection to intellectual creations which would otherwise remain unprotected as a matter of federal law. Both the novelty and the nonobviousness requirements of federal patent law are grounded in the notion that concepts within the public grasp, or those so obvious that they readily could be, are the tools of creation available to all. They provide the baseline of free competition upon which the patent system's incentive to creative effort depends. A state law that substantially interferes with the enjoyment of an unpatented utilitarian or design conception which has been freely disclosed by its author to the public at large impermissibly contravenes the ultimate goal of public disclosure and use which is the centerpiece of federal patent policy. Moreover, through the creation of patent-like rights, the States could essentially redirect inventive efforts away from the careful criteria of patentability developed by Congress over the last 200 years. We understand this to be the reasoning at the core of our decisions in *Sears* and *Compco*, and we reaffirm that reasoning today.

III

We believe that the Florida statute at issue in this case so substantially impedes the public use of the otherwise unprotected design and utilitarian ideas embodied in unpatented boat hulls as to run afoul of the teaching of our decisions in *Sears* and *Compco*. It is readily apparent that the Florida statute does not operate to prohibit "unfair competition" in the usual sense that the term is understood. The law of unfair competition has its roots in the common-law tort of deceit: its general concern is with protecting consumers from confusion as to source. While that concern may result in the creation of "quasi-property rights" in communicative symbols, the focus is on the protection of consumers, not the protection of producers as an incentive to product innovation. Judge Hand captured the distinction well in *Crescent Tool Co. v. Kilborn & Bishop Co.*, 247 F. 299, 301 (2d Cir. 1917), where he wrote:

> [T]he plaintiff has the right not to lose his customers through false representations that those are his wares which in fact are not, but he may not monopolize any design or pattern, however trifling. The defendant, on the other hand, may copy plaintiff's goods slavishly down to the minutest detail: but he may not represent himself as the plaintiff in their sale.

... [T]he common-law tort of unfair competition has been limited to protection against copying of nonfunctional aspects of consumer products which have acquired secondary meaning such that they operate as a designation of source. The "protection" granted a particular design under the law of unfair competition is thus limited to one context where consumer confusion is likely to result; the design "idea" itself may be freely exploited in all other contexts.

In contrast to the operation of unfair competition law, the Florida statute is aimed directly at preventing the exploitation of the design and utilitarian conceptions embodied in the product itself. The sparse legislative history surrounding its enactment indicates that it was intended to create an inducement for the improvement of boat hull designs. To accomplish this goal, the Florida statute endows the original boat hull manufacturer with rights against the world, similar in scope and operation to the rights accorded a federal patentee. Like the patentee, the beneficiary of the Florida statute may prevent a competitor from "making" the product in what is evidently the most efficient manner available and from "selling" the product when it is produced in that fashion. Compare 35 U.S.C. § 154. The Florida scheme offers this protection for an unlimited number of years to all boat hulls and their component parts, without regard to their ornamental or technological merit. Protection is available for subject matter for which patent protection has been denied or has expired, as well as for designs which have been freely revealed to the consuming public by their creators.

In this case, the Bonito 5VBR fiberglass hull has been freely exposed to the public for a period in excess of six years. For purposes of federal law, it stands in the same stead as an item for which a patent has expired or been denied: it is unpatented and unpatentable. Whether because of a determination of unpatentability or other commercial concerns, petitioner chose to expose its hull design to the public in the marketplace, eschewing the bargain held out by the federal patent system of disclosure in exchange for exclusive use. Yet, the Florida statute allows petitioner to reassert a substantial property right in the idea, thereby constricting the spectrum of useful public knowledge. Moreover, it does so without the careful protections of high standards of innovation and limited monopoly contained in the federal scheme. We think it clear that such protection conflicts with the federal policy "that all ideas in general circulation be dedicated to the common good unless they are protected by a valid patent." *Lear, Inc. v. Adkins*, 395 U.S. [653,] 668 [(1969)].

That the Florida statute does not remove all means of reproduction and sale does not eliminate the conflict with the federal scheme. In essence, the Florida law prohibits the entire public from engaging in a form of reverse engineering of a product in the public domain. This is clearly one of the rights vested in the federal patent holder, but has never been a part of state protection under the law of unfair competition or trade secrets. The duplication of boat hulls and their component parts may be an essential part of innovation in the field of hydrodynamic design. Variations as to size and combination of various elements may lead to significant advances in the field. Reverse engineering of chemical and mechanical articles in the public domain often leads to significant advances in technology. If Florida may prohibit this particular method of study and recomposition of an unpatented article, we fail to see the principle that would prohibit a State from banning the use of chromatography in the reconstitution of unpatented chemical compounds, or the use of robotics in the duplication of machinery in the public domain.

Moreover, as we noted in *Kewanee*, the competitive reality of reverse engineering may act as a spur to the inventor, creating an incentive to develop inventions that meet the rigorous requirements of patentability. 416 U.S. at 489-490. The Florida statute substantially reduces this competitive incentive, thus eroding the general rule of free competition upon which the attractiveness of the federal patent bargain depends. . . . The prospect of all 50 States establishing similar protections for preferred industries without the rigorous requirements of patentability prescribed by Congress could pose a substantial threat to the patent system's ability to accomplish its mission of promoting progress in the useful arts.

. . . .

. . . It is difficult to conceive of a more effective method of creating substantial property rights in an intellectual creation than to eliminate the most efficient method for its exploitation. *Sears* and *Compco* protect more than the right of the public to contemplate the abstract beauty of an otherwise unprotected intellectual creation—they assure its efficient reduction to practice and sale in the marketplace.

. . . .

Finally, . . . the federal standards for patentability, at a minimum, express the congressional determination that patent-like protection is unwarranted as to certain classes of intellectual property. The States are simply not free in this regard to offer equivalent protections to ideas which Congress has determined should belong to all. For almost 100 years it has been well established that in the case of an expired patent, the federal patent laws do create a federal right to "copy and to use." *Sears* and *Compco* extended that rule to potentially patentable ideas which are fully exposed to the public. . . .

Our decisions since *Sears* and *Compco* have made it clear that the Patent and Copyright Clauses do not, by their own force or by negative implication, deprive the States of the power to adopt rules for the promotion of intellectual creation within their own jurisdictions. . . .

Nor does the fact that a particular item lies within the subject matter of the federal patent laws necessarily preclude the States from offering limited protection which does not impermissibly interfere with the federal patent scheme. As *Sears* itself makes clear, States may place limited regulations on the use of unpatented designs in order to prevent consumer confusion as to source. In *Kewanee*, we found that state protection of trade secrets, as applied to both patentable and unpatentable subject matter, did not conflict with the federal patent laws. In both situations, state protection was not aimed exclusively at the promotion of invention itself, and the state restrictions on the use of unpatented ideas were limited to those necessary to promote goals outside the contemplation of the federal patent scheme. Both the law of unfair competition and state trade secret law have coexisted harmoniously with federal patent protection for almost 200 years, and Congress has given no indication that their operation is inconsistent with the operation of the federal patent laws.

Indeed, there are affirmative indications from Congress that both the law of unfair competition and trade secret protection are consistent with the balance struck by the patent laws. Section 43(a) of the Lanham Act, 15 U.S.C. § 1125(a), creates a federal remedy for making "a false designation of origin, or any false description or representation, including words or other symbols tending falsely to describe or represent the same * * * ." Congress has thus given federal recognition to many of the concerns that underlie the state tort of unfair competition, and the application of *Sears* and *Compco* to nonfunctional aspects of a product which have been shown to identify source must take account of competing federal policies in this regard. Similarly, as Justice Marshall noted in his concurring opinion in *Kewanee*: "State trade secret laws and the federal patent laws have co-existed for many, many, years. During this time, Congress has repeatedly demonstrated its full awareness of the existence of the trade secret system, without any indication of disapproval. Indeed, Congress has in a number of instances given explicit federal protection to trade secret information provided to federal agencies." *Kewanee*, 416 U.S. at 494 (concurring in result). The case for federal pre-emption is particularly weak where Congress has indicated its awareness of the operation of state law in a field of federal interest, and has nonetheless decided to "stand by both concepts and to tolerate whatever tension there [is] between them." *Silkwood v. Kerr-McGee Corp.*, 464 U. S. 238, 256 (1984). The same cannot be said of the Florida statute at issue here, which offers protection beyond that available under the law of unfair competition or trade secret, without any showing of consumer confusion, or breach of trust or secrecy.

The Florida statute is aimed directly at the promotion of intellectual creation by substantially restricting the public's ability to exploit ideas that the patent system mandates shall be free for all to use. Like the [Seventh Circuit's] interpretation of Illinois unfair competition law in *Sears* and *Compco*, the Florida statute represents a break with the tradition of peaceful coexistence between state market regulation and federal patent policy. The Florida law substantially restricts the public's ability to exploit an unpatented design in general circulation, raising the specter of state-created monopolies in a host of useful shapes and processes for which patent protection has been denied or is otherwise unobtainable. It thus enters a field of regulation which the patent laws have reserved to Congress. The patent statute's careful balance between public right and private monopoly to promote certain creative activity is a "scheme of federal regulation * * * so pervasive as to make reasonable the inference that Congress left no room for the States to supplement it." *Rice v. Santa Fe Elevator Corp.*, 331 U. S. 218, 230 (1947).

Congress has considered extending various forms of limited protection to industrial design either through the copyright laws or by relaxing the restrictions on the availability of design patents. *See generally* Brown, *Design Protection: An Overview*, 34 UCLA L. Rev. 1341 (1987). Congress explicitly refused to take this step [in 1976] in the copyright laws, *see* 17 U.S.C. § 101, and despite sustained criticism for a number of years, it has [to date] declined to alter the patent protections presently available for industrial design. It is for Congress to determine if the present system of design and utility patents is ineffectual in promoting the useful arts in the context of industrial design. By offering patent-like protection for ideas deemed unprotected under the present federal scheme, the Florida statute conflicts with the "strong federal policy favoring free competition in ideas which do not merit patent protection." *Lear, Inc.*, 395 U.S. at 656. We therefore agree with the majority of the Florida Supreme Court that the Florida statute is preempted by the Supremacy Clause, and the judgment of that court is hereby affirmed.

Notes & Questions

1. What, precisely, was the problem with Florida's statute? Was it the type of technology protected under the statute? Was it that the Florida statute did not require novelty or nonobviousness in order for the design to be protected? Was it the remedies that the law provided? Was it all of these aspects of the state law? Or was it something else entirely?

2. The Court references the inducement of "market rewards" as the reason one might disclose a design or a technique. If market rewards create an incentive, why, exactly are patent rights necessary? When might "market rewards" be an insufficient inducement, and which legislature—state, federal, both—gets to make that determination?

3. Could a state ban the creation, sale, and use of 3-d printers, seeking to prohibit the easy replication technology that these devices embody?

4. A decade after *Bonito Boats*, Congress provided protection for creative boat hull designs against the type of slavish copying that the Florida statute had addressed. Specifically, the Vessel Hull Design Protection Act of 1998 ("VHDPA") amended Title 17 of the U.S. Code to protect an "original design of a useful article." The VHDPA defined that phrase by defining both what constitutes an "original design" and what constitutes a "useful article" for purposes of the protection granted by VHDPA:

> 1) A design is "original" if it is the result of the designer's creative endeavor that provides a distinguishable variation over prior work pertaining to similar articles which is more than merely trivial and has not been copied from another source.
> (2) A "useful article" is a vessel hull or deck, including a plug or mold, which in normal use has an intrinsic utilitarian function that is not merely to portray the appearance of the article or to convey information.

17 U.S.C. § 1301(b). The protection granted lasts for ten years from the time the design is first published (or registered with the Copyright Office). *Id.* at §§ 1304, 1305. The protection is copyright-like, rather than patent-like, inasmuch as infringement requires proof of copying, while also requiring proof of knowledge that the copied design was protected under the statute. *Id.* at § 1309(c). As a federal statute, the VHDPA is not subject to challenge under preemption theories. Why is it permissible for Congress to provide this type of protection but not for Florida to do so?

5. The structure of the VHDPA almost invites easy amendment to expand the types of "useful articles" that could gain protection. Congress has, for decades now, considered various bills that would extend protection to industrial design more generally, effectively creating a species of copyright-like protection for the designs of manufactures (with substantially lower thresholds than design patent law). Although design protection of this sort has long been provided in European

countries, and is now the subject of an EU-wide freestanding EU design right, it has never progressed too far in the U.S. Congress. The simple expedient of amending the Vessel Hull Design Protection Act to protect all industrial design remains out of reach for the proponents of such protection.

2. Express Preemption

The Copyright Act includes an explicit preemption provision. It reads, in relevant part, as follows:

(a) On and after January 1, 1978, all legal or equitable rights that are equivalent to any of the exclusive rights within the general scope of copyright as specified by section 106 in works of authorship that are fixed in a tangible medium of expression and come within the subject matter of copyright as specified by sections 102 and 103, whether created before or after that date and whether published or unpublished, are governed exclusively by this title. Thereafter, no person is entitled to any such right or equivalent right in any such work under the common law or statutes of any State.

(b) Nothing in this title annuls or limits any rights or remedies under the common law or statutes of any State with respect to—

(1) subject matter that does not come within the subject matter of copyright as specified by sections 102 and 103, including works of authorship not fixed in any tangible medium of expression; or

(2) any cause of action arising from undertakings commenced before January 1, 1978;

(3) activities violating legal or equitable rights that are not equivalent to any of the exclusive rights within the general scope of copyright as specified by section 106; or

(4) State and local landmarks, historic preservation, zoning, or building codes, relating to architectural works protected under section 102(a)(8).

17 U.S.C. § 301. Note that the two subsections work together. Subsection (a) preempts state law: "all legal or equitable rights that are equivalent" to Copyright Act rights "are governed exclusively by this title." Subsection (b) saves state law: for the listed items, Title 17 does not "annul[] or limit[] any rights or remedies under the common law or statutes of any State." This two-part approach is a common structure for express preemption provisions.

The Supreme Court has never adjudicated a preemption claim applying § 301 of the 1976 Copyright Act.[*] As a result, the lower courts have simply used the general body of preemption cases to apply the strictures of § 301. The next case is a

---

[*] The Court did, however, decide an implied preemption case involving the 1909 Copyright Act and a California state statute that criminalized the production and sale of unauthorized duplicate sound recordings (*e.g.*, cassette tapes recorded from vinyl albums). In that case, *Goldstein v. California*, 412 U.S. 546 (1973), the Supreme Court concluded that the 1909 Copyright Act did not preempt such a state statute. Key to the Court's decision was the fact that, under the 1909 Act, federal copyright law protected only published works; state law continued to provide exclusive rights in unpublished works, showing the states played a partnership role. Moreover, the 1909 Act did not provide any

recent example, using a commonly employed "extra elements" approach to resolving § 301 preemption questions.

Forest Park Pictures v. Universal Television Network, Inc.
683 F.3d 424 (2d Cir. 2012)

WALKER, JUDGE:

This dispute over the concept for a television show presents the question of the extent to which the Copyright Act preempts contract claims involving copyrightable property. Plaintiffs-Appellants Forest Park Pictures, Hayden Christensen, and Tove Christensen (collectively, "Forest Park") developed an idea for a television series and created a writing that embodied it, known in the industry as a "series treatment." Forest Park submitted its idea, first by mail and then in person, to Defendant-Appellee USA Network, a division of Universal Television Network, Inc. ("USA Network"). Forest Park alleges an implied promise by USA Network to pay reasonable compensation if the idea were used. . . .

BACKGROUND

Facts

Because Forest Park appeals from an order dismissing the complaint on the pleadings, we accept as true the facts alleged in the Third Amended Complaint ("Complaint"). In 2005, Forest Park formulated a concept for a television show called "Housecall," in which a doctor, after being expelled from the medical community for treating patients who could not pay, moved to Malibu, California, and became a "concierge" doctor to the rich and famous. Forest Park created a written series treatment for the idea, including character biographies, themes, and storylines. It mailed this written material to Alex Sepiol, who worked for USA Network.

After sending the written materials, Forest Park requested a meeting between its representatives and Sepiol. Sepiol scheduled the meeting "for the express purpose of hearing Plaintiffs pitch" their show. Complaint ¶ 12. At the time, Sepiol and USA Network knew "that writer-creat[o]rs pitch creative ideas to prospective purchasers with the object of selling those ideas for compensation" and "that it was standard in the entertainment industry for ideas to be pitched with the expectation of compensation in the event of use." *Id.* ¶ 9. And, at the meeting, "[i]t was understood that Plaintiffs were pitching those ideas with the object of persuading

---

copyright for a sound recording itself (separate from the underlying musical composition) until a 1971 amendment went into effect in February 1972, after the events in *Goldstein* took place. The Court concluded that the California law created no conflict:

> The California statutory scheme evidences a legislative policy to prohibit "tape piracy" and "record piracy," conduct that may adversely affect the continued production of new recordings, a large industry in California. Accordingly, the State has, by statute, given to recordings the attributes of property. No restraint has been placed on the use of an idea or concept; rather, petitioners and other individuals remain free to record the same compositions in precisely the same manner and with the same personnel as appeared on the original recording.

Id. at 571.

USA Network to purchase those ideas for commercial development." *Id.* ¶ 13. Sepiol said that prior to hearing the idea for "Housecall," he had never heard of "concierge" doctors, or doctors who make house calls for wealthy patients, and "thought it was a fascinating concept for a television show." *Id.* ¶ 15. Over the course of the following week, Sepiol and Forest Park exchanged further communications; however, discussions soon ceased and no further contact between the parties ensued.

A little less than four years later, USA Network produced and aired a television show called "Royal Pains," in which a doctor, after being expelled from the medical community for treating patients who could not pay, became a concierge doctor to the rich and famous in the Hamptons. Forest Park had no prior knowledge of "Royal Pains," did not consent to its production, and received no compensation from USA Network for the use of its idea for the show.

Prior Proceedings

Forest Park Pictures, located in California, and the Christensens, residents of California and Toronto, Canada, brought a diversity action against USA Network and Universal Television Network, a New York corporation, for breach of contract. USA Network moved under Federal Rule of Civil Procedure 12(b)(6) to dismiss the Complaint on the grounds that the Copyright Act preempted the claim and that the contract was too vague to be enforced. . . .

DISCUSSION

This appeal presents two questions: first, whether Forest Park's breach of implied contract claim is preempted by the Copyright Act; and second, if such a claim is not preempted, whether Forest Park adequately pleaded a claim under state law. We hold that Forest Park's claim is not preempted and that the Complaint pleads an enforceable contract under state law that survives a motion to dismiss.

. . . .

I. Preemption

We first turn to USA Network's argument that Forest Park's claim is preempted. Section 301 of the Copyright Act expressly preempts a state law claim only if (i) the work at issue "come[s] within the subject matter of copyright" and (ii) the right being asserted is "equivalent to any of the exclusive rights within the general scope of copyright." 17 U.S.C. § 301.

A. Subject Matter Requirement

In order to be preempted, a claim must involve a work "within the subject matter of copyright." 17 U.S.C. § 301(a). Copyright protection exists for "original works of authorship fixed in any tangible medium of expression," but does not extend to an "idea, * * * regardless of the form in which it is described, explained, illustrated, or embodied." 17 U.S.C. § 102(a), (b). We have held, however, that works may fall within the subject matter of copyright, and thus be subject to preemption, even if they contain material that is uncopyrightable under section 102. *See Nat'l Basketball Ass'n v. Motorola, Inc.* ("*NBA*"), 105 F.3d 841, 849 (2d

Cir. 1997); *Harper & Row, Publishers, Inc. v. Nation Enters.*, 723 F.2d 195, 200 (2d Cir. 1983), *rev'd on other grounds*, 471 U.S. 539 (1985). *See generally* 4 Melville B. Nimmer & David Nimmer, *Nimmer on Copyright* § 19D.03[A][2][b] (2011). In *Harper & Row*, for example, the work at issue, President Ford's memoirs, contained uncopyrightable facts. Nevertheless, we held that the factual content of the book did not remove the work as a whole (indisputably a literary work of authorship, see § 102(a)(1)) from the subject matter of copyright. Similarly, in *Briarpatch Ltd. v. Phoenix Pictures, Inc.*, 373 F.3d 296 (2d Cir. 2004), we held that a novel fell within "the broad ambit of the subject matter categories" listed in section 102(a) despite containing uncopyrightable ideas. *Id.* at 306. The scope of copyright for preemption purposes, then, extends beyond the scope of available copyright protection.

The reason for our broad interpretation of the scope of copyright preemption is that Congress, in enacting section 301, created a regime in which some types of works are copyrightable and others fall into the public domain. *See NBA*, 105 F.3d at 849. In preempting certain state causes of action, Congress deprived the states of the power to "vest exclusive rights in material that Congress intended to be in the public domain." *Id.*; *see also Harper & Row*, 723 F.2d at 200 (recognizing that it would "run directly afoul of one of the Act's central purposes" to allow the states to expand copyright protection to works Congress deemed uncopyrightable). Section 301's preemption scheme functions properly only if the "'subject matter of copyright' includes all works of a type covered by sections 102 and 103, even if federal law does not afford protection to them." *NBA*, 105 F.3d at 850 (quoting *ProCD, Inc. v. Zeidenberg*, 86 F.3d 1447, 1453 (7th Cir. 1996)).

The work at issue in this case is Forest Park's idea for "Housecall," manifested in the series treatment (comprising character biographies, themes, and storylines). This treatment and associated written materials are "works of authorship that are fixed in a tangible medium." 17 U.S.C. § 301(a). Although Forest Park's Complaint does not allege that USA Network took its actual scripts or biographies, the subject matter requirement is met because the Complaint alleges that USA Network used the ideas embodied in those written works. That the work contains within it some uncopyrightable ideas does not remove it from the subject matter of copyright. Moreover, because the ideas that are the subject of the claim were fixed in writing—whether or not the writing itself is at issue—the claim is within the subject matter of copyright. *See NBA*, 105 F.3d at 849; *see also Montz v. Pilgrim Films & Television, Inc.*, 649 F.3d 975, 979 (9th Cir. 2011) (en banc) (holding that an idea for a television show, once fixed in a tangible medium, fell within the subject matter of copyright); *Wrench LLC v. Taco Bell Corp.*, 256 F.3d 446, 455 (6th Cir. 2001) (holding that an idea for a character, conveyed in storyboards, scripts, and drawings, was within the subject matter of copyright). Therefore, the first requirement for preemption is met.

B. Equivalency Requirement

In order to establish preemption, USA Network must also demonstrate that the Complaint seeks to vindicate a "legal or equitable right[] that [is] equivalent to any of the exclusive rights within the general scope of copyright as specified by section 106." 17 U.S.C. § 301(a). Section 106 gives copyright owners the exclusive rights, among other things, to reproduce a copyrighted work, to prepare derivative works, to distribute copies of the work to the public, and to display the work publicly. 17 U.S.C. § 106. A state law right is equivalent to one of the exclusive rights of copyright if it "may be abridged by an act which, in and of itself, would infringe one of the exclusive rights." *Harper & Row*, 723 F.2d at 200. "But if an extra element is required instead of or in addition to the acts of reproduction, performance, distribution or display, in order to constitute a state-created cause of action," there is no preemption. *Computer Assocs. Int'l, Inc. v. Altai, Inc.*, 982 F.2d 693, 716 (2d Cir. 1992).

Applying this "extra element" test, we have held numerous categories of claims to be not preempted, including trade secret claims, in which the plaintiff must show the defendant breached a duty of trust through improper disclosure of confidential material, *id.* at 717; certain "hot news" misappropriation claims, because the plaintiff must show time-sensitive factual information, free-riding by the defendant, and a threat to the very existence of the plaintiff's product, *NBA*, 105 F.3d at 853; and breach of confidential relationship, in which the plaintiff must show an obligation not to disclose ideas revealed in confidence, *Smith v. Weinstein*, 578 F. Supp. 1297, 1307 (S.D.N.Y. 1984), *aff'd without opinion*, 738 F.2d 419 (2d Cir. 1984). By contrast, we have found a state law claim preempted when the extra element changes the scope but not the fundamental nature of the right. *See, e.g.*, *Briarpatch*, 373 F.3d at 306-07 (holding an unjust enrichment claim preempted because, although plaintiff must prove "enrichment," the essential nature of the claim remained the unauthorized use of a work); *Fin. Info., Inc. v. Moody's Investors Serv., Inc.*, 808 F.2d 204, 208 (2d Cir. 1986) (holding that a misappropriation claim was preempted because the element of commercial immorality did not change qualitative nature of the right); *Harper & Row*, 723 F.2d at 201 (holding that a claim of conversion based on unauthorized publication of a work was preempted because it is "coextensive with an exclusive right already safeguarded by the Act").

In this case, the issue is whether a particular breach of contract claim survives preemption. More specifically, Forest Park alleges that it entered into an implied-in-fact agreement with USA Network that required USA Network to pay Forest Park for the use of its idea. There are several qualitative differences between such a contract claim and a copyright violation claim. First, the Copyright Act does not provide an express right for the copyright owner to receive payment for the use of a work. It simply gives the copyright owner the right to prevent distribution, copying, or the creation of derivative works (though, of course, the copyright owner may cede or all part of these rights for payment). See 17 U.S.C. § 106. Second, a plaintiff suing for failure to pay under a contract must prove extra elements beyond use or copying, including mutual assent and valid consideration. Third, a breach of contract claim asserts rights only against the contractual counterparty, not the public at large. As the Seventh Circuit explained in *ProCD*, "A copyright is a right against the world. Contracts, by contrast, generally affect only their parties;

strangers may do as they please, so contracts do not create 'exclusive rights.'" 86 F.3d at 1454.

A number of our sister circuits have accordingly concluded that at least some contract claims involving the subject matter of copyright do not contest rights that are the equivalent of rights under the Copyright Act, and thus are not preempted. Of course, preemption cannot be avoided simply by labeling a claim "breach of contract." A plaintiff must actually allege the elements of an enforceable contract (whether express or implied-in-fact), including offer, acceptance, and consideration, in addition to adequately alleging the defendant's breach of the contract.

As long as the elements of a contract are properly pleaded, there is no difference for preemption purposes between an express contract and an implied-in-fact contract. There is, however, a significant difference for preemption purposes between contracts implied-in-fact and contracts implied-in-law. Theories of implied-in-law contract, quasi-contract, or unjust enrichment differ significantly from breach of contract because the plaintiff need not allege the existence of an actual agreement between the parties. Under these quasi-contractual theories, the plaintiff need only prove that the defendant was unjustly enriched through the use of her idea or work. Such a claim is not materially different from a claim for copyright infringement that requires a plaintiff to prove that the defendant used, reproduced, copied, or displayed a copyrighted work. *See Briarpatch*, 373 F.3d at 306 (finding no extra element in an unjust enrichment claim); *see also Wrench*, 256 F.3d at 459 (noting that there is "a crucial difference" between implied-in-fact contracts and implied-in-law contracts because the latter "depend[] on nothing more than the unauthorized use of the work").

In this case, we need not address whether preemption is precluded whenever there is a contract claim, or only when the contract claim includes a promise to pay. Here the Complaint specifically alleges that the contract includes by implication a promise to pay for the use of Forest Park's idea. The alleged contract does not simply require USA Network to honor Forest Park's exclusive rights under the Copyright Act (assuming the material at issue to be copyrightable); it requires USA Network to pay for the use of Forest Park's ideas. A claim for breach of a contract including a promise to pay is qualitatively different from a suit to vindicate a right included in the Copyright Act and is not subject to preemption.[1]

II. Breach of Contract

. . . .

. . . The remaining question before us thus becomes whether Forest Park has alleged an enforceable contract under California contract law.

---

[1] We need not here consider whether even a promise to pay may be insufficient to avoid preemption in circumstances where, through contracts of adhesion or similar instruments, a plaintiff uses such promises to create a de facto monopoly at odds with federal copyright policy. *See* Arthur R. Miller, *Common Law Protection for Products of the Mind: An "Idea" Whose Time Has Come*, 119 Harv. L. Rev. 703, 768-74 (2006). That is not this case.

B. Implied-in-Fact Contract

California has long recognized that an implied-in-fact contract may be created where the plaintiff submits an idea (the offer) that the defendant subsequently uses (the acceptance) without compensating the plaintiff (the breach). In *Desny v. Wilder*, 299 P.2d 257 (Cal. 1956), the plaintiff, Desny, telephoned Billy Wilder, then a producer and writer for Paramount Pictures, and told Wilder's secretary that he had an idea for a film. At the secretary's request, Desny forwarded to Wilder a brief synopsis of the movie idea and stated that, if the idea were used, he expected to be paid. Faced with the enforceability of such an agreement, the California Supreme Court held that a contract claim based on the submission of an idea could succeed either if the plaintiff received "an express promise to pay" or if "the circumstances preceding and attending disclosure, together with the conduct of the offeree acting with knowledge of the circumstances, show a promise of the type usually referred to as 'implied' or 'implied-in-fact.'" *Id.* at 270. For almost six decades following *Desny*, California courts have continued to recognize contract claims under the authority of that case.

. . . .

Here, although Forest Park does not allege that it expressly conditioned disclosure on a promise of payment, the Complaint alleges facts that, if proven, would establish that USA Network knew or should have known such a condition was implied. Forest Park alleges that it pitched its ideas to USA Network "with the object of persuading USA Network to purchase those ideas for commercial development," and that USA Network and its agent Sepiol "at all relevant times knew (a) that writer-creators pitch creative ideas to prospective purchasers with the object of selling those ideas for compensation; and (b) that it was standard in the entertainment industry for ideas to be pitched with the expectation of compensation in the event of use." Complaint ¶¶ 9, 13. Moreover, the Complaint alleges that USA Network accepted Forest Park's idea when it knew or should have known of that condition by keeping the series treatment Forest Park submitted, scheduling a meeting with Forest Park, allowing Forest Park to pitch its idea uninterrupted, and communicating with Forest Park after the meeting. These allegations are sufficient to plead a *Desny* claim under California law.

USA Network argues that even if Forest Park did allege an implied-in-fact agreement, the agreement would not be enforceable because it lacks a definite price term. California courts, however, enforce contracts without exact price terms as long as the parties' intentions can be ascertained. And California permits custom and usage (among other extrinsic evidence) to supply absent terms. In *Desny* itself, an enforceable contract was found even though the plaintiff stated he expected to be paid "the reasonable value" of his idea. 299 P.2d at 261.

Forest Park alleges that it agreed with USA Network to be paid the industry standard for its idea, which is enough under California law to survive a motion to dismiss. At trial, Forest Park will have to prove that such an industry standard price exists and that both parties implicitly agreed to it. . . . Because Forest Park has alleged an enforceable implied-in-fact contract including a promise of payment for the disclosure of its idea, its claim is not preempted by the Copyright Act and therefore the district court erred in dismissing the Complaint.

. . . .

Notes & Questions

1. Why does the court find that an "idea," something that is not protectable under copyright, nonetheless is "within the subject matter of copyright" for purposes of analyzing whether the plaintiff's claim is preempted by the Copyright Act? If the justification is that Congress meant for certain elements to remain in the public domain, why should it matter that the Plaintiff's idea was "fixed" in a tangible media (*i.e.*, the scripts)? Should § 301 not preempt a claim that a defendant used an idea that had been communicated orally but never written down?

2. Why might Congress have made a distinction in § 301 between state law rights that are "equivalent" to those granted by copyright and those that are not "equivalent," expressly preempting only the former? Recall the Florida plug-molding statute at issue in *Bonito Boats*. Was the problem with that state law equivalency of protection?

Is the extra-element test the best way to determine equivalency between the rights granted by the state statute and the rights granted by the Copyright Act? As the court notes, even state-law claims that have extra elements may not survive a preemption challenge if the extra element does not change "the fundamental nature of the right." Does that mean that the extra element test, in the end, is not actually a sufficient test?

3. Re-read footnote 1. What types of contracts might present preemption problems when a party sought to enforce them through a state-law breach of contract claim? Would it matter if the provision that was allegedly breached was one that was "at odds with federal copyright policy" or would simply having a provision in the contract, breached or not, that is at odds doom the entire contract? Would a provision prohibiting resale of a copy that embodied a copyrighted work be enforceable? How about a provision that prohibited critical reviews of a video game embodied in the ubiquitous click-through End-User License Agreement (EULA)? What about a provision in a EULA that prohibited reverse engineering of a computer software program?

II. ENFORCEMENT LIMITS: THE FIRST-SALE PRINCIPLE

As we observed at the beginning of this chapter, property rights are always limited. What happens when two different property rights collide? Intellectual property rights protect intangible items—*e.g.*, ideas, expressions, business good will. Those intangible items, however, are often embodied in a tangible object—*e.g.*, a pill, a book, a soda can. That tangible object is personal property. Generally, people have an idea that they can do what they want with the personal property that they own. One thing people generally assume they can do with their personal property is give it away or sell it. Indeed, the free alienability of goods is a particularly strong sentiment in the market economy of the United States. Think of all the garage sales, swap meets, or eBay transactions you've participated in, without worrying about whether the items being resold were copyrighted, patented, or bore trademarks. This section addresses how the existence of intellectual property rights accommodates the rights that are normally associated with personal property ownership, and vice versa.

A. COPYRIGHT LAW AND THE FIRST SALE DOCTRINE

In Chapter 4 we discussed the first sale doctrine codified in section 109 of the Copyright Act. As introduced in that chapter, section 109 limits both the public distribution right and the public display right granted to the copyright owner. An important aspect of the first sale principle is the ability to re-sell copies. The next case addresses what happens when the first sale occurs in a foreign country and then that copy is imported to the United States, sometimes referred to as "gray market goods" or parallel importation.

Kirtsaeng v. John Wiley & Sons, Inc.

133 S. Ct. 1351 (2013)

BREYER, JUSTICE:

Section 106 of the Copyright Act grants "the owner of copyright under this title" certain "exclusive rights," including the right "to distribute copies * * * of the copyrighted work to the public by sale or other transfer of ownership." 17 U.S.C. § 106(3). These rights are qualified, however, by the application of various limitations set forth in the next several sections of the Act, §§ 107 through 122. Those sections, typically entitled "Limitations on exclusive rights," include, for example, the principle of "fair use" (§ 107), permission for limited library archival reproduction, (§ 108), and the doctrine at issue here, the "first sale" doctrine (§ 109).

Section 109(a) sets forth the "first sale" doctrine as follows:

> "Notwithstanding the provisions of section 106(3) [the section that grants the owner exclusive distribution rights], the owner of a particular copy or phonorecord *lawfully made under this title* * * * is entitled, without the authority of the copyright owner, to sell or otherwise dispose of the possession of that copy or phonorecord." (Emphasis added.)

Thus, even though § 106(3) forbids distribution of a copy of, say, the copyrighted novel *Herzog* without the copyright owner's permission, § 109(a) adds that, once a copy of *Herzog* has been lawfully sold (or its ownership otherwise lawfully transferred), the buyer of *that copy* and subsequent owners are free to dispose of it as they wish. In copyright jargon, the "first sale" has "exhausted" the copyright owner's § 106(3) exclusive distribution right.

What, however, if the copy of *Herzog* was printed abroad and then initially sold with the copyright owner's permission? Does the "first sale" doctrine still apply? Is the buyer, like the buyer of a domestically manufactured copy, free to bring the copy into the United States and dispose of it as he or she wishes?

To put the matter technically, an "importation" provision, § 602(a)(1), says that

> "[i]mportation into the United States, without the authority of the owner of copyright under this title, of copies * * * of a work that have been acquired outside the United States is an infringement of the exclusive right to distribute copies * * * *under section 106* * * *." 17 U.S.C. § 602(a)(1) (emphasis added).

Thus § 602(a)(1) makes clear that importing a copy without permission violates the owner's exclusive distribution right. But in doing so, § 602(a)(1) refers explicitly to the § 106(3) exclusive distribution right. As we have just said, § 106 is by its terms "[s]ubject to" the various doctrines and principles contained in §§ 107 through 122, including § 109(a)'s "first sale" limitation. Do those same modifications apply—in particular, does the "first sale" modification apply—when considering whether § 602(a)(1) prohibits importing a copy?

In *Quality King Distributors, Inc. v. L'anza Research Int'l, Inc.*, 523 U.S. 135, 145 (1998), we held that § 602(a)(1)'s reference to § 106(3)'s exclusive distribution right incorporates the later subsections' limitations, including, in particular, the "first sale" doctrine of § 109. Thus, it might seem that, § 602(a)(1) notwithstanding, one who buys a copy abroad can freely import that copy into the United States and dispose of it, just as he could had he bought the copy in the United States.

But *Quality King* considered an instance in which the copy, though purchased abroad, was initially manufactured in the United States (and then sent abroad and sold). This case is like *Quality King* but for one important fact. The copies at issue here were manufactured abroad. That fact is important because § 109(a) says that the "first sale" doctrine applies to "a particular copy or phonorecord *lawfully made under this title.*" And we must decide here whether the five words, "lawfully made under this title," make a critical legal difference.

Putting section numbers to the side, we ask whether the "first sale" doctrine applies to protect a buyer or other lawful owner of a copy (of a copyrighted work) lawfully manufactured abroad. Can that buyer bring that copy into the United States (and sell it or give it away) without obtaining permission to do so from the copyright owner? Can, for example, someone who purchases, say at a used bookstore, a book printed abroad subsequently resell it without the copyright owner's permission?

In our view, the answers to these questions are, yes. We hold that the "first sale" doctrine applies to copies of a copyrighted work lawfully made abroad.

I

A

Respondent, John Wiley & Sons, Inc., publishes academic textbooks. Wiley obtains from its authors various foreign and domestic copyright assignments, licenses and permissions—to the point that we can, for present purposes, refer to Wiley as the relevant American copyright owner. See 654 F.3d 210, 213, n. 6 (2d Cir. 2011). Wiley often assigns to its wholly owned foreign subsidiary, John Wiley & Sons (Asia) Pte Ltd., rights to publish, print, and sell Wiley's English language textbooks abroad. Each copy of a Wiley Asia foreign edition will likely contain language making clear that the copy is to be sold only in a particular country or geographical region outside the United States.

For example, a copy of Wiley's American edition says, "Copyright © 2008 John Wiley & Sons, Inc. All rights reserved. * * * Printed in the United States of America." J. Walker, *Fundamentals of Physics*, p. vi (8th ed. 2008). A copy of Wiley Asia's Asian edition of that book says:

> "Copyright © 2008 John Wiley & Sons (Asia) Pte Ltd[.] All rights reserved. This book is authorized for sale in Europe, Asia, Africa, and the Middle East only and may be not exported out of these territories. Exportation from or importation of this book to another region without the Publisher's authorization is illegal and is a violation of the Publisher's rights. The Publisher may take legal action to enforce its rights. * * * Printed in Asia." J. Walker, *Fundamentals of Physics*, p. vi (8th ed. 2008 Wiley Int'l Student ed.).

. . . .

The upshot is that there are two essentially equivalent versions of a Wiley textbook, each version manufactured and sold with Wiley's permission: (1) an American version printed and sold in the United States, and (2) a foreign version manufactured and sold abroad. And Wiley makes certain that copies of the second version state that they are not to be taken (without permission) into the United States.

Petitioner, Supap Kirtsaeng, a citizen of Thailand, moved to the United States in 1997 to study mathematics at Cornell University. He paid for his education with the help of a Thai Government scholarship which required him to teach in Thailand for 10 years on his return. Kirtsaeng successfully completed his undergraduate courses at Cornell, successfully completed a Ph.D. program in mathematics at the University of Southern California, and then, as promised, returned to Thailand to teach. While he was studying in the United States, Kirtsaeng asked his friends and family in Thailand to buy copies of foreign edition English-language textbooks at Thai book shops, where they sold at low prices, and mail them to him in the United States. Kirtsaeng would then sell them, reimburse his family and friends, and keep the profit.

B

In 2008 Wiley brought this federal lawsuit against Kirtsaeng for copyright infringement. Wiley claimed that Kirtsaeng's unauthorized importation of its books and his later resale of those books amounted to an infringement of Wiley's § 106(3) exclusive right to distribute as well as § 602's related import prohibition. 17 U.S.C. §§ 106(3), 602(a). Kirtsaeng replied that the books he had acquired were "'lawfully made'" and that he had acquired them legitimately. Thus, in his view, § 109(a)'s "first sale" doctrine permitted him to resell or otherwise dispose of the books without the copyright owner's further permission.

The District Court held that Kirtsaeng could not assert the "first sale" defense because, in its view, that doctrine does not apply to "foreign-manufactured goods" (even if made abroad with the copyright owner's permission). The jury then found that Kirtsaeng had willfully infringed Wiley's American copyrights by selling and importing without authorization copies of eight of Wiley's copyrighted titles. And it assessed statutory damages of $600,000 ($75,000 per work).

On appeal, a split panel of the Second Circuit agreed with the District Court. It pointed out that § 109(a)'s "first sale" doctrine applies only to "the owner of a particular copy * * * *lawfully made under this title.*" [654 F.3d] at 218–219 (emphasis added). And, in the majority's view, this language means that the "first sale" doctrine does not apply to copies of American copyrighted works manufactured abroad. A dissenting judge thought that the words "lawfully made under this title" do not refer "to a place of manufacture" but rather "focu[s] on whether a particular copy was manufactured lawfully under" America's copyright statute, and that "the lawfulness of the manufacture of a particular copy should be judged by U.S. copyright law." *Id.* at 226 (opinion of Murtha, J.).

We granted Kirtsaeng's petition for certiorari to consider this question in light of different views among the Circuits.

II

We must decide whether the words "lawfully made under this title" restrict the scope of § 109(a)'s "first sale" doctrine geographically. The Second Circuit, the Ninth Circuit, Wiley, and the Solicitor General (as *amicus*) all read those words as imposing a form of *geographical* limitation. The Second Circuit held that they limit the "first sale" doctrine to particular copies "made in territories *in which the Copyright Act is law,*" which (the Circuit says) are copies "manufactured domestically," not "outside of the United States." 654 F.3d at 221–222 (emphasis added). Wiley agrees that those five words limit the "first sale" doctrine "to copies made in conformance with the [United States] Copyright Act *where the Copyright Act is applicable,*" which (Wiley says) means it does not apply to copies made "outside the United States" and at least not to "foreign production of a copy for distribution exclusively abroad." Similarly, the Solicitor General says that those five words limit the "first sale" doctrine's applicability to copies "'*made subject to* and in compliance with [the Copyright Act],'" which (the Solicitor General says) are copies "made in the United States." And the Ninth Circuit has held that those words limit the "first sale" doctrine's applicability (1) to copies lawfully made in the United States, and (2) to copies lawfully made outside the United States but initially sold in the United States with the copyright owner's permission. *Denbicare U.S.A. Inc. v. Toys "R" Us, Inc.*, 84 F.3d 1143, 1149–1150 (1996).

Under any of these geographical interpretations, § 109(a)'s "first sale" doctrine would not apply to the Wiley Asia books at issue here. And, despite an American copyright owner's permission to *make* copies abroad, one who *buys* a copy of any such book or other copyrighted work—whether at a retail store, over the Internet, or at a library sale—could not resell (or otherwise dispose of) that particular copy without further permission.

Kirtsaeng, however, reads the words "lawfully made under this title" as imposing a *non*-geographical limitation. He says that they mean made "in accordance with" or "in compliance with" the Copyright Act. In that case, § 109(a)'s "first sale" doctrine would apply to copyrighted works as long as their manufacture met the requirements of American copyright law. In particular, the doctrine would apply where, as here, copies are manufactured abroad with the permission of the copyright owner.

In our view, § 109(a)'s language, its context, and the common-law history of the "first sale" doctrine, taken together, favor a *non*-geographical interpretation. We also doubt that Congress would have intended to create the practical copyright-related harms with which a geographical interpretation would threaten ordinary scholarly, artistic, commercial, and consumer activities. See Part II–D, *infra*. We consequently conclude that Kirtsaeng's nongeographical reading is the better reading of the Act.

A

The language of § 109(a) read literally favors Kirtsaeng's nongeographical interpretation, namely, that "lawfully made under this title" means made "in accordance with" or "in compliance with" the Copyright Act. The language of § 109(a) says nothing about geography. The word "under" can mean "[i]n accordance with." 18 Oxford English Dictionary 950 (2d ed. 1989). See also Black's Law Dictionary 1525 (6th ed. 1990) ("according to"). And a nongeographical interpretation provides each word of the five-word phrase with a distinct purpose. The first two words of the phrase, "lawfully made," suggest an effort to distinguish those copies that were made lawfully from those that were not, and the last three words, "under this title," set forth the standard of "lawful[ness]." Thus, the nongeographical reading is simple, it promotes a traditional copyright objective (combatting piracy), and it makes word-by-word linguistic sense.

The geographical interpretation, however, bristles with linguistic difficulties. It gives the word "lawfully" little, if any, linguistic work to do. (How could a book be *un*lawfully "made under this title"?) It imports geography into a statutory provision that says nothing explicitly about it. And it is far more complex than may at first appear.

. . . .

B

. . . .

. . . [W]e normally presume that the words "lawfully made under this title" carry the same meaning when they appear in different but related sections. *Department of Revenue of Ore. v. ACF Industries, Inc.,* 510 U.S. 332, 342 (1994). But doing so here produces surprising consequences. Consider:

(1) Section 109(c) says that, despite the copyright owner's exclusive right "to display" a copyrighted work (provided in § 106(5)), the owner of a particular copy "lawfully made under this title" may publicly display it without further authorization. To interpret these words geographically would mean that one who buys a copyrighted work of art, a poster, or even a bumper sticker, in Canada, in Europe, in Asia, could not display it in America without the copyright owner's further authorization.

(2) Section 109(e) specifically provides that the owner of a particular copy of a copyrighted video arcade game "lawfully made under this title" may "publicly perform or display that game in coin-operated equipment" without the authorization of the copyright owner. To interpret these words geographically means that an arcade owner could not ("without the authority of the copyright owner") perform or display arcade games (whether new or used) originally made in Japan.

(3) Section 110(1) says that a teacher, without the copyright owner's authorization, is allowed to perform or display a copyrighted work (say, an audiovisual work) "in the course of face-to-face teaching activities"—unless the teacher knowingly used "a copy that was not lawfully made under this title." To interpret these words geographically would mean that the teacher could not (without further authorization) use a copy of a film during class if the copy was lawfully made in Canada, Mexico, Europe, Africa, or Asia.

(4) In its introductory sentence, § 106 provides the Act's basic exclusive rights to an "owner of a copyright under this title." The last three words cannot support a geographic interpretation.

Wiley basically accepts the first three readings, but argues that Congress intended the restrictive consequences. And it argues that context simply requires that the words of the fourth example receive a different interpretation. Leaving the fourth example to the side, we shall explain in Part II–D, *infra,* why we find it unlikely that Congress would have intended these, and other related consequences.

C

A relevant canon of statutory interpretation favors a nongeographical reading. "[W]hen a statute covers an issue previously governed by the common law," we must presume that "Congress intended to retain the substance of the common law." *Samantar v. Yousuf,* 560 U.S. 305, 320, n. 13 (2010).

The "first sale" doctrine is a common-law doctrine with an impeccable historic pedigree. In the early 17th century Lord Coke explained the common law's refusal to permit restraints on the alienation of chattels. Referring to Littleton, who wrote in the 15th century, Lord Coke wrote:

> "[If] a man be possessed of * * * a horse, or of any other chattell * * * and give or sell his whole interest * * * therein upon condition that the Donee or Vendee shall not alien[ate] the same, the [condition] is voi[d], because his whole interest * * * is out of him, so as he hath no possibilit[y] of a Reverter, and it is against Trade and Traffi[c], and bargaining and contracting betwee[n] man and man: and it is within the reason of our Author that it should ouster him of all power given to him." 1 E. Coke, *Institutes of the Laws of England* § 360, p. 223 (1628).

A law that permits a copyright holder to control the resale or other disposition of a chattel once sold is similarly "against Trade and Traffi[c], and bargaining and contracting." *Id.*

With these last few words, Coke emphasizes the importance of leaving buyers of goods free to compete with each other when reselling or otherwise disposing of those goods. American law too has generally thought that competition, including freedom to resell, can work to the advantage of the consumer.

The "first sale" doctrine also frees courts from the administrative burden of trying to enforce restrictions upon difficult-to-trace, readily movable goods. And it avoids the selective enforcement inherent in any such effort. Thus, it is not surprising that for at least a century the "first sale" doctrine has played an important role in American copyright law. See *Bobbs–Merrill Co. v. Straus,* 210 U.S. 339 (1908).

The common-law doctrine makes no geographical distinctions; nor can we find any in *Bobbs–Merrill* (where this Court first applied the "first sale" doctrine) or in § 109(a)'s predecessor provision, which Congress enacted a year later. Rather, as the Solicitor General acknowledges, "a straightforward application of *Bobbs–Merrill*" would not preclude the "first sale" defense from applying to authorized copies made overseas. And we can find no language, context, purpose, or history that would rebut a "straightforward application" of that doctrine here.

. . . .

D

Associations of libraries, used-book dealers, technology companies, consumer-goods retailers, and museums point to various ways in which a geographical interpretation would fail to further basic constitutional copyright objectives, in particular "promot[ing] the Progress of Science and useful Arts." U.S. Const., Art. I, § 8, cl. 8.

The American Library Association tells us that library collections contain at least 200 million books published abroad (presumably, many were first published in one of the nearly 180 copyright-treaty nations and enjoy American copyright protection under 17 U.S.C. § 104); that many others were first published in the United States but printed abroad because of lower costs; and that a geographical interpretation will likely require the libraries to obtain permission (or at least create significant uncertainty) before circulating or otherwise distributing these books.

How, the American Library Association asks, are the libraries to obtain permission to distribute these millions of books? How can they find, say, the copyright owner of a foreign book, perhaps written decades ago? They may not know the copyright holder's present address. And, even where addresses can be found, the costs of finding them, contacting owners, and negotiating may be high indeed. Are the libraries to stop circulating or distributing or displaying the millions of books in their collections that were printed abroad?

Used-book dealers tell us that, from the time when Benjamin Franklin and Thomas Jefferson built commercial and personal libraries of foreign books, American readers have bought used books published and printed abroad. The dealers say that they have "operat[ed] * * * for centuries" under the assumption that the "first sale" doctrine applies. But under a geographical interpretation a contemporary tourist who buys, say, at Shakespeare and Co. (in Paris), a dozen copies of a foreign book for American friends might find that she had violated the copyright law. The used-book dealers cannot easily predict what the foreign copyright holder may think about a reader's effort to sell a used copy of a novel. And they believe that a geographical interpretation will injure a large portion of the used-book business.

Technology companies tell us that "automobiles, microwaves, calculators, mobile phones, tablets, and personal computers" contain copyrightable software programs or packaging. Many of these items are made abroad with the American copyright holder's permission and then sold and imported (with that permission) to the United States. A geographical interpretation would prevent the resale of, say, a car, without the permission of the holder of each copyright on each piece of copyrighted automobile software. Yet there is no reason to believe that foreign auto manufacturers regularly obtain this kind of permission from their software component suppliers, and Wiley did not indicate to the contrary when asked. Without that permission a foreign car owner could not sell his or her used car.

Retailers tell us that over $2.3 trillion worth of foreign goods were imported in 2011. American retailers buy many of these goods after a first sale abroad. And, many of these items bear, carry, or contain copyrighted "packaging, logos, labels, and product inserts and instructions for [the use of] everyday packaged goods from floor cleaners and health and beauty products to breakfast cereals." [Brief for Retail Litigation Center, Inc., et al. as *Amici Curiae*] at 10–11. The retailers add that American sales of more traditional copyrighted works, "such as books, recorded music, motion pictures, and magazines" likely amount to over $220 billion. *Id.* at 9. A geographical interpretation would subject many, if not all, of them to the disruptive impact of the threat of infringement suits.

Art museum directors ask us to consider their efforts to display foreign-produced works by, say, Cy Twombly, Rene Magritte, Henri Matisse, Pablo Picasso, and others. A geographical interpretation, they say, would require the museums to obtain permission from the copyright owners before they could display the work—even if the copyright owner has already sold or donated the work to a foreign museum. What are the museums to do, they ask, if the artist retained the copyright, if the artist cannot be found, or if a group of heirs is arguing about who owns which copyright?

These examples, and others previously mentioned, help explain *why* Lord Coke considered the "first sale" doctrine necessary to protect "Trade and Traffi[c], and bargaining and contracting," and they help explain *why* American copyright law has long applied that doctrine.

. . . .

III

. . . .

. . . Wiley and the dissent claim that a nongeographical interpretation will make it difficult, perhaps impossible, for publishers (and other copyright holders) to divide foreign and domestic markets. We concede that is so. A publisher may find it more difficult to charge different prices for the same book in different geographic markets. But we do not see how these facts help Wiley, for we can find no basic principle of copyright law that suggests that publishers are especially entitled to such rights.

The Constitution describes the nature of American copyright law by providing Congress with the power to "secur[e]" to "[a]uthors" "for limited [t]imes" the "*exclusive [r]ight to their * * * [w]ritings*." Art. I, § 8, cl. 8. The Founders, too, discussed the need to grant an author a limited right to exclude competition. Compare Letter from Thomas Jefferson to James Madison (July 31, 1788), in 13 Papers of Thomas Jefferson 440, 442–443 (J. Boyd ed. 1956) (arguing against any monopoly) with Letter from James Madison to Thomas Jefferson (Oct. 17, 1788), in 14 *id.* at 16, 21 (arguing for a limited monopoly to secure production). But the Constitution's language nowhere suggests that its limited exclusive right should include a right to divide markets or a concomitant right to charge different purchasers different prices for the same book, say to increase or to maximize gain. Neither, to our knowledge, did any Founder make any such suggestion. We have found no precedent suggesting a legal preference for interpretations of copyright statutes that would provide for market divisions. Cf. Copyright Law Revision, pt. 2, at 194 (statement of Barbara Ringer, Copyright Office) (division of territorial markets was "primarily a matter of private contract").

To the contrary, Congress enacted a copyright law that (through the "first sale" doctrine) limits copyright holders' ability to divide domestic markets. And that limitation is consistent with antitrust laws that ordinarily forbid market divisions. Cf. *Palmer v. BRG of Ga., Inc.,* 498 U.S. 46, 49-50 (1990) (*per curiam*) ("[A]greements between competitors to allocate territories to minimize competition are illegal"). Whether copyright owners should, or should not, have more than ordinary commercial power to divide international markets is a matter for Congress to decide. We do no more here than try to determine what decision Congress has taken.

. . . [T]he dissent and Wiley contend that our decision launches United States copyright law into an unprecedented regime of "international exhaustion." But they point to nothing indicative of congressional intent in 1976. The dissent also claims that it is clear that the United States now opposes adopting such a regime, but the Solicitor General as *amicus* has taken no such position in this case. In fact, when pressed at oral argument, the Solicitor General stated that the consequences of Wiley's reading of the statute (perpetual downstream control) were "worse" than those of Kirtsaeng's reading (restriction of market segmentation). . . .

. . . .

IV

For these reasons we conclude that the considerations supporting Kirtsaeng's nongeographical interpretation of the words "lawfully made under this title" are the more persuasive. The judgment of the Court of Appeals is reversed, and the case is remanded for further proceedings consistent with this opinion.

KAGAN, JUSTICE, CONCURRING (WITH JUSTICE ALITO):

. . . .

. . . [I]f Congress views the shrinking of § 602(a)(1) as a problem, it should recognize *Quality King*—not our decision today—as the culprit. Here, after all, we merely construe § 109(a); *Quality King* is the decision holding that § 109(a) limits § 602(a)(1). Had we come out the opposite way in that case, § 602(a)(1) would allow a copyright owner to restrict the importation of copies irrespective of the first-sale doctrine. That result would enable the copyright owner to divide international markets in the way John Wiley claims Congress intended when enacting § 602(a)(1). But it would do so without imposing downstream liability on those who purchase and resell in the United States copies that happen to have been manufactured abroad. In other words, that outcome would target unauthorized importers alone, and not the "libraries, used-book dealers, technology companies, consumer-goods retailers, and museums" with whom the Court today is rightly concerned. Assuming Congress adopted § 602(a)(1) to permit market segmentation, I suspect that is how Congress thought the provision would work—not by removing first-sale protection from every copy manufactured abroad (as John Wiley urges us to do here), but by enabling the copyright holder to control imports even when the first-sale doctrine applies (as *Quality King* now prevents).[2]

---

[2] Indeed, allowing the copyright owner to restrict imports irrespective of the first-sale doctrine— *i.e.,* reversing *Quality King*—would yield a far more sensible scheme of market segmentation than would adopting John Wiley's argument here. That is because only the former approach turns on the *intended market* for copies; the latter rests instead on their *place of manufacture*. To see the difference, imagine that John Wiley prints all its textbooks in New York, but wants to distribute certain versions only in Thailand. Without *Quality King*, John Wiley could do so—*i.e.,* produce books in New York, ship them to Thailand, and prevent anyone from importing them back into the United States. But with *Quality King*, that course is not open to John Wiley even under its reading of § 109(a): To prevent someone like Kirtsaeng from re-importing the books—and so to segment the Thai market— John Wiley would have to move its printing facilities abroad. I can see no reason why Congress would

. . . .

[JUSTICE GINSBURG'S dissenting opinion is omitted.]

Notes & Questions

1. Have you ever purchased used copies of your textbooks? How do you think the market for used textbooks affects the price charged for new copies? Have you ever resold your textbooks at the end of a semester? Without the first sale doctrine, such sales constitute copyright infringement and, in Mr. Kirtsaeng's case, the jury found him liable for willful infringement.

2. As the majority indicates, the result in *Kirtsaeng* is what is referred to as "international exhaustion." Countries vary in their approach to the question of exhaustion in copyright law. The dissent described the three main approaches to exhaustion:

> One option is a national-exhaustion regime, under which a copyright owner's right to control distribution of a particular copy is exhausted only within the country in which the copy is sold. Another option is a rule of international exhaustion, under which the authorized distribution of a particular copy anywhere in the world exhausts the copyright owner's distribution right everywhere with respect to that copy. The European Union has adopted the intermediate approach of regional exhaustion, under which the sale of a copy anywhere within the European Economic Area exhausts the copyright owner's distribution right throughout the region.

Kirtsaeng v. John Wiley & Sons, Inc., 133 S. Ct. 1351, 1383-84 (2013) (Ginsburg, J., dissenting). In her dissent Justice Ginsburg noted that in recent international trade negotiations the U.S. has consistently argued against international exhaustion, *see id.* at 1383-85, and she worried that the result in the case would "undermine[] the United States' credibility on the world stage" *Id.* at 1385. Should the Court be influenced by the positions taken by the U.S. in trade negotiations?

3. John Wiley & Sons was engaged in classic price discrimination—charging a lower price in one market (Asia) and a higher price in another (U.S.). Sellers desire to engage in price discrimination because varied pricing can expand their customer base and this, ultimately, can increase profits. From a consumer's standpoint, the negative side of price discrimination is that it can seem unfair that a different consumer pays a lower price for the same good, simply because that consumer is located in a different place (or is in a different context, *e.g.*, a drug for humans v. the same drug for pets). However, price discrimination also has a positive side from a consumer standpoint—varied pricing can expand access to a good for those consumers that could not afford a higher, single, price. Price discrimination is effective for producers only to the degree that they can control against arbitrage—a consumer who can afford, and who values the good at, the higher price nonetheless getting access to, and buying, the lower-priced version. Mr. Kirtsaeng

---

have conditioned a copyright owner's power to divide markets on outsourcing its manufacturing to a foreign country.

was an arbitrageur: he provided U.S. consumers with access to the lower-priced versions of the same titles they wanted. Should copyright law block arbitrage to aid those producers who implement a strategy of price discrimination?

4. Another way to block arbitrage is to bind purchasers to a contract that restricts resale. Should these contractual restrictions on resale be enforced? If such a contract existed and the purchaser resold his copy anyway, should any subsequent state-law breach of contract claim succeed? Would it be preempted? What would Lord Coke say about such a restraint on alienation? Setting aside the issue of a potential breach of contract claim, should a re-sale that violates a contractual restriction constitute an act of infringement? What if the contract is styled as a "license," and that license grants permissions as to some copyright rights but does *not* purport to license any aspect of the right to distribute copies to the public, 17 U.S.C. 106(3)? Should the infringement question in this context turn on the precise wording used in the agreement?

5. In *Kirtsaeng* the Court's main task was statutory interpretation. In that sense, the court was engaged in determining whether Congress *intended* to aid copyright owners in dividing up markets and engaging in price discrimination based on geographic markets. The Court concluded that, based on the language used and the structure of the statute, Congress did not intend to do that. As Justice Kagan points out in her concurrence, an amendment to the importation provision, § 602(a)(1), would permit geographic based market segmentation. If you were a member of Congress, would you vote for such an amendment?

B. Patent Law and the Exhaustion Doctrine

In *Kirstsaeng* the Court focused on the precise words Congress used to codify the first sale doctrine. The Patent Act grants a patent owner the "right to exclude others from making, using, offering for sale, or selling the invention," 35 U.S.C. § 154(a)(1), but there is no explicit first-sale provision on a par with § 109(a) of the Copyright Act. The next case is a recent opinion from the Supreme Court concerning the accommodations that patent law makes for personal property rights.

Bowman v. Monsanto

133 S. Ct. 1761 (2013)

Kagan, Justice:

....

Respondent Monsanto invented a genetic modification that enables soybean plants to survive exposure to glyphosate, the active ingredient in many herbicides (including Monsanto's own Roundup). Monsanto markets soybean seed containing this altered genetic material as Roundup Ready seed. Farmers planting that seed can use a glyphosate-based herbicide to kill weeds without damaging their crops. Two patents issued to Monsanto cover various aspects of its Roundup Ready technology, including a seed incorporating the genetic alteration. (U.S. Patent Nos. 5,352,605 and RE39,247E).

Monsanto sells, and allows other companies to sell, Roundup Ready soybean seeds to growers who assent to a special licensing agreement. That agreement permits a grower to plant the purchased seeds in one (and only one) season. He can then consume the resulting crop or sell it as a commodity, usually to a grain elevator or agricultural processor. But under the agreement, the farmer may not save any of the harvested soybeans for replanting, nor may he supply them to anyone else for that purpose. These restrictions reflect the ease of producing new generations of Roundup Ready seed. Because glyphosate resistance comes from the seed's genetic material, that trait is passed on from the planted seed to the harvested soybeans: Indeed, a single Roundup Ready seed can grow a plant containing dozens of genetically identical beans, each of which, if replanted, can grow another such plant—and so on and so on. The agreement's terms prevent the farmer from co-opting that process to produce his own Roundup Ready seeds, forcing him instead to buy from Monsanto each season.

Petitioner Vernon Bowman is a farmer in Indiana who, it is fair to say, appreciates Roundup Ready soybean seed. He purchased Roundup Ready each year, from a company affiliated with Monsanto, for his first crop of the season. In accord with the agreement just described, he used all of that seed for planting, and sold his entire crop to a grain elevator (which typically would resell it to an agricultural processor for human or animal consumption).

Bowman, however, devised a less orthodox approach for his second crop of each season. Because he thought such late-season planting "risky," he did not want to pay the premium price that Monsanto charges for Roundup Ready seed. He therefore went to a grain elevator; purchased "commodity soybeans" intended for human or animal consumption; and planted them in his fields.[1] Those soybeans came from prior harvests of other local farmers. And because most of those farmers also used Roundup Ready seed, Bowman could anticipate that many of the purchased soybeans would contain Monsanto's patented technology. When he applied a glyphosate-based herbicide to his fields, he confirmed that this was so; a significant proportion of the new plants survived the treatment, and produced in their turn a new crop of soybeans with the Roundup Ready trait. Bowman saved seed from that crop to use in his late-season planting the next year—and then the next, and the next, until he had harvested eight crops in that way. Each year, that is, he planted saved seed from the year before (sometimes adding more soybeans bought from the grain elevator), sprayed his fields with glyphosate to kill weeds (and any non-resistant plants), and produced a new crop of glyphosate-resistant—i.e., Roundup Ready—soybeans.

After discovering this practice, Monsanto sued Bowman for infringing its patents on Roundup Ready seed. Bowman raised patent exhaustion as a defense, arguing that Monsanto could not control his use of the soybeans because they were the subject of a prior authorized sale (from local farmers to the grain elevator). The District Court rejected that argument, and awarded damages to Monsanto of

---

[1] Grain elevators . . . purchase grain from farmers and sell it for consumption; under federal and state law, they generally cannot package or market their grain for use as agricultural seed. See 7 U.S.C. § 1571; Ind. Code § 15-15-1-32 (2012). But because soybeans are themselves seeds, nothing (except, as we shall see, the law) prevented Bowman from planting, rather than consuming, the product he bought from the grain elevator.

$84,456. The Federal Circuit affirmed. . . . We granted certiorari to consider the important question of patent law raised in this case and now affirm.

II

The doctrine of patent exhaustion limits a patentee's right to control what others can do with an article embodying or containing an invention. Under the doctrine, "the initial authorized sale of a patented item terminates all patent rights to that item." *Quanta Computer, Inc. v. LG Electronics, Inc.*, 553 U.S. 617, 625 (2008). And by "exhaust[ing] the [patentee's] monopoly" in that item, the sale confers on the purchaser, or any subsequent owner, "the right to use [or] sell" the thing as he sees fit. *United States v. Univis Lens Co.*, 316 U.S. 241, 249-250 (1942). We have explained the basis for the doctrine as follows: "[T]he purpose of the patent law is fulfilled with respect to any particular article when the patentee has received his reward * * * by the sale of the article"; once that "purpose is realized the patent law affords no basis for restraining the use and enjoyment of the thing sold." *Id.* at 251.

Consistent with that rationale, the doctrine restricts a patentee's rights only as to the "particular article" sold; it leaves untouched the patentee's ability to prevent a buyer from making new copies of the patented item. "[T]he purchaser of the [patented] machine * * * does not acquire any right to construct another machine either for his own use or to be vended to another." *Mitchell v. Hawley*, 16 Wall. 544, 548 (1873); *see Wilbur-Ellis Co. v. Kuther*, 377 U.S. 422, 424 (1964) (holding that a purchaser's "reconstruction" of a patented machine "would impinge on the patentee's right '*to exclude others from making*' * * * the article" (quoting 35 U.S.C. § 154)). Rather, "a second creation" of the patented item "call[s] the monopoly, conferred by the patent grant, into play for a second time." *Aro Mfg. Co. v. Convertible Top Replacement Co.*, 365 U.S. 336, 346 (1961). That is because the patent holder has "received his reward" only for the actual article sold, and not for subsequent recreations of it. *Univis*, 316 U.S. at 251. If the purchaser of that article could make and sell endless copies, the patent would effectively protect the invention for just a single sale. Bowman himself disputes none of this analysis as a general matter: He forthrightly acknowledges the "well settled" principle "that the exhaustion doctrine does not extend to the right to 'make' a new product." Brief for Petitioner 37 (citing *Aro*, 365 U.S. at 346).

Unfortunately for Bowman, that principle decides this case against him. Under the patent exhaustion doctrine, Bowman could resell the patented soybeans he purchased from the grain elevator; so too he could consume the beans himself or feed them to his animals. Monsanto, although the patent holder, would have no business interfering in those uses of Roundup Ready beans. But the exhaustion doctrine does not enable Bowman to make additional patented soybeans without Monsanto's permission (either express or implied). And that is precisely what Bowman did. He took the soybeans he purchased home; planted them in his fields at the time he thought best; applied glyphosate to kill weeds (as well as any soy plants lacking the Roundup Ready trait); and finally harvested more (many more) beans than he started with. That is how "to 'make' a new product," to use Bowman's words, when the original product is a seed. [S]ee Webster's Third New International Dictionary 1363 (1961) ("make" means "cause to exist, occur, or appear," or more specifically, "plant and raise (a crop)"). Because Bowman thus reproduced Monsanto's patented invention, the exhaustion doctrine does not protect him.[3]

Were the matter otherwise, Monsanto's patent would provide scant benefit. After inventing the Roundup Ready trait, Monsanto would, to be sure, "receiv[e] [its] reward" for the first seeds it sells. *Univis*, 316 U.S. at 251. But in short order, other seed companies could reproduce the product and market it to growers, thus depriving Monsanto of its monopoly. And farmers themselves need only buy the seed once, whether from Monsanto, a competitor, or (as here) a grain elevator. The grower could multiply his initial purchase, and then multiply that new creation, ad infinitum—each time profiting from the patented seed without compensating its inventor. Bowman's late-season plantings offer a prime illustration. After buying beans for a single harvest, Bowman saved enough seed each year to reduce or eliminate the need for additional purchases. Monsanto still held its patent, but received no gain from Bowman's annual production and sale of Roundup Ready soybeans. The exhaustion doctrine is limited to the "particular item" sold to avoid just such a mismatch between invention and reward.

Our holding today also follows from *J.E.M. Ag Supply, Inc. v. Pioneer Hi-Bred Int'l, Inc.*, 534 U.S. 124 (2001). We considered there whether an inventor could get a patent on a seed or plant, or only a certificate issued under the Plant Variety Protection Act (PVPA), 7 U.S.C. § 2321 et seq. We decided a patent was available, rejecting the claim that the PVPA implicitly repealed the Patent Act's coverage of seeds and plants. On our view, the two statutes established different, but not conflicting schemes: The requirements for getting a patent "are more stringent than those for obtaining a PVP certificate, and the protections afforded" by a patent are correspondingly greater. *J.E.M.*, 534 U.S. at 142. Most notable here, we explained that only a patent holder (not a certificate holder) could prohibit "[a] farmer who legally purchases and plants" a protected seed from saving harvested seed "for replanting." *Id.* at 140; *see id.* at 143 (noting that the Patent Act, unlike the PVPA, contains "no exemptio[n]" for "saving seed"). That statement is inconsistent with applying exhaustion to protect conduct like Bowman's. If a sale cut off the right to control a patented seed's progeny, then (contrary to *J.E.M.*) the patentee could not prevent the buyer from saving harvested seed. Indeed, the patentee could not stop the buyer from selling such seed, which even a PVP certificate

---

[3] This conclusion applies however Bowman acquired Roundup Ready seed: The doctrine of patent exhaustion no more protected Bowman's reproduction of the seed he purchased for his first crop (from a Monsanto-affiliated seed company) than the beans he bought for his second (from a grain elevator). The difference between the two purchases was that the first—but not the second—came with a license from Monsanto to plant the seed and then harvest and market one crop of beans. We do not here confront a case in which Monsanto (or an affiliated seed company) sold Roundup Ready to a farmer without an express license agreement. For reasons we explain below, we think that case unlikely to arise. And in the event it did, the farmer might reasonably claim that the sale came with an implied license to plant and harvest one soybean crop.

owner (who, recall, is supposed to have fewer rights) can usually accomplish. *See* 7 U.S.C. §§ 2541, 2543. Those limitations would turn upside-down the statutory scheme *J.E.M.* described.

Bowman principally argues that exhaustion should apply here because seeds are meant to be planted. The exhaustion doctrine, he reminds us, typically prevents a patentee from controlling the use of a patented product following an authorized sale. And in planting Roundup Ready seeds, Bowman continues, he is merely using them in the normal way farmers do. Bowman thus concludes that allowing Monsanto to interfere with that use would "creat[e] an impermissible exception to the exhaustion doctrine" for patented seeds and other "self-replicating technologies." Brief for Petitioner 16.

But it is really Bowman who is asking for an unprecedented exception—to what he concedes is the "well settled" rule that "the exhaustion doctrine does not extend to the right to 'make' a new product." Reproducing a patented article no doubt "uses" it after a fashion. But as already explained, we have always drawn the boundaries of the exhaustion doctrine to exclude that activity, so that the patentee retains an undiminished right to prohibit others from making the thing his patent protects. *See, e.g., Cotton-Tie Co. v. Simmons*, 106 U.S. 89, 93-94 (1882) (holding that a purchaser could not "use" the buckle from a patented cotton-bale tie to "make" a new tie). That is because, once again, if simple copying were a protected use, a patent would plummet in value after the first sale of the first item containing the invention. The undiluted patent monopoly, it might be said, would extend not for 20 years (as the Patent Act promises), but for only one transaction. And that would result in less incentive for innovation than Congress wanted. Hence our repeated insistence that exhaustion applies only to the particular item sold, and not to reproductions.

Nor do we think that rule will prevent farmers from making appropriate use of the Roundup Ready seed they buy. Bowman himself stands in a peculiarly poor position to assert such a claim. As noted earlier, the commodity soybeans he purchased were intended not for planting, but for consumption. Indeed, Bowman conceded in deposition testimony that he knew of no other farmer who employed beans bought from a grain elevator to grow a new crop. So a non-replicating use of the commodity beans at issue here was not just available, but standard fare. And in the more ordinary case, when a farmer purchases Roundup Ready seed qua seed—that is, seed intended to grow a crop—he will be able to plant it. Monsanto, to be sure, conditions the farmer's ability to reproduce Roundup Ready; but it does not—could not realistically—preclude all planting. No sane farmer, after all, would buy the product without some ability to grow soybeans from it. And so Monsanto, predictably enough, sells Roundup Ready seed to farmers with a license to use it to make a crop. Applying our usual rule in this context therefore will allow farmers to benefit from Roundup Ready, even as it rewards Monsanto for its innovation.

Still, Bowman has another seeds-are-special argument: that soybeans naturally "self-replicate or 'sprout' unless stored in a controlled manner," and thus "it was the planted soybean, not Bowman" himself, that made replicas of Monsanto's patented invention. But we think that blame-the-bean defense tough to credit. Bowman was not a passive observer of his soybeans' multiplication; or put another way, the seeds he purchased (miraculous though they might be in other respects) did not spontaneously create eight successive soybean crops. As we have explained, Bowman devised and executed a novel way to harvest crops from Roundup Ready seeds without paying the usual premium. He purchased beans from a grain elevator anticipating that many would be Roundup Ready; applied a glyphosate-based herbicide in a way that culled any plants without the patented trait; and saved beans from the rest for the next season. He then planted those Roundup Ready beans at a chosen time; tended and treated them, including by exploiting their patented glyphosate-resistance; and harvested many more seeds, which he either marketed or saved to begin the next cycle. In all this, the bean surely figured. But it was Bowman, and not the bean, who controlled the reproduction (unto the eighth generation) of Monsanto's patented invention.

Our holding today is limited—addressing the situation before us, rather than every one involving a self-replicating product. We recognize that such inventions are becoming ever more prevalent, complex, and diverse. In another case, the article's self-replication might occur outside the purchaser's control. Or it might be a necessary but incidental step in using the item for another purpose. *Cf.* 17 U.S.C. § 117(a)(1) ("[I]t is not [a copyright] infringement for the owner of a copy of a computer program to make * * * another copy or adaptation of that computer program provide[d] that such a new copy or adaptation is created as an essential step in the utilization of the computer program"). We need not address here whether or how the doctrine of patent exhaustion would apply in such circumstances. In the case at hand, Bowman planted Monsanto's patented soybeans solely to make and market replicas of them, thus depriving the company of the reward patent law provides for the sale of each article. Patent exhaustion provides no haven for that conduct. We accordingly affirm the judgment of the Court of Appeals for the Federal Circuit.

Notes & Questions

1. What did you think of Bowman's "blame-the-bean" defense? Farmers can purchase soybean seeds that are not "Roundup Ready" for less than the Roundup Ready variety. Why didn't Bowman do that? If he wanted the agricultural advantage that comes from having Roundup Ready seeds, the only way to get that, during the term of the patent, was to pay the higher price.

2. The exhaustion doctrine in patent law does not extend to the right to "make" a new product, but patent law also grants the patent owner the right to exclude others from "using" the patented invention. If you buy a patented tool, don't you expect to be able to "use" that tool for its intended purpose? How does patent law account for that reasonable, and daresay universal, expectation?

3. Recall the discussion of what it means to "use" a trade secret in *Silvaco Data Sys. v. Intel Corp.,* 109 Cal. Rptr. 3d 27 (Cal. Ct. App. 2010). Because trade secret

misappropriation requires "use" of a trade secret, defining "use" to exclude a customer's use of a product that had been made by someone else (who had employed the trade secret to make the product) has a similar effect to the exhaustion doctrine that applies in other areas of IP. In *Silvaco Data Sys.*, the court stated that: "'use' in the ordinary sense is not present when the conduct consists entirely of possessing, and taking advantage of, *something that was made* using the secret." There the court used an analogy to aid in its conclusion that use of a computer program that embodied a trade secret algorithm did not constitute misappropriation: "One who bakes a pie from a recipe certainly engages in the 'use' of the latter; but one who eats the pie does not, by virtue of that act alone, make 'use' of the recipe in any ordinary sense . . . ."

4. Notice that unlike copyright law, the patent exhaustion doctrine is not codified in the statute. Thus the Court in *Bowman* was not engaged in a statutory construction exercise. Instead, it was determining the scope of a judicially created and judicially shaped doctrine. Courts have also permitted owners of products to repair the product as necessary for continued use. There is a line, however, between what constitutes legitimate repair and what constitutes impermissible (*i.e.*, infringing) reconstruction. Reconstruction amounts of a "'a second creation' of the patented item [and] 'call[s] the monopoly, conferred by the patent grant, into play for a second time." *Aro Mfg. Co. v. Convertible Top Replacement Co.*, 365 U.S. 336, 346 (1961). The line between repair and reconstruction can be a difficult one to draw. *Compare Jazz Photo Corp. v. United States International Trade Commission*, 264 F.3d 1094 (Fed. Cir. 2001) (finding refurbished "single-use" film cameras resulted from non-infringing repair), *with Fuji PhotoFilm Co. v. Jazz Photo Corp.*, 394 F.3d 1368, 1376 (Fed. Cir. 2005) (finding some refurbishments of "single-use" film cameras to be infringing reconstructions).

5. As you have seen in other contexts, laws besides intellectual property law may provide protection for innovators. In *Bowman* the Plant Variety Protection Act was used by the court to bolster its conclusion concerning the scope of patent protection. As a federal statute the PVPA does not face pre-emption challenges. However, harmonizing different congressionally adopted schemes of protection with one another is one aspect of the Court's analysis. Did the Court get it right? If a state had adopted a law similar to the PVPA, would it be preempted?

6. Re-read the Court's final paragraph. Besides seeds, what type of self-replicating technologies is the Court concerned with?

7. A key to the exhaustion concept in the most common circumstances is that the patent owner has sold a product without restriction. The sale seems to bring along a promise that the patentee will not interfere with the customer's full enjoyment of that product. But what happens when the product is sold "with restrictions?" Many believed that in granting review in *Bowman* the Court was going to further clarify its holding in *Quanta Computer, Inc. v. LG Electronics, Inc.*, 553 U.S. 617 (2008), concerning the extent to which a patentee can use contractual restrictions to control downstream use and re-sale of goods that embodied patented technology. The facts of *Bowman* however did not ultimately lend themselves to further exploration of that question. Be sure you understand why the

license that Monsanto used in connection with sales of its soybean seeds was not relevant to the Court's decision.

In *Quanta Computer* the Court addressed patents on a computer chip that involved both product and process claims. The Court held that the exhaustion doctrine applies to method claims and that the sale of a product that "substantially embodies" the claimed method exhausts the patent rights. The source of the products at issue in *Quanta*, Intel, was a licensee of the patent owner, LG Electronics. The license permitted Intel to "make, use, [or] sell" the patented invention. The license also required Intel to notify its customers that LG Electronics had not licensed Intel's customers to practice the claimed invention, which involved combining the chip with other elements. When the customers combined the chips with the other elements, LG Electronics sued for infringement. The court held that the sale by Intel was authorized and thus exhausted LG Electronic's rights. The required notification, which Intel had given, did not negate the exhaustion.

8. In its most recent patent-exhaustion decision, the Supreme Court reaffirmed and extended its *Quanta Computer* decision, overturning the Federal Circuit's rejection of an exhaustion defense where the patentee had sold its products—toner cartridges for laser printers—with an explicit "single use, no resale" limitation. In this case, *Impression Products, Inc. v. Lexmark International, Inc.*, 137 S. Ct. __, 2017 WL 2322830 (May 30, 2017), the Court explained that the exhaustion principle is not merely a default rule for construing the basic infringement provision (35 U.S.C. § 271) that patentees and their customers can alter by contract:

> the exhaustion doctrine is not a presumption about the authority that comes along with a sale; it is instead a limit on "the scope of the *patentee's rights*." *United States v. General Elec. Co.*, 272 U.S. 476, 489 (1926) (emphasis added). The right to use, sell, or import an item exists independently of the Patent Act. What a patent adds—and grants exclusively to the patentee—is a limited right to prevent others from engaging in those practices. *See Crown Die & Tool Co. v. Nye Tool & Machine Works*, 261 U.S. 24, 35 (1923). Exhaustion extinguishes that exclusionary power. As a result, the sale transfers the right to use, sell, or import because those are the rights that come along with ownership, and the buyer is free and clear of an infringement lawsuit because there is no exclusionary right left to enforce.

Id. at *11. Critically, "[t]he [patent exhaustion] limit functions automatically: When a patentee chooses to sell an item, that product 'is no longer within the limits of the monopoly' and instead becomes the 'private, individual property' of the purchaser, with the rights and benefits that come along with ownership." *Id.* at *7 (quoting *Bloomer v. McQuewan*, 55 U.S. 539, 549-50). The exhaustion limit also applies fully to authorized sales abroad: "An authorized sale outside the United States, just as one within the United States, exhausts all rights under the Patent Act." *Id.* at *12.

C. Exhaustion under Trademark Law

As with copyright law and patent law, trademark law recognizes the exhaustion of a trademark owner's right upon the sale of a product bearing the mark. The Ninth Circuit recently provided a succinct summary of the basic contours of the doctrine, with examples from prior cases:

> Application of the "first sale" doctrine has generally focused on the likelihood of confusion [as to source] among consumers. In *Sebastian Int'l, Inc. v. Longs Drug Stores Corp.*, 53 F.3d 1073, 1077 (9th Cir. 1995), we held that the "first sale" doctrine protected Longs when it purchased Sebastian hair products from a distributor and sold them in its own store despite Sebastian's efforts to allow only "Sebastian Collective Members" to sell the products. We recognized the principle that "the right of a producer to control distribution of its trademarked product does not extend beyond the first sale of the product." *Id.* at 1074. We emphasized that this rule "preserves an area of competition by limiting the producer's power to control the resale of its product," while ensuring that "the consumer gets exactly what the consumer bargains for, the genuine product of the particular producer." *Id.* at 1075.
>
> We also applied the "first sale" doctrine in *Enesco Corp. v. Price/Costco Inc.*, 146 F.3d 1083, 1084–85 (9th Cir. 1998), in which Costco purchased porcelain figurines manufactured by Enesco, repackaged them in allegedly inferior packaging, and sold them in its own stores. We held that Costco could repackage and sell the Enesco figurines, but that it was required to place labels on the packages that disclosed to the public that Costco had repackaged Enesco's original product. We rejected Enesco's argument that it would be harmed, even with this disclosure, because of the poor quality of the packaging. "The critical issue is whether the public is likely to be confused as a result of the lack of quality control." *Id.* at 1087.
>
> A number of district courts have applied the "first sale" doctrine in cases where the defendants incorporated the trademarked product into a new product. *See, e.g., Alexander Binzel Corp. v. Nu–Tecsys Corp.*, 785 F. Supp. 719 (N.D. Ill. 1992) (defendant manufactured welding gun using some parts bearing trademark of competitor); *Major League Baseball Players Ass'n v. Dad's Kid Corp.*, 806 F. Supp. 458 (S.D.N.Y. 1992) (defendant used three baseball trading cards bearing trademarks to create 3D playing card); *Scarves by Vera, Inc. v. Am. Handbags, Inc.*, 188 F. Supp. 255 (S.D.N.Y. 1960) (defendant used towels bearing plaintiff's trademark to create handbags). In these cases, the courts focused on the possibility of confusion as the dispositive factor. *See Alexander Binzel Corp.*, 785 F. Supp. at 724 (finding that "defendants did all that was required of them to diminish customer confusion by packaging their product with the [defendant's own] name and logo"); *Dad's Kid Corp.*, 806 F. Supp. at 460 (finding "no likelihood that anyone will be confused as to origin"); *Scarves by Vera, Inc.*, 188 F. Supp. at 258 (issuing injunction requiring further labeling because the court found "that the average purchaser would be misled" otherwise).

Au-Tomotive Gold Inc. v. Volkswagen of America, Inc., 603 F.3d 1133, 1136-37 (9th Cir. 2010). The repacking at issue in the *Costco* case and the combined products at issue in other cases illustrate the concern that courts have with altered goods that continue to bear the trademark of the original producer. The key to determining whether the resale of the re-packed or combined goods constitutes trademark infringement is whether consumers are likely to be confused about the source of the trademarked good.

As with patent law and its distinction between permissible repair and impermissible reconstruction, trademark law has a similar dynamic in the market for repaired or refurbished goods.

A separate line of cases further illustrates the central role of the likelihood of confusion, including post-purchase confusion, in trademark infringement claims. In this line of cases, we have held that producers committed trademark infringement by selling refurbished or altered goods under their original trademark. None of these cases directly addressed the "first sale" doctrine, but they establish that activities creating a likelihood of post-purchase confusion, even among non-purchasers, are not protected.

In *Karl Storz Endoscopy–America, Inc. v. Surgical Tech., Inc.* ("Surgi–Tech"), 285 F.3d 848, 852–53 (9th Cir. 2002), Surgi–Tech repaired Storz endoscopes at the request of hospitals that owned them. Surgi–Tech sometimes rebuilt the endoscopes, replacing every part and "retaining only the block element bearing Storz's trademarks." *Id.* at 852. At an earlier time, Surgi–Tech had etched its own mark into rebuilt endoscopes to make clear what it had done, but Surgi–Tech had stopped that practice. *Id.* Storz submitted evidence of confusion on the part of surgeons who were not the purchasers of the endoscopes but who used them and mistakenly blamed Storz when they malfunctioned. *Id.* at 855. We held that there was a triable issue of fact on Storz's trademark infringement claim, even though there was no claim of purchaser confusion. *Id.* at 853–55. We relied entirely on the possibility of confusion among non-purchasers, noting that such confusion "may be no less injurious to the trademark owner's reputation than confusion on the part of the purchaser at the time of sale." *Id.* at 854.

We also relied on the likelihood of non-purchaser confusion in *Rolex Watch, U.S.A., Inc. v. Michel Co.*, 179 F.3d 704 (9th Cir. 1999). The defendant sold used Rolex watches that had been "reconditioned" or "customized" with non-Rolex parts. *Id.* at 707. We agreed with the district court that "retention of the original Rolex marks on altered 'Rolex' watches * * * was deceptive and misleading as to the origin of the non-Rolex parts, and likely to cause confusion to subsequent or downstream purchasers, as well as to persons observing the product." *Id.* at 707.

Au-Tomotive Gold Inc., 603 F.3d at 1137.

Looking beyond the domestic context, to apply the exhaustion principle to gray market goods in trademark law, one must keep in mind that protection for trademarks in the U.S. is based on use *within the U.S.*, whatever may be happening outside the U.S. As a result, different entities may own the exact same trademark for the exact same type of good, so long as those different entities operate in different countries. The different companies may be related; for example, the U.S. company may have a subsidiary company located in Mexico. Or the companies may be wholly unrelated producers that sell the same product with the same mark, albeit in different countries. How, as a matter of trademark law, should one analyze whether a firm that imports a foreign good lawfully bearing mark *M* (when sold in country of origin *O*) infringes the rights of a firm that has been selling goods in the U.S. using the same mark *M*? Is it simply a matter of applying the traditional likelihood-of-confusion factors, or must the analysis take additional matters into account?

Hokto Kinoko Co. v. Concord Farms, Inc.
738 F.3d 1085 (9th Cir. 2013)

Wardlaw, Judge:

In this trademark infringement action, Hokto Kinoko Co. (Hokto USA), a wholly owned subsidiary of Hokuto Co., Ltd. (Hokuto Japan), sued Concord Farms, Inc. (Concord Farms) for violating its rights to marks under which it markets its Certified Organic Mushrooms, which are produced in the United States. Hokto USA claimed that Concord Farms wrongly imported and marketed mushrooms under its marks for Certified Organic Mushrooms, but which were cultivated in Japan by Hokuto Japan under nonorganic standards.... The district court granted summary judgment in favor of Hokto USA and Hokuto Japan on all claims and entered a permanent injunction against Concord Farms....

I. Background

A. Hokuto Japan and Hokto USA

Hokuto Japan is a Japanese corporation that produces mushrooms in Japan. These mushrooms include maitake, white beech (marketed as "Bunapi"), and brown beech (marketed as "Bunashimeji") mushrooms, and are sold in 3.5 ounce packages. Hokuto Japan's mushrooms are grown in nonorganic conditions throughout Japan and sold in Japanese-language packaging.

In 2006, Hokuto Japan incorporated Hokto USA, also a Japanese corporation, to produce and market mushrooms in the United States. Hokto USA is a wholly owned subsidiary of Hokuto Japan. Like Hokuto Japan, Hokto USA produces white beech, brown beech, and maitake mushrooms. Unlike Hokuto Japan's mushrooms, however, Hokto USA's mushrooms are certified organic and produced in a state-of-the-art facility in San Marcos, California. Hokto USA's mushrooms are robotically transported within the facility in plastic bottles, and its entire process is computer controlled. While most mushroom-growing techniques involve manure and compost, Hokto USA uses a sterilized culture medium made of sawdust, corn cob pellets, vegetable protein, and other nutrients. Hokto USA also enforces strict temperature controls and other quality control standards, both in its San Marcos facility and during the transportation and storage of its mushrooms, to ensure that the mushrooms stay fresh for as long as possible.

The production of mushrooms in the United States did not start off quite as smoothly as planned. Although Hokto USA was incorporated in 2006, its San Marcos growing facility was not completed until 2009. While the facility was under construction, Hokto USA resorted to importing mushrooms from Hokuto Japan. Because U.S. consumers have different preferences than Japanese consumers, Hokuto Japan grew mushrooms for Hokto USA in special conditions. Most significantly, Hokuto Japan used a special growing medium that met U.S. Certified Organic standards. Hokuto Japan also worked with Hokto USA to develop English-language packaging for the U.S. market. The packaging identified the mushrooms as "Certified Organic" and provided nutritional information geared toward U.S. consumers.

When the San Marcos facility finally opened in 2009, Hokto USA began producing its own mushrooms and stopped importing Hokuto Japan's mushrooms. But in 2010, there was a shortfall of white beech mushrooms. To meet its customers' demand, Hokto USA imported two shipments of Hokuto Japan's inferior white beech mushrooms, which were produced in Japan and sold in Hokuto Japan's usual Japanese-language packaging. Before selling these mushrooms to U.S. consumers, Hokto USA affixed a white sticker to every package, which clearly identified the mushrooms as a product of Japan and identified the product as "white beech mushrooms." The white stickers also identified the "distributor" as Hokto USA and provided U.S. customer service information.

B. The Trademarks

In 2003, Hokuto Japan acquired Japanese trademark registrations for a series of marks ("Hokto marks"), including variations on its logo and several mushroom-shaped cartoon characters with faces, arms, and legs. These registrations protected Hokuto Japan's rights to use the marks to market a wide variety of goods, [including] mushrooms . . . .

Hokuto Japan also sought U.S. trademark registrations on the same marks . . .
.

The United States Patent and Trademark Office (USPTO) issued registrations for the cartoon-character marks (Reg. Nos. 3182866, 3179700, and 3182867) in December 2006 and for the Hokto logo (Reg. No. 3210268) on February 20, 2007
. . . .

In August 2008, Hokuto Japan granted Hokto USA a license for the exclusive use of the marks in the United States. In 2010, Hokuto Japan assigned all of its rights under the American trademark registrations to Hokto USA. Both the mushrooms sold by Hokuto Japan in Japan and those sold by Hokto USA in the United States are marketed in packaging that prominently features the Hokto marks. . . .

C. Concord Farms

Meanwhile, Concord Farms, a U.S. corporation that grows and imports mushrooms, has been importing Hokuto Japan's mushrooms from Japan since 2003. From 2003 to 2009, Concord Farms imported Hokuto Japan's maitake, brown beech, and white beech mushrooms. Since 2009, it has imported only the maitake mushrooms. Because Concord Farms purchases these products through a series of wholesalers, Hokuto Japan was initially unaware that Concord Farms was importing its mushrooms. The mushrooms Concord Farms imports into the United States are the nonorganic mushrooms that Hokuto Japan produces in Japan for Japanese consumption and are packaged in the Japanese packaging, which features the Hokto marks. Concord Farms's warehouse is not temperature controlled, and Concord Farms does not impose formal limits on how long mushrooms are kept in the warehouse.

In July 2009, Hokto USA learned that Concord Farms imports Hokuto Japan's mushrooms when Hokto USA's representative saw packages of Hokuto Japan's Japanese-packaged, nonorganic maitake mushrooms mixed with packages of Hokto USA's maitake mushrooms in a grocery store display. All of the mushrooms were under a sign that said "organic" and "made in USA," but the Japanese products under the sign were neither. There was too much moisture in the Hokuto Japan packages, and the mushrooms were going bad. The store's manager told Hokto USA's representative that he had purchased the Hokuto Japan mushrooms from Concord Farms. At a produce exposition three months later, Hokto USA's representative requested that Concord Farms refrain from importing, selling, or distributing Hokuto Japan's mushrooms. Concord Farms refused.

Hokto USA filed this trademark action in the United States District Court for the Central District of California. All three parties filed cross-motions for summary judgment. The district court entered judgment in favor of Hokto USA and Hokuto Japan, and permanently enjoined Concord Farms from selling the Hokuto Japan mushrooms in the United States. Concord Farms timely appeals.

. . . .

III. Discussion

A. Gray-Market Goods

The crux of Hokto USA's claim is that when Concord Farms imported mushrooms bearing the Hokto marks from Hokuto Japan and sold those mushrooms in the United States, it infringed Hokto USA's rights to those marks. This case thus implicates the set of trademark principles governing so-called "gray-market goods": goods that are legitimately produced and sold abroad under a particular trademark, and then imported and sold in the United States in competition with the U.S. trademark holder's products.

The Supreme Court has explained that a gray-market good is "a foreign-manufactured good, bearing a valid United States trademark, that is imported without the consent of the United States trademark holder." *K Mart Corp. v. Cartier, Inc.*, 486 U.S. 281, 285 (1988). The mushrooms at issue here fit comfortably within the Supreme Court's definition. Some commentators apply the term "gray market" only where both the trademark owner and the alleged infringer import their product from foreign countries, *see* J. Thomas McCarthy, *McCarthy on Trademarks & Unfair Competition* § 29:46 (4th ed. 2005), or only where the U.S. trademark owner also owns foreign rights in the disputed mark, *see* 1 Jerome Gilson, *Trademark Protection & Practice* § 4.05[6] (2004). Regardless of whether we categorize the mushrooms here as gray-market goods, however, the fundamental nature of the infringement claim is the same as that in gray-market cases: Hokto USA alleges that Concord Farms violated its trademarks by importing legitimately produced goods sold under those same marks. *See American Circuit Breaker Corp. v. Oregon Breakers, Inc.*, 406 F.3d 577, 583-84 (9th Cir. 2005) (discussing ambiguity in definitions of gray-market goods and concluding that "whether this is technically classified as a gray-market case or not does not drive the solution").

B. Genuine Goods

In general, the sale of gray-market goods may infringe on the U.S. trademark holder's rights, subject to the consumer confusion analysis that generally governs trademark infringement claims. An exception to this rule, however, is that trademark law does not extend to the sale of "genuine goods." If the Japanese-produced Hokuto Japan mushrooms that Concord Farms imported were "genuine" Hokto USA goods, then Concord Farms would not be liable for trademark infringement. The district court correctly concluded that the mushrooms were not "genuine goods."

We have approached the "genuine good" inquiry both as a threshold question for the applicability of trademark law, and as part of the test for consumer confusion. *Compare NEC Elecs. v. CAL Circuit Abco*, 810 F.2d 1506, 1509 (9th Cir. 1987) ("Trademark law generally *does not reach* the sale of genuine goods.") (emphasis added), *with American Circuit Breaker*, 406 F.3d at 585 (analyzing genuineness within the discussion of the absence of likelihood of confusion). Here, because we confront a classic gray-market case, we must analyze the genuine goods question as a threshold matter, for if Concord Farms's mushrooms are "genuine," it is not subject to liability for trademark infringement.

1. The No-Material-Difference Requirement

"Genuine," in the trademark context, is a term of art: a gray-market good is "genuine" only if it does not materially differ from the U.S. trademark owner's product. *See, e.g., McCarthy* § 29:51.75 ("[I]f there are material differences between the gray market imports and the authorized imports, then the gray market imports are not 'genuine' goods and can create a likelihood of confusion."); *see also Iberia Foods Corp. v. Romeo*, 150 F.3d 298, 303 (3d Cir. 1998) (explaining that where goods are marketed under "identical marks but are materially different * * * the alleged infringer's goods are considered 'non-genuine' and the sale of the goods constitutes infringement"); *Societe Des Produits Nestle, S.A. v. Casa Helvetia, Inc.*, 982 F.2d 633, 638 (1st Cir. 1992) ("It follows that the Venezuelan chocolates purveyed by Casa Helvetia were not 'genuine' * * * if they (a) were not authorized for sale in the United States and (b) differed materially from the authorized (Italian-made) version.").

We first established that trademark law does not extend to the sale of genuine goods in *NEC Electronics*. There, the question before us was whether a U.S. subsidiary of a foreign manufacturer may sue for trademark infringement where another company "buys the parent's *identical* goods abroad and then sells them here using the parent's true mark." 810 F.2d at 1508-09 (emphasis added). In *American Circuit Breaker*, applying the *NEC Electronics* rule, we explained that a genuine-goods exception "makes good sense and comports with the consumer protection rationale of trademark law" because a consumer who purchases a genuine good receives essentially the product he expected. 406 F.3d at 585. In both *NEC Electronics* and *American Circuit Breaker*, exemption from trademark law turned on whether the allegedly infringing product differed materially from the U.S. trademark holder's product.

Because the likelihood of confusion increases as the differences between products become more subtle, the threshold for determining a material difference is low. The key question is whether a consumer is likely to consider a difference relevant when purchasing a product. Courts have found a wide range of differences "material" in this context. The Second Circuit, for instance, held that Cabbage Patch dolls were not "genuine" when accompanied with fictitious "birth certificates" and "adoption papers" written in a foreign language. *Original Appalachian Artworks, Inc. v. Granada Elecs., Inc.*, 816 F.2d 68, 73 (2d Cir. 1987). The D.C. Circuit held that there were material differences between the British and American versions of dishwasher detergent where the chemical composition of the detergents differed slightly, and the British detergent was labeled "washing up liquid" rather than "dishwashing liquid" and included a "royal emblem." *Lever Bros. v. United States*, 877 F.2d 101, 103 (D.C. Cir. 1989). Along the same lines, a district court in the Central District of California, comparing the Mexican and U.S. versions of Pepsi, held that differences in quality control and the use of Spanish, rather than English, on the soda cans were material differences. *PepsiCo, Inc. v. Reyes*, 70 F. Supp. 2d 1057, 1059 (C.D. Cal.1999). We agree that differences in language, quality control, and packaging may each be sufficiently material to render imported goods not "genuine."

2. Concord Farms's Mushrooms

Whether the mushrooms that Concord Farms imports from Hokuto Japan are genuine goods thus turns on whether they materially differ from Hokto USA's mushrooms. Concord Farms's mushrooms are not organic, are grown under Hokuto Japan's less extensive quality control standards, and are sold in packaging designed for domestic Japanese consumers. The Hokuto Japan packaging is in Japanese, and the packages' weights are measured in grams, not ounces. The Hokuto Japan packaging identifies the mushrooms as the "product of" the specific Japanese prefecture in which they were produced. It also displays Hokuto Japan's website and telephone number. To determine whether these Concord Farms mushrooms are "genuine" Hokto USA goods, we must compare them to the three categories of Hokto USA's mushrooms: (1) the mushrooms that Hokto USA imported from Hokuto Japan prior to the opening of Hokto USA's San Marcos, California plant; (2) the mushrooms that Hokto USA produces at its California plant;

(3) and the white beech mushrooms that Hokto USA imported from Hokuto Japan in May and November 2010 to supplement its supply.

The mushrooms that Hokto USA imported from Hokuto Japan prior to the opening of the San Marcos facility materially differed from Concord Farms's mushrooms both in their production and in their packaging. Hokto USA submitted uncontradicted evidence that certified organic status is more important to American consumers than to Japanese consumers, and that Hokuto Japan used a special growing medium to ensure that the mushrooms intended for sale by Hokto USA in the United States met U.S. Certified Organic standards. In contrast, Hokuto Japan does not use this special growing medium to produce mushrooms intended for Japanese consumption; so, when Concord Farms imported Hokuto Japan's mushrooms for sale in the United States, they did not meet Certified Organic standards. Hokuto Japan and Hokto USA also developed packaging in both English and Japanese, in contrast to the packaging developed for Japanese consumers. The dual-language packaging described in English the mushrooms' recommended serving size, calorie count, and other nutritional information. We agree with the district court that these differences in production and packaging are material, rendering these imports not "genuine."

Concord Farms's mushrooms materially differ even more from the mushrooms that Hokto USA produces in its San Marcos facility. Like the pre-2009 imports, Hokto USA's mushrooms are certified organic and sold in dual-language packaging. At its San Marcos facility, Hokto USA uses a hygienic, computer-controlled cultivation process, which includes the robotic transport of mushrooms in plastic bottles; a sterile culture medium composed of sawdust, corn cob pellets, and other nutrients; and aggressive sterilization and temperature controls to ensure a longer shelf life. The packaging on the domestically produced mushrooms identifies them as a "Product of USA," provides weights in grams and ounces, and displays Hokto USA's website.

Concord Farms's mushrooms are also materially different from the white beech mushrooms that Hokto USA imported from Hokuto Japan in May and November 2010. When Hokto imported Hokuto Japan's mushrooms because of problems at Hokto USA's San Marcos facility, it ensured that a label was affixed to each imported package. The label included the English name for the mushrooms, listed the packages' weights in ounces rather than grams, clearly identified the mushrooms' origin, and provided a U.S. address for customer service inquiries. It is more than likely that consumers would consider these clarifying English-language labels relevant when purchasing the mushrooms. *See Bourdeau Bros. v. Int'l Trade Comm'n*, 444 F.3d 1317, 1323-24 (Fed. Cir. 2006) (explaining that "there need only be one material difference between a domestic and a foreign product" to support the conclusion that the product is not genuine). Concord Farms's mushrooms are therefore not "genuine goods" in relation to any of the three separately sold and packaged Hokto USA products.

. . . .

C. Likelihood of Consumer Confusion

Because Concord Farms's mushrooms are not "genuine" Hokto USA goods, Concord Farms is not exempt from potential liability under trademark law. The *sine qua non* of trademark infringement is consumer confusion. To determine the likelihood of consumer confusion, we apply the long-established factors set forth in *AMF Inc. v. Sleekcraft Boats*, 599 F.2d 341, 348-54 (9th Cir. 1979). The *Sleekcraft* factors include (1) the "similarity of the marks"; (2) the "strength of the mark" that has allegedly been infringed; (3) "evidence of actual confusion"; (4) the relatedness or "proximity" of the goods; (5) the "normal marketing channels" used by both parties; (6) the "type of goods and the degree of care likely to be exercised by the purchaser"; (7) the alleged infringer's "intent in selecting the mark"; and (8) evidence that "either party may expand his business to compete with the other." *Id.* We apply these factors flexibly, and Hokto USA need not demonstrate that every factor weighs in its favor.

Here, the first factor, the similarity of the marks, weighs unequivocally in favor of a finding of consumer confusion because the marks are identical. The second factor, the strength of the mark, also weighs in Hokto USA's favor. A mark is strong if it is particularly unique or memorable, and the more unique a mark is, the greater the trademark protection it is entitled to. "Fanciful" marks, which are words or phrases invented solely to function as trademarks, receive a high level of trademark protection because they are inherently distinctive. The marks here, which consist of a stylized depiction of a fictitious word[2] and cartoon-character mushrooms, are unique, fanciful, and likely to be associated with Hokto USA by U.S. consumers. Hokto USA has adduced no evidence as to the third factor, actual consumer confusion. However, we have specifically recognized that likelihood of confusion may be established absent such evidence. *See Am. Int'l Grp., Inc. v. Am. Int'l Bank*, 926 F.2d 829, 832 (9th Cir. 1991).

The fourth and fifth factors, the relatedness of the goods and the parties' normal marketing channels, also weigh in Hokto USA's favor. Both Hokto USA and Concord Farms sell maitake mushrooms, and both sell them to grocery stores in the United States. Indeed, Hokto USA discovered Concord Farms's infringement because it found Concord Farms's packages of the same variety of mushrooms in the same grocery store in which it sold its own mushrooms. The sixth factor, the type of goods and the degree of care purchasers are likely to exercise, also weighs in Hokto USA's favor. . . . As the district court correctly noted, mushrooms are a "low-cost consumer good," and reasonably prudent purchasers are unlikely to carefully examine the mushrooms' packaging before each purchase.

The seventh factor is the alleged infringer's intent in adopting the marks. When an alleged infringer knowingly adopts a mark identical or similar to another's mark, "courts will presume an intent to deceive the public." *Official Airline Guides, Inc. v. Goss*, 6 F.3d 1385, 1394 (9th Cir. 1993). Here, Concord Farms

---

[2] "Hokto" has no meaning in Japanese, although "Hokuto" means "northern star."

imported mushrooms bearing a mark identical to that of Hokto USA's mushrooms, and it has produced no evidence to negate the presumption that it intended to confuse consumers. Thus, the intent factor also weighs in favor of Hokto USA. The final factor is whether either party is likely to expand its product lines so as to compete directly with the product sold under the allegedly infringing mark. This factor also weighs in Hokto USA's favor, as the companies already directly compete in the relevant market, and indeed sell mushrooms to the same grocery stores.

While Hokto USA failed to introduce evidence of actual confusion, each of the other *Sleekcraft* factors weighs heavily in Hokto USA's favor. The district court thus correctly concluded that there was no genuine dispute of material fact as to whether Concord Farms's importation of Hokuto Japan mushrooms is likely to confuse consumers.

. . . .

Notes & Questions

1. As the First Circuit has stated, "the unauthorized importation and sale of *materially different* merchandise violates [the Lanham Act] because a difference in products bearing the same name confuses consumers and impinges on the local trademark holder's goodwill." *Societe des Produits Nestle v. Casa Helvetia*, 982 F.2d 633, 638 (1st Cir. 1992) (emphasis added). What would constitute a material difference? Consider the First Circuit's explanation of the test:

> There is no mechanical way to determine the point at which a difference becomes "material." Separating wheat from chaff must be done on a case-by-case basis. Bearing in mind the policies and provisions of the Lanham Trade-Mark Act as they apply to gray goods, we can confidently say that the threshold of materiality is always quite low in such cases. . . . [T]he threshold of materiality must be kept low enough to take account of potentially confusing differences–differences that are not blatant enough to make it obvious to the average consumer that the origin of the product differs from his or her expectations.

Id. at 641. Do you think that the removal of the original UPC code on cologne and perfume bottles and packaging should constitute a material difference? *See Zino Davidoff SA v. CVS Corp.*, 571 F.3d 238, 243 (2d Cir. 2009). How about the alteration of sewn-in labels in clothing? *See Abercrombie & Fitch v. Fashion Shops of Ky., Inc.*, 363 F. Supp. 2d 952 (S.D. Ohio 2005). Make sure you understand why Concord Farms was unable to rebut the existence of the required "material difference" between the mushrooms produced for the Japanese market and the mushrooms produced for the U.S. market.

2. The defense of exhaustion cannot be used when the goods are unauthorized. Thus, if the defendant is importing goods from a foreign producer that lacks authority to use the trademark at issue, exhaustion will not excuse otherwise infringing U.S. sales of imported goods bearing the mark. This is true even if the goods

are identical to the U.S. trademark owner's own goods. Those goods may not a have material difference, but they are still not "genuine."

3. Customs Service regulations provide that the restriction on importing trademark-infringing goods will not apply where a label is placed on the product informing the ultimate U.S. purchaser that the "product is not the product authorized by the U.S. trademark holder for importation, and is physically and materially different." 19 C.F.R. § 133.23(b). This disclaimer must be appear "in close proximity to the trademark as it appears in its most prominent location on the article itself or the retail package or container." *Id*. This labeling exception is only available when the foreign manufacturer is either (1) acting under the authority of the U.S. owner, (2) a parent or subsidiary of the U.S. owner, or (3) a party otherwise subject to common ownership or control with the U.S. owner. *Id*.

CHAPTER 7: REMEDIES

Liability and remedy are analytically distinct. The fact that one has demonstrated a misappropriation of trade secrets or an infringement of a patent, copyright, or trademark, and defeated any defenses presented by the defendant, is a necessary step to getting some form of relief for the legal wrong. But it is not sufficient, for liability does not dictate the appropriate remedy to award. Determining the appropriate remedy depends, at least in part, on what goals the law is designed to achieve. First, stopping further infringement or misappropriation might be a paramount goal. A law designed to accomplish that goal would provide confidence in the protection afforded by the intellectual property regime and thus encourage investment in the creation of valuable intangible goods. Second, preventing infringers from profiting from their unlawful acts helps reinforce the policy choice to prohibit that type of free-riding. Providing monetary sanctions in the form of a disgorgement of ill-gotten profits makes infringement far less attractive. Third, compensating intellectual property owners harmed by infringers' past acts comports with an understanding of intellectual property law as protecting an asset that has value. The combination of remedies can deter infringers and bolster the incentives for investment in the creation of valuable information.

This chapter examines the remedies available in intellectual property cases, in terms of both generally applicable principles and domain-specific doctrines. We begin by considering what should be awarded to address the harm that has already occurred. Usually lawyers think of damages for past harm, and indeed this is an important part of the remedy a court may award. Several types of damages awards are possible in intellectual property cases, each identified in the relevant statute. We begin with a category often called "actual damages," which can include both compensation for harm suffered by the plaintiff as well as a disgorgement of the defendant's profits obtained as a result of the infringing activity. Next we examine two different types of proxies that are used as substitutes for evidence of actual damages: a reasonable royalty, in patent and trade secret law, and an amount known as "statutory damages" available in some copyright cases and also in cybersquatting cases. We then consider an intellectual property owner's obligation to provide notice of its rights as a condition precedent to recovering damages at all or receiving a specific types of damages; whether damages awards can be enhanced by a multiplier, and on what proof; and the possibility of, and requirements for, obtaining an award of attorney's fees under the different intellectual property statutes.

A damages award is not the only way to remedy a past harm. In trade secret cases, a certain type of injunction, known as a "head-start" injunction, also addresses a type of past harm. A "head start" injunction is meant to last long enough to ensure that the misappropriator does not benefit from the misappropriation. For example, a court "may delay use of the misappropriated information for the

period of time that it would have taken to reverse engineer or independently develop the secret." Dan L. Burk, *Intellectual Property and the Firm*, 71 U. Chi. L.
Rev. 3, 11 (2004).

After addressing remedies for past harms, we turn to addressing remedies designed to prevent future harm. First, we examine the requirements for obtaining a
permanent injunction and then explore how those standards translate into what a
plaintiff must demonstrate to obtain a preliminary injunction. Future harm may,
in some instances, be addressed by a court mandated royalty payment. Finally we
look at the availability of criminal sanctions for violations of intellectual property
rights.

As you read the cases in this chapter, consider that intellectual property disputes often involve assertions of more than one type of intellectual property violation. For example, a plaintiff may assert that the defendant's actions infringe
both copyright and trademark. In what way might the remedies the plaintiff would
receive differ under these alternative allegations?

I. Looking Back: Redressing Past Harm

A. Actual Damages

Infringement of an intellectual property right is a trespass tort, and "tort
damages generally compensate the plaintiff for loss and return him to the position
he occupied before the injury." *East River Steamship Corp. v. Transamerica Delaval,
Inc.*, 476 U.S. 858, 873 n.9 (1986). Section 504 of the Copyright Act, for example,
sets out the actual damages for which an infringer may be liable:

> (b) Actual Damages and Profits.—The copyright owner is entitled to re-
> cover the actual damages suffered by him or her as a result of the infringe-
> ment, and any profits of the infringer that are attributable to the infringe-
> ment and are not taken into account in computing the actual damages. In
> establishing the infringer's profits, the copyright owner is required to pre-
> sent proof only of the infringer's gross revenue, and the infringer is re-
> quired to prove his or her deductible expenses and the elements of profit
> attributable to factors other than the copyrighted work.

Note that the statute allows for recovery of *both* the copyright owner's actual damages and the infringer's profits. Of course, the statute also cautions courts that are
awarding both types of damages not to permit a double recovery. Considering the
category of actual harm to the copyright owner, awarding "lost profits" is one example where the risk of double recovery is high. Lost profits can occur when the
copyright owner and the accused infringer compete in a goods or services market
and the copyright owner can show it has lost sales to the infringer. If consumers
purchased defendant's infringing book instead of the plaintiff's copyrighted book,
the copyright owner lost the profits from those sales. At the same time, the defendant gained the profits on those sales. To calculate a damages award that included *both* the plaintiff's lost profits and defendant's ill-gotten profits would
amount to a double recovery, absent a showing of a price differential or some other

reason why double recovery is not present. The Copyright statute is also clear about the burden of proof when it comes to proving the relevant elements of a disgoregement award. Watch for these burden of proof issues in the cases throughout this chapter.

The Patent Act does not enumerate the specific categories of actual damages. Instead it makes adequate compensation the key criterion for damages. It also sets a reasonable royalty floor below which damages cannot fall:

> Upon finding for the claimant the court shall award the claimant damages *adequate to compensate* for the infringement, but in no event less than a reasonable royalty for the use made of the invention by the infringer, together with interest and costs as fixed by the court.

35 U.S.C. § 284, ¶ 1 (emphasis added). As the Federal Circuit has emphasized, "the language of the statute is expansive rather than limiting. It affirmatively states that damages must be adequate, while providing only a lower limit and no other limitation." *Rite-Hite Corp. v. Kelley Co.*, 56 F.3d 1538, 1544 (Fed. Cir. 1995) (en banc). Even with that expansive approach, however, it does not provide for disgorgement of the defendants' profits.

Like the Patent Act and the Copyright Act, the Uniform Trade Secrets Act specifies the availability of monetary damages for proven misappropriation: "Damages can include both the actual loss caused by misappropriation and the unjust enrichment caused by misappropriation that is not taken into account in computing actual loss." UTSA § 3(a).

1. Lost Profits and Causation

Recovering lost profits under any theory of intellectual property protection requires proof of *causation*. It is not enough to show that the defendant sold an infringing item. To warrant *lost* profits, the plaintiff must show that it *lost* the sale that the defendant made. The cases below—one a trade secret case, and the other a patent case—focus on this causation question.

Pioneer Hi-Bred Int'l v. Holden Found. Seeds, Inc.

35 F.3d 1226 (8th Cir. 1994)

GIBSON, JUDGE:

This case involves a dispute between competing breeders of corn seed. The district court awarded $46,702,230 to Pioneer Hi-Bred International based on Holden Foundation Seeds, Inc.'s misappropriation of the genetic make-up of certain seed corn. Holden contests the district court's liability determination and damage award on numerous grounds. . . . We affirm.

. . . .

The sale of hybrid seed corn is a multi-billon dollar industry. Pioneer is a vertically integrated seed corn company. It conducts a breeding program, develops parent seed, and produces hybrid[2] seed corn for the retail market. The superior performance of its products, obtained in part through millions of dollars spent annually on corn research and development, has enabled it to gain a sizable portion of the retail seed corn market. Its marketed seed corns include Pioneer hybrids 3780 (Pioneer's leader in sales for several years) and 3541—which share a common parent, designated H3H.

Holden indirectly competes with Pioneer. Holden is a foundation seed company that develops inbred parent seed lines and sells these lines to its customers, also seed corn companies, which use them to produce hybrid seed in competition with Pioneer. During the 1980s, Holden's LH38, LH39, and LH40 ("LH38-39-40") were among its most popular parent lines.

Pioneer sued Holden on a number of legal theories, claiming that Holden developed LH38-39-40 from misappropriated Pioneer H3H or H43SZ7—protected trade secrets of Pioneer ("H3H/H43SZ7"). During discovery, the district court decided, at Pioneer's request, that the nature of this dispute required that Holden "freeze in" a particular story regarding the development of LH38-39-40. The court did so to prevent Holden from altering its story to conform to the scientific evidence eventually introduced. Holden claimed that although LH38-39-40 demonstrated some similarity to Pioneer's seed lines, Holden developed LH38-39-40 from an internal line, designated L120. The district court held that the genetic make-up of H3H/H43SZ7 is a protected trade secret of Pioneer. The court determined that although LH38-39-40 differed in some respects from Pioneer's seed, the lines had nonetheless been derived from misappropriated Pioneer material. The district court based its decision on expert testimony that LH38-39-40 derived from H3H/H43SZ7 and Holden's inability to offer adequate evidence of its "L120 story." . . .

     . . . .

The district court bifurcated the liability and damage trials. After considering the various submissions of the parties in the separate damage trial, the court awarded Pioneer $46,703,230 in lost profits.[49] The court further determined that,

---

[2] "Hybrid" corn seed is produced by planting two inbred parents together and allowing pollen from one inbred (used as the male parent) to fertilize silks on the other inbred (used as the female parent). In corn, inbred lines are lines developed by self-pollination . . . . Inbred lines may be "public" if developed and released by a public university, or "private" if developed by a private entity.

[49] The various damage estimates presented include:

Pioneer:			
	Lost profits range:	$198,467,255	- 276,493,736
	Unjust enrichment range:	31,702,877	- 44,658,469
Holden:			
	Unjust enrichment range:	$ 7,748,065	- 14,840,473
	(after 80% reduction)	1,549,613	- 2,968,096
	Royalty:	2,406,382	- 3,104,062
Court:			
	Lost profits awarded:	$46,703,230	
	Lost profits range:	46,703,230	- 70,054,845
	Alt. unjust enrichment	21,174,913	

in the alternative, the proper damages under an unjust enrichment theory would be $21,174,913. The court also enjoined Holden from distributing or disposing of 177 inbred lines containing LH38-39-40 or their progeny, pending a determination of whether such lines should be turned over to Iowa State University.

. . . .

Courts have used a wide variety of methods to ensure damages in trade secret cases, including lost profits, unjust enrichment and reasonable royalty. Of these, the reasonable royalty theory is least plausible on our facts. Although occasionally applied in trade secret cases, *see* 11 A.L.R.4th § 33[a], at 94-100 (collecting cases), this theory is most appropriate when the other theories would result in no recovery or when parties actually had or contemplated a royalty arrangement. The district court's assessment of over 40 million dollars for lost profits and 20 million dollars for unjust enrichment suggests that this is not a case where a royalty provides the only possible basis for recovery. Nor is there any evidence that Pioneer ever licensed or considered licensing its trade secrets to any direct competitor or foundation seed company.

In general, both lost profits and unjust enrichment theories have been widely accepted. *See* 11 A.L.R.4th §§ 4, 11-12. Courts and commentators alike have suggested that in choosing among competing theories, courts typically select the measure "which affords the plaintiff the greatest recovery." 11 A.L.R.4th § 2[a], at 21.

The district court determined that Pioneer's lost profits were the appropriate measure for damages. The Iowa Supreme Court has recognized lost profits as an appropriate measure of damages. *Basic Chemicals*[*, Inc. v. Benson*], 251 N.W.2d [220,] 233 [(Iowa 1977)]. The selection of lost profits as the appropriate measure in this case, however, presented formidable problems stemming from the fact that Holden does not directly compete with Pioneer. Rather, Holden provides inbred lines which its customers use to develop hybrids which directly compete with Pioneer's hybrids.[51]

In choosing a particular method to quantify Pioneer's lost profits, the district court adopted the basic methodology of Pioneer's expert, Michael Wagner.[52]

---

[51] This explains why Holden's profits might not adequately capture the full extent of Pioneer's losses. The evidence showed that Holden sold infringing seed to as many as 200-300 competitors of Pioneer.

[52] The following is a summary of Wagner's analysis.

Step 1. Take Holden's sales of misappropriated inbreds and companion inbreds (i.e., related seed tied to sale of LH38-39-40) to its customers.

Step 2. Calculate how many bags of hybrids those customers produced and sold to farmers.

Step 3. Take the total market share occupied by look-alikes and determine what share of that market Pioneer would have obtained if it had not been forced to compete with seed derived from its own inbreds.

Step 4. Deduct Pioneer's costs from predicted sales and arrive at net lost profits.

Steps 1, 2, and 4 were based on relatively solid data. The court used Holden's sales and production estimates for steps 1 and 2. Step 4 relied on Pioneer's actual profitability experience. The dispute centers upon step 3.

Wagner sought to quantify the amount Pioneer would have made "but for" Holden's misappropriation. Wagner began by considering the sales of "look-alikes"[53] from 1979 to 1989. Then, he assumed that, absent Holden's distribution of LH38-39-40, Pioneer would have obtained the same percentage of these sales as its market share in all other sales (36% average). From this Wagner computed lost profits of $140,109,691. The court, however, reduced this amount by two-thirds reflecting its determination that Pioneer would not have obtained the full 36% market penetration in that portion of the market occupied by the "look-alikes."[54]

Holden attacks the court's analysis at several points. First, it claims the district court lacked an adequate basis for its assessment of damages. Although Holden blurs the two, there are two distinct steps to such damage inquiry. *Basic Chemicals* provides the governing test:

> If it is speculative and uncertain whether damages have been sustained, re-covery is denied. If the uncertainty lies only in the amount of damages, re-covery may be had if there is proof of reasonable basis from which the amount can be inferred or approximated.

251 N.W.2d at 233. Thus, under Iowa law, a court must first determine if damage occurred, and then consider whether there is a reasonable basis for awarding a specific damage amount.

The district court found that Pioneer lost profits, and that Holden's expert had so conceded. This finding is amply supported by the record. Hybrids developed from LH38-39-40 were used for several years in direct competition with Pioneer. As demonstrated by the growouts, some of these hybrids were quite similar to Pi-oneer's hybrids. The availability of these hybrids (which Holden contends are ac-tually superior to those of Pioneer) resulted in fewer Pioneer sales at perhaps lower prices. Thus, the only relevant question is the second part of the *Basic Chemicals* analysis—whether there was a "reasonable basis" from which the court could award a particular amount.

We believe the district court had a reasonable basis for its award of $46,703,230. The court relied on Holden's actual sales figures, the known pro-ductive capacity of Pioneer's parent lines, Pioneer's profitability history and a rea-sonable estimate of Pioneer's lost share of the "look-alike" market. This meets the "reasonable basis" requirement of *Basic Chemicals*.

---

[53] The term "look-alikes" describes the hybrids created by Holden's customers from LH38-39-40. These hybrids, according to the court, were similar to Pioneer's hybrids. Thus, farmers desiring a particular type of corn could, after Holden developed LH38-39-40, select either a Pioneer hybrid or a "look-alike" where they earlier could have chosen only the Pioneer product.

[54] The court's reasons for reducing the award included: Pioneer's increasing market share during the 1980s, Holden's customers' inability to match Pioneer's yield per female acre, the existence of "loyal" Holden customers, and Pioneer's repeated problems with inadequate seed supplies. None-theless, the court stated that "there is absolutely no doubt that [Pioneer's lost profits] would amount to a minimum average of approximately 12%." Indeed, "the court [was] persuaded that [the loss] may well average 18%."

Holden also contends that the entire "but for" rationale is misguided. However, the court's rationale appears to be straightforward and not uncommon approach to determining lost profits. *See King Instrument Corp. v. Otari Corp.,* 767 F2d. 853, 863 (Fed. Cir. 1985) (lost profits award "requires (1) a showing that the patent owner would have made the sale but for the infringement, *i.e.*, causation existed, and (2) proper evidence of the computation on the loss of profits"). Holden's attack on this point . . . seems largely over the verbiage the court should have employed. The court's result, we believe, would be the same under any lost profits verbiage. Accordingly, we find no error in the court's lost profits methodology and conclude the district court did not err in awarding Pioneer $46,703,230.

. . . .

Notes & Questions

1. As this case shows, one's choice of how to measure damages in a case can lead to drastically different awards. Re-examine footnote 49 of the court's opinion. What was the maximum amount that Holden argued the court should have awarded? What was the problem the court identified with the unjust enrichment approach?

If Holden had taken laboratory equipment belonging to Pioneer Hi-Bred, the damages awarded for the conversion of that property likely would have been the replacement value of the equipment. Instead, Holden "took" Pioneer Hi-Bred's trade secrets. Why is measuring damages in the trade secret context so difficult? Unlike a tangible asset, the non-rivalrous consumption characteristic of information makes establishing a value for that asset more difficult. In selecting a damages model for a case, how should the court set a value that is genuinely compensatory?

2. In *Pioneer Hi-Bred,* the Eighth Circuit notes that "courts typically select the measure 'which affords the plaintiff the greatest recovery.'" *Pioneer Hi-Bred,* 35 F.3d 1226, 1244 (quoting 11 A.L.R.4th §2[a] at 21). Why might it be appropriate to favor the approach that results in the largest award?

3. In affirming the district court's award, the Eighth Circuit notes that the plaintiff's productive capacity was factored into determining the appropriate award. Why would the trade secret owner's productive capacity to satisfy market demand be relevant to the damages that the defendant should have to pay? Would productive capacity be equally relevant under each of the different approaches to measuring damages?

Grain Processing Corp. v. American Maize-Products Co.
185 F.3d 1341 (Fed. Cir. 1999)

RADER, JUDGE:

. . . .

This appeal culminates the lengthy and complex history of this case, spanning more than eighteen years and eight prior judicial opinions, three by this court. The patent featured in this infringement suit [U.S. Patent No. 3,849,194 (the '194 patent)] involves maltodextrins, a versatile family of food additives made from starch. Commercial food manufacturers purchase hundreds of millions of pounds of maltodextrins annually from producers such as Grain Processing and American Maize.

Maltodextrins serve well as food additives because they are bland in taste and clear in solution. They do not affect the natural taste or color of other ingredients in food products. Maltodextrins also improve the structure or behavior of food products. For instance, they inhibit crystal growth, add body, improve binding and viscosity, and preserve food properties in low temperatures. Consequently, food manufacturers use maltodextrins in a wide variety of products such as frostings, syrups, drinks, cereals, and frozen foods.

Maltodextrins belong to a category of chemical products known as "starch hydrolysates." Producers make starch hydrolysates by putting starch through hydrolysis, a chemical reaction with water. . . . [Grain Processing had sold a line of maltodextrins under the "Maltrin" brand name that did not practice the claimed invention; they were not made from a "waxy-starch," a requirement of the patent. American Maize made a waxy-starch maltodextrin that it sold under the name "Lo-Dex 10."]

. . . American Maize sold Lo-Dex 10 . . . during the entire time Grain Processing owned the '194 patent rights, from 1979 until the patent expired in 1991. During this time, however, American Maize used four different processes for producing Lo-Dex 10. The changes in American Maize's production processes, and the slight chemical differences in the Lo-Dex 10 from each process, are central to the lost profits issue in this appeal.

American Maize used a first process (Process I) from June 1974 to July 1982. In Process I, American Maize used a single enzyme (an alpha amylase) to facilitate starch hydrolysis. [Grain Processing sued for infringement of the '194 patent on May 12, 1981. In August 1982, while the suit was pending, American Maize changed to a different process (Process II) to lower its production costs. It made Lo-Dex 10 with Process II from August 1982 to February 1988. American Maize contended that Lo-Dex 10 made by Process I or Process II did not infringe because it did not have a characteristic required by the claim language as measured by the Lane-Eynon test (an accepted test), while Grain Process disagreed based on measures it took using the Schoorl test (another accepted test). Each test yielded slightly different results (one within the claim requirements and one outside the claim requirements). After a bench trial determining no infringement, an appeal reversing that determination, and a subsequent bench trial determination of infringement, the district court issued an injunction. In response to the injunction American Maize developed another process for producing Lo-Dex 10, Process III, using it from March 1988 to April 1991. Grain Processing, after testing the Process III Lo-Dex with the Schoorl test and finding the claim requirement present, filed a contempt motion. After an appeal determined that the prosecution history showed a preference for use of the Schoorl test the Process III Lo-Dex 10 was found to be an infringement.]

American Maize then adopted a fourth process (Process IV) for producing Lo-Dex 10. In Process IV, American Maize added a second enzyme, glucoamylase, to the reaction. [This addition resulted in the absence of a claim requirement.]

From the time American Maize began experimenting with the glucoamylase-alpha amylase combination, or the "dual enzyme method," it took only two weeks to perfect the reaction and begin mass producing Lo-Dex 10 using Process IV. According to the finding of the district court, this two-week development and production time is "practically instantaneous" for large-scale production. American Maize simply experimented with different combinations of glucoamylase and alpha amylase, along with pH, heat, and time of the reaction. American Maize did not change any equipment, source starches, or other ingredients from Process III. Glucoamylase has been commercially available and its effect in starch hydrolysis widely known since the early 1970's, before the '194 patent issued. American Maize had not used Process IV to produce Lo-Dex earlier because the high cost of glucoamylase makes Process IV more expensive than the other processes.

The parties agree that Process IV yielded only noninfringing Lo-Dex 10 and that consumers discerned no difference between Process IV Lo-Dex 10 and Lo-Dex 10 made by Processes I-III. American Maize used Process IV exclusively to produce Lo-Dex 10 from April 1991 until the '194 patent expired in November 1991, and then switched back to the cheaper Process III.

The district court commenced the damages portion of the trial on July 10, 1995. Grain Processing claimed lost profits in the form of lost sales of Maltrin M100, price erosion, and American Maize's accelerated market entry after the patent expired. Grain Processing further claimed that, for any of American Maize's infringing sales not covered by a lost profits award, Grain Processing should receive a 28% royalty. After a three day bench trial, the district court denied lost profits and determined that a 3% reasonable royalty was adequate to compensate Grain Processing. The royalty applies to all of American Maize's Lo-Dex 10 sales from May 12, 1981 (when Grain Processing filed suit)[3] to April 1991 (when American Maize converted to Process IV, thereby producing a noninfringing product).

The trial court determined that Grain Processing could not establish causation for lost profits, because American Maize "could have produced" a noninfringing substitute 10 D.E. maltodextrin using Process IV. "With infringing Lo-Dex 10 banned, the customers' substitute is non-infringing Lo-Dex 10." American Maize did not actually produce and sell this noninfringing substitute until April 1991, seven months before the '194 patent expired, but the district court nevertheless found that its availability "scotches [Grain Processing's] request for lost-profits damages."

---

[3] Grain Processing was not entitled to damages before this date because neither Grain Processing nor its predecessor in interest had marked the [patented] products [it sold] with the patent number pursuant to 35 U.S.C. § 287(a).

The district court also found that American Maize's production cost difference between infringing and noninfringing Lo-Dex 10 effectively capped the reasonable royalty award. American Maize showed that it cost only 2.3% more to make noninfringing Process IV products than it did to make infringing Process I-III products. The district court also found that "buyers viewed as equivalent" the Process I-III and Process IV output: "Lo-Dex 10 made by Process IV had a lower D.R. [which is what makes it noninfringing] * * * but no one argues that any customer cared a whit about the product's descriptive ratio." The district court concluded that under these facts, American Maize, when faced with a hypothetical offer to license the '194 patent in 1974 (or to renegotiate the rate in 1979, when Grain Processing acquired the patent rights and its ability to collect damages began), would not have paid more than a 3% royalty rate. The court reasoned that this rate would reflect the cost difference between Processes I-III and Process IV, while also taking into account possible cost fluctuations (due to fluctuating enzyme prices) and the elimination of American Maize's risk of producing an infringing product, despite its best efforts. The court concluded that if Grain Processing had insisted on a rate greater than 3% in the hypothetical negotiations, American Maize instead would have chosen to invest in producing noninfringing Lo-Dex 10 with Process IV.

Grain Processing appealed the district court's denial of lost profits . . . . This court reversed and remanded . . . observ[ing] that "[t]he [district] court denied [Grain Processing's] request for lost profits because [American Maize] developed a new process of producing Lo-Dex 10 in 1991 [after years of infringement] that did not infringe the '194 patent." This court noted, however, that the mere fact of "switching to a noninfringing product years after the period of infringement [does] not establish the presence of a noninfringing substitute during the period of infringement." This court noted that a product or process must be "available or on the market at the time of infringement" to qualify as an acceptable noninfringing substitute.

On remand, the district court again denied Grain Processing lost profits. The district court found that Process IV was "available" throughout the period of infringement. This factual finding, the district court explained, was not based merely on "the simple fact of switching [to Process IV]" but rather on several subsidiary factual findings regarding the technology of enzyme-assisted starch hydrolysis and the price and market structure for the patentee's and accused infringer's products. The trial court found that American Maize could obtain all of the materials needed for Process IV, including the glucoamylase enzyme, before 1979, and that the effects of the enzymes in starch hydrolysis were well known in the field by that time. American Maize also had all of the necessary equipment, know-how, and experience to implement Process IV whenever it chose to do so during the time of infringement. "The sole reason [American Maize did not use Process IV to produce Lo-Dex 10 prior to 1991] was economic: glucoamalyse is more expensive than the alpha amylase enzyme that [American Maize] had been using." American Maize did not make the substitution sooner because its test results using the Lane-Eynon method convinced it that it was not infringing.

The district court concluded that "the profit lost from infringement is the cost and market price difference attributable to using glucoamylase." The court did not further address the amount of damages, having already found in *Grain Processing VI* [the district court's earlier opinion in this case,] that the infringement did not affect the market price of Lo-Dex 10, and having figured the 2.3% cost increase into the 3% royalty award.

. . . .

II.

. . . .

Upon proof of infringement, Title 35, Section 284 provides that "the court shall award [the patent owner] damages adequate to compensate for the infringement but in no event less than a reasonable royalty for the use made of the invention by the infringer." 35 U.S.C. § 284. The phrase "damages adequate to compensate" means "full compensation for 'any damages' [the patent owner] suffered as a result of the infringement." *General Motors Corp. v. Devex Corp.*, 461 U.S. 648, 654 (1983). Full compensation includes any foreseeable lost profits the patent owner can prove.

To recover lost profits, the patent owner must show "causation in fact," establishing that "but for" the infringement, he would have made additional profits. *See King Instruments Corp. v. Perego*, 65 F.3d 941, 952 (Fed. Cir. 1995). When basing the alleged lost profits on lost sales, the patent owner has an initial burden to show a reasonable probability that he would have made the asserted sales "but for" the infringement. *See id.* Once the patent owner establishes a reasonable probability of "but for" causation, "the burden then shifts to the accused infringer to show that [the patent owner's "but for" causation claim] is unreasonable for some or all of the lost sales." [*Rite-Hite Corp. v. Kelley Co.*, 56 F.3d 1538, 1544 (Fed. Cir. 1995) (en banc).]

At trial, American Maize proved that Grain Processing's lost sales assertions were unreasonable. The district court adopted Grain Processing's initial premise that, because Grain Processing and American Maize competed head-to-head as the only significant suppliers of 10 D.E. maltodextrins, consumers logically would purchase Maltrin 100 if Lo-Dex 10 were not available. *See Lam, Inc. v. Johns-Manville Corp.*, 718 F.2d 1056, 1065 (Fed. Cir. 1983) (holding that the patent owner may satisfy his initial burden by inference in a two-supplier market). However, the district court found that American Maize proved that Process IV was available and that Process IV Lo-Dex 10 was an acceptable substitute for the claimed invention. In the face of this noninfringing substitute, Grain Processing could not prove lost profits.

American Maize concedes that it did not make or sell Lo-Dex 10 from Process IV until 1991, after the period of infringement. However, an alleged substitute not "on the market" or "for sale" during the infringement can figure prominently in determining whether a patentee would have made additional profits "but for" the infringement. . . .

In *Aro Manufacturing*, the Supreme Court stated that the statutory measure of "damages" is "the difference between [the patent owner's] pecuniary condition after the infringement, and what his condition would have been if the infringement had not occurred." *Aro Mfg Co. v. Convertible Top Replacement Co.*, 377 U.S. 476, 507 (1964) (plurality opinion). The determinative question, the Supreme Court stated, is: "had the Infringer not infringed, what would the Patent Holder-Licensee have made?" *Aro*, 377 U.S. at 507. The "but for" inquiry therefore requires a reconstruction of the market, as it would have developed absent the infringing product, to determine what the patentee "would . . . have made."

Reconstructing the market, by definition a hypothetical enterprise, requires the patentee to project economic results that did not occur. To prevent the hypothetical from lapsing into pure speculation, this court requires sound economic proof of the nature of the market and likely outcomes with infringement factored out of the economic picture. Within this framework, trial courts, with this court's approval, consistently permit patentees to present market reconstruction theories showing all of the ways in which they would have been better off in the "but for world," and accordingly to recover lost profits in a wide variety of forms. *See, e.g.*, *King Instrument*, 65 F.3d at 953 (upholding award for lost sales of patentee's unpatented goods that compete with the infringing goods); *Rite-Hite*, 56 F.3d 1550 (holding that a patentee may recover lost profits on components that have a functional relationship with the patented invention); *Minnesota Mining & Mfg. Co. v. Johnson & Johnson Orthopaedics, Inc.*, 976 F.2d 1559, 1579 (Fed. Cir. 1992) (upholding award for price erosion due to infringing sales). In sum, courts have given patentees significant latitude to prove and recover lost profits for a wide variety of foreseeable economic effects of the infringement.

By the same token, a fair and accurate reconstruction of the "but for" market also must take into account, where relevant, alternative actions the infringer foreseeably would have undertaken had he not infringed. Without the infringing product, a rational would-be infringer is likely to offer an acceptable noninfringing alternative, if available, to compete with the patent owner rather than leave the market altogether. The competitor in the "but for" marketplace is hardly likely to surrender its complete market share when faced with a patent, if it can compete in some other lawful manner. Moreover, only by comparing the patented invention to its next-best available alternative(s)—regardless of whether the alternative(s) were actually produced and sold during the infringement—can the court discern the market value of the patent owner's exclusive right, and therefore his expected profit or reward, had the infringer's activities not prevented him from taking full economic advantage of this right. Thus, an accurate reconstruction of the hypothetical "but for" market takes into account any alternatives available to the infringer.

Accordingly, this court in *Slimfold Manufacturing Co. v. Kinkead Industries, Inc.* held that an available technology not on the market during the infringement can constitute a noninfringing alternative. 932 F.2d 1453 (Fed. Cir. 1991). In *Slimfold*, the patent owner (Slimfold) claimed lost profits on its bi-fold doors with a patented pivot and guide rod assembly. This court noted, however, that Slimfold did not show "that the alleged infringer [Kinkead] would not have made a substantial portion or the same number of sales *had it continued with its old hardware* or with the hardware utilized by any of the other companies." Id. at 1458 (emphasis added). On the basis of this noninfringing substitute, which was not on the market at the time of infringement, this court affirmed the district court's denial of lost profits. This court determined that the record supported the district court's finding that this noninfringing "old hardware" was available to Kinkead at the time of the infringement. Specifically, Kinkead and others had used the substitute technology on other doors before the period of infringement. Furthermore, consumers considered Kinkead's noninfringing alternative an acceptable substitute for the infringing doors. Therefore, this court upheld the district court's award of a "small" royalty, rather than lost profits. *Id.* at 1458-59.

. . . .

Grain Processing asserts that permitting the infringer to show substitute availability without market sales, thereby avoiding lost profits, undercompensates for infringement. Section 284, however, sets the floor for "damages adequate to compensate for the infringement" as "a reasonable royalty." 35 U.S.C. § 284. Thus, the statute specifically envisions a reasonable royalty as a form of adequate compensation. While "damages adequate to compensate" means "full compensation," *General Motors*, 461 U.S. at 654, "full compensation" does not entitle Grain Processing to lost profits in the absence of "but for" causation. Moreover, although Grain Processing stresses that American Maize should not reap the benefit of its "choice" to infringe rather than use the more expensive Process IV, Grain Processing does not allege willful infringement and the record shows none. To the extent that Grain Processing feels undercompensated, it must point out a reversible error in the district court's fact-finding, reasoning, or legal basis for denying lost profits or in its reasonable royalty determination.

III.

This court next turns to the district court's findings that Process IV was in fact "available" to American Maize for producing Lo-Dex 10 no later than October, 1979, and that consumers would consider Process IV Lo-Dex 10 an acceptable substitute. This court reviews these factual findings for clear error.

The critical time period for determining availability of an alternative is the period of infringement for which the patent owner claims damages, i.e., the "accounting period." Switching to a noninfringing substitute after the accounting period does not alone show availability of the noninfringing substitute during this critical time. When an alleged alternative is not on the market during the accounting period, a trial court may reasonably infer that it was not available as a noninfringing substitute at that time. The accused infringer then has the burden to overcome this inference by showing that the substitute was available during the accounting period. Mere speculation or conclusory assertions will not suffice to overcome the inference. After all, the infringer chose to produce the infringing, rather than noninfringing, product. Thus, the trial court must proceed with caution in assessing proof of the availability of substitutes not actually sold during the period of infringement. Acceptable substitutes that the infringer proves were available during the accounting period can preclude or limit lost profits; substitutes only theoretically possible will not.

In this case, the district court did not base its finding that Process IV was available no later than October 1979 on speculation or possibilities, but rather on several specific, concrete factual findings, none of which Grain Processing challenges on appeal. The district court found that American Maize could readily obtain all of the materials needed for Process IV, including the glucoamylase enzyme, before 1979. The court also found that the effects of the enzymes in starch hydrolysis were well known in the field at that time. Furthermore, the court found that American Maize had all of the necessary equipment, know-how, and experience to use Process IV to make Lo-Dex 10, whenever it chose to do so during the time it was instead using Processes I, II or III. American Maize "did not have to 'invent around' the patent," the district court observed; "all it had to do was use a glucoamaylase enzyme in its production process."

The trial court also explained that "the sole reason [American Maize did not use Process IV prior to 1991] was economic: glucoamylase is more expensive than the alpha amylase enzyme American Maize had been using," and American Maize reasonably believed it had a noninfringing product. While the high cost of a necessary material can conceivably render a substitute "unavailable," the facts of this case show that glucoamylase was not prohibitively expensive to American Maize. The district court found that American Maize's "substantial profit margins" on Lo-Dex 10 were sufficient for it to absorb the 2.3% cost increase using glucoamylase.

. . . .

. . . Accordingly, this court holds that the district court did not clearly err in finding that Process IV Lo-Dex 10 was an available alternative throughout the accounting period.

Whether and to what extent American Maize's alleged alternative prevents Grain Processing from showing lost sales of Maltrin 100 depends not only on whether and when the alternative was available, but also on whether and to what extent it was acceptable as a substitute in the relevant market. Consumer demand defines the relevant market and relative substitutability among products therein. Important factors shaping demand may include consumers' intended use for the patentee's product, similarity of physical and functional attributes of the patentee's product to alleged competing products, and price. *See Fonar Corp. v. General Elec. Co.*, 107 F.3d 1543, 1553 (Fed. Cir. 1997). Where the alleged substitute differs from the patentee's product in one or more of these respects, the patentee often must adduce economic data supporting its theory of the relevant market in order to show "but for" causation.

In this case, the parties vigorously dispute the precise scope of the relevant market. The district court's uncontroverted factual findings, however, render this dispute moot. In the eyes of consumers, according to the district court, Process IV Lo-Dex 10 was the same product, for the same price, from the same supplier as Lo-Dex 10 made by other processes. Process IV Lo-Dex 10 was a perfect substitute for previous versions, and therefore Grain Processing's efforts to show a distinct 10 D.E. maltodextrin market do not assist its lost profits case.

Market evidence in the record supports the district court's uncontroverted findings and conclusions on acceptability. First, for example, American Maize's high profit margin on Lo-Dex 10 and the consumers' sensitivity to price changes support the conclusion that American Maize would not have raised the price of Process IV Lo-Dex 10 to offset the cost of glucoamylase. Further, American Maize's sales records showed no significant changes when it introduced Process IV Lo-Dex 10 at the same price as previous versions, indicating that consumers considered its important properties to be effectively identical to previous versions. Witness testimony supported this market data. Thus, this court discerns no clear error in the district court's finding that Process IV Lo-Dex 10 was an acceptable substitute in the marketplace.

It follows from the district court's findings on availability and acceptability that Grain Processing's theory of "but for" causation fails. As the district court correctly noted, "[a]n [American Maize] using the dual-enzyme method between 1979 and 1991 * * * would have sold the same product, for the same price, as the actual [American Maize] did * * * "and consequently would have retained its Lo-Dex 10 sales. Grain Processing did not present any other evidence of lost profits, such as individual lost transactions . . . . Thus, the district court properly determined that, absent infringing Lo-Dex 10, Grain Processing would have sold no more and no less Maltrin 100 than it actually did.

. . . .

Notes & Questions

1. Make a list of everything the courts requires the defendant to prove in order to defeat an otherwise proper award of lost profits. The Federal Circuit seems quite concerned about it becoming too easy for an accused infringer to defeat an otherwise meritorious lost profits award with a story about a noninfringing product or process it could have switched to, but didn't. The court stresses that "[s]witching to a noninfringing substitute after the accounting period does not alone show availability of the noninfringing substitute during" the accounting period, and that "the trial court must proceed with caution in assessing proof of the availability of substitutes not actually sold during the period of infringement." If it were to become much easier for an accused infringer to defeat a lost profits award in this way, how might that affect license negotiations that take place before any lawsuit is filed? How might it affect the way a person might weigh taking a bigger risk with an unlicensed product or process?

2. Is there a potential hazard here for the accused infringer? If one stresses how easy it would have been to switch to an available noninfringing alternative that one *didn't* use, could it leave the factfinder annoyed that the accused infringer didn't simply "do the right thing," *i.e.*, avoid infringement? Put differently, is it more than a coincidence that American Maize pushed its theory, successfully, in a bench trial rather than before a jury? And not just any bench trial: Seventh Circuit

Judge Frank Easterbrook sat by designation as the trial judge in the latter stages of the case. Many identify Judge Easterbrook as a leading judicial proponent of the economic analysis of law. Additionally, note the impact that the bifurcation of the trial may have on different strategy options available to the parties.

3. The court notes that Claim 12 of the '194 patent did not cover Grain Processing's own maltodextrin products. (When Grain Processing bought the company that owned the '194 patent, it discontinued the maltodextrins covered by the '194 patent.) Should it matter, in determining a damages award, whether the patentee is exploiting the patent? For example, what if the patentee seeks lost profits where its competitor sells an infringing product that purportedly diverts sales but the patentee sells only a product that *doesn't* embody the claimed invention? Four years before *Grain Processing*, the Federal Circuit held that a patentee *can* recover lost profits in these circumstances. *See Rite-Hite Corp. v. Kelley Co.*, 56 F.3d 1538, 1544 (Fed. Cir. 1995) (en banc). The limiting principle is foreseeability: "If a particular injury was or should have been reasonably foreseeable by an infringing competitor in the relevant market, broadly defined, that injury is generally compensable absent a persuasive reason to the contrary." *Id.* at 1546. Examine the other cases that the court cites in *Grain Processing*. In addition to lost profits on plaintiff's products that do not practice the claimed invention, what else might be recoverable?

4. How did American Maize's profits on the products it sold factor into the damages the court ultimately awarded? The Patent Act does not provide a disgorgement remedy. Thus, seeking to obtain an award based on the defendant's ill-gotten gains is not expressly permitted for patent infringement.

2. Determining Defendant's Profits

Frank Music Corp. v. Metro-Goldwyn-Mayer, Inc.
772 F.2d 505 (9th Cir. 1985)

FLETCHER, JUDGE:

This copyright infringement suit arises out of defendants' use of five songs from plaintiffs' dramatico-musical play *Kismet* in a musical revue staged at defendant MGM Grand Hotel in 1974-76. . . .

. . . .

On April 26, 1974, defendant MGM Grand Hotel premiered a musical revue entitled *Hallelujah Hollywood* in the hotel's Ziegfeld Theatre. The show . . . featured ten acts of singing, dancing, and variety performances. Of the ten acts, four were labeled as "tributes" to MGM motion pictures of the past, and one was a tribute to the "Ziegfeld Follies." The remaining acts were variety numbers, which included performances by a live tiger, a juggler, and the magicians, Siegfried and Roy.

. . . .

Act IV of *Hallelujah Hollywood,* the subject of this lawsuit, was entitled "Kismet," and was billed as a tribute to the MGM movie of that name. Comprised of four scenes, it was approximately eleven and one-half minutes in length. It was set in ancient Baghdad, as was plaintiffs' play, and the characters were called by the same or similar names to those used in plaintiffs' play. Five songs were taken in whole or in part from plaintiffs' play. No dialogue was spoken during the act, and, in all, it contained approximately six minutes of music taken directly from plaintiffs' play.

The total running time of *Hallelujah Hollywood* was approximately 100 minutes, except on Saturday nights when two acts were deleted, shortening the show to 75 minutes. The show was performed three times on Saturday evenings, twice on the other evenings of the week.

. . . .

[When plaintiffs informed MGM Grand that they considered *Hallelujah Hollywood* to infringe their rights in *Kismet,* MGM Grand responded that it believed its use of plaintiffs' music was covered by its blanket license agreement with the American Society of Composers, Authors and Publishers ("ASCAP"). The Plaintiffs disagreed and filed this lawsuit. The district court concluded that the ASCAP license did not cover the dramatic performance of the songs. After a bench trial, the district court found infringement and awarded the plaintiffs $22,000 as a share of defendants' profits. Plaintiffs appealed and defendants cross-appealed.]

1. Actual Damages

. . . .

The district court declined to award actual damages. The court stated that it was "unconvinced that the market value of plaintiffs' work was in any way diminished as a result of defendant's infringement." We are obliged to sustain this finding unless we conclude it is clearly erroneous.

. . . .

In a copyright action, a trial court is entitled to reject a proffered measure of damages if it is too speculative. Although uncertainty as to the amount of damages will not preclude recovery, uncertainty as to the fact of damages may. It was the *fact* of damages that concerned the district court. The court found that plaintiffs "failed to establish *any* damages attributable to the infringement." (emphasis in original). This finding is not clearly erroneous.

Plaintiffs offered no disinterested testimony showing that *Hallelujah Hollywood* precluded plaintiffs from presenting *Kismet* at some other hotel in Las Vegas. It is not implausible to conclude, as the court below apparently did, that a production presenting six minutes of music from *Kismet,* without telling any of the story of the play, would not significantly impair the prospects for presenting a full production of that play. Based on the record presented, the district court was not clearly erroneous in finding that plaintiffs' theory of damages was uncertain and speculative.

2. Infringer's Profits

As an alternative to actual damages, a prevailing plaintiff in an infringement action is entitled to recover the infringer's profits to the extent they are attributa-

ble to the infringement. In establishing the infringer's profits, the plaintiff is required to prove only the defendant's sales; the burden then shifts to the defendant
to prove the elements of costs to be deducted from sales in arriving at profit. Any
doubt as to the computation of costs or profits is to be resolved in favor of the
plaintiff. If the infringing defendant does not meet its burden of proving costs, the
gross figure stands as the defendant's profits.

The district court, following this approach, found that the gross revenue MGM
Grand earned from the presentation of *Hallelujah Hollywood* during the relevant
time period was $24,191,690. From that figure, the court deducted direct costs of
$18,060,084 and indirect costs (overhead) of $3,641,960, thus arriving at a net
profit of $2,489,646.

. . . .

A portion of an infringer's overhead properly may be deducted from gross revenues to arrive at profits, at least where the infringement was not willful, conscious, or deliberate. *Kamar International, Inc. v. Russ Berrie & Co.*, 752 F.2d 1326,
1331 (9th Cir. 1984); *Sammons v. Colonial Press, Inc.*, 126 F.2d 341, 351 (1st Cir.
1942). Plaintiffs argue that the infringement here was conscious and deliberate,
but the district court found to the contrary. The court's finding is not clearly erroneous. Defendants believed their use of *Kismet* was protected under MGM
Grand's ASCAP license. Although their contention ultimately proved to be
wrong, it was not implausible. Defendants reasonably could have believed that
their production was not infringing plaintiffs' copyrights, and, therefore, the district court was not clearly erroneous in finding that their conduct was not willful.

We find more merit in plaintiffs' second challenge to the deduction of overhead costs. They argue that defendants failed to show that each item of claimed
overhead assisted in the production of the infringement. The evidence defendants
introduced at trial segregated overhead expenses into general categories, such as
general and administrative costs, sales and advertising, and engineering and
maintenance. Defendants then allocated a portion of these costs to the production
of *Hallelujah Hollywood* based on a ratio of the revenues from that production as
compared to MGM Grand's total revenues. The district court adopted this approach.

We do not disagree with the district court's acceptance of the defendants'
method of allocation, based on gross revenues. Because a theoretically perfect allocation is impossible, we require only a "reasonably acceptable formula." *Sammons,* 126 F.2d at 349. We find, as did the district court, that defendants' method
of allocation is reasonably acceptable.

We disagree with the district court, however, to the extent it concluded the
defendants adequately showed that the claimed overhead expenses actually contributed to the production of *Hallelujah Hollywood*. Recently, in *Kamar International,* we stated that a deduction for overhead should be allowed "only when the
infringer can demonstrate that [the overhead expense] was of actual assistance in
the production, distribution or sale of the infringing product." 752 F.2d at 1332.
We do not take this to mean that an infringer must prove his overhead expenses

and their relationship to the infringing production in minute detail. Nonetheless, the defendant bears the burden of explaining, at least in general terms, how claimed overhead actually contributed to the production of the infringing work. *Taylor v. Meirick,* 712 F.2d 1112, 1121-22 (7th Cir. 1983) ("It is too much to ask a plaintiff who has proved infringement also to do the defendant's cost accounting.").

. . . .

Plaintiffs next challenge the district court's failure to consider MGM Grand's earnings on hotel and gaming operations in arriving at the amount of profits attributable to the infringement. The district court received evidence concerning MGM Grand's total net profit during the relevant time period, totaling approximately $395,000,000, but its memorandum decision does not mention these indirect profits and computes recovery based solely on the revenues and profits earned on the production of *Hallelujah Hollywood* (approximately $24,000,000 and $2,500,000 respectively). We surmise from this that the district court determined plaintiffs were not entitled to recover indirect profits, but we have no hint as to the district court's reasons.

Whether a copyright proprietor may recover "indirect profits" is one of first impression in this circuit. We conclude that under the 1909 Act indirect profits may be recovered.

The 1909 Act provided that a copyright proprietor is entitled to "all the profits which the infringer shall have made from such infringement * * * ." 17 U.S.C. § 101(b). The language of the statute is broad enough to permit recovery of indirect as well as direct profits. At the same time, a court may deny recovery of a defendant's profits if they are only remotely or speculatively attributable to the infringement.

. . . .

. . . Defendants maintain that they endeavor to earn profits on all their operations and that *Hallelujah Hollywood* was a profit center. However, that fact does not detract from the promotional purposes of the show—to draw people to the hotel and the gaming tables. MGM's 1976 annual report states that "[t]he hotel and gaming operations of the MGM Grand—Las Vegas continue to be materially enhanced by the popularity of the hotel's entertainment[, including] 'Hallelujah Hollywood,' the spectacularly successful production revue * * * ." Given the promotional nature of *Hallelujah Hollywood,* we conclude indirect profits from the hotel and gaming operations, as well as direct profits from the show itself, are recoverable if ascertainable.

3. Apportionment of Profits

How to apportion profits between the infringers and the plaintiffs is a complex issue in this case. Apportionment of direct profits from the production as well as indirect profits from the hotel and casino operations are involved here, although the district court addressed only the former at the first trial.

When an infringer's profits are attributable to factors in addition to use of plaintiff's work, an apportionment of profits is proper. *Sheldon v. Metro-Goldwyn Pictures, Inc.*, 309 U.S. 390, 405-06 (1939) (*Sheldon II*). The burden of proving apportionment, (i.e., the contribution to profits of elements other than the infringed property), is the defendant's. We will not reverse a district court's findings regarding apportionment unless they are clearly erroneous.

After finding that the net profit earned by *Hallelujah Hollywood* was approximately $2,500,000, the district court offered the following explanation of apportionment:

> While no precise mathematical formula can be applied, the court concludes in light of the evidence presented at trial and the entire record in this case, a fair approximation of the profits of Act IV attributable to the infringement is $22,000.

The district court was correct that mathematical exactness is not required. However, a reasonable and just apportionment of profits is required.

Arriving at a proper method of apportionment and determining a specific amount to award is largely a factual exercise. Defendants understandably argue that the facts support the district court's award. They claim that the infringing material, six minutes of music in Act IV, was an unimportant part of the whole show, that the unique features of the Ziegfield Theater contributed more to the show's success than any other factor. This is proved, they argue, by the fact that when the music from *Kismet* was removed from *Hallelujah Hollywood* in 1976, the show suffered no decline in attendance and the hotel received no complaints.

Other evidence contradicts defendants' position. For instance, defendant Donn Arden testified that *Kismet* was "a very important part of the show" and "[he] hated to see it go." Moreover, while other acts were deleted from the shortened Saturday night versions of the show, Act IV "Kismet" never was.

We reject defendants' contention that the relative unimportance of the *Kismet* music was proved by its omission and the show's continued success thereafter. *Hallelujah Hollywood* was a revue, comprised of many different entertainment elements. Each element contributed significantly to the show's success, but no one element was the sole or overriding reason for that success. Just because one element could be omitted and the show goes on does not prove that the element was not important in the first instance and did not contribute to establishing the show's initial popularity.

The difficulty in this case is that the district court has not provided us with any reasoned explanation of or formula for its apportionment. We know only the district court's bottom line: that the plaintiffs are entitled to $22,000. Given the nature of the infringement, the character of the infringed property, the success of defendants' show, and the magnitude of the defendants' profits, the amount seems to be grossly inadequate. It amounts to less than one percent of MGM

Grand's profits from the show, or roughly $13 for each of the 1700 infringing performances.[11]

On remand, the district court should reconsider its apportionment of profits, and should fully explain on the record its reasons and the resulting method of apportionment it uses. Apportionment of indirect profits may be a part of the calculus. . . .

Notes & Questions

1. As the court notes, the Copyright Act allows for recovery of both actual harm and defendant's profits (so long as there is no double recovery). What was the problem with the plaintiffs' evidence of actual damages in *Frank Music*?

2. When the disgorgement of defendant's profits is warranted, the court must first determine what constitutes "profit"—which expenses can be deducted from gross revenue. Here the Copyright Act is clear that the defendant bears the burden of proof. In *Frank Music* the defendant provided evidence of both direct and indirect, or "overhead," expenses. Direct expenses would include items such as the performers' salaries and the cost of set production. How should a court evaluate the different overhead expenses proffered by the defendant? The Second Circuit indicates that overhead may be deducted so long as the infringement is not willful. This rule is not in the statute, but rather is judge made. Does this rule make sense?

3. Once the amount of profit is established, the next inquiry is one of apportionment—what *portion* of that profit is attributable to the infringement. What percentages might you offer as a rational basis for apportionment if you were representing Frank Music? What percentages would you offer if you were representing MGM? While *Frank Music* was decided under the 1909 Act, it is an important and often-cited precedent for the appropriate approach to determining monetary awards.

4. What do you think of the court's decision that an award of a portion of indirect profits is permissible? Is such an award similar to an award in a patent case, discussed in *Grain Processing*, of lost profits on component parts of a patented product?

---

[11] The apportionment percentages in similar cases are markedly higher. *See, e.g., Universal Pictures Co. v. Harold Lloyd Corp.,* 162 F.2d at 377 (infringing use of one comedy sketch in motion picture; court affirmed award of 20% of infringing movie's profits); *MCA, Inc. v. Wilson,* 677 F.2d 180, 181-82 (2d Cir. 1981) (defendants copied substantial portion plaintiff's song, "Boogie Woogie Bugle Boy", substituted "dirty" lyrics, and performed the song as a portion of an erotic nude show; court affirmed special master's award of approximately $244,000 representing 5% of defendants' total profits from the show); *Lottie Joplin Thomas Trust v. Crown Publishers, Inc.,* 592 F.2d at 657 (infringing songs filled one side of five-record set; court affirmed award of 50% of profits because inclusion of infringing songs made record set the only "complete" collection of Scott Joplin's works); *Abkco Music, Inc. v. Harrisongs Music, Ltd.,* 508 F. Supp. 798, 800-801 (S.D.N.Y. 1981) (infringing song reproduced on one side of single record, "flip side" contained noninfringing song, court awarded 70% of profits from sales of the single because infringing song was more popular than noninfringing song; similarly, court awarded 50% of profits for reproduction of same song on album containing twenty-one other songs).

5. The Patent Act does not provide for a disgorgement remedy. What, if anything, might justify the difference in remedies available for copyright infringement? Patent infringement and copyright infringement are both strict liability offenses. However, copyright infringement, unlike patent infringement, requires proof of copying-in-fact. Does that fully justify the availability of the disgorgement remedy? Should the Patent Act be amended to provide for a disgorgement remedy? Would it make sense to limit disgorgement in patent law to instances where the defendant could not show that it did not copy from the patent or the patent owner's patented product?

3. The equitable nature of trademark damages

Under the Lanham Act, a successful infringement or plaintiff is entitled "*subject to the principles of equity*, to recover (1) defendant's profits, (2) any damages sustained by the plaintiff, and (3) the costs of the action." 15 U.S.C. § 1117(a) (emphasis added). The reference to equity in the context of damages—which we typically think of as the classic remedy at law—is a lasting sign of trademark litigation's history, from the era when equity courts were separate from law courts. As the *Restatement (3rd)* explains:

> [b]ecause of the difficulty of establishing the fact and extent of loss attributable to a competitor's unfair competition and the threat of continuing harm, the equitable action for injunctive relief became the preferred remedy. Courts of equity, in order to avoid the need for a separate action at law, sometimes awarded damages or an accounting of the defendant's profits in addition to injunctive relief. These monetary awards were subject to traditional equitable principles such as laches and unclean hands. The modern rules governing the recovery of monetary relief in actions for trademark infringement and unfair competition reflect this history.

Rest. (3rd) Unfair Competition § 36, cmnt. b (1995). Courts thus remain attuned, in the trademark context, to providing a damages remedy that most closely responds to all the facts, and the fairness, of the individual case.

Damages sustained by the mark owner typically include lost sales and other damage to goodwill (if proven), as well as the cost of corrective advertising to restore the value of the mark. Older cases required a finding of willful infringement or bad faith before awarding an accounting of the infringer's profits, *see, e.g., Banff, Ltd. v. Colberts, Inc.,* 996 F.2d 33 (2d Cir. 1993), but courts have interpreted a 1999 statutory change to signal that willfulness is not required for an accounting. *See, e.g., Banjo Buddies, Inc. v. Renosky,* 399 F.3d 168, 174-75 (3d Cir. 2005). However, for a plaintiff to recover damages for a claim of trademark dilution, the diluter must have "willfully intended to" dilute. 15 U.S.C. § 1125(c)(5)(A), (B).

Plaintiffs often prefer to pursue the disgorgement of the defendant's profits, both because this remedy does not entail proving any lost plaintiff sales, and because the Lanham Act, similar to the Copyright Act, expressly gives the plaintiff a strategic advantage in proving this damages theory: "In assessing profits the plaintiff shall be required to prove defendant's sales only; defendant must prove all elements of cost or deduction claimed." 15 U.S.C. § 1117(a).

Tamko Roofing Prods. v. Ideal Roofing Co.
282 F.3d 23 (1st Cir. 2002)

LYNCH, JUDGE:

Tamko Roofing Products, Inc. won its trademark infringement case against Ideal Roofing Company, Ltd. after a six-day jury trial. The district court awarded Ideal's profits to Tamko, ordered Ideal to pay Tamko's attorneys' fees, which amounted to a sum larger than the profits, and issued a permanent injunction. Ideal now appeals each of these district court actions.

We affirm. . . .

I.

The facts are described "as a jury might have found them, consistent with the record but in the light most favorable to the verdict." *Grajales-Romero v. Am. Airlines, Inc.,* 194 F.3d 288, 292 (1st Cir. 1999).

Tamko and Ideal each manufacture and sell roofing products. Tamko manufactures and sells asphalt roofing products, including shingles, in the United States and Canada. Ideal is based in Ottawa, Canada, and manufactures metal roofing and siding products, which it sells in Canada and the United States.

Since 1975, Tamko has been using the trademark "Heritage" in its roofing products business. By 1997, when Ideal began to use the Heritage mark, Tamko had registered ten marks in the Heritage family with the USPTO, including "The American Heritage Series" mark, and two Heritage family trademarks in Canada. Tamko has vigorously defended the Heritage marks, and has successfully enforced its trademark rights.

In April 1997, Ideal selected the trademark "Heritage Series" for hidden fastener metal roofing panels, a new product it introduced to the market later that year. Ideal's "Heritage Series" mark used very similar cursive script to Tamko's "The American Heritage Series" mark. Ideal made the selection through a four-member executive committee: Marcel Laplante (President), René Laplante (Vice President), Pierre Tessier (Sales Manager), and Mark Lebreque (Quebec City Office Manager).

Before Ideal adopted the Heritage Series mark, Tessier attended several roofing trade shows where Tamko prominently displayed its Heritage mark. Ideal hired an advertising agency, Innovacom, to help in the selection and marketing of the new mark. Although the agency usually recommends a trademark search to its clients before they adopt a new mark, René Laplante of Ideal decided against conducting such a search through the agency, an attorney, or Ideal itself. Two other trademarks considered by Ideal were "Carriage" and "Royal Albert," both of which are similar to marks owned by other manufacturers in the roofing industry: Certain-Teed uses the mark "Carriage House," and IKO uses "Royal Victorian."

Although Tamko and Ideal produce and sell different types of roofing products, their products—asphalt and metal roofing respectively—are both appropriate for steep-slope roofs. They compete directly in the roofing industry market, particularly in the northeastern United States. For example, Ideal belongs to the

Metal Roofing Alliance, which, among other things, attempts to persuade home-owners to install metal roofing instead of asphalt shingles. Ideal also tried to per-suade consumers to use metal roofing rather than asphalt shingles in its brochure called "The Smartest Looking House."

When Tamko discovered that Ideal was using the Heritage mark for its new product line, its president, David Humphreys, wrote to Marcel Laplante on March 9, 1999. In the letter, Humphreys discussed the importance of the mark to Tamko, expressed his concern that Ideal's use of the mark would cause "confusion in the marketplace," and asked Ideal to "cease and desist all use of HERITAGE in con-nection with its building products." When Humphreys did not receive a response, he sent another letter to Ideal on March 26, 1999, demanding a response and warn-ing Ideal that if Tamko did not receive a response, it would have "no choice but to seek legal help to resolve this matter." Ideal responded to the second letter, but the companies could not negotiate a mutually agreeable phase out period in which Ideal would stop using the Heritage mark. Ideal wanted a two-year period, while Tamko claimed that a few months would be sufficient.

Tamko gave Ideal notice that it was going to file a suit against it, and that the USPTO had previously rejected another metal roofing manufacturer's application for the Heritage mark. In response, Ideal suggested a one-year phase out as a com-promise.

In August 1999, Tamko filed suit against Ideal for trademark infringement in violation of section 32(1)(a) of the Lanham Act. On November 3, 1999, Tamko filed a motion for a preliminary injunction to enjoin Ideal from using the Heritage mark until the trial resolved the infringement issue. The district court granted the preliminary injunction on February 29, 2000, adopting the report and recommen-dation from the magistrate judge, who was briefed and held an evidentiary hearing on the issue.

Despite the preliminary injunction, Ideal continued to use the Heritage mark in its brochures and on its web site. Ideal distributed brochures containing the Heritage mark at two trade shows which took place in March 2000 in the United States. Ideal also did not modify its web site which contained several references to the Heritage mark. As a result, on March 16, 2000, Tamko moved for contempt. After a hearing, the magistrate judge issued another report and recommendation that "Ideal should be held in contempt," finding Ideal distributed brochures that contained the Heritage mark at a trade show two weeks after the preliminary in-junction issued, and "intentionally kept the 'Heritage' mark on its web site" after the injunction issued. The district court adopted the magistrate judge's report and recommendation and held Ideal in contempt on May 26, 2000. The contempt or-der provided that Ideal would be fined $200 for each day of noncompliance, start-ing on May 29, 2000. Ideal was fined $3,000 for its failure to comply with the contempt order until June 13, 2000.

. . . .

At the end of the trial, the district court ruled that Tamko's Heritage trade-marks were valid. The jury found: (i) "by a preponderance of the evidence that

Ideal infringed Tamko's trademarks"; (ii) "by clear and convincing evidence that Ideal acted willfully in infringing Tamko's trademarks"; and (iii) "by a preponderance of the evidence that the roofing product[s] of Ideal and Tamko directly competed with each other."

After the close of the trial, the judge requested briefing on the issues of an accounting of defendant's profits and attorneys' fees. On August 21, 2000, the district court issued an order that Tamko was entitled to both an accounting of Ideal's profits and attorneys' fees. On October 19, 2000, the court issued an order awarding $201,385.60 in profits. On August 30, 2000, the district court permanently enjoined Ideal from "using the term Heritage, Heritage Series, H Series, or any name or mark confusingly similar to Heritage."

On appeal, Ideal is represented by new counsel. It does not contest the jury's findings, but disputes the district court's rulings. . . . Ideal argues that the district court should not have awarded Tamko an accounting of 100% of Ideal's profits because the two companies did not compete in 100% of their markets, and that the court erred in setting the amount of profits. . . .

II.

. . . .

B. Award to Tamko of Ideal's Profits on the Heritage Series Products

The district court awarded Tamko Ideal's profits of $201,385.60, a sum calculated on the basis of conservative estimates of Ideal's actual profits from the Heritage Series products between November 1997 and February 2000. Ideal argues that no profits should be awarded and the district court committed errors of law; and if any award was justified, this award was too high.

We review de novo the legal standard by which an award of defendant's profits is calculated, and for clear error the factual findings supporting the award.

1. Standard for an Award of Defendant's Profits

The jury found that Tamko and Ideal were in direct competition. Ideal does not argue that the factual finding was unsupported, but does argue that the district court was nonetheless obligated to inquire further as to the percent of direct market overlap in which such competition took place before it could award an accounting of profits. At most, Ideal says, the two companies competed in 20-30% of their business, so it was error to award 100% of Ideal's profits. Ideal's argument is based on product differentiation. That is, Ideal sold only metal roofing, while Tamko sold only asphalt roofing. Only customers with residential steep-slope roofs would consider buying each of the two types of roofing. Ideal says that only 20% of its sales are in this residential market; its remaining sales are in the commercial and agricultural buildings market, where the products do not compete.

The thrust of the argument is essentially that most of the products Ideal sold under the infringing mark should be considered to be noncompeting products and so it is inequitable to award 100% of the profits to Tamko. Under circuit precedent, there may be infringement, as well as an accounting of defendant's profits, even when most of the products are not in competition, if there is evidence, as there was here, of likelihood of confusion. An accounting of defendant's profits may be awarded in a trademark infringement action "subject to the principles of equity." 15 U.S.C. § 1117(a). Here, Tamko did not seek its actual damages, but did seek an accounting as well as injunctive relief. If injunctive relief provides a complete and adequate remedy, then the equities of the case may not require an accounting of profits. For example, injunctive relief may be adequate if there has been no fraud or palming off and there is little likelihood of actual damage to the plaintiff or profit to the defendant.

Trying to fit itself into these shoes, Ideal suggests that injunctive relief should suffice, as it engaged in neither fraud nor palming off. Ideal's argument is misplaced. The presence of injunctive relief does not preclude an accounting here. There was adequate evidence that Tamko did suffer actual damages and that Ideal did benefit from its infringement. To boot, Ideal's contumacious behavior also raises a question of the adequacy of injunctive relief alone as a remedy.

. . . .

This circuit and others have articulated three justifications for awarding to plaintiff an accounting of the defendant's profits: (1) as a rough measure of the harm to plaintiff; (2) to avoid unjust enrichment of the defendant; or (3) if necessary to protect the plaintiff by deterring a willful infringer from further infringement.

Ideal's most cogent argument is that it did not directly compete against Tamko in all the markets in which it profited from use of the mark, and so all of its profits should not go to Tamko. Ideal points to the articulated justification for the . . . rule [that where the products directly compete an accounting of defendant's profits does not require fraud, bad faith, or palming off], which is that the defendant acts as a "trustee" of profits that would otherwise belong to plaintiff. From this, Ideal argues . . . [for] a "but for" rule: a defendant may be deemed to have acted as "trustee" for the plaintiff's profits, which, but for the infringement, would have been made by plaintiff. It follows then, Ideal argues, that it cannot be deemed to have acted as "trustee" for the plaintiff's profits as to the 70-80% of the market where, it says, the two companies were not in direct competition; that is, where asphalt roofing and metal roofing did not compete.

We reject any such limitation for an accounting of profits award, once there has been a finding of direct competition, for three reasons, each articulated in the Lanham Act, 15 U.S.C. § 1117(a).

First, we think once plaintiff has shown direct competition and infringement, the statute places the burden on the infringer to show the limits of the direct competition: "In assessing profits the plaintiff shall be required to prove defendant's sales only; defendant must prove all elements of cost or deduction claimed." 15 U.S.C. § 1117(a). Even before the Lanham Act, the Supreme Court had squarely placed the burden on the infringer "to prove that his infringement had no cash value in sales made by him. If he does not do so, the profits made on sales of goods bearing the infringing mark properly belong to the owner of the mark." *Mishawaka Rubber & Woolen Mfg. Co. v. S.S. Kresge Co.,* 316 U.S. 203, 206-07 (1942). Here, the plaintiff proved the amount of defendant's sales; we consider the defense of only partial direct competition to be very similar to an element of cost or deduction claimed, which defendant has the burden of showing. At trial, Ideal did not ask for a jury finding on the percentage of market overlap. Nor did it present the issue and evidence to the trial judge when he requested briefing from both parties on the accounting of profits. The trial judge was under no obligation to raise or resolve the issue sua sponte. In fact, Tamko disputes Ideal's assertions made on appeal about the percentage of direct competition in the market, and says there was 100% overlap. The place for resolution of this issue was the trial court.

Second, even ignoring momentarily Ideal's waiver at the district court level, the argument is inconsistent with the first rationale for providing an accounting of profits—recompense to plaintiff for the harms it has suffered. Congress recognized that the defendant's profits may be an inexact proxy for the detriment suffered by plaintiffs. Toward this end, the Act also provides:

> If the court shall find that the amount of the recovery based on profits is either inadequate or excessive the court may in its discretion enter judgment for such sum as the court shall find to be just, according to the circumstances of the case. Such sum in * * * the above circumstances shall constitute compensation and not a penalty.

15 U.S.C. § 1117(a). Here, Ideal provided little basis for the district court to conclude that an award of all of defendant's profits was excessive. In addition to its own loss of profits, a plaintiff may, for example, suffer harm to the goodwill associated with its mark. But beyond that, the district court, so long as the sum awarded was not a penalty, was entitled to consider two other policy objectives once it found that defendant's conduct was inequitable: awarding defendant's profits based on unjust enrichment to the defendant, or based on a deterrence theory. *AB Electrolux*, 999 F.2d at 5.

Even assuming that Tamko and Ideal directly compete as to only a portion of Ideal's sales, and even if we were to give Ideal the benefit of plain error review, we could not say that there was an abuse of discretion in awarding defendant's profits in order to avoid unjust enrichment where the infringement was willful. The award itself was conservative. Further, the evidence is that the infringement was willful, intended to divert customers from Tamko, and, importantly, to trade on the goodwill Tamko had established, nurtured, and assiduously guarded in its Heritage mark. There was also evidence of customer confusion. Thus, even in the absence of palming off, it is reasonable to conclude that Ideal was unjustly enriched by trading on Tamko's goodwill beyond the two companies' areas of direct competition.

In cases of at least some direct competition and willfulness, some role may exist for deterrence in an award of an accounting of profits. The role of deterrence must be carefully weighed in light of the statutory prohibition on the imposition of penalties. In an analogous case, one circuit revised an award of 20% of defendant's profits and directed an award of 100% of the profits because it believed that 20% was "clearly inadequate to ensure that similar conduct will not reoccur in the future." *Truck Equip. Serv. Co. v. Fruehauf Corp.,* 536 F.2d 1210, 1223 (8th Cir. 1976). The rule that where there is willful infringement, an accounting of profits is not necessarily restricted to the particular area of direct competition is reinforced by the intention of the Lanham Act to provide *nationwide* protection of a mark, in contrast to the more limited geographic protection afforded by the common law. Tamko had an unusually strong interest in deterrence, given Ideal's track record, and it would not be an abuse of discretion to conclude that the accounting of profits should reflect some recognition of that interest. Nonetheless, the deterrence rationale is primarily served by the attorneys' fees award, and should not be the primary reason for an accounting of profits.

Our third reason for rejecting Ideal's limitation on profits awards is that it is inconsistent with the equitable nature of the court's remedial power. It may well be equitable for a court to include in the damages calculation an award of less than the defendant's complete profits in light of less than complete direct competition. *See, e.g., Truck Equip. Serv. Co.,* 536 F.2d at 1221-22 (awarding defendant's profits only from geographical areas where parties directly competed). On other facts it may be inequitable to give defendants such a benefit. Mechanical rules are of little aid in this analysis.

Equity must take account of the purposes served by the Lanham Act:

> One is to protect the public so it may be confident that, in purchasing a product bearing a particular trade-mark which it favorably knows, it will get the product which it asks for and wants to get. Secondly, where the owner of a trade-mark has spent energy, time, and money in presenting to the public the product, he is protected in his investment from its misappropriation by pirates and cheats.

S. Rep. No. 1333 (1946), *reprinted in* 1946 U.S.C.C.S. 1274, 1274. As another circuit cogently observed in a case raising the issue of less than 100% competition, either through product differentiation or geographical separation: "We think it doubtful whether even the second of these purposes, protection of the trademark owner, is adequately served by a rule which would allow accountings only where the parties directly compete." *Monsanto Chem. Co. v. Perfect Fit Prods. Mfg. Co.,* 349 F.2d 389, 395 (2d Cir. 1965).

The award of Ideal's entire profits was correct.

2. Amount of Award

This still leaves Ideal's attack on the amount of the profits award. This attack is also without merit. The calculation of the award is up to the trial court's discretion, and we will not disturb it unless it rests on clearly erroneous findings of fact, incorrect legal standards, or a meaningful error in judgment.

The court awarded Tamko $201,385.60 as Ideal's profits from the sales of its Heritage Series products. To calculate this amount, the court accepted the $449,522 figure for Ideal's sales of the Heritage Series product in the United States between January 1, 1998 and January 31, 2000, which was provided by René

Laplante (Ideal's Vice President) in response to an interrogatory. The court pro-rated this amount "to account for sales made in November and December of 1997 and February 2000" to arrive at $503,464. The court then subtracted 60% from that amount because "LaPlante previously testified that Ideal's profit margin on sales of its 'Heritage Series' product [was] 40%." Thus, the court arrived at its final figure.

Ideal argues that the court used the wrong numbers for the amount of the costs. The defendant has the burden of producing evidence as to its costs. 15 U.S.C. § 1117(a). Ideal only provided the district court with a conclusory earnings statement which included a number for costs. The court decided that this statement was unreliable without supporting documentation. Instead, it relied on a statement made by Ideal at trial as to its profit margin on the Heritage Series products. This was not an abuse of discretion, and the amount of the accounting award is affirmed.

. . . .

Notes & Questions

1. How did the equitable nature of trademark damages affect the award in *Tamako Roofing*?

2. Why did the court reject the defendant's argument concerning the effect that the lack of competitive overlap between defendant's product and plaintiff's product should have on a disgorgement-of-profits award? Was defendant's problem with the evidence it offered or with the fundamentals of the argument? Did the equitable nature of the remedy affect the court's determination of the proper standard to employ?

B. PROXY MEASURES OF DAMAGES

As the cases above demonstrate, difficulties abound in proving causation and presenting sufficient evidence from which a court can calculate an award of actual damages. All four of the major intellectual property statutes, however, provide alternatives to the type of actual damages explored in the previous subsection.

Both the Patent Act and the Uniform Trade Secrets Act provide a minimum recovery of a reasonable royalty. On the one hand, a reasonable royalty might be seen as a type of "actual damages," in the form of one's best estimate of the amount that the defendant failed to pay the plaintiff for the activity in which the defendant engaged. On the other hand, the plaintiff may not have been interested in permitting the defendant to engage in that use at all, in which case a reasonable royalty is really just a proxy for the harm caused by the defendant's actions.

The Copyright Act, by contrast, allows for certain plaintiffs to recover "statutory damages" as an alternative to actual damages or the infringer's profits. The Act provides a dollar-value range that a court may award as damages, with no requirement to prove that the defendant's actions caused any harm, let alone the amount of the harm caused. With limited exceptions, for a plaintiff to be eligible to elect to receive statutory damages, the plaintiff must have registered her work prior to the infringement commencing. The details of the registration requirement are explored in subsection C, below. Trademark law also uses the statutory damages mechanism, but only for a successful claim of cybersquatting. 15 U.S.C. § 1117(d).

1. A Reasonable Royalty

For reasonable royalty determinations, the courts have long used the construct of a hypothetical negotiation between a willing patentee and a willing licensee, from which the parties would determine the royalty. "A reasonable royalty is the amount that a person, desiring to manufacture, use, or sell a patented article, as a business proposition, would be willing to pay as a royalty and yet be able to make, use, or sell the patented article, in the market, at a reasonable profit." *Trans-World Mfg. v. Al Nyman & Sons, Inc.*, 750 F.2d 1552, 1568 (Fed. Cir. 1984) (internal quotations and alterations omitted).

Although a damages expert can range far and wide over a large set of factual considerations to justify a given royalty rate in her testimony,[*] the courts recognize two touchstones. First, where there is an established royalty rate, it carries great weight. Indeed, where it exists, it "is usually the best measure of a 'reasonable' royalty for a given use of an invention because it removes the need to guess at the terms to which parties would hypothetically agree." *Monsanto Co. v. McFarling*, 488 F.3d 973, 979 (Fed. Cir. 2007). Second, where the accused infringer can establish the cost of using the next-cheapest noninfringing substitute, the cost difference between it and the infringing technology functions as something of a soft cap on the likely royalty rate the parties would have determined in the hypothetical negotiation. Recall that in *Grain Processing, supra*, the availability of a non-infringing substitute resulted in the court's refusal to base the damages award on plaintiff's lost profits. Instead, the court awarded a reasonable royalty and factored the cost of switching to a non-infringing substitute into the royalty rate.

The Federal Circuit has categorically *rejected* what, until recently, was a widely used rule of thumb among expert witnesses on patent damages—the "25% rule"—according to which a hypothetical would-be licensee would agree to pay the patentee 25% of its expected profits from using the patented technology. Canvassing widespread criticism of the rule and its overuse among licensing witnesses, the Federal Circuit held that "as a matter of Federal Circuit law . . . the 25 percent rule of thumb is a fundamentally flawed tool for determining a baseline royalty rate in a hypothetical negotiation. Evidence relying on the 25 percent rule of thumb is thus inadmissible under [the] *Daubert* [case, which governs expert testimony in federal court,] and the Federal Rules of Evidence, because it fails to tie a reasonable royalty base to the facts of the case at issue." *Uniloc USA, Inc. v. Microsoft Corp.*, 632 F.3d 1292, 1315 (Fed. Cir. 2011).

---

[*] The conventional approach—using the so-called *Georgia-Pacific* factors, from *Georgia-Pacific Corp. v. United States Plywood Corp.,* 318 F. Supp. 1116 (S.D.N.Y. 1970)—embraces 15 factors.

The Uniform Trade Secrets Act also permits a recovery of a "reaonsable royalty." The Fifth Circuit has articulated an approach to determining whether a reasonable royalty award is appropriate:

> In *University Computing* [*v. Lykes-Youngstown Corp.*, 504 F.2d 518, 535-36 (5th Cir. 1974)], this court recognized that a reasonable royalty method provides a means of measuring the benefit to the defendant, which is the appropriate measure of damages where the secret has not been destroyed, where the plaintiff is unable to prove specific injury, and where the defendant has gained no actual profits by which to value the worth to the defendant of what it misappropriated. 504 F.2d at 536. In calculating what a reasonable royalty would have been had the parties agreed, the trier of fact should consider the following factors: (1) the resulting and foreseeable changes in the parties' competitive posture; (2) prices paid by licensees in the past; (3) the total value of the secret to the plaintiff, including the plaintiff's development cost and the importance of the secret to the plaintiff's business; (4) the nature and extent of the use the defendant intended for the secret; and (5) whatever other unique factors in the particular case might have been affected by the parties' agreement, such as the ready availability of alternative process. *Metallurgical Indus.,* [*Inc. v. Fourtek, Inc.*, 790 F.2d 1195, 1208 (5th Cir.1986)].

Carbo Ceramics, Inc. v. Keefe, 166 Fed. Appx. 714, 723 (5th Cir. 2006).

Bohnsack v. Varco, L.P.
668 F.3d 262 (5th Cir. 2012)

STEWART, JUDGE:

. . . .

This dispute is between Clyde Bohnsack, a drilling fluids engineer, and Varco, a company that cleans drilling fluids. Drawing on several decades of experience in the industry, Bohnsack invented the "Pit Bull," a machine intended to make the process of cleaning drilling fluids more efficient. Bohnsack and Varco negotiated over the right to manufacture the Pit Bull for several years. After Varco pulled out of these discussions, Bohnsack sued Varco for . . . misappropriation of trade secrets. A jury found for Bohnsack . . . and awarded him compensatory damages [of $600,000. Varco appealed.] . . .

. . . .

. . . After Varco drafted [a] business plan, negotiation of terms for the use of [the] Pit Bull began in earnest. In response to Bohnsack's request for a meeting, [Bill Crabbe, Varco's vice president of marketing,] offered to meet "in mid-March to make a formal offer on a go forward plan of action between [Varco] and yourself." On March 20 and 21, Bohnsack met with several Varco representatives to discuss the financial arrangements that would allow Varco to use the Pit Bull.

An exchange of proposals and counterproposals followed the March meeting. Shortly after the meeting, Crabbe outlined a fifteen-point proposal for financial arrangements between Varco and Bohnsack for Varco's use of the Pit Bull. Bohnsack responded to Crabbe's proposed agreement with a counterproposal on March 28. Crabbe replied to Bohnsack on April 17 with another counterproposal: $150,000 to Bohnsack up front, 15% of revenues to Bohnsack, a $25,000 consulting fee to Bohnsack, and a $100,000 annual payment to Bohnsack for Bohnsack's IP rights until Bohnsack receives $450,000 or until Varco discontinues the product. These payments were capped at $2,750,000. Apparently believing an agreement was close, Crabbe added that once the parties have "agreed in principle," they would send the agreement to Varco's legal department so it could draw up the formal documents. Bohnsack responded with yet another counterproposal, this one very similar to the offer sent by Crabbe. The differences were a lower total cap of $2,450,000 and more specific formulations of several of the terms: Bohnsack proposed that the 15% of revenue belonging to him be calculated "prior to expenses and any tax consequences"; he added "justified travel expenses" to the consulting fee; and he specified the exact dates of the $100,000 annual payments, though it was unclear whether he had agreed to Varco's proposal that those annual payments be contingent on Varco's business decisions. Bohnsack also wrote that his counsel would have to review any formal agreement.

On June 11, Bohnsack asked Crabbe . . . for the status of the Pit Bull negotiations. Bohnsack wrote that he believed they were waiting to receive a document from Varco's lawyers. Crabbe responded that same day, indicating that he had received a draft contract from Varco's legal department that he would review the next day and then send to [Mark Lapeyrouse, then Varco's Vice President for U.S. Operations] for approval. Once Lapeyrouse approved, Crabbe said, Bohnsack would receive the document for final approval. The next day, Crabbe wrote to Bohnsack to tell him that he was not pleased with the draft he had received from the legal department, so he would "get [Bohnsack] a proper document" the following week.

Before Bohnsack received a formal document from Varco memorializing their "agree[ment] in principle,"[10] however, Varco abruptly pulled out of the discussions.

       . . . .

Varco . . . argues that Bohnsack did not present evidence of damages caused by Varco's use of the Pit Bull. Damages in misappropriation cases can take several forms: the value of plaintiff's lost profits, the defendant's actual profits from the use of the secret, *Elcor Chem. Corp. v. Agri–Sul, Inc.*, 494 S.W.2d 204, 214 (Tex. Civ. App. 1973); the value that a reasonably prudent investor would have paid for the trade secret, the development costs the defendant avoided incurring through misappropriation, *Univ. Computing Co. v. Lykes–Youngstown Corp.*, 504 F.2d 518,

---

[10] While Crabbe testified that Varco had never "agreed in principle," the e-mail he wrote in which he said that he would send terms to Varco's legal department once Varco and Bohnsack "agreed in principle" was sufficient evidence for the jury to conclude that, since Crabbe had sent terms to Varco's legal department, an agreement in principle had been reached.

535–36 (5th Cir. 1974) (applying Georgia law);[14] and a "reasonable royalty," *Elcor Chem. Corp.*, 494 S.W.2d at 214. This variety of approaches demonstrates the "flexible" approach used to calculate damages for claims of misappropriation of trade secrets. *See Univ. Computing*, 504 F.2d at 535.

Bohnsack has sufficiently proven that he is entitled to $600,000 in damages for the misappropriation of trade secrets verdict. Varco argues that Bohnsack must prove his precise damages to recover for misappropriation. This is incorrect. *See id.* at 539 ("Where the damages are uncertain, however, we do not feel that the uncertainty should preclude recovery."). A jury need only have sufficient evidence to determine the value a reasonably prudent investor would pay for the trade secret. Here, the final terms negotiated between Varco and Bohnsack are sufficient evidence to prove the value of the Pit Bull to a reasonably prudent investor. Those terms demonstrated Varco's willingness to pay at least $600,000, and possibly much more, for the Pit Bull. The terms were the result of a long, careful process involving significant testing of the Pit Bull and years of negotiation. Further, even if the final terms did not represent a contract, one senior officer at Varco effectively deemed these terms to represent an "agree[ment] in principle." Thus, the jury had sufficient evidence to infer that a reasonably prudent investor would have been willing to pay at least $600,000 for the rights to use the Pit Bull. The district court did not err when it denied judgment as a matter of law on Bohnsack's claim for misappropriation of trade secrets.

. . . .

Notes & Questions

1. An important part of the value of a trade secret or a patent is the competitive advantage it gives the owner. If the owner is interested in fully exploiting that value, it is not going to agree to license the technology to a competitor. In a very real sense, a 'reasonable royalty" award in that situation is entirely fictional. Or is it? Is there a price at which the owner would, nonetheless, being willing to license the technology? If you were counsel for the plaintiff, what type of evidence might you present in order for the court to set an appropriately high royalty rate? Some courts have indicated that the amount a trade secret owner would ask a competitor to pay as a means of ensuring the competitor refuses the offer and thus does not enter the market is not an appropriate measure. *See, e.g., MGE UPS Systems, Inc. v. GE Consumer and Indus., Inc.*, 622 F.3d 361 (5th Cir. 2010).

2. Review *Pioneer Hi-Bred Int'l v. Holden Found. Seeds, Inc., supra*. Why did the court reject the use of a reasonable royalty in that case?

3. One problem with an award of a reasonable royalty is that it may not create enough of a deterrent against wrongful conduct: "Since the imposition of a reasonable royalty requires the defendant to pay only the amount it would have paid

---

[14] Although *University Computing* was applying Georgia law, we have previously described the case as persuasive authority for interpreting Texas law because the misappropriation of trade secrets doctrine in both Georgia and Texas are based on the Restatement of Torts. *See Carbo Ceramics, Inc. v. Keefe*, 166 Fed. Appx. 714, 722 n. 4 (5th Cir. 2006) (unpublished).

had it fairly bargained for use of the plaintiff's secret, it may not adequately discourage the appropriation of trade secrets." *Russo v. Ballard Medical Products*, 550 F.3d 1004 (10th Cir. 2008). How might a legislature address this concern? We explore the possibility of enhanced damages and attorney fees award below in Section D.

2. Statutory Damages

As an alternative to actual damages/lost profits, the Copyright Act provides for a different type of monetary award: statutory damages.

(c) Statutory Damages.—
(1) Except as provided by clause (2) of this subsection, the copyright owner may elect, at any time before final judgment is rendered, to recover, instead of actual damages and profits, an award of statutory damages for all infringements involved in the action, with respect to any one work, for which any one infringer is liable individually, or for which any two or more infringers are liable jointly and severally, in a sum of not less than $750 or more than $30,000 as the court considers just. For the purposes of this subsection, all the parts of a compilation or derivative work constitute one work.
(2) In a case where the copyright owner sustains the burden of proving, and the court finds, that infringement was committed willfully, the court in its discretion may increase the award of statutory damages to a sum of not more than $150,000. In a case where the infringer sustains the burden of proving, and the court finds, that such infringer was not aware and had no reason to believe that his or her acts constituted an infringement of copyright, the court in its discretion may reduce the award of statutory damages to a sum of not less than $200.

17 U.S.C. § 504.

In a cybersquatting case under § 43(d) of the Lanham Act, a plaintiff also has the option to pursue statutory damages rather than prove actual damages. Specifically, "the plaintiff may elect, at any time before final judgment is rendered by the trial court, to recover, instead of actual damages and profits, an award of statutory damages in the amount of not less than $1,000 and not more than $100,000 per domain name, as the court considers just." 15 U.S.C. § 1117(d).

The difficulty in proving damages—both actual damages and the infringer's profits—can often make an award of statutory damages an attractive option. Additionally, statutory damages can sometimes result in a higher monetary award, depending on the facts of the case. How should a court determine the "right" amount of statutory damages to award?

Bryant v. Media Right Productions, Inc.
603 F.3d 135 (2d Cir. 2010)

WOOD, JUDGE:

. . . .

Appellants Anne Bryant and Ellen Bernfeld are songwriters who own a record label, Appellant Gloryvision Ltd (collectively with Bryant and Bernfeld, "Appellants"). In the late 1990s, Appellants created and produced two albums, *Songs for Dogs* and *Songs for Cats* (the "Albums"). They registered the Albums with the United States Copyright Office. They also separately registered at least some of the twenty songs on the Albums.

On February 24, 2000, Appellants entered into an agreement with Media Right ("Media Right Agreement"), which authorized [Appellee] Media Right to market the Albums in exchange for twenty percent of the proceeds from any sales. The Agreement did not grant Media Right permission to make copies of the Albums. If Media Right needed more copies of the Albums, Appellants would provide them.

The Media Right Agreement resulted from conversations between Appellant Ellen Bernfeld ("Bernfeld") and Appellee Douglas Maxwell ("Maxwell"), President of Media Right, during which Maxwell told Bernfeld that Media Right would be distributing music through [Appellee] Orchard [Enterprises, Inc.], a music wholesaler.

Media Right entered into an agreement with Orchard on February 1, 2000 ("Orchard Agreement"). The Orchard Agreement authorized Orchard to distribute on Media Right's behalf eleven albums listed in the Agreement, two of which were the Albums (apparently in anticipation of the Media Right Agreement). The Orchard Agreement provided, in relevant part, that:

> [Media Right] grant[s] [Orchard] * * * non-exclusive rights to sell, distribute and otherwise exploit * * * [Media Right's albums] by any and all means and media (whether now known or existing in the future), including * * * throughout E-stores including * * * those via the Internet, as well as all digital storage, download and transmission rights, whether now known or existing in the future.

In the Orchard Agreement, Media Right warranted that Orchard's use of the Albums would not infringe any copyrights. Maxwell gave Orchard physical copies of the Albums, which bore copyright notices stating that the copyrights for the Albums were held by Appellants.

When Media Right entered into the Orchard Agreement in 2000, Orchard sold only physical copies of recordings. In about April 2004, however, Orchard began making digital copies of the Albums to sell through internet-based music retailers such as iTunes. Internet customers were able to purchase and download digital copies of the Albums and individual songs on the Albums. Orchard did not inform Media Right or Appellants that it was selling digital copies of the Albums and individual songs on the Albums.

From April 1, 2002 to April 8, 2008, Orchard generated $12.14 in revenues from sales of physical copies of the Albums, and $578.91 from downloads of digital copies of the Albums and of individual songs. Media Right's share of these revenues was $413.82, of which $331.06 should have been forwarded to Appellants pursuant to the Media Right Agreement. Because the $413.82 was aggregated with other monies Orchard paid to Media Right, Media Right overlooked that it owed a portion of the payments to Appellants. Media Right, therefore, did not pay Appellants the $331.06 to which they were entitled.

In 2006, Appellants discovered that digital copies of the Albums were available online. On April 16, 2007, Appellants filed a complaint against Appellees in the Southern District of New York, alleging direct and contributory copyright infringement, and seeking statutory damages.

In 2008, Appellants and Appellees both moved for summary judgment in the case. They agreed to permit the District Court to treat the motions as a case stated. The Court conducted two evidentiary hearings before issuing its order. The Court held, in relevant part, that Appellees had committed direct copyright infringement by making and selling digital copies of the Albums and the individual songs on the Albums.

The Court awarded Appellants statutory damages in the total amount of $2400, pursuant to Section 504 of the Copyright Act of 1976 (the "Act"). 17 U.S.C. § 504(c)....

The District Court made the following three rulings regarding damages, all of which Appellants contest on appeal.

First, the Court held that the Albums were compilations, and thus that each Appellee was liable for only one award of statutory damages per Album, rather than one award per song, as Appellants had sought.

Second, the Court found that Orchard had proven that its infringement was innocent, and thus ordered Orchard to pay only minimal statutory damages of $200 per Album, for a total of $400.

Third, the Court found that Maxwell and Media Right had failed to prove that their infringement was innocent, but that Appellants had failed to prove that Maxwell and Media Right's infringement was willful. The Court found that because neither side had met its burden of proof, and because Appellees' revenues from the Albums were very low, Media Right and Maxwell were jointly and severally liable for an award of only $1000 per Album, for a total of $2000.

The Court did not award Appellants attorneys fees. Accordingly, the total award to Appellants was $2400. This appeal followed.

II. *Discussion*

Appellants argue that we should vacate the District Court's statutory damage award, contending that: (1) the Court erred in refusing to grant a separate statutory damage award for each song on the Albums; (2) the Court erred in its findings on intent; and (3) the Court erred in determining the amount of damages. Appellants also argue that the Court abused its discretion by refusing to award them attorneys' fees. We address each of these arguments in turn.

A. *The District Court's Decision to Award Statutory Damages on a Per-Album Basis*

Appellants contend that the District Court erred in holding that the Albums were compilations, and thus limiting statutory damages to one award for each Album. Appellants argue that each song on the Albums qualifies as a separate work because, according to Appellants, each song is separately copyrighted, and because Orchard sold the songs individually.

. . . .

The Copyright Act allows only one award of statutory damages for any "work" infringed. 17 U.S.C. § 504(c)(1). It states that "all the parts of a compilation ... constitute one work." *Id*. § 504(c)(1). It defines a "compilation" as "a work formed by the collection and assembling of preexisting materials or of data that are selected, coordinated, or arranged in such a way that the resulting work as a whole constitutes an original work of authorship." *Id*. § 101. The term compilation includes collected works, which are defined as works "in which a number of contributions, constituting separate and independent works in themselves, are assembled into a collective work." *Id*. The Conference Report that accompanied the Act and explains many of its provisions, states that a "compilation" "results from a process of selecting, bringing together, organizing, and arranging previously existing material of all kinds, *regardless of whether * * * the individual items in the material have been or ever could have been subject to copyright.*" H.R. Rep. No. 1476, 94th Cong., 2d Sess. 162, *reprinted in* 1976 U.S.C.C.A.N. 5659 (emphasis added).

An album falls within the Act's expansive definition of compilation. An album is a collection of preexisting materials–songs–that are selected and arranged by the author in a way that results in an original work of authorship–the album. Based on a plain reading of the statute, therefore, infringement of an album should result in only one statutory damage award. The fact that each song may have received a separate copyright is irrelevant to this analysis. *See* H.R. Rep. No. 1476, 94th Cong., 2d Sess. 162, *reprinted in* 1976 U.S.C.C.A.N. 5659.

We have addressed in two previous decisions the issue of what constitutes a compilation subject to Section 504(c)(1)'s one-award restriction. *See Twin Peaks Prods., Inc. v. Publ'ns. Int'l Ltd.,* 996 F.2d 1366, 1381 (2d Cir. 1993); *WB Music Corp. v. RTV Comm. Group, Inc.,* 445 F.3d 538, 541 (2d Cir. 2006). In both decisions, we focused on whether the plaintiff—the copyright holder—issued its works separately, or together as a unit.

In *Twin Peaks,* the plaintiff issued each episode of a television series sequentially, each at a different time. The *defendant* printed eight teleplays from the series in one book. 996 F.2d at 1381. We held that the plaintiff could receive a separate award of statutory damages for each of the eight teleplays because the *plaintiff* had issued the works separately, as independent television episodes.[5] In *WB Music Corp.*, the plaintiff had separately issued each of thirteen songs. 445 F.3d at 541. It was the *defendant* who issued the songs in album form. We held that the plaintiff could receive a separate statutory damage award for each song, because there was "no evidence * * * that any of the separately copyrighted works were included in a compilation authorized by the [*plaintiff*]." *Id.* (emphasis added).

Here, it is the copyright holders who issued their works as "compilations"; they chose to issue Albums. In this situation, the plain language of the Copyright Act limits the copyright holders' statutory damage award to one for each Album.[6]

Appellants argue that the District Court should have allowed a statutory damage award for each song, because each song has "independent economic value": internet customers could listen to and purchase copies of each song, each of which Appellants claim was independently copyrighted. Plaintiffs point to a decision from the First Circuit, *Gamma Audio,* in which the Court held that a work that is part of a multi-part product can constitute a separate work for the purposes of statutory damages if it has "independent economic value and * * * is viable." 11 F.3d at 1116-17. Applying what that court described as a "functional" test, the court held that each episode of a television show, al-though released on videotape as part of a complete series, could be the subject of a separate statutory damage award because each episode *could* be rented and viewed separately. *Id.* at 1117-18. At least three other circuits have adopted the "independent economic value" test, although to date none has applied the test to an album of music. *See MCA Television Ltd. v. Feltner,* 89 F.3d 766, 769 (11th Cir. 1996) (holding that each episode of a television show can be the subject of a separate statutory damage award because each episode has independent economic value); *Walt Disney Co. v. Powell,* 897 F.2d 565, 569 (D.C. Cir. 1990) (holding that plaintiff could not receive a separate statutory damage award for each, separate picture of Mickey Mouse and Minnie Mouse in different poses, because each picture did not have independent economic value). Appellants argue that it is particularly appropriate to apply the "independent economic value" test to music albums, because music is increasingly

---

[5] We also relied on the facts that the episodes were separately written and separately produced. *Twin Peaks,* 996 F.2d at 1381. We held that, although the episodes taken together had a common plot line ("Who killed Laura Palmer?"), that did not suffice to render the episodes a "compilation."

[6] The few district courts that have considered whether a compilation is subject to only one statutory damage award have reached the same conclusion. *See UMG Recordings, Inc. v. MP3.COM, Inc.,* 109 F. Supp. 2d 223, 225 (S.D.N.Y. 2000) (Rakoff, J.) (finding that where the infringed works were albums issued by the plaintiff, statutory damages should be awarded on a per-album basis); *Country Road Music, Inc. v. MP3.com, Inc.,* 279 F. Supp. 325, 332 (S.D.N.Y. 2003); *Arista Records, Inc. v. Flea World, Inc.,* Civ. No. 03-2670, 2006 WL 842883, at *21 (D.N.J. Mar. 31, 2006); *see also Xoom, Inc. v. Imageline, Inc.,* 323 F.3d 279, 285 (4th Cir. 2003) (holding that plaintiff could receive only one statutory damage award for its computer clip art software, which contained many individual pieces of clip art, because plaintiff had packaged and sold the clip art in one piece of software, and thus the software constituted a compilation); *Stokes Seeds Ltd. v. Geo. W. Park Seed Co.,* 783 F. Supp. 104, 106 (W.D.N.Y. 1991) (holding that catalog containing many separately copyrighted photographs of plant seedlings constituted a compilation because plaintiff had assembled the photographs into a compilation (the catalog)).

available in digital form, which has made it easier for infringers to break apart albums and sell the album's songs individually, as Appellees did here.

This Court has never adopted the independent economic value test, and we decline to do so in this case. The Act specifically states that all parts of a compilation must be treated as one work for the purpose of calculating statutory damages. This language provides no exception for a part of a compilation that has independent economic value, and the Court will not create such an exception. *See UMG Recordings, Inc.,* 109 F. Supp. 2d at 225 (stating that to award statutory damages on a per-song basis would "make a total mockery of Congress' express mandate that all parts of a compilation must be treated as a single 'work' for purposes of computing statutory damages"). We cannot disregard the statutory language simply because digital music has made it easier for infringers to make parts of an album available separately. . . .

. . . .

B. *The District Court's Decision on Intent and the Amount of Damages*

Appellants contend that the District Court erred in finding that Orchard proved that its conduct was innocent, and that Appellants failed to prove that Appellees' conduct amounted to willful infringement.

Pursuant to Section 504(c)(2) of the Copyright Act, an infringer's intent can affect the amount of statutory damages awarded: only a minimal award may be warranted where the infringement is innocent; a higher award may be warranted where the infringer acted willfully.

We review the district court's findings on intent for clear error. The burden of proving innocence is on the alleged infringer. The burden of proving willfulness is on the copyright holder. *See* 17 U.S.C. § 504(c).

1. *Innocence*

The District Court found that Orchard acted innocently because Orchard, in making digital copies, reasonably relied on two provisions of the Orchard Agreement: (1) a provision permitting Orchard to distribute the Albums "by any and all means and media * * * including digital storage, download and transmission"; and (2) a provision warranting that Orchard's use of the Albums in accordance with the Agreement would not infringe any copyrights.

We hold that it was not clear error for the District Court to find that it was reasonable for Orchard to believe that it had received the right to copy the Albums.

2. *Willfulness*

A copyright holder seeking to prove that a copier's infringement was willful must show that the infringer "had knowledge that its conduct represented infringement or * * * recklessly disregarded the possibility." *Twin Peaks,* 996 F.2d at 1382.

The District Court found that Appellees did not prove that Maxwell and Media Right acted willfully in infringing Appellees' copyright. The District Court found that it was not unreasonable for Maxwell not to have anticipated that Orchard would distribute digital copies of the Albums, notwithstanding that the Orchard Agreement granted Orchard the right to do so, because Orchard did not distribute digital music in 2000, when the Orchard Agreement was signed. The District Court also found credible Maxwell's testimony at the evidentiary hearing that he had never before marketed recordings that were not his own, and that, in allowing Orchard broad distribution rights, he focused only on his belief that Appellants wanted him to do everything possible to market their Albums. This testimony shows that Maxwell did not have experience marketing music owned by a third party; that he did not fully understand the rights he had obtained under the Media Right Agreement; and that his focus was on maximizing sales of the Albums.

We hold that it was not clear error for the District Court to find that Maxwell and Media Right's infringement was not willful.

C. *The District Court's Calculation of Statutory Damages*

Appellants also argue that the statutory damages awarded by the District Court were too low. District courts "enjoy wide discretion * * * in setting the amount of statutory damages." *Fitzgerald Pbl'g Co. [v. Baylor Pbl'g Co.*, 807 F.2d 1110,] 1116 [(2d Cir. 1986)]. We review for clear error the District Court's factual findings supporting its determination of the appropriate level of statutory damages, and we review an award of those damages for abuse of discretion.

When determining the amount of statutory damages to award for copyright infringement, courts consider: (1) the infringer's state of mind; (2) the expenses saved, and profits earned, by the infringer; (3) the revenue lost by the copyright holder; (4) the deterrent effect on the infringer and third parties; (5) the infringer's cooperation in providing evidence concerning the value of the infringing material; and (6) the conduct and attitude of the parties. *See N.A.S. Impor. Corp. v. Chenson Enter., Inc.,* 968 F.2d 250, 252-53 (2d Cir. 1992).

The District Court awarded a total of $2400 in statutory damages, based on its finding that Appellees' profits from infringing sales of the Albums and songs were meager, and that the award did not need to be higher to achieve deterrence, because deterrence was effectuated here by Appellees having to pay their own attorneys fees. We hold that the District Court did not abuse its discretion in calculating statutory damages.

. . . .

III. *Conclusion*

For the reasons stated above, the order of the District Court is AFFIRMED.

Notes & Questions

1. The Copyright Act is clear: plaintiffs who satisfy the registration requirement may elect to recover either actual damages or statutory damages. Why did the plaintiff in *Bryant* pursue statutory damages?

2. Despite the statute's language directing *courts* to award statutory damages, there is a right to a jury trial under the Seventh Amendment on the issue of statutory damages. *Feltner v. Columbia Pictures Television, Inc.,* 523 U.S. 340 (1998). Why did the parties not seek a jury trial in this case?

3. Copyright infringement is a strict liability offense; there is no "mental state" requirement to be liable for infringement. Proof of innocent infringement can, however, have consequences for an award of statutory damages, and the statute indicates that innocence might also affect an award of actual damages. *See* 17 U.S.C. §§ 401, 402. The maximum statutory damage award for willful infringement is $150,000 per work infringed. What test did the court employ to determine whether the defendants were willful infringers? We return to the issue of willful infringement in Subsection D below. What state of mind does a defendant have if he is neither innocent nor willful?

4. A court must determine what constitutes one work for purposes of statutory damages, but it also must pick a damages number within the statutory range of permissible damages. A recent example demonstrates the wide range of possibilities. The RIAA sued Jamie Thomas for her peer-to-peer file sharing activities. A jury found her liable for infringement of 24 sound recordings and determined that her infringement was willful. It awarded $9,250 for each work, for a total of $222,000. When the court granted a new trial based on an erroneous jury instruction related to the issue of liability, a second jury found she had willfully infringed and awarded $80,000 per song, for a total of $1.9 million. When the court granted a remittitur, conditionally reducing the award to $2,250 per song, the plaintiffs exercised their right to reject the remittitur in favor of a new trial, and a third trial was held. The jury in that trial awarded $1.5 million. *Capitol Records, Inc. v. Thomas-Rasset,* 692 F.3d 899 (8th Cir. 2012) (reinstating the award amount of $222,000 from the first trial pursuant to the plaintiff's request).

5. As explored more fully below, Subsection C, in most instances the Copyright Act requires registration of the copyright prior to the infringement in order for a plaintiff to be entitled to elect statutory damages. In this way, statutory damages are an important carrot to promptly registering a copyright. The proxy award of a "reasonable royalty" is available in all patent and trade secret cases. Does this difference make sense? What happens when actual damages are extremely difficult or impossible to prove, the plaintiff did not timely register his copyright, and the defendant infringed? What should the measure of damages be in that situation? The next case confronts that scenario.

Davis v. The Gap, Inc.
246 F.3d 152 (2d Cir. 2001)

LEVAL, JUDGE:

    . . . .

Davis is the creator and designer of nonfunctional jewelry worn over the eyes in the manner of eyeglasses. The Gap, Inc. is a major international retailer of clothing and accessories marketed largely to a youthful customer base with annual revenues of several billions of dollars. It operates several chains of retail stores, some under the name "Gap." It is undisputed that the Gap, without Davis's permission, used a photograph of an individual wearing Davis's copyrighted eyewear in an advertisement for the stores operating under the "Gap" trademark that was widely displayed throughout the United States.

Davis brought this action seeking a declaratory judgment of infringement and damages, including $2,500,000 in unpaid licensing fees, a percentage of the Gap's profits, punitive damages of $10,000,000, and attorney's fees. The district court granted summary judgment for the Gap. . . .

BACKGROUND

Davis has created at least fifteen different designs of eye jewelry, which he markets under the name "Onoculii Designs." Davis describes Onoculii eyewear as "sculptured metallic ornamental wearable art." Am. Compl. ¶ 7. Each piece is made of gold, silver, or brass, and is constructed in a manner similar to eyeglasses (a frame hinged to templates that hook over the ears), but with very different effect. The frames support decorative, perforated metallic discs or plates in the place that would be occupied by the lenses of a pair of eyeglasses. The discs effectively conceal the wearer's eyes, although the perforations permit the wearer to see through them. Some of Davis's designs are of flowery or abstract filagree shapes, some are crescents with protruding spokes or wings. The particular piece that gives rise to this action consists of a horizontal bar at the level of the eyebrows from which are suspended a pair of slightly convex, circular discs of polished metal covering the eyes, perforated with dozens of tiny pinprick holes. Davis registered his copyright for the design at issue, effective May 16, 1997.

. . . .

In May 1996, prior to Davis's registration of his copyright, the defendant created a series of advertisements showing photographs of people of various lifestyles wearing Gap clothing. The campaign was designed to promote the concept that Gap merchandise is worn by people of all kinds. The ad in question, which bears the caption "fast" emblazoned in red (the "fast" ad), depicts a group of seven young people probably in their twenties, of Asian appearance, standing in a loose V formation staring at the camera with a sultry, pouty, provocative look. The group projects the image of funky intimates of a lively afterhours rock music club. They are dressed primarily in black, exhibiting bare arms and partly bare chests, goatees (accompanied in one case by bleached, streaked hair), large-brimmed, Western-style hats, and distinctive eye shades, worn either over their eyes, on their hats, or cocked over the top of their heads. The central figure, at the apex of the V formation, is wearing Davis's highly distinctive Onoculii eyewear; he peers over the metal disks directly into the camera lens.

The "fast" photograph was taken by the Gap in May 1996 during a photo shoot in the Tribeca area of Manhattan. The defendant provided the subjects with Gap apparel to wear for the shoot, and a trailer in which to change. The Gap claims that it did not furnish eyewear to any of the subjects, and that the subjects were told to wear their own eyewear, wristwatches, earrings, nose-rings or other incidental items, thereby "permitting each person to project accurately his or her own personal image and appearance."

The Gap's "fast" advertisement was published in a variety of magazines, including *W, Vanity Fair, Spin, Details*, and *Entertainment Weekly*. Davis claims that the total circulation of these magazines was over 2,500,000. For five weeks during August and September of 1996, the advertisement was displayed on the sides of buses in New York, Boston, Chicago, San Francisco, Atlanta, Washington, D.C., and Seattle. The advertisement may also have been displayed on bus shelters. According to Davis, when used on buses the photograph was cropped so that only the heads and shoulders of the subjects were shown.

. . . [*Eds. Note*: Later in the opinion, the court rejects the defendant's infringement defenses of "de minimis" use and fair use.]

DISCUSSION

. . . .

B. *Compensatory Damages*

17 U.S.C. § 504 imposes two categories of compensatory damages. Taking care to specify that double recovery is not permitted where the two categories overlap, the statute provides for the recovery of both the infringer's profits and the copyright owner's "actual damages." It is important that these two categories of compensation have different justifications and are based on different financial data. The award of the infringer's profits examines the facts only from the infringer's point of view. If the infringer has earned a profit, this award makes him disgorge the profit to insure that he not benefit from his wrongdoing. The award of the owner's actual damages looks at the facts from the point of view of they copyright owner; it undertakes to compensate the owner for any harm he suffered by reason of the infringer's illegal act.

. . . [T]he district court believed Davis failed to show any causal connection between the infringement and the defendant's profits. With respect to Davis's claim of entitlement to "actual damages" based on the license fee he should have been paid for the Gap's unauthorized use of his copyrighted material, the district court believed that his evidence was too speculative . . . .

We agree with the district court as to the defendant's profits, but not as to Davis's claim for damages based on the Gap's failure to pay him a reasonable license fee.

1. *Infringer's profits*

Davis submitted evidence that, during and shortly after the Gap's advertising campaign featuring the "fast" ad, the corporate parent of the Gap stores realized net sales of $1.668 billion, an increase of $146 million over the revenues earned in the same period of the preceding year. The district court considered this evidence inadequate to sustain a judgment in the plaintiff's favor because the overall revenues of the Gap, Inc. had no reasonable relationship to the act of alleged infringement. Because the ad infringed only with respect to Gap label stores and eyewear, we agree with the district court that it was incumbent on Davis to submit evidence at least limited to the gross revenues of the Gap label stores, and perhaps also limited to eyewear or accessories. Had he done so, the burden would then have shifted to the defendant under the terms of § 504(b) to prove its deductible expenses and elements of profits from those revenues attributable to factors other than the copyrighted work.

It is true that a highly literal interpretation of the statute would favor Davis. It says that "the copyright owner is required to present proof only of the infringer's gross revenue," 17 U.S.C. § 504(b), leaving it to the infringer to prove what portions of its revenue are not attributable to the infringement. Nonetheless we think the term "gross revenue" under the statute means gross revenue reasonably related to the infringement, not unrelated revenues.

Thus, if a publisher published an anthology of poetry which contained a poem covered by the plaintiff's copyright, we do not think the plaintiff's statutory burden would be discharged by submitting the publisher's gross revenue resulting from its publication of hundreds of titles, including trade books, textbooks, cookbooks, etc. In our view, the owner's burden would require evidence of the revenues realized from the sale of the anthology containing the infringing poem. The publisher would then bear the burden of proving its costs attributable to the anthology and the extent to which its profits from the sale of the anthology were attributable to factors other than the infringing poem, including particularly the other poems contained in the volume. The point would be clearer still if the defendant publisher were part of a conglomerate corporation that also received income from agriculture, canning, shipping, and real estate development. While the burden-shifting statute undoubtedly intended to ease plaintiff's burden in proving the defendant's profits, we do not believe it would shift the burden so far as to permit a plaintiff in such a case to satisfy his burden by showing gross revenues from agriculture, canning, shipping and real estate where the infringement consisted of the unauthorized publication of a poem. The facts of this case are less extreme; nonetheless, the point remains the same: the statutory term "infringer's gross revenue" should not be construed so broadly as to include revenue from lines of business that were unrelated to the act of infringement.

. . . .

2. The copyright owner's actual damages: Davis's failure to receive a reasonable licensing fee

Among the elements Davis sought to prove as damages was the failure to receive a reasonable license fee from the Gap for its use of his copyrighted eyewear. The complaint asserted an entitlement to a $2.5 million licensing fee. . . .

a. *Was Davis's evidence too speculative?*

While there was no evidence to support Davis's wildly inflated claim of entitlement to $2.5 million, in our view his evidence did support a much more modest claim of a fair market value for a license to use his design in the ad. In addition to his evidence of numerous instances in which rock music stars wore Onoculii eyewear in photographs exhibited in music publications, Davis testified that on one occasion he was paid a royalty of $50 for the publication by *Vibe* magazine of a photo of the deceased musician Sun Ra wearing Davis's eyewear.

On the basis of this evidence, a jury could reasonably find that Davis established a fair market value of at least $50 as a fee for the use of an image of his copyrighted design. This evidence was sufficiently concrete to support a finding of fair market value of $50 for the type of use made by *Vibe*. And if Davis could show at trial that the Gap used the image in a wider circulation than *Vibe*, that might justify a finding that the market value for the Gap's use of the eyewear was higher than $50. Therefore, to the extent the district court dismissed the case because Davis's evidence of the market value of a license fee was too speculative, we believe this was error.

. . . .

. . . The Gap was not seeking . . . to surreptitiously steal material owned by a competitor. There is no reason to suppose that the Gap's use of Davis's copyrighted eyewear without first receiving his permission was attributable to anything other than oversight or mistake. To the contrary, the facts of this case support the view that the Gap and Davis could have happily discussed the payment of a fee, and that Davis's consent, if sought, could have been had for very little money, since significant advantages might flow to him from having his eyewear displayed in the Gap's ad. Alternatively, if Davis's demands had been excessive, the Gap would in all likelihood have simply eliminated Davis's eyewear from the photograph. . . .

c. *Actual damages under § 504(a) and (b)*

. . . [W]e proceed to consider whether that measure of damages is permissible under the statute. The question is as follows: Assume that the copyright owner proves that the defendant has infringed his work. He proves also that a license to make such use of the work has a fair market value, but does not show that the infringement caused him lost sales, lost opportunities to license, or diminution in the value of the copyright. The only proven loss lies in the owner's failure to receive payment by the infringer of the fair market value of the use illegally appropriated. Should the owner's claim for "actual damages" under § 504(b) be dismissed? Or should the court award damages corresponding to the fair market value of the use appropriated by the infringer?

Neither answer is entirely satisfactory. If the court dismisses the claim by reason of the owner's failure to prove that the act of infringement cause economic harm, the infringer will get his illegal taking for free, and the owner will be left uncompensated for the illegal taking of something of value. On the other hand, an award of damages might be seen as a windfall for an owner who received no less than he would have if the infringer had refrained from the illegal taking. In our view, the more reasonable approach is to allow such an award in appropriate circumstances.

. . . .

If a copier of protected work, instead of obtaining permission and paying the fee, proceeds without permission and without compensating the owner, it seems entirely reasonable to conclude that the owner has suffered damages to the extent of the infringer's taking without paying what the owner was legally entitled to exact a fee for. We can see no reason why, as an abstract matter, the statutory term "actual damages" should not cover the owner's failure to obtain the market value of the fee the owner was entitled to charge for such use.

The problem is roughly analogous to illegal takings or uses in other contexts outside the realm of copyright. For example:

(a) D, who lives on property adjacent to P, without authorization regularly swims and canoes in P's lake and uses a road crossing P's land because it provides more direct access to town. The right to use P's property for such purposes has a fair market value. P proves neither harm to his property nor loss of opportunity to license others to use the property for such recreation. Nonetheless P has lost the revenue he would have recovered if D had paid the fair market value of what he took.

(b) P, the owner of a baseball stadium charges $50 for admission to games. D, through a corrupt arrangement with a stadium usher, sneaks in free on days when the stadium has excess seating capacity. P shows no economic injury inflicted by D's free entrance, other than the hypothetical loss of the revenue for the tickets D did not purchase.

(c) P Telephone Company charges set rates for calls on its lines. D devises an electronic "black box" that emits signals mimicking the telephone company's codes, enabling D to place calls without charge. D places free calls at late-night, offpeak hours when the telephone lines are underutilized, and P is unable to prove any loss inflicted by the unauthorized use, other than its failure to collect its regular rates.

(d) P is a public transportation system. D, a subway rider, jumps the turnstile and avoids paying the normal $1.50 fee.

(e) P, a manufacturing company in reorganization, owns (or leases) warehousing space in a warehouse facility. Because of P's financial and legal difficulties, the space has been unused for some time. D company, which leases adjacent warehouse space, noting that P's space is not being used, secretly uses P's space to warehouse its own inventories.

In each of these cases, the defendant has surreptitiously taken a valuable right, for which plaintiff could have charged a reasonable fee. Plaintiff's revenue is thus smaller than it would have been if defendant had paid for what he took. On the other hand, plaintiff's revenue is no less than it would have been if the defendant had refrained from the taking. In our view, as between leaving the victim of the illegal taking with nothing, and charging the illegal taker with the reasonable cost of what he took, the latter, at least in some circumstances, is the preferable solution.

It is important to note that under the terms of § 504(b), unless such a foregone payment can be considered "actual damages," in some circumstances victims of infringement will go uncompensated. If the infringer's venture turned out to be unprofitable, the owner can receive no recovery based on the statutory award of the "infringer's profits." And in some instances, there will be no harm to the market value of the copyrighted work. The owner may be incapable of showing a loss of either sales or licenses to third parties. To rule that the owner's loss of the fair market value of the license fees he might have exacted of the defendant do not constitute "actual damages," would mean that in such circumstances an infringer may steal with impunity. We see no reason why this should be so. Of course, if the terms of the statute compelled that result, our perception of inequity would make no difference; the statute would control. But in our view, the statutory term "actual damages" is broad enough to cover this form of deprivation suffered by infringed owners.

We recognize that awarding the copyright owner the lost license fee can risk abuse. Once the defendant has infringed, the owner may claim unreasonable amounts as the license fee—to wit Davis's demand for an award of $2.5 million. The law therefore exacts that the amount of damages may not be based on "undue speculation." The question is not what the owner would have charged, but rather what is the fair market value. In order to make out his claim that he has suffered actual damage because of the infringer's failure to pay the fee, the owner must show that the thing taken had a fair market value. But if the plaintiff owner has done so, and the defendant is thus protected against an unrealistically exaggerated claim, we can see little reason not to consider the market value of the uncollected license fee as an element of "actual damages" under § 504(b).[5]

We recognize also that finding the fair market value of a reasonable license fee may involve some uncertainty. But that is not sufficient reason to refuse to consider this as an eligible measure of actual damages. Many of the accepted methods of calculating copyright damages require the court to make uncertain estimates in the realm of contrary to fact. A classic element of the plaintiff's copyright damages is the profits the plaintiff would have earned from third parties, were it not for the infringement. This measure requires the court to explore the counterfactual hypothesis of the contracts and licenses the plaintiff would have made absent the infringement and the costs associated with them. A second accepted method, focusing on the "infringer's profits," similarly requires the court to explore circumstances that are counterfactual. The owner's entitlement to the infringer's profits is limited to the profits "attributable to the infringement." 17 U.S.C. § 504(b). The court, therefore, must compare the defendant's actual profits to what they would have been without the infringement, awarding the plaintiff the difference. Neither of these approaches is necessarily any less speculative than the approach that requires the court to find the market value of the license fee for what the infringer took. Indeed, it may be far less so. Many copyright owners are represented by agents who have established rates that are regularly paid by licensees. In such cases, establishing the fair market value of the license fee of which the owner

---

[5] Furthermore, the fair market value to be determined is not of the highest use for which plaintiff might license but the use the infringer made. Thus, assuming the defendant made infringing use of a Mickey Mouse image for a single performance of a school play before schoolchildren, teachers and parents with tickets at $3, the fair market value would not be the same as the fee customarily charged by the owner to license the use of this image in a commercial production.

was deprived is no more speculative than determining the damages in the case of a stolen cargo of lumber or potatoes. Given our long-held view that in assessing copyright damages "courts must necessarily engage in some degree of speculation," some difficulty in quantifying the damages attributable to infringement should not bar recovery.

. . . .

Notes & Questions

1. Unlike the Patent Act, the Copyright Act does not mention the possibility of an award of a reasonable royalty as a minimum measure of damages. Should that omission from the statute bar the use of "the fair market value of a reasonable license fee" as an aspect of "actual damages"? In determining the amount to award, should it matter whether the plaintiff would have objected to licensing the inclusion of his work in a particular advertisement campaign?

2. Clearly in *Davis* the Second Circuit was concerned with leaving the plaintiff whose work has been infringed with no remedy. Statutory damages require no proof of a damages amount, but require the copyright to have been registered before the accused infringement began. Thus, Davis was not eligible to receive an award of statutory damages. In a portion of the opinion not reproduced above, the Second Circuit rejected the possibility that Congress, as a way to encourage prompt registration, chose to leave those who delay in registering their copyrights with no remedy if they can't prove actual harm or the infringer's ill-gotten gains. Such a result would certainly provide an incentive for prompt registration. Registering a claim to an entitlement such as a copyright generates the benefit of creating a clear registry that provides information about who is interested in protecting their rights along with information concerning who to contact for obtaining a license to use the work. Was the Second Circuit right to reject a result that would have led to no remedy for the copyright owner? Does the requirement of showing "that the thing taken had a fair market value" provide sufficient protection for the plaintiff? *See Dash v. Mayweather*, 731 F.3d 303, 333 n.6 (4th Cir. 2013) (rejecting an award of a lost license fee award because plaintiff had failed to present "sufficient nonspeculative evidence to show that [plaintiffs' infringed work] had a fair market value").

3. Given the problems of proof in *Davis v. The Gap*, if Davis had been eligible for statutory damages, what would the maximum recovery have been for a statutory damages award?

4. Should the court in *Frank Music* have considered awarding a lost license fee, similar to what the court approved in *Davis*? Why did the court in *Davis* need to resort to an award of a lost license fee?

5. If an award of actual damages is meant to make the plaintiff whole, would an award of Davis' $50 license fee really do that? What about the attorney's fees he paid to file the lawsuit and bring it to judgment? The Copyright Act does contain a fee shifting provision, but for a prevailing copyright owner to be eligible for an award of attorney's fees, the copyright must be registered prior to infringement

commencing. Would the "carrot" of the ability to recover attorney fees, alone, be a sufficient incentive to register?

C. NOTICE, REGISTRATION, AND LIMITATIONS PERIODS

1. Marking & Notice

The Patent Act precludes damages for product-claim infringement during the time before which the patentee put the accused infringer on notice. Specifically,

> [p]atentees . . . may give notice to the public that [any patented article] is patented, either by fixing thereon the word "patent" or the abbreviation "pat.", together with the number of the patent, or by fixing thereon the word "patent" or the abbreviation "pat." together with an address of a posting on the Internet, accessible to the public without charge for accessing the address, that associates the patented article with the number of the patent, or when, from the character of the article, this can not be done, by fixing to it, or to the package wherein one or more of them is contained, a label containing a like notice. *In the event of failure so to mark, no damages shall be recovered by the patentee in any action for infringement, except on proof that the infringer was notified of the infringement and continued to infringe thereafter, in which event damages may be recovered only for infringement occurring after such notice.* Filing of an action for infringement shall constitute such notice.

35 U.S.C. § 287(a) (emphasis added). While a patent application is pending an applicant can mark its product with "patent pending," although such marking is not legally required and has no legal effect. Falsely marking something as patented, or false use of phrases such as "patent pending" or "patent applied for" with "the purpose of deceiving the public," is subject to a $500 fine that can be imposed by the United States. 35 U.S.C. § 292(a).

The need for notice is more complex as to process claims. Where a patent contains only process claims, and thus a patentee can assert only process claims, no notice or marking is required. *State Contracting & Eng'g Corp. v. Condotte America, Inc.*, 346 F.3d 1057, 1073-74 (Fed. Cir. 2003). Where a patent contains both process and product claims, but only the process claims are asserted, no notice is required. *Hanson v. Alpine Valley Ski Area, Inc.*, 718 F.2d 1075, 1082-83 (Fed. Cir. 1983). Where, however, a patent contains both process and product claims, the patentee alleges infringement of both claim types, and the patentee sells products of its own to which the method claims pertain, the Federal Circuit has held that notice under § 287 is a condition for recovering damages. *American Medical Sys. v. Medical Eng'g Corp.*, 6 F.3d 1523, 1538-39 (Fed. Cir. 1993).

Note that in *Grain Processing Corp. v. American Maize-Products Co.*, 185 F.3d 1341 (Fed. Cir. 1999), reproduced in Subsection A.1, the patentee could recover damages only for activity that had occurred *after* it had filed suit. This was because the patentee had failed to properly mark its own products. The evidence demonstrated that the defendant was, nonetheless, aware of the plaintiff's patent. Should this matter? Another way to put people on notice of a patent is to send a letter, but a letter notice doesn't start the damages clock ticking unless it charges infringement: "For purposes of section 287(a), notice must be of 'the infringement,' not merely notice of the patent's existence or ownership. Actual notice requires the affirmative communication of a specific charge of infringement by a specific accused product or device." *Amsted Indus. v. Buckeye Steel Castings Co.*, 24 F.3d 178, 187 (Fed. Cir. 1994). Moreover, "[i]t is irrelevant . . . whether the defendant [already] knew of the patent or knew of his own infringement. The correct approach to determining notice under section 287 must focus on the action of the patentee, not the knowledge or understanding of the infringer." *Id.* Is the purpose behind the notice requirement to prevent a trap for the unaware infringer, or is it to place a burden on the patentee to positively assert her claims?

In the case of registered trademarks, the Lanham Act limits damages with a notice requirement: a registrant must provide either constructive notice, akin to that required by the patent marking statute; or actual notice of the registration to a defendant. 15 U.S.C. § 1111. The registrant gives constructive notice "by displaying with the mark the words 'Registered in U.S. Patent and Trademark Office' or 'Reg. U.S. Pat. & Tm. Off.' or the letter R enclosed within a circle, thus ®." *Id.* In the absence of notice, "no profits and no damages shall be recovered." *Id.* However, monetary awards for claims brought under § 43(a) are not subject to this marking requirement, even if brought by a registered mark owner. Thus, in practice, the trademark marking requirement typically has no effect on the award of damages.

2. Registration

Under the Copyright Act statutory damages and attorney fees are available only if the copyright owner of a published work registered the work *prior* to the infringement commencing, although the statute does provide for a grace period of three months following publication to seek registration and retain the ability to elect statutory damages. 17 U.S.C. § 412. If the work is unpublished there is no grace period—the registration must have been made prior to the infringement commencing. Section 412 contains a complicated exception for works that have followed a "preregistration" process that only applies to certain types of works, like motion pictures, sound recordings, and video games, that are infringed prior to their release date. Regulations issued by the Copyright Office contain the details of the pre-registration process. *See* 37 C.F.R. § 202.16. One type of claim does not require registration to be eligible to elect statutory damages and seek attorney fees: claims for violations of §106A, added by the Visual Artists Rights Act to protect moral rights for works of visual art. Why might it make sense not to impose the registration requirement for these types of claims?

3. Limitations Periods

Damages are also limited in time. Specifically, the Patent Act provides that "no recovery shall be had for any infringement committed more than six years prior to the filing of the complaint or counterclaim for infringement." 35 U.S.C. § 286. Damages generally accrue for conduct occurring after the PTO formally issues the patent. In addition, under a "provisional rights" system Congress created in 1999, an issued patent gives a patentee the right to recover against someone who practiced the invention *before* the patent's issue date. The patentee's provisional rights are, however, subject to the following constraints: (a) the accused infringer must have had "actual notice of the published patent application"; (b) the allegedly infringed application claim must be "substantially identical to" a claim in the issued patent; and (c) the patentee must bring the action within six years of the patent's issue date. 35 U.S.C. § 154(d).

Section 507(b) of the Copyright Act provides that copyright claims must be "commenced within three years after the claim accrued." 17 U.S.C. § 507(b). The general rule is that "[a] plaintiff's right to damages is limited to those suffered during the statutory period for bringing claims . . . ." *Los Angeles News Serv. v. Reuters Television Int'l, Ltd.,* 149 F.3d 987, 992 (9th Cir. 1998). Most circuits employ a "rolling" statute of limitations: "in a case of continuing copyright infringements, an action may be brought for all acts that accrued within the three years preceding the filing of suit." *Roley v. New World Pictures, Ltd.,* 19 F.3d 479, 481 (9th Cir. 1994). The copyright plaintiff cannot, however, reach back beyond the three-year limit and sue for damages or other relief for infringing acts that he knew about at the time but did not pursue. *Petrella v. Metro-Goldwyn-Mayer, Inc.,* 134 S. Ct. 1962 (2014).

"[T]he Lanham Act does not contain a statute of limitations. Rather, courts use the doctrine of laches," an equitable doctrine, "to determine whether a suit should be barred." *Audi AG v. D'Amato,* 469 F.3d 534, 545 (6th Cir. 2006). Courts differ in the nuances with which they define the laches bar. It is fair to say, however, that what the doctrine amounts to is this—"Laches consists of two elements: (1) inexcusable delay in bringing suit, and (2) prejudice to the defendant as a result of the delay." *Santana Products, Inc. v. Bobrick Washroom Equipment, Inc.,* 401 F.3d 123, 138 (3d Cir. 2005). In the Lanham Act context, many regional Courts of Appeals establish a presumption of laches, *i.e.*, a presumption of unreasonable delay and prejudice to the defendant, if a plaintiff delays filing suit for more time than the period set by borrowing the most closely analogous state limitations statute. *See, e.g., Miller v. Glen Miller Productions, Inc.,* 454 F.3d 975, 997 (9th Cir. 2006).

D. Enhanced Damages and Attorney Fee Awards

What if the defendants conduct was particularly bad? Are there other tools that a court may use to enhance the award to the plaintiff, to punish the defendant, or to deter future infringers (or to accomplish all three goals)? In *Bryant v. Media Right Productions, Inc.*, reproduced above in Section B.2, the plaintiff had asserted that the defendant's copyright infringement was willful. Demonstrating willful copyright infringement can result in a higher statutory damages award. But the Copyright Act does not contain any provision for enhanced *actual* damages, no matter how egregious the defendant's conduct. By contrast, both the Patent Act and the Lanham Act allow courts to treble the amount of the damages award. 35 U.S.C. § 284; 15 U.S.C. § 1117(a).

Litigation can also involve fee-shifting. The default rule in the U.S. is that each party, win or lose, pays its own attorney fees: "Under the American Rule, the prevailing litigant is ordinarily not entitled to collect a reasonable attorneys' fee from the loser. This default rule can, of course, be overcome by statute." *Travelers Casualty & Surety Co. v. Pacific Gas & Elec. Co.*, 549 U.S. 443, 448 (2007). For example, the Patent Act expressly states that a court "in exceptional cases may award reasonable attorney fees to the prevailing party." 35 U.S.C. § 285. The Lanham Act uses the same formulation: "The court in exceptional cases may award reasonable attorney fees to the prevailing party." 15 U.S.C. § 1117. Neither statute defines what makes a case "exceptional," and thus fee-shifting eligible. The Copyright Act also contains a fee shifting provision, permitting the court, in its discretion, to "award a reasonable attorney's fee to the prevailing party." 17 U.S.C. § 505. As discussed in the previous section, for a prevailing copyright owner to be eligible for a fee award, they must have timely registered the work (registering prior to the infringement commencing, or within the first three months after publication). See 17 U.S.C. § 412. Finally, the Uniform Trade Secrets Act § 4 provides that attorney fees may be awarded in the case of "willful and malicious misappropriation."

What type of conduct will lead to damages enhancement or an attorney fee award?

1. The Role of Willfulness in Enhanced Damages and Fee Awards

Zomba Enterprises, Inc. v. Panorama Records, Inc.
491 F.3d 574 (6th Cir. 2007)

MOORE, JUDGE:

From Japan to the United States and beyond, karaoke is wildly popular. Countless people have lined up at various venues to perform their favorite songs with, and in front of, their friends. But few participants (with the possible exception of IP lawyers) ever stop to consider the intellectual property regime governing karaoke.

Panorama Records, Inc. ("Panorama"), a purveyor of karaoke discs, resembles the majority of these participants. It entered the business of recording and selling karaoke discs without considering whether doing so infringed the intellectual property rights of others. Before long, this lack of foresight caught up with Panorama.

. . . .

I. FACTS AND PROCEDURE

Since 1998, Panorama has been in the business of manufacturing and selling karaoke compact discs. It issues a new disc monthly in each of a variety of musical genres, including country, pop, rock, and R&B. Each installment (or "karaoke package") contains the top hits in that genre for the relevant month. Laurindo Santos is one of Panorama's four shareholders, and at all times relevant to this case he was the decision-maker regarding the release of products.

The individual discs that Panorama makes and sells are in the CD+G format—shorthand for "compact disc plus graphics." As Panorama explains, "[t]hese are compact discs on which musicians that are hired by Panorama record a musical composition of a work which at some time may have been made popular by another artist. The CD+G contains a graphic element and is designed to be viewed when played on a karaoke machine." (Resp. to Req. for Admis. #1). The graphic element consists of the text of each song's lyrics, and it scrolls across a screen as the music (sans vocals) plays, permitting karaoke participants to read the lyrics as they sing along. Each of Panorama's karaoke packages contained nine or ten songs, with two tracks for each song, one track released with audible lyrics and one without.

Zomba Enterprises, Inc. and Zomba Songs, Inc. (collectively, "Zomba") are music publishing corporations often identified by the trade name Zomba Music Publishing. Zomba "is in the business of exploiting musical compositions for commercial gain." Toward this purpose, and at all times relevant to this case, Zomba held and administered the copyrights to a variety of musical compositions, including songs performed by pop music performers such as 98 Degrees, Backstreet Boys, *NSYNC, and Britney Spears.

Without Anna Music ("Without Anna") is another music publishing company, but is not a party to this action. In 2000, Without Anna discovered that some of the songs to which it owned copyrights appeared on Panorama's karaoke packages. In response, attorney Linda Edell Howard sent a cease-and-desist letter to Panorama on Without Anna's behalf, demanding that Panorama quit selling unlicensed copies of Without Anna's songs. When Panorama received this letter from Howard in 2000, it had not acquired licenses from the copyright owners of any of the songs it had released in its karaoke packages. Panorama then hired Vincent Castalucci, a licensing agent, and began negotiating licenses. Eventually, Panorama obtained license agreements from Without Anna.

On February 28, 2002, Howard sent another cease-and-desist letter to Panorama, this time on behalf of Zomba. Like Without Anna, Zomba had discovered that Panorama's karaoke packages contained copies of songs it owned. Zomba's cease-and-desist letter specified the terms upon which Zomba would be willing to grant a license: a $250 fixing fee for each Zomba-owned song on each package,[3] plus royalties of $0.16 per song per CD+G sold for the first half of the five-year license term, and $0.19 per song per CD+G sold for the second-half of the term. Santos and Castalucci contacted Howard in response to this letter, but Panorama did not stop selling CD+Gs with Zomba's songs on them, nor did it obtain any licenses.

On April 12, 2002, Howard sent a follow-up cease-and-desist letter on Zomba's behalf. Again, Santos and Castalucci responded to the letter. And again,

---

[3] Howard testified that a "fixing fee" "is basically a permission to fix the copyrighted musical work with the technology to allow you to visually see the lyric[s]. And it is a one-time fee . . . ." Such permission generally comes in the form of a "synch license." *ABKCO Music*, 96 F.3d at 62-63 & n.4.

Panorama failed both to obtain licenses to Zomba's songs and to cease selling CD+Gs containing them.

C. Procedural History

On January 13, 2003, Zomba filed its complaint, asserting thirty counts of copyright infringement—one count for each Zomba-owned musical composition that Panorama recorded and sold in its karaoke packages. Panorama answered, asserting no affirmative defenses other than estoppel, laches, waiver, and acquiescence. On April 22, 2003, the parties entered into a consent order in which Panorama agreed "to be restrained from distributing, releasing or otherwise exploiting any karaoke package containing compositions owned or administered by" Zomba. (4/22/03 Dist. Ct. Order). Within a week of entering this consent order, Panorama breached its agreement and resumed selling CD+Gs containing Zomba's copyrighted work. This conduct continued, and a year later, Zomba moved for sanctions on this basis.

After the parties filed cross-motions for summary judgment but before the district court ruled, Panorama's counsel withdrew on May 10, 2004. On June 18, 2004, the district court granted Zomba's, and denied Panorama's, motion for summary judgment on the issue of copyright infringement, rejecting Panorama's fair-use defense.

To determine damages, the district court scheduled a bench trial for August 10, 2004.

Panorama was unable to obtain new counsel and consequently failed to file any of the required pretrial documents or to appear at the pretrial conference held on July 29, 2004. The district court accordingly entered a default against Panorama on the issue of damages. . . .

. . . .

On November 5, 2005, the district court held a hearing to determine the amount of damages. A month later, the district court issued findings of fact and conclusions of law. The district court concluded that Panorama's infringement was willful, and accordingly awarded Zomba $31,000 for each of the twenty-six infringements at issue, for a total of $806,000. On January 23, 2006, the district court also awarded Zomba $76,456.16 in attorney fees and $1058.91 in costs.

. . . .

III. ANALYSIS

On appeal, Panorama raises a series of challenges. First, it objects to the district court's rejection of its fair-use defense and the district court's correlative conclusion that Panorama infringed Zomba's copyrights. Next, it disputes the district court's statutory-damage calculation, arguing both that any infringement was not willful and that the $806,000 damage award is unconstitutionally high. Third, it argues that the district court erred by refusing to grant its motion to transfer venue. Finally, Panorama appeals the district court's award of attorney fees to Zomba.

A. Copyright Infringement

. . . .

Panorama's core argument is that its copying should be considered "fair use" under the Copyright Act. . . .

. . . .

a. Purpose and Character of Use

. . . .

As an initial matter, Panorama's use of the compositions is only minimally, if at all, transformative. Although Panorama created its own recordings of these songs, Santos admitted that the hired musicians did not "change the words or music."

Unlike a parody, *see Campbell*, 510 U.S. at 579-80, a facsimile recording of a copyrighted composition adds nothing new to the original and accordingly has virtually no transformative value.

. . . .

The crux of Panorama's fair-use argument is its assertion that its use was transformative because its karaoke packages are used for "teaching." Notably, this focus is newly found. . . .

More importantly though, the end-user's utilization of the product is largely irrelevant; instead, the focus is on whether alleged infringer's use is transformative and/or commercial. . . .

. . . Zomba does not challenge karaoke crooners' renditions (atrocious or otherwise) of the relevant compositions, but rather Panorama's decision to copy these songs onto CD+Gs and then distribute them without paying royalties. . . . Panorama's manufacturing and selling the karaoke packages at issue "was performed on a profit-making basis by a commercial enterprise." *Id.* at 1389. Accordingly, Panorama's use is commercial in nature, a fact militating against its fair-use defense.

b. The Nature of the Copyrighted Work

The second of the § 107 factors "calls for recognition that some works are closer to the core of intended copyright protection than others," *Campbell*, 510 U.S. at 586, and accordingly are entitled to stronger protection. Like the musical composition in *Campbell* (Roy Orbison's "Oh, Pretty Woman"), the compositions of pop songs here at issue "fall[] within the core of the copyright's protective purposes." *Id.* Accordingly, this factor militates against a finding of fair use.

c. The Amount and Substantiality of the Portion Used

. . . Panorama acknowledges that it copied the entire compositions. It hired studio musicians to play the songs as closely as possible to the original performers, and distributed copies of their efforts. Additionally, it copied the lyrics, in both an auditory (on the tracks containing vocals) and a visual (on the graphics display) fashion. Because Panorama copied the relevant compositions in their entirety, this factor, too, cuts against Panorama's fair-use defense.

d. Effect on the Potential Market for the Copyrighted Work

. . . Panorama has failed to sustain its burden of proving that its copying does not adversely affect the market value of Zomba's copyrights.

The thrust of Panorama's argument—that Zomba and Panorama operate in different markets—is factually inaccurate, as the record illustrates that Zomba has previously licensed (and continues to license) its musical compositions to purveyors of karaoke products. It follows, then, that market harm is a given, as Panorama's unlicensed copying deprived Zomba of licensing revenues it otherwise would have received. Further, there can be no doubt that Panorama's practices, if they became widespread throughout the karaoke industry, would have a deleterious effect on the potential market for licenses to Zomba's songs. Accordingly, Panorama cannot show that its copying passes the market-harm test, and this factor, like the other three, militates in favor of rejecting its fair-use defense.

Because all four fair-use factors indicate that Panorama's copying was not a fair use, we conclude that the district court correctly rejected this defense and concluded that Panorama infringed Zomba's copyrights.

B. Willfulness

Next, Panorama argues that even if it infringed Zomba's copyrights, the district court erred by concluding that the infringement was willful and thus was subject to enhanced statutory damages. According to Panorama, any infringement was innocent, and certainly not willful. We disagree.

For infringement to be "willful," it must be done "with knowledge that [one's] conduct constitutes copyright infringement." *Princeton Univ. Press* [*v. Mich. Doc. Servs., Inc.*], 99 F.3d [1381,] 1392 [(6th Cir. 1996) (en banc)]. Accordingly, "one who has been notified that his conduct constitutes copyright infringement, but who reasonably and in good faith believes the contrary, is not 'willful' for these purposes." *Id.* This belief must be both (1) reasonable *and* (2) held in good faith. *See id.*

Panorama argues that it held a good-faith belief that the copying here at issue was a fair use, and contends that even if ultimately erroneous, this belief precludes a finding of willfulness. . . . [T]he issue is not so much whether Panorama held in good faith its belief that its copying was fair use (although we have serious misgivings on this matter),[9] but whether Panorama *reasonably* believed that its conduct did not amount to copyright infringement. To decide this issue, we must determine "whether the copyright law supported the plaintiffs' position so clearly that the defendants must be deemed as a matter of law to have exhibited a reckless disregard of the plaintiffs' property rights." *Id.* . . .

---

[9] The reasons for these misgivings are numerous. For instance, we note that Panorama did not mention fair use in its initial licensing discussions with Howard. Instead, Panorama waited to raise its fair-use defense until this litigation was over a year old. This delay calls into serious question whether Panorama sincerely believed that its copying was a fair use before it began recording and distributing the karaoke packages. Instead, it appears that the fair-use defense was merely a post-hoc rationalization concocted to skirt liability. Further, we note that the record is bereft of evidence that Panorama ever discussed the fair-use doctrine with an attorney. . . .

Here, we conclude that Panorama exhibited a reckless disregard for Zomba's rights, and accordingly, that Panorama's reliance on its fair-use defense was objectively unreasonable. The fact most crucial to this inquiry is that Panorama continued to sell karaoke packages containing copies of each of the relevant compositions *after* the district court entered its April 22, 2003, consent order forbidding Panorama to do so.

In copyright cases, it is unreasonable to rely on a defense to infringement after a court rejects it on the merits. This principle does not dispose fully of Panorama's position because the April 22, 2003, consent order did not resolve any issues on the merits. To the contrary, the order stated that it "shall not be construed as a finding of any fact by the [District] Court or a finding by the [District] Court that Plaintiff has established . . . any . . . substantive element of its case." (4/22/03 Dist. Ct. Order).

This caveat, however, does not justify Panorama's continued reliance on its fair-use defense. By entering into the consent decree, Panorama agreed to cease infringing Zomba's copyrights. Thus, it implicitly agreed to suspend its reliance on the fair-use defense at least temporarily, and this agreement was reduced to an order of the court. Because an order entered by a court of competent jurisdiction must be obeyed even if it is erroneously issued, Panorama lacked any legal justification for continuing to distribute copies of Zomba's copyrighted works after April 22, 2003. Without expressing any opinion regarding the reasonableness of Panorama's position before that date, we are certain that from April 22, 2003, on, the law applicable to this case "supported [Zomba's] position so clearly that [Panorama] must be deemed as a matter of law to have exhibited a reckless disregard of [Zomba's] property rights." *Princeton Univ. Press*, 99 F.3d at 1392. On this basis, we agree with the district court's conclusion that Panorama's infringements were willful and accordingly justified enhanced statutory damages.

C. Amount of Statutory-Damage Award

. . . .

1. Abuse of Discretion

Panorama contends that the district court believed that, after making a finding of willfulness, it lacked discretion to award statutory damages of less than $30,000 per infringement. On this basis, Panorama maintains that the district court abused its discretion. The record does not support this argument.

In its conclusions of law, the district court recognized that it had "wide discretion in determining the amount of statutory damages to be awarded, constrained only by the maximum and minimum amounts." (12/5/05 Dist. Ct. Op. at 7). It found "that the maximum statutory amount of $30,000 per work for 'innocent' infringement is not sufficient in this case because of the clearly willful nature of Defendant's conduct," but that the maximum award of $150,000 per infringement was excessive, given the dollar amounts involved in the case. (*id.* at 7- 8).

Nowhere did the district court indicate that it believed that it lacked discretion to award statutory damages of less than $30,000 per infringement. To the contrary, Panorama's willfulness prompted the district court to conclude that the maximum penalty for nonwillful infringement was *not sufficient* given Panorama's conduct. We therefore conclude that Panorama has not shown that the district court abused its discretion by setting the statutory damage award at $31,000 per infringement.

2. Eighth Amendment

Panorama next argues that such a high award of statutory damages, in light of the relatively low actual damages,[10] renders the district court's award an "excessive fine" under the Eighth Amendment. However, the Supreme Court has explained that the word "fine" in this context means a payment to the *government*. *United States v. Bajakajian*, 524 U.S. 321, 327 (1998). Consequently, the Court has held that the Excessive Fines Clause "does not constrain an award of money damages in a civil suit when the government neither has prosecuted the action nor has any right to receive a share of the damages awarded." *Browning-Ferris Indus. of Vermont, Inc. v. Kelco Disposal, Inc.*, 492 U.S. 257, 264 (1989). Panorama's Eighth Amendment argument thus fails.

3. Due Process

Panorama argues, based upon *BMW of North America, Inc. v. Gore*, 517 U.S. 559 (1996), and *State Farm Mutual Automobile Insurance Co. v. Campbell*, 538 U.S. 408 (2003), that an award of statutory damages that (it alleges) is thirty-seven times the actual damages violates its right to due process. Again, we disagree.

We note at the outset that both *Gore* and *Campbell* addressed due-process challenges to *punitive*-damages awards. In both cases, the award was greater than one hundred times the amount of compensatory damages awarded. In *Gore*, the Court concluded that the award in question (which amounted to 500 times the compensatory-damages award) was "grossly excessive," 517 U.S. at 574, after considering three "guideposts": (1) "the degree of reprehensibility of the" defendant's conduct; (2) "the disparity between the harm or potential harm suffered by [the plaintiff] and [the] punitive damages award"; and (3) "the difference between this remedy and the civil penalties authorized or imposed in comparable cases," *id.* at 575. In *Campbell*, the Court considered the same three guidepost factors and concluded that the punitive-damages award (which amounted to 145 times the compensatory-damages award) "was an irrational and arbitrary deprivation of the property of the defendant." 538 U.S. at 429. Regarding the second guidepost, neither case created a "concrete constitutional limit[]" to the punitive-to-compensatory damages ratio. *Id.* at 424. Instead, the *Campbell* Court explicitly "decline[d]

---

[10] The district court concluded that it was unable to calculate the actual damages based on the record before it. Panorama asserted that it sold a total of 74,734 copies of the twenty-six infringed compositions. Based upon this figure, Zomba lost approximately $11,957.92 in royalties, plus an additional $6500 in fixing fees, for a total of $18,457.92. Additionally, Panorama asserted that its net profit attributable to the twenty-six compositions at issue was $9693.86, although Panorama never substantiated its expenses to the district court. Assuming arguendo that this profit figure is accurate, Zomba would have been entitled to $28,151.78 if it had not elected to pursue statutory damages. *See* 17 U.S.C. § 504(b).

... to impose a bright-line ratio which a punitive damages award cannot exceed," although it expressed a general preference for single-digit ratios. *Id.* at 425.

The Supreme Court has not indicated whether *Gore* and *Campbell* apply to awards of *statutory* damages. We know of no case invalidating such an award of statutory damages under *Gore* or *Campbell*, although we note that some courts have suggested in dicta that these precedents may apply to statutory-damage awards. *See, e.g., Parker v. Time Warner Entm't Co.*, 331 F.3d 13, 22 (2d Cir. 2003) (suggesting that "in a sufficiently serious case," due process may require courts to reduce a statutory-damage award in a class action, and citing both *Campbell* and *Gore*); *Leiber v. Bertelsmann AG (In re Napster, Inc. Copyright Litig.)*, 2005 WL 1287611, at *10 (N.D. Cal. June 1, 2005) (unpublished) (citing *Gore* and *Campbell*); *DIRECTV v. Gonzalez*, 2004 WL 1875046, at * 4 (W.D. Tex. Aug. 23, 2004) (unpublished) (citing *Campbell*); *but see Lowry's Reports, Inc. v. Legg Mason Inc.*, 302 F. Supp. 2d 455, 459-60 (D. Md. 2004) (concluding that *Gore* and *Campbell* do not limit statutory damages in copyright cases).

Regardless of the uncertainty regarding the application of *Gore* and *Campbell* to statutory damage awards, we may review such awards under *St. Louis, I.M. & S. Ry. Co. v. Williams*, 251 U.S. 63, 66-67 (1919), to ensure they comport with due process. In such cases, we inquire whether the awards are "so severe and oppressive as to be wholly disproportioned to the offense and obviously unreasonable." *Id.* at 67. This review, however, is extraordinarily deferential—even more so than in cases applying abuse-of-discretion review.

Williams is instructive, and leads us to conclude that the statutory-damage award against Panorama was not sufficiently oppressive to constitute a deprivation of due process. In that case, a railroad charged two sisters sixty-six cents apiece more than the maximum rate permissible by regulation. 251 U.S. at 64. A state statute sought to deter such overcharges by providing for statutory damages of between $50 and $350 when a railroad charged more than the permissible rate. *Id.* The sisters sued separately, and received statutory damage awards of $75 apiece— over 113 times the amount they were overcharged. *Id.* Before the Supreme Court, the railroad argued that the penalty was so disproportionate to the harm sustained that it violated due process. Rejecting this argument, the Court concluded that the award "properly cannot be said to be so severe and oppressive as to be wholly disproportioned to the offense or obviously unreasonable." *Id.* at 67.

If the Supreme Court countenanced a 113:1 ratio in *Williams*, we cannot conclude that a 44:1 ratio[11] is unacceptable here. We acknowledge the Supreme Court's preference for a lower punitive to-compensatory ratio, as stated in *Campbell*, but emphasize that this case does not involve a punitive-damages award. Until

---

[11] Dividing the statutory-damage award ($806,000) by the lost licensing fees, as calculated by Panorama, ($18,458) yields a ratio of about 44:1. The 37:1 figure that Panorama advances in its brief includes in the denominator Panorama's unsubstantiated net profits figure of $9693.86. This sum is separate from the "actual damages," as defined by 17 U.S.C. § 504(b), which permits a copyright plaintiff to elect to receive "the actual damages suffered by him or her as a result of the infringement, and any profits of the infringer that are attributable to the infringement and not taken into account in computing the actual damages."

the Supreme Court applies *Campbell* to an award of statutory damages, we conclude that *Williams* controls, not *Campbell*, and accordingly reject Panorama's due process argument.[12]

. . . .

E. Attorney fees

Lastly, Panorama argues that the district court abused its discretion by awarding attorney fees to Zomba. The Copyright Act provides that "the [district] court in its discretion may allow the recovery of full costs . . . [and] may also award a reasonable attorney's fee to the prevailing party as part of the costs." 17 U.S.C. § 505. In *Fogerty v. Fantasy, Inc.*, 510 U.S. 517 (1994), the Supreme Court endorsed in dicta a series of factors "to guide courts' discretion" in awarding attorney fees. The factors include "frivolousness, motivation, objective unreasonableness (both in the factual and in the legal components of the case) and the need in particular circumstances to advance considerations of compensation and deterrence." *Id.*

According to Panorama, the district court abused its discretion by failing to consider the *Fogerty* factors. However, as Zomba notes, the parties provided the district court with briefing focused on the factors, so even if the district court did not precisely recount its assessment of the factors, it was alerted to them. Further, given the unreasonableness of Panorama's positions and the need to deter such conduct, it is difficult to see how the imposition of attorney fees here could qualify as an abuse of discretion.

Notes & Questions

1. Is the standard for finding willful infringement the same in the Sixth Circuit and Second Circuit?

2. In *Zomba* the Sixth Circuit rejects a due process challenge to the statutory damages award. Both the Second Circuit and the Eighth Circuit have also rejected Due Process challenges. *Capitol Records, Inc. v. Thomas-Rasset*, 692 F.3d 899 (8th Cir. 2012); *Sony BMG Music Entertainment v. Tenenbaum,* 660 F.3d 487 (1st Cir. 2011). Without some sense of the basis on which a court determines the size of an award, does the range in the Copyright Act comply with a notion of "fair notice" of the potential liability one might face for violations of the law?

3. The Patent Act provides that a "court may increase the damages up to three times the [damages] amount found." 35 U.S.C. § 284. The Act says nothing, however, about when such an enhancement is warranted. The jurisprudence the Act codifies, however, makes clear that "enhanced damages are punitive, not compen-

---

[12] The Supreme Court's recent decision in *Philip Morris USA v. Williams* , 127 S. Ct. 1057 (2007), which was issued after the parties submitted their briefs but before oral argument, does not alter our analysis. There, the Supreme Court considered only "whether the Constitution's Due Process Clause permits a jury to base [a punitive-damage] award in part upon its desire to *punish* the defendant for harming persons who are not before the court." *Id.* at 1060. As there is absolutely no indication that the district court based its award of statutory damages on Panorama's conduct vis-a-vis copyright holders other than Zomba, *Philip Morris USA* is inapposite.

satory. Enhancement is not a substitute for perceived inadequacies in the calculation of actual damages, but depends on a showing of willful infringement or other indicium of bad faith warranting punitive damages." *Sensonics, Inc. v. Aerosonic Corp.*, 81 F.3d 1566, 1574 (Fed. Cir. 1996). Importantly, "a finding of willful infringement merely *authorizes*, but does not *mandate*, an award of increased damages." *Modine Mfg. v. Allen Group, Inc.*, 917 F.2d 538, 543 (Fed. Cir. 1990) (emphasis in original). The Federal Circuit has approved the following list of factors for determining whether to enhance damages upon a willfulness finding:

> (1) whether the infringer deliberately copied the ideas or design of another, (2) whether the infringer, when he knew of the other's patent protection, investigated the scope of the patent and formed a good-faith belief that it was invalid or that it was not infringed, (3) the infringer's behavior as a party to the litigation, (4) the infringer's size and financial condition, (5) the closeness of the case, (6) the duration of the infringer's misconduct, (7) any remedial action by the infringer, (8) the infringer's motivation for harm, and (9) whether the infringer attempted to conceal its misconduct.

Johns Hopkins University v. CellPro, Inc., 152 F.3d 1342, 1352 n.16 (Fed. Cir. 1998).

In 2016, the Supreme Court held that "[a]wards of enhanced damages under the Patent Act over the past 180 years establish that they are not to be meted out in a typical infringement case, but are instead designed as a 'punitive' or 'vindictive' sanction for egregious infringement behavior. The sort of conduct warranting enhanced damages has been variously described in our cases as willful, wanton, malicious, bad-faith, deliberate, consciously wrongful, flagrant, or—indeed— characteristic of a pirate." *Halo Elecs. v. Pulse Elecs.*, 136 S. Ct. 1923, 1932 (2016). In other words, in patent cases "such damages are generally reserved for egregious cases of culpable behavior." *Id.* How does this compare with the standard applied in *Zomba*? Should the Supreme Court's willfulness standard in *Halo* prompt the Federal Circuit to reconsider the factors it used in the *Johns Hopkins* case to decide whether to enhance damages?

4. Once a defendant's subjective state of mind is relevant, a well-grounded, well-reasoned opinion of counsel can be used to demonstrate that one's belief that one's conduct is reasonable, or at least held in good faith. And this is so even though the counsel's opinion turned out to be wrong, *i.e.*, one has been found to have infringed a valid ip claim. (If infringement hasn't been found, the willfulness of the conduct is irrelevant.) Lawyers can be asked to provide opinion letters that address the assertions of infringement that intellectual property owners have made. An opinion letter may address the validity of the intellectual property asserted (*e.g.*, whether the patent may be vulnerable to an obviousness challenge), the strength of the infringement claim (*e.g.*, whether the client's work is, in fact, substantially similar to the ip owner's alleged copyrighted work), or the viability of certain defenses (*e.g.*, fair use or prior use). Reliance on an opinion of counsel, however, produces an important practical litigation problem: potential waiver of attorney-client privilege. For this reason, the counsel that provides the opinion letter should be different from the lawyer that is engaged to represent the client in

litigation. *In re Seagate Tech.*, 497 F.3d 1360, 1371 (Fed. Cir. 2007) (en banc) (holding that "as a general proposition . . . asserting the advice of counsel defense and disclosing opinions of opinion counsel do not constitute waiver of the attorney-client privilege for communications with trial counsel").

The America Invents Act of 2011 introduced into the Patent Act, for the first time, a provision expressly addressing the connection between legal advice on another's patent rights and the question of willful infringement:

> The failure of an infringer to obtain the advice of counsel with respect to any allegedly infringed patent, or the failure of the infringer to present such advice to the court or jury, may not be used to prove that the accused infringer willfully infringed the patent or that the infringer intended to induce infringement of the patent.

35 U.S.C. § 298. How this new provision will translate into practice remains to be seen.

5. Enhanced damages are not meant to be compensatory; they are punitive. The literature on the law and economics of punitive damages is vast. One recurring concept in this literature is that punitive damages are best thought of as a means to address rights violations that are harder to detect or to proceed against. *See, e.g.*, Thomas F. Cotter, *An Economic Analysis of Enhanced Damages and Attorney's Fees for Willful Patent Infringement*, 14 FED. CIR. B.J. 291 (2004).

2. Fee Shifting Under the Copyright Act

Kirtsaeng v. John Wiley & Sons, Inc.
136 S. Ct. 1979 (2016)

KAGAN, JUSTICE :

Section 505 of the Copyright Act provides that a district court "may * * * award a reasonable attorney's fee to the prevailing party." 17 U.S.C. § 505. The question presented here is whether a court, in exercising that authority, should give substantial weight to the objective reasonableness of the losing party's position. The answer, as both decisions below held, is yes—the court should. But the court must also give due consideration to all other circumstances relevant to granting fees; and it retains discretion, in light of those factors, to make an award even when the losing party advanced a reasonable claim or defense. . . .

I

Petitioner Supap Kirtsaeng, a citizen of Thailand, came to the United States 20 years ago to study math at Cornell University. He quickly figured out that respondent John Wiley & Sons, an academic publishing company, sold virtually identical English-language textbooks in the two countries—but for far less in Thailand than in the United States. Seeing a ripe opportunity for arbitrage, Kirtsaeng asked family and friends to buy the foreign editions in Thai bookstores and ship them to him in New York. He then resold the textbooks to American students, reimbursed his Thai suppliers, and pocketed a tidy profit.

Wiley sued Kirtsaeng for copyright infringement, claiming that his activities violated its exclusive right to distribute the textbooks. Kirtsaeng invoked the 'first-sale doctrine' as a defense. That doctrine typically enables the lawful owner of a book (or other work) to resell or otherwise dispose of it as he wishes. See § 109(a). But Wiley contended that the first-sale doctrine did not apply when a book (like those Kirtsaeng sold) was manufactured abroad.

At the time, courts were in conflict on that issue. . . . In this case, the District Court sided with Wiley; so too did a divided panel of the Court of Appeals for the Second Circuit. To settle the continuing conflict, this Court granted Kirtsaeng's petition for certiorari and reversed the Second Circuit in a 6-to-3 decision, thus establishing that the first-sale doctrine allows the resale of foreign-made books, just as it does domestic ones. See *Kirtsaeng v. John Wiley & Sons, Inc.*, 133 S. Ct. 1351 (2013).

Returning victorious to the District Court, Kirtsaeng invoked § 505 to seek more than $2 million in attorney's fees from Wiley. The court denied his motion. Relying on Second Circuit precedent, the court gave "substantial weight" to the "objective reasonableness" of Wiley's infringement claim. In explanation of that approach, the court stated that "the imposition of a fee award against a copyright holder with an objectively reasonable"—although unsuccessful—"litigation posi-tion will generally not promote the purposes of the Copyright Act." Here, Wiley's position was reasonable: After all, several Courts of Appeals and three Justices of the Supreme Court had agreed with it. And according to the District Court, no other circumstance "overr[o]de" that objective reasonableness, so as to warrant fee-shifting. The Court of Appeals affirmed, concluding in a brief summary order that "the district court properly placed 'substantial weight' on the reasonableness of [Wiley's] position" and committed no abuse of discretion in deciding that other "factors did not outweigh" the reasonableness finding.

We granted certiorari to resolve disagreement in the lower courts about how to address an application for attorney's fees in a copyright case.

II

Section 505 states that a district court "may * * * award a reasonable attorney's fee to the prevailing party." It thus authorizes fee-shifting, but without specifying standards that courts should adopt, or guideposts they should use, in determining when such awards are appropriate.

In *Fogerty v. Fantasy, Inc.*, 510 U.S. 517 (1994), this Court recognized the broad leeway § 505 gives to district courts—but also established several principles and criteria to guide their decisions. The statutory language, we stated, "clearly con-notes discretion," and eschews any "precise rule or formula" for awarding fees. *Id.* at 533, 534. Still, we established a pair of restrictions. First, a district court may not "award [ ] attorney's fees as a matter of course"; rather, a court must make a more particularized, case-by-case assessment. *Id.* at 533. Second, a court may not treat prevailing plaintiffs and prevailing defendants any differently; defendants should be "encouraged to litigate [meritorious copyright defenses] to the same ex-tent that plaintiffs are encouraged to litigate meritorious claims of infringement."

Id. at 527. In addition, we noted with approval "several nonexclusive factors" to inform a court's fee-shifting decisions: "frivolousness, motivation, objective unreasonableness[,] and the need in particular circumstances to advance considerations of compensation and deterrence." *Id.* at 534, n.19. And we left open the possibility of providing further guidance in the future, in response to (and grounded on) lower courts' evolving experience. *See id.* at 534-35; *Martin v. Franklin Capital Corp.*, 546 U.S. 132, 140 n.* (2005).

The parties here, though sharing some common ground, now dispute what else we should say to district courts. Both Kirtsaeng and Wiley agree—as they must—that § 505 grants courts wide latitude to award attorney's fees based on the totality of circumstances in a case. Yet both reject the position . . . that *Fogerty* spelled out the only appropriate limits on judicial discretion—in other words, that each district court should otherwise proceed as it sees fit, assigning whatever weight to whatever factors it chooses. Rather, Kirtsaeng and Wiley both call, in almost identical language, for "[c]hanneling district court discretion towards the purposes of the Copyright Act." Brief for Petitioner 16; see Brief for Respondent 21. But at that point, the two part ways. Wiley argues that giving substantial weight to the reasonableness of a losing party's position will best serve the Act's objectives. By contrast, Kirtsaeng favors giving special consideration to whether a lawsuit resolved an important and close legal issue and thus "meaningfully clarifie[d]" copyright law.

We join both parties in seeing a need for some additional guidance respecting the application of § 505. In addressing other open-ended fee-shifting statutes, this Court has emphasized that "in a system of laws discretion is rarely without limits." *Flight Attendants v. Zipes*, 491 U. S. 754, 758 (1989); *see also Halo Elecs. v. Pulse Elecs.*, 136 S. Ct. 1923 (2016). Without governing standards or principles, such provisions threaten to condone judicial "whim" or predilection. *Martin*, 546 U.S. at 139. At the least, utterly freewheeling inquiries often deprive litigants of "the basic principle of justice that like cases should be decided alike," [*id.*]—as when, for example, one judge thinks the parties' "motivation[s]" determinative and another believes the need for "compensation" trumps all else. *Fogerty*, 510 U.S. at 534, n.19. And so too, such unconstrained discretion prevents individuals from predicting how fee decisions will turn out, and thus from making properly informed judgments about whether to litigate. For those reasons, when applying fee-shifting laws with "no explicit limit or condition," *Halo* at 1931, we have nonetheless "found limits" in them—and we have done so, just as both parties urge, by looking to "the large objectives of the relevant Act," *Zipes*, 491 U.S. at 759.

In accord with such precedents, we must consider if either Wiley's or Kirtsaeng's proposal well advances the Copyright Act's goals. Those objectives are well settled. As *Fogerty* explained, "copyright law ultimately serves the purpose of enriching the general public through access to creative works." 510 U.S. at 527; *see* U.S. Const., Art. I, §8, cl. 8 ("To promote the Progress of Science and useful Arts"). The statute achieves that end by striking a balance between two subsidiary aims: encouraging and rewarding authors' creations while also enabling others to build on that work. Accordingly, fee awards under § 505 should encourage the types of lawsuits that promote those purposes. (That is why, for example, *Fogerty* insisted on treating prevailing plaintiffs and prevailing defendants alike—because the one could "further the policies of the Copyright Act every bit as much as" the other. 510 U.S. at 527.) On that much, both parties agree. The contested issue is whether giving substantial weight to the objective (un)reasonableness of a losing party's litigating position—or, alternatively, to a lawsuit's role in settling significant and uncertain legal issues—will predictably encourage such useful copyright litigation.

The objective-reasonableness approach that Wiley favors passes that test because it both encourages parties with strong legal positions to stand on their rights and deters those with weak ones from proceeding with litigation. When a litigant—whether plaintiff or defendant—is clearly correct, the likelihood that he will recover fees from the opposing (*i.e.,* unreasonable) party gives him an incentive to litigate the case all the way to the end. The holder of a copyright that has obviously been infringed has good reason to bring and maintain a suit even if the damages at stake are small; and likewise, a person defending against a patently meritless copyright claim has every incentive to keep fighting, no matter that attorney's fees in a protracted suit might be as or more costly than a settlement. Conversely, when a person (again, whether plaintiff or defendant) has an unreasonable litigating position, the likelihood that he will have to pay two sets of fees discourages legal action. The copyright holder with no reasonable infringement claim has good reason not to bring suit in the first instance (knowing he cannot force a settlement and will have to proceed to judgment); and the infringer with no reasonable defense has every reason to give in quickly, before each side's litigation costs mount. All of those results promote the Copyright Act's purposes, by enhancing the probability that both creators and users (*i.e.,* potential plaintiffs and defendants) will enjoy the substantive rights the statute provides.

By contrast, Kirtsaeng's proposal would not produce any sure benefits. We accept his premise that litigation of close cases can help ensure that "the boundaries of copyright law [are] demarcated as clearly as possible," thus advancing the public interest in creative work. Brief for Petitioner 19 (quoting *Fogerty*, 510 U.S. at 527). But we cannot agree that fee-shifting will necessarily, or even usually, encourage parties to litigate those cases to judgment. Fee awards are a double-edged sword: They increase the reward for a victory—but also enhance the penalty for a defeat. And the hallmark of hard cases is that no party can be confident if he will win or lose. That means Kirtsaeng's approach could just as easily discourage as encourage parties to pursue the kinds of suits that "meaningfully clarif[y]" copyright law. It would (by definition) raise the stakes of such suits; but whether those higher stakes would provide an incentive—or instead a disincentive—to litigate hinges on a party's attitude toward risk. Is the person risk-preferring or risk-averse—a high-roller or a penny-ante type? Only the former would litigate more in Kirtsaeng's world. *See* Posner, *An Economic Approach to Legal Procedure and Judicial Administration*, 2 J. Legal Studies 399, 428 (1973) (fees "make[ ] the expected value of litigation less for risk-averse litigants, which will encourage [them to] settle[ ]"). And Kirtsaeng offers no reason to think that serious gamblers predominate. So the value of his standard, unlike Wiley's, is entirely speculative.[2]

What is more, Wiley's approach is more administrable than Kirtsaeng's. A district court that has ruled on the merits of a copyright case can easily assess whether the losing party advanced an unreasonable claim or defense. That is closely related to what the court has already done: In deciding any case, a judge cannot help but consider the strength and weakness of each side's arguments. By contrast, a judge may not know at the conclusion of a suit whether a newly decided issue will have, as Kirtsaeng thinks critical, broad legal significance. The precedent-setting, law-clarifying value of a decision may become apparent only in retrospect—sometimes, not until many years later. And so too a decision's practical impact (to the extent Kirtsaeng would have courts separately consider that factor). District courts are not accustomed to evaluating in real time either the jurisprudential or the on-the-ground import of their rulings. Exactly how they would do so is uncertain (Kirtsaeng points to no other context in which courts undertake such an analysis), but we fear that the inquiry would implicate our oft-stated concern that an application for attorney's fees "should not result in a second major litigation." *Zipes*, 491 U.S. at 766. And we suspect that even at the end of that post-lawsuit lawsuit, the results would typically reflect little more than educated guesses.

Contrary to Kirtsaeng's view, placing substantial weight on objective reasonableness also treats plaintiffs and defendants even-handedly, as *Fogerty* commands. No matter which side wins a case, the court must assess whether the other side's position was (un)reasonable. And of course, both plaintiffs and defendants can (and sometimes do) make unreasonable arguments. Kirtsaeng claims that the reasonableness inquiry systematically favors plaintiffs because a losing defendant "will virtually *always* be found to have done something culpable." But that conflates two different questions: whether a defendant in fact infringed a copyright and whether he made serious arguments in defense of his conduct. Courts every day see reasonable defenses that ultimately fail (just as they see reasonable claims that come to nothing); in this context, as in any other, they are capable of distinguishing between those defenses (or claims) and the objectively unreasonable variety. And if some court confuses the issue of liability with that of reasonableness, its fee award should be reversed for abuse of discretion.

---

[2] This case serves as a good illustration. Imagine you are Kirtsaeng at a key moment in his case—say, when deciding whether to petition this Court for certiorari. And suppose (as Kirtsaeng now wishes) that the prevailing party in a hard and important case—like this one—will probably get a fee award. Does that make you more likely to file, because you will recoup your own fees if you win? Or less likely to file, because you will foot Wiley's bills if you lose? Here are some answers to choose from (recalling that you cannot confidently predict which way the Court will rule): (A) Six of one, half a dozen of the other. (B) Depends if I'm feeling lucky that day. (C) Less likely—this is getting scary; who knows how much money Wiley will spend on Supreme Court lawyers? (D) More likely—the higher the stakes, the greater the rush. Only if lots of people answer (D) will Kirtsaeng's standard work in the way advertised. Maybe. But then again, maybe not.

All of that said, objective reasonableness can be only an important factor in assessing fee applications—not the controlling one. As we recognized in *Fogerty*, § 505 confers broad discretion on district courts and, in deciding whether to fee-shift, they must take into account a range of considerations beyond the reasonableness of litigating positions. That means in any given case a court may award fees even though the losing party offered reasonable arguments (or, conversely, deny fees even though the losing party made unreasonable ones). For example, a court may order fee-shifting because of a party's litigation misconduct, whatever the reasonableness of his claims or defenses. Or a court may do so to deter repeated instances of copyright infringement or overaggressive assertions of copyright claims, again even if the losing position was reasonable in a particular case. Although objective reasonableness carries significant weight, courts must view all the circumstances of a case on their own terms, in light of the Copyright Act's essential goals.

And on that score, Kirtsaeng has raised serious questions about how fee-shifting actually operates in the Second Circuit. To be sure, the Court of Appeals' framing of the inquiry resembles our own: It calls for a district court to give "substantial weight" to the reasonableness of a losing party's litigating positions while also considering other relevant circumstances. But the Court of Appeals' language at times suggests that a finding of reasonableness raises a presumption against granting fees—and that goes too far in cabining how a district court must structure its analysis and what it may conclude from its review of relevant factors. Still more, district courts in the Second Circuit appear to have overly learned the Court of Appeals' lesson, turning "substantial" into more nearly "dispositive" weight. . . . For these reasons, we vacate the decision below so that the District Court can take another look at Kirtsaeng's fee application. In sending back the case for this purpose, we do not at all intimate that the District Court should reach a different conclusion. Rather, we merely ensure that the court will evaluate the motion consistent with the analysis we have set out—giving substantial weight to the reasonableness of Wiley's litigating position, but also taking into account all other relevant factors.

Notes & Questions

1. The default rule in U.S. civil litigation is that each side pays for its own attorneys. Fee-shifting provisions such as § 505 of the Copyright Act change that default to a loser pays rule. A district court can order the losing side to pay the winning side's attorney's fees. What do you make of the Supreme Court's assessment of the fight-or-settle calculus that a loser-pays rule injects into copyright litigation? In light of that calculus, does the Supreme Court's approach to limiting a district court's discretion under § 505 make for good copyright policy?

2. Review the portion of the *Zomba Enterprises* opinion in Section I.D.1 concerning the award of attorney's fees. After the Supreme Court's *Kirtsaeng* decision should the result be different? Under the *Kirtsaeng* standard, would it have been an abuse of discretion for the court to award attorney's fees? Would it have been an abuse of discretion if the court refused to award fees in that case? In *Bryant v.*

Media Rights, excerpted above in Section I.B.2, the plaintiff prevailed, but the district court refused to award fees. The Court of Appeals affirmed. After *Kirtsaeng* should the result be different?

3. You are selling the copyright in your memoir manuscript to a publishing company. The contract includes an indemnity clause, wherein you promise to pay the publisher's costs and fees in the event the publication of your book results in a copyright infringement suit against the publisher. How does *Kirtsaeng* affect your thinking about that contract term? Is it more agreeable to you if the contract also gives you some say over how the publishing company conducts that copyright litigation?

3. Fee Shifting In Trademark and Patent Litigation: The "Exceptional Case" Requirement

Both the Patent Act and the Lanham Act use the idea of an "exceptional case" as the threshold condition for awarding attorney fees to the prevailing party. 35 U.S.C. § 285; 15 U.S.C. § 1117(a). The Supreme Court recently addressed the meaning of exceptionality in patent cases and, in doing so, relied on a notable D.C. Circuit decision applying the trademark statute. The opinion in the D.C. Circuit case was written by then-Circuit Judge, now-Justice, Ruth Bader Ginsburg, and joined by then-Circuit Judge, now-Justice, Antonin Scalia.

Octane Fitness, LLC v. ICON Health & Fitness, Inc.

134 S. Ct. 1749 (2014)

SOTOMAYOR, JUSTICE:

. . . .

Prior to 1946, the Patent Act did not authorize the awarding of attorney's fees to the prevailing party in patent litigation. Rather, the "American Rule" governed: "[E]ach litigant pa[id] his own attorney's fees, win or lose * * * ." *Marx v. General Revenue Corp.*, 133 S. Ct. 1166, 1175 (2013). In 1946, Congress amended the Patent Act to add a discretionary fee-shifting provision, then codified in § 70, which stated that a court "may in its discretion award reasonable attorney's fees to the prevailing party upon the entry of judgment in any patent case." 35 U.S.C. § 70 (1946 ed.).

Courts did not award fees under § 70 as a matter of course. They viewed the award of fees not "as a penalty for failure to win a patent infringement suit," but as appropriate "only in extraordinary circumstances." *Park-In-Theatres, Inc. v. Perkins*, 190 F.2d 137, 142 (9th Cir. 1951). The provision enabled them to address "unfairness or bad faith in the conduct of the losing party, or some other equitable consideration of similar force," which made a case so unusual as to warrant fee-shifting. *Id.*; *see also Pennsylvania Crusher Co. v. Bethlehem Steel Co.*, 193 F.2d 445, 451 (3d Cir. 1951) (listing as "adequate justification[s]" for fee awards "fraud practiced on the Patent Office or vexatious or unjustified litigation").

Six years later, Congress amended the fee-shifting provision and recodified it as § 285. . . . [T]he revised language of § 285 (which remains in force today) provides that "[t]he court in exceptional cases may award reasonable attorney fees to the prevailing party." We have observed, in interpreting the damages provision of the Patent Act, that the addition of the phrase "exceptional cases" to § 285 was "for purposes of clarification only." *General Motors Corp. v. Devex Corp.*, 461 U.S. 648, 653, n.8 (1983). And the parties agree that the recodification did not substantively alter the meaning of the statute.

For three decades after the enactment of § 285, courts applied it—as they had applied §70—in a discretionary manner, assessing various factors to determine whether a given case was sufficiently "exceptional" to warrant a fee award.

In 1982, Congress created the Federal Circuit and vested it with exclusive appellate jurisdiction in patent cases. 28 U.S.C. § 1295. In the two decades that followed, the Federal Circuit, like the regional circuits before it, instructed district courts to consider the totality of the circumstances when making fee determinations under § 285. *See, e.g., Rohm & Haas Co. v. Crystal Chemical Co.*, 736 F.2d 688, 691 (Fed. Cir. 1984) ("Cases decided under § 285 have noted that 'the substitution of the phrase "in exceptional cases" has not done away with the discretionary feature'").

In 2005, however, the Federal Circuit abandoned that holistic, equitable approach in favor of a more rigid and mechanical formulation. In *Brooks Furniture Mfg. v. Dutailier Int'l, Inc.*, 393 F.3d 1378 (Fed. Cir. 2005), the court held that a case is "exceptional" under § 285 only "when there has been some material inappropriate conduct related to the matter in litigation, such as willful infringement, fraud or inequitable conduct in procuring the patent, misconduct during litigation, vexatious or unjustified litigation, conduct that violates Fed. R. Civ. P. 11, or like infractions." *Id.* at 1381. "Absent misconduct in conduct of the litigation or in securing the patent," the Federal Circuit continued, fees "may be imposed against the patentee only if both (1) the litigation is brought in subjective bad faith, and (2) the litigation is objectively baseless." *Id.* The Federal Circuit subsequently clarified that litigation is objectively baseless only if it is "so unreasonable that no reasonable litigant could believe it would succeed," *iLOR, LLC v. Google, Inc.*, 631 F.3d 1372, 1378 (Fed. Cir. 2011), and that litigation is brought in subjective bad faith only if the plaintiff "actually know[s]" that it is objectively baseless, *id.*, at 1377.

Finally, *Brooks Furniture* held that because "[t]here is a presumption that the assertion of infringement of a duly granted patent is made in good faith[,] * * * the underlying improper conduct and the characterization of the case as exceptional must be established by clear and convincing evidence." 393 F.3d at 1382.

B

The parties to this litigation are manufacturers of exercise equipment. The respondent, ICON Health & Fitness, Inc., owns U.S. Patent No. 6,019,710 ('710 patent), which discloses an elliptical exercise machine that allows for adjustments to fit the individual stride paths of users. ICON is a major manufacturer of exercise equipment, but it has never commercially sold the machine disclosed in the '710 patent. The petitioner, Octane Fitness, also manufactures exercise equipment, including elliptical machines known as the Q45 and Q47.

ICON sued Octane, alleging that the Q45 and Q47 infringed several claims of the '710 patent. The District Court granted Octane's motion for summary judgment, concluding that Octane's machines did not infringe ICON's patent. Octane then moved for attorney's fees under § 285. Applying the *Brooks Furniture* standard, the District Court denied Octane's motion. It determined that Octane could show neither that ICON's claim was objectively baseless nor that ICON had brought it in subjective bad faith. . . .

ICON appealed the judgment of noninfringement, and Octane cross-appealed the denial of attorney's fees. The Federal Circuit affirmed both orders. 496 Fed. Appx. 57 (Fed. Cir. 2012). In upholding the denial of attorney's fees, it rejected Octane's argument that the District Court had "applied an overly restrictive standard in refusing to find the case exceptional under § 285." *Id.* at 65. The Federal Circuit declined to "revisit the settled standard for exceptionality." *Id.*

We granted certiorari and now reverse.

II

The framework established by the Federal Circuit in *Brooks Furniture* is unduly rigid, and it impermissibly encumbers the statutory grant of discretion to district courts.

A

Our analysis begins and ends with the text of § 285: "The court in exceptional cases may award reasonable attorney fees to the prevailing party." This text is patently clear. It imposes one and only one constraint on district courts' discretion to award attorney's fees in patent litigation: The power is reserved for "exceptional" cases.

The Patent Act does not define "exceptional," so we construe it "'in accordance with [its] ordinary meaning.'" *Sebelius v. Cloer*, 133 S. Ct. 1886, 1893 (2013); *see also Bilski v. Kappos*, 130 S.Ct. 3218, 3226 (2010) ("In patent law, as in all statutory construction, unless otherwise defined, words will be interpreted as taking their ordinary, contemporary, common meaning."). In 1952, when Congress used the word in § 285 (and today, for that matter), "[e]xceptional" meant "uncommon," "rare," or "not ordinary." *Webster's New International Dictionary* 889 (2d ed. 1934); *see also* 3 *Oxford English Dictionary* 374 (1933) (defining "exceptional" as "out of the ordinary course," "unusual," or "special"); *Merriam-Webster's Collegiate Dictionary* 435 (11th ed. 2008) (defining "exceptional" as "rare"); *Noxell Corp. v. Firehouse No. 1 Bar-B-Que Restaurant*, 771 F.2d 521, 526 (D.C. Cir. 1985) (R.B. Ginsburg, J., joined by Scalia, J.) (interpreting the term "exceptional" in the Lanham Act's identical fee-shifting provision, 15 U. S. C. § 1117(a), to mean "uncommon" or "not run-of-the-mill").

We hold, then, that an "exceptional" case is simply one that stands out from others with respect to the substantive strength of a party's litigating position (considering both the governing law and the facts of the case) or the unreasonable manner in which the case was litigated. District courts may determine whether a case is "exceptional" in the case-by-case exercise of their discretion, considering the totality of the circumstances.[6] As in the comparable context of the Copyright Act, "there is no precise rule or formula for making these determinations, but instead equitable discretion should be exercised in light of the considerations we have identified." *Fogerty v. Fantasy, Inc.*, 510 U.S. 517, 534 (1994).

B

1

The Federal Circuit's formulation is overly rigid. Under the standard crafted in *Brooks Furniture*, a case is "exceptional" only if a district court either finds litigation-related misconduct of an independently sanctionable magnitude or determines that the litigation was both "brought in subjective bad faith" and "objectively baseless." 393 F.3d at 1381. This formulation superimposes an inflexible framework onto statutory text that is inherently flexible.

For one thing, the first category of cases in which the Federal Circuit allows fee awards—those involving litigation misconduct or certain other misconduct—appears to extend largely to independently sanctionable conduct. *See id.* (defining litigation-related misconduct to include "willful infringement, fraud or inequitable conduct in procuring the patent, misconduct during litigation, vexatious or unjustified litigation, conduct that violates Fed. R. Civ. P. 11, or like infractions"). But sanctionable conduct is not the appropriate benchmark. Under the standard announced today, a district court may award fees in the rare case in which a party's unreasonable conduct—while not necessarily independently sanctionable—is nonetheless so "exceptional" as to justify an award of fees.

The second category of cases in which the Federal Circuit allows fee awards is also too restrictive. In order for a case to fall within this second category, a district court must determine both that the litigation is objectively baseless and that the plaintiff brought it in subjective bad faith. But a case presenting either subjective bad faith or exceptionally meritless claims may sufficiently set itself apart from mine-run cases to warrant a fee award. *Cf. Noxell*, 771 F.2d at 526 ("[W]e think it fair to assume that Congress did not intend rigidly to limit recovery of fees by a [Lanham Act] defendant to the rare case in which a court finds that the plaintiff acted in bad faith, vexatiously, wantonly, or for oppressive reasons * * *. Something less than bad faith, we believe, suffices to mark a case as 'exceptional'").

. . . .

---

[6] In *Fogerty v. Fantasy, Inc.*, 510 U.S. 517 (1994), for example, we explained that in determining whether to award fees under a similar provision in the Copyright Act, district courts could consider a "nonexclusive" list of "factors," including "frivolousness, motivation, objective unreasonableness (both in the factual and legal components of the case) and the need in particular circumstances to advance considerations of compensation and deterrence." *Id.* at 534, n.19.

2

We reject *Brooks Furniture* for another reason: It is so demanding that it would appear to render § 285 largely superfluous. We have long recognized a common-law exception to the general "American rule" against fee-shifting—an exception, "inherent" in the "power [of] the courts" that applies for "willful disobedience of a court order" or "when the losing party has acted in bad faith, vexatiously, wantonly, or for oppressive reasons * * * ." *Alyeska Pipeline Service Co. v. Wilderness Society*, 421 U.S. 240, 258-259 (1975). We have twice declined to construe fee-shifting provisions narrowly on the basis that doing so would render them superfluous, given the background exception to the American rule, see *Christiansburg Garment Co. v. EEOC*, 434 U.S. 412, 419 (1978); *Newman v. Piggie Park Enterprises, Inc.*, 390 U.S. 400, 402 (1968), and we again decline to do so here.

3

Finally, we reject the Federal Circuit's requirement that patent litigants establish their entitlement to fees under § 285 by "clear and convincing evidence," *Brooks Furniture*, 393 F.3d at 1382. We have not interpreted comparable fee-shifting statutes to require proof of entitlement to fees by clear and convincing evidence. *See, e.g., Fogerty*, 510 U.S. at 519; *Cooter & Gell v. Hartmarx Corp.*, 496 U.S. 384 (1990); *Pierce v. Underwood*, 487 U.S. 552, 558 (1988). And nothing in § 285 justifies such a high standard of proof. Section 285 demands a simple discretionary inquiry; it imposes no specific evidentiary burden, much less such a high one. Indeed, patent-infringement litigation has always been governed by a preponderance of the evidence standard, *see, e.g., Béné v. Jeantet*, 129 U.S. 683, 688 (1889), and that is the "standard generally applicable in civil actions," because it "allows both parties to 'share the risk of error in roughly equal fashion,'" *Herman & MacLean v. Huddleston*, 459 U.S. 375, 390 (1983).

*   *   *

For the foregoing reasons, the judgment of the United States Court of Appeals for the Federal Circuit is reversed, and the case is remanded for further proceedings consistent with this opinion.

Highmark Inc. v. Allcare Health Management Sys.
134 S. Ct. 1744 (2014)

SOTOMAYOR, JUSTICE:

Section 285 of the Patent Act provides: "The court in exceptional cases may award reasonable attorney fees to the prevailing party." 35 U.S.C. § 285. . . . We granted certiorari to determine whether an appellate court should accord deference to a district court's determination that litigation is "objectively baseless." On the basis of our opinion in *Octane Fitness, LLC v. ICON Health & Fitness, Inc.*, argued together with this case and also issued today, we hold that an appellate court should review all aspects of a district court's § 285 determination for abuse of discretion.

I

Allcare Health Management System, Inc., owns U. S. Patent No. 5,301,105 ('105 patent), which covers "utilization review" in "managed health care systems."[1] 687 F.3d 1300, 1306 (Fed. Cir. 2012). Highmark Inc., a health insurance company, sued Allcare seeking a declaratory judgment that the '105 patent was invalid and unenforceable and that, to the extent it was valid, Highmark's actions were not infringing it. Allcare counterclaimed for patent infringement. Both parties filed motions for summary judgment, and the District Court entered a final judgment of noninfringement in favor of Highmark. The Federal Circuit affirmed. 329 Fed. Appx. 280 (Fed. Cir. 2009).

Highmark then moved for fees under § 285. The District Court granted Highmark's motion. The court reasoned that Allcare had engaged in a pattern of "vexatious" and "deceitful" conduct throughout the litigation. Specifically, it found that Allcare had "pursued this suit as part of a bigger plan to identify companies potentially infringing the '105 patent under the guise of an informational survey, and then to force those companies to purchase a license of the '105 patent under threat of litigation." And it found that Allcare had "maintained infringement claims [against Highmark] well after such claims had been shown by its own experts to be without merit" and had "asserted defenses it and its attorneys knew to be frivolous." In a subsequent opinion, the District Court fixed the amount of the award at $4,694,727.40 in attorney's fees and $209,626.56 in expenses, in addition to $375,400.05 in expert fees.

The Federal Circuit affirmed in part and reversed in part. 687 F.3d 1300. It affirmed the District Court's exceptional-case determination with respect to the allegations that Highmark's system infringed one claim of the '105 patent but reversed the determination with respect to another claim of the patent. In reversing the exceptional-case determination as to one claim, the court reviewed it *de novo*. The court held that because the question whether litigation is "objectively baseless" under *Brooks Furniture* "is a question of law based on underlying mixed questions of law and fact," an objective-baselessness determination is reviewed on appeal "*de novo*" and "without deference." It then determined, contrary to the judgment of the District Court, that "Allcare's argument" as to claim construction "was not so unreasonable that no reasonable litigant could believe it would succeed." The court further found that none of Allcare's conduct warranted an award of fees under the litigation-misconduct prong of *Brooks Furniture*.

. . . .

We granted certiorari and now vacate and remand.

II

Our opinion in *Octane Fitness, LLC v. ICON Health & Fitness, Inc.*, rejects the *Brooks Furniture* framework as unduly rigid and inconsistent with the text of § 285. It holds, instead, that the word "exceptional" in § 285 should be interpreted in

---

[1] "'Utilization review' is the process of determining whether a health insurer should approve a particular treatment for a patient." 687 F.3d at 1306.

accordance with its ordinary meaning . . . [and] that "[d]istrict courts may determine whether a case is 'exceptional' in the case-by-case exercise of their discretion, considering the totality of the circumstances." Our holding in *Octane* settles this case: Because § 285 commits the determination whether a case is "exceptional" to the discretion of the district court, that decision is to be reviewed on appeal for abuse of discretion.

Traditionally, decisions on "questions of law" are "reviewable *de novo*," decisions on "questions of fact" are "reviewable for clear error," and decisions on "matters of discretion" are "reviewable for 'abuse of discretion.'" *Pierce v. Underwood*, 487 U.S. 552, 558 (1988). For reasons we explain in *Octane*, the determination whether a case is "exceptional" under §285 is a matter of discretion. And as in our prior cases involving similar determinations, the exceptional-case determination is to be reviewed only for abuse of discretion.[2] *See Pierce*, 487 U.S. at 559 (determinations whether a litigating position is "substantially justified" for purposes of fee-shifting under the Equal Access to Justice Act are to be reviewed for abuse of discretion); *Cooter & Gell v. Hartmarx Corp.*, 496 U.S. 384, 405 (1990) (sanctions under Federal Rule of Civil Procedure 11 are to be reviewed for abuse of discretion).

As in *Pierce*, the text of the statute "emphasizes the fact that the determination is for the district court," which "suggests some deference to the district court upon appeal," 487 U.S. at 559. As in *Pierce*, "as a matter of the sound administration of justice," the district court "is better positioned" to decide whether a case is exceptional, *id.* at 559-560, because it lives with the case over a prolonged period of time. And as in *Pierce*, the question is "multifarious and novel," not susceptible to "useful generalization" of the sort that de novo review provides, and "likely to profit from the experience that an abuse-of-discretion rule will permit to develop," *id.* at 562.

We therefore hold that an appellate court should apply an abuse-of-discretion standard in reviewing all aspects of a district court's § 285 determination. Although questions of law may in some cases be relevant to the § 285 inquiry, that inquiry generally is, at heart, "rooted in factual determinations," *Cooter*, 496 U.S. at 401.

. . . .

Notes & Questions

1. Does the Supreme Court's rejection of the dual requirement of both subjective bad faith and objective baselessness in the context of attorney fee awards make sense? Recall that the Federal Circuit has adopted a similar dual requirement for a finding of willful infringement. *See In re Seagate Tech.*, 497 F.3d 1360, 1371 (Fed.

---

[2] The abuse-of-discretion standard does not preclude an appellate court's correction of a district court's legal or factual error: "A district court would necessarily abuse its discretion if it based its ruling on an erroneous view of the law or on a clearly erroneous assessment of the evidence." *Cooter & Gell v. Hartmarx Corp.*, 496 U.S. 384, 405 (1990).

Cir. 2007) (en banc) (requiring both and "objectively high likelihood that defendant's actions constituted infringement and that this was either known or so obvious that it should have been known to the accused infringer). Does the Supreme Court's *Octane Fitness* decision cast doubt on the requirement of a dual standard for determining wilfullness? Or is an "exceptional" case different from a case of "willful infringement"? If they *are* different, which one should be harder to prove: willfulness or exceptionality?

2. The *Octane Fitness* decision makes it easier for district courts to award attorney fees in patent infringement cases. Some believe that a greater willingness to award fees against plaintiffs who assert patent claims inappropriately will help alleviate the problem with patent "trolls." Do you agree?

3. In *Octane Fitness* the court references the fee shifting provision in the Copyright Act and the Court's earlier decision in *Fogerty v. Fantasy, Inc.*, 510 U.S. 517 (1994). That case involved a lower court determination of no infringement. The fee shifting provisions in all of the federal intellectual property statutes grant courts the discretion to award attorney fees to either the plaintiff or the defendant—whoever is the "prevailing party." The Supreme Court provided the following guidance on when it would be appropriate to award fees to a prevailing defendant:

> Because copyright law ultimately serves the purpose of enriching the general public through access to creative works, it is peculiarly important that the boundaries of copyright law be demarcated as clearly as possible. To that end, defendants who seek to advance a variety of meritorious copyright defenses should be encouraged to litigate them to the same extent that plaintiffs are encouraged to litigate meritorious claims of infringement. . . . [A] successful defense of a copyright infringement action may further the policies of the Copyright Act every bit as much as a successful prosecution of an infringement claim by the holder of a copyright.

Id. at 527 (rejecting a "dual standard" under which prevailing defendants had to prove that the claim was frivolous or lacking in good faith in order to recover fees). On remand the district court granted the defendant $1.3 million in attorney fees, finding that the defendant was an author; the defendant had "prevailed on the merits rather than on a technical defense, such as the statute of limitations, laches, or the copyright registration requirements[;]" "the benefit conferred by Fogerty's successful defense was not slight or insubstantial relative to the costs of litigation[; and] the fee award [would not] have too great a chilling effect or impose an inequitable burden on Fantasy, which was not an impecunious plaintiff." *Fantasy, Inc. v. Fogerty*, 94 F.3d 553, 556 (9th Cir. 1996)

The UTSA provides that a court may award attorney fees to a defendant if a plaintiff has made a misappropriation claim in bad faith. What policy goal is served by permitting an award of attorney fees to a defendant in a trade secret case? Does it make sense to limit awards of attorney fees to defendant in trade secret cases to only those cases in which the plaintiff has asserted a claim in bad faith?

II: Looking Forward: Preventing Future Harm

A. Permanent Injunctive Relief

We begin with the Supreme Court's most recent decision concerning injunctions in intellectual property cases. The case involves the Patent Act which expressly gives the federal courts broad discretion to enjoin patent infringement: "The several courts having jurisdiction of cases under this title may grant injunctions in accordance with the principles of equity to prevent the violation of any right secured by patent, on such terms as the court deems reasonable." 35 U.S.C. § 283. Prior to the *eBay* case, the Federal Circuit caselaw mandated a permanent injunction in a prevailing patentee's favor in all but the most unusual circumstances. Statements like the following reflected the Federal Circuit's injunction standards: "Infringement having been established, it is contrary to the laws of property, of which the patent law partakes, to deny the patentee's right to exclude others from use of his property." *Richardson v. Suzuki Motor Co.*, 868 F.2d 1226, 1246-47 (Fed. Cir. 1989). Courts also routinely granted permanent injunctive relief once copyright infringement had been demonstrated. For example, the Ninth Circuit stated that "[a]s a general rule, a permanent injunction will be granted when liability has been established and there is a threat of continuing violations." *MAI Sys. Corp. v. Peak Computer, Inc.*, 991 F.2d 511, 520 (9th Cir. 1993). A presumption for granting a permanent injunction was also pervasive in trademark cases. The following decision significantly changed the proof required to obtain injunctive relief.

eBay Inc. v. MercExchange, L.L.C.

547 U.S. 388 (2006)

Thomas, Justice:

Ordinarily, a federal court considering whether to award permanent injunctive relief to a prevailing plaintiff applies the four-factor test historically employed by courts of equity. Petitioners eBay Inc. and Half.com, Inc., argue that this traditional test applies to disputes arising under the Patent Act. We agree and, accordingly, vacate the judgment of the Court of Appeals.

I

Petitioner eBay operates a popular Internet Web site that allows private sellers to list goods they wish to sell, either through an auction or at a fixed price. Petitioner Half.com, now a wholly owned subsidiary of eBay, operates a similar Web site. Respondent MercExchange holds a number of patents, including a business method patent for an electronic market designed to facilitate the sale of goods between private individuals by establishing a central authority to promote trust among participants. See U.S. Patent No. 5,845,265. MercExchange sought to license its patent to eBay and Half.com, as it had previously done with other companies, but the parties failed to reach an agreement. MercExchange subsequently filed a patent infringement suit against eBay and Half.com in the United States

District Court for the Eastern District of Virginia. A jury found that Mer-cExchange's patent was valid, that eBay and Half.com had infringed that patent, and that an award of damages was appropriate.

Following the jury verdict, the District Court denied MercExchange's motion for permanent injunctive relief. The Court of Appeals for the Federal Circuit reversed, applying its "general rule that courts will issue permanent injunctions against patent infringement absent exceptional circumstances." 401 F.3d 1323, 1339 (2005). We granted certiorari to determine the appropriateness of this general rule.

II

According to well-established principles of equity, a plaintiff seeking a permanent injunction must satisfy a four-factor test before a court may grant such relief. A plaintiff must demonstrate: (1) that it has suffered an irreparable injury; (2) that remedies available at law, such as monetary damages, are inadequate to compensate for that injury; (3) that, considering the balance of hardships between the plaintiff and defendant, a remedy in equity is warranted; and (4) that the public interest would not be disserved by a permanent injunction. See, *e.g.*, *Weinberger* v. *Romero-Barcelo,* 456 U.S. 305, 311–313 (1982); *Amoco Production Co.* v. *Gambell,* 480 U.S. 531, 542 (1987). The decision to grant or deny permanent injunctive relief is an act of equitable discretion by the district court, reviewable on appeal for abuse of discretion.

These familiar principles apply with equal force to disputes arising under the Patent Act. As this Court has long recognized, "a major departure from the long tradition of equity practice should not be lightly implied." *Id.*; see also *Amoco, supra*, at 542. Nothing in the Patent Act indicates that Congress intended such a departure. To the contrary, the Patent Act expressly provides that injunctions "may" issue "in accordance with the principles of equity." 35 U.S.C. § 283.

To be sure, the Patent Act also declares that "patents shall have the attributes of personal property," § 261, including "the right to exclude others from making, using, offering for sale, or selling the invention," § 154(a)(1). According to the Court of Appeals, this statutory right to exclude alone justifies its general rule in favor of permanent injunctive relief. But the creation of a right is distinct from the provision of remedies for violations of that right. Indeed, the Patent Act itself indicates that patents shall have the attributes of personal property "[s]ubject to the provisions of this title," § 261, including, presumably, the provision that injunctive relief "may" issue only "in accordance with the principles of equity," § 283.

This approach is consistent with our treatment of injunctions under the Copyright Act. Like a patent owner, a copyright holder possesses "the right to exclude others from using his property." *Fox Film Corp.* v. *Doyal,* 286 U. S. 123, 127 (1932); see also *id.* at 127–128 ("A copyright, like a patent, is at once the equivalent given by the public for benefits bestowed by the genius and meditations and skill of individuals, and the incentive to further efforts for the same important objects" (internal quotation marks omitted)). Like the Patent Act, the Copyright Act provides that courts "may" grant injunctive relief "on such terms as it may deem reasonable to prevent or restrain infringement of a copyright." 17 U.S.C. § 502(a). And as in our decision today, this Court has consistently rejected invitations to replace traditional equitable considerations with a rule that an injunction automatically follows a determination that a copyright has been infringed. See, *e.g.*, *New York Times Co.* v. *Tasini,* 533 U.S. 483, 505 (2001) (citing *Campbell* v. *Acuff-Rose Music, Inc.,* 510 U.S. 569, 578, n. 10 (1994)); *Dun* v. *Lumbermen's Credit Assn.,* 209 U.S. 20, 23–24 (1908).

Neither the District Court nor the Court of Appeals below fairly applied these traditional equitable principles in deciding respondent's motion for a permanent injunction. Although the District Court recited the traditional four-factor test, it appeared to adopt certain expansive principles suggesting that injunctive relief could not issue in a broad swath of cases. Most notably, it concluded that a "plaintiff's willingness to license its patents" and "its lack of commercial activity in practicing the patents" would be sufficient to establish that the patent holder would not suffer irreparable harm if an injunction did not issue. But traditional equitable principles do not permit such broad classifications. For example, some patent holders, such as university researchers or self-made inventors, might reasonably prefer to license their patents, rather than undertake efforts to secure the financing necessary to bring their works to market themselves. Such patent holders may be able to satisfy the traditional four-factor test, and we see no basis for categorically denying them the opportunity to do so. To the extent that the District Court adopted such a categorical rule, then, its analysis cannot be squared with the principles of equity adopted by Congress. The court's categorical rule is also in tension with *Continental Paper Bag Co.* v. *Eastern Paper Bag Co.,* 210 U.S. 405, 422–430 (1908), which rejected the contention that a court of equity has no jurisdiction to grant injunctive relief to a patent holder who has unreasonably declined to use the patent.

In reversing the District Court, the Court of Appeals departed in the opposite direction from the four-factor test. The court articulated a "general rule," unique to patent disputes, "that a permanent injunction will issue once infringement and validity have been adjudged." 401 F.3d at 1338. The court further indicated that injunctions should be denied only in the "unusual" case, under "exceptional circumstances" and "in rare instances * * * to protect the public interest." *Id.* at 1338–1339. Just as the District Court erred in its categorical denial of injunctive relief, the Court of Appeals erred in its categorical grant of such relief.

Because we conclude that neither court below correctly applied the traditional four-factor framework that governs the award of injunctive relief, we vacate the judgment of the Court of Appeals, so that the District Court may apply that framework in the first instance. In doing so, we take no position on whether permanent injunctive relief should or should not issue in this particular case, or indeed in any number of other disputes arising under the Patent Act. We hold only that the decision whether to grant or deny injunctive relief rests within the equitable discretion of the district courts, and that such discretion must be exercised consistent with traditional principles of equity, in patent disputes no less than in other cases governed by such standards.

Accordingly, we vacate the judgment of the Court of Appeals, and remand for further proceedings consistent with this opinion.

ROBERTS, CHIEF JUSTICE, concurring (with Justices Scalia & Ginsburg):

I agree with the Court's holding that "the decision whether to grant or deny injunctive relief rests within the equitable discretion of the district courts, and that such discretion must be exercised consistent with traditional principles of equity, in patent disputes no less than in other cases governed by such standards," and I join the opinion of the Court. That opinion rightly rests on the proposition that "a major departure from the long tradition of equity practice should not be lightly implied." *Weinberger* v. *Romero-Barcelo,* 456 U.S. 305, 320 (1982).

From at least the early 19th century, courts have granted injunctive relief upon a finding of infringement in the vast majority of patent cases. This "long tradition of equity practice" is not surprising, given the difficulty of protecting a right to *exclude* through monetary remedies that allow an infringer to *use* an invention against the patentee's wishes—a difficulty that often implicates the first two factors of the traditional four-factor test. This historical practice, as the Court holds, does not *entitle* a patentee to a permanent injunction or justify a *general rule* that such injunctions should issue. . . . At the same time, there is a difference between exercising equitable discretion pursuant to the established four-factor test and writing on an entirely clean slate. "Discretion is not whim, and limiting discretion according to legal standards helps promote the basic principle of justice that like cases should be decided alike." *Martin* v. *Franklin Capital Corp.,* 126 S. Ct. 704, 710 (2005). When it comes to discerning and applying those standards, in this area as others, "a page of history is worth a volume of logic." *New York Trust Co.* v. *Eisner,* 256 U.S. 345, 349 (1921) (Holmes, J.).

KENNEDY, JUSTICE, concurring (with Justices Stevens, Souter, & Breyer):

The Court is correct, in my view, to hold that courts should apply the well-established, four-factor test—without resort to categorical rules—in deciding whether to grant injunctive relief in patent cases. The Chief Justice is also correct that history may be instructive in applying this test. The traditional practice of issuing injunctions against patent infringers, however, does not seem to rest on "the difficulty of protecting a right to *exclude* through monetary remedies that allow an infringer to *use* an invention against the patentee's wishes." (Roberts, C. J., concurring). Both the terms of the Patent Act and the traditional view of injunctive relief accept that the existence of a right to exclude does not dictate the remedy for a violation of that right. To the extent earlier cases establish a pattern of granting an injunction against patent infringers almost as a matter of course, this pattern simply illustrates the result of the four-factor test in the contexts then prevalent. The lesson of the historical practice, therefore, is most helpful and instructive when the circumstances of a case bear substantial parallels to litigation the courts have confronted before.

In cases now arising trial courts should bear in mind that in many instances the nature of the patent being enforced and the economic function of the patent holder present considerations quite unlike earlier cases. An industry has developed in which firms use patents not as a basis for producing and selling goods but, instead, primarily for obtaining licensing fees. See Federal Trade Commission, *To Promote Innovation: The Proper Balance of Competition and Patent Law and Policy*, ch. 3, pp. 38–39 (Oct. 2003), available at http://www.ftc.gov/os/2003/10/innovationrpt.pdf. For these firms, an injunction, and the potentially serious sanctions arising from its violation, can be employed as a bargaining tool to charge exorbitant fees to companies that seek to buy licenses to practice the patent. When the patented invention is but a small component of the product the companies seek to produce and the threat of an injunction is employed simply for undue leverage in negotiations, legal damages may well be sufficient to compensate for the infringement and an injunction may not serve the public interest. In addition injunctive relief may have different consequences for the burgeoning number of patents over business methods, which were not of much economic and legal significance in earlier times. The potential vagueness and suspect validity of some of these patents may affect the calculus under the four-factor test.

The equitable discretion over injunctions, granted by the Patent Act, is well suited to allow courts to adapt to the rapid technological and legal developments in the patent system. For these reasons it should be recognized that district courts must determine whether past practice fits the circumstances of the cases before them. With these observations, I join the opinion of the Court.

Notes & Questions

1. On remand, the trial court denied MercExchange's renewed motion for a permanent injunction. 500 F. Supp. 2d 556 (E.D. Va. 2007). The parties then settled, with eBay agreeing to "buy three patents from MercExchange needed to run its web search, online auctions and fixed price sales, as well as other assets, for an undisclosed sum." Reuters, *eBay, MercExchange Settle Long Patent Legal Fight* (Feb. 28, 2008), available at http://www.reuters.com/article/company-News/idUSWNAS295020080228.

2. How should a trial court decide whether the infringer's conduct, if it continues, would cause the type of harm a patent injunction is intended to prevent?

3. Justice Kennedy appears concerned that we keep the standards for granting equitable remedies responsive to changing circumstances. Is his concern in *eBay* in tension with his caution in *Festo*, the doctrine-of-equivalents case, that patent courts should guard against disturbing patentees' settled expectations? What is the difference, if any, between a settled expectation about the scope of one's right and a settled expectation about the remedies available for infringements of that right?

As noted at the beginning of this section, prior to the *eBay* decision, many courts had routinely granted permanent injunctive relief once copyright infringement had been demonstrated. After *eBay* the standard has changed. For example, upon remand from the Supreme Court in the *Grokster* case concerning inducement to infringement that you read in Section VII of Chapter 4, the district court struggled with what evidence is sufficient to prove irreparable harm, the first prong of the four-prong test articulated by the Supreme Court:

> Irreparable harm cannot be established solely on the fact of past infringement. Additionally, it must also be true that the mere likelihood of future infringement by a defendant does not by itself allow for an inference of irreparable harm. As to the latter, future copyright infringement can always be redressed via damages, whether actual or statutory. . . .
>
> "[I]rreparable harm may not be presumed[, but] [i]n run-of-the-mill copyright litigation, such proof should not be difficult to establish * * * ." 6 [William F.] Patry, [*Patry on Copyrights,*] § 22:74. Thus, Plaintiffs may establish an irreparable harm stemming from the infringement (e.g., loss of market share, reputational harm). It is also possible that some qualitative feature about the infringement itself, such as its peculiar nature, could elevate its status into the realm of "irreparable harm."
>
> . . . .
>
> In *eBay*, Chief Justice Roberts indicated that irreparable harm can result from the infringement itself, depending upon the circumstances of the case:

> > From at least the early 19th century, courts have granted injunctive relief upon a finding of infringement in the vast majority of patent cases. This "long tradition of equity practice" is not surprising, given the difficulty of protecting a right to exclude through monetary remedies that allow an infringer to use an invention against the patentee's wishes-a difficulty that often implicates the first two factors of the traditional four-factor test.

> 126 S. Ct. at 1841 (Roberts, C.J., concurring). And "[l]ike a patent owner, a copyright holder possesses 'the right to exclude others from using his property.'" *eBay*, at 1840. . . .
>
> This Court also recognizes that a competing *eBay* concurrence took issue with Chief Justice Roberts's "right to exclude" language. Justice Kennedy explained his view that "the existence of a right to exclude does not dictate the remedy for a violation of that right." *eBay*, 126 S.Ct. at 1842 (Kennedy, J., concurring). This Court agrees, since a contrary conclusion would come close to permitting a presumption of irreparable harm. This Court also observes that Justice Kennedy's statement was made primarily in the context of certain recent developments in the patent field that are wholly inapplicable to this lawsuit. For example, this is simply not a case in which the copyright infringement represents "but a small component of the

product the companies seek to produce," such that "legal damages may well be sufficient to compensate for the infringement." *Id*. As this Court previously held, StreamCast's entire business was built around the fundamental premise that Morpheus would be utilized to infringe copyrights, including those owned by Plaintiffs. Furthermore, Justice Kennedy emphasized that "[t]he equitable discretion over injunctions . . . is well suited to allow courts to adapt to the rapid technological and legal developments. . . ." *Id*. Given the technological aspects of the infringement induced by StreamCast, and the flexibility conferred by the Copyright Act, this Court is persuaded that its bases for finding irreparable harm, *infra*, are supported by both Chief Justice Roberts's and Justice Kennedy's concurrences.

In light of this authority, the Court concludes that certain qualities pertaining to the nature of StreamCast's inducement of infringement are relevant to a finding of irreparable harm. . . .

The irreparable harm analysis centers on two basic themes: (1) StreamCast has and will continue to induce far more infringement than it could ever possibly redress with damages; and (2) Plaintiffs' copyrights (especially those of popular works) have and will be rendered particularly vulnerable to continuing infringement on an enormous scale due to StreamCast's inducement. The Court agrees with both arguments, and each is independently sufficient to support of finding of irreparable harm in this case.

Metro-Goldwyn-Mayer Studios v. Grokster, 518 F. Supp. 2d 1197, 1214-18 (N.D. Cal. 2007).

The *Restatement (3rd) of Unfair Competition* explains the tilt toward injunctive relief in Lanham Act cases this way:

> In cases of deceptive marketing, trademark infringement, or trademark dilution, a prevailing plaintiff is ordinarily awarded injunctive relief to protect both the plaintiff and the public from the likelihood of future harm.
> . . .
> . . . The plaintiff's interest in protecting the good will symbolized by its trademark . . . is unlikely to be adequately secured by monetary relief, and the equities thus normally favor the award of an injunction. The public interest in preventing confusion and deception also typically weigh in favor of an injunction. . . .

Rest. (3rd) Unfair Competition § 35, cmnt. b (1995).

Protecting a secret from threatened disclosure or from use by a competitor is often the top priority in any new trade secret litigation. The UTSA, Section 2 provides:

> (a) Actual or threatened misappropriation may be enjoined. Upon application to the court, an injunction shall be terminated when the trade secret has ceased to exist, but the injunction may be continued for an additional reasonable period of time in order to eliminate commercial advantage that otherwise would be derived from the misappropriation.

(b) In exceptional circumstances, an injunction may condition future use upon payment of a reasonable royalty for no longer than the period of time for which use could have been prohibited. Exceptional circumstances include, but are not limited to, a material and prejudicial change of position prior to acquiring knowledge or reason to know of misappropriation that renders a prohibitive injunction inequitable.

(c) In appropriate circumstances, affirmative acts to protect a trade secret may be compelled by court order.

When litigating any particular trade secret case, the applicable state law must be consulted. Even when litigating in one of the many jurisdictions that have adopted the UTSA, the enacted version of the UTSA in that jurisdiction must be examined, as variations from the uniform act do exist.

Once a court uses its equitable power to issue an injunction, it retains jurisdiction over that injunction and can modify or vacate it at any time, although such action is usually based on a motion from one or more of the parties. A court uses its power of contempt to enforce compliance with any injunction it issues. Sanctions for contempt can be serious, ranging from fines to even jail time. 18 U.S.C. § 401(3). *See also Young v. United States ex rel. Vuitton et Fils, S.A.,* 481 U.S. 787 (1987); Earl C. Dudley, Jr., *Getting Beyond the Civil/Criminal Distinction: a New Approach to the Regulation of Indirect Contempts,* 79 Va. L. Rev. 1025 (1993).

Notes & Questions

1. Keep in mind that irreparable harm is only one of the four factors that a court must consider in determining whether to exercise its discretion to grant injunctive relief. If an injunction is not granted and the defendant continues to infringe, under the standards for willfulness explored earlier in this Section, what remedies would an intellectual property owner be entitled to in a second lawsuit against the same defendant? Is that enough to discourage the defendant from continuing to infringe?

2. When a right is protected solely by a damages remedy we call that a "liability rule." When, on the other hand, injunctive relief is available, the term that applies is "property rule." *eBay* has weakened the nature of the property rule in intellectual property cases. No longer is an injunction automatic upon proof of infringement. If the court denies an injunction should it grant a running royalty for future infringement? If it does so, it would be creating a compulsory license— in effect forcing the intellectual property owner to grant a license to use the intellectual property at a price set by the court. The Supreme Court has indicated in dicta several times that such a course of action may be appropriate, despite no express statutory authority for such a remedy. *See, e.g., Campbell v. Acuff-Rose,* 510 U.S. 569, 578 n. 10 (*reproduced supra* Chapter 4); *see also* Tomás Gómez Arostegui, *Prospective Compensation in Lieu of a Final Injunction in Patent and Copyright Cases,* 78 Fordham L. Rev. 1661 (2010). When might such a remedy be appropriate? If the reasonable royalty used as a floor in calculating damages is intended to approximate what a willing licensor and licensee would agree to, is there any problem in using this measure for ongoing use of the patented technology?

3. Throughout the chapters on copyright and trademark law you have seen the importance of balancing the intellectual property protections with the protections of the First Amendment. Injunctions in these types of cases constitute state action aimed at a type of expressive activity. Thus, in the fourth consideration, the public interest, it is appropriate to take into account the speech interests that may be affected by the injunction that the copyright owner seeks. Mark Lemley and Eugene Volokh, *Freedom of Speech and Injunctions in Intellectual Property Cases,* 48 Duke L.J. 147 (1998). What other interests might courts consider?

4. The Supreme Court's decision in *eBay* has refocused court attention on the equitable requirement of proof of irreparable injury before granting injunctive relief in patent, copyright and trademark cases. Back when law and equity courts were separate, before granting an injunction, an equity court would generally require the plaintiff to show that there was no adequate remedy at law. In trademark cases, what is the strongest argument to be made for a presumption of irreparable harm, *i.e.*, a presumption that the defendant's actions are causing a harm that an award of damages cannot repair?

When entering any injunction, the court must craft language to specify what, exactly, the defendant is prohibited from doing.

Tamko Roofing Prods. v. Ideal Roofing Co.
282 F.3d 23 (1st Cir. 2002)

Lynch, Judge:

[Review the facts of this case, reproduced above in Section I.A.3.]

. . . .

D. The Permanent Injunction Against Ideal's Use of "H-Series"

The district court issued a permanent injunction against Ideal on August 30, 2000, barring Ideal "from using the term Heritage, Heritage Series, H Series, or any name or mark confusingly similar to Heritage in connection with the sale, offer to sell, promotion, marketing, or advertising of any roofing product or service in the United States."

Ideal argues that the injunction is overbroad in barring use of "H-Series" by Ideal, because H-Series is not one of Tamko's registered trademarks. "[I]njunctive relief should be no more burdensome to the defendant than necessary to provide complete relief to plaintiffs," *Califano v. Yamasaki,* 442 U.S. 682, 702 (1979), and courts must "closely tailor injunctions to the harm that they address," *ALPO Petfoods[, Inc. v. Ralston Purina Co.],* 913 F.2d [958,] 972 [(D.C. Cir. 1990)]. While generally the issuance of injunctive relief is reviewed for abuse of discretion, we review underlying factual determinations for clear error. *I.P. Lund Trading ApS v. Kohler Co.,* 163 F.3d 27, 33 (1st Cir. 1998).

On different facts, we might have more sympathy for a claim that an injunction against the use of a mark not registered to plaintiff is overbroad. Not here. This case is a perfect example of the need for the "safe distance rule," which counsels that "an infringer, once caught, must expect some fencing in. * * * Thus, a court can frame an injunction which will keep a proven infringer safely away from the perimeter of future infringement." 5 *McCarthy on Trademarks, supra,* § 30:4, at 30-12. Indeed, it was after Ideal faced contempt charges that it came up with "H-Series," effectively dropping the "eritage" of "Heritage Series." The district court, during the contempt proceedings, heard evidence that Ideal's use of "H-Series" would cause confusion: Tamko representatives and their customers used "H" as an abbreviated reference for Tamko Heritage products, and Tamko's "Heritage 25" product is often referred to as H25. There are circumstances in which abbreviations of trademarks may be protectable as independent marks. Whether or not that is the case here, the evidence that Ideal's use of "H-Series" would cause confusion is sufficient to justify the injunction requiring Ideal to steer clear of this similar abbreviation of the mark.

III.

Although Ideal's counsel on appeal have striven mightily, the trial record dooms the appeal. The judgment is *affirmed.* . . .

Notes & Questions

1. Imagine the district court in this case had, pursuant to the safe distance rule, barred Ideal from *any* use of the word "Series" in a mark for its metal roofing products. Would that be improperly overbroad, or merely a good way to keep Ideal safely away from Tamko's marks? What argument could Ideal have made, at the infringement stage, that its use of the word "Series" could not be the predicate for a likelihood of confusion finding?

2. The Lanham Act also empowers the courts—in cases where the court finds a defendant to have violated either basic infringement provision, §§ 32 and 43(a), *or* to have willfully violated the antidilution provision, § 43(c)—to impound and destroy infringing items. Section 1118 states, in relevant part, as follows:

> the court may order that all labels, signs, prints, packages, wrappers, receptacles, and advertisements in the possession of the defendant, bearing the . . . mark or . . . the word, term, name, symbol, device, combination thereof, designation, description, or representation that is the subject of the violation, or any reproduction, counterfeit, copy, or colorable imitation thereof, and all plates, molds, matrices, and other means of making the same, shall be delivered up and destroyed.

15 U.S.C. § 1118. As you can see, the statute aims to be comprehensive.

3. The Copyright Act provides that courts may "grant temporary and final injunctions on such terms as it may deem reasonable to prevent or restrain infringement of a copyright." 17 U.S.C. § 502. In addition to furnishing the general power to grant injunctions, the statute allows courts to order the impoundment and destruction of copies or phonorecords "made or used in violation of the copyright owner's exclusive rights, and of all plates, molds, matrices, masters, tapes, film negatives, or other articles by means of which such copies or phonorecords may be reproduced." 17 U.S.C. § 503.

B. PRELIMINARY INJUNCTIONS

The Supreme Court's *eBay* decision affected not only the standard for permanent injunctions, it also changed the way courts approach granting preliminary injunctions. The hearing in which a plaintiff is granted the preliminary injunction is often referred to as a "show cause" hearing because the plaintiff moves for the defendant to "show cause why the preliminary injunctive relief should not issue." The burden, however, remains on the plaintiff.

Salinger v. Colting
607 F.3d 68 (2d Cir. 2010)

CALABRESI, JUDGE:

[J.D. Salinger wrote the coming-of-age story about disaffected sixteen-year-old Holden Caulfield, *The Catcher in the Rye*, in 1951. In 2009, without Salinger's permission, the defendant wrote *60 Years Later: Coming Through the Rye*, published through his own publishing company. Defendant's book tells the story of a 76-year-old Holden Caulfield in a world that includes a "fictionalized Salinger" character. The district court granted Salinger's motion for a preliminary injunction. The defendant appealed, arguing that the Second Circuit's standard for granting preliminary injunctions was an unconstitutional prior restraint on speech and that it was in conflict with the Supreme Court's decision in *eBay*.]

. . . .

This Court's pre-*eBay* standard for when preliminary injunctions may issue in copyright cases is inconsistent with the principles of equity set forth in *eBay*. The Supreme Court's decision in *Winter* [*v. Natural Resources Defense Counsel*, 129 S. Ct. 365 (2008)] tells us that, at minimum, we must consider whether "irreparable injury is likely in the absence of an injunction," we must "'balance the competing claims of injury,'" and we must "'pay particular regard for the public consequences in employing the extraordinary remedy of injunction.'" 129 S. Ct. at 375-77 (quoting *Amoco [Production Co. v. Village of Gambell*, 480 U.S. 531, 542 (1987)]; *Weinberger [v. Romero-Barcelo*, 456 U.S. 305, 312 (1982)]). Therefore, in light of *Winter* and *eBay*, we hold that a district court must undertake the following inquiry in determining whether to grant a plaintiff's motion for a preliminary injunction in a copyright case. First, as in most other kinds of cases in our Circuit, a court may issue a preliminary injunction in a copyright case only if the plaintiff has demonstrated "either (a) a likelihood of success on the merits or (b) sufficiently serious questions going to the merits to make them a fair ground for litigation and a balance of hardships tipping decidedly in the [plaintiff]'s favor." *NXIVM Corp. [v. Ross Inst.*, 364 F.3d 471, 476 (2d Cir. 2004)]. Second, the court may issue the injunction only if the plaintiff has demonstrated "that he is likely to suffer irreparable injury in the absence of an injunction." *Winter*, 129 S. Ct. at 374. The court must not adopt a "categorical" or "general" rule or presume that the plaintiff will suffer irreparable harm (unless such a "departure from the long tradition of equity practice" was intended by Congress). *eBay*, 547 U.S. at 391, 393-94. Instead, the court must actually consider the injury the plaintiff will suffer if he or she loses on the preliminary injunction but ultimately prevails on the merits, paying particular attention to whether the "remedies available at law, such as monetary damages, are inadequate to compensate for that injury." *eBay*, 547 U.S. at 391. Third, a court must consider the balance of hardships between the plaintiff and defendant and issue the injunction only if the balance of hardships tips in the plaintiff's favor. *Winter*, 129 S. Ct. at 374; *eBay*, 547 U.S. at 391. Finally, the court must ensure that the "public interest would not be disserved" by the issuance of a preliminary injunction. *eBay*, 547 U.S. at 391; *accord Winter,* 129 S. Ct. at 374.

A.

The first consideration in the preliminary injunction analysis is the probability of success on the merits. In gauging this, we emphasize that courts should be particularly cognizant of the difficulty of predicting the merits of a copyright claim at a preliminary injunction hearing. *See* Lemley & Volokh, [*Freedom of Speech and Injunctions in Intellectual Property Cases*, 48 Duke L.J. 147, 201-02 (1998)] ("[When deciding whether to grant a TRO or a preliminary injunction,] the judge has limited time for contemplation. The parties have limited time for briefing. Preparation for a typical copyright trial, even a bench trial, generally takes many months; the arguments about why one work isn't substantially similar in its expression to another, or about why it's a fair use of another, are often sophisticated and fact-intensive, and must be crafted with a good deal of thought and effort."). This difficulty is compounded significantly when a defendant raises a colorable fair use defense. "Whether [a] taking[] will pass the fair use test is difficult to predict. It depends on widely varying perceptions held by different judges." Pierre N. Leval, *Toward a Fair Use Standard*, 103 Harv. L. Rev. 1105, 1132 (1990); *see also Campbell*, 510 U.S. at 578 n.10 (noting that "the fair use enquiry often requires close questions of judgment").

B.

Next, the court must consider whether the plaintiff will suffer irreparable harm in the absence of a preliminary injunction, and the court must assess the balance of hardships between the plaintiff and defendant. Those two items, both of which consider the harm to the parties, are related. The relevant harm is the harm that (a) occurs to the parties' legal[9] interests and (b) cannot be remedied after a final adjudication, whether by damages or a permanent injunction. The plaintiff's interest is, principally, a property interest in the copyrighted material. But as the Supreme Court has suggested, a copyright holder might also have a First Amendment interest in *not* speaking. *See Harper & Row Publishers, Inc. v. Nation Enters.*, 471 U.S. 539, 559 (1985). The defendant to a copyright suit likewise has a property interest in his or her work to the extent that work does not infringe the plaintiff's

---

[9] As Judge Leval noted in *New Era Publications International, ApS v. Henry Holt & Co.*, "the justification of the copyright law is the protection of the *commercial* interest of the artist/author. It is not to coddle artistic vanity or to protect secrecy, but to stimulate creation by protecting its rewards." 695 F. Supp. 1493, 1526 (S.D.N.Y. 1988).

copyright. And a defendant also has a core First Amendment interest in the freedom to express him- or herself, so long as that expression does not infringe the plaintiff's copyright.

But the above-identified interests are relevant only to the extent that they are not remediable after a final adjudication. Harm might be irremediable, or irreparable, for many reasons, including that a loss is difficult to replace or difficult to measure, or that it is a loss that one should not be expected to suffer. In the context of copyright infringement cases, the harm to the plaintiff's property interest has often been characterized as irreparable in light of possible market confusion. And courts have tended to issue injunctions in this context because "to prove the loss of sales due to infringement is . . . notoriously difficult." *Omega Importing Corp. v. Petri-Kine Camera Co.*, 451 F.2d 1190, 1195 (2d Cir. 1971) (Friendly, C.J.). Additionally, "'[t]he loss of First Amendment freedoms," and hence infringement of the right *not* to speak, "for even minimal periods of time, unquestionably constitutes irreparable injury." *Elrod v. Burns*, 427 U.S. 347, 373 (1976).

After *eBay*, however, courts must not simply presume irreparable harm. Rather, plaintiffs must show that, on the facts of their case, the failure to issue an injunction would actually cause irreparable harm. This is not to say that most copyright plaintiffs who have shown a likelihood of success on the merits would not be irreparably harmed absent preliminary injunctive relief. As an empirical matter, that may well be the case, and the historical tendency to issue preliminary injunctions readily in copyright cases may reflect just that. *See* H. Tomás Gómez-Arostegui, *What History Teaches Us About Copyright Injunctions and the Inadequate-Remedy-at-Law Requirement*, 81 S. Cal. L. Rev. 1197, 1201 (2008) (concluding, after a thorough historical analysis, that "the historical record suggests that in copyright cases, legal remedies were deemed categorically inadequate"). As Chief Justice Roberts noted, concurring in *eBay*:

> From at least the early 19th century, courts have granted injunctive relief upon a finding of infringement in the vast majority of patent cases. This "long tradition of equity practice" is not surprising, given the difficulty of protecting a right to *exclude* through monetary remedies. * * * This historical practice, as the Court holds, does not *entitle* a patentee to [an] * * * injunction or justify a *general rule* that such injunctions should issue. * * * At the same time, there is a difference between exercising equitable discretion pursuant to the established four-factor test and writing on an entirely clean slate. * * * When it comes to discerning and applying those standards, in this area as others, a page of history is worth a volume of logic.

547 U.S. at 395 (quotation marks omitted).

But by anchoring the injunction standard to equitable principles, albeit with one eye on historical tendencies, courts are able to keep pace with innovation in this rapidly changing technological area. Justice Kennedy, responding to Justice Roberts, made this very point as to patent injunctions in his *eBay* concurrence. Although the "lesson of the historical practice * * * is most helpful and instructive when the circumstances of a case bear substantial parallels to litigation the courts have confronted before[,] * * * in many instances the nature of the patent being enforced and the economic function of the patent holder present considerations quite unlike earlier cases." *Id.* at 396. Justice Kennedy concluded that changes in the way parties use patents may now mean that "legal damages [are] sufficient to compensate for the infringement." *Id.*

C.

Finally, courts must consider the public's interest. The object of copyright law is to promote the store of knowledge available to the public. But to the extent it accomplishes this end by providing individuals a financial incentive to contribute to the store of knowledge, the public's interest may well be already accounted for by the plaintiff's interest.

The public's interest in free expression, however, is significant and is distinct from the parties' speech interests. *See Pac. Gas & Elec. Co. v. Pub. Utils. Comm'n of Cal.*, 475 U.S. 1, 8 (1986). "By protecting those who wish to enter the market-place of ideas from government attack, the First Amendment protects the public's interest in receiving information." *Id.* Every injunction issued before a final adjudication on the merits risks enjoining speech protected by the First Amendment. Some uses, however, will so patently infringe another's copyright, without giving rise to an even colorable fair use defense, that the likely First Amendment value in the use is virtually nonexistent.

IV.

Because the District Court considered only the first of the four factors that, under *eBay* and our holding today, must be considered before issuing a preliminary injunction, we vacate and remand the case. But in the interest of judicial economy, we note that there is no reason to disturb the District Court's conclusion as to the factor it did consider—namely, that Salinger is likely to succeed on the merits of his copyright infringement claim.

. . . .

Notes & Questions

1. How difficult should it be to obtain a preliminary injunction once a likelihood of success on the merits is shown? In some ways, the standard for granting injunctive relief is fundamentally about the nature of the right being violated. What is the nature of the right granted by federal copyright law, and what role does an injunction play in providing adequate protection against an invasion of that right?

2. Does the Second Circuit treat the first two prongs of the four-prong test separately? What is the difference between showing "irreparable harm" and showing that one does not have an "adequate remedy at law"?

3. If a plaintiff is interested in obtaining a preliminary injunction, it is important to request one very early in the litigation.

It has been held that any presumption of irreparable harm that may arise upon a finding of likelihood of success on the merits of a trademark infringement claim "is inoperative if the plaintiff has delayed either in bringing suit or in moving for preliminary injunctive relief." *Tough Traveler, Ltd.*

v. Outbound Prods., 60 F.3d 964, 968 (2d Cir. 1995). The reasoning behind this principle is that "the 'failure to act sooner undercuts the sense of urgency that ordinarily accompanies a motion for preliminary relief and suggests that there is, in fact, no irreparable injury.'" *Id.* (quoting *Citibank, N.A. v. Citytrust*, 756 F.2d 273, 277 (2d Cir. 1985)). However, it has also been held that the aforementioned presumption may still operate where "the delay was caused by the plaintiff's ignorance of the defendant's competing product or the plaintiff's making good faith efforts to investigate the alleged infringement." *Id.*

Voice of the Arab World, Inc. v. MDTV Medical News Now, Inc., 645 F.3d 26, 35 (1st Cir. 2011).

4. How strong is the public interest in a case such as *Salinger*, where the work alleged to be infringed is widely assigned in high school and college literature classes? Is the possibility of a fair use defense a sufficient safeguard for the public's interest? Does it matter whether Salinger himself ever planned to write a derivative work based on *Catcher*? In a portion of the opinion not reproduced above, the Second Circuit noted its agreement with the District Court's conclusion that the plaintiff was likely to succeed on the merits of its claim, which included a conclusion that the defendant was not likely to prevail on its fair use defense. *Salinger*, 607 F.3d at 83.

Is the potential irreparable harm from trademark infringement qualitatively different, or substantially more likely, such that a presumption of irreparable harm, even after *eBay*, is appropriate? Consider the following case.

Herb Reed Enterprises, LLC v. Florida Entertainment Management, Inc.
736 F.3d 1239 (9th Cir. 2013)

McKEOWN, CIRCUIT JUDGE:

"The Platters"—the legendary name of one of the most successful vocal performing groups of the 1950s—lives on. With 40 singles on the Billboard Hot 100 List, the names of The Platters' hits ironically foreshadowed decades of litigation—"Great Pretender," "Smoke Gets In Your Eyes," "Only You," and "To Each His Own." Larry Marshak and his company Florida Entertainment Management, Inc. (collectively "Marshak") challenge the district court's preliminary injunction in favor of Herb Reed Enterprises ("HRE"), enjoining Marshak from using the "The Platters" mark in connection with any vocal group with narrow exceptions. We consider an issue of first impression in our circuit: whether the likelihood of irreparable harm must be established—rather than presumed, as under prior Ninth Circuit precedent—by a plaintiff seeking injunctive relief in the trademark context. In light of Supreme Court precedent, the answer is yes, and we reverse the district court's order granting the preliminary injunction.

BACKGROUND

The Platters vocal group was formed in 1953, with Herb Reed as one of its founders. Paul Robi, David Lynch, Zola Taylor, and Tony Williams, though not founders, have come to be recognized as the other "original" band members. The group became a "global sensation" during the latter half of the 1950s, then broke up in the 1960s as the original members left one by one. After the break up, each member continued to perform under some derivation of the name "The Platters." *Marshak v. Reed,* 2001 WL 92225, at *4 (E.D.N.Y. and S.D.N.Y.) (*"Marshak I"*).

Litigation has been the byproduct of the band's dissolution; there have been multiple legal disputes among the original members and their current and former managers over ownership of "The Platters" mark. Much of the litigation stemmed from employment contracts executed in 1956 between the original members and Five Platters, Inc. ("FPI"), the company belonging to Buck Ram, who became the group's manager in 1954. As part of the contracts, each member assigned to FPI any rights in the name "The Platters" in exchange for shares of FPI stock. *Marshak I,* 2001 WL 92225, at *3. According to Marshak, FPI later transferred its rights to the mark to Live Gold, Inc., which in turn transferred the rights to Marshak in 2009. Litigation over the validity of the contracts and ownership of the mark left a trail of conflicting decisions in various jurisdictions, which provide the backdrop for the present controversy. . . . [The court then summarized the "tangled web multi jurisdictional litigation that spans more than four decades."]

. . . .

Last year brought yet another lawsuit. HRE commenced the present litigation in 2012 against Marshak in the District of Nevada, alleging trademark infringement and seeking a preliminary injunction against Marshak's continued use of "The Platters" mark. . . . The district court found that HRE had established a likelihood of success on the merits, a likelihood of irreparable harm, a balance of hardships in its favor, and that a preliminary injunction would serve public interest. Accordingly, the district court granted the preliminary injunction and set the bond at $10,000. Marshak now appeals from the preliminary injunction.

ANALYSIS

. . . .

III. PRELIMINARY INJUNCTION

To obtain a preliminary injunction, HRE "must establish that [it] is likely to succeed on the merits, that [it] is likely to suffer irreparable harm in the absence of preliminary relief, that the balance of equities tips in [its] favor, and that an injunction is in the public interest." *Winter v. Natural Res. Def. Council, Inc.,* 555 U.S. 7, 20 (2008). We review a district court's preliminary injunction for abuse of discretion, a standard of review that is "limited and deferential." *Johnson v. Couturier,* 572 F.3d 1067, 1078 (9th Cir. 2009). If the district court "identified and applied the correct legal rule to the relief requested," we will reverse only if the court's decision "resulted from a factual finding that was illogical, implausible, or without support in inferences that may be drawn from the facts in the record." *United States v. Hinkson,* 585 F.3d 1247, 1263 (9th Cir. 2009) (en banc).

Marshak's key arguments are that the district court erred in concluding that HRE had established a likelihood of success on the merits because Reed abandoned "The Platters" mark and that the district court erred in finding a likelihood of irreparable harm.

. . . [The Court of Appeals concluded] that the record supports the district court's determination that HRE did not abandon "The Platters" mark.

B. LIKELIHOOD OF IRREPARABLE HARM

We next address the likelihood of irreparable harm. As the district court acknowledged, two recent Supreme Court cases have cast doubt on the validity of this court's previous rule that the likelihood of "irreparable injury may be *presumed* from a showing of likelihood of success on the merits of a trademark infringement claim." *Brookfield Commc'ns, Inc. v. W. Coast Entm't Corp.*, 174 F.3d 1036, 1066 (9th Cir. 1999) (emphasis added). Since *Brookfield*, the landscape for benchmarking irreparable harm has changed with the Supreme Court's decisions in *eBay Inc. v. MercExchange, L.L.C.*, 547 U.S. 388, in 2006, and *Winter* in 2008.

In *eBay*, the Court held that the traditional four-factor test employed by courts of equity, including the requirement that the plaintiff must establish irreparable injury in seeking a permanent injunction, applies in the patent context. 547 U.S. at 391. Likening injunctions in patent cases to injunctions under the Copyright Act, the Court explained that it "has consistently rejected * * * a rule that an injunction automatically follows a determination that a copyright has been infringed," and emphasized that a departure from the traditional principles of equity "should not be lightly implied." *Id.* at 391–93 (citations omitted). The same principle applies to trademark infringement under the Lanham Act. Just as "[n]othing in the Patent Act indicates that Congress intended such a departure," so too nothing in the Lanham Act indicates that Congress intended a departure for trademark infringement cases. *Id.* at 391–92. Both statutes provide that injunctions may be granted in accordance with "the principles of equity." 35 U.S.C. § 283; 15 U.S.C. § 1116(a).

In *Winter*, the Court underscored the requirement that the plaintiff seeking a preliminary injunction "demonstrate that irreparable injury is *likely* in the absence of an injunction." 555 U.S. at 22 (emphasis in original) (citations omitted). The Court reversed a preliminary injunction because it was based only on a "possibility" of irreparable harm, a standard that is "too lenient." *Id. Winter*'s admonition that irreparable harm must be shown to be likely in the absence of a preliminary injunction also forecloses the presumption of irreparable harm here.

Following *eBay* and *Winter*, we held that likely irreparable harm must be demonstrated to obtain a preliminary injunction in a copyright infringement case and that actual irreparable harm must be demonstrated to obtain a permanent injunction in a trademark infringement action. *Flexible Lifeline Sys. v. Precision Lift, Inc.*, 654 F.3d 989, 998 (9th Cir. 2011). Our imposition of the irreparable harm requirement for a permanent injunction in a trademark case applies with equal force in the preliminary injunction context. *Amoco Prod. Co. v. Village of Gambell, AK*, 480 U.S. 531, 546 n. 12 (1987) (explaining that the standard for a preliminary injunction is essentially the same as for a permanent injunction except that "likelihood of" is replaced with "actual"). We now join other circuits in holding that the *eBay* principle—that a plaintiff must establish irreparable harm—applies to a preliminary injunction in a trademark infringement case.

Having anticipated that the Supreme Court's decisions in *eBay* and *Winter* signaled a shift away from the presumption of irreparable harm, the district court examined irreparable harm in its own right, explaining that HRE must "establish that remedies available at law, such as monetary damages, are inadequate to compensate" for the injury arising from Marshak's continuing allegedly infringing use of the mark. *HRE,* 2012 WL 3020039, at *15. Although the district court identified the correct legal principle, we conclude that the record does not support a determination of the likelihood of irreparable harm.

Marshak asserts that the district court abused its discretion by relying on "unsupported and conclusory statements regarding harm [HRE] *might* suffer." We agree.

The district court's analysis of irreparable harm is cursory and conclusory, rather than being grounded in any evidence or showing offered by HRE. To begin, the court noted that it "cannot condone trademark infringement simply because it has been occurring for a long time and may continue to occur." The court went on to note that to do so "could encourage wide-scale infringement on the part of persons hoping to tread on the goodwill and fame of vintage music groups." Fair enough. Evidence of loss of control over business reputation and damage to goodwill could constitute irreparable harm. *See, e.g., Stuhlbarg Int'l Sales Co., Inc. v. John D. Brush and Co., Inc.,* 240 F.3d 832, 841 (9th Cir. 2001) (holding that evidence of loss of customer goodwill supports finding of irreparable harm). Here, however, the court's pronouncements are grounded in platitudes rather than evidence, and relate neither to whether "irreparable injury is *likely* in the absence of an injunction," *Winter,* 555 U.S. at 22, nor to whether legal remedies, such as money damages, are inadequate in this case. It may be that HRE could establish the likelihood of irreparable harm. But missing from this record is any such evidence.

In concluding its analysis, the district court simply cited to [one of the many different lawsuits between the parties relating to the use of "The Platters"] in Nevada "with a substantially similar claim" in which the court found that "the harm to Reed's reputation caused by a different unauthorized Platters group warranted a preliminary injunction." *HRE,* 2012 WL 3020039, at *15–16. As with its speculation on future harm, citation to a different case with a different record does not meet the standard of showing "likely" irreparable harm.

Even if we comb the record for support or inferences of irreparable harm, the strongest evidence, albeit evidence not cited by the district court, is an email from a potential customer complaining to Marshak's booking agent that the customer wanted Herb Reed's band rather than another tribute band. This evidence, however, simply underscores customer confusion, not irreparable harm.[5]

The practical effect of the district court's conclusions, which included no factual findings, is to reinsert the now-rejected presumption of irreparable harm based solely on a strong case of trademark infringement. Gone are the days when "[o]nce the plaintiff in an infringement action has established a likelihood of confusion, it is ordinarily presumed that the plaintiff will suffer irreparable harm if injunctive relief does not issue." *Rodeo Collection, Ltd. v. W. Seventh*, 812 F.2d 1215, 1220 (9th Cir. 1987) (citing *Apple Computer, Inc. v. Formula International Inc.*, 725 F.2d 521, 526 (9th Cir. 1984)). This approach collapses the likelihood of success and the irreparable harm factors. Those seeking injunctive relief must proffer evidence sufficient to establish a likelihood of irreparable harm. As in *Flexible Lifeline*, 654 F.3d at 1000, the fact that the "district court made no factual findings that would support a likelihood of irreparable harm," while not necessarily establishing a lack of irreparable harm, leads us to reverse the preliminary injunction and remand to the district court.

In light of our determination that the record fails to support a finding of likely irreparable harm, we need not address the balance of equities and public interest factors.

WALLACE, SENIOR CIRCUIT JUDGE, concurring:

I agree that the district court's preliminary injunction should be reversed. However, I write separately to emphasize that we are solely reviewing a *preliminary* injunction, and that we thus can express no view on issues arising after a trial dealing with a permanent injunction.

Notes & Questions

1. The Seventh Circuit recently noted that "irreparable harm is especially likely in a trademark case because of the difficulty of quantifying the likely effect on a brand of a nontrivial period of consumer confusion (and the interval between the filing of a trademark infringement complaint and final judgment is sure not to be trivial)." *Kraft Foods Group Brands LLC v. Cracker Barrel Old Country Store, Inc.*, 735 F.3d 735, 741 (7th Cir. 2013). Given the nature of the harm that stems from trademark infringement coupled with the need to provide evidence showing a likelihood of irreparable harm, what types of evidence could a plaintiff present?

2. In a later case, the Ninth Circuit repeated the rule that irreparable harm may not be presumed based on a likelihood of success in a trademark action, but—citing *Herb Reed*—held that "[e]vidence of loss of control over business reputation and damage to goodwill c[an] constitute irreparable harm." *2DIE4KOURT v. Hillair Capital Mgmt., LLC*, 2017 WL 2304376, at *2 (9th Cir. May 26, 2017). The

---

[5] In assessing the evidence with respect to irreparable harm, we reject Marshak's assertion that the district court may rely only on admissible evidence to support its finding of irreparable harm. Not so. Due to the urgency of obtaining a preliminary injunction at a point when there has been limited factual development, the rules of evidence do not apply strictly to preliminary injunction proceedings.

court then noted that there was evidence showing that the defendant, a licensee of the plaintiffs' trademark, had used the plaintiffs' trademarks in connection with the release an unapproved line of cosmetics. In affirming the district court's grant of a preliminary injunction, the Ninth Circuit held that "[t]his is enough to support a finding, at this early stage, that the [plaintiffs] likely will lose some measure of control over their business reputation in the absence of injunctive relief." How is that different than a presumption of irreparable harm upon a showing of likelihood of success on the merits of the claim of trademark infringement?

III. Criminal Infringement

Congress has established criminal sanctions to punish or deter violations of intellectual property rights in three of the four main intellectual property areas: copyright, trademarks, and trade secrets. State law also plays a role in creating criminal sanctions for certain types of trade secret violations.

A. Copyright

The Copyright Act identifies certain types of infringement as worthy of criminal sanctions. Specifically, the statute provides:

> Any person who willfully infringes a copyright shall be punished as provided under section 2319 of title 18,[*] if the infringement was committed—
> (A) for purposes of commercial advantage or private financial gain;
> (B) by the reproduction or distribution, including by electronic means, during any 180–day period, of 1 or more copies or phonorecords of 1 or more copyrighted works, which have a total retail value of more than $1,000; or
> (C) by the distribution of a work being prepared for commercial distribution, by making it available on a computer network accessible to members of the public, if such person knew or should have known that the work was intended for commercial distribution.

17 U.S.C. § 506. Note that the three different types of criminal infringement activity are joined by an "or"—any one of the types of infringing activity identify in (A)-(C) qualifies for criminal sanctions. Note also that for all three, the infringement must be committed "willfully."

Congress first added criminal sanctions to the Copyright Act in 1897, but only for limited types of willful infringement and only if engaged in for profit. In 1997 Congress added the text that is currently subsection (B) to § 506, with the passage of the No Electronic Theft (NET) Act. While the Net Act was passed to combat digital piracy, the statutory language is not so circumscribed. Additionally, the Net Act added a definition of "financial gain" to include the "receipt, or expectation of receipt, of anything of value, including the receipt of other copyrighted works." 17 U.S.C. § 101. Thus, "sharing" digital files with an expectation that others will

---

[*] This provision specifies the criminal sanctions imposed, ranging from 1 to 5 years in prison for a first offense and up to 10 years for a second or subsequent offense.

share their files with you could fit into the definition of financial gain. Eric Gold-man, *A Road to No Warez: The No Electronic Theft Act and Criminal Copyright Infringement*, 82 Or. L. Rev. 369 (2003) (discussing criminal liability in the context of peer-to-peer filesharing). Congress added subsection (C) as part of the Family Entertainment and Copyright Act of 2005.

When one considers how easy it is to reach the $1,000 threshold for criminal liability over any six-month period, determining what constitutes willfulness takes on increased urgency. In the criminal context "willful" takes on a somewhat different meaning from that encountered in cases determining whether enhanced civil damages should be awarded. *See* Lydia Pallas Loren, *Digitization, Commodification, Criminalization: The Evolution of Criminal Copyright Infringement and the Importance of the Willfulness Requirement*, 77 WASH. U. L.Q. 835, 853 (1999). In the criminal context demonstrating willfulness requires showing a "voluntary, intentional violation of a known legal duty." *United States v. Moran*, 757 F. Supp. 1046 (D. Neb. 1991) (quoting *Cheek v. United States*, 498 U.S. 192 (1991)). In *Cheek* the Supreme Court held that showing willfulness requires proof that the defendant was aware of the duty at issue, and also requires negating a defendant's claim of either ignorance of the law or a good-faith belief in the lawfulness of the activity at issue. This negation is part of the government's burden "because one cannot be aware that the law imposes a duty upon him and yet . . . believe that the duty does not exist." *Cheek*, 498 U.S. at 202. The defendant's belief that his conduct is lawful is not to be judged by an objective standard but rather a subjective, good-faith belief in the lawfulness of the activity. Of course, the more unreasonable that belief is, the more difficult it will be for the finder of fact to believe the credibility of an assertion of a good-faith belief, but such credibility determination is to be left to the trier of fact. *Moran* 757 F. Supp. at 1051.

B. TRADEMARK

In the trademark area, trafficking in counterfeit goods and labels carries the possibility of criminal prosecution:

> (a) Whoever intentionally —
>> (1) traffics in goods or services and knowingly uses a counterfeit mark on or in connection with such goods or services,
>> (2) traffics in labels, patches, stickers, wrappers, badges, emblems, medallions, charms, boxes, containers, cans, cases, hangtags, documentation, or packaging of any type or nature, knowing that a counterfeit mark has been applied thereto, the use of which is likely to cause confusion, to cause mistake, or to deceive, . . .
>
> or attempts or conspires to violate any of paragraphs (1) through (4) shall be punished as provided in subsection (b).
>
> (b) Penalties. —
>> (1) In general. — Whoever commits an offense under subsection (a) —
>>> (A) if an individual, shall be fined not more than $2,000,000 or imprisoned not more than 10 years, or both, and, if a person

other than an individual, shall be fined not more than $5,000,000; and

(B) for a second or subsequent offense under subsection (a), if an individual, shall be fined not more than $5,000,000 or imprisoned not more than 20 years, or both, and if other than an individual, shall be fined not more than $15,000,000. . . .

18 U.S.C § 2320. The statute defines a "counterfeit mark" as:

(A) a spurious mark—

(i) that is used in connection with trafficking in any goods, services, labels, patches, stickers, wrappers, badges, emblems, medallions, charms, boxes, containers, cans, cases, hangtags, documentation, or packaging of any type or nature;

(ii) that is identical with, or substantially indistinguishable from, a mark registered on the principal register in the United States Patent and Trademark Office and in use, whether or not the defendant knew such mark was so registered;

(iii) that is applied to or used in connection with the goods or services for which the mark is registered with the United States Patent and Trademark Office, or is applied to or consists of a label, patch, sticker, wrapper, badge, emblem, medallion, charm, box, container, can, case, hangtag, documentation, or packaging of any type or nature that is designed, marketed, or otherwise intended to be used on or in connection with the goods or services for which the mark is registered in the United States Patent and Trademark Office; and

(iv) the use of which is likely to cause confusion, to cause mistake, or to deceive. . . .

Id. In a prosecution, the government must prove that defendant knew that the trademark was spurious, that it was used in connection with trafficking in goods or services, that it was identical to or virtually indistinguishable from another mark, and that it was likely to cause confusion, mistake or to deceive. The government need not prove that the defendant was aware of the fact that the mark was registered with the Patent Office.

C. TRADE SECRET

Trade secret misappropriation had long been, for the most part, a civil law matter between private parties—the purported secret owner and the alleged misappropriator. Beginning in the 1960s, however, legislatures began to criminalize trade secret theft, treating it as a matter of greater public concern. Today, about half the states have laws that expressly criminalize trade secret theft. Geraldine Szott Moohr, *The Problematic Role of Criminal Law in Regulating Use of Information: The Case of the Economic Espionage Act*, 80 N.C. L. Rev. 853, 875 (2004).[*]

---

[*] To learn more about a given state's trade secret theft law, one might begin with Brian M. Malsberger, *Trade Secrets: A State-by-State Survey* (3d ed. 2006).

These statutes overcame what, before their enactment, had been judicial concerns about applying antitheft statutes focused on the physical taking of tangible property to the misappropriation of intangible information. Daniel D. Fetterly, *Historical Perspectives on Criminal Laws Relating to the Theft of Trade Secrets*, 25 BUS. LAW. 1535, 1537 (1970).

The California Penal Code's trade secret theft law is illustrative. Its substantive core provides as follows:

> Every person is guilty of theft who, with intent to deprive or withhold the control of a trade secret from its owner, or with an intent to appropriate a trade secret to his or her own use or to the use of another, does any of the following:
>
> (1) Steals, takes, carries away, or uses without authorization, a trade secret.
>
> (2) Fraudulently appropriates any article representing a trade secret entrusted to him or her.
>
> (3) Having unlawfully obtained access to the article, without authority makes or causes to be made a copy of any article representing a trade secret.
>
> (4) Having obtained access to the article through a relationship of trust and confidence, without authority and in breach of the obligations created by that relationship, makes or causes to be made, directly from and in the presence of the article, a copy of any article representing a trade secret.

Cal. Penal Code § 499c(b). Subsection (a) of the statute provides definitions of such key terms as "access," "article," and "trade secret."

Congress also made trade secret theft a federal crime when it enacted the Economic Espionage Act in 1996. The Act is codified at 18 U.S.C. §§ 1831-1839. Section 1831 criminalizes trade secret theft that is undertaken "intending or knowing that the offense will benefit any foreign government, foreign instrumentality, or foreign agent." 18 U.S.C. § 1831(a). Indeed, the congressional witnesses supporting the Act focused on foreign industrial espionage. *See* Moohr, *supra*, at 864-65. Section 1832, however, criminalizes theft as to any trade secret "that is related to or included in a product that is produced for or placed in interstate or foreign commerce." 18 U.S.C. § 1832(a). In other words, this act makes *domestic* trade secret theft a federal felony as well. Attempt and conspiracy are also felonies. 18 U.S.C. §§ 1832(a)(4), (5).

Many of the prosecutions in the first decade following passage of the EEA involved purely domestic trade secret theft and often involved an employee with access to trade secret information. *See, e.g., United States v. Lange,* 312 F.3d 263 (7th Cir. 2002) (former employee charged with selling technical specifications for aircraft parts); *United States v. Genovese,* 409 F. Supp. 2d 253 (S.D.N.Y. 2005) (defendant charged with selling portions of Microsoft source code); *United States v. Krumrei,* 258 F.3d 535 (6th Cir. 2001) (defendant charged with selling technical information to a competitor); *United States v. Martin,* 228 F.3d 1 (1st Cir. 2000) (employee charged with passing information to competitor). The initial prosecutions helped establish the broad sweep of the federal law. *See also United States v. Hsu*, 155 F.3d 189 (3d Cir. 1998). The EEA and the use of criminal prosecutions in addition to civil liability raise many interesting legal issues that are beyond the scope of a survey course in intellectual property.

Notes & Questions

1. When a private individual or entity has detected or suspects infringement of an intellectual property right that is subject to criminal sanctions, it can notify the proper law enforcement authorities. Why might a trade secret owner not be eager to report suspected trade secret theft to law enforcement? Whether to pursue prosecution is a decision that is left to the discretion of the prosecutor. While the assistance of the intellectual property owner may be important to pursuing the case, a private individual cannot bring an action under the criminal provisions. What types of intellectual property crimes might be worthy of criminal prosecution?

2. Article 61 of the Agreement on Trade-Related Aspects of Intellectual Property Rights (TRIPs) requires that signatory countries establish criminal procedures and penalties in cases of "willful trademark counterfeiting or copyright piracy on a commercial scale." The United States is a signatory to the TRIPs Agreement. Why are there no criminal sanctions for patent infringement?

3. The Department of Justice Task Force on Intellectual Property monitors and coordinates overall intellectual property enforcement efforts at the federal level, including a focus on the international aspects of IP enforcement and the links between IP crime and international organized crime. *See* http://www.justice.gov/dag/iptaskforce/.

4. In addition to the criminal provisions identified above, other criminal statutes may come into play, especially where some or all of the targeted activity took place on computer networks. An important statute, the Computer Fraud and Abuse Act, 18 U.S.C. § 1030, is beyond the scope of this book. *See* James Grimmelmann, *Internet Law: Cases and Problems* Ch. 5, Part B (6th ed. 2016).

Made in the USA
Middletown, DE
10 January 2020